The Geometry and Petrogenesis of Dolomite Hydrocarbon Reservoirs

Special Publication reviewing procedures

The Society makes every effort to ensure that the scientific and production quality of its books matches that of its journals. Since 1997, all book proposals have been refereed by specialist reviewers as well as by the Society's Books Editorial Committee. If the referees identify weaknesses in the proposal, these must be addressed before the proposal is accepted.

Once the book is accepted, the Society has a team of Book Editors (listed above) who ensure that the volume editors follow strict guidelines on refereeing and quality control. We insist that individual papers can only be accepted after satisfactory review by two independent referees. The questions on the review forms are similar to those for *Journal of the Geological Society*. The referees' forms and comments must be available to the Society's Book Editors on request.

Although many of the books result from meetings, the editors are expected to commission papers that were not presented at the meeting to ensure that the book provides a balanced coverage of the subject. Being accepted for presentation at the meeting does not guarantee inclusion in the book.

Geological Society Special Publications are included in the ISI Index of Scientific Book Contents, but they do not have an impact factor, the latter being applicable only to journals.

More information about submitting a proposal and producing a Special Publication can be found on the Society's web site: www.geolsoc.org.uk.

It is recommended that reference to all or part of this book should be made in one of the following ways:

Braithwaite, C. J. R., Rizzi, G. & Darke, G. (eds) 2004. *The Geometry and Petrogenesis of Dolomite Hydrocarbon Reservoirs.* Geological Society, London, Special Publications, **235**.

Whitaker, F. F., Smart, P. L. & Jones, G. D. 2004. Dolomitization: from conceptual to numerical models. *In*: Braithwaite, C. J. R., Rizzi, G. & Darke, G. (eds) *The Geometry and Petrogenesis of Dolomite Hydrocarbon Reservoirs.* Geological Society, London, Special Publications, **235**, 99–139.

GEOLOGICAL SOCIETY SPECIAL PUBLICATION NO. 235

The Geometry and Petrogenesis of Dolomite Hydrocarbon Reservoirs

EDITED BY

C. J. R. BRAITHWAITE
Division of Earth Sciences, University of Glasgow, UK

G. RIZZI
Trask Geoscience Ltd, Aberdeen, UK

and

G. DARKE
Statoil, Stavanger, Norway

2004
Published by
The Geological Society
London

THE GEOLOGICAL SOCIETY

The Geological Society of London (GSL) was founded in 1807. It is the oldest national geological society in the world and the largest in Europe. It was incorporated under Royal Charter in 1825 and is Registered Charity 210161.

The Society is the UK national learned and professional society for geology with a worldwide Fellowship (FGS) of 9000. The Society has the power to confer Chartered status on suitably qualified Fellows, and about 2000 of the Fellowship carry the title (CGeol). Chartered Geologists may also obtain the equivalent European title, European Geologist (EurGeol). One fifth of the Society's fellowship resides outside the UK. To find out more about the Society, log on to www.geolsoc.org.uk.

The Geological Society Publishing House (Bath, UK) produces the Society's international journals and books, and acts as European distributor for selected publications of the American Association of Petroleum Geologists (AAPG), the American Geological Institute (AGI), the Indonesian Petroleum Association (IPA), the Geological Society of America (GSA), the Society for Sedimentary Geology (SEPM) and the Geologists' Association (GA). Joint marketing agreements ensure that GSL Fellows may purchase these societies' publications at a discount. The Society's online bookshop (accessible from www.geolsoc.org.uk) offers secure book purchasing with your credit or debit card.

To find out about joining the Society and benefiting from substantial discounts on publications of GSL and other societies worldwide, consult www.geolsoc.org.uk, or contact the Fellowship Department at: The Geological Society, Burlington House, Piccadilly, London W1J 0BG: Tel. +44 (0)20 7434 9944; Fax +44 (0)20 7439 8975; E-mail: enquiries@geolsoc.org.uk.

For information about the Society's meetings, consult *Events* on www.geolsoc.org.uk. To find out more about the Society's Corporate Affiliates Scheme, write to enquiries@geolsoc.org.uk.

Published by The Geological Society from:
The Geological Society Publishing House
Unit 7, Brassmill Enterprise Centre
Brassmill Lane
Bath BA1 3JN,
UK
(*Orders*: Tel. +44 (0)1225 445046
Fax +44 (0)1225 442836)
Online bookshop: http://bookshop.geolsoc.org.uk

British Library Cataloguing in Publication Data
A catalogue record for this book is available from the British Library.

ISBN 1-86239-166-1

Typeset by Type Study, Scarborough, UK
Printed by MPG, Bodmin, UK

Distributors
USA
AAPG Bookstore
PO Box 979
Tulsa
OK 74101-0979
USA
Orders: Tel. +1 918 584-2555
Fax +1 918 560-2652
E-mail bookstore@aapg.org

India
Affiliated East–West Press PVT Ltd
G-1/16 Ansari Road, Darya Ganj,
New Delhi 110 002
India
Orders: Tel. +91 11 2327-9113/2326-4180
Fax +91 11 2326-0538
E-mail affiliat@nda.vsnl.com

Japan
Kanda Book Trading Company
Cityhouse Tama 204
Tsurumaki 1-3-10
Tama-shi
Tokyo 206-0034
Japan
Orders: Tel. +81 (0)423 57-7650
Fax +81 (0)423 57-7651
E-mail geokanda@ma.kcom.ne.jp

Contents

The geometry and petrogenesis of dolomite hydrocarbon reservoirs: introduction

COLIN J. R. BRAITHWAITE[1], GIANCARLO RIZZI[2] & GILLIAN DARKE[3]
[1]*Division of Earth Sciences, The University of Glasgow, Glasgow G12 8QQ, UK*
[2]*Trask Geoscience Ltd, Aberdeen Offshore Technology Park, Exploration Drive, Bridge of Don, Aberdeen AB23 8GX, UK*
[3]*Statoil, Forusbeen 50, N-4035, Stavanger, Norway*

The present volume represents a series of papers arising from a conference on 'The Geometry and Petrogenesis of Dolomite Hydrocarbon Reservoirs' held at the Geological Society of London on 3–4 December 2002. More than 70 dolomite enthusiasts gathered in the largest specialist meeting since the bicentennial conference, in Ortisei in 1991, honouring the first description of the mineral by the French Engineer Déodat de Dolomieu (1750–1801). The proceedings of the latter meeting, sponsored jointly by the International Association of Sedimentologists (IAS) and the Society of Economic Paleontologists and Mineralogists (SEPM), were subsequently published in IAS Special Publication 21 edited by Purser *et al.* (1994).

As the title indicates, the focus of the meeting reported here was on the implications of the various processes and products of dolomitization for hydrocarbon exploration. This is by no means a trivial issue, where an estimated 80% of North American and many Middle and Far Eastern reservoirs are in rocks affected by dolomitization.

As in any meeting of this kind, the weight of emphasis lies with the contributors. Sessions were devoted to Early Dolomites, Burial Dolomitization, Fractures and Reservoir Quality, the Geometry of Dolomite Bodies, Petrophysics and a variety of case histories. Although a polite accord was maintained throughout the conference discussion raised the blood pressures of at least some delegates. In keeping with the diversity of views on the origins of dolomites, there was a schizoid disconnection between those driven by curiosity; 'art for arts sake', and those with clear economic objectives, with fanatics and sceptics in both groups drawn together by a common interest in the origins of this mysterious rock.

Since its discovery, dolomite has been something of an enigma and an embarrassment for sedimentologists. Dolomite rocks do not behave as respectable sediments; they neither consist of aggregations of detrital or bioclastic grains nor, for the most part, have they crystallized from solution. In one sense they are sedimentological cuckoos, stealing the resources and progressively replacing the calcite and aragonite of existing limestones, and, perhaps worse, apparently neglecting the rules of stratigraphic superposition. The 'dolomite question', how they form, is clear and unambiguous, but the contributions here indicate that the answers (there really are more than one) have been (and for many still are) elusive.

The basic crystal structure of the mineral was not known until the work of Wyckoff & Merwyn (1924) and an understanding of the atomic parameters waited a further 30 years until X-ray powder diffraction data became available (Howie & Broadhurst 1958). Reviews and refinements of this knowledge continued into the 1980s with the work of Reeder (1983) and Wenk *et al.* (1983) among others. Laboratory synthesis of dolomite at low temperatures has proved difficult because reaction rates are relatively slow, but optimistically we can see a steady progression in our understanding of how features of the atomic structure function in issues such as solubility and replacement.

The issue of origins cannot be viewed in this way and its history can reasonably be described as a Gadarene rush after a series of bandwagons. It is difficult to estimate the number of 'universal' models that have come and gone in the last century. The early reviews by van Tuyl (1914) and by Fairbridge (1957) provided important milestones, and Fairbridge in particular was already aware of at least 10 alternative geometries of deposition. As **Machel**'s paper indicates, more have been added since. However, many of the originators of these ideas seem to have been convinced that the 'dolomite question' had a unique answer. Reason has prevailed and there is now something of a concensus, following Land (1985) and Hardie

From: BRAITHWAITE, C. J. R., RIZZI, G. & DARKE, G. (eds) 2004. *The Geometry and Petrogenesis of Dolomite Hydrocarbon Reservoirs*. Geological Society, London, Special Publications, **235**, 1–6. 0305-8719/$15.00

(1987) among others, that although the magnesium required for large-scale replacement is most commonly derived from seawater the required physicochemical conditions can be met in a number of ways.

A further problem lay in the assumption that the process of replacement was inexorable and that, once started, it would proceed to completion. To some degree this was borne out by the observation that dolomitic rocks tended either to contain rather less than 20% dolomite or to be wholly dolomite (eg. Sperber *et al.* 1984). It was subsequently realized not only that the process commonly includes multiple phases (eg. Machel & Anderson 1989), but also that these may be punctuated by periods of dissolution (Theriault & Hutcheon 1987) or replacement by calcite (Shearman *et al.* 1961).

It has long been appreciated that dolomite is not uniformly distributed in the geological record. Given & Wilkinson (1987) showed that it did not, as had commonly been supposed, increase in older rocks, but that there were peaks and troughs. It was relatively common in the Precambrian, in the Lower Ordovician and Middle Silurian, and there was a minor peak in the Middle Devonian, falling to a minimum in the Permo-Carboniferous, before rising to a maximum (around 70%) in the Lower Cretaceous. It is, however, important to remember that that the dolomite is always younger, and sometimes significantly so, than the host rocks. The greater proportions generally correspond with major periods of high sea level, as defined by Vail *et al.* (1977) and Hallam (1984). This observation is especially important because of the implicit link between the origins of most dolomitization and seawater.

Work by Dickson (2002) indicated that this distribution might be linked to changes in ocean chemistry that may be preserved in records of temporal changes in the magnesium/calcium ratio of the world ocean. Kerr (2002) linked these to the activity of the ocean ridges. Greater activity or longer ridges are reflected, it is argued, in greater flows of heated waters along ridge axes. The likely connection between large-scale dolomitization and the waters generated at ocean ridges had earlier been recognized by Given & Wilkinson (1987), but at the time there were few data to provide direct support. Work by Kelley *et al.* (2001) has shown that where waters emerge from serpentized ultramafic rocks they are warm rather than hot, and are alkaline (pH *c.* 9.8). They precipitate calcite, aragonite and brucite, but not dolomite. The key issue here is that magnesium concentrations are very much lower than in typical seawater (9–19 mmol kg^{-1} compared to 54 mmol kg^{-1}). There is thus apparently an inverse relationship between periods of high sea-level when oceans are depleted in magnesium and periods when most dolomite is produced, an enigma yet to be resolved.

As indicated, conference presentations were grouped under a number of themes, but contributions were not uniformly distributed. This may have been a reflection of differences in commercial significance or of general interest but is repeated in the present offerings. A number of contributors chose not to provide formal accounts of their data. We have therefore only attempted to separate papers into two sections. The first is concerned with what can be regarded as general analyses of particular problems relating to dolomites. These include both 'state-of-the-art' reviews and theoretical modelling of specific issues. The second group provides examples of reservoirs or potential reservoirs in dolomites. Some offer solutions to issues raised in the reviews, while others present evidence apparently supporting more contentious views of origins abandoned elsewhere.

Machel provides a fitting hors d'oeuvre to the body of the contributions here. He sets out to review the development of ideas on the origins of dolomite, emphasizing the requirement for an understanding of both geochemical and hydrological parameters. As an opening to a menu it also serves a substantial main course! It reviews both the history of investigation and terminology; thermodynamic, kinetic and mass-balance constraints; dolomite textures resulting from both replacement and cementation; porosity and permeability; dolomite geochemistry and environments; models of dolomitization; and secular distribution of dolostones. To those new to the field it provides an invaluable perspective not only on what has gone before but where we stand now. This is not a dry listing of past ideas but a critique and statement of reasoned opinion: 'This notion is clearly incorrect because. . .'. Some conclusions may not please everyone but they challenge dissenters to produce evidence.

Only one paper deals at any length with what might be termed a biogenic origin for dolomites. In an important contribution **Wright & Wacey**, revisiting the sediments of the Coorong, present the results of experiments that support the involvement of sulphate-reducing bacteria in the precipitation of poorly ordered dolomite. These results shed further light on the widely stated view that sulphates act to inhibit dolomite formation. The experiments themselves are convincing and circumstantial evidence has previously supported similar views, but conference speakers from the floor made clear their scepticism that the model can be applied to large-scale platform deposits. Do we need more experiments or more field observations?

The paper by **Wright *et al.*** stands out, as it is devoted to geochemical modelling of the waters responsible for generating replacive planar dolomite in the Lower Carboniferous Waulsortian Limestones of Ireland. Systems using oxygen, carbon and strontium isotopes, as well as bulk calcium and carbon, are modelled and closely argued. The conclusions reached are that the waters responsible had compositions similar to those of seawater that was only slightly modified and were active at low temperatures, but also that the general model is applicable on a global scale. The driving mechanism for circulation is considered to have been similar to the Kohout model (Kohout 1967).

Although others allude to it, only one paper, by **Whitaker *et al.***, addresses the issue of numerical modelling of the fluid flow associated with dolomitization. A variety of conceptual models are examined, addressing both fluid flow and water–rock interaction. Three principle drives are identified: topographic head of meteoric or seawater; gradients in fluid density, reflecting variations in salinity, temperature, pressure or some combination of these; or as a result of compaction or tectonic loading, but it is recognized that these commonly act in concert, albeit with differing emphases. The balance between these variables changes over time in response to changes in the underlying palaeogeography and sea-level. Thus, simplistic assumptions in models can only provide simplistic answers. Dolomitization reflects a delicate interplay between mass transport and reaction kinetics. It is kinetically favoured at the higher temperatures associated with greater depth, but here reactions are slower because permeability and therefore flow are typically lower. Flow rates are likely to be higher near surfaces, but here reaction rates are limited by lower temperatures. Lithological variations impose additional constraints. The paper by **Whitaker *et al.*** implies that there is a golden mean, a fairway in which dolomitization is more effective, and supports the belief that the diagenetic patterns in dolostones provide valuable analogues that aid understanding of the distribution of many other diagenetic features.

The principal theme of **Lucia** is that much of the porosity in dolomites is not derived via the process of dolomitization but is inherited from precursor limestones. There has been a long-cherished belief, arising from calculations by de Beaumont (1837), that the mole-for-mole conversion of calcite to dolomite necessarily results in a 12% increase in porosity, reflecting the difference in density. However, drawing on his experience of Palaeozoic, Mesozoic and Tertiary examples, Lucia's contention is that dolomitization is unlikely in dense tight limestones because there is little possibility of the passage of large volumes of fluids. In young rocks in particular, high primary porosities are conserved during the dolomitization process and are only significantly reduced by the later growth of pore-filling dolomite cements. Drawing on the Whitaker *et al.* statements regarding transport and kinetics, one can reasonably conclude that if limestones are to be dolomitized and retain porosity they had better do it while they are young and near the surface.

Gregg also draws on years of personal experience together with that of his many students to analyse in particular the sequential changes in limestones and dolomites affecting porosity and permeability. Although this is largely from the perspective of mineral exploration, the common association of base metals with both dolomites and hydrocarbons in Mississippi Valley-type deposits points to common factors in generation and thus there is much to be learned that can be applied to reservoir analysis. The passage of evolving basinal fluids through carbonate rocks is responsible for dolomitization but also for subsequent neomorphism, dissolution and growth of cements. The resulting micro- and mesoporosity is dominated by intra- and intercrystal pores and vugs, with progressive changes in texture increasing pore tortuosity. As in many oil fields, large-scale porosity commonly reflects telogenetic processes and the influence of karst features.

Pores and the passage of fluids are also central to the paper by **Gale *et al.*** However, this stands apart from others in that the focus is on the important issue of fractures. In particular, it argues that the characteristics of fractures on a microscopic scale, including both basic geometry and diagenetic features, can be used to predict those on the macroscopic scale. The data presented suggest that over several orders of magnitude fracture sets follow the same power-law distributions. Case examples demonstrate the potential of the method in reservoir analysis. The timing of fracture generation and resolution of the question whether fractures provide permeable pathways or effective barriers are additional significant issues.

Petrophysics is addressed in more detail in

the comprehensive study of the Arab Carbonates of the Al Rayan Field, Qatar, by **Clark *et al.*** This details depositional environment and early diagenetic history, with an interplay between cement growth and leaching phases, and subsequent dolomitization. Qatar is of particular interest as it was initially regarded as uneconomic, although good porosity and oil were clearly present locally. Careful placing of horizontal wells through horizons controlled by both depositional and diagenetic characteristics outlined in the paper have proved highly productive. Details of porosity, permeability, capillary pressure curves, water-saturation states and general reservoir characteristics complete this thorough account.

Remaining papers are largely devoted to studies that provide feet-on-the-ground examples of regional patterns of distribution, and differing views of fluid movements. As in the contributions listed above, there is no shortage of ideas. What we need is evidence! It is of particular interest to compare the views expressed in the discussions of the following papers (and it is important to note that they are not always the views of the authors) with the sometimes theoretical arguments in the papers outlined above.

Eherenberg provides one of the few examples of relatively recent dolomites in his important study of Miocene carbonate platforms of the Marion Plateau, off NE Australia. As might be expected, these reflect circulation of normal–slightly modified seawater, but variations in the style of replacement and the degree of fabric destruction with depth, leading to metre-scale heterogeneity, are thought to be the result of cycles of sea-level change. Given the environments in which most precursor limestones are thought to have formed, these should be an issue in many dolomitizing environments yet are seldom emphasized; why not?

The paper by **Carnell & Wilson** provides a regional overview and an extensive bibliography of the distribution, characteristics and economic significance of various dolomites in SE Asia, with particular (and economically justified) emphasis on those of the Neogene. It is interesting to see in this area a common emphasis on mechanisms involving waters expressed from compacting shales, a model subject to some criticism elsewhere. However, the authors highlight, and are careful to stress, the need for further studies of dolomitization in the region. These will undoubtedly also focus, in part, on the origins of the exceptional volumes of CO_2 present in many reservoirs in the area, making 'otherwise attractive targets uneconomic'.

Some support for Wright's observations on sulphate-reducing bacteria is provided by **Kirkham**'s contribution on the occurrence of so-called patterned dolomites in the Kimmeridgian Arab Formation of the Arabian Gulf. Interbedded with stromatolites and thin carbonates containing anhydrite nodules, the dark mottled beds are interpreted to have formed as by-products of the activities of sulphate-reducing bacteria and to reflect the presence of microcrystalline iron sulphide. The shallow salinas in which the dolomites are thought to have formed were depleted in sulphate, and the sulphide formed penecontemporaneously with the dolomite. The facies patterns derived from this interpretation offer one explanation for the known reservoir heterogeneity.

The paper by **Saller** provides a clear and comprehensive overview of reservoirs in the Permian Basin of Texas and New Mexico. This has held a particular fascination since the publication in 1953 of the seminal book by Newall *et al.* (1953) on the Permian Reef Complex. After more than 50 years it remains an important producing area and has been favourable not only for the repeated deposition of both source rocks and reservoirs but also for the recurrent occurrence of dolomite over time. Field-scale (kilometre) variations in porosity are generally related to position in the original dolomitizing system, and in platform dolomites generally increase in a basin-ward direction. Although most dolomite fields are structural traps, there is commonly stratigraphic closure associated with up-dip loss in porosity reflecting evaporite precipitation and compactional drape.

Bouch *et al.* offer a detailed account of reservoir analogues in the Lake District of England. The references suggest that the study was driven in part by investigations related to planning for a nuclear-waste repository, and fluid movements were clearly the key issue. Lower Carboniferous basin-margin limestones within the region have been dolomitized, fractured and mineralized by saline fluids, but these have migrated from contrasting sources. The important conclusion here is that differing relationships between fluid movements, dolomite development and fracture initiation have apparently had similar net results on porosity and permeability.

In a paper dealing with the Lower Carboniferous Pekisko Formation of western Canada **Hopkins** draws attention to the important potential of dolomudstones as reservoirs. These may begin as mudstones interbedded and sometimes encased by grainstones that have an apparently higher oil-bearing potential.

However, because of the relative ease with which the coarser lithologies become tightly cemented the muddy components remain accessible to dolomitizing fluids and may ultimately develop significantly higher porosities (again contrary to the views of Lucia). In the Pekisko example porosity was further enhanced as the formation was locally exposed at several Jurassic and Cretaceous unconformities during which times both sandstones and shales were deposited in solution cavities.

In a group of papers dealing with dolomitization in the Irish Lower Carboniferous (Mississippian) **Nagy *et al.*** outline patterns of migration of fluids dolomitizing shallow-marine and peritidal sequences in the Irish Midlands. Fine-grained dolomites developed on the margins of the Leinster Massif reflect some evaporitic influence, indicated by stable isotope compositions, but have been modified by later diagenesis by slightly modified seawater. Following alteration, base-metal mineralization was associated with the apparently vertical migration of brines enriched in chlorides above levels that might be expected from the evaporation of seawater alone. The resulting rocks are not oil-bearing, but parallels can be drawn and the migration paths are considered to provide an analogue for hydrocarbon migration elsewhere.

Mulhall & Sevastopulo are also engaged with the Carboniferous of Ireland driven, like others, by the relationship of dolomitization to what is quoted as being, per square kilometre, one of the richest ore fields in the world. Whereas others have emphasized models (or perhaps fashions) in the dolomitization mechanism, here this issue is bypassed to some degree and the focus is on the state of the limestones at the time of dolomitization. It seems that the original porosity of the limestones was occluded within several hundred metres of burial prior to the onset of dolomitization (compare this with the views of Lucia and of Hopkins). Contrary to the interpretations of some other contributions, temperatures are thought to have been relatively high and therefore this process is believed to have occurred at relatively greater depth, perhaps 2.5 km. In some areas dolomite formation increased permeability on a regional scale, but elsewhere rising fluids may have exploited steep fracture zones.

Conclusions

The diverse papers presented at the conference reflect the current distribution of interests in the various issues associated with dolomites, and dolostones and their functions as reservoirs and host rocks for base-metal mineralization. The lack of any obvious balance between the various issues addressed seems to imply that they are all to some degree 'open' with no concensus view that problems are 'solved'. There are, indeed, contrasting views. However, it is clear from several contributions that much progress has been made and we can look forward to the next decade and increasing degrees of sophistication. Where might this lead us? It is inevitable that there will remain a need to generate ever more detailed field investigations. It is sometimes argued that there are already sufficient numbers of these and we merely have to assimilate what they tell us, but chance determines what will be found in any particular study and some new component or new technique may appear that provides the key to an outstanding problem. Petrography remains a likely productive area. Coupled with laser ablation and mass spectrometry, the analysis of radiogenic and stable isotopes, together with trace elements, allows the history of growth of phases to be mapped in ever more detail. The increasing resolution of scanning transmission electron microscopy (STEM) offers the possibility of determining the mechanisms of replacement at finer and finer scales. Numerical models may also be expected to make greater advances, although there is a risk in these that as the number of variables that can be accommodated in calculations increases it will be difficult to generalize sufficiently to relate models to field examples where boundary conditions will, for the most part, remain unknowable. There seem to be few areas of investigation where we can claim to have closed the file and it is likely that dolomites and dolostones will provide a stimulus for creative research for *at least* another decade.

References

DE BEAUMONT, E. 1837. Application du calcul a l'hypothèse de la formation par epigenie des anhydrites, des gypses et des dolomies. *Société Géologique de France, Bulletin*, **8**, 174–177.

DICKSON, J.A.D. 2002. Fossil echinoderms as monitor of the Mg/Ca ratio of Phanerozoic oceans. *Science*, **298**, 1222–1224.

FAIRBRIDGE, R.W. 1957. The dolomite question. *In*: LE BLANC, R.J. & BREEDING, J.G. (eds) *Regional Aspects of Carbonate Deposition*. Society of Economic Paleontologists and Mineralogists, Special Publications, **5**, 123–178.

GIVEN, R.K. & WILKINSON, B.H. 1987. Dolomite abundance and stratigraphic age: constraints on rates and mechanisms of Phanerozoic dolostone formation. *Journal of Sedimentary Petrology*, **57**, 1068–1078.

HALLAM, A. 1984. Pre-Quaternary sea level changes. *Annual Review of Earth and Planetary Sciences*, **12**, 205–243.

HARDIE, L.A. 1987. Perspectives: dolomitization, a critical review of some current views. *Journal of Sedimentary Petrology*, **57**, 166–183.

HOWIE, R.A. & BROADHURST, F.M. 1958. X-Ray data for dolomite and ankerite. *American Mineralogist*, **43**, 1210–1214.

KELLEY, D.S., KARSON, J.A. *ET AL*., & THE AT3-60 SHIPBOARD PARTY. 2001. An off-axis hydrothermal vent field near the Mid-Atlantic Ridge at 30°N. *Nature*, **412**, 145–149.

KERR, R.A. 2002. Inconstant ancient seas and life's path. *Science*, **298**, 1165–1166.

KOHOUT, F.A. 1967. Groundwater flow and the geothermal regime of the Floridan Plateau. *Gulf Coast Association of Geological Societies, Transactions*, **17**, 339–354.

LAND, L.S. 1985. The origin of massive dolostones. *Journal of Geological Education*, **33**, 112–125.

MACHEL, H.G. & ANDERSON, J.H. 1989. Pervasive subsurface dolomitization of the Nisku Formation in Central Alberta. *Journal of Sedimentary Petrology*, **59**, 891–911.

NEWELL, N.D., RIGBY, J.K., FISCHER, A.G., WHITEMAN, A.J., HICKOX, J.E. & BRADLEY, J.S. 1953. *The Permian Reef Complex of the Guadalupe Mountains Region, Texas and New Mexico*. Freeman, San Francisco, CA.

PURSER, B.H., TUCKER, M.E. & ZENGER, D.H. (eds). 1994. *Dolomites: A Volume in Honour of Dolomieu*. International Association of Sedimentologists, Special Publications, **21**.

REEDER, R.J. 1983. Crystal chemistry of the rhombohedral carbonates. *In*: REEDER, R.J. (ed.) *Carbonates: Mineralogy and Chemistry*. Mineralogical Society of America, Reviews in Mineralogy, **11**, 1–47.

SHEARMAN, D.J., KHOURI, J. & TAHA, S. 1961. On the replacement of dolomite by calcite in some Mesozoic limestones from the French Jura. *Proceedings of the Geologists Association*, **72**, 1–12.

SPERBER, C.M., WILKINSON, B.H. & PEACOR, D.R. 1984. Rock composition, dolomite stoichiometry, and water/rock interactions in dolomite carbonate rocks. *Journal of Geology*, **92**, 609–622.

THERIAULT, F. & HUTCHEON, I. 1987. Dolomitization and calcitization of the Devonian Grosmont Formation, northwestern Alberta. *Journal of Sedimentary Petrology*, **57**, 955–966.

VAIL, P.R., MITCHUM, R.M., JR., *ET AL*. 1977. *Seismic Stratigraphy – Application to Hydrocarbon Exploration*. American Association of Petroleum Geologists, Memoir, **26**, 49–212.

VAN TUYL, F.M. 1914. *The Origin of Dolomite*. Annual Report AR-25C, Iowa Geological Survey, **25**, 251–422.

WENK, H.-R., BARBER, D.J. & REEDER, R.J. 1983. Microstructures in carbonates. *In*: REEDER, R.J. (ed.) *Carbonates: Mineralogy and Chemistry*. Mineralogical Society of America, Reviews in Mineralogy, **11**, 301–367.

WYCKOFF, W.G. & MERWYN, H.S. 1924. The crystal structure of dolomite. *American Journal of Science*, **8**, 447–461.

Concepts and models of dolomitization: a critical reappraisal

HANS G. MACHEL

Department of Earth and Atmospheric Sciences, University of Alberta, Edmonton, Alberta, Canada T6G 2E3 (e-mail: hans.machel@ualberta.ca)

Abstract: Despite intensive research over more than 200 years, the origin of dolomite, the mineral and the rock, remains subject to considerable controversy. This is partly because some of the chemical and/or hydrological conditions of dolomite formation are poorly understood, and because petrographic and geochemical data commonly permit more than one genetic interpretation. This paper is a summary and critical appraisal of the state of the art in dolomite research, highlighting its major advances and controversies, especially over the last 20–25 years.

The thermodynamic conditions of dolomite formation have been known quite well since the 1970s, and the latest experimental studies essentially confirm earlier results. The kinetics of dolomite formation are still relatively poorly understood, however. The role of sulphate as an inhibitor to dolomite formation has been overrated. Sulphate appears to be an inhibitor only in relatively low-sulphate aqueous solutions, and probably only indirectly. In sulphate-rich solutions it may actually promote dolomite formation.

Mass-balance calculations show that large water/rock ratios are required for extensive dolomitization and the formation of massive dolostones. This constraint necessitates advection, which is why all models for the genesis of massive dolostones are essentially hydrological models. The exceptions are environments where carbonate muds or limestones can be dolomitized via diffusion of magnesium from seawater rather than by advection.

Replacement of shallow-water limestones, the most common form of dolomitization, results in a series of distinctive textures that form in a sequential manner with progressive degrees of dolomitization, i.e. matrix-selective replacement, overdolomitization, formation of vugs and moulds, emplacement of up to 20 vol% calcium sulphate in the case of seawater dolomitization, formation of two dolomite populations, and – in the case of advanced burial – formation of saddle dolomite. In addition, dolomite dissolution, including karstification, is to be expected in cases of influx of formation waters that are dilute, acidic, or both.

Many dolostones, especially at greater depths, have higher porosities than limestones, and this may be the result of several processes, i.e. mole-per-mole replacement, dissolution of unreplaced calcite as part of the dolomitization process, dissolution of dolomite due to acidification of the pore waters, fluid mixing (mischungskorrosion), and thermochemical sulphate reduction. There also are several processes that destroy porosity, most commonly dolomite and calcium sulphate cementation. These processes vary in importance from place to place. For this reason, generalizations about the porosity and permeability development of dolostones are difficult, and these parameters have to be investigated on a case-by-case basis.

A wide range of geochemical methods may be used to characterize dolomites and dolostones, and to decipher their origin. The most widely used methods are the analysis and interpretation of stable isotopes (O, C), Sr isotopes, trace elements, and fluid inclusions. Under favourable circumstances some of these parameters can be used to determine the direction of fluid flow during dolomitization.

The extent of recrystallization in dolomites and dolostones is much disputed, yet extremely important for geochemical interpretations. Dolomites that originally form very close to the surface and from evaporitic brines tend to recrystallize with time and during burial. Those dolomites that originally form at several hundred to a few thousand metres depth commonly show little or no evidence of recrystallization.

Traditionally, dolomitization models in near-surface and shallow diagenetic settings are defined and/or based on water chemistry, but on hydrology in burial diagenetic settings. In this paper, however, the various dolomite models are placed into appropriate diagenetic settings.

Penecontemporaneous dolomites form almost syndepositionally as a normal consequence of the geochemical conditions prevailing in the environment of deposition. There are many such settings, and most commonly they form only a few per cent of microcrystalline dolomite(s). Many, if not most, penecontemporaneous dolomites appear to have formed through the mediation of microbes.

Virtually all volumetrically large, replacive dolostone bodies are post-depositional and formed during some degree of burial. The viability of the many models for dolomitization

From: BRAITHWAITE, C. J. R., RIZZI, G. & DARKE, G. (eds) 2004. *The Geometry and Petrogenesis of Dolomite Hydrocarbon Reservoirs*. Geological Society, London, Special Publications, **235**, 7–63. 0305-8719/$15.00

in such settings is variable. Massive dolomitization by freshwater–seawater mixing is a myth. Mixing zones tend to form caves without or, at best, with very small amounts of dolomite. The role of coastal mixing zones with respect to dolomitization may be that of a hydrological pump for seawater dolomitization. Reflux dolomitization, most commonly by mesohaline brines that originated from seawater evaporation, is capable of pervasively dolomitizing entire carbonate platforms. However, the extent of dolomitization varies strongly with the extent and duration of evaporation and flooding, and with the subsurface permeability distribution. Complete dolomitization of carbonate platforms appears possible only under favourable circumstances. Similarly, thermal convection in open half-cells (Kohout convection), most commonly by seawater or slightly modified seawater, can form massive dolostones under favourable circumstances, whereas thermal convection in closed cells cannot. Compaction flow cannot form massive dolostones, unless it is funnelled, which may be more common than generally recognized. Neither topography driven flow nor tectonically induced ('squeegee-type') flow is likely to form massive dolostones, except under unusual circumstances. Hydrothermal dolomitization may occur in a variety of subsurface diagenetic settings, but has been significantly overrated. It commonly forms massive dolostones that are localized around faults, but regional or basin-wide dolomitization is not hydrothermal.

The regionally extensive dolostones of the Bahamas (Cenozoic), western Canada and Ireland (Palaeozoic), and Israel (Mesozoic) probably formed from seawater that was 'pumped' through these sequences by thermal convection, reflux, funnelled compaction, or a combination thereof. For such platform settings flushed with seawater, geochemical data and numerical modelling suggest that most dolomites form(ed) at temperatures around 50–80 °C commensurate with depths of 500 to a maximum of 2000 m. The resulting dolostones can be classified both as seawater dolomites and as burial dolomites. This ambiguity is a consequence of the historical evolution of dolomite research.

More than 200 years ago Déodat de Dolomieu was the first to provide a description of a rock consisting chiefly of the mineral dolomite (Dolomieu 1791). Dolomite, the mineral and the rock, has been found in almost all diagenetic settings, and in rocks that range in age from the Precambrian to the Recent. Yet, despite its widespread occurrence, and after more than 200 years of research, several aspects of the mineral dolomite, especially the origin of massive dolostones, are still much debated, as highlighted by the conference 'The Geometry and Petrogenesis of Dolomite Hydrocarbon Reservoirs', held by the Geological Society in London on 3–4 December 2002. This paper is an outgrowth of the author's presentation and discussions at that conference, with the objective of providing a critical summary of the state of the art of the most important aspects of dolomite, dolostones, and dolomitization. An article with these same objectives was published some 20 years ago under the title 'Chemistry and environments of dolomitization – a reappraisal' (Machel & Mountjoy 1986). This paper was a reappraisal of the state of the art of dolomite research at that time, as represented in the Society of Economic Paleontologists and Mineralogists (SEPM) Special Publication No. 28 (Zenger *et al.* 1980), entitled *Concepts and Models of Dolomitization*. The title used here is a deliberate hybrid of Machel & Mountjoy's (1986) title and that of the SEPM Special Publication, reflecting a reappraisal of dolomite research over the last 20–25 years. The sources underlying the arguments made below are hundreds of articles published mainly over the last 50 years, while some notable sources are much older, as well as studies by the author and his research group, including collaborating researchers.

Scanning the scientific literature published after World War II (WW II), one gets the impression that not much was learnt about dolomite and dolomitization for the first 130–150 years after its discovery. This impression, however, is incorrect. The first comprehensive review paper on dolomite and dolomitization was written by van Tuyl (1914) some 90 years ago. van Tuyl (1914, p. 257) began his long paper with the following sentences: 'Bischof has well said: "No rock has attracted greater attention than dolomite". The problem of the origin of this rock has long occupied the minds of geologists and many theories have been advanced for its formation'. van Tuyl (1914) then documented a large array of studies with knowledge, data, interpretations, theories, and hypotheses that had been published since the discovery of dolomite. van Tuyl's (1914) insights and those of the researchers he cited are positively astounding to the reader even today, considering what is being discussed at present in the dolomite research community. For example, the first viable experimental studies and a whole series of dolomitization models, some of them remarkably similar to modern ones, date from well over 100 years ago. Most of van Tuyl's (1914) paper rings true even today, in the early part of the 21st century, and only very few aspects are outdated. This writer salutes our scientific forefathers for their diligent work and ingenuity, and encourages everyone to read van Tuyl's (1914) most enlightening contribution. Several aspects of his work and the sources cited therein are included below where appropriate. Most of the numerous sources published in the 1800s could not be obtained, however, and reference is made to them as 'cited in van Tuyl (1914)'.

Between van Tuyl (1914) and the end of World War II not much was added to the knowledge on dolomite and dolomitization, probably because dolomite was of no particular significance to civilization, and possibly because of the geopolitical developments during that period. In the 1920s the first hydrocarbon reservoirs in dolomitized carbonates were discovered, most notably the giant PJWDM oil field in the Permian of west Texas (e.g. Major *et al.* 1988) and the Turner Valley gas condensate field in the Mississippian of Alberta, western Canada (Gray 1970; Stenson 1992). Although these discoveries shifted petroleum exploration in the USA and Canada in new directions, throughout the 1920s–1940s most oil companies were primarily interested in finding oil in the easiest and cheapest way possible, for which one needed only a structure map and a drilling rig. Hence, the fact that some hydrocarbon reservoir rocks were dolomitized received little more than passing attention, and academically carbonate geology was focused on palaeontology and biostratigraphy. In those years, to quote James Lee Wilson, dolomite was considered 'a disease' (Lucia pers. comm. 2003). Reservoir work was focused on dolomite porosity and petrophysics.

All this changed in 1947, when the Leduc No. 1 drillhole in Alberta, western Canada, found oil in a dolomitized Devonian reef that had exceptionally high porosities and permeabilities. This discovery ushered in the modern oil and gas era in Canada, and essentially provided the foundation for Canada's petroleum industry (Gray 1970; Stenson 1992). In rapid succession many more oil pools were discovered in dolostone reservoir rocks, not only in Alberta but also in the United States. Coincidentally, it was found that many dolostone reservoirs had higher porosities and permeabilities, and thus had better reservoir properties, than limestone reservoirs. This finding, together with the advent of relatively modern investigative techniques (certain downhole logging tools, X-ray diffractometry, porosimetry, thin-section petrography, isotope geochemistry), led to a dramatic increase in the intensity of dolomite research in the 1950s. Suddenly dozens of case studies were published each year on every aspect of dolomitization imaginable. An SEPM symposium in 1964, subsequently published as SEPM Special Publication No. 13 (edited by Pray & Murray 1965), acted as a further important catalyst for dolomite research. In this volume, dolomitization was covered with three discoveries of penecontemporaneous dolomites (i.e. Illing *et al.*'s study in the Persian Gulf; Shinn *et al.*'s study on Andros Island; and Deffeyes *et al.*'s study on Bonaire). Major compilations of research papers on concepts and models of dolomitization followed as SEPM Special Publication No. 28 (Zenger *et al.* 1980), then SEPM Special Publication No. 43 (Shukla & Baker 1988), and the International Association of Sedimentologists (IAS) Special Publication No. 21 (Purser *et al.* 1994). These books attained a status similar to that of seminal textbooks of their times because they provided comprehensive state-of-the-art views of dolomite research, albeit in somewhat fragmented form, as many authors contributed to each volume. In addition, a few notable individual review articles were published by Morrow (1982*a*, *b*, 1999), Machel & Mountjoy (1986, 1987), Hardie (1987), Last (1990), Budd (1997), Mazzullo (2000), and Warren (2000), who succinctly summarized dolomite research or aspects thereof.

Judging from the literature it is clear that the last 50 years, and especially the last 20–25 years, have resulted in major advances of our understanding of dolomite and dolomitization. Nevertheless, several aspects of the so-called 'dolomite problem' remain unresolved and controversial, and these are one of the reasons for writing this reappraisal. Collectively, four aspects make-up the 'dolomite problem': (a) dolomites occur in many different sedimentary and/or diagenetic settings; (b) in many cases the available data permit more than one genetic interpretation; (c) dolomite is fairly rare in Holocene environments and sediments, yet very abundant in older rocks; and (d) well-ordered, stoichiometric dolomite has never been successfully grown inorganically in laboratory experiments at near-surface conditions of 20–30 °C and 1 atm pressure. The fourth aspect necessitates that geochemical parameters that are needed for back-calculating the composition(s) of the dolomitizing fluid(s), such as the equilibrium oxygen isotope fractionation and trace-element partition coefficients, have to be extrapolated from high-temperature experiments. This renders them notoriously inaccurate and often leads to ambiguous genetic interpretations in studies of dolomitization. This paper addresses all of these aspects to one degree or another.

The most controversial aspects of dolomite research are the various models for dolomitization, and whether or not they can explain the origin of massive dolostones. Many researchers applied these models as soon as they had been published, in spite of the fact that many model(s) had not been sufficiently tested or verified. Thus, almost every new model became a bandwagon until it proved to be insufficient in some way. By the time of the bicentennial conference in honour of Déodat de Dolomieu in Ortisei 1991, a 'seawater model' had emerged that was prominently featured in IAS Special Publication No. 21 (Purser *et al.* 1994) that arose from the conference. At the same time, the gamut of dolomite models had been enlarged to include new variants of the compaction model, dolomitization by tectonically expelled fluids, and various forms of dolomitization by thermal convection, along with other more exotic alternatives. Some of these new(er) models, such as hydrothermal dolomitization (Machel & Lonnee 2002), have since attained bandwagon status, at least in certain parts of the world.

This paper summarizes the major advances and current controversies in dolomite research. Following Machel & Mountjoy (1986), it begins with a brief review of the chemical (thermodynamic and kinetic) conditions that favour dolomitization, including mass-balance considerations for the generation of massive dolostones. Classifications for dolomite textures and pore spaces in dolostones are presented, along with a series of photographs of representative textural types and development. These parts provide the basis for a discussion of porosity evolution during, and as a result of, dolomitization, one of the more controversial aspects of dolomitization, yet of great practical importance to the petroleum industry. The following section provides a brief overview of the geochemical methods used in dolomite studies, emphasizing the role of recrystallization, another controversial subject. A major part of this paper deals with the various dolomitization models, starting with a rigorous definition of the term 'model'. The final section briefly covers secular variations in dolomite abundance.

Basic facts and terminology

Ideal, ordered *dolomite* has a formula of $CaMg(CO_3)_2$ and consists of alternating layers of Ca^{2+}–CO_3^{2-}–Mg^{2+}–CO_3^{2-}–Ca^{2+}, etc., perpendicular to the crystallographic *c*-axis. Most natural dolomite has up to a few per cent Ca-surplus (and a corresponding Mg-deficit), as well as less than ideal ordering. *Protodolomite* has about 55–60% Ca, is poorly ordered, i.e. the alternating cation layer structure is poorly developed, and is common as a metastable precursor of well-ordered, nearly stoichiometric dolomite in both laboratory experiments and in nature (Graf & Goldsmith 1956; Gaines 1977). Good arguments have been made to abandon the term protodolomite (e.g. Land 1980) or to restrict it to laboratory products (Gidman 1978), yet the term is useful to describe metastable precursors of dolomite in nature. The term *dolostone*, introduced by Shrock (1948) and semantically equivalent to limestone, refers to a rock that consists largely (>75%) of the mineral dolomite. The term dolostone has been rejected (e.g. Vatan 1958), mainly because the term *dolomite* has historical priority for the rock (Dolomieu 1791 discussed the rock, not the mineral). However, *dolostone* has gained wide acceptance during the last 20 years, probably because of the confusion arising from the word dolomite referring to both a mineral and a rock. The term *dolomites* is used to refer to types of dolomite that vary in texture, composition, genesis, or a combination thereof.

Two types of *dolomite formation* are common: *dolomitization*, the replacement of $CaCO_3$ by $CaMg(CO_3)_2$; and *dolomite cementation (precipitation)*, the precipitation of dolomite from aqueous solution as a cement in primary or secondary pore spaces. Contrary to common practice, the term dolomitization should not be applied to dolomite cementation. Similarly, the term dolomitization should not be applied to cases where a hot or hydrothermal fluid leads to recrystallization of pre-existing dolomites. Dolomites and dolostones that originate via replacement of $CaCO_3$ are called *replacement dolomites* or *secondary dolomites*, especially in the older literature. A third type, called *primary dolomites*, originates by direct precipitation from aqueous solution to form sedimentary deposits.

Genetically, all natural dolomites can be placed in two major families, *penecontemporaneous* dolomites and *post-depositional* dolomites (*sensu* Budd 1997). Penecontemporaneous dolomites may also be called *syndepositional* dolomites. They form while carbonate sediment or limestone resides in the original environment of deposition as a result of the geochemical conditions that are 'normal' for that environment. Such dolomites are also called *primary* or *early diagenetic*, although these terms are not strictly synonymous with 'penecontemporaneous'. For example, dolomites that formed syndepositionally may well be secondary and formed by replacement of $CaCO_3$. Also, dolomite may form after hundreds of thousands or even millions of years in limestones that have resided in their original environment of deposition throughout this time or that returned to this environment after some burial or exposure. In the latter case the term 'early diagenetic' is not justified. In any event, true penecontemporaneous dolomites appear to be relatively rare, most are of Holocene age, and are restricted to certain evaporitic lagoonal and/or lacustrine settings. It is possible, however, that such dolomites are more common in the geological record than is presently known, but their presence is hard to prove because of diagenetic overprinting.

Post-depositional dolomites may also be called *post-sedimentary*. They form after carbonate sediment has been deposited and removed from the active zone of sedimentation. This may happen through progradation of the sedimentary surface, burial and subsidence, uplift and emergence, eustatic sea-level fluctuations, or any combination of these. Such dolomites and dolostones are commonly called *late diagenetic*, although this term is not synonymous with post-depositional. Carbonate sediment may be rapidly removed from its site of deposition, i.e. within a few hundreds to thousands of years, and dolomite formed during this time would be early diagenetic compared with the truly 'late diagenetic' phases that formed millions to tens of millions of years later. Almost all examples of massive, regionally extensive dolostones are post-depositional, and they are the main topic of this paper.

One aspect that transcends the above genetic grouping is that of hydrology. Whether syndepositional or post-depositional, the formation of large amounts of dolomite requires advection, i.e. fluid flow, because of chemical mass-balance constraints (discussed below). On the other hand, small amounts of dolomite can be formed without advection. In such cases the Mg needed for dolomite formation is locally derived and redistributed, or it is supplied via (slow) diffusion. Examples include dolomite formed from Mg contained in (high-) Mg calcite, adsorbed to the surfaces of minerals, organic substances or biogenic silica, or contained in older primary or secondary dolomites (e.g. Lyons *et al.* 1984; Baker & Burns 1985). Dolomites formed in these ways are mentioned only briefly or in connection with other, volumetrically much more important, types.

Thermodynamic and kinetic constraints

The chemical conditions of dolomite formation have been under investigation ever since the rock and the mineral were defined. van Tuyl (1914, p. 306) elegantly summarized the first 100 years of experimental and circumstantial evidence, and found that 'experiments have failed to indicate the conditions under which dolomite can be precipitated directly at ordinary temperatures and pressures', a statement that is still (almost) true today. However, even back in the 1800s, several researchers had managed to grow dolomite at elevated temperatures, either directly from solution or as replacements of calcium carbonate. Similar experiments were carried out by others, most notably after World War II.

As a result, the thermodynamic conditions of dolomite formation have been known quite well, at least since the 1970s (see summary in Carpenter 1980). The only notable new experimental studies in this context are those of Usdowski (1994), Land (1998), and Arvidson & MacKenzie (1999). Usdowski (1994) ran his experiments for up to 7 years, and Land (1998) ran his for 32 years, the longest run laboratory experiments to form dolomite to date. Land did not manage to grow dolomite in his particular set-up (dilute solution, 25 °C, 1000-fold supersaturation with respect to dolomite), whereas both Usdowski's and Arvidson & MacKenzie's studies essentially confirmed the results of earlier experimental studies. However, the kinetics, i.e. catalysts and inhibitors of dolomite formation, continue to be a source of controversy.

According to present knowledge, dolomite formation is thought to be favoured chemically, that is thermodynamically and/or kinetically, under the following conditions: low Ca^{2+}/Mg^{2+}-ratios; low Ca^{2+}/CO_3^{2-} ratios (i.e. high carbonate alkalinity); high temperatures; salinities substantially lower or higher than that of seawater; and where fluids suddenly release CO_2 (Carpenter 1980; Morrow 1982*a*; Machel & Mountjoy 1986; Leach *et al.* 1991; Usdowski 1994; Arvidson & MacKenzie 1999) (Figs 1 and 2). These constraints translate into four

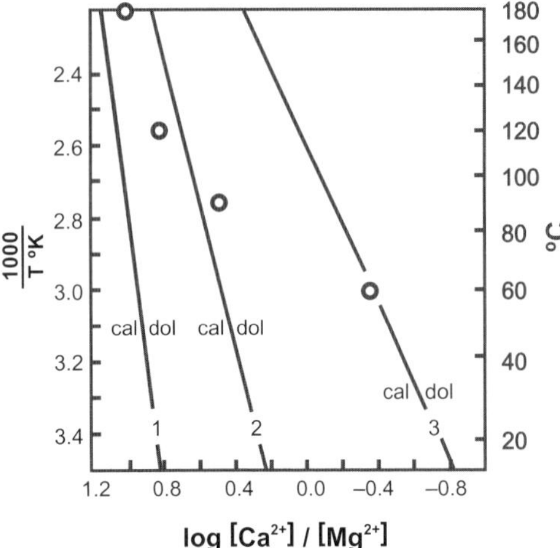

Fig. 1. Bivariate thermodynamic stability diagram for the system calcite–dolomite–water. Square brackets denote activities. Lines are calculated from experimental data: line 1, calcite + ideal, fully ordered dolomite; line 2, calcite + ordered dolomite with slight Ca-surplus; line 3, calcite + fully disordered protodolomite. The four open circles denote the experimental results of Usdowski (1994), whose up to 7 year-long runs represent the lowest-temperature experimental dolomite formation performed to date. Usdowski's (1994) data for 90, 120 and 180 °C plot close to line 2, but his data for 60 °C plot on line 1, which probably reflects the fact that protodolomite rather than dolomite formed at 60 °C. Data from natural aquifers (not shown) cluster close to line 2, which can be considered representative of most natural dolomite. This figure is modified from Carpenter (1980).

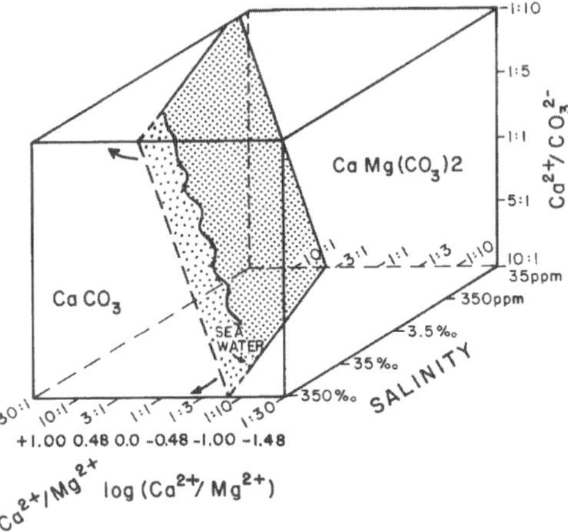

Fig. 2. Trivariate kinetic stability diagram for the system calcite–dolomite–water. The ionic ratios are molar ratios. Seawater plots just in the calcite field. The stippled field boundary is bent towards higher Ca/Mg ratios at salinities greater than 35‰. The figure is reproduced with permission from Machel & Mountjoy (1986).

common conditions to form dolostones in natural settings.

(1) *settings with a sufficient supply of Mg^{2+} and CO_3^{2-}*; this condition favours marine and burial-diagenetic settings with pore fluids of marine parentage because seawater is the only common Mg-rich natural fluid in such settings;
(2) *settings with a long-lasting and efficient delivery system for Mg^{2+} and/or CO_3^{2-}* (and also exporting Ca^{2+} in the case of calcite replacement); this favours settings with an active and long-lasting hydrologic drive;
(3) *carbonate depositional settings* or limestones that can be replaced, i.e. abundant calcium carbonate must be available to be replaced;
(4) *from hydrothermal solutions* that ascend rapidly through fault systems.

Considering that the above chemical constraints allow dolomite formation in almost the entire range of surface and subsurface diagenetic settings, the question arises as to why there are so many undolomitized limestones. The likely conditions for the lack of dolomitization are:

(5) *ion pair formation* (especially hydration), inactivating much of the Mg^{2+} and CO_3^{2-} in solution;
(6) *insufficient flow* because of the lack of a persistent hydraulic head, too small a hydrologic head, or insufficient diffusion, resulting in insufficient magnesium and/or carbonate ion supply;
(7) *the limestones are cemented and not permeable enough*, inhibiting or prohibiting the throughput of Mg-rich waters;
(8) *the diagenetic fluids are incapable of forming dolomite because of kinetic inhibition*, e.g. because the environment is too cold; most kinetic inhibitors of dolomite nucleation and growth are rather potent at temperatures below about 50 °C, and the Ca^{2+}/Mg^{2+} ratio of many relatively cold diagenetic fluids is not low enough for dolomitization;
(9) *the conditions conducive to dolomite formation do not last long enough to overcome the induction period* (discussed below).

The last point leads to the kinetic factors, many of which are relatively poorly understood. Three aspects deserve special mention. First, almost all researchers agree that most kinetic inhibitors that lower the nucleation rate and growth rate of dolomite are especially potent at temperatures below about 50 °C (see summaries in Morrow 1982*a*; Machel & Mountjoy 1986). Hence, dolomite formation is easier at higher temperatures. Secondly, it is also generally acknowledged that dolomite forms via metastable precursors, but the significance of this phenomenon for studies of massive dolomitization is not clear and much debated. Thirdly, the role of sulphate in dolomitization is highly controversial.

Regarding the second point, it has been known for a long time that quartz, metal sulphides and other minerals do not form directly from aqueous solution. Rather, the first phases to form during the overall replacement process are metastable phases with a similar, but not identical, composition and/or ordering to the final replacement product, such as opal-A or mackinawite that transform (often in multiple steps – a process commonly referred to as Ostwald's step rule) to the actual and final replacement products of quartz and pyrite, respectively (e.g. Morse & Casey 1988). It only stands to reason that dolomite could also form via some metastable precursor(s). This was elegantly demonstrated in a series of hydrothermal experiments by Sibley (1990), Nordeng & Sibley (1994) and Sibley *et al.* (1994) (see also Gaines 1974; Katz & Matthews 1977), who showed that dolomite forms after an induction period, during which no detectable products form, via so-called VHMC (very-high-Mg calcite with about 36 mole% Mg), then VHMC plus non-stoichiometric dolomite, then stoichiometric dolomite. The induction period can be very long and is one, and perhaps the best, explanation for the apparent lack of dolomite in recent and geologically relatively young marine carbonate environments (Nordeng & Sibley 2003). The transformations from one metastable phase to another are a form of recrystallization, and they take place very fast, i.e. within hours–days in hydrothermal experiments. By analogy, metastable precursors to dolomite (often referred to as protodolomite) are common in a variety of Holocene sediments but nearly absent in older settings. It appears, therefore, that Nature 'performs' the various steps of recrystallization from the VHMC nucleus to dolomite very fast, i.e. commonly within a few hundreds to a few thousands of years in typical low-temperature diagenetic settings, and obviously even faster in deep-burial and hydrothermal settings. Hence, these transitions are pretty much irrelevant for the investigation of ancient (older Cenozoic, Mesozoic and Palaeozoic) dolomites, except for some exceptional cases where the fluid chemistry has changed dramatically within the short time frame of these transitions. This topic is further discussed in the section on recrystallization below.

The role of sulphate as a potential kinetic inhibitor to dolomitization deserves special mention and is much debated. Following the hydrothermal–experimental study by Baker & Kastner (1981), which suggested that dissolved sulphate inhibits dolomite formation and that lowered sulphate concentrations can enhance the rate of dolomite formation, a number of studies have been published that proposed a positive correlation between (bacterial) sulphate reduction and dolomitization, or they claimed that sulphate reduction is necessary for dolomite formation. However, Morrow & Rickets (1986) and Morrow & Abercrombie (1994) have shown through further experiments and geochemical modelling that the amount of dissolved sulphate has no influence on the rate of dolomitization under relatively low-temperature diagenetic (<80 °C) conditions. On the other hand, they also showed that dissolved sulphate does appear to reduce dolomite formation at relatively high-temperature diagenetic conditions (*c.* 100–200 °C), but only indirectly, because the degree of calcite undersaturation correlates inversely with the sulphate concentration. This leads to higher calcite dissolution rates, and these enhance the rate of dolomite formation when the sulphate concentration is reduced. But even this effect is probably negligible in most natural environments (Morrow & Abercrombie 1994).

Brady *et al.* (1996) went further and suggested, on the basis of field relationships and their own experimental data, that one path for dolomite growth is through the adsorption of Mg-sulphate complexes, which at the very least provides a mechanistic explanation for dolomite formation in sulphate-rich fluids. Thus, where this path is taken, sulphate actually promotes dolomitization. In Brady *et al.*'s (1996, p. 730) words: 'We argue that massive dolostone sequences formed from evaporatively modified seawater due, in part, to the attendant high sulfate levels', and '. . . in a kinetic sense, sulfate does not hinder dolomite growth in evaporitic environments but, rather, accelerates it'. For sulphate-poor solutions, Brady *et al.*'s (1996) results support those of Baker & Kastner (1981).

These considerations suggest that the role of sulphate as an inhibitor for dolomitization has been significantly overrated. On the one hand, there is little doubt that sulphate reduction can significantly enhance or even trigger dolomitization in pelagic environments and in at least some lacustrine settings that are relatively rich in organic matter, as suggested by the formation of dolomite with negative (organogenic) carbon isotope ratios in such settings (e.g. Mazzullo 2000). However, the generalization that low sulphate concentrations and/or bacterial sulphate reduction always enhance or even trigger dolomite formation is unjustified. This notion is supported by three direct or circumstantial lines of evidence: (a) gypsum and anhydrite appear to be common by-products of dolomitization from seawater (discussed below); (b) ancient examples show that dolomitization can happen in or from mesohaline–penesaline seawater, i.e. seawater evaporated between normal salinity and gypsum saturation, and therefore with considerable amounts of dissolved sulphate (e.g. Qing *et al.* 2001; Melim & Scholle 2002); and (c) there are many modern evaporitic environments rich in dissolved sulphate that form dolomite (Friedman 1980; Brady *et al.* 1996).

Mass-balance constraints

Within the chemical constraints outlined in the previous section, the amount of dolomite that can be formed in a given diagenetic setting depends on the stoichiometry of the reaction, the temperature, and the fluid composition (Morrow 1982*a*; Land 1985; Machel & Mountjoy 1986; Machel *et al.* 1996*b*). Dolomitization can be represented by two equations:

$$2CaCO_3\ (s) + Mg^{2+}(aq) \rightarrow CaMg(CO_3)_2\ (s) + Ca^{2+}(aq) \qquad \text{(reaction 1)}$$

(where s = solid and aq = aqueous) or by

$$CaCO_3\ (s) + Mg^{2+}(aq) + CO_3^{2-}(aq) \rightarrow CaMg(CO_3)_2\ (s). \qquad \text{(reaction 2)}$$

Reactions 1 and 2 are end members of a range of possible reaction stoichiometries, *i.e.*

$$(2-x)\ CaCO_3\ (s) + Mg^{2+}(aq) + xCO_3^{2-}(aq) \rightarrow CaMg(CO_3)_2\ (s) + (1-x)\ Ca^{2+}(aq) \qquad \text{(reaction 3)}$$

(Lippman 1973; Morrow 1982*a*; Machel & Mountjoy 1986 – these authors discussed several other mass-balance reactions for dolomitization). Reaction 3 can be used to represent dolomitization in general, as it 'contains' reactions 1 and 2. For $x = 0$, reaction 3 becomes reaction 1, and for $x = 1$ reaction 3 becomes reaction 2. Magnesium has to be imported to the reaction site and calcium has to be exported from it in the case of reaction 1, whereas there is no export of calcium in reaction 2. Intermediate cases of reaction stoichiometry are represented by values of x between 0 and 1.

Dolomite cementation is most simplistically represented by:

$$Ca^{2+}(aq) + Mg^{2+}(aq) + 2CO_3^{2-}(aq) \rightarrow CaMg(CO_3)_2\ (s) \qquad \text{(reaction 4)}$$

If dolomitization proceeds via reaction 1, and if the dolomitizing solution is average (normal) seawater, about 650 m^3 of solution are needed to dolomitize 1 m^3 of limestone with 40% initial porosity at 25 °C (Land 1985). However, dolomitization may not take place with 100% efficiency, and some Mg in excess of that required for saturation is carried away by the dolomitizing solution. In such cases, larger water/rock ratios are needed for complete dolomitization. If seawater is diluted to 10% of its original concentration, as is the case in a typical seawater–freshwater mixing zone, 10 times as much water is needed. By contrast, only about 30 m^3 of halite-saturated brine are needed per m^3 of limestone at 100% dolomitization efficiency. The role of increasing temperature in the underlying thermodynamic calculations is to reduce the amount of magnesium necessary for dolomitization because the equilibrium constant (and hence the equilibrium Ca/Mg-ratio) is temperature-dependent (Fig. 1). For example, at 50 °C only about 450 m^3 of seawater are needed for complete dolomitization of 1 m^3 of limestone with 40% initial porosity at 100% efficiency. The amounts of dilute and hypersaline waters change accordingly.

These calculations have two major implications. First, large water/rock ratios are required for complete dolomitization, whereby the more dilute the solution, the larger the water/rock ratio. This necessitates advection for extensive and pervasive dolomitization, and this is why all models for the genesis of massive dolostones are essentially hydrogeological models. The exceptions are natural environments where carbonate muds or limestones are dolomitized via diffusion of magnesium from seawater rather than by advection. Secondly, variable reaction stoichiometries result in variable porosity development during dolomite formation (discussed below).

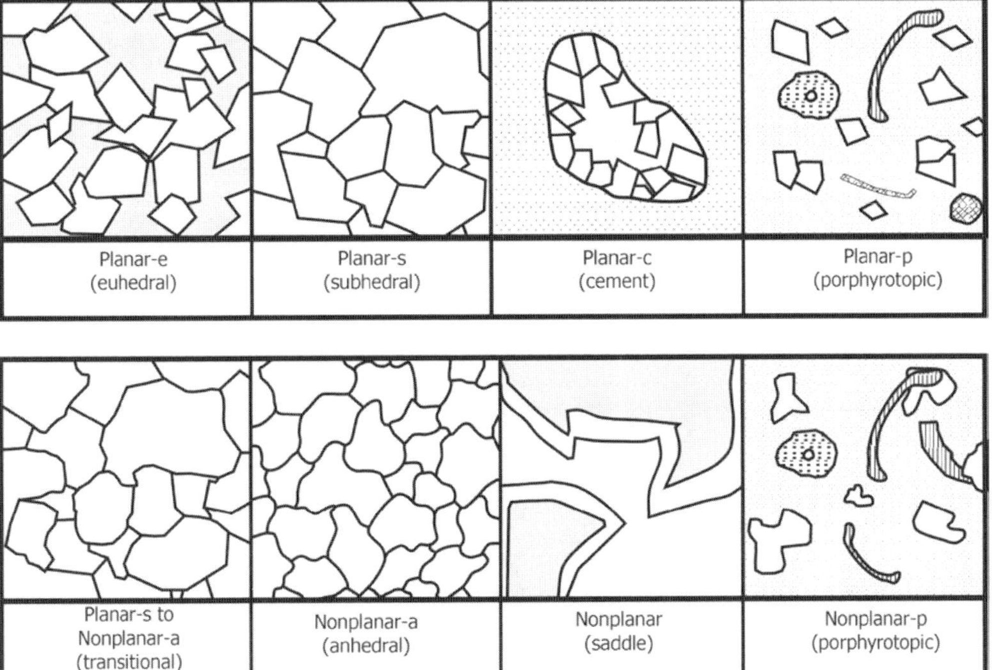

Fig. 3. Dolomite textural classification combined from Gregg & Sibley (1984) and Sibley & Gregg (1997), supplemented by a 'transitional' form. The figure is reproduced with permission from Wright (2001).

Textures

Rock classification

The most widely used classification of dolomite/dolostone textures is that proposed by Sibley & Gregg (1987), based on Gregg & Sibley (1984) (Fig. 3). This classification is popular because it is simple and largely descriptive. However, it carries some genetic implications, and it is restricted to the microscopic scale (in this aspect it differs from the classification(s) for porosity – see below). Crystal size distributions are classified as 'unimodal' or 'polymodal', whereas crystal shapes are classified as 'planar-e' (euhedral), 'planar-s' (subhedral), and 'nonplanar-a' (anhedral). Using this semantic scheme, almost all dolomite texture types can be named, e.g. planar-c (cement), planar-p and nonplanar-p (both porphyrotopic). Saddle dolomite, with its distinctive warped crystal faces, is categorized as nonplanar or nonplanar-c (when it is a cement). Wright (2001) defined one additional texture type, i.e. planar-s to nonplanar-a (transitional), in which planar and nonplanar crystals occur side by side (Fig. 3). A complete textural description includes recognizable allochems or biochems, matrix and void fillings. Particles and cements may be unreplaced, partially replaced or completely replaced. Replacement may be mimetic or non-mimetic, depending mainly on crystal size (Bullen & Sibley 1984; Sibley 2003), and qualifying terms such as 'unimodal, non-mimetic planar-s dolomite' can be added to a rock description.

Many authors use the time-honoured terms euhedral, subhedral and anhedral, as well as the equivalent terms idiomorphic, hypidiomorphic and xenomorphic. While these terms are accurate in principle for characterizing dolomite crystals that display free growth surfaces, they are inappropriate for describing interlocking crystals in thin sections. This is because the identification of crystal–crystal boundaries as growth faces requires use of the universal stage (Sibley 2003).

Pore classification(s)

Pores in dolostones are commonly addressed using the pore classification for limestones proposed by Choquette & Pray (1970). This classification is texturally descriptive, i.e. it discriminates between pore types (such as mouldic, vuggy, shelter, etc.), while it is also

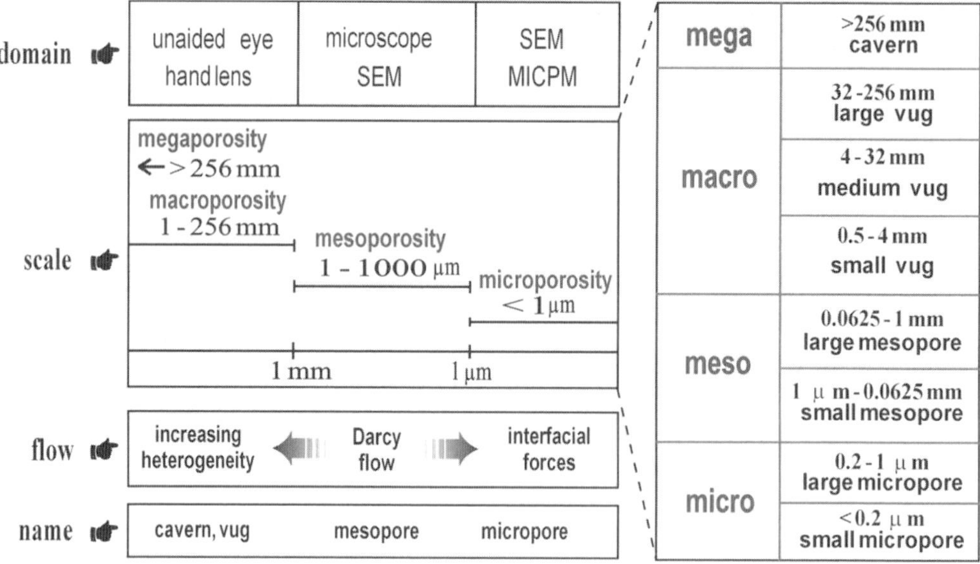

Fig. 4. Pore size classification for carbonates. Measurements under 'scale' refer to pore diameters. The figure is reproduced with permission from Luo & Machel (1995).

genetic (primary v. secondary), yet it is independent of pore size. The petroleum industry, however, is also, and often more so, interested in pore sizes and overall porosity. For this reason, Luo & Machel (1995) established a carbonate pore size classification (Fig. 4) that is applicable to both limestones and dolostones. This classification is based on the textural and petrophysical classifications of Archie (1952), Choquette & Pray (1970), and Pittman (1979, 1992), as well as Luo & Machel's (1995) investigations of dolostone reservoir rocks. The categories contained in this classification range in size/magnitude from the very smallest to the very largest, i.e. from MICPM (mercury injection capillary measurements) and SEM (scanning electron microscopy) to karst caverns, respectively.

At about the same time as the classification scheme by Luo & Machel (1995) appeared in print, Lucia (1995) published his 'geological and petrophysical classification of carbonate interparticle pore space', which resembles Luo & Machel's classification in several aspects. At the time of writing this article, it remains to be seen whether one or the other of these classifications will take hold.

It is important to remember that both Luo & Machel's (1995) and Lucia's (1995) classifications are essentially of pore types, not of porosity, which is the sum of all pore spaces relative to the total rock volume. Both classification schemes contain elements of permeability, which is commonly positively correlated with porosity. These aspects are discussed below.

Textural evolution

The textures and hydrocarbon reservoir characteristics of dolostones are highly variable. On the microscopic scale, unimodal size distribution generally results from a single nucleation event and/or a unimodal primary (pre-dolomite) size distribution of the substrate. Polymodal size distributions indicate multiple nucleation events, differential nucleation on an originally polymodal substrate, or both. Planar crystal boundaries tend to develop up to the so-called 'critical roughening temperature', which appears to be about 50–60 °C for dolomites (Gregg & Sibley 1983, 1984), whereas nonplanar boundaries tend to develop at higher temperatures and/or high degrees of supersaturation. It is not clear, however, whether a critical roughening temperature really exists for dolomites that are replacive (Braithwaite 1991). On a macroscopic scale, there is a distinctive difference in the textures resulting from 'low-temperature' v. 'high-temperature' dolomitization of limestones, i.e. fabric-retentive v. fabric-obliterative. Within this framework, observations from many dolostone occurrences show that dolomitization often proceeds in a certain sequence of steps that correspond to

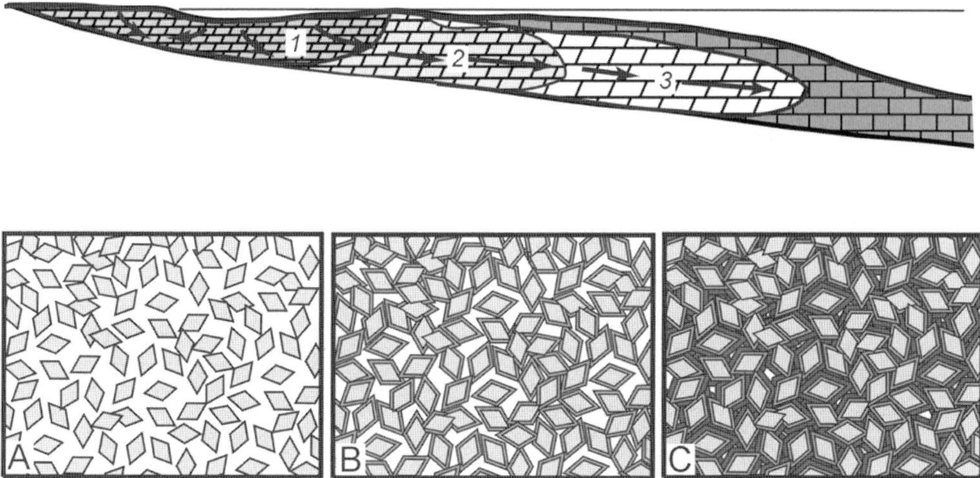

Fig. 5. Schematic model of reflux dolomitization where the dolomitizing brines form on tidal flats and/or in an evaporative lagoon. The textural evolution shown in the bottom row is representative for all situations of matrix-selective dolomitization, not just by evaporative reflux. Numbers denote areas/zones of dolomitization, whereby the locus of dolomitization moves progressively basinward with time. Dolomitization starts in zone 1. Once the dolomitizing fluids have exhausted their Mg, they keep flowing basinward without causing further dolomitization. New increments of the dolomitizing fluids through zone 1 will lead to dolomite cementation, i.e. overgrowths around the earlier formed dolomite crystals, as there is no more calcite to replace, and excess Mg will form new dolomite downflow. Successive time steps in zone 1 are shown in (**A**)–(**C**). Initial replacive dolomitization would form a loose meshwork of dolomite crystals, but porosity would decrease over time via progressive dolomite cementation. This diagram is modified from Saller & Henderson (2001).

specific textural types on the macroscopic scale. Within limits, these steps also reflect particular types of dolomitizing fluids, especially seawater and its derivatives, or meteoric water incursion.

Matrix-selective dolomitization. Most commonly, dolomitization begins as a selective replacement of the matrix, probably as the result of three interacting and reinforcing factors: (a) the matrix contains or consists of thermodynamically metastable carbonates (aragonite and/or high-Mg-calcite), which have higher solubilities than low-Mg-calcite; (b) the matrix has much smaller grain sizes and, thus, a higher surface area per grain than the larger biochems, allochems or cement crystals formed prior to dolomitization; and (c) the matrix has a higher permeability than the larger, more massive particles or cements. The nuclei that are commonly scattered in the matrix become more abundant and grow to form microscopic crystals over time. Soon they form a loose meshwork of crystals that ultimately coalesce to an interconnected mosaic (Fig. 5, bottom row; the textural evolution shown is applicable to replacement dolomitization in general, although this figure was designed to represent replacement dolomitization by evaporative reflux). If the fluid composition does not change significantly over the course of the dolomitization process, the crystals will be unzoned on the scale of thin-section microscopy. The zoning shown in Figure 5B is included to illustrate successive overgrowths on the originally smaller nuclei, and does not mark a change in composition. If dolomitization begins or proceeds below the critical roughening temperature, these crystals tend to be planar-e and/or planar-s.

In the early stages of such matrix-selective dolomitization, large and relatively much less soluble (more stable) biochems and allochems remain unreplaced. This type of texture is shown in Figure 6A. In this particular example, a Devonian reef rock from western Canada, the outer walls and even the most delicate intraskeletal platelets of fasciculate corals are preserved as calcite, despite the fact that these corals are completely uncemented. More commonly, the internal pores of biochems are partially or completely cemented prior to dolomitization, rendering them almost impermeable to the dolomitizing fluids. The corals shown in Figure 6A survived initial dolomitization because of their mineralogical stability

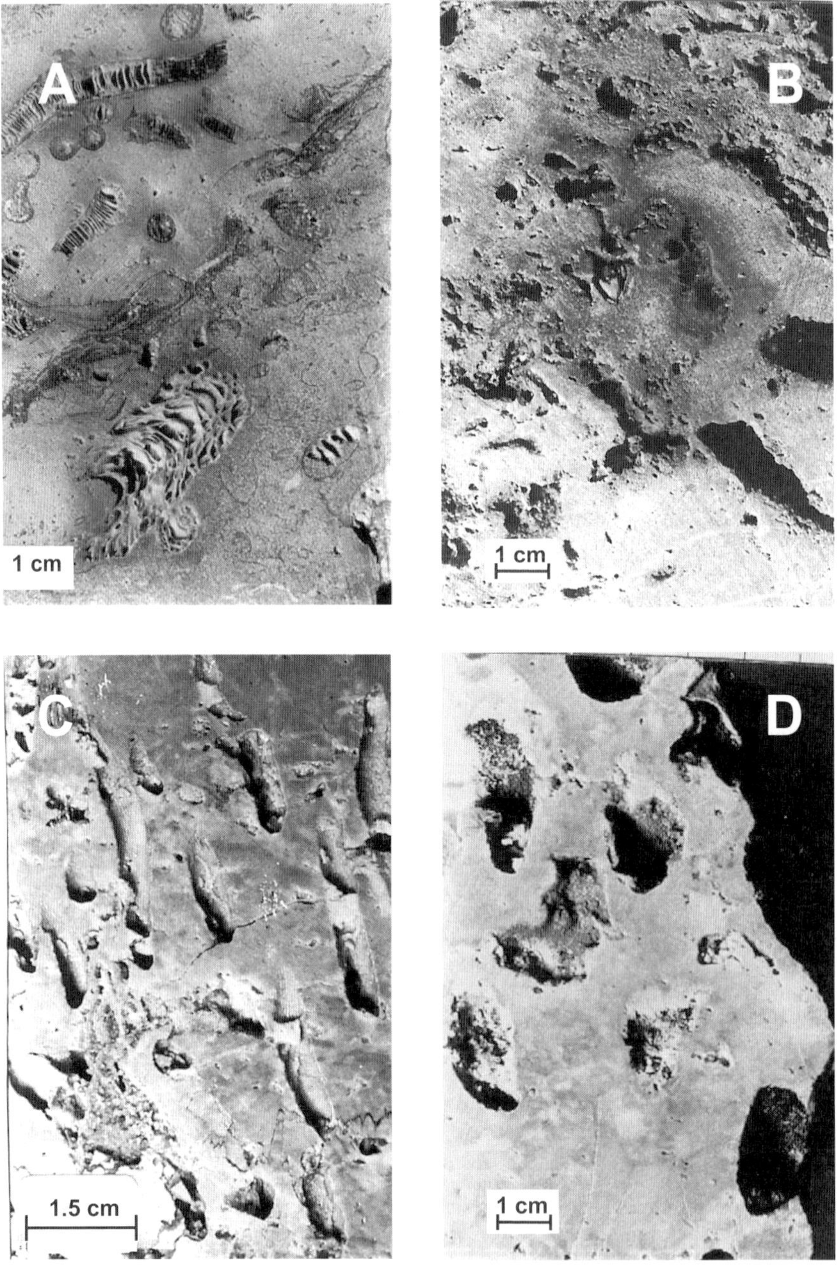

Fig. 6. Successive steps during matrix-selective dolomitization with subsequent dissolution of unreplaced calcite. All samples are from the Upper Devonian Nisku Formation, Alberta, Canada. (**A**) Uncemented *Smithiphyllum* and *Phacelophyllum* with calcite preservation of the delicate chamber walls (trabeculae) in partially dolomitized matrix. (**B**) Vuggy dolostone that resulted from (macro-) dissolution of unreplaced calcite matrix and fossils similar to the sample shown in (A). Connection of pores is intercrystalline-pervasive. (**C**) Coral-mouldic porosity in tight matrix dolomite. The moulds originated from dissolution of corals such as the slender *Smithiphyllum* shown in (A). Connection of macropores is mainly via hairline fractures (centre). (**D**) Coral-mouldic porosity in tight matrix dolomite. The moulds originated from dissolution of corals such as the large *Phacelophyllum* shown in (A). Connection of macropores is mainly via fractures (top right margin of sample).

(low-Mg calcite), not because of low permeability.

In this context the resistance to dissolution of metastable calcium carbonates is governed by an interplay of three factors: permeability and an interplay of microstructure and thermodynamic solubility (Walter 1985). Hence, one should expect certain biochems or allochems to be more susceptible to dissolution and dolomitization than others.

Dolomite cementation ('overdolomitization'). The replacement of lime mud or limestone by dolomite may generate up to 13% interparticle or intercrystal porosity as a function of reaction stoichiometry (discussed below). If so, this newly formed porosity may not survive the continued influx of the dolomitizing solution, which, being supersaturated with respect to dolomite, will tend to form dolomite cement as overgrowths on the earlier formed dolomite crystals, as shown in Figure 5 for reflux dolomitization. Lucia (2002, 2004) calls this process 'overdolomitization'. As pointed out by Saller & Henderson (2001) and Lucia (2002, 2004), this process could severely reduce the porosity and permeability in dolostones, at least in situations of evaporative reflux where the refluxing brines tend to have very high degrees of supersaturation, and where fluxes tend to be high.

Vugs and moulds. Vugs and moulds, the two most common forms of secondary macroporosity, develop without or during dolomitization. The best-known example of the first possibility is meteoric dissolution of biochems or allochems consisting of relatively unstable calcium carbonate, such as aragonite or high-Mg calcite fossils, or grains with a very high surface area (e.g. Tucker & Wright 1990). If a limestone with vugs or moulds is dolomitized, these secondary pores tend to remain open during the dolomitization process.

The advanced stages of dolomitization are characterized by two possibilities: dolomitization of unreplaced biochems and allochems; or their dissolution. The first alternative commonly results in mimetic replacement to some variable degree, depending on dolomite crystal size. Vugs and moulds result from dissolution during advanced stages of dolomitization that appears to be an integral part of the replacement process. This is indicated by numerous cases of limestones that did not contain secondary macroporosity before penetration by dolomitizing fluids, yet where vugs and moulds appeared once the percentage of dolomite replacement exceeded about 70–80 vol%. Examples are shown in Figure 6B–D. Where the matrix was not entirely dolomitized, dissolution removed both the remnants of unreplaced matrix and larger unreplaced particles, generating highly permeable dolostones with vuggy porosity (Fig. 6B). Where matrix dolomitization had gone to completion, only the larger allochems or biochems are removed, as they are the only undolomitized particles left, leaving moulds in a fairly tight matrix. This is shown in Figure 6C & D, with moulds of elongate and more equant corals, such as those in Figure 6A, connected only via fractures in a relatively tight matrix. By comparison, the rock shown in Figure 6B is a much better reservoir rock than those in Figure 6C & D, mainly because the larger voids are connected through pervasively distributed intercrystal pores. The development of mouldic porosity can be spatially highly restricted. This was recognized long ago in various Palaeozoic sequences of the United Stated (e.g. Landes 1946 and references therein), then also in much younger carbonates, including the sub-recent (less than 2200 years old) reflux dolomites from Bonaire, where shells and pellets are dissolved in dolomite crusts but not in the adjacent limestone layers (Deffeyes *et al.* 1965).

Vuggy dolostones are commonly interpreted to result from dissolution that took place in a completely matrix-dolomitized rock, whereby dissolution started in the unreplaced larger allochems and biochems and then proceeded beyond the margins of the moulds into the already dolomitized matrix. This interpretation, however, not only requires that the solution had stopped 'making' dolomite, but also two unlikely circumstances, namely that the solution had attained undersaturation with respect to dolomite, and that the principle of Occam's Razor was violated. The thermodynamically and kinetically much more likely, and thus more plausible, explanation of vuggy dolostones is that the matrix was incompletely replaced before the vugs formed.

The cause of the development of vuggy and mouldic porosity is probably an interplay of fluid supply, composition, and reaction kinetics. One possibility is that the dolomitizing flow system continues to supply fluid, yet the Mg available for dolomitization (in excess of saturation) is used up while undersaturation with respect to calcium carbonate is maintained or acquired, as suggested by geochemical modelling (Sun 1992; Morrow 2001). Another possibility is that the fluid, while approaching dolomite–calcite–water three-phase equilibrium through continued Mg-loss and/or Ca-gain, passes a kinetic threshold below which dolomite formation is severely limited or inhibited while calcite dissolution remains relatively rapid. Thermodynamic and kinetic experiments and modelling are needed to evaluate these alternatives.

Calcium sulphate. The next step in textural evolution during dolomitization is the emplacement of calcium sulphate, either as gypsum or as anhydrite. This step is typical for, and largely restricted to, cases where the dolomitizing solution is seawater or a derivative of seawater (evaporated or slightly changed by water–rock interaction). This reflects the fact that seawater and its derivatives are relatively rich in dissolved sulphate and, as Butler (1970) and others have pointed out, calcium sulphate is a common by-product of dolomitization by such fluids. Curiously, most geochemical modelling of dolomitization has ignored the formation of calcium sulphate during seawater dolomitization, with the notable exceptions of Wilson *et al.* (2001) and Whitaker *et al.* (2002, 2004). The textural development and attendant geochemical (isotopic and trace-element) characteristics of dolomitization with concomitant calcium sulphate formation are extensively discussed in Machel (1985, 1986) yet largely overlooked.

Reefs and platform carbonates of the Upper Devonian Nisku Formation in the West Pembina area of western Canada, presently at depths of about 2600–4600 m, were dolomitized by chemically slightly modified seawater (Anderson 1985; Machel 1985; Machel & Anderson 1989). Anhydrite is absent in undolomitized and in little to moderately (up to about 70 vol%) dolomitized parts of the Nisku Formation in the up-dip part of the reef trend. Down-dip, however, where the Nisku is largely to completely dolomitized, anhydrite is abundant and comprises up to about 20 vol% over any 20–40 m-core interval.

Three questions arise. (1) Did the anhydrite replace calcite or dolomite? (2) In the case of calcite replacement, did anhydrite form before or after the dolomite? Lastly, (3) did the anhydrite form as anhydrite or originally as gypsum? All of these questions can be answered using petrography.

In hand specimen, the most conspicuous features are that most of the anhydrite (more than 90 vol%) appears both as a cement in mouldic pores and also as a partial replacement of the matrix dolomite (Fig. 7). It is clear that these textures did not originate from calcium sulphate replacing calcite because of the combined occurrence of the following features: (a) the moulds originated from the removal of calcite (Fig. 6C & D); (b) the dissimilarity in the appearance of vuggy pores and the replacive anhydrite (compare Figs 6B and 7C: the anhydrite forms much more massive areas/volumes than the calcite that had survived dolomitization and that is now represented by the vugs); (c) the common occurrence of 'islands' of dolomite floating in anhydrite, both on the macroscopic (Fig. 7C) and on the microscopic scale (Fig. 7B); (d) marginal, partial replacement of individual dolomite crystals; and (e) the complete absence of calcite remnants (small, 'undigested' calcite islands or crystals are common when anhydrite replaces calcite).

The microscopic textures further reveal whether the original calcium sulphate mineral was gypsum or anhydrite. Specifically: (a) the textures shown in Figure 7 are absent from primary anhydrites, yet (b) similar textures occur in salt dome cap rocks that have undergone repeated gypsum–anhydrite transformations (Goldman 1952); (c) the presence of large porphyroblastic/porphyrotopic anhydrite crystals floating in a relatively fine-crystalline, partially felted, anhydrite matrix that appears to marginally corrode some of the large crystals (Fig. 7D & E); (d) large porphyroblasts impinge upon one another, generating stress twins and crystal breakage (Fig. 7E). These features indicate that the calcium sulphate was originally emplaced as gypsum that dewatered to anhydrite during further burial, thereby generating the 'corroblastic' or 'corrotopic' textures shown in Figure 7D & E. Apparently the porphyroblasts grew floating in a dewatering gypsum mush until they ran out of room, at which point the remainder of the gypsum mush converted to the finer-crystalline, partially felted anhydrite matrix. Machel (1985, 1986) called this type of anhydrite 'corroblastic' or 'corrotopic' because, in thin section, the most striking features are the corroded porphyroblasts/porphyrotopes.

These observations suggest that anhydritization took place during progressive burial in an interval that overlaps with the gypsum–anhydrite transformation depth. This depth depends on temperature, pressure, and the composition of the fluid (Cruft & Chao 1970), and is normally of the order of 600–800 m, but somewhat greater in overpressured regions. This depth range coincides with the later stages of matrix dolomitization in the Nisku Formation, the total range of which is interpreted to be about 300–1000 m (Machel & Anderson 1989). There also are minor amounts (commonly less than 10 vol%) of anhydrite cement and replacive anhydrite with 'pile-of brick' textures (Carozzi 1960) that appear to be

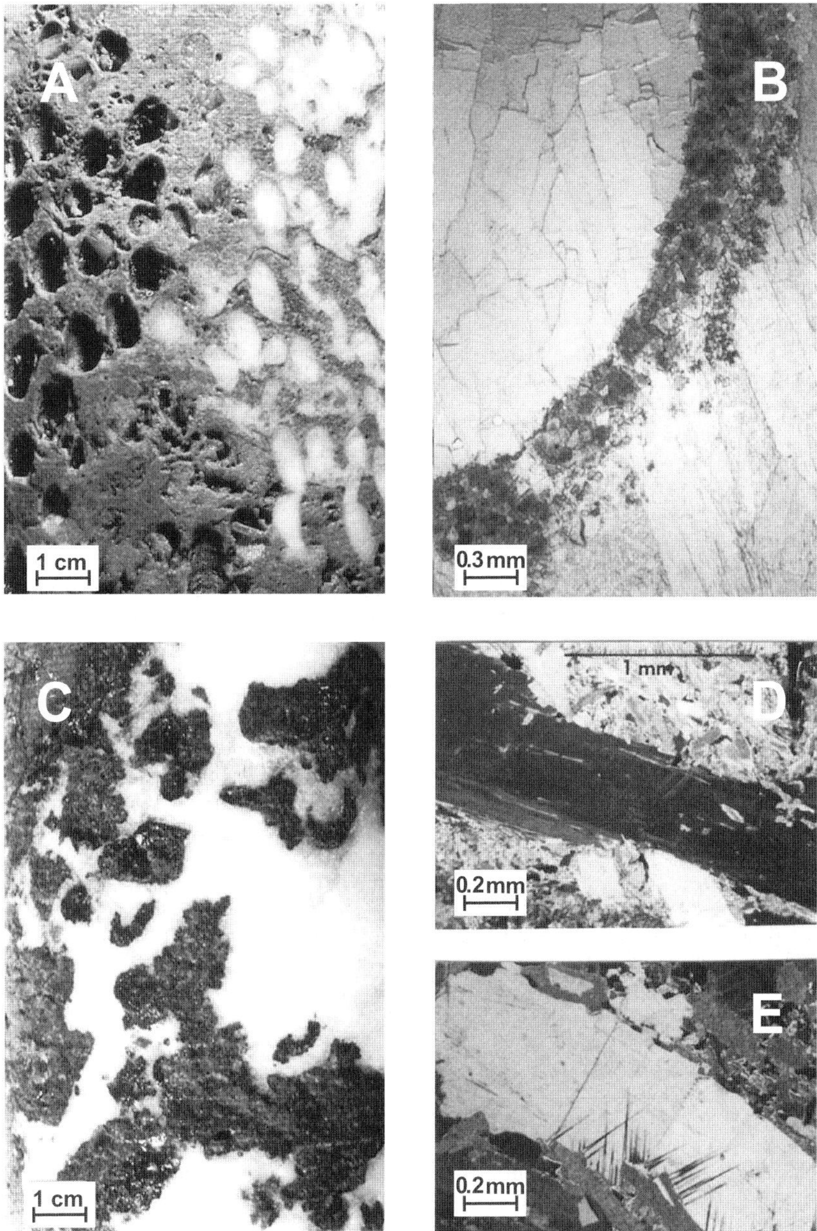

Fig. 7. Calcium sulphate formed as a by-product of dolomitization from (chemically slightly modified) seawater. All samples are from the Upper Devonian Nisku Formation, Alberta, Canada, from depths of 3300–4300 m. (**A**) White anhydrite partially as a replacement (top right) and partially as a cement in coral-mouldic porosity (centre and bottom right). (**B**) Thin-section photomicrograph of top right of sample shown in (A). The interior of the coral contains sparry anhydrite cement, while outside of the coral the anhydrite is replacive/corrotopic. (**C**) White anhydrite that is replacive after dolomite, as it contains 'islands' of undigested dolostone. (**D**) and (**E**) Thin-section photomicrographs, crossed polars, of anhydrite shown in (**A**)–(**C**). Note corroblastic/corrotopic fabrics, i.e. the anhydrite consists of large porphyroblastic/porphyrotopic crystals that float in a medium- to fine-crystalline 'felted' anhydrite matrix. Some of the large crystals appear corroded (top of D), giving the name to this type of anhydrite (Machel 1985, 1986). The large crystal in (E) has strain twins that originate from the point near the bottom where another large crystal impinges.

'primary', i.e. they formed without a gypsum precursor. In combination, these observations indicate that anhydritization started at depths of about 600–800 m (perhaps up to 1000 m if the strata were somewhat overpressured) during advanced phases of dolomitization. During progressive burial, the gypsum transformed to corroblastic/corrotopic anhydrite, and some anhydrite formed directly as anhydrite during the deepest, most advanced, phases of anhydritization.

The emplacement of gypsum–anhydrite during the latter stages of dolomitization from seawater or chemically altered seawater is probably a general phenomenon. It is widespread in the Devonian of western Canada (Mountjoy *et al.* 1999). Conversely, neither gypsum nor anhydrite emplacement should be expected from relatively sulphate-poor dolomitizing solutions, such as in dilute mixing zones, hydrothermal brines, etc.

Two dolomite populations. A common phenomenon in massive dolostones is the occurrence of two crystal populations with different sizes and shapes, and with differing pore types and degrees of pore interconnection. A typical example is shown in Figure 8 in various magnifications. Two aspects of these textures are especially noteworthy. First, crystals of the smaller sized population are commonly 'cloudy' with or without clear rims (overgrowths), whereby the rims appear to be similar to the larger sized population. Secondly, the domains with the coarser crystal size population have a much higher intercrystal porosity and permeability than those with the finer crystal size. In such rocks most petroleum is stored in and flows through the coarser domains, which have a much higher permeability.

These types of textures are genetically ambiguous, whereby the crystals with cloudy centres and clear rims are most easily understood. Initially dolomitization most probably forms short-lived, metastable phases, and large(r) rhombs overgrow the initial phases. This has been shown in hydrothermal experiments, in which the very irregular interiors of larger crystals (equivalent to the cloudy interiors of natural dolomites) apparently underwent intracrystalline recrystallization during or after formation of the stable overgrowths (e.g. Sibley 1990).

The two size populations may result from one dolomitization event or from dolomite recrystallization. In the first alternative, the two dolomite types may reflect textural differences in the precursor limestone(s). One example would be a lime mud or a limestone that consisted of irregularly shaped domains of two populations of matrix, such as lime mud and lime silt or sand. Another example would be a more or less homogeneous lime sediment or limestone that had undergone heterogenous lithification before dolomitization, such that some patches were better cemented or recrystallized than others. In both cases dolomitization could result in the textures shown in Figure 8. The geochemical compositions of both textural types of dolomite would be identical or nearly so, possibly with very small differences inherited from the precursor substrate(s), if the geochemical system is not entirely water-dominated. An example of this type of situation is the Devonian Nisku Formation in central Alberta, Canada, where two populations of matrix dolomite are geochemically indistinguishable within the margin of analytical error (Machel & Anderson 1989).

The second alternative, recrystallization, would be expected where dolomites formed very near the surface and/or in evaporitic environments. Such dolomites commonly form as metastable 'protodolomites' that are prone to recrystallization during burial (discussed below). It is conceivable that recrystallization proceeds in a spatially heterogeneous manner, or at different times within a given volume of rock, or both, governed by heterogeneities in permeability, mineralogy, and corresponding reaction kinetics. As a result, some rock domains may recrystallize to a coarser crystal mosaic than others, possibly including the development of zonation in one domain but not in another. In such cases, there may be a marked difference in the geochemical compositions of the two dolomite populations. An example of this type of situation is the reflux dolomites in the Upper Devonian Grosmont Formation in eastern Alberta, Canada (Huebscher 1996; Machel & Huebscher 2000).

Dolomite dissolution. Textures that resemble those shown in Figure 8 in hand specimen may also result from dolomite dissolution. Great care must be taken to differentiate this alternative from the two discussed in the previous section.

Figure 9 shows samples from the karst-modified part of the Upper Devonian Grosmont Formation in eastern Alberta, Canada. An undolomitized sample (Fig. 9A) would become a dolostone with mouldic porosity (Figure 9B) if matrix-selective dolomitization was accompanied or followed by the dissolution of biochems that had survived replacement. Closer

Fig. 8. Dolostones consisting of domains of relatively tight, light- to medium-grey dolomite intergrown with domains of highly porous, brownish dolomite. The porous domains originated either from replacement of coarser matrix, or from recrystallization of the tight dolomite type. All samples are from the Upper Devonian Nisku Formation, Alberta, Canada. (**A**) Hand specimen. (**B**) Thin-section photomicrograph of sample shown in (A), from the boundary region between the two dolomite types. (**C**) SEM of tight dolomite domain (left in B). Most crystals are planar-s. The pore throats are rather tight yet lined with small 'roundish' calcite crystals that look like rubble in morphological depressions. Permeability is through these parts of the rock, i.e. along crystal boundaries. (**D**) SEM of porous dolomite domain (right in B). The crystals are also planar-s but much larger than those in (C), and there is significant intercrystalline porosity and permeability.

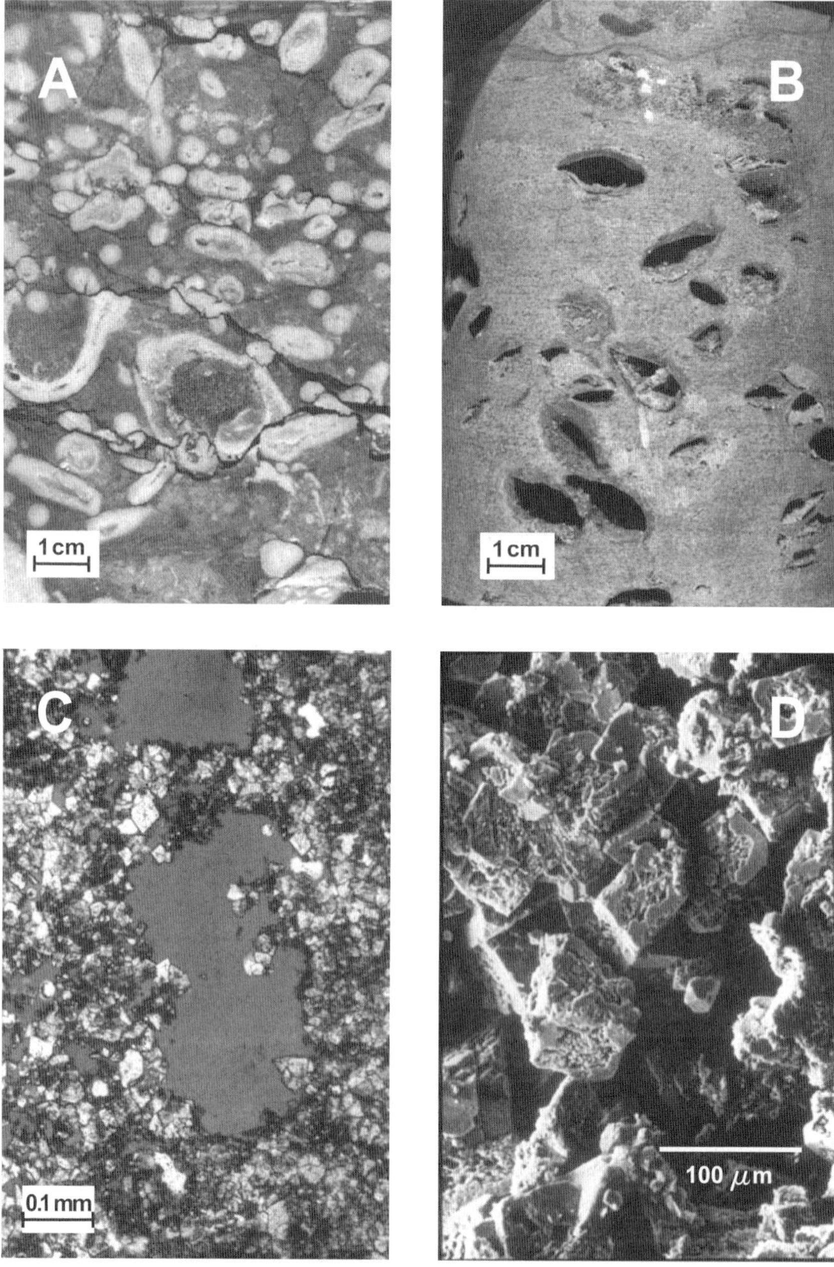

Fig. 9. Development of matrix-selective dolomitization, subsequent dissolution of unreplaced biochems (mainly brachiopods), marginal recrystallization around moulds, and subsequent partial dissolution of dolomite. All samples are from the Upper Devonian Grosmont Formation, Alberta, Canada. (**A**) Hand specimen of (rare) limestone with partial matrix dolomitization. (**B**) Hand specimen of rock type shown in (A) after complete matrix dolomitization, development of mouldic porosity from dissolution of biochems (mainly brachiopods), as well as enhanced porosity around moulds (dark fringes). The domains with enhanced porosity originated from a combination of dolomite dissolution and recrystallization. (**C**) Thin-section photomicrograph of sample shown in (B), illustrating oversized dissolution pores from dark, porous fringes around moulds. (**D**) SEM photomicrograph of sample shown in (B), illustrating intra-crystal dissolution porosity from dark, porous fringes around moulds.

inspection shows that the matrix immediately surrounding the moulds is more coarsely crystalline, as well as more porous and more permeable, than the bulk of the rock. Similar textural differences occur in larger patches in other core intervals, resembling those shown in Figure 8A. However, thin section and SEM images of the Grosmont samples reveal the presence of oversized pores (Fig. 9C) and abundant dissolution pits in the dolomite crystals (Fig. 9D), which indicate that these textures originated from dolomite dissolution. In the Grosmont Formation dissolution was facilitated by meteoric water (Machel & Huebscher 2000), but other mechanisms are possible, as discussed below.

Saddle dolomite. Saddle dolomite, also called baroque dolomite, pearl spar and other names, is a distinctive type of dolomite. Its crystallographic, geochemical and paragenetic characteristics suggest a special type of genesis with respect to the crystal growth mechanism and diagenetic setting(s) (Radke & Mathis 1980; Machel 1987; Searl 1989; Kostecka 1995; Spötl & Pitman 1998).

Saddle dolomite is almost invariably coarse crystalline and milky-white or pink in hand specimens, with a pearly lustre and a distinctively distorted crystal structure that is macroscopically expressed as warped crystal faces and cleavage planes, and microscopically as sweeping extinction. The crystal faces, although well developed, are often faceted like a pavement. Fluid-inclusion homogenization temperatures commonly range between about 80 and 150 °C, in some places up to about 300 °C. There are no confirmed cases of saddle dolomite formation below 60–80 °C; hence, this temperature may be taken as a minimum for the formation of this phase (Spötl & Pitman 1998).

The association of saddle dolomite is also distinctive. It commonly occurs as a gangue mineral in Mississippi Valley-type (MVT) metal sulphide deposits, including features characteristic of thermochemical sulphate reduction (solid bitumen, elemental sulphur, depleted carbon isotope ratios, etc.: Machel *et al.* 1995; Machel 2001). Saddle dolomite is, however, also common as cement in dolostones without any association to MVT-sulphides or hydrocarbons.

Figure 10 illustrates typical occurrences and textural associations of saddle dolomite. As shown in Figure 10A, matrix-selective dolomitization generated grey medium-crystalline dolomite, and later calcite dissolution created elongate moulds after fasciculate corals (compare to Fig. 6C). These moulds were probably filled during stylolitization of the matrix dolomite with milky-white saddle dolomite cement derived from matrix dolomite dissolved along the stylolite(s). Similar moulds are empty or open about 50 cm above and below this core sample, indicating that the material for saddle dolomite was derived locally rather than by advection, a relatively common process in deeply buried dolostones. In Figure 10B, most of the saddle dolomite is replacive (nonplanar-a, using the classification shown in Fig. 3), but some of it is cement and intergrown with galena. Prior to saddle dolomite formation the host rock was medium-grey matrix dolomite, some of which is still visible as dark bands. In some locations, such bands alternate with bands of white saddle dolomite, a texture commonly referred to as 'zebra dolomite'. The origin of zebra dolomite is much debated and probably involves more than one mechanism, including repeated fracturing or some type of geochemical self-organization (e.g. Krug *et al.* 1996). Figure 10C & D illustrate the common association of solid bitumen with saddle dolomite. In these samples, the saddle dolomite is the latest paragenetic phase, and the crystals grew into large voids that were already partially coated with oil. The crystals nucleated on small dolomite crystals of the wall rock that protruded through the oil coating, which may or may not have been solidified at the time of dolomite formation.

The peculiar crystal structure and features associated with saddle dolomite require special conditions during crystal growth. The crystallographic and geochemical characteristics indicate that the crystals most probably grow very fast from highly supersaturated solutions, together with or under conditions where surface-related activation energy barriers are much reduced (Searl 1989). The former can be expected in hydrothermal solutions that cool or depressurize and de-gas rapidly. The reason for the occurrence of the latter condition is not obvious. Considering, however, that saddle dolomite usually forms at temperatures in excess of about 60–80 °C, it seems logical to assume that elevated temperatures are a key factor in reducing the surface energy barriers. On a macroscopic scale, studies have shown that saddle dolomite can be formed as a cement or as a replacement in at least three ways: from advection (commonly, but not necessarily, by hydrothermal fluids), from local redistribution of older dolomite during stylolitization, and as a by-product of thermochemical sulphate reduction in a closed or semi-closed system (Radke & Mathis 1980; Machel 1987; Machel & Lonnee 2002).

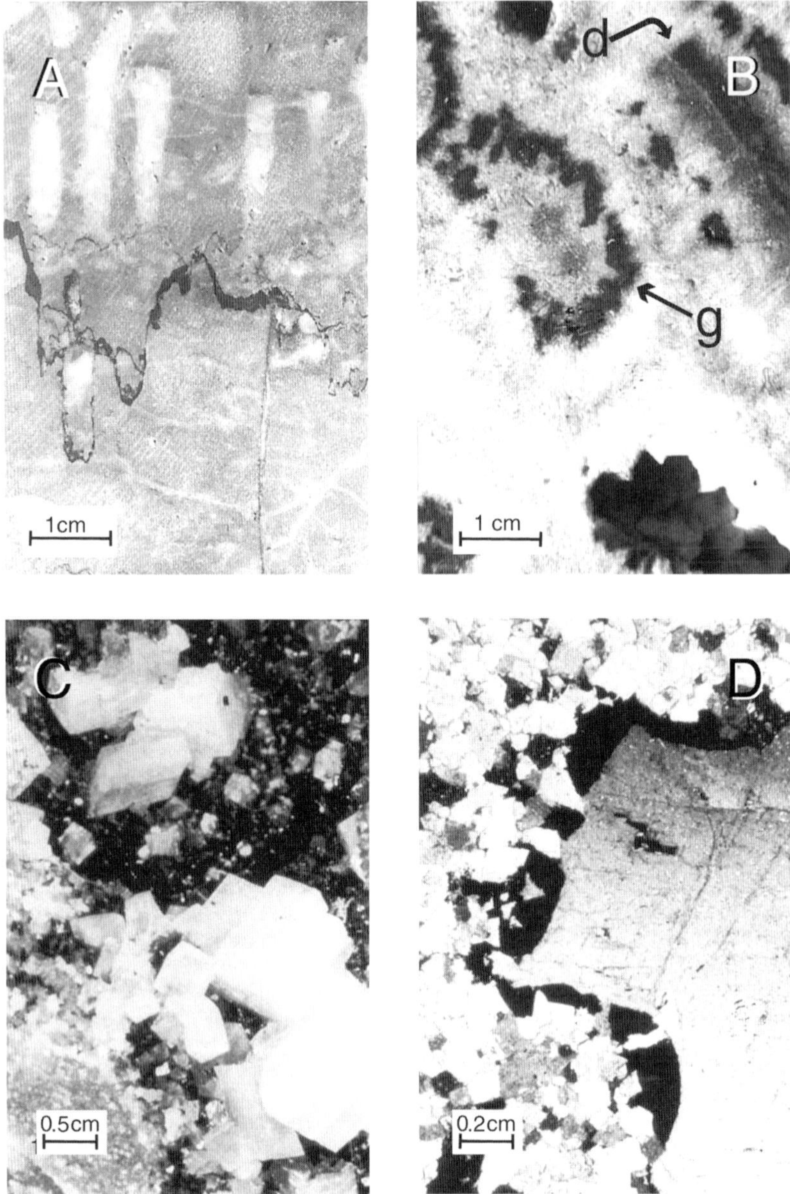

Fig. 10. Saddle dolomite. All samples are from western Canada. Sample (A) is from the Upper Devonian Nisku Formation; (B) is Middle Devonian Presqu'ile dolomite from the MVT mine site at Pine Point; (C) and (D) are from the Upper Devonian Leduc Formation. (**A**) Hand specimen of stylolitized dolostone with milky-white saddle dolomite cement in coral moulds (compare with Fig. 6C). (**B**) Hand specimen of milky-white, coarse-crystalline saddle dolomite intergrown with bands of galena (g). Typical MVT-paragenesis. Most of the saddle dolomite replaced grey matrix dolomite, which is preserved in some thin bands (d). Saddle dolomite forms subhedral cement fringes around some large vugs (lower right). (**C**) Core specimen of milky-white saddle dolomite cement in vug that is coated with solid bitumen ('dead oil'). Host rock is grey matrix dolomite. Note than the saddle dolomite postdates the emplacement of the bitumen. (**D**) Thin-section photomicrograph, transmitted light with crossed polarizers, of sample shown in (C). Most bitumen displays round convex surfaces toward the centre of the pore, mimicking formerly liquid oil droplets that clung to the margins of this pore. This bitumen may have formed as a byproduct of thermal cracking or from thermochemical sulphate reduction. Saddle dolomite appears as large crystals in the centre and lower right, with undulouse extinction.

'Low-temperature' v. 'high-temperature' dolomitization. Outcrop evidence shows that there may be a distinct difference in the textures resulting from 'low-temperature' v. 'high-temperature' dolomitization of limestones. Empirical evidence suggests that the range of 50–80 °C marks the approximate boundary between these two temperature realms, but this aspect remains to be investigated.

In low-temperature settings dolomitization is commonly matrix-selective and at least partially fabric-retentive, as discussed earlier, whereas in high-temperature settings it tends to be fabric-destructive. Figure 11 shows examples from outcrops of Carboniferous carbonates in Cantabria, Spain, where high-temperature solutions dolomitized shallow-marine limestones that constitute the wall rocks around the dolomitized domains. Based on uncorrected fluid inclusion homogenization temperatures, these dolomites formed at a minimum of 130–140 °C, i.e. they are high-temperature dolomites. Furthermore, they are hydrothermal because they formed at temperatures significantly higher than the wall rocks that, based on Conodont Alteration Index (CAI) values, experienced maximum burial temperatures of about 70–95 °C (Gasparrini 2003). Hence, the temperature differential between the wall rock and the intruding hydrothermal solutions was 35–70 °C. The limestone–dolostone boundaries are sharp, and sedimentary, as well as diagenetic, features of the limestone are largely or completely obliterated in the dolomitized portions of the outcrops. Similar textures are common in Devonian reef carbonates of Germany (Machel 1990; Grobe & Machel 1996, 1997; Grobe 1999), and in Cambrian carbonates of the Rocky Mountains, Canada (Moore 1994; Yao & Demicco 1995; Spencer & Hutcheon 1999) where high-temperature, hydrothermal solutions dolomitized limestones.

Figure 12 is a schematic comparison of low-temperature and high-temperature dolomitization. In low-temperature settings dolomitization proceeds essentially as documented in the previous sections and in Figures 6–9 (bottom part of Fig. 12). In high-temperature settings it tends to proceed via a sharp, straight–irregular front behind which sedimentary and diagenetic textures are obliterated (top part of Fig. 12). However, there are counter-examples. Dolomitization of fossils in hydrothermal bombs produced fabric-retentive dolomite (Bullen & Sibley 1984), whereas low-temperature dolomitization in Eniwetok atoll is fabric-destructive, at least in part (Saller 1984). In some locations, such as western Canada, hydrothermal fabric-obliterative dolomites grade laterally into fabric-retentive dolomites, suggesting that both types of replacement were generated by a single fluid pulse, and that the textural gradient reflects a decrease in temperature, temperature differential, and supersaturation away from the faults from which the fluids emanated (Spencer & Hutcheon 1999). It seems plausible that fabric-obliterative replacement may be favoured by an overall high temperature (wall rock plus intruding dolomitizing solution); by a high temperature differential between the wall rock and an intruding dolomitizing solution (which would be hydrothermal); by high supersaturation of the intruding solution with respect to dolomite; or by a combination of these factors. A delicate interaction may be at play.

Porosity and permeability

It has long been claimed that most dolostones are more porous and more permeable than limestones (e.g. Blatt *et al.* 1972), a circumstance of obvious importance for the petroleum industry. A related aspect is that dolostones commonly form aquifers and preferential migration pathways for hydrocarbons, or both. A striking example is the dolomitized margin of the Devonian Cooking Lake platform in Alberta, Canada, that has acted as a water and hydrocarbon migration pathway over a distance of several hundred kilometres (e.g. Amthor *et al.* 1993, 1994).

The claim that most dolostones are more porous and permeable than limestones is contentious. van Tuyl (1914, p. 259) stated that 'Some dolomites are very compact, but most of them are vesicular and porous', a view held for many years. Schmoker & Halley (1982) and Halley & Schmoker (1983), however, demonstrated with porosity–depth profiles of Cenozoic carbonates in southern Florida that many dolostones that have not been buried too deeply (less than about 1 km) have porosities equal to or less than those of adjacent limestones. Budd (2001) showed that the same is true for permeability in these particular rocks, where many dolostones have permeabilities equal to or less than those of adjacent limestones. On a larger scale, Schmoker *et al.* (1985) compared thousands of limestones and dolostones from across the USA and found that dolostone reservoirs commonly have lower matrix porosities and permeabilities, yet higher fracture porosities and permeabilities, than limestones. However, Amthor *et al.* (1994) in a study of 31 wells from Devonian reservoirs in Alberta that span a depth range of several thousand metres, found

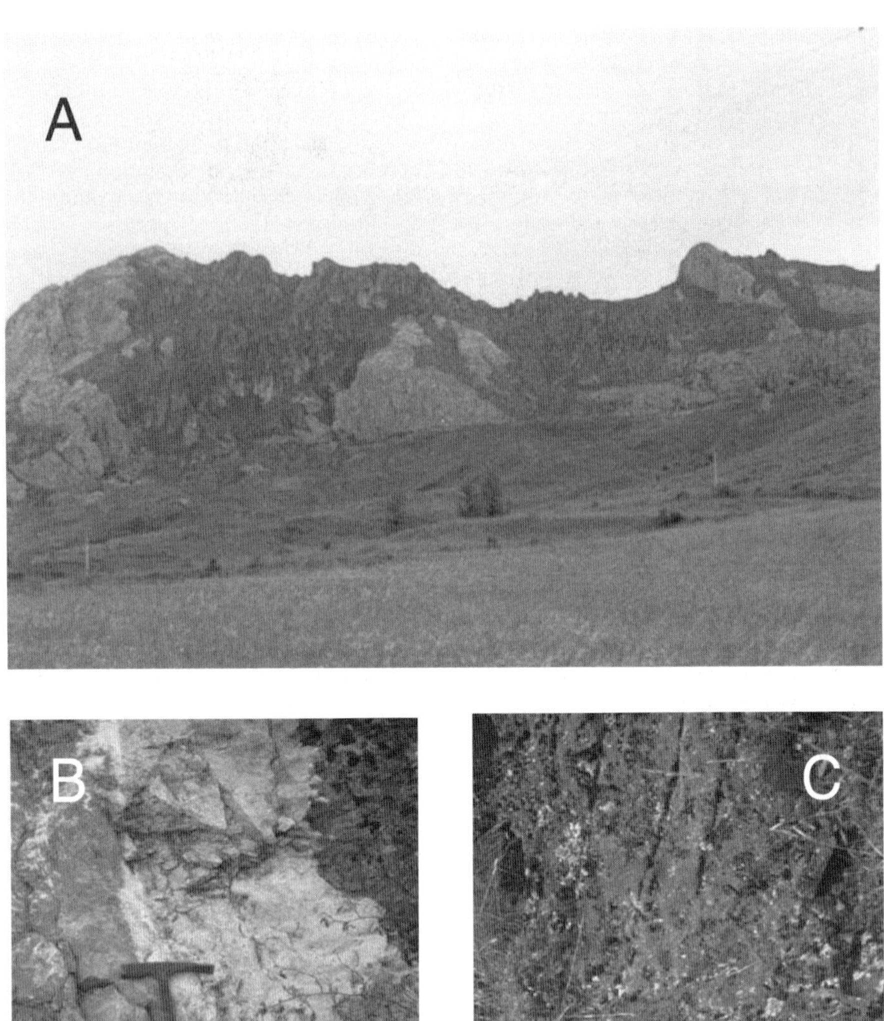

Fig. 11. Outcrop photographs of Upper Carboniferous carbonates from the SW Cantabrian zone, Spain: hydrothermal and high-temperature dolomitization of limestones. Photographs courtesy of M. Gasparrini. (**A**) Cliff face showing sharp, irregular contacts and irregular distribution of limestone (light) v. dolomite (dark). In this location the dolomite appears dark because of the lichen cover (the lichen grows only on the dolostone). The dolomitizing fluids ascended via faults. (**B**) Close-up of limestone–dolostone contact such as shown in (A) from a location nearby. The dolostone appears dark where covered with lichen (upper right corner) yet light beige where cleaned of lichen (centre). The limestone (left) has a medium-grey colour. Note the sharp yet irregular contact between the limestone and dolostone. Sedimentary and diagenetic textures visible in the limestone are obliterated in the dolostone. The hammer is for scale. (**C**) Rock face similar to that shown in (B), rotated by 90° (top to the left). Note that the limestone (dark, top) is well bedded, whereas the bedding is obliterated in the dolostone (light, bottom; no lichen cover). The hammer is for scale.

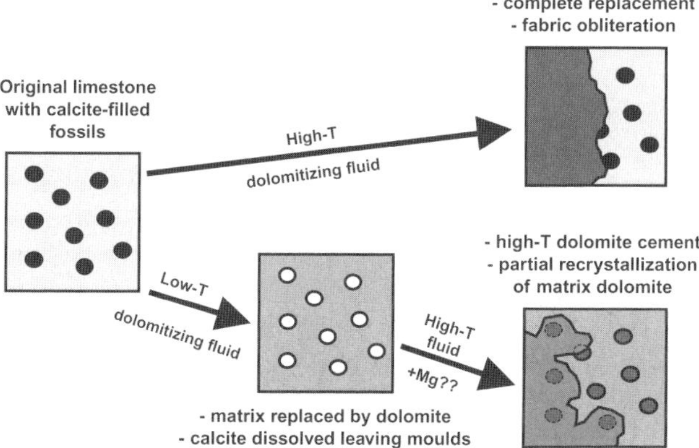

Fig. 12. Schematic illustration of limestone replacement by low-temperature and high-temperature dolomitizing solutions. See text for further explanation.

that there is a distinct dependence on depth. If considered irrespective of depth, limestones and dolomitic limestones are more porous than dolostones, whereas at burial depths of greater than 2000 m dolostones are significantly more porous and permeable than limestones. There are also notable examples of very young, near-surface dolomites that are tight and apparently devoid of porosity generated during the replacement process, as in the Plio-Pleistocene carbonates of Bonaire (Lucia & Major 1994).

Porosity

The theory that dolostones have higher porosities than limestones originated with the classic work by Elie de Beaumont in 1836 (cited by van Tuyl 1914), who proposed that 'molecular replacement' of limestone by dolomite would result in a volume loss of 12.1% (this is now called 'mole-per-mole' replacement, and the percentage commonly cited is 13%: discussed below). This view, however, is far too simplistic. Several other processes are involved, summarized diagrammatically in Figure 13. Some of these processes, such as excess calcite dissolution over dolomite, were recognized fairly early (Landes 1946; Murray 1960), and one relatively recent article provided an overview of several of the processes involved with reference to the ages and types of dolomite reservoirs (Sun 1995). Figure 13 is designed to represent the porosity and permeability evolution in both limestones and dolostones. The only aspect specific to dolostones is the effect of variable reaction stoichiometry during replacement, as represented by reactions 1–4 (earlier). Reaction stoichiometry can be added to all parts of the circle in Figure 13 because it can reduce, leave unchanged, or enhance porosity and/or permeability. The porosity and permeability distribution and evolution in dolomites and dolostones should be discussed in this context. Some of the processes are illustrated in Figures 6–10.

Six processes appear to be responsible for this phenomenon: (a) mole-per-mole replacement; (b) dissolution of unreplaced calcite (the solution is undersaturated for calcite after all Mg in excess of dolomite saturation is exhausted); (c) dissolution of dolomite (without externally controlled acidification); (d) acidification of pore waters (via decarboxylation, clay mineral diagenesis, etc.); (e) fluid mixing (mischungskorrosion); and (f) thermochemical sulphate reduction, which may generate porosity under certain circumstances (Machel 2001). The porosity v. depth compilations of Schmoker & Halley (1982), Halley & Schmoker (1983), and Amthor *et al.* (1994) did not separate these possibilities or recognize the lack of porosity destruction (porosity preservation) with depth. In fact, most workers have made no attempt to discriminate between these alternatives. The wide scatter and lack of systematic relationships between the porosity of limestones and dolostones observed in Florida and Alberta probably reflects locally and regionally heterogeneous interplays between the various processes that generate, preserve, or destroy porosity (Fig. 13). Clearly, it appears unwise to make generalizations about the porosity development of dolostones, which should be evaluated individually regarding their porosity development.

Perhaps the best known and/or most widely

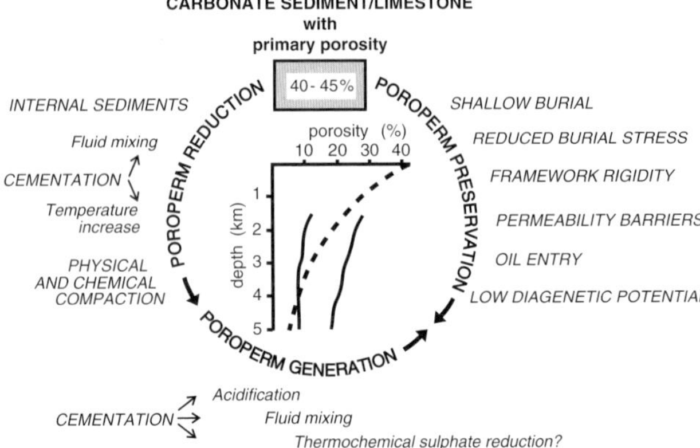

Fig. 13. Major processes of porosity and permeability ('poroperm') generation, preservation and reduction in carbonates. The inset contains averaged porosity–depth data from Mesozoic and Cenozoic limestones and dolostones in south Florida (stippled trend, from Schmoker & Halley 1982) and of the Jurassic Smackover oolite carbonate reservoirs in the southern United States (solid trends, which envelope the measured maximum and minimum values below depths of about 1.5 km, from Scholle & Halley 1985; Heydari 1997). The Florida trend can be considered typical for most carbonates elsewhere. The large variations in the Smackover carbonates at any given depth reflect highly variable degrees of porosity generation, preservation and reduction due to competing diagenetic processes. The figure is reproduced with permission from Machel (1999).

recognized mode of porosity gain during dolomitization is the replacement process *sensu stricto*. Comparison of the molar volumes of calcite and dolomite reveals that about 13% of porosity is generated in the so-called 'mole-per-mole' replacement of calcite by dolomite according to reaction 1 (section on 'Mass-balance Constraints': 2 moles of calcite are replaced by 1 mole of dolomite). If, for example, a limestone has 40% initial porosity, mole-per-mole replacement will generate a dolostone with about 45% porosity. More generally, porosity gains or losses can be represented by the values of x in reaction 3 ('Mass-balance Constraints'). Volume-per-volume replacement is represented by the special cases of $x = 0.11$ and $x = 0.25$ (for aragonite and calcite, respectively), when there is no volume loss or gain (Morrow 1982*a*). It is not clear how Nature 'chooses' one reaction stoichiometry over another. Geochemical modelling suggests that an interplay of the degree of evaporation and flow rate determines the relative saturation states of dolomite to calcite and aragonite through space and time (Sun 1992; Morrow 2001). This also controls the rates of calcite/aragonite dissolution relative to dolomite formation, and where and when macrodissolution of calcium carbonate (formation of moulds and vugs) happens along the flow path of the dolomitizing solution.

Lucia (2002, 2004) claimed that dolomitization does not normally result in an increase in porosity, arguing against the notion that the commonly observed higher porosity of dolostones compared to limestones is the result of the dolomitizing process. Rather, he suggests that most dolostones have lower porosities than limestones due to 'overdolomitization', i.e. dolomite cementation following matrix replacement and reducing pore sizes (Fig. 5), as well as permeability. However, Lucia's argument, although correct, is an incomplete explanation of those dolostones that have lower porosities than corresponding limestones. Where dolomitization is only partial, mole-per-mole replacement, if it takes place, will generate porosity. Where dolomitization is complete, mole-per-mole replacement, if it takes place, will generate porosity only if the supply of the dolomitizing solution ends roughly at the time of dolomitization approaching completion. If, however, there is a continued supply of dolomitizing solution, then 'overdolomitization' may indeed obliterate much or most of the porosity previously generated. It remains to be seen just how common 'overdolomitization' really is.

Dissolution of unreplaced calcite has the potential of generating much more than the theoretical maximum of 13% porosity in the mole-per-mole replacement process. This potential appears to be realized quite frequently (Landes 1946; Amthor *et al.* 1994). In addition, the fact that dolostones are more porous than

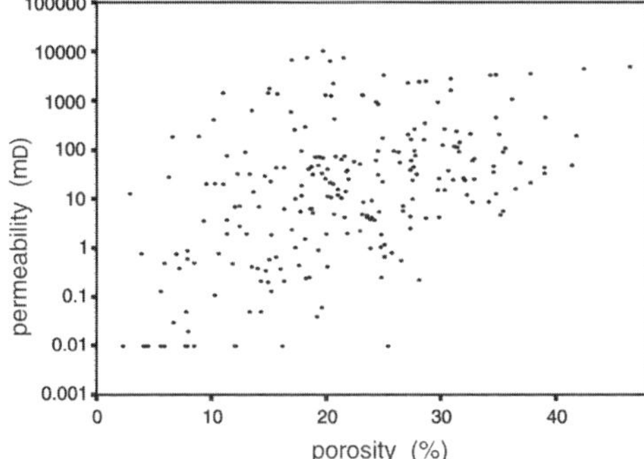

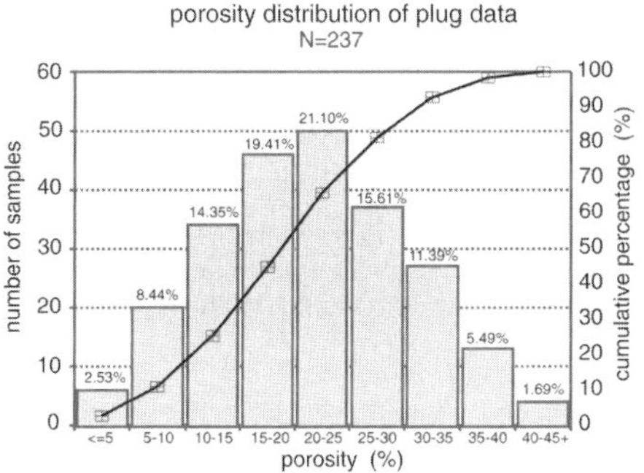

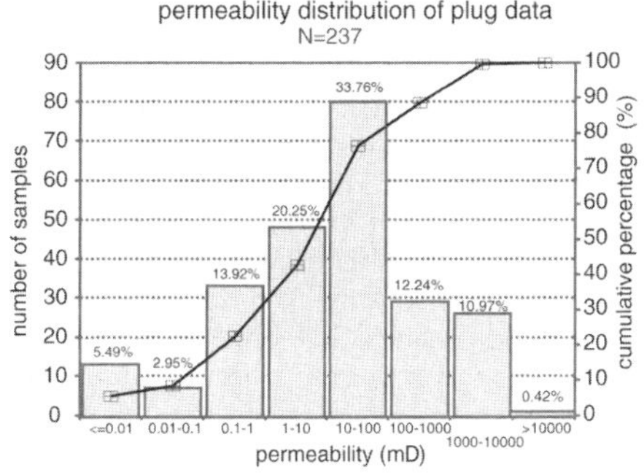

Fig. 14. Cross-plot of porosity and permeability data from 237 core plugs from 24 wells in the Upper Devonian Grosmont Formation, Alberta, Canada, with histograms of porosity and permeability. Despite considerable scatter, these data show a general positive correlation between porosity and permeability, and also attest to the excellent reservoir quality of the Grosmont Formation with modes around 20–25% and 10–100 mD, respectively (where 1D = 0.9868 × 10^{-12} m^2). The figure is reproduced with permission from Luo *et al.* (1994).

limestones at depths greater than 2000 m in some basins, such as in Alberta, is caused at least in part by the greater extent of stylolitization of the limestones, which have a higher pressure-solubility than dolostones at any given depth (Amthor *et al.* 1994).

The textural developments discussed previously illustrate some common possibilities of porosity development during, or as a result of, dolomitization; that is, porosity gain through 'excess' calcite dissolution (Fig. 6) and through dolomite dissolution (Fig. 9). On the other hand, porosity loss may occur through gypsum and/or anhydrite emplacement (Fig. 7), as well as cementation with saddle dolomite (Fig. 10), or with base-metal sulphides. The completely dolomitized Cambrian Bonneterre Formation in the USA is a case in point. Thin-section petrography, gas porosimetry and point counting show that the porosity of these rocks was reduced from about 19% prior to mineralization to less than 4% in several steps by successive cement generations of dolomites, quartz and sulphides, whereas some porosity was re-established during sulphide mineralization with concurrent dolomite dissolution (Gregg *et al.* 1993).

In this context, a few generalizations can be made regarding textural development. In cases of mole-per-mole replacement, the fabrics of the original limestone must be at least partially obliterated in order to account for the volume change during the replacement process. On the other hand, limestones dolomitized in a volume-per-volume replacement should not contain secondary intercrystal pores or dolomite cements, and the primary textures may be partially or largely, even mimetically (if the crystal size is very small), preserved. Partial or complete obliteration of primary textures can occur even in a volume-per-volume replacement, however, if there is a marked change in crystal size (usually an increase, due to Ostwald ripening), with or without porosity redistribution.

Permeability

Dolomitization almost invariably involves the reorganization of permeability pathways. Commonly, permeability increases along with porosity, and vice versa. This is documented through studies of examples such as the Upper Devonian Grosmont Formation in eastern Alberta, which hosts a giant heavy-oil reservoir (Luo *et al.* 1994; Luo & Machel 1995; Machel & Huebscher 2000). A comparison of porosity and permeability data from 237 core plugs reveals an overall positive correlation, despite considerable scatter (Fig. 14). This correlation is also expressed in the displacement pressures from mercury injection capillary measurements that

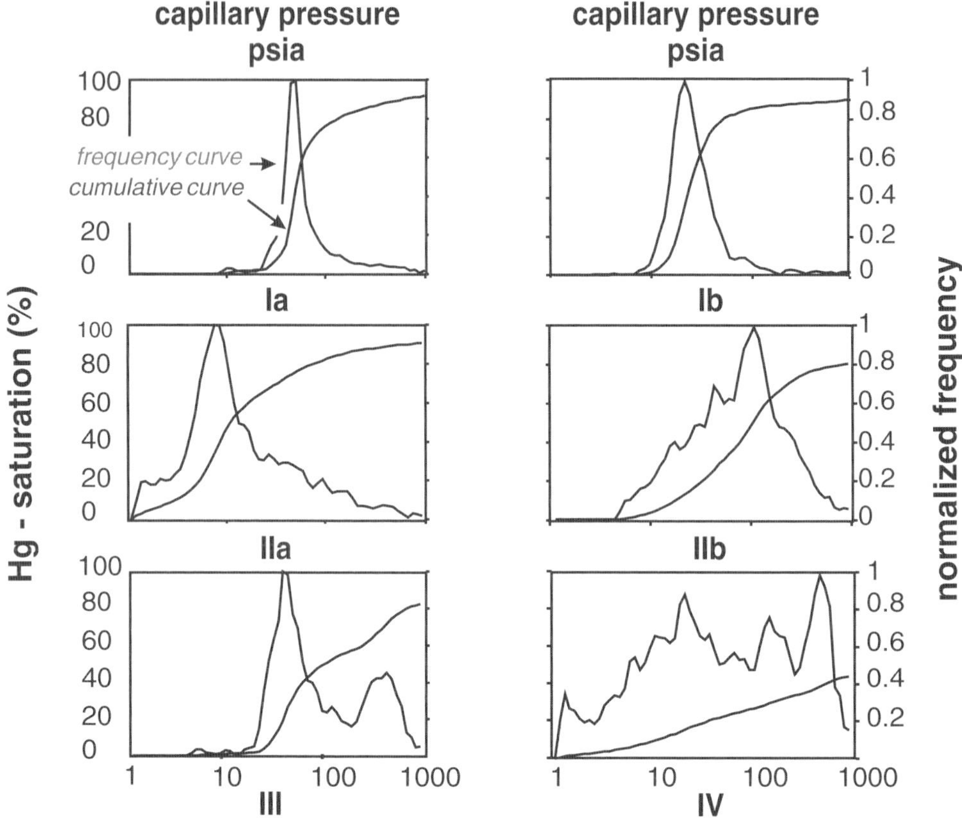

Fig. 15. Mercury injection capillary pressure measurement curve types that represent six groups of samples out of a total of 38 samples from the Upper Devonian Grosmont Formation of Alberta, Canada, which is a heavy oil reservoir. (**Ia**) Symmetrical frequency curve, representing porosity enhanced by pervasive dissolution after dolomitization. (**Ib**) Symmetrical curve, representing intercrystalline porosity generated during dolomitization. (**IIa**) Finely-skewed curve. (**IIb**) Coarsely skewed curve. (**III**) Bimodal curve. (**IV**) Non-sorting curve. The various frequence curves (IIa)–(IV) reflect a complex interplay of porosity–permeability generating processes (see Fig. 13). The figure is modified from Luo & Machel (1995).

permit the identification of four major and two minor dolomite reservoir rock types (Fig. 15). Types Ia, IIb and III have relatively high displacement pressures that correspond to the lowest porosities and permeabilities (Luo *et al.* 1994; Luo & Machel 1995). Using a similar approach (thin-section and SEM petrography, combined with helium porosimetry and mercury injection capillary measurements) Woody *et al.* (1996) documented positive and statistically significant correlations between porosity and permeability for planar dolomites in the Cambrian–Ordovician Bonneterre Formation of Missouri, USA, host to one of the world's largest MVT-sulphide deposits. Woody *et al.* (1996) further found that the planar-e dolomites have the highest porosities and permeabilities, the latter caused by well-connected pore systems with low pore to throat size ratios (as indicated by mercury injection curves); in planar-s dolomite the permeabilities do not increase as rapidly with increasing porosity, corresponding to relatively large pore to throat size ratios; and nonplanar dolomites have a statistically insignificant porosity–permeability relationship, whereby the pore systems have a high tortuosity and large pore to throat size ratios (see also Gregg 2004).

Other authors have contended that there is no systematic correlation between porosity and permeability in dolostones, or that these two petrophysical parameters are enhanced in dolostones relative to limestones. Halley & Schmoker (1983), in the absence of reliable or sufficient permeability data, attempted to assess the permeability of carbonate rocks from porosity data. They found that carbonate aquifers and carbonate aquicludes cannot be distinguished on the basis of porosity. Lucia (2002, 2004) claimed that '. . . there is no relationship between porosity and permeability in dolostones . . . and dolomite crystal size and the precursor fabric are key elements in predicting permeability', and 'Dolomitization of grain-dominated limestones usually does not change porosity–permeability relationships. Instead, the precursor fabric controls pore-size distribution'. While this may be so in some, perhaps many, cases, the Grosmont and the Bonneterre examples clearly show that there is a relationship between porosity and permeability in at least some major and economically important dolostone sequences. The cause(s) for this relationship are not just the dolomitization process itself but an interplay of various diagenetic processes. If these processes can be quantified, the diagenetic evolution of a dolomitized rock unit could be used as a predictor for the petrophysical properties of the resulting dolostone reservoir unit (Woody *et al.* 1996).

Dolomite geochemistry

A wide range of geochemical methods may be used to characterize dolomites and dolostones, and to decipher their origins. The most extensively applied are the analysis and interpretation of stable isotopes (O, C), Sr-isotopes, trace elements, and fluid inclusions, along with less common methods such as palaeomagnetics and others (e.g. Land 1980; Tucker & Wright 1990; Allen & Wiggins 1993; see also various case studies in Purser *et al.* 1994). This paper cannot discuss all of these possibilities, most of which are adequately covered in the references cited. The focus here is on two aspects of particular interest, the determination of the type of the dolomitizing fluid(s) (marine, evaporitic, subsurface brine, etc.), and the identification of the direction of fluid flow during dolomitization. The latter can commonly be determined by mapping a gradient in dolomite abundance, from complete dolomitization near the upflow direction to decreasing abundance downflow. However, this approach necessarily fails where dolomitization is 'complete' or where exposure and/or core material are insufficient. In such cases the geochemical compositions of dolomites can be used, within limits, to determine the flow direction.

Stable isotopes and fluid inclusions

Oxygen and carbon isotope ratios ($\delta^{18}O$ and $\delta^{13}C$) are the most widely applied and probably the best understood geochemical parameters in dolomite research. In brief, $\delta^{18}O$ values of carbonates can be used, within limits, to determine the $\delta^{18}O$ value and/or temperature of the fluid present during crystallization, providing a possible distinction between meteoric, marine and/or evaporitic waters.

Fluid-inclusion homogenization temperatures are arguably the best means of determining the temperature of formation of dolomites (or any other minerals), in addition to the highly desirable information on fluid compositions that can be gained from freezing experiments (e.g. McLimans 1987). Unfortunately, the vast majority of fluid inclusions in dolomites are too small for standard heating–freezing runs, and thus phase transitions within the inclusions are not observable. This is especially true of matrix-selective, replacive dolomites. On the other hand, sparry saddle dolomite cements found in late-diagenetic dissolution vugs, but also as a replacement, commonly yield excellent fluid-inclusion data.

Where possible, fluid-inclusion homogenization temperatures are used in conjunction with $\delta^{18}O$ values to further characterize the conditions of dolomite formation. This type of analysis can reveal the direction(s) and temperature gradient(s) of the dolomitizing fluid flow on a local (a few kilometres: Wilson *et al.* 1990) or on a regional scale (over several hundred kilometres: Qing & Mountjoy 1992, 1994). These latter two studies are special, in that mapping and contouring of the oxygen isotope and/or fluid-inclusion homogenization temperatures display clear, spatially resolved gradients. Unfortunately, such gradients do not appear to be particularly common.

The $\delta^{13}C$ values of the carbonates can be used to identify whether meteoric water (carrying soil CO_2) was involved, whether thermogenic or biogenic CH_4 was oxidized, whether CO_2 from microbial processes or organic matter maturation was available, or whether thermochemical sulphate reduction (TSR) contributed carbon to the system (e.g. Hudson 1977; Machel *et al.* 1995). There also is a secular carbon isotope trend that may be used for dating marine dolostones, but only under very favourable circumstances (Veizer *et al.* 1999).

Sr-isotopes

Radiogenic isotopes are less commonly used in studies of carbonate diagenesis, mainly because their analysis is much more expensive. Yet, strontium isotopic compositions (usually quoted as $^{87}Sr/^{86}Sr$ ratios) are an excellent parameter to deduce compositional changes and especially

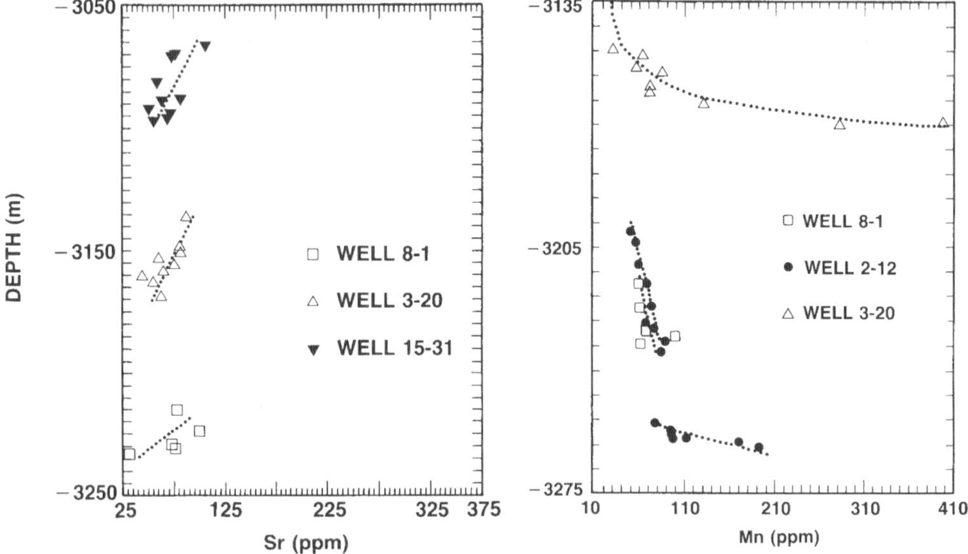

Fig. 16. Sr v. depth and Mn v. depth cross-plots of matrix dolomites from four Upper Devonian Nisku drill cores in two traverses down the structural dip of the Nisku reef trend. Well 15–31 is structurally the shallowest and 2–12 is structurally the deepest. The rocks cored in these wells were nearly at the same depth during dolomitization (very minor structural dip at that time). Sr increases upwards in each well, whereas Mn decreases upwards in each of these wells that penetrate two facies types. These trends demonstrate that dolomitization was post-depositional and that the dolomitizing fluid flow was upwards. Figures are reproduced with permission from Machel (1988).

flow directions of the fluids from which diagenetic carbonates have formed. This is because Sr isotopes, unlike the more commonly used stable isotopes of oxygen and carbon, are not fractionated by pressure, temperature and microbial processes (e.g. Faure & Powell 1972), and $^{87}Sr/^{86}Sr$ ratios display a distinctive secular trend (Smalley *et al.* 1994; Veizer *et al.* 1999).

Examples of how $^{87}Sr/^{86}Sr$ ratios can be used to decipher palaeofluid flow direction (and origin) are provided by Machel & Cavell (1999) and Buschkuehle & Machel (2002). In these studies, the spatial distribution of the $^{87}Sr/^{86}Sr$ ratios in sparry calcite cements, and to a minor degree in sparry dolomite cements, suggests a general W–E flow pattern through an Upper Devonian carbonate complex, with $^{87}Sr/^{86}Sr$ ratios decreasing eastward as a result of dilution and increasing water–rock interaction.

Trace elements

The direction of fluid flow can also be determined using trace elements. This is especially attractive because trace-element analysis is the cheapest of all the common geochemical methods. To this end, Machel (1988) developed a mathematical model applying a variant of the Heterogeneous Distribution Law. Mechanistically, dolomitization is assumed to take place in a manner analogous to the textural evolution shown in Figure 5, and the resulting trace-element compositions are obtained by drilling out powders that sample batches of several tens to hundreds of crystals, rather than individual crystals. The model predicts that systematic trace-element trends indicating fluid flow direction(s) can result during dolomitization. Unfortunately, any transition is possible between: (a) a large up-flow and down-flow trace-element difference stretched over a large flow distance, and (b) a small up-flow and down-flow trace-element difference over a negligible flow distance, depending on the interplay of all involved parameters. In the case of dolomite cementation, trace elements with distribution coefficients smaller than 1 increase, and those with a distribution coefficient larger than 1 decrease, in the down-flow direction. Trace-element trends have been documented in several Phanerozoic dolostone sequences (Machel 1988), and one example is shown in Figure 16. Such trends may be more common than previously recognized. Their absence in other dolostone sequences may be real in a percentage of cases, because dolomitization does not necessarily yield trace-element trends.

Recrystallization

For all practical applications, such as the determination of fluid composition and flow direction, the absence, presence and/or degree of recrystallization is important. The degree of recrystallization in dolomites and dolostones is much in dispute (Mazzullo 1992; Machel 1997). Some authors claim that all dolomites and dolostones are recrystallized, and that recrystallization commonly proceeds by multiple steps (Land 1992). Implicit in Land's contention is that all dolostones form near the surface from seawater (see also Land 1985) and that such seawater dolomites are thermodynamically unstable during burial. Others claim that 'early' near-surface and shallow subsurface dolomites commonly (but not always) recrystallize during burial, but that burial dolomites often do not recrystallize because they have little if any thermodynamic drive to do so (e.g. Machel *et al.* 1994). A resolution of this problem is of utmost importance for genetic interpretations of dolomites and dolostones. This has led to broadening of the definition of the term recrystallization, and to the introduction of the concept of 'significant recrystallization' (Machel 1997), which is of great use in genetic interpretations of dolomites and dolostones.

As pointed out earlier, it is well known from hydrothermal experiments that dolomite forms in stages via so-called VHMC (very-high-Mg calcite with about 36 mole% Mg), then VHMC plus non-stoichiometric dolomite, then stoichiometric dolomite. These recrystallization steps commonly appear to take place very fast, i.e. within a few hundreds to at most a few thousands of years in low-temperature diagenetic settings, but are even faster in high-temperature settings. Hence, these transitions are pretty much irrelevant for the investigation of ancient (older Cenozoic, Mesozoic and Palaeozoic) dolomites, except for some exceptional cases where the fluid chemistry has changed dramatically within this time frame. One example would be hypersaline protodolomites that may recrystallize in meteoric or brackish water relatively soon after their formation. Furthermore, additional recrystallization may (and often does) happen after many thousands to millions of years, especially during deep burial, and it is these later recrystallization(s) that are of concern.

If changes in texture, structure, composition and/or palaeomagnetic properties through recrystallization are so small that the total range of properties after recrystallization is the same as when the dolomite first formed, a dolomite/dolostone is said to be 'insignificantly recrystallized' (Fig. 17, top), and its properties are still representative of the fluid and environment of dolomitization. On the other hand, if these changes result in ranges of properties that are larger than the original ones, a dolomite/dolostone is said to be 'significantly recrystallized' (Fig. 17, bottom), and its properties are no longer representative of the fluid and environment of dolomitization. In this case, the measured properties are reset and characterize the last event of recrystallization. Not all measurable properties are necessarily reset during recrystallization. For a dolomite to be recognized as 'significantly recrystallized' only one of the measurable properties has to be modified to a range larger than that in the original crystals. In this case, the inherited properties may still represent the event of dolomitization, whereas the reset properties represent recrystallization.

At present it is not clear how common significant recrystallization is in dolomites and dolostones. There are unequivocal examples of the lack of recrystallization and of insignificant recrystallization (e.g. Tan & Hudson 1971; Packard 1992), and there also are convincing cases of significant recrystallization, especially in geologically young dolomites of evaporative origin, but also in some ancient examples (e.g. Gregg *et al.* 1992, 2001; Montañez & Reid 1992*a*; Malone *et al.* 1994; Durocher & Al-Aasm 1997). The progressive and stepwise recrystallization proposed by Land (1992) has been found in only a few geologically young dolomites, most notably in the hemipelagic Miocene Monterey Formation, California (Malone *et al.* 1994). In many, if not most, other cases the evidence is ambiguous.

The Carboniferous dolostones of the Dunvegan gas field in Alberta, Canada, provide a striking, albeit unusual, example of a lack of significant recrystallization. The Dunvegan gas field is a trend about 24 km long, 5 km wide and 35 m thick that was buried for about 300 Ma to depths of up to 4000 m. Yet, the Dunvegan dolomites are texturally and geochemically (stable isotopes, stoichiometry, ordering) virtually identical to the Recent dolomites of the Abu Dhabi sabkha (Packard 1992). This is highly unusual because sabkha dolomites tend to recrystallize fairly easily and early in the burial history, due to the fact that they usually form as metastable protodolomites. A case in

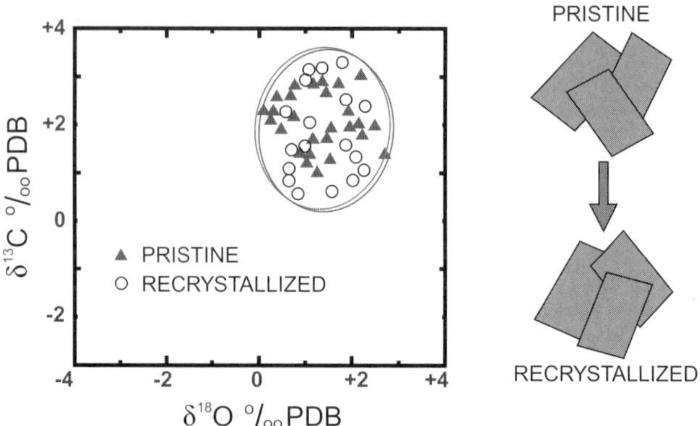

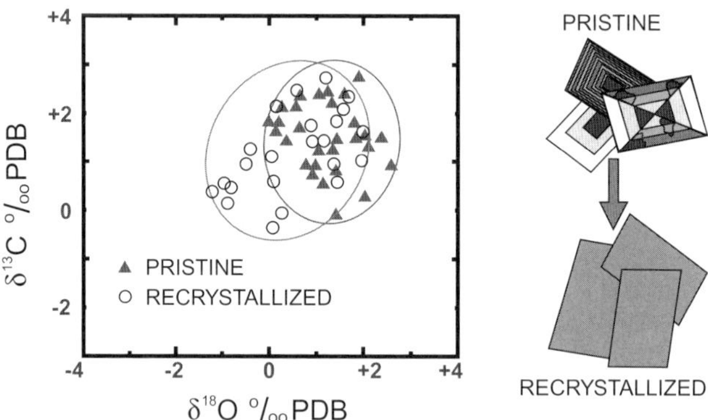

Fig. 17. Schematic illustration of insignificant and significant recrystallization. For the properties shown, $\delta^{13}C$ and $\delta^{18}O$ values, crystal sizes and luminescence, the ranges in the pristine and recrystallized samples are identical for insignificant recrystallization but different, despite some overlap, for significant recrystallization, where at least some isotope values of the recrystallized samples fall outside the range of the pristine samples. The crystals have also increased in size and lost their zonation. The figure is reproduced with permission from Machel (1997).

point is the genetically similar dolomites of the Ordovician Knox Group in the Appalachians, USA, that are significantly dolomitized (Montañez & Read 1992*a*).

There also are examples where, up to now, it has not been possible to determine the extent of recrystallization, or where the evidence shows a lack of significant recrystallization, as in most of the massive dolostones of the Devonian of western Canada. A particularly instructive example is the famous Rimbey–Meadowbrook reef trend that extends through the subsurface of Alberta for several hundred kilometres, forming a structural homocline with burial depths near 200 m at its NNE end and about 6 km near its SSW end. The reefs are located on top of the Cooking Lake platform, and both the platform margin and the overlying reefs have been replaced by matrix-replacive, commonly fine- to medium-crystalline dolomites. Multiple lines of evidence (facies, structure, petrography, and geochemistry) taken together suggest that the reef trend and the underlying platform margin were dolomitized by chemically modified seawater at depths of about 500–1500 m (Amthor *et al.* 1993; Machel *et al.* 1994; Mountjoy *et al.* 1999). Considering that the Rimbey–Meadowbrook reef trend is so long and now varies widely in depth from one end to the other, this trend is an ideal place to test the hypothesis of Land's (1992) 'quantum theory of dolomite stabilization'. If there is any place where a stepwise progression in recrystallization with increasing temperature and pressure (depth) and time is developed, this is it. However, there is no evidence of systematic significant recrystallization in this reef trend. Plots of textural and geochemical data form clusters (with a few outliers), and there are no stepwise offsets down-dip (Amthor *et al.* 1993; Machel *et al.* 1994; Horrigan 1996; Drivet & Mountjoy 1997; Mountjoy *et al.* 1999). Only the early, peritidal–supratidal dolostones of the Grosmont formation at the shallow end of the trend (Huebscher 1996; Machel & Huebscher 2000), and the most deeply buried dolostones close to the Rocky Mountain deformed belt presently at depths in excess of about 4500 m and formerly buried by up to 2000 m more, are significantly recrystallized (Machel *et al.* 1996*b*; Drivet & Mountjoy 1997).

Taken together the data suggest the following generalizations. Most dolomites that originally form very close to the surface and/or from evaporitic brines tend to recrystallize with time and burial because they form as metastable protodolomite phases and become thermodynamically highly unstable as a result of increasing temperature and pressure, and changing fluid composition. A perhaps typical example is the Monterey Formation, yet there are exceptions, the Dunvegan gas field being particularly striking. By contrast, dolomites that form at several hundred to a few thousand metres depth are either not or hardly prone to recrystallization because they tend to form as rather stable (nearly stoichiometric, well-ordered) phases, the stability of which does not change much during further burial and with increasing time. A conspicuous example in this regard is the replacive matrix dolomites of the Rimbey–Meadowbrook reef trend, except for its most deeply buried part.

Environments and models of dolomitization

One of the most striking developments in dolomite research after World War II was the rapid evolution of a series of models of dolomitization that started in the 1950s and is continuing to this day. These models were designed to explain the origin of the various types of dolomite, and especially of massive dolostones. Interestingly, many of them, some long forgotten, had first appeared in the 100 years after the discovery of dolomite. van Tuyl (1914), in a section entitled 'Historical Review' discussed in great detail eight models, then called 'theories of the origin of dolomite', which he presented in three groups:

I. Primary deposition theories:
   IA: The chemical theory
   IB: The organic theory
   IC: The clastic theory.
II. Alteration theories:
   IIA: The marine alteration theory
   IIB: The groundwater alteration theory
   IIC: The pneumatolytic alteration theory.
III. Leaching theories:
   IIIA: The marine leaching theory
   IIIB: The surface leaching theory.

Some of these theories/models are outdated in the light of present knowledge, but several form the basis of current models of dolomitization, as will be pointed out below.

Traditionally, dolomitization models have been defined or based on water chemistry in near-surface and shallow diagenetic settings, but on hydrology in burial diagenetic settings (e.g. Morrow 1982*b*). This poses an obvious dilemma where a near-surface diagenetic fluid moves into the deeper subsurface, or where a deep(er) subsurface fluid ascends into shallow diagenetic settings. Research over the last 15–20 years has revealed several such 'crossovers' or 'overlaps' between models that have resulted in ambiguities in semantics and classification. This problem became bothersome about 15 years ago and surfaced at two major international conferences held in the early 1990s, i.e. the 1991 Dolomieu conference in Ortisei, Italy, and the 1992 National Conference of Earth Science in Banff, Canada. As the most striking example, there was an intense debate at the Banff conference regarding the meaning of the term 'burial dolomite', which meant 3000+ m for some, yet encompassed a much wider range from a few hundred to a few thousand metres of burial for others. If nothing else, this debate highlighted the need to establish clear definitions of burial diagenetic settings, and these were eventually published by Machel (1999) (Fig. 18). Additional research has led to considerable refinements of some established models and to a small number of new ones. For all these reasons, the various models currently in use are discussed in new categories or groups and are placed into an unambiguous, clearly defined context of diagenetic settings (Fig. 18).

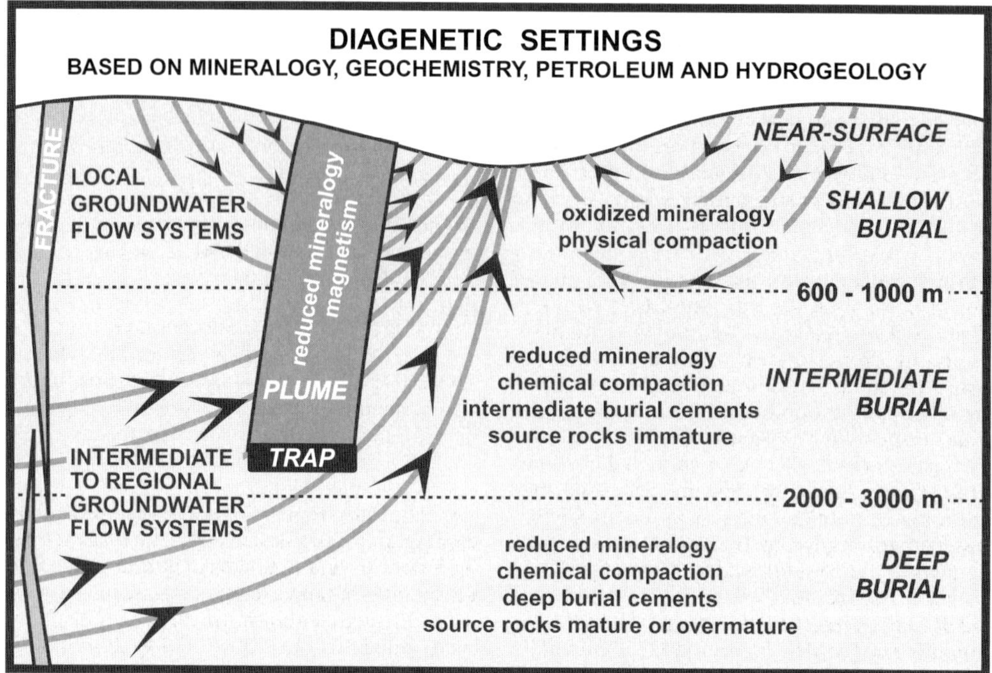

Fig. 18. Classification of diagenetic settings on the basis of mineralogy, petroleum, hydrogeochemistry, and hydrogeology. For illustrative simplicity, the geological section is assumed to be isotropic and homogeneous, with idealized groundwater flow lines. The hydrocarbon-contaminated plume is slightly deflected by the local and regional groundwater flow systems. The depth limits separating the burial diagenetic settings are approximate and based on geological phenomena that are easily recognizable. Near-surface settings may be meteoric, brackish, marine, or hypersaline. The figure is reproduced with permission from Machel (1999).

However, there is certainly more than one viable way to group dolomitization models.

A further problem in dolomite research is a commonly unwitting and certainly needless obfuscation: the widespread practice of calling any interpretation a 'model'. A model is not the same as an interpretation. Rather, a model is a complex concept that is based on a set of criteria, one of which is an interpretation. Unfortunately, this is often ignored, so it appears necessary to define what a model is, what it consists of, and what it can or cannot do. In a general sense, a model is:

> a working hypothesis or precise simulation, by means of description, statistical data, or analogy, of a phenomenon or process that cannot be observed directly or that is difficult to observe directly. Models can be derived by various methods, e.g. by computer, from stereoscopic photographs, or from scaled experiments (AGI 1999).

Walker (1992), using stratigraphic models as an example, elegantly summarized the general criteria for a model, which must act as:

(1) a norm for purposes of comparison;
(2) a framework and guide for future observations;
(3) a predictor in new geological situations;
(4) an integral basis for interpretation of the environment or system that it represents.

Several so-called models do not fulfill these criteria but are merely interpretations. In addition, dolomite models must fulfill three specific criteria, i.e.

(5) *thermodynamic*: there must be supersaturation for dolomite, with variable saturation states for calcite and aragonite; replacement dolomite (dolomitization *sensu stricto*) requires undersaturation with respect to calcium carbonate; otherwise there will be dolomite cementation;

(6) *kinetic*: the rate of dolomite formation must be equal to or greater than the rate of calcium carbonate dissolution, otherwise there will be significant dissolution porosity up to the scale of megascopic karst;
(7) *hydrologic*: there must be long-lasting pore-water flow, preferentially with high Mg-content (an exception is Mg-supply via diffusion).

Several models have been published that fail one or more of these criteria specific for dolomitization. Some examples are discussed below.

The following sections critically evaluate the major dolomitization models, with emphasis on those dealing with the origin of the massive dolostones that commonly form hydrocarbon reservoir rocks and/or regional aquifers. Some of the less important models that deal with small amounts of dolomite formation, and are insignificant regarding the formation of reservoir rocks, are mentioned only briefly. These include the relatively well-researched and academically interesting lacustrine Coorong dolomite(s) (von der Borch 1976; Muir *et al.* 1980; Rosen *et al.* 1989), the microbial/organogenic dolomites, and most other penecontemporaneous dolomites. For more information on these types of dolomite see the summary articles by Last (1990), Budd (1997), and Mazzullo (2000).

This section contains a series of illustrations of flow mechanism and domains, and the resulting dolomite/dolostone bodies (Figs 19–23). Whitaker *et al.* (2003, 2004) recently emphasized that it may be misleading to conceptualize individual flow mechanisms in isolation, and that fluid flow in a geologic situation may be the product of a number of different drives acting simultaneously or consecutively. As a result, the dolomite bodies resulting from a single hydrologic drive, such as those shown in Figure 19, may also be misleading. This is correct in principle and should be kept in mind at all times. However, illustrations of amalgamated flow regimes and the resulting dolomitization would be confusing, and most probably there are dolostone bodies that originated from one predominant flow mechanism. Hence, the examples shown in the following figures are schematic illustrations of individual flow regimes and resulting dolomitization, as they would be expected from field data and circumstantial evidence, including the most relevant numerical models. The reader is referred to Whitaker *et al.* (2004) for a detailed evaluation of the various analytical and numerical models used to predict the patterns of groundwater flow, the rate and distribution of dolomitization resulting from groundwater flow, and the limitations of these types of models.

Penecontemporaneous dolomites and the microbial/organogenic model

Penecontemporaneous (synsedimentary) dolomites form very shortly after deposition, i.e. within a few years to tens of years, as a normal by-product of the geochemical conditions at the site of deposition. There are two preferred settings for penecontemporaneous dolomite formation, shallow marine to supratidal and hemipelagic to pelagic. In terms of diagenetic environments, these are near-surface settings (Fig. 18).

In shallow-marine to supratidal environments, penecontemporaneous dolomites commonly form in quantities of <5 vol%, mostly as Ca-rich and poorly ordered, microcrystalline to fine-crystalline cements, or directly from aqueous solution (see summary in Budd 1997). These occurrences include lithified supratidal crusts (e.g. Andros Island; Sugarloaf Key; Ambergris Cay); thin layers in salinas (e.g. Bonaire; West Caicos Island) and evaporative lagoons/lakes (e.g. the Coorong); fine-crystalline cements and replacements in peritidal sediments (e.g. Florida Bay; Andros Island). The dolomite-forming fluid is normal seawater and/or evaporated seawater, in some cases with admixtures of evaporated groundwater. There also are two examples of penecontemporaneous dolomite formation in association with volcanic activity: dolomite as a fine-crystalline supratidal weathering product of basic rocks (Capo *et al.* 2000), and hydrothermal dolomite forming at submarine vents (Pichler & Humphrey 2001). These cases have in common that the amount of dolomite formed is very small, and that it is well ordered and nearly stoichiometric.

One especially important type of penecontemporaneous dolomite forms lenses and layers of up to 100 vol% in sabkhas. Genetically, these dolomites belong to the family discussed in this section. However, they are considered separately below because of their historical significance, and because they have a genetic affinity to reflux dolomitization where sabkhas grade into evaporative lagoons.

Penecontemporaneous dolomites in hemipelagic to pelagic settings commonly form in very small quantities as microcrystalline protodolomite, generally less than 1 wt% (Lumsden 1988). However, under favourable circumstances the amount of dolomite locally

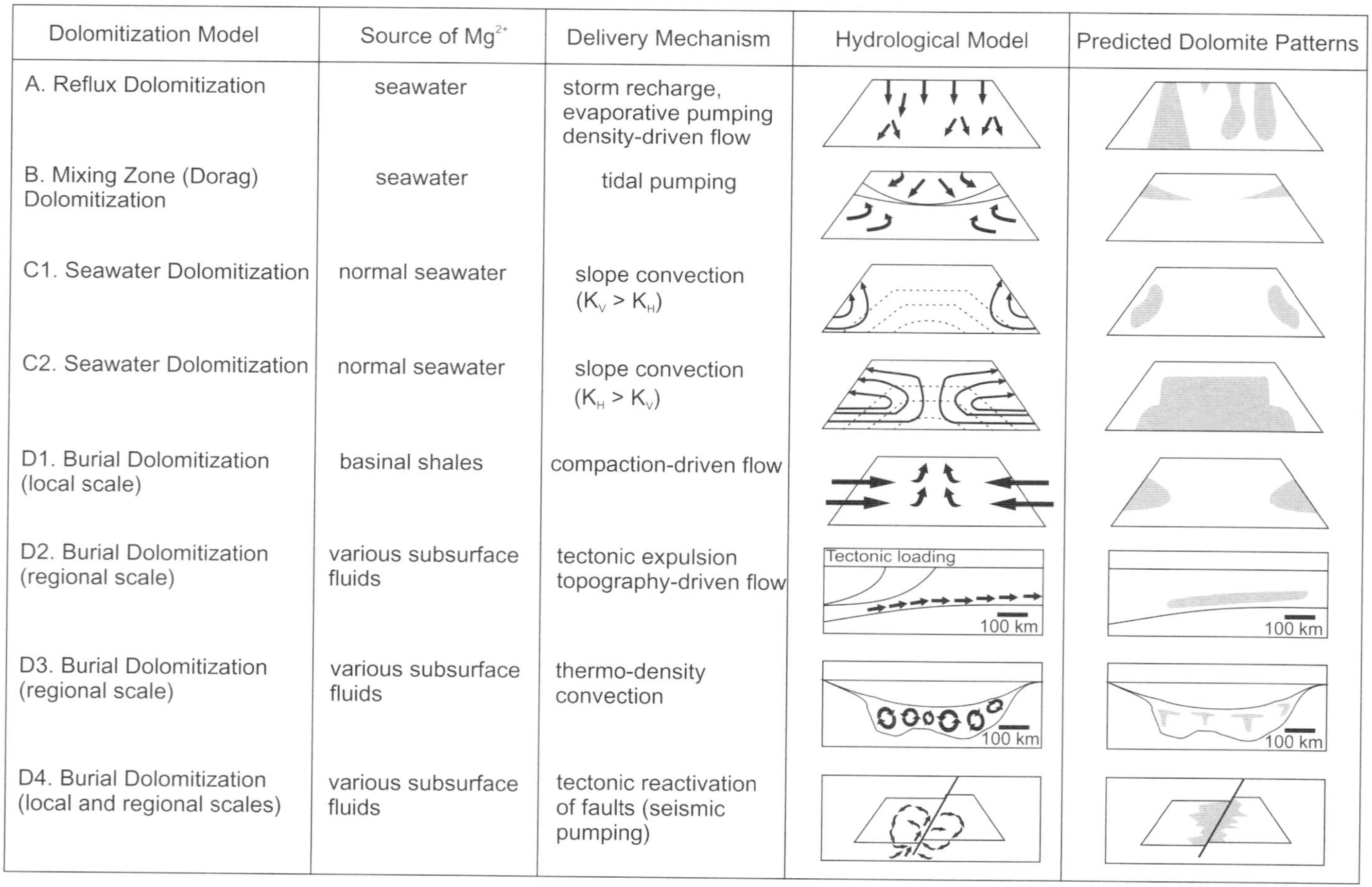

Dolomitization Model	Source of Mg^{2+}	Delivery Mechanism	Hydrological Model	Predicted Dolomite Patterns
A. Reflux Dolomitization	seawater	storm recharge, evaporative pumping density-driven flow		
B. Mixing Zone (Dorag) Dolomitization	seawater	tidal pumping		
C1. Seawater Dolomitization	normal seawater	slope convection ($K_V > K_H$)		
C2. Seawater Dolomitization	normal seawater	slope convection ($K_H > K_V$)		
D1. Burial Dolomitization (local scale)	basinal shales	compaction-driven flow		
D2. Burial Dolomitization (regional scale)	various subsurface fluids	tectonic expulsion topography-driven flow	Tectonic loading; 100 km	100 km
D3. Burial Dolomitization (regional scale)	various subsurface fluids	thermo-density convection	100 km	100 km
D4. Burial Dolomitization (local and regional scales)	various subsurface fluids	tectonic reactivation of faults (seismic pumping)		

Fig. 19. Selected models of dolomitization, illustrated as groundwater flow systems and predicted dolomitization patterns. Examples are of incomplete dolomitization of carbonate platforms or reefs, i.e. they represent early phases of dolomitization. Arrows denote flow directions; dashed lines show isotherms. Predicted dolomitization patterns are shaded. Models A–D1 and D4 are kilometre-scale; models D2 and D3 are basin-scale. The figure is modified from Amthor *et al.* (1993).

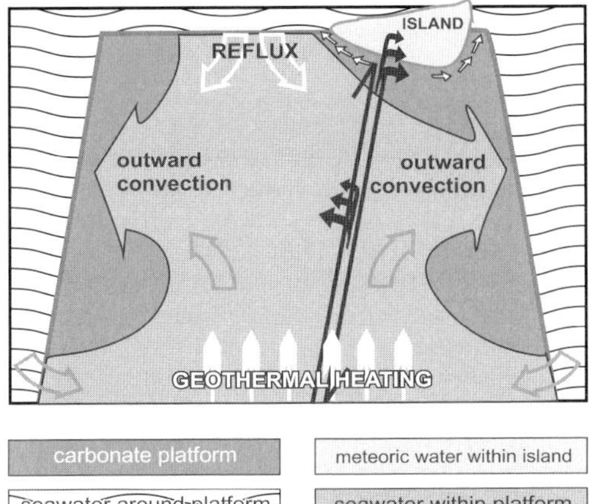

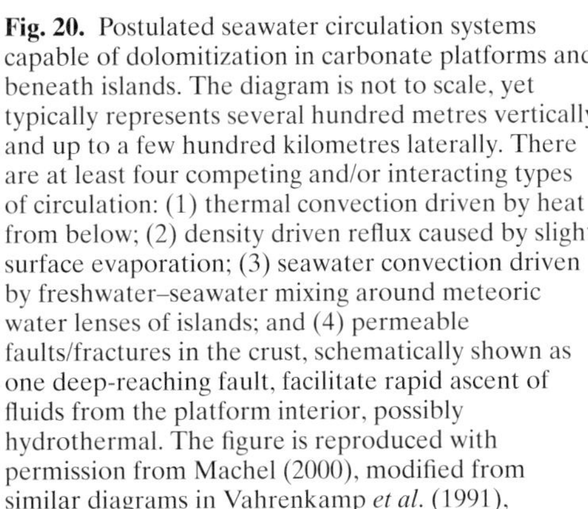

Fig. 20. Postulated seawater circulation systems capable of dolomitization in carbonate platforms and beneath islands. The diagram is not to scale, yet typically represents several hundred metres vertically and up to a few hundred kilometres laterally. There are at least four competing and/or interacting types of circulation: (1) thermal convection driven by heat from below; (2) density driven reflux caused by slight surface evaporation; (3) seawater convection driven by freshwater–seawater mixing around meteoric water lenses of islands; and (4) permeable faults/fractures in the crust, schematically shown as one deep-reaching fault, facilitate rapid ascent of fluids from the platform interior, possibly hydrothermal. The figure is reproduced with permission from Machel (2000), modified from similar diagrams in Vahrenkamp *et al.* (1991), Vahrenkamp & Swart (1994) and Whitaker *et al.* (1994).

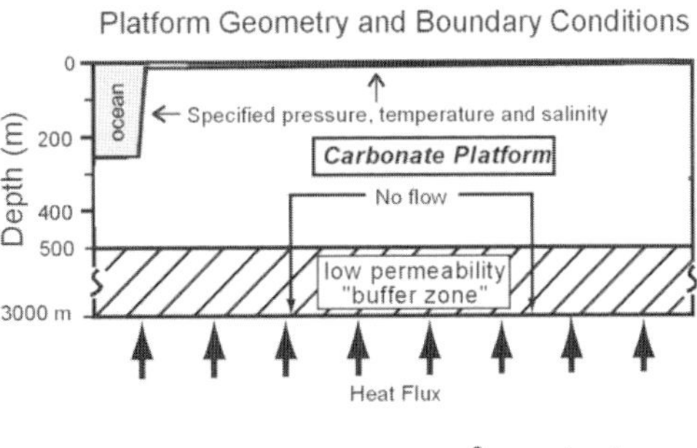

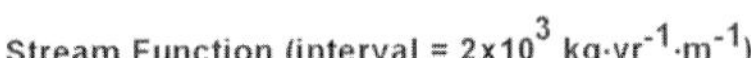

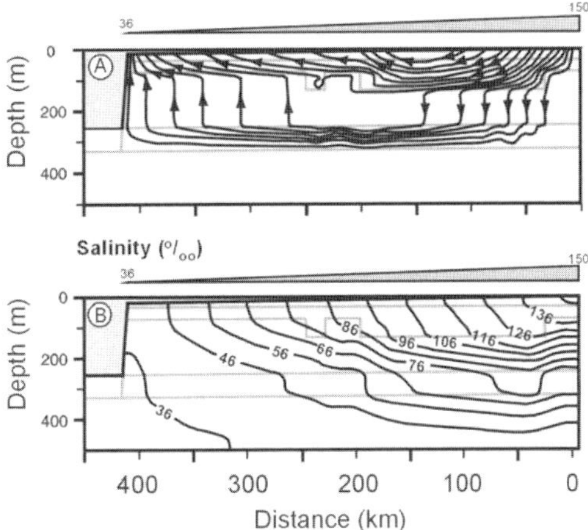

Fig. 21. Simulation boundary conditions (top) and modelling results (A and B) of reflux in a carbonate platform. The salinity at the top of the flooded platform ranges from marine (36 g l^{-1}) near the platform margin to mesohaline–hypersaline (150 g l^{-1}) near the shoreline. The platform contains a weak aquitard near its centre and has a strong aquitard as its base ('buffer zone'; note the change in vertical scale within this zone). (**A**) Fluid flux as stream function; (**B**) salinity distribution (g l^{-1}), after 500 000 years. The weak aquitard shows as a zone with little horizontal flow but is breached vertically. Most of the flow is within the upper and lower, highly permeable, parts of the platform. The diagram is modified from Jones *et al.* (2003).

reaches up to 100%. For example, Miocene hemipelagic carbonate sediments from the margin of the Great Bahama Bank are partially to completely dolomitized over a depth range of about 50–500 m subsea. In this setting, dolomite forms as a primary void-filling cement and by replacing micritic sediments, red calcareous algae and echinoderm grains (Swart & Melim 2000). Dolomites in these settings are prone to recrystallization because they tend to form as metastable protodolomites (e.g. Baker & Burns 1985; Malone *et al.* 1994; Mazzullo 2000).

Both settings of penecontemporaneous dolomite formation appear to be linked to the 'microbial' or 'organogenic' model of dolomitization (Vasconcelos & McKenzie 1997; Burns *et al.* 2000; Mazzullo 2000) that has its roots in the 'Organic theory' based on studies by Forchhammer (1850), Damour (1851), Ludwig & Theobold (1982), and Doelter & Hoernes (1875) that are reported by van Tuyl (1914). According to this model, dolomite may be formed syndepositionally or early post-depositionally and at depths of a few centimetres to a few hundred metres under the influence of, or promoted by, bacterial sulphate reduction and/or methanogenesis. The latter is commonly indicated by

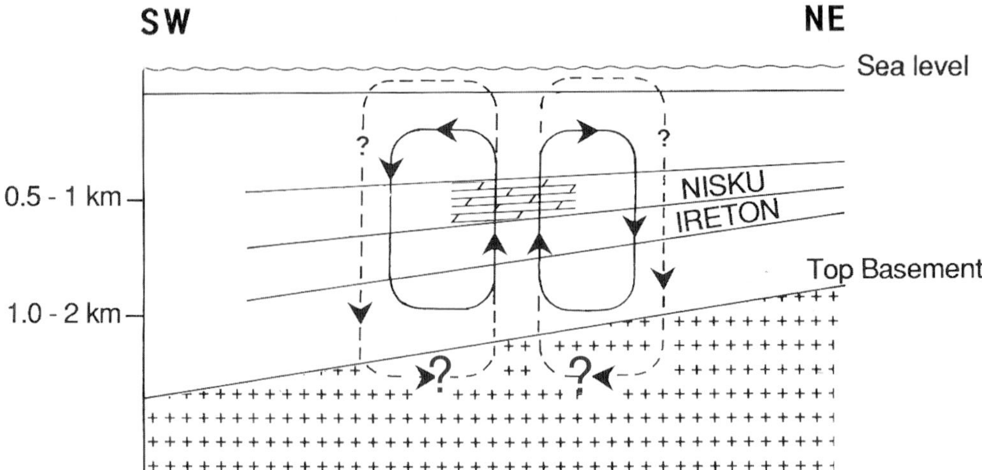

Fig. 22. Thermal convection half-cells that are open to seawater recharge only at the top. Note that thermal convection is assumed to take place only to a depth maximum of 2 km where the strata are still relatively permeable, and where the dolomitized part of the sequence (Nisku) is at shallower depths of ± 1 km, consistent with the recent modelling by Wilson *et al.* (2001). The figure is reproduced with permission from Machel & Anderson (1989); this paper speculated that the convection cells may have penetrated the uppermost parts of the basement, where it was permeable.

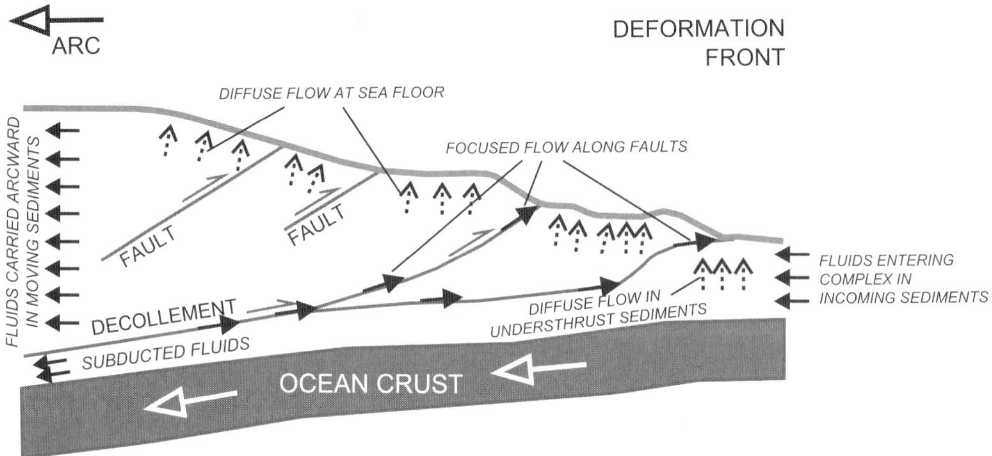

Fig. 23. Schematic cross-section of an accretionary prism/wedge and potential fluid pathways. Flow within the décollement and other faults may be spatially focused in dilated networks, temporally transient, or both. The figure is reproduced with permission from Machel (2000).

depleted $\delta^{13}C$ values. The exact role of microbial activity in reducing the notorious kinetic barriers to dolomitization is unknown, although it seems likely that a reduction of Mg- and Ca-hydration barriers or an increase in alkalinity, or a change in pH, is involved. Most microbial/organogenic dolomites are cements, but some are replacive, typically fine-crystalline–microcrystalline (less than 10 μm), calcic and poorly ordered protodolomites. The chief modes of Mg-supply are diffusion from the overlying seawater or release from Mg-calcites and clay minerals, and these place severe limits on the amounts of dolomite that can be formed.

Microbial/organogenic dolomites may act as nuclei for later, more pervasive, dolomitization during burial.

Hyposaline environments and the mixing zone model

Hyposaline environments are those with salinities below that of normal seawater (35–36 g l^{-1}). These environments include coastal and inland freshwater–seawater mixing zones, marshes, rivers, lakes, and caves. Post-depositional dolomite has been found to form in all of these environments, but only in small amounts and commonly as cements. Virtually all hyposaline environments are near-surface to shallow-burial diagenetic settings at depths of less than about 600–1000 m (Figs 16, 17A, and 18).

One hyposaline environment, the coastal freshwater–seawater mixing zone (often simply called the mixing zone) has given rise to one of the oldest and most popular models, the 'mixing zone model' for dolomitization. Dolomitization by brackish water in a freshwater–seawater mixing zones was first proposed by Hanshaw *et al.* (1971) on the basis of their study of a Tertiary carbonate aquifer in Florida. This concept was expanded to the status of a model by Badiozamani (1973), who coined the term 'Dorag model' and advocated that dolomite should form in massive amounts in those parts of mixing zones where the waters have much less than 50% seawater salinity down to about 5%. His rationale was a thermodynamic calculation of saturation states for dolomite and calcite, whereby the mixing waters were found to be supersaturated for dolomite yet undersaturated for calcite in the said salinity range. At the same time, Land (1973) proposed a mixing model on the basis of his study of dolomites in the Pleistocene Hope Gate Formation, Jamaica, advocating dolomitization by 'high P_{CO_2} meteoric waters and small amounts of seawater . . .' (Land 1973, p. 86).

For several years many authors invoked the mixing model to explain pervasive dolomitization of entire carbonate platforms of several hundreds to thousands of square kilometres in extent (e.g. Choquette & Steinen 1980; Dunham & Olson 1980; Xun & Fairchild 1987). In addition, modelling by Humphrey & Quinn (1989) suggested that coastal mixing zones may form thick sections of dolomite in platform-margin settings, and that such dolostones may be common in the geological record. Their model, however, was based on several incorrect assumptions (see discussion by Machel & Mountjoy 1990).

The mixing model has been highly overrated with regard to its potential to form massive dolostones. Not a single location in the world has been shown to be extensively dolomitized in a freshwater–seawater mixing zone, in recent or in ancient carbonates, and many lines of evidence indicate that massive dolomitization in mixing zones is so unlikely as to be virtually impossible (Hardie 1987; Smart *et al.* 1988; Machel & Mountjoy 1990; Melim *et al.* 2003). Although the waters in many mixing zones are thermodynamically supersaturated with respect to dolomite in at least a part of the mixing range (commonly between about 10 and 50% seawater), these waters also tend to be supersaturated with respect to calcite and/or aragonite in the same salinity range. Thus, the 'salinity window' of dolomitization is much smaller or does not exist, and model criterion (5) (above) is not fulfilled. Moreover, where the waters are supersaturated with respect to dolomite and undersaturated with respect to calcium carbonate, the dissolution rate of calcium carbonate is many times higher than the nucleation and growth rate of dolomite, hence model criterion (6) (above) is also not fulfilled.

The dominant diagenetic process in most typical freshwater–seawater mixing zones is extensive dissolution of calcium carbonate, often up to the dimensions of caves. This has been shown in many studies, especially from Florida and Yucatan (Back *et al.* 1986; Smith *et al.* 2002; Smart & Whitaker 2003; Whitaker *et al.* 2004), and was previously indicated by geochemical modelling (Sanford & Konikow 1989). Also, most coastal mixing zones are only a few hundreds of metres wide and the waters pass relatively quickly through the rocks in response to eustatic sea-level fluctuations and subsidence. This prevents a long-lasting supply of Mg, and model criterion (7) is not fulfilled. Even where mixing zones are capable of forming dolomite, the dolomitized rock volume tends to be relatively small and restricted to the platform margin(s) (Figs 19B and 20). If dolomite forms at all, it is commonly in comparatively minuscule amounts (a few vol%) that form in the more saline parts, i.e. more than 70% seawater, as thin cement fringes, replacements, or both.

Most mixing zone dolomites are petrologically and geochemically distinct. The crystals tend to be relatively clear, planar-e or planar-s, stoichiometric, well-ordered rhombs, although some mixing zone dolomite is non-stoichiometric and poorly ordered. Crystal sizes commonly range from 1 to 100 μm, but reach several millimetres in some cases. Most mixing-zone dolomite occurs as cement in microscopic interstices and macroscopic voids, moulds, vugs and caverns, and subordinately as a replacement. Alternating generations or growth zones of calcite and dolomite are common in coastal mixing zones with rapid cyclical changes of salinity (Ward & Halley 1985).

The main role of coastal mixing zones in dolomitization might be that of hydrologic pumps for seawater dolomitization, rather than that of a geochemical environment favourable for dolomitization (Machel & Mountjoy 1990). Seawater is driven through the sediments both by the hydrologic action of freshwater–seawater mixing at the seaward margin of a coastal mixing zone, and by tidal pumping. Hence, over time, substantial amounts of seawater may pass through the sediments. Dolomitization would be facilitated by seawater in which the kinetic barriers to dolomitization have been sufficiently lowered, and extensive dissolution is absent. However, the mixing model, as originally proposed with waters of about 10–30% seawater, would form caves with very small amounts of dolomite.

Hypersaline environments and the reflux and sabkha models

Hypersaline environments have salinities greater than that of normal seawater and are widespread at latitudes of less than about 30°, although some occur at higher latitudes. Hypersaline environments thus defined include the so-called mesohaline (also called penehaline: Adams & Rhodes 1960) environments, which are mildly hypersaline, i.e. between normal seawater salinity (35–36 g l^{-1}) and that of gypsum saturation (about 120 g l^{-1}). In all these environments dolomite is formed from water in which the salinity is controlled by surface evaporation, that is, in near-surface and shallow-burial diagenetic settings (Fig. 18). However, the latest numerical modelling suggests that reflux dolomites may also form at intermediate burial depths, as discussed below.

Whereas most dolomites formed from evaporated seawater are post-depositional and form via reflux, a few are penecontemporaneous and form in sabkhas. The latter are discussed in this section because of the geochemical affinity of the two settings, and because sabkhas often grade into evaporative lagoons.

Reflux model. The (evaporative) reflux model, illustrated in various forms in Figures 5, 19B, 20 and 21, was originally proposed by Adams & Rhodes (1960) for seawater evaporated beyond gypsum saturation in lagoonal and shallow-marine settings on a carbonate platform behind a barrier, such as a reef. Surface-water circulation on such a platform is severely restricted because of the barrier, leading to evaporation and a landward salinity gradient. The evaporated seawater flows downward into and seaward through the platform sediments because of its increased density (i.e. active reflux), thereby dolomitizing the penetrated sediments.

This model was first applied to stratiform dolostones that extend over several hundred square kilometres in the Permian Basin of west Texas and New Mexico (Adams & Rhodes 1960). A few years later, this type of reflux was found 'in action' in the Pekelmeer, a lagoon on the island of Bonaire (Deffeyes *et al.* 1965), albeit on a much smaller scale of only a few square kilometres. The reflux model has since become one of the most popular and enduring models of dolomitization, often invoked to explain pervasive dolomitization of entire carbonate platforms and, on an even larger scale, of entire sedimentary basins (Shields & Brady 1995; Potma *et al.* 2001).

Early numerical modelling by Simms (1984) and Kaufman (1994) showed that mesohaline reflux is possible in principle, and is capable of forming dolomite. The latest numerical modelling has reinforced the viability and enlarged the scope of the model, while also placing limits on the possible extent of reflux dolomitization and the amounts of dolomite formed. Jones & Rostron (2000) and Jones *et al.* (2002, 2003, 2004) modelled evaporative reflux with concomitant dolomitization in a carbonate platform of several hundred kilometres width and about 3 km thickness. In the model the platform is flooded by seawater that increases in salinity from normal at the platform edge to 150 g l^{-1} at the coastline, similar to the conditions in the original reflux model of Adams & Rhodes (1960). Figure 21 illustrates these boundary conditions, along with representative distributions of stream lines and salinity contours within the platform. The distributions shown are established after 500 000 years and are representative of future time steps. Under the chosen conditions, the platform is penetrated by mesohaline 'active reflux' at depths that were not anticipated in the original reflux model, down to several hundred metres. Jones *et al.* (2002) also recognized a hitherto unknown type of flow that they termed 'latent reflux'. In their model, latent reflux is predicted to occur following the cessation of brine generation at the platform top after flooding of the platform with seawater of normal salinity, such as after a significant rise in sea-level. Latent reflux is driven by the greater density of the earlier generated subsurface brines of reflux origin that continue to sink and disperse laterally. At the same time, seawater is entrained (sucked in from above) through the platform top. Latent reflux, like active reflux during brine generation, has the potential to form dolomite, albeit in much smaller amounts. This is because the brine and the entrained seawater together move more slowly and contain less Mg than a pure brine reflux system.

One might think that active plus latent reflux would dolomitize any carbonate platform rapidly and completely. Using realistic assumptions for repeated eustatic sea-level fluctuations that flooded the Devonian Grosmont platform episodically over a period of 1.6 Ma, near the maximum time available for reflux, Jones *et al.* (2003) found that the combined action of active and latent reflux could only form discrete layers of dolostone that alternate with undolomitized limestone. A platform can only be dolomitized completely if it has very high permeabilities and does not contain effective aquitards (such as shale or evaporite layers), and if reflux is permitted to persist for a relatively long time (Jones *et al.* 2003, 2004). At present it is not clear just how commonly such conditions are or have been realized in Nature. Also, it is to be expected that gypsum and/or anhydrite layers would form close to or at the sediment–water interface if the brines were evaporated past 120 g l^{-1} salinity. Such layers of calcium sulphate would tend to be effective aquitards and would thus supress deeply penetrating reflux and 'sucking in' of seawater during times of latent reflux, leading to near-surface brine and seawater runoff that would effectively inhibit dolomitization at greater depths (Machel *et al.* 1996*b*). Alternatively, evaporite aquitards could act to focus reflux at carbonate–evaporite interfaces, as proposed by Adams & Rhodes (1960) and observed on a small scale in the MacLeod Evaporite Basin (Logan 1987).

There are several examples of localities that were probably dolomitized by evaporative reflux, including the type location of the reflux model, the Permian carbonates of west Texas and New Mexico (Adams & Rhodes 1960). In these locations reflux was responsible for dolomitization of lagoonal carbonate sediments. These dolomites are fine to medium crystalline and matrix-selective, commonly with good–excellent fabric preservation, as illustrated in Figures 5 and 9. In addition, they may be intergrown with abundant gypsum and anhydrite in layers and nodules that appear to be cogenetic. This shows that the brines must have been evaporated past gypsum saturation, at least episodically. Recent case studies have shown that reflux can also form dolomite that is free of calcium sulphates, if the brines are mesohaline/penehaline. One example is the peritidal Jurassic carbonates of Gibraltar (Qing *et al.* 2001), formed in a situation corresponding to the model illustrated in Figure 21. Melim & Scholle (2002) investigated another example of apparent mesohaline reflux dolomitization, albeit much less extensive and restricted to fractures in the Capitan Reef, which forms a barrier at the platform margin of the Permian reflux model type location. Such dolomitization can be expected where the lagoonal, platform-interior sediments have been dolomitized almost completely near the surface, and where the dolomitizing fluids have either lost all their calcium sulphate, forming layers or nodules (as suggested above), or never reached gypsum saturation in the first place (Jones *et al.* 2003).

Whether active or latent, all refluxing brines exit at or near the platform margin (Machel *et al.* 1996*b*, 2002; Jones & Rostron 2000; Jones *et al.* 2002, 2003, 2004), and this confines reflux dolomitization to the platform. This recognition is important because some authors have invoked evaporative reflux beyond the platform margin and/or on a basin-wide scale (Shields & Brady 1995; Potma *et al.* 2001). This notion is clearly incorrect because it is physically and hydrologically impossible. The brines simply do not have enough energy to flow through the sediments and rocks beyond the platform margin (Machel *et al.* 1996*b*, 2002; Jones & Rostron 2000; Jones *et al.* 2002, 2003, 2004).

Sabkha model. The sabkha model is hydrologically and hydrochemically related to the reflux model yet differs in several important aspects. Sabkhas are intertidal–supratidal deflation surfaces that are episodically flooded. The sabkha of the Trucial Coast of Abu Dhabi is the type location of the sabkha dolomitization model. It is probably the best researched recent hypersaline intertidal–supratidal flat (Butler 1970; McKenzie *et al.* 1980; Patterson & Kinsman 1982; Müller *et al.* 1990; Baltzer *et al.* 1994), and is also representative of prolific reflux dolomite formation, as most sabkhas elsewhere produce much less dolomite.

In the Abu Dhabi sabkha, the Mg for dolomitization is supplied synsedimentarily (penecontemporaneously) by seawater that is propelled periodically onto the lower supratidal zone and along remnant tidal channels by strong onshore winds. The seawater has normal to slightly elevated salinity (up to about 38 g l^{-1}) but becomes significantly evaporated beyond gypsum saturation on and within the supratidal flats, through which it refluxes via its increased density, similar to flow in the reflux model. Sabkhas undergo hydrological and hydrochemical cycles as a result of their episodic flooding. A full cycle consists of three phases (McKenzie *et al.* 1980): storm-driven flooding of the near-coastal supratidal flats (and tidal channels); capillary evaporation; and evaporative pumping. The hydrogeological characteristics of the Abu Dhabi sabkha have been confirmed in the study by Müller *et al.* (1990), which prompted these authors to rename the sabkha model the 'flood recharge–evaporative pumping model'. The last part of the cycle, evaporative pumping, briefly gained the status of an independent model (Hsü & Siegenthaler 1969; Hsü & Schneider 1973), but most researchers soon abandoned this independent status.

Sabkha dolomite appears to form via evaporative pumping in a narrow (1–1.5 km) fringe next to the strandline, and in flooded tidal channels that extend farther landward. The distribution of dolomite is uneven. In the Abu Dhabi sabkha, the best dolomitized parts contain from 5 to about 65 wt% protodolomite. Dolomite forms as a cement and aragonite is replaced, but lithification does not occur, or only partially. Dolomitization is restricted to the upper 1–2 m of the sediments and appears to be most intense where the pore waters become chemically reducing, leading to enhanced carbonate alkalinity via sulphate reduction and/or microbial methanogenesis. In this respect, sabkha dolomitization is related to the organogenic/microbial model of dolomitization (see above). Not surprisingly, therefore, sabkha dolomites are texturally and geochemically similar to organogenic dolomites in some respects; they tend to form as protodolomite and may have reduced carbon isotope ratios. However, the oxygen isotope ratios of sabkha dolomites tend to be enriched because of evaporation. Another difference is their association with gypsum and anhydrite, common in sabkhas but missing in hemipelagic and pelagic settings. Sabkha sulphates are formed as by-products of dolomitization in texturally distinctive varieties (nodules, chicken-wire). As a result of repeated eustatic and/or relative sea-level changes, sabkhas commonly form distinctive shallowing-upward cycles that consist of undolomitized shallow-marine or lagoonal sediments at the base, overlain by dolomitized intertidal algal mats that grade up into dolomitized supratidal sediments that contain sulphates (Butler 1970; McKenzie *et al.* 1980).

In most respects, the Abu Dhabi sabkha appears to be a good recent analogue for dolomitization in many ancient intertidal–supratidal flats, such as landward of the famous Permian Capitan Reef complex in Texas and New Mexico. Rather than forming reservoir rocks, these dolostones – including the associated evaporites – generally form tight seals for underlying hydrocarbon reservoirs (e.g. Major *et al.* 1988; Harris & Walker 1990; Machel & Longacre 2000). More generally, sabkhas and similar intertidal–supratidal depositional systems in more humid climates typically form small quantities of fine crystalline protodolomite in thin beds, crusts or nodules, either within the upper 1–2 m of sediment or at the sediment surface. Repeated transgressions and regressions may stack such sequences upon one another to cumulative thicknesses of several tens of metres.

Two ancient examples shed further light on the dynamics and potency of sabkha dolomitization. One is the aforementioned Permian carbonates of Texas and New Mexico. Mutti & Simo (1994) found that the efficiency of sabkha dolomitization was variable during transgression and regression, and suggested that dolomitization during transgressive cycles affected only the intertidal and supratidal facies, whereas during regression it affected supra-, inter- and subtidal facies. These authors also speculated that tidal pumping may have aided in supplying Mg. The other example is the Ordovician Knox Group in the Appalachians, USA, one of the best-documented cases of sabkha dolomitization in ancient carbonates. The Knox carbonates consists of multiple metre-scale dolomitized cycles that formed as a result of fourth- and fifth-order eustatic sea-level changes (Montañez & Reid 1992*a*, *b*; Montañez 1997). These authors found that the transgressive cycles were not dolomitized, whereas the facies within regressive cycles were almost completely replaced by tight fine-crystalline sabkha dolomite. Thus, the timing of dolomitization relative to sea-level fluctuations appears to differ from that in the Permian carbonates of Texas and New Mexico.

Seawater dolomitization

Post-depositional formation of massive dolostones can also be attributed to the 'seawater dolomitization model' or 'seawater dolomitization' (Purser *et al.* 1994). These terms have been in common use for only about 10 years. Strictly speaking, they do not constitute or identify an independent dolomitization model. Rather, 'seawater dolomitization' refers to a group of models whose common denominator is seawater as the principle dolomitizing fluid, and that differ in hydrology and/or depth and timing of dolomitization. All dolomites in this group are post-depositional, and the diagenetic settings range in depth from shallow to intermediate burial (Fig. 18). Penecontemporaneous dolomites that formed in and from seawater are not part of this group of models but belong to the microbial/organogenic model discussed above.

Dolomitization by seawater appears to be a relatively recent addition to the array of dolomitization models. In modern times, Land (1985) was the first to advocate the notion that post-depositional dolomitization by seawater should be common in the geological record. In the 1980s this idea was unusual and other models were very much in vogue, as it was common and uncontested knowledge that the vast majority of modern marine environments are devoid of dolomite, suggesting that dolomite does not normally form from seawater because of kinetic inhibition. Thus, dolomitization from seawater was disregarded as a viable process by most authors, except for the formation of traces of penecontemporaneous, microcrystalline (proto-) dolomites that were known to form in some hemipelagic settings. However, Land (1985) recognized that seawater is by far the most common natural Mg-rich fluid, and that there had to be mechanisms to pump seawater through carbonates at considerable depths long after deposition, whereby the kinetic barriers to dolomitization are somehow reduced. This advance opened new avenues in dolomite research, now grouped as 'seawater dolomitization model(s)'.

The main credit for these 'models' of dolomitization must go to Dana (1843, again cited by van Tuyl 1914), who was the first to advocate alteration of calcite to dolomite by seawater for the dolomitic reef rock of the coral islands of the Pacific. In 1852, in discussing the origin of a dolomitic coral limestone from the Island of Metia, Dana stated (van Tuyl 1914, p. 275): 'We cannot account for the supply of magnesia except by referring to the magnesium salts of the ocean. It is an instance of dolomitization during consolidation of the rock beneath seawater'. In 1872 Dana extended his 'marine alteration theory' to evaporated seawater in lagoons, and in his *Manual of Geology* (1895) wrote: 'If this is the true theory of dolomite-making, then great shallow areas or basins of salt-pan character must have existed in past time over various parts of the continental area and have been the result of oscillation of the water level... The frequent alternation of calcite and dolomite strata would indicate alternations between the clear water and salt-pan conditions' (cited in van Tuyl 1914, p. 275). From today's point of view, Dana's insights are positively remarkable, as he was clearly almost 100 years ahead of his time. In the early 1900s the 'marine alteration theory' and its variants proposed by a number of other authors was the most popular of the then-existing models for the formation of dolomite (van Tuyl 1914). Skeats (1903, cited in van Tuyl 1914, p. 282) even provided a list of conditions favourable for the formation of dolomite masses that is almost identical to the list in use today, both in case studies and in numerical modelling:

(1) shallow water, between 0 and 150 feet in depth, and corresponding to a pressure of 1–5 atm;
(2) the presence of carbon dioxide in abundance, causing the partial solution of the limestones and the possibility of chemical interchange with the magnesium salts in seawater;
(3) porosity of the limestones, allowing percolation of seawater through the mass of the rocks;
(4) sufficiently slow subsidence of elevation to render the change from calcite to dolomite complete.

Curiously, dolomitization by seawater or evaporated seawater then went out of fashion, only to be rediscovered and embellished long after World War II in the models involving seawater that we recognize today.

The Cenozoic dolostones of the Bahamas platform, often used as an analogue for older dolomitized carbonate platforms elsewhere, can be considered the type location for seawater dolomitization. Petrographic and geochemical data indicate that seawater and/or chemically slightly modified seawater was the principle agent of dolomitization at shallow–intermediate depths and commensurate temperatures. The compositional modifications were caused by slight evaporation and/or water–rock interaction (Dawans & Swart 1988; Vahrenkamp *et al.* 1991; Vahrenkamp & Swart 1994). The hydrology of seawater during dolomitization is still very much contested. Various flow systems (summarized in Fig. 20) have been invoked to drive the large amounts of seawater needed for pervasive dolomitization through the Bahamas platform: thermal convection (Sanford *et al.* 1998); a combination of thermal seawater convection and reflux of slightly evaporated seawater derived from above (Whitaker *et al.* 1994); or seawater driven by an overlying freshwater–seawater mixing zone during partial platform exposure (Vahrenkamp & Swart 1994), possibly layer-by-layer in several episodes (Vahrenkamp *et al.* 1991). Thermal convection is discussed below under 'burial models', as it necessarily occurs under considerable (at least intermediate: Fig. 18) burial, unless the heat source is a local hot spot, such as an igneous intrusion.

The Bahamas dolostones represent a hybrid with respect to the traditional, conventional classifications of models. Petrographic and geochemical data indicate seawater as the principal dolomitizing agent, yet thermal convection, as a hydrologic system and drive for dolomitization, is better classified under the burial (subsurface) models discussed below. Analogously, the regionally extensive Devonian dolostones in Alberta, western Canada, are also a hybrid. These dolostones probably formed at depths of 300–1500 m at temperatures of about 50–80 °C from chemically slightly modified seawater, and have been classified as burial dolostones (Amthor *et al.* 1993; Machel *et al.* 1994; Mountjoy & Amthor 1994; Mountjoy *et al.* 1999). The regionally extensive dolostones of the Carboniferous of Ireland that are petrographically and geochemically very similar to the Devonian dolostones of Alberta, and whose genesis has been interpreted in an analogous manner (Gregg *et al.* 2001), are another Palaeozoic example. In both cases, the hydrology that facilitated dolomitization is unclear, with thermal convection, reflux, compaction, tectonic expulsion, or a combination thereof as theoretically viable alternatives. Mesozoic examples of this type of dolomitization are the regionally extensive dolostones of the Cretaceous Soreq Formation in Israel investigated by Sass & Katz (1982). All of these Palaeozoic and Mesozoic dolostones can be (re-) classified along with the Cenozoic Bahamas dolostones as 'seawater dolomites'. This classification dilemma arises from the historical evolution of our understanding of these dolostones. This conflict does not invalidate the earlier 'burial' interpretations, which were and are correct.

van Tuyl (1914, p. 334) summarized the various examples of 'alteration theories': 'It is not possible to say in all cases whether the dolomitization took place while the limestone was still beneath the sea, through the agency of seawater, or after its emergence through the agency of ground water'. He was right.

Intermediate–deep burial (subsurface) environments and models

Burial (subsurface) environments are those removed from active sedimentation by burial, and in which the pore-fluid chemistry is no longer entirely governed by surface processes, i.e. where water–rock interaction has modified the original pore waters to a significant degree, or where the fluid chemistry is dominated by subsurface diagenetic processes. Such environments are found in intermediate–deep burial settings (Fig. 18) and are characterized by chemically reducing conditions. These are reflected in the mineralogy of redox-sensitive compounds, such as ferroan carbonates rather than non-ferroan carbonates, iron sulphides rather than iron oxides, and hydrocarbons in fluid inclusions, stylolites, etc. (Machel 1999).

The textures, porosities and permeabilities of dolostones formed in intermediate- and deep-burial settings vary. Except for dolomite dissolution textures (Fig. 9), which appear to form largely in near-surface and shallow-burial settings permeated by meteoric water, matrix-selective dolomitization and related textures (Figs 6–8) are as common as in shallower diagenetic settings. Hence, these textures alone are not indicators of depth of burial. Three specific characteristics may be used:

(1) dolomites cross-cut by stylolites suggest burial of at least 600 m; stylolites in dolostones appear to require at least 600 m of burial, as implied by the studies of Lind (1993) and Fabricius (2000);
(2) development of nonplanar crystal textures and coarse planar textures at temperatures in excess of about 60 °C;
(3) the presence of saddle dolomite suggests temperatures of formation in excess of about 80 °C.

All burial (subsurface) models for dolomitization are essentially hydrological models. They differ mainly in the nature of the drives and direction(s) of fluid flow (e.g. Morrow 1982*b*, 1999). Four main types of fluid flow take place in subsurface diagenetic settings: (1) compaction flow; (2) thermal convection; (3) topography driven flow; and (4) tectonically driven flow. Combinations of these flow regimes and fluids are possible under certain circumstances. In addition, hydrothermal or hydrofrigid fluids may be injected into any burial setting where fractures open up (Fig. 18).

Compaction model. The oldest burial model of dolomitization is the compaction model (Illing 1959; Jodry 1969). According to this model (Fig. 19D, showing only one variant of subsurface compaction flow), seawater or its subsurface derivative(s) buried along with the sediments are pumped through the rocks at several tens to several hundreds of metres as a result of compaction dewatering.

The compaction model in its original form was never especially popular because it rapidly became clear that burial compaction could generate only rather limited amounts of dolostone due to the limited amounts of compaction water (Morrow 1982*b*; Land 1985; mass-balance calculations for specific cases: Machel & Anderson 1989; Amthor *et al.* 1993). However, despite the mass-balance constraint, the compaction model remains a viable alternative for burial/subsurface dolomitization where focusing (funnelling) of the compaction waters through relatively small volumes of limestones is possible. This may happen on a local as well as on a regional–basinal scale. Jodry (1969) was the first to recognize the necessity of focusing and advocated this process for the dolomitization of Silurian reefs in Michigan that are encased in aquitards. Similarly, Machel & Anderson (1989) advocated focused compaction flow as one of two viable alternatives for dolomitization of Upper Devonian reefs in Alberta. On a regional–basinal scale, compaction flow in typical asymmetrical basins is mainly 'up and out' laterally through aquifers, and is cumulative along the flow path (e.g. Garven & Freeze 1984). Very high amounts of Mg can be supplied for dolomitization on this scale, especially where only a part of a geological unit is permeable, such as a carbonate platform margin into which the waters are focused (represented in Fig. 19D2). This may have happened in the several hundred kilometres long Cooking Lake platform margin with the overlying Rimbey–Meadowbrook reef trend in Alberta, Canada. The available data are consistent with focusing of compaction-driven seawater and/or longer-range expulsion of formation fluids from far down-dip where the foreland basin of the Antler Orogen was located (Amthor *et al.* 1993, 1994; Machel *et al.* 1994; Mountjoy & Amthor 1994; Mountjoy *et al.* 1999). This process may be much more common than generally recognized.

Thermal convection models. Thermal convection is driven by spatial variations in temperature that result in changes in pore-water density and thus effective hydraulic head. Variations in temperature may be due to elevated heat flux in the vicinity of igneous intrusions (Wilson *et al.* 1990), the lateral contrast between warm platform waters and cold ocean waters (Kohout *et al.* 1977), or lithology controlled variations in thermal conductivity, for example where carbonates are overlain by thick evaporites (Combarnous and Bories 1975; Wood and Hewett 1982; Phillips 1991; Jones *et al.* 2004). Thermal convection is classified as 'open', 'closed', or 'mixed' (Raffensberger & Vlassopoulos 1999).

Open convection cells (also called half-cells) may form in carbonate platforms that are open to seawater recharge and discharge laterally and at the top (Figs 19C1 & C2 and 20). This type of convection was first recognized in carbonate platforms by Kohout *et al.* (1977), and thus was named Kohout convection. It has been numerically modelled by Simms (1984), Kaufman (1994), Sanford *et al.* (1998), Wilson *et al.* (2001), and Whitaker *et al.* (2002, 2003, 2004). The most recent studies modelled dolomitization and/or calcium sulphate formation resulting from this type of thermal convection in reactive-transport simulations, and also considered platforms that are open to seawater recharge on one side only (platform geometry similar to that shown in Fig. 21). These studies found that the magnitude and distribution of permeability are the most important parameters governing flow and dolomitization, and that Kohout convection is active to a depth of about 2–3 km, provided that the sequence does not contain effective aquitards, such as (overpressured) shales or evaporites. One result of this modelling is especially noteworthy. Dolomitization is most favoured at the depth where the ambient temperature is around 50–60 °C, i.e. 0.5–2 km, depending on the geothermal gradient. Above this thermal regime the formation of dolomite is severely limited because of the low ambient temperatures, whereas at greater depths it is equally limited by very low permeabilities. In other words, below 2–3 km depth compaction has reduced porosity and permeability to levels that are too low to sustain viable convection cells. The models by Whitaker *et al.* (2002, 2003, 2004) also indicate that even at a moderate width of only 40 km, complete dolomitization in a 2 km-thick sequence takes about 30–60 Ma, much longer than most carbonate platforms remain laterally open to seawater recharge. Hence, most carbonate platforms, even if subjected to thermal convection by seawater, would at best get only partially dolomitized during the time that they are open to seawater recharge.

Furthermore, numerical modelling by Jones *et al.* (2002, 2003, 2004) has shown that thermal convection is overpowered rather easily by reflux, where the latter is established due to surface evaporation (see Fig. 21 and the discussion above). Thus, whether or not open-cell thermal convection is established or leads to dolomitization depends mainly on the permeability distribution within the platform (especially the presence of effective aquitards), but also on the presence or absence of reflux, and the amounts of dolomite formed are constrained by the time the platform is open to seawater circulation.

Kohout convection, as defined above, is enhanced by basin to platform relief (Sanford *et al.* 1998), as shown in Figures 19 and 21, but is also predicted to occur in ramp situations with relatively shallow basins, like the Persian Gulf (Jones 2000). Open convection may occur in the absence of such topographic relief, i.e. far in the interior of an epeiric-scale platform. In this situation, the carbonate strata are open to recharge (and discharge) only at the platform top. Given significant variations in groundwater temperatures and sufficient permeability, it is conceivable that thermal convection could take place in such a situation, at least during shallow–intermediate burial (about 500–1500 m) while the strata are still highly permeable (Fig. 22). Based on circumstantial evidence, this type of open convection was advocated as an alternative for dolomitization of the Nisku reefs near the centre of the Alberta Basin (Machel & Anderson 1989). However, the conditions necessary for this type of open thermal convection await confirmation by appropriate modelling.

In this context, it is worth recalling that the four cases of regional platform and reef dolomitization noted previously (Bahamas, western Canada, Ireland, Israel) have similar petrographical and geochemical characteristics that point to seawater dolomitization of the bulk of the rocks at depths of about 500 m to at most 1–2 km, and temperatures of about 50–80 °C. These depths and temperatures happen to coincide with the depth and temperature range within which Kohout convection is most conducive to dolomitization (see above). The most sophisticated, recent modelling results thus confirm what has been inferred from petrographic–geochemical case studies of massive dolomitization for more than 15 years, i.e. a favourable environment exists for extensive dolomitization by seawater at temperatures of 50–60 (80) °C and commensurate depths of 0.5 to a maximum of 2 km. Hence, it appears likely that these carbonate platforms and reefs were indeed dolomitized by open thermal convection of seawater in one of the scenarios discussed above.

Significantly higher temperatures would favour higher convective fluxes, provided the rocks are permeable enough (Combarnous & Bories 1975; Wood & Hewett 1982; Phillips 1991). Thermal convection half-cells would thus be especially vigorous when a platform is underlain or penetrated by an igneous intrusion, which should result in especially fast and pervasive dolomitization. This appears to have happened in the Triassic Latemar reef in the Italian Alps (Wilson *et al.* 1990). The Latemar reef is dolomitized in a mushroom-shaped body in which oxygen isotope ratios and fluid-inclusion temperatures can be contoured to an underlying igneous intrusion, and in which replacive dolomitization appears to have occurred rapidly from seawater heated to about 200 °C.

Thermal convection can also occur in closed cells, referred to as 'free convection' by some authors (Fig. 19D3). In principle, this can happen in any sedimentary basin over tens to hundreds of metres thickness, provided that the temperature gradient is high enough relative to the permeability of the strata. As a rule of thumb, however, such convection cells will only be established and be capable of dolomitizing a carbonate sequence of interest if this sequence is of substantial thickness (several hundred metres), highly permeable and not interbedded with aquitards (Combarnous & Bories 1975; Wood & Hewett 1982; Bjørlykke *et al.* 1988; Phillips 1991). Such conditions are very rare in typical sedimentary basins, most of which contain effective aquitards. Furthermore, even if closed convection cells are established the amounts of dolomite that can be formed are severely limited, to an even greater extent than in compaction flow, by the pre-convection Mg-content, as no new Mg is supplied to the system. It appears, therefore, that extensive, pervasive dolomitization by closed-cell thermal convection is highly unlikely (Jones *et al.* 2003, 2004). Nevertheless, thermal convection of this type has been suggested, at least in principle, for pervasive dolomitization in carbonate platforms and regional carbonate aquifers (e.g. Morrow 1999).

Mixed convection is a variant of thermal convection and occurs when flow driven by an external hydraulic gradient interacts with thermal convection cells (Raffensberger &

Vlassopoulos 1999). Under such conditions Mg can be supplied to otherwise closed convection cells, thus increasing the potential for dolomitization (see Whitaker *et al.* 2004).

Topography driven model. Topography driven flow takes place in all uplifted sedimentary basins that are exposed to meteoric recharge on scales from a few tens of kilometres to that of whole basins (Tóth 1988; Garven 1995). Pore-water flow geometrically resembles the pattern shown in Figure 18 (allowing for the assumption, made here for illustrative simplicity, that the subsurface is hydrologically isotropic and homogeneous). With time, topography can drive enormous quantities of meteoric water through a basin, commonly concentrating it by water–rock interaction (especially salt dissolution), and preferentially focusing it through aquifers. However, volumetrically significant dolomitization can take place only where meteoric water dissolves enough Mg en route before encountering limestones. This does not appear to be common. At present there are no proven cases of extensive dolomitization via topography driven flow, with the possible exceptions of Cambrian carbonates in Missouri (Gregg 1985) and Cambrian–Ordovicain carbonates in the southern Canadian Rocky Mountains (Yao & Demicco 1995). It appears that these strata may have been affected by vigorous topography driven flow, but insufficient evidence is available to demonstrate that the flow system(s) contained enough Mg for regional dolomitization.

Tectonic (squeegee) model. Another type of flow that has been suggested to result in pervasive dolomitization is tectonically driven squeegee-type flow (Oliver 1986). In this type of flow system, metamorphic fluids are expelled from crustal sections affected by tectonic loading so that basinal fluids are driven towards the basin margin (Fig. 19D2). Such fluids could be injected into compactional and/or topography driven flow, with attendant fluid mixing. Tectonically driven flow was invoked by several authors to explain extensive dolomitization. These include Dorobek (1989) for the Siluro-Devonian Helderberg Group, USA, and Drivet & Mountjoy (1997) for Devonian reefs in western Canada. Similarly, Montañez (1994) invoked extensive burial dolomitization via tectonic loading in Ordovician carbonates in several thrust sheets of the southern Appalachians. However, it is unlikely that tectonically-induced flow, or the related fluid mixing, form massive replacive dolostones. Modelling studies have shown that squeegee-type flow systems have low fluxes that are short lived (e.g. Deming *et al.* 1990), which has been affirmed by diagenetic studies (Machel & Cavell 1999; Machel *et al.* 2000; Buschkuehle & Machel 2002). This effectively precludes extensive dolomitization via squeegee flow. On the other hand, if the squeegee fluids are hot and flow relatively fast, and if they encounter highly porous pre-existing dolostones, the latter may significantly recrystallize, such that the textures and geochemistry reflect the hot recrystallization event rather than the original dolomitization event. This appears to have happened in the Ordovician carbonates discussed by Montañez (1994), where squeegee fluids, originally undersaturated with respect to dolomite, invaded the Knox dolostones that were partially dissolved and then recrystallized.

The same argument must apply to a new variant of the squeegee model proposed by Machel (2000) for accretionary prisms (Fig. 23). It has been shown that there is substantial pore-water flow in such geological settings, as indicated in Figure 23, and that some isotopically distinct dolomite can be formed in this way (Machel 2000). However, the amount of dolomite that can be produced is limited by the same mass-balance constraints calculated for the compaction model (see above).

High-temperature and hydrothermal dolomitization. Convection cells invariably have rising limbs that penetrate the overlying and cooler strata, linking thermal convection to hydrothermal dolomitization. However, as in dolomitization by seawater, hydrothermal dolomitization is not a model in its own right because hydrothermal conditions may occur in a variety of situations in all types of diagenetic settings from near surface to deep burial, especially where fractures transgress more than one burial-diagenetic zone (Fig. 18, left).

The possibility of hydrothermal dolomitization dates back to the earliest days of dolomite research. In 1779, 12 years before the mineral dolomite was properly defined, Arduino (cited in van Tuyl 1914, p. 288) mentioned a magnesian limestone that he believed to have been formed by alteration of ordinary limestones by volcanic activity. Then Heim (1894) and several others (cited in van Tuyl 1914, pp. 289–290) advocated dolomitization by volcanic vapours or water-bearing magnesia, and these types of dolomitization were summarized under a 'pneumatolytic alteration theory' (van Tuyl 1914). In all cases thus classified at the time, volcanics were near the sites of dolomitization.

It is not clear when the term 'hydrothermal' was first used in the context of dolomitization, but it is frequently misused, giving rise to the relatively new bandwagon of 'hydrothermal dolomitization' on the scale of entire sedimentary basins. Most commonly, dolomites are called hydrothermal on the basis of two observations: (a) the dolomite is saddle dolomite (e.g. Davies 1997, 2002); or (b) the dolomite, whatever its texture, is associated with base metal mineralization (e.g. Auajjar & Boulègue 2002). Both observations are insufficient for a viable identification of hydrothermal activity.

Using White's (1957) time-honoured definition of hydrothermal, including Stearns *et al.*'s (1935) requirement for a 'significant' temperature difference of at least 5–10 °C, a mineral can be described as '*hydrothermal*' only if it is demonstrated to have formed at a temperature that was 5–10 °C higher than the temperature of the surrounding strata, regardless of fluid source or drive. If a mineral was formed at or near the same temperature as the surrounding rocks (within 5–10 °C), it can be described as 'geo*thermal*', whatever the geothermal gradient. The qualifier 'geothermal' may be omitted, unless special emphasis needs to be placed on the geothermal nature of a particular mineralization event. Minerals formed at temperatures significantly lower than ambient (by >5–10 °C) can be described as '*hydrofrigid*', even if they formed at a rather high temperature (Machel & Lonnee 2002).

There are well-documented examples of hydrothermal dolomite on a local as well as on a regional scale. Most are rather small and restricted to the vicinity of faults and fractures and/or localized heat sources (Fig. 19D4). One striking case of this type is the Pb–Zn-mineralized Navan dolomite plume in Ireland (Braithwaite & Rizzi 1997), another is the dolomitized plume of the Latemar build-up in the Italian Alps (Wilson *et al.* 1990). There are also examples of larger scale, even regionally extensive, hydrothermal dolomitization (e.g. Spencer-Cervato & Mullis 1992; Qing & Mountjoy 1992, 1994; Duggan *et al.* 2001) (Fig. 24).

Unfortunately, examples of misinterpretations of hydrothermal dolomitization abound. Dolomite is commonly syngenetic with base-metal minerals in hydrothermal systems, but this is no justification for assuming that all dolomite(s) formed in association with base-metal mineralization are hydrothermal. A study by Auajjar & Boulègue (2002) of dolomites in Liassic rocks of Morocco illustrates this issue. Three paragenetically different dolomites were interpreted as hydrothermal, but only their dolomite type 2 is associated and probably syngenetic with Pb–Zn sulphides. The other two 'hydrothermal dolomites' are not associated with sulphides, and unfortunately no palaeotemperature data have been provided for these dolomites or for their host rocks. On a larger scale, Davies (1997, p. 59) asserted that an 'HTD (hydrothermal dolomite) overprint in Devonian carbonates in Alberta ... often attributed to burial "matrix" processes may be the product of hydrothermal fluid migration'. There is no credible evidence for most of these dolomites being hydrothermal in origin and/or having a hydrothermal 'overprint', except for isolated cases, such as in the Wabamun Group and the Keg River Formation, which are, respectively, in the upper and lower of the four stratigraphic levels of the Devonian (Machel & Lonnee 2002).

Furthermore, hydrothermal dolomitization must be separated from hydrothermal alteration of pre-existing dolomites. Hot and/or hydrothermal fluids often ascend via faults in geologically short time spans and with relatively low fluxes (compared to the overall rock volume outside of the fractures). If such fluids are undersaturated with respect to dolomite, which is not uncommon, they will lead to dissolution and recrystallization of pre-existing dolostones that make up the wall rocks, analogous to the case of deep burial dissolution and recrystallization in the squeegee system discussed by Montañez (1994). A good example of this type of hydrothermal alteration originating from fault systems is in Devonian carbonates of western Canada (Lonnee & Machel 2004).

Texturally, most high-temperature and hydrothermal dolomites are distinct. First, most dolomite replacing limestone at temperatures in excess of about 60 °C is medium–coarse crystalline, nonplanar and/or planar with a relatively narrow size distribution. As Sibley & Gregg (1987) and Gregg (2004) pointed out, similar textures and crystal size distributions also result from recrystallization of older, low-temperature dolomites. Hence, these textures by themselves are not indicative of limestone replacement or dolomite recrystallization. However, textures such as those shown in Figures 11 and 12 may be taken as evidence for high-temperature dolomitization, rather than dolomite recrystallization. This is because high-temperature replacement of limestone, whether hydrothermal or not, tends to be fabric-obliterative, as discussed previously. Furthermore, much if not most high-temperature dolomite, especially when grown as

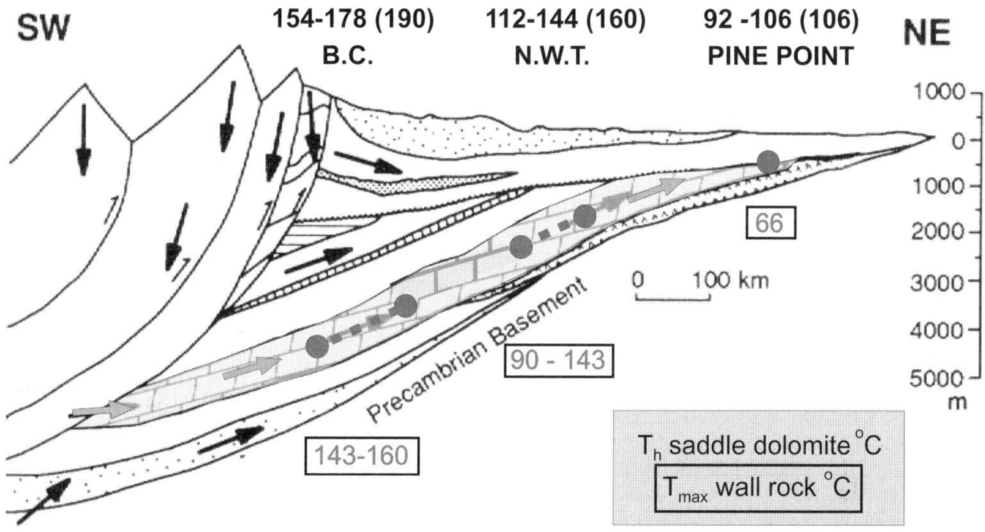

Fig. 24. Hydrothermal Presqu'ile saddle dolomite aquifer. The figure is modified from Qing & Mountjoy (1992, 1994).

cement into macropores, is saddle dolomite. This does not mean, however, that all saddle dolomite is hydrothermal. It can be formed in at least three ways: i.e. from advection (fluid flow); local redistribution of older dolomite during stylolitization; and as a by-product of thermochemical sulphate reduction in a closed or semi-closed system, as discussed above. Only the first and the last of these possibilities have a chance of being hydrothermal.

Reimer & Teare (1992) and Reimer *et al.* (2001) proposed that breccias cemented with saddle dolomite encased in limestone formed in a so-called 'HTD-furnace' ('hydrothermal dolomite furnace'), and that thermochemical sulphate reduction (TSR) initiated and promoted such dolomitization. This 'TSR–HTD model' is partially based on the notion that TSR is exothermic (Reimer & Teare 1992; Reimer *et al.* 2001). However, such saddle dolomite bodies are likely to be hydrofrigid where associated with TSR, and TSR did not initiate such dolomitization. First, it is not justified to assume that all or even most TSR settings are hydrothermal. Simpson *et al.* (1996) and Simpson (1999) have shown that TSR is probably endothermic in many, if not most, cases. Secondly, most TSR settings are closed or nearly closed hydrodynamically (e.g. Machel 2001), whereas dolomitization requires an open system because of the requirement to deliver Mg. At best, TSR may coincide with dolomitization in such a setting and add some oxidized carbon to the saddle dolomite. On the other hand, where brecciated dolomite bodies such as those discussed by Reimer & Teare (1992), Reimer *et al.* (2001), and seen in many MVT occurrences (Gregg 2004), formed without the involvement of TSR, and where dolomitization was caused by fluid flow ascending through faults, saddle dolomite bodies are commonly hydrothermal.

Secular distribution of dolostones

The relative abundance of dolostones that originated by the replacement of marine limestones appears to have varied cyclically through time. Early data suggested that dolomite was most abundant in rocks of the early Palaeozoic systems and decreased in abundance with time (van Tuyl 1914, table 1, with a reference to Daly 1909). Relatively recent reassessments of the dolomite distribution throughout time reveal discrete maxima of 'significant early' dolomite formation, i.e. massive early diagenetic replacement of marine limestones, during the Early Ordovician–Middle Silurian and the Early Cretaceous (Given & Wilkinson 1987). Furthermore, it is well known from geologically young carbonate platforms, such as the Bahamas Bank, that marine carbonate rocks younger than about the late Pliocene are almost devoid of dolomite.

The reason(s) for the secular variations in dolostone abundance throughout the Phanerozoic are much debated. Various explanations have been proposed, including periods of enhanced 'early' dolomite formation related to or controlled by plate tectonics that changed the compositions of the atmosphere and seawater, such as increased atmospheric CO_2 levels, high eustatic sea levels, low saturation states of seawater with respect to calcite, changes in the marine Mg/Ca ratio or low atmospheric O_2 levels that coincided with enhanced rates of bacterial sulphate reduction (see discussions in Stanley & Hardie 1999; Burns *et al.* 2000). It appears possible that a combination of two or more of these factors were involved. Perhaps the most elegant explanation was recently provided by Nordeng & Sibley (2003) who interpreted the notable absence of dolomite in carbonates younger than late Pliocene in the Bahamas Bank as a result of the lengthy induction period for dolomite formation. According to Nordeng & Sibley (2003), Bahamas carbonates older than late Pliocene are dolomitized because they remained in contact with the dolomitizing solution (seawater) long enough to exceed the induction period. On the other hand, carbonates younger than late Pliocene have not been in contact with seawater long enough, and any metastable precursors to dolomite that may have formed were readily destroyed by freshwater diagenesis during several intervening periods of exposure. This interpretation, if true, may also explain the secular variations in dolomite in the earlier Phanerozoic. Mesozoic and Palaeozoic variations in marine dolomitization may simply reflect periods of seawater contact longer or shorter than the induction period.

Summary and conclusions

(1) The thermodynamic conditions of dolomite formation are well known. The kinetics of dolomite formation are relatively poorly understood, although it is clear that there are significant kinetic barriers to formation below about 50 °C.

(2) Mass-balance calculations necessitate advection for extensive dolomitization, and this is why all models for the genesis of massive dolostones are essentially hydrological models. The exceptions are natural environments where carbonate muds or limestones can be dolomitized by diffusion of magnesium from seawater rather than by advection.

(3) The replacement of shallow-water limestones, the most common form of dolomitization, results in a series of distinctive textures that often form in a sequential manner with progressive degrees of dolomitization.

(4) Many dolostones have higher porosities than limestones, and this may be the result of several processes. There also are several processes that destroy porosity and which vary in importance from place to place. The evolution of permeability during dolomitization is also variable. Generalizations are difficult.

(5) A wide range of geochemical methods may be used to characterize dolomites and dolostones, and to decipher their origins. Of particular interest are those methods that can be used to identify the direction of fluid flow during dolomitization.

(6) Dolomites that originally form very close to the surface and/or from evaporitic brines tend to recrystallize with time and during burial. On the other hand, those that form at several hundred to a few thousand metres depth are not, or hardly, prone to recrystallization.

(7) Penecontemporaneous dolomites commonly form only small amounts (a few per cent) of microcrystalline dolomite. Many, if not most, penecontemporaneous dolomites appear to form through the mediation of microbes.

(8) Virtually all volumetrically large, replacive dolostones are post-depositional and form during some degree of burial.

(9) In its original form the mixing model of dolomitization does not provide a viable explanation for the formation of massive dolostones.

(10) Dolomitization can occur in hypersaline environments and below, either via reflux in subtidal environments (reflux model) or via reflux and/or evaporative pumping in intertidal–supratidal environments (sabkha model).

(11) Seawater dolomitization is not an independent model. Rather, the various possibilities of dolomitization by seawater form a group of models that have seawater in common as the principle source of Mg.

(12) Thermal convection in open half-cells (Kohout convection) can form massive dolostones only under favourable circumstances. Thermal convection in closed cells cannot form massive dolostones.

(13) Compaction flow cannot form massive

dolostones, unless it is funnelled. The latter may be more common than is generally recognized.

(14) Neither topography driven nor tectonically induced flow (squeegee-type) are likely to form massive dolostones, except under highly unusual circumstances.

(15) The regionally extensive dolostones of the Bahamas (Cenozoic), Israel (Mesozoic), western Canada and Ireland (Palaeozoic), can be classified both as seawater dolomites and as burial dolomites. This apparent ambiguity is a consequence of the historical evolution of dolomite research.

(16) Hydrothermal dolomitization is not an independent model. Rather, hydrothermal fluids may occur in a variety of overlapping diagenetic settings.

(17) The secular distribution of dolostones that replaced shallow-marine limestones is uneven throughout the Phanerozoic. The reasons for this phenomenon are not clear.

I am indebted to dozens of colleagues, too many to enumerate, who have helped shape my understanding of dolomites over the last 20 years. The constructive manuscript reviews by C. Braithwaite, G. Jones, J. Lonnee and, especially, J. Gregg greatly improved this paper. Financial support has been provided by the Natural Sciences and Engineering Research Council of Canada (NSERC) and the Alexander von Humboldt Foundation (AvH).

References

ADAMS, J.E. & RHODES, M.L. 1960. Dolomitization by seepage refluxion. *AAPG Bulletin*, **44**, 1912–1920.

AGI. 1999. *Illustrated Dictionary of Earth Science.* American Geological Institute. CD-ROM edition.

ALLEN, J.R. & WIGGINS, W.D. 1993. *Dolomite Reservoirs – Geochemical Techniques for Evaluating Origin and Distribution.* American Association of Petroleum Geologists, Continuing Education Course Notes, **36**.

AMTHOR, J.E., MOUNTJOY, E.W. & MACHEL, H.G. 1993. Subsurface dolomites in Upper Devonian Leduc Formation buildups, central part of Rimbey–Meadowbrook reef trend, Alberta, Canada. *Bulletin of Canadian Petroleum Geology*, **41**, 164–185.

AMTHOR, J.E., MOUNTJOY, E.W. & MACHEL, H.G. 1994. Regional-scale porosity and permeability variations in Upper Devonian Leduc buildups: implications for reservoir development and prediction in carbonates. *AAPG Bulletin*, **78**, 1541–1559.

ANDERSON, J.H. 1985. *Depositional facies and carbonate diagenesis of the downslope reefs in the Nisku Formation (U. Devonian), central Alberta, Canada.* PhD thesis, University of Austin, Texas.

ARCHIE, G.E. 1952. Classification of carbonate reservoir rocks and petrophysical considerations. *AAPG Bulletin*, **36**, 278–298.

ARVIDSON, R.S. & MACKENZIE, F.T. 1999. The dolomite problem: control of precipitation kinetics by temperature and saturation state. *American Journal of Science*, **299**, 257–288.

AUAJJAR, J. & BOULÈGUE, J. 2002. Dolomites in the Tazekka Pb–Zn district, eastern Morocco: polyphase origin of hydrothermal fluids. *Terra Nova*, **14**, 175–182.

BACK, W., HANSHAW, D.B., HERMAN, J.S. & VAN DRIEL, J.N. 1986. Differential dissolution of a Pleistocene reef in the groundwater-mixing zone of coastal Yucatan, Mexico. *Geology*, **14**, 137–140.

BADIOZAMANI, K. 1973. The Dorag dolomitization model – application to the Middle Ordovician of Wisconsin. *Journal of Sedimentary Petrology*, **43**, 965–984.

BAKER, P.A. & BURNS, S.J. 1985. Occurrence and formation of dolomite in organic-rich continental margin sediments. *AAPG Bulletin*, **69**, 1917–1930.

BAKER, P.A. & KASTNER, M. 1981. Constraints on the formation of sedimentary dolomite. *Science*, **213**, 214–216.

BALTZER, F., KENIG, F., BOICHARD, R., PLAZIAT, J.-C. & PURSER, B.H. 1994. Organic matter distribution, water circulation and dolomitization beneath the Abu Dhabi sabhka (United Arab Emirates). *In*: PURSER, B.H., TUCKER, M.E. & ZENGER, D.H. (eds) *Dolomites – A Volume in Honour of Dolomieu.* International Association of Sedimentologists, Special Publications, **21**, 409–428.

BJØRLYKKE, K., MO, A. & PALM, E. 1988. Modelling of thermal convection in sedimentary basins and its relevance to diagenetic reactions. *Marine and Petroleum Geology*, **5**, 338–351.

BLATT, H., MIDDLETON, G. & MURRAY, R. 1972 *Origin of Sedimentary Rocks.* Prentice-Hall, Englewood Cliffs, NJ.

BRADY, P.V., KRUMHANSL, J.L. & PAPENGUTH, H.W. 1996. Surface complexation clues to dolomite growth. *Geochimica et Cosmochimia Acta*, **60**, 727–731.

BRAITHWAITE, C.J.R. 1991. Dolomites, a review of origins, geometry and textures. *Transaction of the Royal Society of Edinburgh, Earth Sciences*, **82**, 99–112.

BRAITHWAITE, C.J.R. & RIZZI, G. 1997. The geometry and petrogenesis of hydrothermal dolomite at Navan, Ireland. *Sedimentology*, **44**, 421–440.

BUDD, D.A. 1997. Cenozoic dolomites of carbonate islands: their attributes and origin. *Earth Science Reviews*, **42**, 1–47.

BUDD, D.A. 2001. Permeability loss with depth in the Cenozoic carbonate platform of west-central Florida. *AAPG Bulletin*, **85**, 1253–1277.

BULLEN, S.A. & SIBLEY, D.F. 1984. Dolomite selectivity and mimetic replacement. *Geology*, **12**, 655–658.

BURNS, S.J., MCKENZIE, J.A. & VASCONCELOS, C. 2000. Dolomite formation and biochemical cycles in the Phanerozoic. *Sedimentology*, **47**, 49–61.

BUSCHKUEHLE, B.E. & MACHEL, H.G. 2002. Diagenesis and paleofluid flow in the Devonian Southesk–Cairn carbonate complex in Alberta, Canada. *Marine and Petroleum Geology*, **19**, 219–227.

BUTLER, G.P. 1970. Holocene gypsum and anhydrite of the Abu Dhabi sabkha, Trucial Coast: an alternative explanation of origin. *In*: RAU, J.L. & DELLWIG, L.F. (eds) *Third Symposium on Salt, Volume 1*. Northern Ohio Geological Society, Cleveland, OH, 120–152.

CAPO, R.C., WHIPKEY, C.E., BLACHÈRE, J.R. & CHADWICK, O.A. 2000. Pedogenic origin of dolomite in a basaltic weathering profile, Kohala Peninsula, Hawaii. *Geology*, **28**, 271–274.

CAROZZI, A.V. 1960. *Microscopic Sedimentary Petrography*. Wiley, New York.

CARPENTER, A.B. 1980. The chemistry of dolomite formation I: the stability of dolomite. *In*: ZENGER, D.H., DUNHAM, J.B. & ETHINGTON, R.L. (eds) *Concepts and Models of Dolomitization*. Society of Economic Paleontologists and Mineralogists, Special Publications, **28**, 111–121.

CHOQUETTE, P.W. & PRAY, L.C. 1970. Geologic nomenclature and classification of porosity in sedimentary carbonates. *AAPG Bulletin*, **54**, 207–250.

CHOQUETTE, P.W. & STEINEN, R.P. 1980. Mississippian non-supratidal dolomite, Ste. Genevieve Limestone, Illinois Basin: evidence for mixed-water dolomitization. *In*: ZENGER, D.H., DUNHAM, J.B. & ETHINGTON, R.L. (eds) *Concepts and Models of Dolomitization*. Society of Economic Paleontologists and Mineralogists, Special Publications, **28**, 163–196.

COMBARNOUS, M.A. & BORIES, S.A. 1975. Hydrothermal convection in saturated porous media. *Advances in Hydroscience*, **10**, 231–307.

CRUFT, E.F. & CHAO, P.C. 1970, Nucleation kinetics of the gypsum–anhydrite system. *In*: RAU, J.L. & DELLWIG, L.F. (eds) *Third Syposium on Salt*, Volume 1. Northern Ohio Geological Society, Cleveland, OH, 109–118.

DANA, J.D. 1895. *Manual of Geology: Treating the Principles of the Science with Special Reference to American Geological History, for the Use of Colleges, Academies and Schools*, 4th edn. Ivison Blakeman, Taylor and Co., New York.

DAVIES, G.R. 1997. *Hydrothermal Dolomite (HTD) Reservoir Facies: Global Perspectives on Tectonic–Structural and Temporal Linkage Between MVT and Sedex Pb–Zn Ore Bodies, and Subsurface HTD Reservoir Facies*. Canadian Society of Petroleum Geologists, Short Course Notes.

DAVIES, G.R. 2002. Thermobaric dolomitization: transient fault-controlled pressure-driven processes and the role of boiling/effervescence. *In*: *Diamond Jubilee Convention of the Canadian Society of Petroleum Geologists*, Calgary, 3–7 June 2002, Program & Abstracts, 105.

DAWANS, J.M. & SWART, P.K. 1988. Textural and geochemical alternations in Late Cenozoic Bahamian dolomites. *Sedimentology*, **35**, 385–404.

DE DOLOMIEU, D. 1791. Sur un genre de Pierres calcaires très-peu effervescentes avec les Acides, & phosphorescentes par la collision. *Journal de Physique*, **39**, 3–10. Translation of Dolomieu's paper with notes reporting his discovery of dolomite by CAROZZI, A.V. & ZENGER, D.H. 1981. *Journal of Geological Education*, **29**, 4–10.

DEFFEYES, K.S., LUCIA, F.J. & WEYL, P.K. 1965. Dolomitization of recent and Plio-Pleistocene sediments by marine evaporite waters on Bonaire, Netherlands Antilles. *In*: PRAY, L.C. & MURRAY, R.C. (eds) *Dolomitization and Limestone Diagenesis*. Society of Economic Paleontologists and Mineralogists, Special Publications, **13**, 71–88.

DEMING, D., NUNN, J.A. & EVANS, D.G. 1990. Thermal effects of compaction-driven groundwater flow from overthrust belts. *Journal of Geophysical Research*, **95**, 6669–6683.

DOROBEK, S. 1989. Migration of orogenic fluids through the Siluro-Devonian Helderberg Group during late Paleozoic deformation: constraints on fluid sources and implications for thermal histories of sedimentary basins. *Tectonophysics*, **159**, 25–45.

DRIVET, E. & MOUNTJOY, E.W. 1997. Dolomitization of the Leduc Formation (Upper Devonian), southern Rimbey–Meadowbrook reef trends, Alberta. *Journal of Sedimentary Research*, **67**, 411–423.

DUGGAN, J.P., MOUNTJOY, E.W. & STASIUK, L.D. 2001. Fault-controlled dolomitization at Swan Hills Simonette oil field (Devonian), deep basin west-central Alberta, Canada. *Sedimentology*, **48**, 301–323.

DUNHAM, J.B. & OLSON, E.R. 1980. Shallow subsurface dolomitization of subtidally deposited carbonate sediments in the Hanson Creek Formation (Ordovician–Silurian) of central Nevada. *In*: ZENGER, D.H., DUNHAM, J.B. & ETHINGTON, R.L. (eds) *Concepts and Models of Dolomitization*. Society of Economic Paleontologists and Mineralogists, Special Publications, **28**, 139–161.

DUROCHER, S. & AL-AASM, I.S. 1997. Dolomitization and neomorphism of Mississippian (Visean) Upper Debolt Formation, Blueberry Field, Northeastern British Columbia: geologic, petrologic, and chemical evidence. *AAPG Bulletin*, **81**, 954–977.

FABRICIUS, I.L. 2000. 10. Interpretation of burial history and rebound from loading experiments and occurrence of microstylolites in mixed sediments of Caribbean Sites 999 and 1001. *In*: LECKIE, R.M., SIGURDSON, H., ACTON, G.D. & DRAPER, G. (eds) *Proceedings of the Ocean Drilling Program, Scientific Results*, **165**. Ocean Drilling Program, College Station, TX, 177–190.

FAURE, G. & POWELL, J.L. 1972. *Strontium Isotope Geology*. Springer, Berlin.

FRIEDMAN, G.M. 1980. Dolomite is an evaporite mineral: evidence from the rock record and from sea-marginal ponds of the Red Sea. In: ZENGER, D.H., DUNHAM, J.B. & ETHINGTON, R.L. (eds)

Concepts and Models of Dolomitization. Society of Economic Paleontologists and Mineralogists, Special Publications, **28**, 69–80.

GAINES, A.M. 1974. Protodolomite synthesis at 100 °C and atmospheric pressure. *Science*, **183**, 518–520.

GAINES, A.M. 1977. Protodolomite redefined. *Journal of Sedimentary Petrology*, **47**, 543–546.

GARVEN, G. 1995. Continental-scale groundwater flow and geological processes. *Annual Review of Earth and Planetary Sciences*, **23**, 89–117.

GARVEN, G. & FREEZE, R. 1984. Theoretical analysis of the role of groundwater flow in the genesis of stratabound ore deposits. *American Journal of Science*, **284**, 1085–1174.

GASPARRINI, M. 2003. *Large-scale hydrothermal dolomitization in the southwestern Cantabrian Zone (NW Spain): causes and controls of the process and origin of the dolomitizing fluids*. PhD thesis, Ruprecht-Karls-Universität Heidelberg.

GIDMAN, J. 1978. Discussion: Protodolomite redefined. *Journal of Sedimentary Petrology*, **47**, 1007–1008.

GIVEN, R.K. & WILKINSON, B.H. 1987. Dolomite abundance and stratigraphic age constraints on rates and mechanisms of Phanerozoic dolostone formation. *Journal of Sedimentary Petrology*, **57**, 1068–1078.

GOLDMAN, M.I. 1952. *Deformation, Metamorphism, and Mineralization in Gypsum–Anhydrite Cap Rock, Sulphur Salt Dome, Louisiana*. Geological Society of America, Memoir, **50**.

GRAF, D.L. & GOLDSMITH, J.R. 1956. Some hydrothermal synthesis of dolomite and protodolomite. *Journal of Geology*, **64**, 173–186.

GRAY, E. 1970. *The Great Canadian Oil Patch*. Maclean-Hunter.

GREGG, J.M. 1985. Regional epigenetic dolomitization in the Bonneterre dolomite (Cambrian), southeastern Missouri. *Geology*, **13**, 503–506.

GREGG, J.M. 2004. Basin fluid flow, base-metal sulphide mineralization and the development of dolomite petroleum reservoirs. *In*: BRAITHWAITE, C.J.R., RIZZI, G. & DARKE, G. (eds) *The Geometry and Petrogenesis of Dolomite Hydrocarbon Reservoirs*. Geological Society, London, Speical Publications, **235**, 157–175.

GREGG, J.M. & SIBLEY, D.F. 1983. Xenotopic dolomite texture – Implications for dolomite neomorphism and late diagenetic dolomitization in the Galena Group (Ordovician) Wisconsin and Iowa. *In*: DELGADO, D.J. (ed.) *Ordovician Galena Group of the Upper Mississippi Valley – Deposition, Diagenesis, and Paleoecology*. Guidebook for the 13th Annual Field Conference, Great Lakes Section. Society of Economic Paleontologists and Mineralogists, G1–G15.

GREGG, J.M. & SIBLEY, D.F. 1984. Epigenetic dolomitization and the origin of xenotopic dolomite texture. *Journal of Sedimentary Petrology*, **54**, 908–931.

GREGG, J.M., HOWARD, S.A. & MAZULLO, S.J. 1992. Early-diagenetic recrystallization of Holocene (<3000 years old) peritidal dolomites, Ambergris Cay, Belize. *Sedimentology*, **39**, 143–160.

GREGG, J.M., LAUDON, P.R., WOODY, R.E. & SHELTON, K.L. 1993. Porosity evolution of the Cambrian Bonneterre dolomite, south-eastern Missouri, USA. *Sedimentology*, **40**, 1153–1169.

GREGG, J.M. SHELTON, K.L., JOHNSON, A.W., Somerville, I.S. & WRIGHT, W.R. 2001. Dolomitization of the Waulsortian Limestone (Lower Carboniferous) in the Irish Midlands. *Sedimentology*, **48**, 745–766.

GROBE, M. 1999. *Origin and formation of dolostone in the Devonian Brilon Reef Complex, northeastern Rhenish Schiefergebirge, Germany*. PhD thesis, University of Alberta.

GROBE, M. & MACHEL, H.G. 1996. Postvariszische Dolomitisierung des Devonischen Briloner Riffkomplexes, nordöstliches Rheinisches Schiefergebirge, Deutschland. *Zentralblatt für Geologie und Paläontologie, Teil I*, **1995**, 131–143.

GROBE, M. & MACHEL, H.G. 1997. Petrographie und stabile Isotopendaten (C und O) für störungskontrollierte hydrothermale Mineralisation im devonischen Briloner Riffkomplex. *Zentralblatt für Geologie und Paläontologie, Teil I*, **1996**, 397–413.

HALLEY, R.B. & SCHMOKER, J.W. 1983. High-porosity Cenozoic carbonate rocks of South Florida: progressive loss of porosity with depth. *American AAPG Bulletin*, **67**, 191–200.

HANSHAW, B.B., BACK, W. & DEIKE, R.G. 1971. A geochemical hypothesis for dolomitization by ground water. *Economic Geology*, **66**, 710–724.

HARDIE, L.A. 1987. Dolomitization: a critical view of some current views. *Journal of Sedimentary Petrology*, **57**, 166–183.

HARRIS, P.M. & WALKER, S.D. 1990. McElroy Field–U.S.A., Central Basin Platform, Permian Basin, Texas. *In*: BEAUMONT, E.A. & FOSTER, N.H. (eds) *Stratigraphic Traps 1*. Treatise of Petroleum Geology, Atlas of Oil and Gas Fields. AAPG, 195–227.

HEYDARI, E. 1997. Hydrotectonic models of burial diagenesis in platform carbonates based on formation water geochemistry in North American sedimentary basins. *In*: MONTAÑEZ, I.P., GREGG, J.M. & SHELTON, K.L. (eds) *Basin-wide Diagenetic Patterns: Integrated Petrologic, Geochemical, and Hydrologic Considerations*. Society of Economic Paleontologists and Mineralogists, Special Publications, **57**, 53–79.

HORRIGAN, E.K. 1996. *Extent of dolomite recrystallization along the Rimbey–Meadowbrook reef trend, Western Canada Sedimentary Basin, Alberta, Canada*. MSc thesis University of Alberta.

HSÜ, K.J. & SCHNEIDER, J. 1973. Progress report on dolomitization hydrology of Abu Dhabi Sabkhas: Arabian Gulf. *In*: PURSER, B.H. (ed.) *The Persian Gulf*. Springer, Berlin, 409–422.

HSÜ, K.J. & SIEGENTHALER, C. 1969. Preliminary experiments on hydrodynamic movements induced by evaporation and their bearing on the dolomite problem. *Sedimentology*, **12**, 11–25.

HUDSON, J.D. 1977. Stable isotopes and limestone lithification. *Journal of the Geological Society, London*, **133**, 637–660.

HUEBSCHER, H. 1996. *Regional controls on the stratigraphic and diagenetic evolution of Woodbend Group carbonates, north-central Alberta.* PhD thesis, University of Alberta.

HUMPHREY, J.D. & QUINN, T.M. 1989. Coastal mixing zone dolomite, forward modeling, and massive dolomitization of platform-margin carbonates. *Journal of Sedimentary Petrology*, **59**, 438–454.

ILLING, L.V. 1959. Deposition and diagenesis of some upper Paleozoic carbonate sediments in western Canada. *In*: *Fifth World Petroleum Congress New York*, Proceedings Section 1, 23–52.

JODRY, R.L. 1969. Growth and dolomitization of Silurian Reefs, St. Clair County, Michigan. *AAPG Bulletin*, **53**, 957–981.

JONES, G.D. 2000. *Numerical modelling of saline groundwater circulation in carbonate platforms.* Unpublished PhD thesis, University of Bristol.

JONES, G.D. & ROSTRON, B.J. 2000. Analysis of flow constraints in regional-scale reflux dolomitization: constant versus variable-flux hydrological models. *Bulletin of Canadian Petroleum Geology*, **48**, 230–245.

JONES, G.D., WHITAKER, F.F., SMART, P.L. & SANFORD, W.E. 2002. Fate of reflux brines in carbonate platforms. *Geology*, **30**, 371–374.

JONES, G.D., SMART, P.L., WHITAKER, F.F., ROSTRON, B.J. & MACHEL, H.G. 2003. Numerical modeling of reflux dolomitization in the Grosmont platform complex (Upper Devonian), Western Canada Sedimentary Basin. *AAPG Bulletin*, **87**, 1273–1298.

JONES, G.D., WHITAKER, F.F., SMART, P.L. & SANFORD, W.E. 2004. Numerical analysis of seawater circulation in carbonate platforms: II. The dynamic interaction between geothermal and brine reflux circulation. *American Journal of Science*, **304**, 250–284.

KATZ, A. & MATTHEWS, A. 1977. The dolomitization of $CaCO_3$: an experimental study at 225–295°C. *Geochimica et Cosmochimica Acta*, **41**, 297–308.

KAUFMAN, J. 1994. Numerical models of fluid flow in carbonate platforms. *Journal of Sedimentary Research*, **A64**, 128–139.

KOHOUT, F.A., HENRY, H.R. & BANKS, J.E. 1977. Hydrogeology related to geothermal conditions of the Floridan Plateau. *In*: SMITH, K.L. & GRIFFIN, G.M. (eds) *The Geothermal Nature of the Floridan Plateau.* Florida Department of Natural Resources Bureau, Geology Special Publications, **21**, 1–34.

KOSTEKA, A. 1995. A model of orientation of subcrystals in saddle dolomite. *Journal of Sedimentary Research*, **A65**, 332–336.

KRUG, H.-J., BRANDSTÄDTER, H. & JACOB, K.H. 1996. Morphological instabilities in pattern formation by precipitation and crystallization processes. *Geologische Rundschau*, **85**, 19–28.

LAND, L.S. 1973. Contemporaneous dolomitization of middle Pleistocene reefs meteoric water, North Jamaica. *Bulletin of Marine Science*, **23**, 64–92.

LAND, L.S. 1980. The isotopic and trace element geochemistry of dolomite: the state of the art. *In*: ZENGER, D.H., DUNHAM, J.B. & ETHINGTON, R.L. (eds) *Concepts and Models of Dolomitization.* Society of Economic Paleontologists and Mineralogists, Special Publications, **28**, 87–110.

LAND, L.S. 1985. The origin of massive dolomite. *Journal of Geological Education*, **33**, 112–125.

LAND, L.S. 1992. The quantum theory of dolomite stabilization: does dolomite stabilize by 'Ostwald steps'? *In*: *Dolomite – From Process and Models to Porosity and Reservoirs.* Canadian Society of Petroleum Geologists and Faculty of Extension of the University of Alberta (sponsors): 1992 National Conference of Earth Science, 13–18 September, Banff, Alberta.

LAND, L.S. 1998. Failure to precipitate dolomite at 25°C from dilute solution despite 1000-fold oversaturation after 32 years. *Aquatic Geochemistry*, **4**, 361–368.

LANDES, K.K. 1946. Porosity through dolomitization. *AAPG Bulletin*, **30**, 305–318.

LAST, W.M. 1990. Lacustrine dolomite – an overview of modern, Holocene, and Pleistocene occurrences. *Earth Science Reviews*, **27**, 221–263.

LEACH, D.L., PLUMLEE, G.S., HOFSTRA, A.H., LANDIS, G.P., ROWAN, E.L. & VIETS, J.G. 1991. Origin of late dolomite cement by CO_2-saturated deep basin brines: evidence from the Ozark region, central United States. *Geology*, **19**, 348–351.

LIND, I.L. 1993. Stylolites in chalk from Leg 130, Ontong Java Plateau. *In*: BERGER, W.H., KROENKE, J.W. & MAYER, L.A. (eds) *Proceedings of the Ocean Drilling Program, Scientific Results*, **130**. Ocean Drilling Program, College Station, TX, 445–451.

LIPPMAN, F. 1973. *Sedimentary Carbonate Minerals.* Springer, New York.

LOGAN, B.W. 1987. The MacLeod Evaporite Basin, Western Australia. *AAPG Memoir*, **44**.

LONNEE, J. & MACHEL, H.G. 2004. Dolomitization by halite-saturated brine and subsequent hydrothermal alteration in the Devonian Slave Point Formation, Clark Lake gas field, British Columbia. CSPG-CHOA-CWLS Joint Conference Abstracts, S013.

LUCIA, F.J. 1995. Rock-fabric petrophysical classification of carbonate pore space for reservoir characterization. *AAPG Bulletin*, **79**, 1275–1300.

LUCIA, F.J. 2002. Origin and petrophysics of dolostone pore space. *In*: RIZZI, G., DARKE, G. & BRAITHWAITE, C.J.R. (convenors) *The Geometry and Petrogenesis of Dolomite Hydrocarbon Reservoirs.* Final Programme and Abstracts. Geological Society Petroleum Group, London.

LUCIA, F.J. 2004. Origin and petrophysics of dolostone pore space. *In*: BRAITHWAITE, C.J.R., RIZZI, G. & DARKE, G. (eds) *The Geometry and Petrogenesis of Dolomite Hydrocarbon Reservoirs*. Geological Society, London, Special Publications, **235**, 141–155.

LUCIA, F.J. & MAJOR, R.P. 1994. Porosity evolution through hypersaline reflux dolomitization. *In*: PURSER, B.H., TUCKER M.E. & ZENGER, D.H. (eds) *Dolomites – A Volume in Honour of Dolomieu.* International Association of Sedimentologists, Special Publications, **21**, 325–341.

LUMSDEN, D.N. 1988. Characteristics of deep-marine dolomites. *Journal of Sedimentary Petrology*, **58**, 1023–1031.

LUMSDEN, D.N. & CAUDLE, G.C. 2001. Origin of massive dolostone: the Upper Knox model. *Journal of Sedimentary Research*, **71**, 400–409.

LUO, P. & MACHEL, H.G. 1995. Pore size and pore-throat types in a heterogeneous Dolostone reservoir, Devonian Grosmont Formation, Western Canada Sedimentary Basin. *AAPG Bulletin*, **79**, 1698–1720.

LUO, P., MACHEL, H.G. & SHAW, J. 1994. Petrophysical properties of matrix blocks of a heterogeneous dolostone reservoir – the Upper Devonian Grosmont Formation, Alberta, Canada. *Bulletin of Canadian Petroleum Geology*, **42**, 465–481.

LYONS, W.B., HINES, M.E. & GAUDETTE, H.E. 1984. Major and minor element pore water geochemistry of modern marine sabkhas: the influence of cyanobacterial mats. *In*: COHEN, Y., CASTENHOLZ, R.W. & HALVORSON, H.O. (eds) *Microbial Mats*. Liss, New York, 411–423.

MACHEL, H.G. 1985. *Facies and diagenesis of the Upper Devonian Nisku Formation in the subsurface of central Alberta*. PhD thesis, McGill University, Montreal.

MACHEL, H.G. 1986. Early lithification, dolomitization, and anhydritization of Upper Devonian Nisku buildups, subsurface of Alberta, Canada. *In*: SCHROEDER, J.H. & PURSER, B.H. (eds) *Reef Diagenesis*. Springer, Berlin, 336–356.

MACHEL, H.G. 1987. Saddle dolomite as a by-product of chemical compaction and thermochemical sulphate reduction. *Geology*, **15**, 936–940.

MACHEL, H.G. 1988. Fluid flow direction during dolomite formation as deduced from trace element trends. *In*: SHUKLA, V. & BAKER, P.A. (eds) *Sedimentology and Geochemistry of Dolostones*. Society of Economic Paleontologists and Mineralogists, Special Publications, **43**, 115–125.

MACHEL, H.G. 1990. Submarine Frühdiagenese, Spaltenbildungen, und prätektogenetische Spätdiagenese des Briloner Riffs. *Geologisches Jahrbuch*, **D95**, 85–137.

MACHEL, H.G. 1997. Recrystallization versus neomorphism, and the concept of 'significant recrystallization' in dolomite research. *Sedimentary Geology*, **113**, 161–168.

MACHEL, H.G. 1999. Effects of groundwater flow on mineral diagenesis, with emphasis on carbonate aquifers. *Hydrogeology Journal*, **7**, 94–107.

MACHEL, H.G. 2000. Dolomite formation in Caribbean islands – driven by plate tectonics?! *Journal of Sedimentary Research*, **70**, 977–984.

MACHEL, H.G. 2001. Bacterial and thermochemical sulphate reduction in diagenetic settings. *Sedimentary Geology*, **140**, 143–175.

MACHEL, H.G. & ANDERSON, J.H. 1989. Pervasive subsurface dolomitization of the Nisku Formation in central Alberta. *Journal of Sedimentary Petrology*, **59**, 891–911.

MACHEL, H.G. & CAVELL, P.A. 1999. Low-flux, tectonically induced squeegee fluid flow ('hot flash') into the Rocky Mountain Foreland Basin. *Bulletin of Canadian Petroleum Geology*, **47**, 510–533.

MACHEL, H.G. & HUEBSCHER, H. 2000. The Devonian Grosmont heavy oil reservoir in Alberta, Canada. *Zentralblatt für Geologie und Paläontologie*, Teil I, Heft 1/2, 55–84.

MACHEL, H.G. & LONGACRE, S.A. 2000. Die permische McElroy Öllagerstätte in Texas, U.S.A. – Fazies, Sequenzstratigraphie, und Diagenese. *Zentralblatt für Geologie und Paläontologie*, Teil I, Heft 1/2, 85–115.

MACHEL, H.G. & LONNEE, J. 2002. Hydrothermal dolomite – a product of poor definition and imagination. *Sedimentary Geology*, **152**, 163–171.

MACHEL, H.G. & MOUNTJOY, E.W. 1986. Chemistry and environments of dolomitization – a reappraisal. *Earth Science Reviews*, **23**, 175–222.

MACHEL, H.G. & MOUNTJOY, E.W. 1987. General constraints on extensive pervasive dolomitization – and their application to the Devonian carbonates of western Canada. *Bulletin of Canadian Petroleum Geology*, **35**, 143–158.

MACHEL, H.G. & MOUNTJOY, E.W. 1990. Coastal mixing zone dolomite, forward modeling, and massive dolomitization of platform-margin carbonates – Discussion. *Journal of Sedimentary Petrology*, **60**, 1008–1012.

MACHEL, H.G., MOUNTJOY, E.W. & AMTHOR, J.E. 1994. Dolomitisierung von devonischen Riff- und Plattformkarbonaten in West-Kanada. *Zentralblatt für Geologie und Paläontologie*, Teil I, **1993**, 941–957.

MACHEL, H.G., KROUSE, H.R. & SASSEN, R. 1995. Products and distinguishing criteria of bacterial and thermochemical sulphate reduction. *Applied Geochemistry*, **10**, 373–389.

MACHEL, H.G., CAVELL, P.A., & PATEY, K.S. 1996*a*. Isotopic evidence for carbonate cementation and recrystallization, and for tectonic expulsion of fluids into the Western Canada Sedimentary Basin. *Geological Society of America Bulletin*, **108**, 1108–1119.

MACHEL, H.G., MOUNTJOY, E.W. & AMTHOR, J.E. 1996*b*. Mass balance and fluid flow constraints on regional-scale dolomitization, Late Devonian, Western Canada Sedimentary Basin. *Bulletin of Canadian Petroleum Geology*, **44**, 566–571.

MACHEL, H.G., CAVELL, P.A., BUSCHKUEHLE, B.E. & MICHAEL, K. 2000. Tectonically induced fluid flow in Devonian carbonate aquifers of the Western Canada Sedimentary Basin. *Journal of Geochemical Exploration*, **69–70**, 213–217.

MACHEL, H.G., MOUNTJOY, E.W., ROSTRON, B.J. & JONES, G.D. 2002. Toward a sequence stratigraphic framework for the Frasnian of the Western Canada Basin. *Bulletin of Canadian Petroleum Geology*, **50**, 332–338.

MAJOR, R.P., BEBOUT, D.G. & LUCIA, F.J. 1988. Depositional facies and porosity distribution, Permian (Guadalupian) San Andres and Grayburg Formations, PJWDM Field Complex, Central Basin Platform, West Texas. *In*: LOMANDO, A.J. & HARRIS, P.M. (eds) *Giant Oil and Gas Fields – A*

Core Workshop. Society of Economic Paleontologists and Mineralogists, Core Workshop, **12**, 615–648.

MALONE, M.L., BAKER, P.A. & BURNS, S.J. 1994. Recrystallization of dolomite: evidence from the Monterey Formation (Miocene), California. *Sedimentology*, **41**, 1223–1239.

MAZZULLO, S.J. 1992. Geochemical and neomorphic alteration of dolomite: a review. *Carbonates and Evaporites*, **6**, 21–37.

MAZZULLO, S.J. 2000. Organogenic dolomitization in peritidal to deep sea sediments. *Journal of Sedimentary Research*, **70**, 10–23.

MCKENZIE, J.A., HSÜ, K.J. & SCHNEIDER, J.F. 1980. Movement of subsurface waters under the sabkha, Abu Dhabi, UAE, and its relation to evaporative dolomite genesis. *In*: ZENGER, D.H., DUNHAM, J.B. & ETHINGTON, R.L. (eds) *Concepts and Models of Dolomitization*. Society of Economic Paleontologists and Mineralogists, Special Publications, **28**, 11–30.

MCLIMANS, R.K. 1987. The application of fluid inclusions to migration of oil and diagenesis in petroleum reservoirs. *Applied Geochemistry*, **2**, 585–603.

MELIM, L.A. & SCHOLLE, P.A. 2002. Dolomitization of the Capitan Formation forereef facies (Permian, West Texas and New Mexico): seepage reflux revisited. *Sedimentology*, **49**, 1207–1227.

MELIM, L.A., SWART, P.K. & EBERLI, G.P. 2003. Mixing-zone diagenesis: separating fact from theory. *In*: *12th Bathurst Meeting – International Conference of Carbonate Sedimentologists*, 8–10 July 2003, Durham, 68.

MONTAÑEZ, I.P. 1994. Late diagenetic dolomitization of Lower Ordovician, Upper Know carbonates: a record of hydrodynamic evolution of the southern Appalachian Basin. *AAPG Bulletin*, **78**, 1210–1239.

MONTAÑEZ, I.P. 1997. Secondary porosity and late diagenetic cements in the Upper Knox Group, central Tennessee region: a temporal and spatial history of fluid flow conduit development within the Knox regional aquifer. *In*: MONTAÑEZ, I.P., GREGG, J.M. & SHELTON, K.L. (eds) *Basin-wide Diagenetic Patterns: Integrated Petrologic, Geochemical, and Hydrologic Considerations*. Society of Economic Palentologists and Mineralogists, Special Publications, **57**, 101–117.

MONTAÑEZ, I.P. & READ, J.F. 1992*a*. Fluid–rock interaction history during stabilization of early dolomites, Upper Knox Group (Lower Ordovician), U.S. Appalachians. *Journal of Sedimentary Petrology*, **62**, 753–778.

MONTAÑEZ, I.P. & READ, J.F. 1992*b*. Eustatic control on early dolomitization of cyclic peritidal carbonates: evidence from the Early Ordovician Upper Knox Group, Appalachians. *Geological Society of America Bulletin*, **104**, 872–886.

MOORE, S.L.O. 1994. *The origin of dolomite of the Middle Cambrian Eldon and Pika Formations in the Yoho Glacier area, Yoho National Park, British Columbia*. PhD thesis, University of Calgary.

MORROW, D.W. 1982*a*. Diagenesis 1. Dolomite – Part 1: The chemistry of dolomitization and dolomite precipitation. *Geoscience Canada*, **9**, 5–13.

MORROW, D.W. 1982*b*. Diagenesis 2. Dolomite – Part 2: Dolomitization models and ancient dolostones. *Geoscience Canada*, **9**, 95–107.

MORROW, D.W. 1999. Regional subsurface dolomitization: models and constraints. *Geoscience Canada*, **25**, 57–70.

MORROW, D.W. 2001. Distribution of porosity and permeability in platform dolomites: insight from the Permian of west Texas: Discussion. *American AAPG Bulletin*, **85**, 525–529.

MORROW, D.W. & ABERCROMBIE, H.J. 1994. Rates of dolomitization: the influence of dissolved sulphate. *In*: PURSER, B.H., TUCKER, M.E. & ZENGER, D.H. (eds) *Dolomites – A Volume in Honour of Dolomieu*. International Association of Sedimentologists, Special Publications, **21**, 377–386.

MORROW, D.W. & RICKETS, B.D. 1986. Chemical controls on the precipitation of mineral analogues of dolomite: the sulphate enigma. *Geology*, **14**, 408–410.

MORSE, J.W. & CASEY, W.H. 1988. Ostwald process and mineral paragenesis in sediments. *American Journal of Science*, **288**, 537–560.

MOUNTJOY, E.W. & AMTHOR, J.E. 1994. Has burial dolomitization come of age? Some answers from the Western Canada Sedimentary Basin. *In*: PURSER, B.H., TUCKER, M.E. & ZENGER, D.H. (eds) *Dolomites – A Volume in Honour of Dolomieu*. International Association of Sedimentologists, Special Publications, **21**, 203–230.

MOUNTJOY, E.W., MACHEL, H.G., GREEN, D., DUGGAN, J. & WILLIAMS-JONES, A.E. 1999. Devonian matrix dolomites and deep burial carbonate cements: a comparison between the Rimbey–Meadowbrook reef trend and the deep basin of west-central Alberta. *Bulletin of Canadian Petroleum Geology*, **47**, 487–509.

MUIR, M., LOCK, D. & VON DER BORCH, C.C. 1980. The Coorong model for penecontemporaneous dolomite formation in the Middle Proterozoic McArthur Group, Northern Territory, Australia. *In*: ZENGER, D.H., DUNHAM, J.B. & ETHINGTON, R.L. (eds) *Concepts and Models of Dolomitization*. Society of Economic Paleontologists and Mineralogists, Special Publications, **28**, 51–67.

MÜLLER, D.W., MCKENZIE, J.A & MUELLER, P.A. 1990. Abu Dhabi sabkha, Persian Gulf, revisited: application of strontium isotopes to test an early dolomitization model. *Geology*, **18**, 618–621.

MURRAY, R.C. 1960. Origin of porosity in carbonate rocks. *Journal of Sedimentary Petrology*, **30**, 59–84.

MUTTI, M. & SIMO, J.A. 1994. Distribution, petrograhy and geochemistry of early dolomite in cyclic shelf facies, Yates Formation (Guadalupian), Capitan Reef Complex, USA. *In*: PURSER, B.H., TUCKER M.E. & ZENGER, D.H. (eds) *Dolomites – A Volume in Honour of Dolomieu*. International Association of Sedimentologists, Special Publications, **21**, 91–107.

NORDENG, S.H. & SIBLEY, D.F. 1994. Dolomite stoichiometry and Ostwald's step rule. *Geochimica et Cosmochimica Acta*, **58**, 191–196.

NORDENG, S.H. & SIBLEY, D.F. 2003. The induction period and dolomite kinetics in experimental and natural systems. *In*: *12th Bathurst Meeting – International Conference of Carbonate Sedimentologists*, 8–10 July 2003, Durham, 74.

OLIVER, J. 1986. Fluids expelled tectonically from orogenic belts: their role in hydrocarbon migration and other geologic phenomena. *Geology*, **14**, 99–102.

PACKARD, J.J. 1992. Early formed reservoir dolomites of the Mississippian Upper Debolt Fm.: Dunvegan Gas Field: A highly stratified reservoir hosting 1.2 TCF of sweet gas in Mississippian sabkha and contiguous subtidal sediments, NW Alberta, CA. *In*: *Dolomite – From Process and Models to Porosity and Reservoirs*. Canadian Society of Petroleum Geologists and Faculty of Extension of the University of Alberta (sponsors): 1992 National Conference of Earth Science, 13–18 September, Banff, Alberta.

PATTERSON, R.J. & KINSMAN, D.J. 1982. Formation of diagenetic dolomite in coastal sabkha along the Arabian (Persian) Gulf. *AAPG Bulletin*, **66**, 28–43.

PHILLIPS, O.M. 1991. *Flow and Reactions in Permeable Rocks*. Cambridge University Press, Cambridge.

PICHLER, T. & HUMPHREY, J.D. 2001. Formation of dolomite in recent island-arc sediments due to gas–seawater sediment interaction. *Journal of Sedimentary Research*, **71**, 394–399.

PITTMAN, E.D. 1979. *Porosity, Diagenesis and Productive Capability of Sandstone Reservoir*. Society of Economic Paleontologists and Mineralogists, Reprint Series, **26**, 159–173.

PITTMAN, E.D. 1992. Relationship of porosity and permeability to various parameters derived from mercury injection-capillary pressure curves for Sandstone. *AAPG Bulletin*, **75**, 191–198.

POTMA, K., WEISSENBERGER, J.A.W., WONG, P.K. & GILHOOLY, M.G. 2001. Toward a sequence stratigraphic framework for the Frasnian of the Western Canada Basin. *Bulletin of Canadian Petroleum Geology*, **49**, 37–85.

PRAY, L.C. & MURRAY, R.C. (eds). 1965. *Dolomitization and Limestone Diagenesis – A Symposium*. Society of Economic Paleontologists and Mineralogists, Special Publications, **13**.

PURSER, B.H., TUCKER, M.E. & ZENGER, D.H. (eds). 1994. *Dolomites – A Volume in Honour of Dolomieu*. International Association of Sedimentologists, Special Publications, **21**.

QING, H. & MOUNTJOY, E.W. 1992. Large-scale fluid flow in the Middle Devonian Presqu'ile Barrier, Western Canada Sedimentary Basin. *Geology*, **20**, 903–906.

QING, H. & MOUNTJOY, E.W. 1994. Formation of coarsely crystalline, hydrothermal dolomite reservoirs in the Presqu'ile Barrier, Western Canada Sedimentary Basin. *AAPG Bulletin*, **78**, 55–77.

QING, H., BOSENCE, D.W.J. & ROSE, P.F. 2001. Dolomitization by penesaline seawater in Early Jurassic pertidal platform carbonates, Gibraltar, western Mediterranean. *Sedimentology*, **48**, 153–163.

RADKE, B.M. & MATHIS, R.L. 1980. On the formation and occurrence of saddle dolomite. *Journal of Sedimentary Petrology*, **50**, 1149–1168.

RAFFENSBERGER, J.P. & VLASSOPOULOS, D. 1999. The potential for free and mixed convection in sedimentary basins. *Journal of Hydrogeology*, **7**, 505–520.

REIMER, J.D & TEARE, M.R. 1992. Thermalorganic sulphate reduction and hydrothermal dolomitization. *In*: *Dolomite – From Process and Models to Porosity and Reservoirs*. Canadian Society of Petroleum Geologists and Faculty of Extension of the University of Alberta (sponsors): 1992 National Conference of Earth Science, 13–18 September, Banff, Alberta.

REIMER, J.D., HUDEMA, T. & VIAU, C. 2001. TSR-HTD ten years later: an exploration update, with examples from western and eastern Canada. *Canadian Society of Petroleum Geologists Annual Convention Abstracts*, 276–279.

ROSEN, M.E., MISER, D.E., STARCHER, M.A. & WARREN, J.K. 1989. Formation of dolomite in the Coorong region, South Australia. *Geochimica et Cosmochimica Acta*, **53**, 661–669.

SALLER, A.H. 1984. Petrologic and geochemical constraints on the origin of subsurface dolomite, Enewetak Atoll: an example of dolomitization by normal seawater. *Geology*, **12**, 217–220.

SALLER, A.H. & HENDERSON, N. 2001. Distribution of porosity and permeability in platform dolomites: Insight from the Permian of west Texas: Reply. *AAPG Bulletin*, **85**, 530–532.

SANFORD, W.E. & KONIKOW, L.F. 1989. Porosity development in coastal carbonate aquifers. *Geology*, **17**, 249–252.

SANFORD, W.E., WHITAKER, F.F., SMART, P.L. & JONES, G. 1998. Numerical analysis of seawater circulation in carbonate platforms: I. Geothermal convection. *American Journal of Science*, **298**, 801–828.

SASS, E. & KATZ, A. 1982. The origin of platform dolomites. *American Journal of Science*, **282**, 1184–1213.

SCHMOKER, J.W. & HALLEY, R.B. 1982. Carbonate porosity versus depth: a predictable relation for south Florida. *AAPG Bulletin*, **66**, 2561–2570.

SCHMOKER, J.W., KRYSTINICK, K.B. & HALLEY, R.B. 1985. Selected characteristics of limestone and dolomite reservoirs in the United States. *AAPG Bulletin*, **69**, 733–741.

SCHOLLE, P.A. & HALLEY, R.B. 1985. Burial diagenesis: out of sight, out of mind! *In*: SCHNEIDERMANN, N. & HARRIS, P.M. (eds) *Carbonate Cements*. Society of Economic Paleontologists and Mineralogists, Special Publications, **36**, 309–334.

SEARL, A. 1989. Saddle dolomite: a new view of its nature and origin. *Mineralogical Magazine*, **53**, 547–555.

SHIELDS, M.J. & BRADY, P.V. 1995. Mass Balance and fluid flow constraints on regional scale dolomitization, Late Devonian, Western Canada Sedimentary Basin. *Bulletin of Canadian Petroleum Geology*, **43**, 371–392.

SHROCK, R.R. 1948. A classification of sedimentary rocks. *Journal of Geology*, **56**, 118–129.

SHUKLA, V. & BAKER, P.A. (eds) 1988. *Sedimentology and Geochemistry of Dolostones*. Society of Economic Paleontologists and Mineralogists, Special Publications, **43**.

SIBLEY, D.F. 1990. Unstable to stable transformations during dolomitization. *Journal of Geology*, **98**, 739–748.

SIBLEY, D.F. 2003. Dolomite textures. *In*: MIDDLETON, G.M., CHURCH, M.J., CONIGLIO, M., HARDIE, L.A. & LONGSTAFFE, F.S. (eds) *Encyclopedia of Sediments and Sedimentary Rocks*. Kluwer, Dordrecht, 231–234.

SIBLEY, D.F. & GREGG, J.M. 1987. Classification of dolomite rock textures. *Journal of Sedimentary Petrology*, **57**, 967–975.

SIBLEY, D.F., NORDENG, S.H. & BARKOWSKI, M.L. 1994. Dolomitization kinetics in hydrothermal bombs and natural settings. *Journal of Sedimentary Research*, **64A**, 630–637.

SIMMS, M. 1984. Dolomitization by ground water-flow systems in carbonate platforms. *Transactions of the Gulf Coast Association of Geological Societies*, **XXXIV**, 411–470.

SIMPSON, G.P. 1999. *Sulfate reduction and fluid chemistry of the Devonian Leduc and Nisku Formations in south-central Alberta.* PhD thesis University of Calgary.

SIMPSON, G.P., YANG, C. & HUTCHEON, I. 1996. Thermochemical sulphate reduction: a local process that does not generate thermal anomalies. *In*: ROSS, G.M. (compiler) *1995 Alberta Basement Transects Workshop*. Lithoprobe Report, **51**. Lithoprobe Secretariat, University of British Columbia, 241–245.

SMALLEY, P.C., HIGGINS, A.C., HOWARTH, R.J., NICHOLSON, H., JONES, C.E., SWINBURNE, N.H.M. & BESSA, J. 1994. Seawater Sr isotope variations through time: a procedure for constructing a reference curve to date and correlate marine sedimentary rocks. *Geology*, **22**, 431–434.

SMART, P.L. & WHITAKER, F.F. 2003. Are models of collapsed paleocave breccias based on continental caves adequate? *In*: *12th Bathurst Meeting – International Conference of Carbonate Sedimentologists*, 8–10 July 2003, Durham, 102.

SMART, P.L., DAWANS, J.M. & WHITAKER, F.F 1988. Carbonate dissolution in a modern mixing zone. *Nature*, **335**, 811–813.

SMITH, S.L., WHITAKER, F.F. & PARKES, R.J. 2002. Dolomitization by saline groundwaters in the Yucatan Peninsula. *In*: RIZZI, G., DARKE, G. & BRAITHWAITE, C.J.R. (convenors) *The Geometry and Petrogenesis of Dolomite Hydrocarbon Reservoirs*. Final Programme and Abstracts. Geological Society Petroleum Group, London.

SPENCER, R.J. & HUTCHEON, I. 1999. *Diagenesis and Fluid Flow*. Canadian Society of Petroleum Geologists, Hydrogeology Division Field Trip, 11 September 1999. Canadian Society of Petroleum Geologists.

SPENCER-CERVATO, C. & MULLIS, J. 1992. Chemical study of tectonically controlled hydrothermal dolomitization: an example from the Lessini Mountains, Italy. *Geologische Rundschau*, **81/82**, 347–370.

SPÖTL, C. & PITMAN, J.K. 1998. Saddle (baroque) dolomite in carbonates and sandstones: a reappraisal of the burial-diagenetic concept. *In*: MORAD, S. (ed.) *Carbonate Cementation in Sandstones*. International Association of Sedimentologists, Special Publications, **26**, 437–460.

STANLEY, S.M. & HARDIE, L.A. 1999. Hypercalcification: Paleontology links plate tectonics and geochemistry to sedimentology. *GSA Today*, **9**(2), 2–7.

STEARNS, N.D., STEARNS, H.T. & WARING, G.A. 1935. *Thermal Springs in the United States*. US Geological Survey Water Supply Paper, **679-B**, 59–191.

STENSON, F. 1992. *Waste to Wealth – A History of Gas Processing in Canada*. Friesen Printers.

SUN, S.Q. 1992. Skeletal aragonite dissolution from hypersaline seawater: a hypothesis. *Sedimentary Geology*, **77**, 249–257.

SUN, S.Q. 1995. Dolomite reservoirs: porosity evolution and reservoir characteristics. *AAPG Bulletin*, **79**, 186–204.

SWART, P.K. & MELIM, L.A. 2000. The origin of dolomites in Tertiary sediments from the margin of Great Bahama Bank. *Journal of Sedimentary Research*, **70**, 738–748.

TAN, F.C. & HUDSON, J.D. 1971. Carbon and oxygen isotope relationships of dolomites and co-existing calcites, Great Estuarine Series (Jurassic), Scotland. *Geochimica et Cosmochimica Acta*, **35**, 755–767.

TUCKER, M.E. & WRIGHT, V.P. 1990. *Carbonate Sedimentology*: Blackwell Scientific, Oxford.

TÓTH, J. 1988. Groundwater and hydrocarbon migration. *In*: BACK, W., ROSENHEIN, J.S. & SAEBER, P.R. (eds) *Hydrogeology.* The Geology of North America, **0-2**. Geological Society of America, Boulder, CO, 485–502.

USDOWSKI, E. 1994. Synthesis of dolomite and geochemical implications. *In*: PURSER, B.H., TUCKER, M.E. & ZENGER, D.H. (eds) *Dolomites – A Volume in Honour of Dolomieu*. International Association of Sedimentologists, Special Publications, **21**, 345–360.

VAHRENKAMP, V.C. & SWART, P.K 1994. Late Cenozoic dolomites of the Bahamas: metastable analogues for the genesis of ancient platform dolomites. *In*: PURSER, B.H., TUCKER, M.E. & ZENGER, D.H. (eds) *Dolomites – A Volume in Honour of Dolomieu.* International Association of Sedimentologists, Special Publications, **21**, 133–153.

VAHRENKAMP, V.C., SWART, P.K. & RUIZ, J. 1991. Episodic dolomitization of Late Cenozoic carbonates in the Bahamas: evidence from strontium isotopes. *Journal of Sedimentary Petrology*, **61**, 1002–1014.

VON DER BORCH, C.C. 1976. Stratigraphy and formation of Holocene dolomitic carbonate deposits in the Coorong area, South Australia. *Journal of Sedimentary Petrology*, **46**, 952–966.

VAN TUYL, F.M. 1914. *The Origin of Dolomite*. Annual Report 1914, Iowa Geological Survey, **XXV**, 257–421.

VASCONCELOS, C. & MCKENZIE, J.A. 1997. Microbial mediation of modern dolomite precipitation and diagenesis under anoxic conditions (Lagoa Vermelha, Rio de Janeiro, Brazil). *Journal of Sedimentary Research*, **67**, 378–390.

VATAN, A. 1958. Discussion: 'Dolostone'. *Journal of Sedimentary Petrology*, **28**, 514.

VEIZER, J., ALA, D. ET AL. 1999. $^{87}Sr/^{86}Sr$, $\delta^{13}C$ and $\delta^{18}O$ evolution of Phanerozoic seawater. *Chemical Geology*, **161**, 59–88.

WALKER, R.G. 1992. Facies, facies models and modern stratigraphic concepts. *In*: WALKER, R.G. & JAMES, N.P. (eds) *Facies Models – Response to Sea Level Changes*, 3rd edn. Geoscience Canada, Reprint Series, **1**, 1–14.

WALTER, L.M. 1985. Relative reactivity of skeletal carbonate during dissolution: implications for diagenesis. *In*: SCHNEIDERMANN, N. & HARRIS, P.M. (eds) *Carbonate Cements*. Society of Economic Paleontologists and Mineralogists, Special Publications, **36**, 3–16.

WARD, W.C. & HALLEY, R.B. 1985. Dolomitization in a mixing zone of near-seawater composition, Late Pleistocene, Northeastern Yucatan Peninsula. *Journal of Sedimentary Petrology*, **55**, 407–420.

WARREN, J. 2000. Dolomite: occurrence, evolution and economically important associations. *Earth Science Reviews*, **52**, 1–81.

WHITAKER, F.F., JONES, G. & SMART, P. 2003. From conceptual to numerical models of dolomitization. *In*: *12th Bathurst Meeting – International Conference of Carbonate Sedimentologists*, 8–10 July 2003, Durham, 114.

WHITAKER, F.F., SMART, P.L. & JONES, G. 2004. Dolomitization: from conceptual to numerical models. *In*: BRAITHWAITE, C.J.R., RIZZI, G. & DARKE, G. (eds) *The Geometry and Petrogenesis of Dolomite Hydrocarbon Reservoirs*. Geological Society, London, Special Publications, **235**, 99–139.

WHITAKER, F.F., SMART, P.L., VAHRENKAMP, V.C., NICHOLSON, H. & WOGELIUS, R.A. 1994. Dolomitization by near-normal seawater? Field evidence from the Bahamas. *In*: PURSER, B.H., TUCKER, M.E. & ZENGER, D.H. (eds) *Dolomites – A Volume in Honour of Dolomieu*. International Association of Sedimentologists, Special Publications, **21**, 111–132.

WHITAKER, F.F., WILSON, A.M., SANFORD, W.E. & SMART, P.L. 2002. Spatial patterns and rates of dolomitization and anhydritization during geothermal convection in carbonate platforms. *In*: RIZZI, G., DARKE, G. & BRAITHWAITE, C.J.R. (convenors) *The Geometry and Petrogenesis of Dolomite Hydrocarbon Reservoirs*. Final Programme and Abstracts. Geological Society Petroleum Group, London.

WHITE, D.E. 1957. Thermal waters of volcanic origin. *Geological Society of America, Bulletin*, **68**, 1637–1658.

WILSON, A.M., SANFORD, W.E., WHITAKER, F.F. & SMART, P.L. 2001. Spatial patterns of diagenesis during geothermal circulation in carbonate platforms. *American Journal of Science*, **301**, 727–752.

WILSON, E.N., HARDIE, L.A. & PHILLIPS, O.M. 1990. Dolomitization front geometry, fluid flow patterns, and the origin of massive dolomite: the Triassic Latemar buildup, northern Italy. *American Journal of Science*, **290**, 741–796.

WOOD, J.R. & HEWETT, T.A. 1982. Fluid convection and mass transfer in porous sandstones – a theroretical approach. *Geochimica et Cosmochimica Acta*, **46**, 1707–1713.

WOODY, R.E., GREGG, J.M. & KOEDERITZ, L.F. 1996. Effect of texture on petrophysical properties of dolomite: evidence from the Cambrian–Ordovician of southeastern Missouri. *AAPG Bulletin*, **80**, 119–132.

WRIGHT, W.R. 2001. *Dolomitization, fluid-flow and mineralization of the Lower Carboniferous rocks of the Irish Midlands and Dublin Basin*. PhD thesis, University College Dublin, Belfield, Ireland.

XUN, Z. & FAIRCHILD, I.J. 1987. Mixing zone dolomitization of Devonian carbonates, Guangxi, South China. *In*: MARSHALL, J.D. (ed.) *Diagenesis in Sedimentary Sequences*. Geological Society, London, Special Publications, **36**, 157–170.

YAO, Q. & DEMICCO, R.V. 1995. Paleoflow patterns of dolomitizing fluids and paleohydroogeology of the southern Canadian Rockie Mountains: evidence from dolomite geometry and numerical modeling. *Geology*, **23**, 791–794.

ZENGER, D.H., DUNHAM, J.B. & ETHINGTON, R.L. (eds) 1980. *Concepts and Models of Dolomitization*. Society of Economic Paleontologists and Mineralogists, Special Publications, **28**.

Sedimentary dolomite: a reality check

DAVID T. WRIGHT[1] & DAVID WACEY[2]

[1]*Department of Geology, University of Leicester, Leicester LE1 7RH, UK*

[2]*Department of Earth Sciences, University of Oxford, Oxford OX1 3PR, UK*

Abstract: The failure to precipitate dolomite experimentally at low temperatures or from seawater in which it is both supersaturated and the most thermodynamically favoured carbonate phase, together with its unequal distribution through geological time relative to limestone, are all aspects of the 'dolomite problem', a subject of continuing controversy. A plethora of physicochemical models has been invoked to explain sedimentary dolomite formation, none of which satisfactorily addresses the basic problem of how kinetic barriers are overcome. These barriers are related to the disproportionate distribution of the component ions of dolomite, cation hydration and ion complexing in seawater. Competing claims for the effectiveness of sulphate as an inhibitor to dolomite formation further confuse the debate, although there are many reports of modern dolomite associated with bacterial sulphate reduction. The uppermost sediments in some lakes of the Coorong region of South Australia comprise almost 100% dolomite, and afford an ideal opportunity to study this association.

Samples of lake waters taken during late evaporative stages of several shallow hypersaline dolomitic lakes showed high initial sulphate concentrations, high pH and high carbonate alkalinities. Pore waters from unlithified lake sediment cores directly below the lake-water sample sites showed a substantial and progressive decrease in sulphate concentrations with depth, coupled with an exponential increase in carbonate concentrations, through the sulphate-reduction zone. By the end of the evaporative cycle, sulphate was entirely removed. High bacterial counts on cultures from the sediment cores, and sulphur isotope values consistent with 'bacterial' fractionation in lake waters, indicate that the chemical changes in ambient water chemistry can be related to active bacterial sulphate reduction. Laboratory experiments using sulphate reducers cultured from the lake sediments and simulating the anoxic microbiogeochemical environment of the lakes, have resulted in the precipitation of dolomite, demonstrating that bacterial sulphate reduction in the Coorong lakes modifies lake-water and pore-water chemistry so that dolomite precipitation is kinetically favoured.

Given the wide spatial and temporal distribution of sulphate-reducing bacteria, and their frequent association, both past and present, with cyanobacteria, it is likely that this process was more widespread in the geological past when dolomite was found in far greater abundance than limestone. Bacterial sulphate reduction may thus have played an important role in dolomite formation throughout the geological record.

In this paper, the term '*sedimentary dolomite*' means primary, diagenetic or replacive dolomite, formed as or within sediment at sediment-surface temperatures and pressures. This distinguishes it from burial and hydrothermally formed dolomite. 'Normal' sedimentary conditions are the range of temperatures and pressures expected at or within a few metres of the sediment surface.

Numerous physicochemical investigations have failed to reveal the precise mechanism of sedimentary dolomite formation, with many observed chemical trends not following thermodynamic predictions (e.g. Kelleher & Redfern 2002). Precipitation of dolomite from a fluid, or by replacement of a $CaCO_3$ precursor, has only been achieved experimentally at or near hydrothermal conditions (e.g. Gaines 1980; Lumsden *et al.* 1995; Tribble *et al.* 1995; Land 1998). Recently, certain studies have emphasized the kinetic control of ordering and growth by mineral surface chemistry, for example Brady *et al.* (1996) who used a synthetic phase produced at 70 °C that 'approximated to fully disordered dolomite'. The absence of an ordering peak, however, means that the precipitate was not a true dolomite. Where Brady *et al.* (1996) studied adsorption of metal ions onto dolomite surfaces using pure dolomite slurry as a starting ingredient, Mg^{2+} and Ca^{2+} ions were adsorbed to the dolomite surface in near-stoichiometric proportions. They did not, however, precipitate the initial dolomite. The synthesis of large amounts of experimental data led Arvidson & Mackenzie (1999) to propose a global rate law based, in part, on the equilibrium

From: BRAITHWAITE, C. J. R., RIZZI, G. & DARKE, G. (eds) 2004. *The Geometry and Petrogenesis of Dolomite Hydrocarbon Reservoirs*. Geological Society, London, Special Publications, **235**, 65–74. 0305-8719/$15.00

saturation state, which is poorly known. The use of transmission electron microscopy for imaging of microstructures in carbonates may give clues to growth mechanisms in calcite and dolomite, but the problem remains that precipitation has not been achieved in physicochemical experiments attempting to simulate normal sedimentary conditions.

Clearly, there is a barrier to precipitation of dolomite from supersaturated seawater solution under Earth-surface conditions, and this has been attributed to kinetic inhibition. These kinetic barriers have been identified as: (1) the high hydration energy of the Mg^{2+} ion (Lippmann 1973; Dasent 1982; Slaughter & Hill 1991); (2) the extremely low concentration and activity of CO_3^{2-} (e.g. Garrels & Thompson 1962; Lippmann 1973); and (3) the presence of even very low concentrations of sulphate (Usdowski 1967; Baker & Kastner 1981; Kastner 1984), related to ion complexing, i.e. the formation of strongly bonded neutral $MgSO_4^0$ ion pairs (Slaughter & Hill 1991). Both (1) and (3) significantly increase the solubility of the Mg^{2+} ion, which in experiments is always the last to precipitate out from evaporated brines (e.g. Borchert & Muir 1964).

The requirement to overcome these inhibitors has led some researchers to investigate settings where modern sedimentary dolomite is forming, specifically to understand how the kinetic barriers might be overcome in the natural environment. A genetic link between bacterial sulphate reduction, raised alkalinity, removal of sulphate, and dolomite formation has frequently been reported in both modern and ancient sediments (e.g. Gieskes *et al.* 1982; Compton 1988; Wright 1993, 1994, 1997, 1998, 1999, 2000; Vasconcelos *et al.* 1995; Vasconcelos & McKenzie 1997). Some researchers considered that further investigations of the microbiogeochemical effects of sulphate-reducing bacteria (SRB) were merited in the search for a solution to the dolomite problem.

This approach has met with success, and the reality is that only in the case of recent models of microbial mediation driven by sulphate reduction has theory, in the form of an organogenic model, been substantiated by empirical proof (e.g. Vasconcelos *et al.* 1995; Vasconcelos & McKenzie 1997; Warthmann *et al.* 2000; Wacey 2003). Experiments duplicating microbially mediated, anaerobic, saline conditions occurring in Lagoa Vermelha, Brazil, and in the distal Coorong lakes of South Australia, have successfully resulted in the precipitation of dolomite at low temperatures, but only through the metabolic activities of SRB. This paper briefly examines the role of bacterial sulphate reduction in dolomite formation.

The role of sulphate and bacterial sulphate reduction

There are competing claims in the literature for the effectiveness of sulphate in inhibiting dolomite (and calcite) precipitation (Usdowski 1967; Baker & Kastner 1981; Kastner 1984; Hardie 1987; Morrow & Ricketts 1988; Slaughter & Hill 1991; Morrow & Abercrombie 1994; Wright 1999). Sulphate is an abundant component of seawater, and forms neutral ion pairs with metal cations, enhancing their solubility; moreover the proportion of ion pairs increases with ionic concentration. The work of Walter (1986), among others, has shown that sulphate can inhibit calcite precipitation, and this is entirely consistent with dolomite inhibition – the same kinetics are involved, with sulphate forming neutral ion pairs with Ca^{2+}, but the hydration shell surrounding the Ca^{2+} ion is less tightly bound than that surrounding the Mg^{2+} ion.

Brady *et al.* (1996) argued that sulphate acts as a catalyst to dolomite precipitation, using observations that dolomite occurs today in waters with high sulphate levels, and that many ancient dolomites are associated with evaporites. However, the seasonally evaporitic, SO_4^{2-}-rich, dolomitic Coorong lakes contain no solid sulphate sediments, but support abundant microbial populations with a high proportion of SRB (reaching 3.73×10^6 ml^{-1}: Wacey 2003), which by consuming SO_4^{2-} significantly modify ambient water chemistry (see below). Endo-evaporitic microbial communities in gypsum in salterns in Eilat comprise consortia of cyanobacteria, sulphur bacteria and SRB (among others) living in stratified communities in hypersaline conditions (Sørensen *et al.* 2003). A lowermost, black, organic-rich layer of the gypsum contains SRB, so that microbiogeochemical transformations within intraevaporite micropore fluids can be expected to favour carbonate precipitation by removing kinetic inhibitors. These and similar evaporites are associated with carbonates, which may thus be related to sulphate reduction in 'solid sulphate'. It is highly likely that SRB colonized similar environments in the past, in addition to marine environments where microbes dominated the environment in the absence of Metazoa, as evidenced by, for example, microbialitic rocks. SRB are universally distributed in microbial communities in modern marine sediments, microbial mats and stromatolites (Teske *et al.* 1998; Reid *et al.* 2000).

Hardie (1987) used the presence of sulphate in certain lake waters to argue against the effectiveness of sulphate as an inhibitor to dolomite formation. However, the experiments of Usdowski (1967), Baker & Kastner (1981) and Kastner (1984) provide compelling evidence for sulphate inhibition. Sulphate is, of course, present in seawater, but in modern deep-sea organic-rich sediments, where continual diffusion of seawater SO_4^{2-} feeds and controls SRB populations, dolomite forms in sulphate-free anoxic interstitial waters immediately beneath the zone of bacterial sulphate reduction (e.g. Irwin 1980; Baker & Burns 1985; Compton 1988). Overlying sediment pore waters show increased concentrations of ammonia and carbonate alkalinity associated with bacterial sulphate reduction, but where sulphate reduction was absent in deep-sea samples, no dolomite was reported (Gieskes *et al.* 1982). This provides strong evidence that the removal of sulphate by SRB from a solution in which it was originally present may lead to dolomitization if other kinetic inhibitors are also overcome. The origin of nodular dolomite in the Miocene Drakes Bay and Monterey formations, California, is attributed to biochemical changes in interstitial waters driven by SRB (Burns *et al.* 1988; Compton 1988). Although SRB and methanogens are often closely associated key components of anaerobic microbial consortia, it is doubtful that methanogens can actually drive dolomite formation in the presence of sulphate because SRB can out-compete methanogens for common growth substrates and energy sources, such as acetate and hydrogen (Lovley *et al.* 1982; Kristjansson *et al.* 1982; Schonheit *et al.* 1982; Claypool & Kvenvolden 1983). Moreover, high levels of sulphide can strongly inhibit methanogens (Mizuno *et al.* 1995; Scholten *et al.* 2002).

Observations and data reported in Wright (1999, 2000) and Wacey (2003), together with experiments described here, demonstrate that dolomite forms where sulphate reduction has removed sulphate in the presence of organic matter. For example, high counts of SRB recorded from Coorong lake sediments (in excess of 1×10^6 ml^{-1} in dolomitic lakes), and enrichment of ^{34}S in residual lake waters, indicate flourishing microbial populations and bacterial fractionation of sulphate during sulphate reduction. Sulphate concentrations were extremely high (>20 000 mg^{-1}) in dolomitic lake-water samples, but declined dramatically and progressively with depth through the sulphate-reduction zone in underlying pore-water samples. By the end of the evaporative cycle, sulphate was entirely removed. In aquatic systems SRB can also overcome other kinetic inhibitors to dolomite formation by reducing the solubility of Mg^{2+} and Ca^{2+} ions, and raising carbonate alkalinity through organic degradation (Slaughter & Hill 1991; Wright 2000). In the Coorong, on-site titrations of lake-water and sediment pore-water samples were used for calculating carbonate concentrations: these show exponential rises with depth in the sulphate-reduction zone (Fig. 1), reaching >1400 mg^{-1} in some lakes. These observations, and data and experiments, indicate that it is bacterial sulphate reduction (which requires the presence of sulphate) and the consequent biochemical interactions, not the presence of sulphate itself, that drives dolomite formation.

Organogenic dolomite is generally distinguished by having a significant proportion of its carbon derived directly from organic sources, and is characterized by a widely variable stable carbon isotope signature reflecting changes in $\delta^{13}C$ of dissolved CO_2 caused by sulphate reduction and methanogenesis. However, carbon isotopes alone are not a sufficiently precise tool for determining the role of organic matter in dolomitization (Hill 1990), nor indeed for tracking other microbially driven carbonate precipitation. The degree of ^{13}C depletion through sulphate reduction, leading to a 'light' $\delta^{13}C$ signal, is controlled by the carbon source, the extent of organic diagenesis and the contribution from the ambient water reservoir (Mazzullo 2000; Wright 2000).

The $\delta^{13}C$ values for Coorong dolomite (ranging from –1.19 to +3.22‰, Wacey 2003) fall in the 'normal marine' field. This can be explained by the fact that inorganic 'marine' HCO_3^- is abundant in the lake-water reservoir. In response to increased alkalinity derived from the release of ammonia during anoxic organic degradation, driven by bacterial sulphate reduction (in pore waters) and buffering, HCO_3^- dissociates to form CO_3^{2-} and H_2O (Berner 1980; Durand 1980; Slaughter & Hill 1991):

$$2NH_3 + CO_2 + H_2O \rightarrow 2NH_4^+ + CO_3^{2-} + OH^-$$
$$NH_3 + H_2O \rightarrow NH_4^+ + OH^-$$
$$OH^- + HCO_3^- \rightarrow H_2O + CO_3^{2-}$$

Thus, the lake-water-supplied pore waters provide the largest reservoir of CO_3^{2-} available for dolomite formation, significantly in excess of the proportion of isotopically light carbon derived from organic diagenesis. If the carbon in the dolomite were derived solely from inorganic species in equilibrium with dissolved

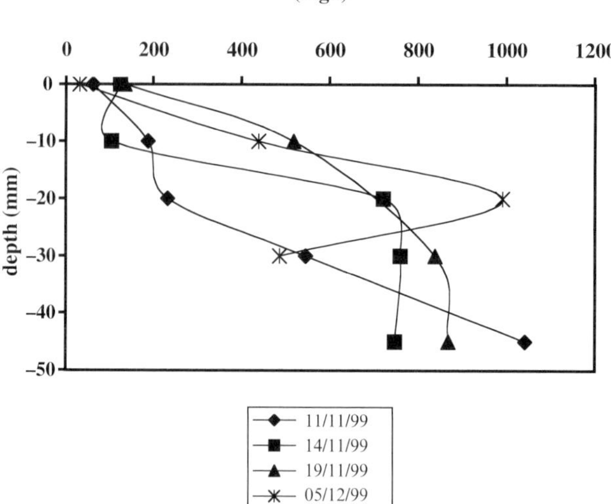

Fig. 1. Carbonate ion concentrations in lake and pore waters from Mini-Dolomite Lake, Coorong. Titrations were made on four different dates, in 1999, towards the end of the lake evaporative cycle, at 1 cm-intervals to a depth of 4.5 cm. Concentrations in lake-water samples (given at zero on the vertical scale) are always higher than seawater averages (16.2 mg^{-1}) but increase dramatically in pore waters during sulphate reduction, helping to overcome this kinetic barrier to dolomite formation. Variations in the concentration trends with time and depth are attributed to dynamic microbiogeochemical interactions within migrating, depth-stratified benthic microbial communities within the sediments.

CO_2 in the lake water, then the $\delta^{13}C$ values should be around +5‰, reflecting equilibrium fractionation from atmospheric CO_2. It is clear that the majority of lakes are not precipitating carbonate in equilibrium with atmospheric CO_2 and that some input from organically-derived, isotopically light carbon is involved.

Dolomite from experiments simulating the microbiogeochemistry of dolomitic Coorong lakes

We have precipitated dolomite in experiments undertaken at normal Earth-surface conditions that simulate those of the Coorong dolomitic lakes. These experiments have shown, without doubt, that sulphate reduction is essential for the precipitation of dolomite under microbiogeochemical conditions that simulate those of the dolomitic Coorong lakes.

Methods

Total bacteria were counted using acridine orange direct counts (AODC) and the number of sulphate reducers isolated from each sediment sample was calculated by the most probable number (MPN) statistical technique (Hurley & Roscoe 1983). Cultures of SRB were obtained from Coorong lake sediment cores at the Geomicrobiology Unit, Bristol University, UK. A Postgate Media B culture medium for SRB was used, and a 1:5 dilution series was prepared in triplicate in vials with 13 stages of dilution. Vials were incubated at 22 °C for 5 weeks, after which positive vials were recognized by a black pyritiferous precipitate.

Experiments were then initiated (week 1) using microbial populations cultured from the study lakes in order to replicate as closely as possible the microbiogeochemical conditions prevailing in the lake sediments. Twenty-one unused vials were labelled and partly filled with sterile mineral grains (quartz sand, aragonite, glass beads), so that there were seven vials of each. These vials were filled with Postgate Culture Media B and placed in an anaerobic cabinet, then injected with the cultured bacteria, except for control vials 1, 2 and 3 (each with a different grain-fill) that remained sterile. A drop in sulphate concentration levels in active vials after 2 months indicated a sustainable population of SRB had been achieved.

Results

At the beginning of week 8, the vials, except control vials 1 and 10, were injected with additional components thought to be necessary

Table 1. *Components of the 21 experimental vials, all of which contain a Postgate B culture medium. Each vial is partially filled with a different mineral seed (grain base) – glass beads, quartz or aragonite. Vials 1–3 are controls. Vials 4–21 are all injected with bacterial cultures from Coorong dolomitic lake sediments; additional components were added as shown*

Vial	Grain base	Bacteria (ml)	Conc. seawater (ml)	$MgSO_4$ (ml)	Organic matter (ml)
1	Quartz	X	X	X	X
2	Glass beads	X	X	X	X
3	Aragonite	X	X	X	X
4	Quartz	1	X	X	X
5	Quartz	1	2	1	1
6	Quartz	1	X	1	1
7	Quartz	2	X	6	X
8	Quartz	2	2	1	X
9	Quartz	2	X	X	1
10	Glass beads	1	X	X	X
11	Glass beads	1	X	1	1
12	Glass beads	1	2	X	1
13	Glass beads	2	2	X	X
14	Glass beads	2	X	X	1
15	Glass beads	2	X	6	X
16	Aragonite	6	X	X	X
17	Aragonite	6	2	1	1
18	Aragonite	3	2	X	1
19	Aragonite	2	X	1	1
20	Aragonite	2	X	X	1
21	Aragonite	2	X	1	1

X, not added.

for dolomite formation (brine, $MgSO_4$, organic matter – details are shown in Table 1). The vials were monitored for chemical changes. The slight fall in sulphate concentration in control vials 1 and 10 is attributed to background sulphate mineral precipitation and the difficulty of extracting all of the dissolved sulphate from the vial for measuring concentrations. The rise in sulphate concentration in the active vials as $MgSO_4$ was added is shown in Figure 2 for representative vials 6, 7, 8, 15 and 21. The subsequent rapid fall in sulphate concentrations in all active vials is interpreted as the result of bacterial sulphate reduction, supported by liberation of H_2S and no precipitation of sulphate minerals.

After week 12, injections of $MgSO_4$ were added to vials 7 and 15 on two separate occasions in order to simulate reactions in the lake sediment pore waters caused by SRB metabolism, and to monitor changes in pH. A similar trend was observed in each vial (Fig. 3). The initial drop from pH values >8 is attributed to the pH of 7.35 for the added $MgSO_4$; the subsequent rise in pH in the days following injection is attributable to increased carbonate alkalinity of water chemistry through SRB metabolism.

Within 2 weeks of the injection of the additional components a distinct layer of precipitate was observed in some vials. Vial 6 was left for 2 more weeks and then opened. Representative vials were opened a further month later and the sediment analysed using scanning electronic microscopy (SEM) and X-ray diffraction (XRD). Two of the vials represented here (6 and 21) produced dolomite within 2 months of injection, both having received additional $MgSO_4$ and organic matter.

In vial 6 and vial 21, 90% of the added sulphate was consumed within 3–4 weeks, followed by the precipitation of dolomite (Fig. 4a & b), as determined by XRD (Fig. 5). SEM analysis shows that the dolomite was in the form of subspherical nanocrystals, identical in morphology and size to those in the sediment of the dolomitic lakes (Wright 1999). The submicron grains are spherical–elliptical in shape, and many tend to be darker in the centre than around their edges suggesting that microbes, perhaps nanobacteria, acted as nuclei for dolomite precipitation.

The results demonstrate that SRB metabolism can remove kinetic inhibitors to dolomite formation that operate in aqueous solutions at Earth-surface temperatures. Experiments

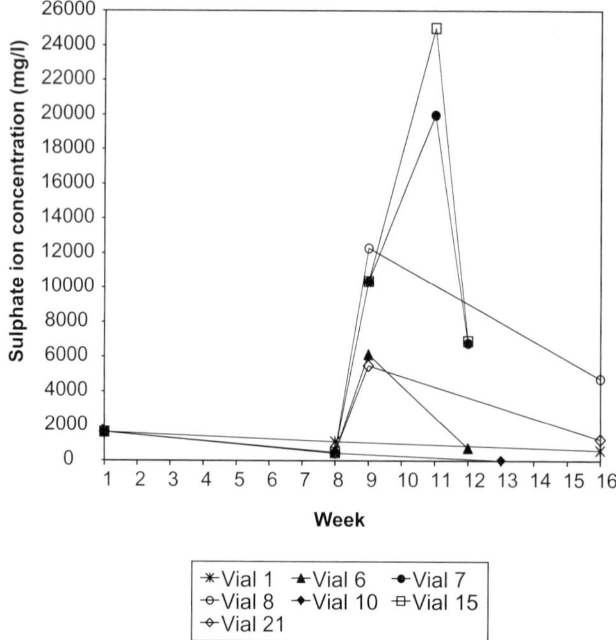

Fig. 2. Changes in sulphate concentrations in a number of experimental vials filled with Postgate Culture Medium B and cultured microbes from Coorong dolomitic lake samples. Selected vials in an anaerobic cabinet were injected at week 8 with additional $MgSO_4$; subsequent falls in sulphate concentrations were attributed to bacterial sulphate reduction. The varying values for samples are related to the different proportions of injected materials, but where the added amounts were identical in different vials (e.g. vials 7 and 15), very similar trends are seen. Only vials 6 and 21 produced dolomite, indicating that sulphate and organic matter are both necessary for dolomite to form. Vials 6 and 21 were both injected with 1 ml additional $MgSO_4$ and 1 ml organic matter; both show comparable trends and both produced dolomite. Vials 1 and 10 were controls; vial 1 was completely sterile, vial 10 contained SRB but received no injected material during the experiment. No precipitates were found. Vials 7 and 15 were both injected with 6 ml additional $MgSO_4$ but no organic matter; both show comparable trends and produced no dolomite. Vial 8 was injected with 1 ml additional $MgSO_4$ but no organic matter, and produced no dolomite.

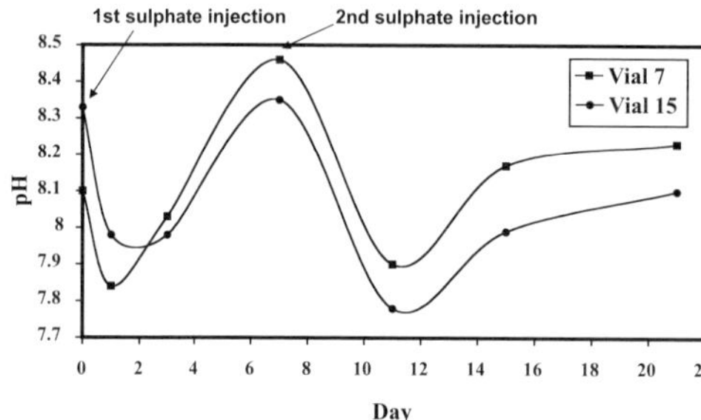

Fig. 3. Changes in the pH of vials 7 and 15 after injections of $MgSO_4$ on two separate occasions. $MgSO_4$ has a pH of 7.35, explaining the initial drop from pH values >8 in the vial liquid when introduced. The subsequent rise in pH after injection is attributable to the release of soluble ammonium compounds from organic matter diagenesis by bacterial sulphate reduction, leading to an increase in carbonate ion concentrations.

performed by Judy McKenzie and her team (Vasconcelos *et al.* 1995; Vasconcelos & McKenzie 1997; Warthmann *et al.* 2000) reached similar conclusions on the critical role of SRB in dolomite formation.

'Conventional' models of sedimentary dolomite formation

When interpreting dolomite formation in the context of conventional sedimentary models, consideration of fundamental chemical constraints has often been avoided or neglected. For example, it is generally held that massive dolomitization in carbonate platform sequences requires the long-term addition of magnesium by the circulation of saline dolomitizing fluids (the 'magnesium pump') derived from seawater or by the circulation or tidal

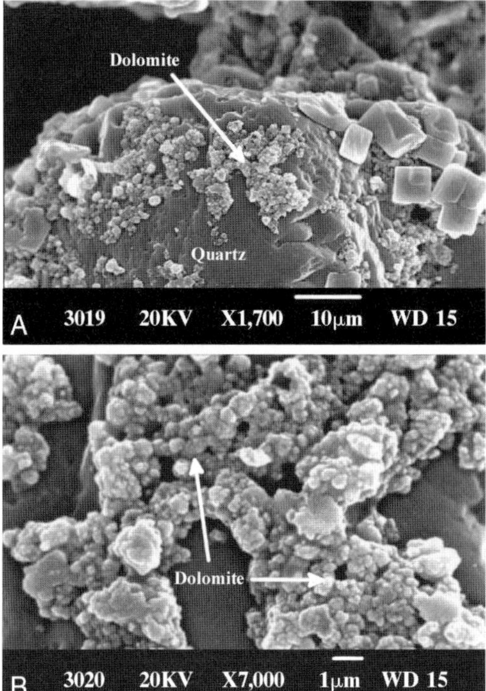

Fig. 4. SEM photomicrographs of sediment deposited after 8 weeks in experimental vial 6. (**A**) Subspherical grains of dolomite about 1 μm in size were precipitated on the surface of a quartz grain; the dolomite preceded precipitation of overlying halite cubes seen in the centre right of the view. (**B**) The precipitated dolomite seen under higher magnification is identical in size and morphology to the dolomite sediment from the distal lakes. Mineralogy was confirmed using SEM and XRD analysis.

pumping of normal seawater through carbonate platforms (e.g. Land 1991). However, it seems most unlikely that unaltered seawater is the dolomitizing solution because of the high enthalpy of hydration of the magnesium ion, low carbonate ion activity, and the presence of sulphate – all effective inhibitors to dolomite formation at normal temperatures and pressures. Furthermore, in many cases there is little evidence to support the inherent assumption that thick carbonate platforms were built up long before being dolomitized. The common observation of dolomicritic lamination and cyanobacterial filaments in ancient, thick carbonate sequences, such as the Cambrian Eilean Dubh Formation of NE Scotland (Wright 1997) and the Neoarchaean Campbellrand Supergroup (Wright & Altermann 2000), suggests the preservation of fabrics of penecontemporaneous dolomite.

The 'Dorag' model (Badiozamani 1973) invokes mixing marine and meteoric waters to produce a solution both undersaturated in calcite and supersaturated in dolomite. However, subsequent high Mg:Ca ratios are unlikely to lead to dolomitization because there is no mechanism to breach the high hydration barrier of the magnesium ion, nor to elevate carbonate ion activities. This model predicts that dolomitization would be concomitant with the dissolution of calcite, but as Hardie (1987) reports, this does not occur in, nor is there consensus that dolomite forms in, modern mixing zones. New evidence from a mixing zone in the Yucatan Peninsula (Smith *et al.* 2002), supported by laboratory simulations, indicates that SRB may stimulate the primary dolomite formation.

Overriding kinetic problems are also inherent in models invoking an evaporative mechanism, including seepage reflux (Adams & Rhodes 1960) and evaporative pumping (Hsü & Siegenthaler 1969). Lippmann (1973) showed that evaporation could not favour dolomitization despite consequent high Mg:Ca ratios because of the accompanying decrease in the activity of the CO_3^{2-} free ion. The already low activity of CO_3^{2-} ions would be further suppressed due to more frequent formation of neutral ion pairs, giving $MgCO_3^0$, and Mg^{2+} would also complex with sulphate, as $MgSO_4^0$. Deffeyes *et al.* (1965) attempted to apply the seepage reflux model to the sediments beneath the Pekelmeer Lake on Bonaire, but reality refused to recognize the model – subsequent investigations showed no dolomite beneath the lake! Where dolomite occurs beneath the Abu Dhabi sabkha it does not cross-cut facies, as might be expected from seepage reflux or evaporative pumping, but is located within horizontal beds associated with cyanobacterial mats (McKenzie *et al.* 1980; Wright 2000) – suggesting that microbial mediation may be involved. Experimental evaporation of Coorong lake brines in beakers did not produce dolomite in the sequence of precipitated minerals (personal observation), clearly indicating that simple evaporation does not produce dolomite.

Conclusion

The study of modern and ancient *sedimentary* dolomite formation has traditionally been approached as an inorganic geochemical problem, but although dolomite should precipitate spontaneously from supersaturated

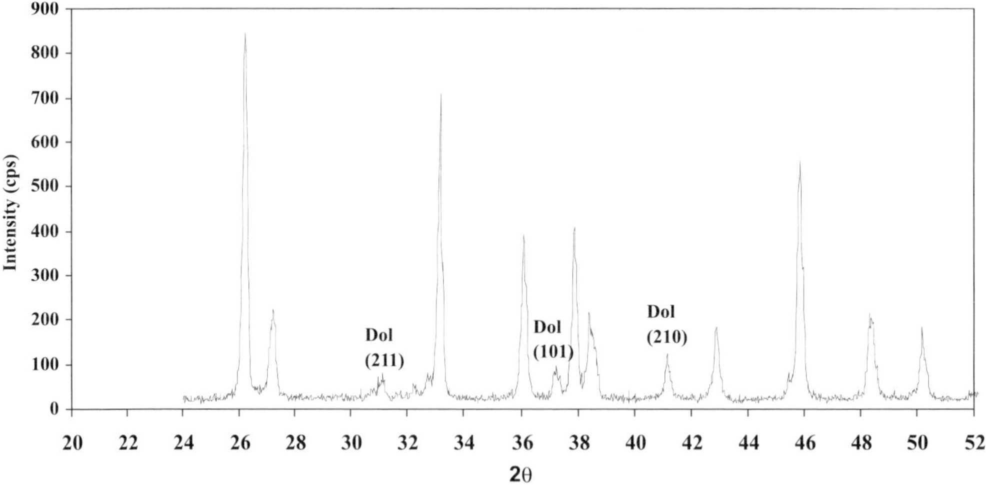

Fig. 5. X-Ray diffractogram trace of dolomite precipitated in experimental vial 21 after just 8 weeks in a medium containing sulphate reducing bacteria, organic matter and magnesium sulphate. The major dolomite peak is present at 30.99 2θ, and a number of subsidiary peaks can also be identified, including weak ordering peaks at 37.4 and 41.2 2θ, confirming the mineralogy as dolomite. Remaining peaks are from aragonite grains in the vial.

solution in seawater, it does not, and the predicted thermodynamic transformations fail to materialize.

Microbially mediated experiments have indicated that the solution to the long-standing problem of the origin of sedimentary dolomite hinges on understanding the kinetic constraints to carbonate precipitation that operate in saline waters, and how these are overcome in the natural environment. The successful precipitation of dolomite within a few weeks in reproducible experiments at 'normal' temperatures and pressures using microbial cultures from lake sediments has provided empirical proof that bacterial sulphate reduction can drive dolomite formation in Lago Vermelha and the Coorong distal lakes by overcoming those constraints. The microbiogeochemistry of the lakes reflects conditions prevailing at various times in the geological past whenever microbes dominated the ecosystem, and in particular during the Precambrian. It is clear that further research into microbial mediation offers a realistic route to understanding sedimentary dolomite formation in both modern and ancient environments.

We would like to place on record our sincere thanks for the facilities and support freely given by J. Parkes and B. Cragg at Bristol University, without whose help much of this work would not have been possible. We also acknowledge with gratitude the support and advice of C. von der Borch and N. McClure of Flinders University, South Australia. For technical assistance, we wish to thank A. Boyce at SUERC, K. Sharkey at Leicester University and S. Wyatt at Oxford University.

References

Adams, J.E. & Rhodes, M.L. 1960. Dolomitization by seepage refluxion. *AAPG Bulletin*, **44**, 1912–1921.

Arvidson, R.S. & Mackenzie, F.T. 1999. The dolomite problem: Control of precipitation kinetics by temperature and saturation state. *American Journal of Science*, **299**, 257–288.

Badiozamani, K. 1973. The Dorag dolomitization model – application to the Middle Ordovician of Wisconsin. *Journal of Sedimentary Petrology*, **43**, 965–984.

Baker, P. & Burns, S. 1985. Occurrence and formation of dolomite. *AAPG Bulletin*, **69**, 1917–1930.

Baker, P. & Kastner, M. 1981. Constraints on the formation of sedimentary dolomite. *Science*, **213**, 214–216.

Berner, R.A. 1980. *Early Diagenesis: A Theoretical Approach*. Princeton University Press, Princeton, NJ.

Borchert H. & Muir R.O. 1964. *Salt Deposits*. Van Nostrand Reinhold, London.

Brady, P.V., Krumhansl, J.L. & Papenguth, H.W. 1996. Surface complexation clues to dolomite growth. *Geochimica et Cosmochimica Acta*, **60**, 727–731.

Burns, S.J., Baker, P.A. & Showers, W.J. 1988. The factors controlling the formation and chemistry of dolomite in organic-rich sediments: Miocene Drakes Bay Formation, California. *In*: SHUKLA, V. & BAKER, P.A. (eds) *Sedimentology and Geochemistry of Dolostones*. Society of Economic Paleontologists and Mineralogists, Special Publications, **43**, 1–52.

CLAYPOOL, G.E. & KVENVOLDEN, K.A. 1983. Methane and other hydrocarbon gases in marine sediment. *Annual Review of Earth and Planetary Sciences*, **11**, 229–327.

COMPTON, J.S. 1988. Sediment composition and precipitation of dolomite and pyrite in the Neogene Monterey and Sisquoc formations, Santa Maria Basin area, California. *In*: SHUKLA, V. & BAKER, P.A. (eds) *Sedimentology and Geochemistry of Dolostones*. Society of Economic Paleontologists and Mineralogists, Special Publications, **43**, 53–64.

DASENT, W.E. 1982. *Inorganic Energetics*, 2nd edn. Cambridge University Press, New York.

DEFFEYES, K.S., LUCIA, F.J. & WEYL, P.K. 1965. Dolomitization of Recent and Plio-Pleistocene sediments by marine evaporite waters on Bonaire, Netherlands Antilles. *In*: PRAY, L.C. & MURRAY, R.C. (eds) *Dolomitization and Limestone Diagenesis*. Society of Economic Paleontologists and Mineralogists, Special Publications, **13**, 71–88.

DURAND, B. 1980. Sedimentary organic matter and kerogen. Definition and quantitative importance of kerogen. *In*: DURAND, B. (ed.) *Kerogen*. Technip, Paris, 13–34.

GAINES, A.M. 1980. Dolomitisation kinetics: Recent experimental studies. *In*: ZENGER, D.H., DUNHAM, J.B. & ETHINGTON, R.L. (eds) *Concepts and Models of Dolomitisation*. Society of Economic Paleontologists and Mineralogists, Special Publications, **28**, 81–86.

GARRELS, R.M. & THOMPSON, M.E. 1962. A chemical model for seawater at 25°C and one atmosphere total pressure. *American Journal of Science*, **260**, 57–66.

GIESKES, J.M., ELDERFIELD, H., LAWRENCE, J.R., JOHNSON, J., MEYERS, B. & CAMPBELL, A. 1982. Geochemistry of interstitial waters and sediments, Leg 64, Gulf of California. *In*: CURRAY, J.R., MOORE, D.G. *ET AL.* (eds) *Initial Reports of the Deep Sea Drilling Project*, **64**. U.S. Government Printing Office, Washington, DC, 675–694.

HARDIE, L.A. 1987. Dolomitization: a critical review of some current views. *Journal of Sedimentary Petrology*, **57**, 166–183.

HILL, R.J. 1990. *Field evidence for the role of organic matter in dolomitization*. Master's abstract, Colorado School of Mines.

HSÜ, K.J. & SIEGENTHALER, C. 1969. Preliminary experiments and hydrodynamic movement induced by evaporation and their bearing on the dolomite problem. *Sedimentology*, **12**, 11–25.

HURLEY, M.A. & ROSCOE, M.E. 1983. Automated statistical analysis of microbial enumeration by dilution series. *Journal of Applied Bacteriology*, **55**, 159–164.

IRWIN, H. 1980. Early diagenetic carbonate precipitation and pore fluid migration in the Kimmeridge Clay of Dorset, England. *Sedimentology*, **27**, 577–591.

KASTNER, M. 1984. Control of dolomite formation. *Nature*, **311**, 410–411.

KELLEHER, I.J. & REDFERN, S.A.T. 2002. Hydrous calcium magnesium carbonate, a possible precursor to the formation of sedimentary dolomite. *Molecular Simulation*, **28**, 557–572.

KRISTJANSSON, J., SCHONHEIT, P. & THAUER, R. 1982. Different K values for hydrogen of methanogenic bacteria and sulfate reducing bacteria: An explanation for the apparent inhibition of methanogenesis by sulfate. *Archives of Microbiology*, **131**, 278–282.

LAND, L.S. 1991. Dolomitization models – seawater mixing zones. (Abs.) *In*: *Dolomieu Conference on Carbonate Platforms and Dolomitization*, 144. Karo-Drucle, Eppan, Italy.

LAND, L.S. 1998. Failure to precipitate dolomite at 25°C from dilute solution despite 1000-fold oversaturation after 32 years. *Aquatic Geochemistry*, **4**, 361–368.

LIPPMANN, F. 1973. *Sedimentary Carbonate Minerals*. Springer, Berlin.

LUMSDEN, D.N., MORRISON, J.W. & LLOYD, R. 1995. Role of iron and Mg/Ca ratio in dolomite synthesis at 192°C. *Journal of Geology*, **103**, 51–61.

LOVLEY, D.R., DWYER, D.F. & KLUG, M.J. 1982. Kinetic analysis of competition between sulfate reducers and methanogens for hydrogen in sediments. *Applied Environmental Microbiology*, **43**, 1373–1379.

MAZZULLO, S.J. 2000. Organogenic dolomitization in peritidal to deep-sea sediments. *Journal of Sedimentary Research*, **70**, 10–23.

MCKENZIE, J.A., HSÜ, K.J. & SCHNEIDER, J.F. 1980. Movement of subsurface waters under the sabkha, Abu Dhabi, UAE, and its relation to evaporative dolomite genesis. *In*: ZENGER, D.H., DUNHAM J.B. & ETHINGTON, R.E. (eds) *Concepts and Models of Dolomitisation*. Society of Economic Paleontologists and Mineralogists, Special Publications, **28**, 11–30.

MIZUNO, O., LI, Y.Y. & NOIKE, T. 1994. Effects of sulfate concentration and sludge retention time on the interaction between methane production and sulfate reduction for butyrate. *Water Science & Technology*, **30**, 45–54.

MORROW, D.W. & ABERCROMBIE, H.J. 1994. Rates of dolomitization: the influence of dissolved sulphate. *In*: PURSER, B.H., TUCKER, M.E. & ZENGER, D.H. (eds) *Dolomites – A volume in honour of Dolomieu*. International Association of Sedimentologists, Special Publications, **21**, 377–386.

MORROW, D.W. & RICKETTS, B.D. 1986. Chemical controls on the precipitation of mineral analogues of dolomite: the sulfate enigma. *Geology*, **14**, 408–410.

MORROW, D.W. & RICKETTS, B.D. 1988. Experimental investigation of sulfate inhibition of dolomite and its mineral analogues. *In*: SHUKLA, V. & BAKER, P.A. (eds) *Proceedings of Sedimentology and*

Geochemistry of Dolostones. Raleigh, North Carolina, 26–28 September 1986. Society of Economic Paleontologists and Mineralogists, Special Publications, **43**, 25–38.

REID, R.P., VISSCHER, P.T. *ET AL*. 2000. The role of microbes in accretion, lamination and early lithification of modern marine stromatolites. *Nature*, **406**, 989–992.

SCHOLTEN, J.C.M., VAN BODEGOM, P.M., VOGELAAR, J., VAN ITTERSUM, A., HORDIJK, K., ROELOFSEN, W.W. & STAMS, A.J.M. 2002. Effect of sulfate and nitrate on acetate conversion by anaerobic microorganisms in a freshwater sediment. *Fems Microbiology Ecology*, **42**, 375–385.

SCHONHEIT, P., KRISTJANSSON, J.K. & THAUER, R.K. 1982. Kinetic mechanism for the ability of sulfate reducers to out-compete methanogens for acetate. *Archives of Microbiology*, **132**, 285–288.

SLAUGHTER, M. & HILL, R.J. 1991. The influence of organic matter in organogenic dolomitization. *Journal of Sedimentary Petrology*, **61**, 296–303.

SMITH, S.L., WHITAKER, F.F. & PARKES, R.J. 2002. Dolomitisation by saline groundwaters in the Yucatan Peninsula. *In*: *The Geometry and Petrogenesis of Dolomite Hydrocarbon Reservoirs*. Final Programme and Abstracts 3–4 December 2002, Geological Society of London.

SØRENSEN, K.B., CANFIELD, D.E., TESKE, A. & OREN, A. 2003. A microbial garden within a gypsum crust: a biogeochemical and phylogenetic study of an endoevaporitic, stratified bacterial community from a solar saltern. (Abs.) *In*: *Aquatic Science Meeting* ASLO 2003, Salt Lake City, Utah. http://www.sgmeet.com/aslo/slc2003/viewabstract2.asp?AbstractID=591&SessionID=SS4.06

TESKE, A., RAMSING, N.B., HABICHT, K.S., FUKUI, M., KUVER, J., JORGENSEN, B.B. & COHEN, Y. 1998. Sulfate-reducing bacteria and their activities in cyanobacterial mats of Solar Lake (Sinai, Egypt). *Applied and Environmental Microbiology*, **64**, 2943–2951.

TRIBBLE, J.S., ARVIDSON, R.S., LANE, M., IV. & MACKENZIE, F.T. 1995. Crystal chemistry, and thermodynamic and kinetic properties of calcite, dolomite, apatite, and biogenic silica: applications to petrologic problems. *Sedimentary Geology*, **95**, 11–37.

USDOWSKI, H.E. 1967. The formation of dolomite in sediments. *In*: MULLER, G. & FRIEDMAN, G.M. (eds) *Recent Developments in Carbonate Sedimentology in Central Europe*. Springer, Heidelberg, 21–32.

VASCONCELOS, C. & MCKENZIE, J.A. 1997. Microbial mediation of modern dolomite precipitation and diagenesis under anoxic conditions (Lagoa Vermelha, Rio de Janeiro, Brazil). *Journal of Sedimentary Research*, **67**, 378–390.

VASCONCELOS, C., MCKENZIE, J.A., BERNASCONI, S., GRUJIC, D. & TIEN, A.J. 1995. Microbial mediation as a possible mechanism for natural dolomite formation at low temperatures. *Nature*, **377**, 220–222.

WACEY, D. 2003. *Microbial mediation of dolomite formation: geochemical investigations in the Coorong region of South Australia*. DPhil thesis, University of Oxford.

WALTER, L.M. 1986. Relative efficiency of carbonate dissolution and precipitation during diagenesis: a progress report on the role of solution chemistry. *In*: GAUTIER, D.L. (ed.) *Roles of Organic Matter in Sediment Diagenesis*. Society of Economic Paleontologists and Mineralogists, Special Publications, **38**, 1–11.

WARTHMANN, R., VAN LITH, Y., VASCONCELOS, C., MCKENZIE, J.A. & KARPOFF, A.M. 2000. Bacterially induced dolomite precipitation in anoxic culture experiments. *Geology*, **28**, 1091–1094.

WRIGHT, D.T., 1993. *Studies of the Cambrian Eilean Dubh Formation of northwest Scotland*. DPhil thesis, University of Oxford.

WRIGHT, D.T. 1994. The role of benthic microbial communities in widespread dolomite formation. (Abs.) *In*: AWRAMIK, S.M. (ed.) *Death Valley International Stromatolite Symposium*, Laughlin, Nevada, USA, 95.

WRIGHT, D.T. 1997. An organogenic origin for widespread dolomite in the Cambrian Eilean Dubh Formation, north western Scotland. *Journal of Sedimentary Research*, **67**, 54–64.

WRIGHT, D.T. 1998. Origin of carbonate in marine stromatolites of the Eilean Dubh Formation, north-western Scotland. *Journal of the Open University Geological Society*, **25**, 3–12 (Anniversary Issue).

WRIGHT, D.T. 1999. The role of sulphate-reducing bacteria and cyanobacteria in dolomite formation in distal ephemeral lakes of the Coorong region, South Australia. *Sedimentary Geology*, **126**, 147–157.

WRIGHT, D.T. 2000. Benthic microbial communities and dolomite formation in marine and lacustrine environments – a new dolomite model. *In*: GLENN, C.R., LUCAS, J. & PREVOT-LUCAS, L. (eds) *Marine Authigenesis: From Global to Microbial*. Society of Economic Paleontologists and Mineralogists, Special Publications, **66**, 7–14.

WRIGHT, D.T. & ALTERMANN, W. 2000. Microfacies development in late Archaean stromatolites and ooids of the Ghaap Group, Republic of South Africa. *In*: INSALACO, E., SKELTON, P.W. & PALMER, T.J. (eds) *Carbonate Platform Systems: Components and Interactions*. Geological Society, London, Special Publications, **178**, 51–70.

Irish Lower Carboniferous replacement dolomite: isotopic modelling evidence for a diagenetic origin involving low-temperature modified seawater

WAYNE R. WRIGHT[1,2], I. D. SOMERVILLE[3], J. M. GREGG[4], K. L. SHELTON[5] & A. W. JOHNSON[5]

[1]*Department of Geology, University College Dublin, Belfield, Dublin 4, Ireland*

[2]*Present address: Robertson Research International Ltd, Llandudno LL30 1SA, UK (e-mail: wrw@robresint.co.uk)*

[3]*Department of Geology, University College Dublin, Belfield, Dublin 4, Ireland*

[4]*Department of Geology and Geophysics, University of Missouri-Rolla, Rolla, MO 65409, USA*

[5]*Department of Geological Sciences, University of Missouri-Columbia, Columbia, MO 65211, USA*

Abstract: Irish Lower Carboniferous (Dinantian Subsystem) carbonate rocks are extensively replaced by planar dolomite. This dolomitization is unrestricted in lithology replaced, age of host rock and geographical occurrence. This paper presents geochemical ($\delta^{18}O$, $\delta^{13}C$, $^{87}Sr/^{86}Sr$, Sr conc.) modelling evidence, and discussion supporting and extending the theory that replacement planar dolomitization formed in the Waulsortian, and other host rocks under shallow burial conditions via interaction with a low-temperature (*c.* 50–70 °C) slightly modified seawater. The probable mechanism for transporting the fluid into the carbonates appears to be a variant of Kohout convection, driven by an elevated geothermal gradient. As seawater was drawn inwards, it encountered the units beneath the Waulsortian and scavenged radiogenic Sr. This warmer fluid then migrated upwards and up-slope into overlying Waulsortian and Supra-Waulsortian platform carbonates still undergoing early diagenesis. Calcium in pore fluids, provided by dissolution–precipitation reactions of calcite, was probably incorporated into the modified, and slightly warmer, seawater resulting in the variability noted in the planar replacement dolomite Sr concentrations. The early low-temperature dolomitization of the carbonate host rocks provided a crucial preparation event by creating/redistributing and preserving porosity and permeability. Younger regionally migrating high-temperature fluids directly related to the Zn–Pb mineralization in Ireland exploited these dolomitized units as aquifers. The models and methodologies presented for understanding dolomite genesis in the Lower Carboniferous rocks of Ireland can be applied to any dolomitized reservoir.

Dolomite is the principal secondary carbonate within the Irish Lower Carboniferous rocks. Volumetrically, planar dolomite dominates (*c.* 95%), whereas nonplanar dolomite is subordinate (*c.* 5%). The occurrence of both morphologies is extensive, both distally and proximally to Zn–Pb mineralization. The intimate association of dolomite (planar replacement and nonplanar open-space filling cements) and ore dictates that understanding the paragenesis and diagenesis of dolomitization is fundamental for creating valid theories for fluid flow and Zn–Pb mineralization. The geochemical fluid–rock interaction modelling presented illustrates the fluid types, sources and rock interactions active during replacement dolomitization of the Irish Lower Carboniferous units.

Fluid–rock models and diagnostic interaction trends on specific geochemical covariation diagrams enable the question of dolomite genesis to be addressed. (1) Is it reflected by a simple, single, open-system fluid–rock interaction environment? (2) Or are multiple fluids and rocks interacting? (3) Are the fluid(s) compositions evolving within the system? (4) Is there a mixing of fluids? (5) Do the data reflect physically mixed mineral end members? (Banner & Hanson 1990; Banner 1995).

Previous applications of fluid–rock interaction modelling to the genesis of Irish Lower Carboniferous carbonates (calcite and

From: BRAITHWAITE, C. J. R., RIZZI, G. & DARKE, G. (eds) 2004. *The Geometry and Petrogenesis of Dolomite Hydrocarbon Reservoirs*. Geological Society, London, Special Publications, **235**, 75–97. 0305-8719/$15.00

dolomite) and ore deposits are limited to studies by Samson & Russell (1987), Dixon (1990) and Everett (1999). In all of these studies only the $\delta^{13}C$ and $\delta^{18}O$ systems were modelled. Although modelling has been sparingly applied to the Irish system, extensive geochemical data sets have been amassed. These data are commonly spatially associated with mineralization, and comprised of $\delta^{13}C$ and $\delta^{18}O$ values.

The models and diagrams in this paper are largely restricted to the Waulsortian limestones/dolomites. However, both sampling and modelling was performed on other older and younger intervals/units (see Wright *et al.* 2000*a*, *b*, 2001; Gregg *et al.* 2001). Detailed descriptions of the stratigraphy, facies and intervals sampled is provided in Gregg *et al.* (2001) and Wright *et al.* (2001, 2003). A detailed diagenetic history and paragenesis of Waulsortian limestones and dolomite is available in Gregg *et al.* (2001).

Previous work suggested that the regional replacement dolomitization of the Waulsortian occurred under early diagenetic conditions (Gregg *et al.* 2001; Wright *et al.* 1999, 2000*a*, *b*; Wright 2001). The modelling and discussion in this paper seeks to address the validity of this assertion.

Methods

The Sr concentration, $^{87}Sr/^{86}Sr$ ratio and $\delta^{13}C$, $\delta^{18}O$ values were the parameters modelled in this study. Both fluid–rock interaction and fluid-mixing models were generated in an effort to understand the complex dolomitizing system. However, here we focus on the results of the fluid–rock interaction modelling for the genesis of the planar replacement dolomite. On many of the diagrams that follow there are dolomite morphologies referred to as either planar neomorphosed and/or transitional. The dolomite morphotypes are identified and discussed in Wright *et al.* (2003), and these distinctions are not critical to the present discussions and/or conclusions. However, it is noteworthy that there is planar dolomite thought to closely reflect low-temperature replacement of calcite, as well as outwardly similar appearing generations that reflect being neomorphosed by later, higher-temperature fluids (Wright *et al.* 2003).

Fluid–rock interaction models

The fluid–rock interaction models are iterative calculations using mass balance, in which successive increments of fluid are passed through a given rock volume (Banner & Hanson 1990). Each fluid increment reacts until isotopic and elemental equilibrium is reached and is then displaced by the next addition of fluid (Banner & Hanson 1990).

Several sets of equations have been developed to model fluid–rock interactions, those used here are from Banner *et al.* (1988), Banner & Hanson (1990) and Banner (1995). They were chosen because they have been applied previously to dolomitization, using trace elements and stable/radiogenic isotopes and, importantly, use additional model parameters such as: *porosity*, *single element and bulk distribution coefficients*, and *salinity* in the overall evolution of the system.

Only a cursory explanation of some parameters is included, and detailed explanations and derivations of the model parameters and equations are given in Banner *et al.* (1988), Banner & Hanson (1990) and Banner (1995). The effects of carbon speciation and temperature dependence are additional parameters modelled in this study. The 'fixed' parameters during the modelling experiments are listed in Table 1, with some general definitions.

Porosity for all the models was fixed at 28% for consistency. This value comes from a porosity study of the Waulsortian in Gregg *et al.* (2001) and represents an 'average' value for this lithology *prior* to most diagenetic cementation. Porosity of this magnitude may be considered too high for *post*-planar dolomitization, but the effect of decreased porosity on the models can be conceptualized and thus removes the need to model all of the different permutations at different porosities. Rocks with high porosity require higher fluid:rock ratios (N) to reach the equivalent values of extent of fluid–rock interaction (EQ) or isotopic EQ (IEQ) (see Table 1) (Banner & Hanson 1990). The N value required for a solid to reach the initial and final equilibrium stages for trace-element substitution is a function of the Ca content of the fluid and *not* of either the trace-element (e.g. Sr) concentration of the fluid or of the rock (Banner & Hanson 1990) (see Table 1).

Changes of $\delta^{18}O$ values are strongly dependent on porosity, but Banner & Hanson (1990) assert that the stoichiometry of the fluid and rock constrain the low N (<10) values at which the limestone/dolomite will reach equilibrium. Because of this, the N value does not vary as a function of the fluid or rock composition in $\delta^{18}O$ models.

In all of the models, the large differences in proportions of the elements (e.g. Sr and O) in

Table 1. *The fixed parameters and general definitions used in performing the isotopic and trace element models*

Parameters	Value (if applied)
K_D^{I-Ca} (Exchange reaction between the distribution coefficients for element I and, in this case, Ca)	0.03 for Sr (average of values from several studies)
D^I (Single element distribution coefficient)	Stoiciometric substitution for Ca
$\delta^{18}O$ (Temperature-dependent fractionation equation)	$10^3 \ln \alpha_{dolomite-water} = 2.78 \times 10^6 T^{-2} + 0.91$ (Land 1985; Machel & Burton 1994)
Porosity (Estimated)	28% (Gregg *et al.* 2001)
$^{87}Sr/^{86}Sr$ (Approximate ratio of Ivorian to Chadian seawater)	(*c.* 0.7079) (Bruckschen *et al.* 1995)
FIP (Fluid Interaction Path)	NA
EQ (Extent of fluid–rock interaction for a concentration of an element)	NA
IEQ (Isotopic equivalent to EQ)	NA
Undolomitized Waulsortian limestone (Mean values)	$^{87}Sr/^{86}Sr$ ratio of 0.70765 200 ppm Sr $\delta^{18}O$ PDB = –2.37‰ PDB
Dissolved carbon concentration	10 000 ppm
EMD (Estimated Marine Lower Carboniferous Dolomite @ 25 °C)	$\delta^{18}O$ PDB value of 1.24 to –1.67‰ $\delta^{13}C$ PDB value of 2–5‰ Sr concentration (average) 100 ppm $^{87}Sr/^{86}Sr$ ratio of 0.7079

the fluid and the rock result in a plot with a characteristic 'right-angle' bend (e.g. Fig. 3). This 'L-shaped' plot is generated for any pair of elements or isotopes in which the carbonate nears equilibration with the fluid for one member of the pair while the other retains nearly its initial value (Banner & Hanson 1990; Banner 1995). In all cases of modelling that follow, it is important to remember that the calculated N values on the pathways illustrated are less significant than the comparison of relative N values for the different processes and fluids modelled.

Limited data describe the unaltered background geochemical composition of the Waulsortian Formation in Ireland. Studies include those of Douthit *et al.* (1993), Gregg *et al.* (2001), and unpublished data in Wright (2001) and Daly (pers. comm.). The Bruckschen *et al.* (1995) and Bruckstein & Veizer (1997) data set was used to constrain the isotopic compositions of the late Ivorian–early Chadian equatorial oceans.

Modelling types

Modelling of the geochemical evolution of the fluids that produced the replacement dolomite in the Waulsortian limestones indicates that multiple chemical systems must be evaluated simultaneously because individual systems (e.g. $\delta^{18}O$ v. $\delta^{13}C$) give only part of the history. Only the precipitation and geochemical conditions that can satisfy all of the models will bear any resemblance to the reality of the system and will still be, at best, approximations.

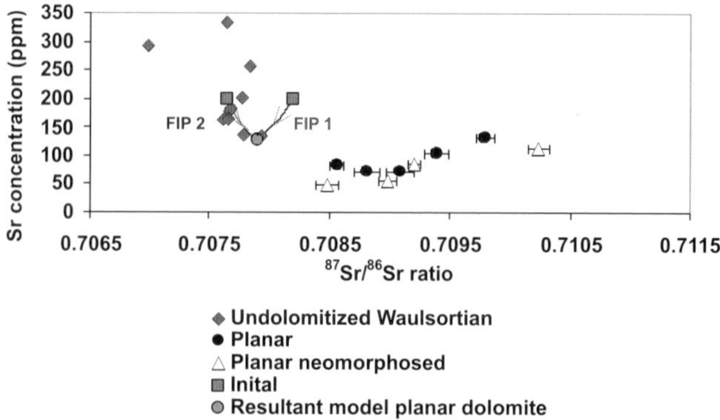

Fig. 1. Sr concentration v. $^{87}Sr/^{86}Sr$ ratio cross-plot for planar dolomite, planar neomorphosed dolomite and undolomitized Waulsortian data (from Douthit *et al.* 1993). Also illustrated are FIPs 1 and 2 of seawater interacting with two possible Waulsortian host compositions.

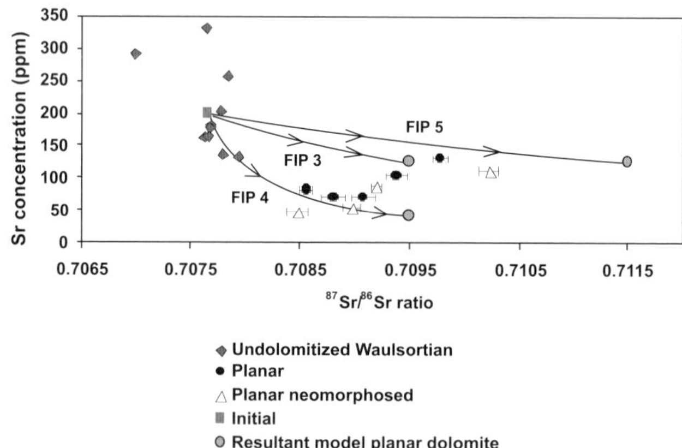

Fig. 2. Sr concentration v. $^{87}Sr/^{86}Sr$ ratio cross-plot for all dolomite data and undolomitized Waulsortian (from Douthit *et al.* 1993). Also illustrated are FIPs 3, 4 and 5 of more radiogenic fluids interacting with the Waulsortian.

Sr concentration v. $^{87}Sr/^{86}Sr$ ratio modelling

Detailed petrography associated with this study and that illustrated in Gregg *et al.* (2001) attests that planar dolomite formed as an early diagenetic, low-temperature (<50 °C) replacement of Waulsortian limestone. This assertion can be tested using the outlined models.

If the fluid (seawater) is constrained to low Sr (8 ppm) and low Ca (415 ppm) concentrations, and has the $^{87}Sr/^{86}Sr$ ratio considered to be present in late Ivorian–early Chadian oceans (*c.* 0.7079, Bruckschen *et al.* 1995), and if Waulsortian rock values are set at 0.7082 $^{87}Sr/^{86}Sr$ and *c.* 200 ppm Sr (Daly pers. comm.), then fluid–rock interaction would produce dolomites with approximately 125 ppm Sr, but with a $^{87}Sr/^{86}Sr$ ratio similar to the seawater ('fluid interaction path', FIP 1, Fig. 1). An increased $^{87}Sr/^{86}Sr$ ratio for the Waulsortian (and other units) of 0.7082 results in $^{87}Sr/^{86}Sr$ values for the limestone similar to those of the planar dolomite, but elevated relative to the standard marine value. However, limestones commonly reflect decreased Sr concentrations and more radiogenic $^{87}Sr/^{86}Sr$ values after only limited diagenesis (Banner & Hanson 1990).

All of the planar replacement dolomite analysed is more radiogenic than either seawater of equivalent age or Waulsortian rock values ($^{87}Sr/^{86}Sr$ up to 0.7098) with average values of 0.7091 and 96 ppm Sr. The fluid–rock interaction pathway (FIP 1) proceeds in the 'wrong' direction to produce dolomite with our measured values (Fig. 1). Therefore, if a 'standard' Lower Carboniferous seawater

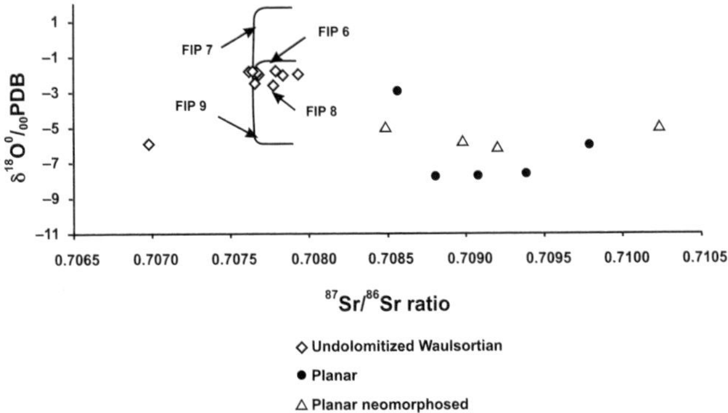

Fig. 3. $\delta^{18}O$ v. $^{87}Sr/^{86}Sr$ ratio cross-plot for planar and planar neomorphosed dolomite data and undolomitized Waulsortian (from Douthit *et al.* 1993). Also illustrated are FIPs 6–9 of seawater interacting with undolomitized Waulsortian. Note FIP 6 and 8 have a $\delta^{18}O$ fluid value of 0‰ SMOW, and FIP 7 and 9 have a $\delta^{18}O$ fluid value of –3‰ SMOW.

cannot produce dolomite with the required values then a modification to the fluid, the environment (host rock) or the modelling system is required.

By decreasing the proposed original $^{87}Sr/^{86}Sr$ values for the Waulsortian to the Douthit *et al.* (1993) values of 0.7076–0.7079, the fluid–rock interaction pathway proceeds toward the analysed values for the planar replacement dolomite, but still falls well short of the measured values (FIP 2, Fig. 1). Changes to the composition of the precursor limestone are therefore inadequate and a fluid different from the 'standard seawater' must be invoked. Drastic changes can be assumed for the original isotopic composition of the Waulsortian and the extent of diagenesis, prior to introduction of the dolomitizing fluids, resulting in $^{87}Sr/^{86}Sr$ values similar to those of the planar dolomites. Although using more diagenetically altered starting compositions for the 'host' rock is not unrealistic, there still is a problem with the fluid itself. Using the seawater composition outlined at the beginning of the section still causes the fluid–rock interaction pathway to migrate toward equilibrium, *away* from the measured planar dolomite values.

Thus, it appears that a fluid equal to or more radiogenic than the planar replacement dolomite is needed to interact with the precursor limestones prior to dolomitization. This fluid would need to have a maximum $^{87}Sr/^{86}Sr$ ratio of 0.7098 to produce the analysed values of the planar dolomite (Fig. 1). In the iterative modelling of the Sr isotope system, a fluid with the same Sr and Ca concentrations as 'standard' seawater and with a $^{87}Sr/^{86}Sr$ ratio of 0.7095 will interact with the limestone and, depending on the exact initial isotopic value, produce values identical to those determined for planar dolomite analysed (FIPs 3 and 4, Fig. 2).

Sensitivity evaluations of the K_D^{Sr-Ca} used in the model indicate that changing from $K_D^{Sr-Ca} = 0.03$ (FIP 3) to $K_D^{Sr-Ca} = 0.01$ (FIP 4) will result in equilibrated Sr concentrations in the resultant dolomite being approximately 84 ppm lower in FIP 4 (42 ppm Sr) than in in FIP 3 (Fig. 2). Therefore, even though the model accurately describes the system needed to produce the values noted in the replacement dolomite, changes to some of the variables have a profound effect on the composition of the resultant dolomite.

$\delta^{18}O$ values v. $^{87}Sr/^{86}Sr$ ratio modelling

Modelling of the relationship of the isotopic ratio of oxygen to that of strontium allows us potentially to define the temperatures of the dolomitizing fluids. Comparison of the $\delta^{18}O$ v. $^{87}Sr/^{86}Sr$ ratio enables us to simultaneously model two geochemical variables in a fluid affected by very different parameters that commonly react at very different *N* values. In Figure 3, four fluid interaction paths (FIPs 6–9) are modelled with respect to the generation of planar dolomite via replacement of undolomitized Waulsortian limestone.

In all instances, the starting solid composition of each FIP is the mean value for the undolomitized Waulsortian limestone (see Table 1). The $^{87}Sr/^{86}Sr$ ratio, and Sr and Ca concentrations are the same for all four of the FIP fluids, 0.7079, 8 ppm and 415 ppm, respectively. These are the

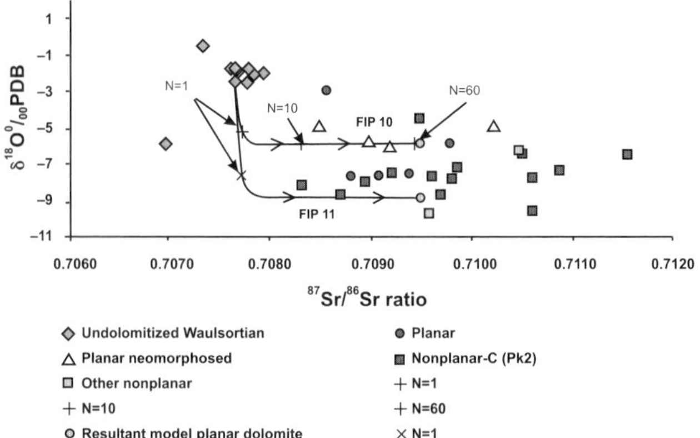

Fig. 4. $\delta^{18}O$ v. $^{87}Sr/^{86}Sr$ ratio cross-plot for all dolomite data and undolomitized Waulsortian (from Douthit *et al.* 1993). Also illustrated are FIPs 10 and 11 of radiogenic seawater interacting with undolomitized Waulsortian. Note FIP 10 and 11 have a $\delta^{18}O$ fluid value of –3‰ SMOW, and temperatures of 50 and 70 °C, respectively. Note the *N* values of 10 and 60 on FIP 11 are in the same spot as those of FIP 10.

same values used in the earlier section on 'Sr concentration v. $^{87}Sr/^{86}Sr$ ratio modelling' to model the interaction of seawater with the undolomitized Waulsortian (Fig. 1). The values used for the range of $\delta^{18}O$ values for seawater are from the study by Bruckschen *et al.* (1999).

The differences apparent in Figure 3, between FIPs 6 and 9, are functions of the fluid $\delta^{18}O$ value and the temperature of the fluid. In FIP 6, the $\delta^{18}O$ value of the seawater is 0‰ SMOW (Standard Mean Ocean Water) and is at a temperature of 25 °C. In FIP 7, the $\delta^{18}O$ value of the seawater is –3‰ SMOW, and is also at a temperature of 25 °C. FIPs 1 and 2 both produce dolomite that has more positive $\delta^{18}O$ values than the undolomitized Waulsortian, but is slightly more radiogenic (Fig. 3). Comparing these modelled values (FIPs 6 and 7) with our data, indicates this fluid (seawater), at 25 °C, is not viable for producing dolomite with values similar to our analysed samples.

A temperature increase to 50 °C will produce markedly different pathways for the fluid–rock interaction (FIPs 8 and 9). This is a direct function of the temperature dependency of oxygen isotope fractionation. FIPs 8 and 9 progress toward the analysed values of the dolomite, but still fall well short due to lack of a more radiogenic $^{87}Sr/^{86}Sr$ ratio in the fluid. It should be noted, however, that if the temperature of the 0‰ SMOW seawater were increased to 70 °C, its FIP would be almost identical to that of FIP 9 (50 °C, –3‰ SMOW fluid). A similar increase in temperature to the –3‰ SMOW fluid would cause the path of the fluid–rock interaction to migrate to more negative $\delta^{18}O$ values producing dolomite with $\delta^{18}O$ values as low as –8.89‰ PDB (Peedee Belemnite Formation standard), similar to those of the analysed dolomite.

Using a different initial Waulsortian value results in the same pathway, as in the first section of 'Modelling types', that diverges and migrates toward lower $^{87}Sr/^{86}Sr$ ratios of the fluid, i.e. away from the analysed data.

The modelling predicts that for a fluid with a $\delta^{18}O$ value of –3‰ SMOW, *all of the planar dolomites* (analysed values) could precipitate from this fluid–rock interaction if the fluid were slightly more radiogenic, 0.70979 or greater, and the temperature varied between approximately 33 and 61 °C (Fig. 3). However, the majority of the data will fall into a smaller range as discussed below.

Modelling the interaction of this more radiogenic seawater with the undolomitized Waulsortian is illustrated in Figure 4. FIP 10 has the same parameters as FIP 9 in Figure 3, but has a fluid $^{87}Sr/^{86}Sr$ ratio of 0.7095. FIP 11 has the same parameters as FIP 10, but is modelled at a temperature of 70 °C.

Although the model constraints and fluids appear realistic and encompass the data, a similar range of values and shapes of the FIPs could be reproduced by fixing the temperature and varying the $\delta^{18}O$ value for the fluid, if the fluid has a radiogenic enough $^{87}Sr/^{86}Sr$ ratio. For example, at 50 °C, a fluid value variation of –1 to –5‰ SMOW $\delta^{18}O$ would encompass the data. However, it is not thought, given the proposed

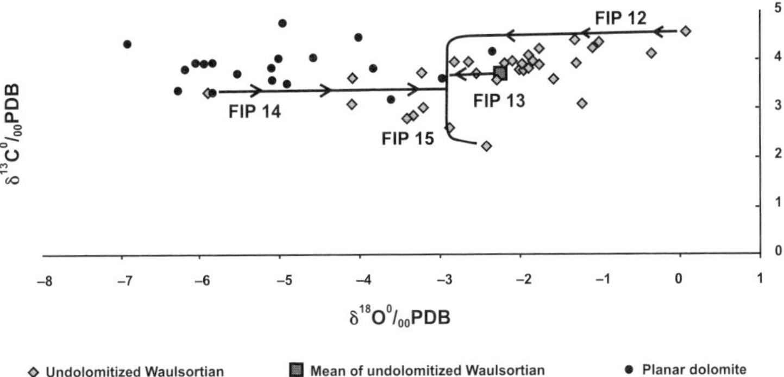

Fig. 5. $\delta^{18}O$ v. $\delta^{13}C$ PDB cross-plot of undolomitized Waulsortian and planar dolomite. Also illustrated are the FIPs 12–15 of seawater interacting with different initial compositions of the undolomitized Waulsortian. See text for parameters of the FIPs.

sedimentological environment and diagenetic information, that the fluid varied 4–6‰ in its $\delta^{18}O$ value. If, as discussed previously, the mean $\delta^{18}O$ value of the planar dolomite (–4.99 PDB) was represented on Figure 4, the lower temperature 30–50 °C FIPs would overlap the data.

$\delta^{18}O$ v. $\delta^{13}C$ value modelling

The modelling of $\delta^{18}O$ versus $\delta^{13}C$ values follows the same method as other models. The equations used for modelling $\delta^{13}C$ values have two parameters that were not used in previous models: (1) the total dissolved carbon concentration in the fluids; and (2) the species of carbon, either CO_2 or HCO_3^-, dominant in the fluid. The concentration of the carbon *will not* affect the overall shape of a fluid–interaction pathway (FIP) but will affect the N values needed to reach a particular $\delta^{13}C$ value. The fluid–rock ratio (N) is inversely proportional to the dissolved carbon content of the fluid. Changing the species of carbon in the fluid has a variable but profound effect on the shape of the FIP, and this stems largely from the fact that the dolomite–HCO_3^- and dolomite–CO_2 fractionation curves are very different, depending on the temperature. At low temperatures, a CO_2-rich fluid will have a higher α (fractionation factor) and therefore tends to have less effect on the solid it is interacting with, but at higher temperatures (above *c.* 150 °C) a CO_2-rich fluid has a lower α than a fluid dominated by HCO_3^- and the reverse relationship is present.

In the following modelling diagrams the data points representing the undolomitized Waulsortian are a composite of the data from Douthit *et al.* (1993) and Gregg *et al.* (2001), and therefore the area encompassed by the field is different from that illustrated previously.

Figure 5 illustrates several fluid interaction pathways for a 50 °C $\delta^{18}O$ SMOW = 0‰ and $\delta^{13}C$ PDB = 0‰ seawater interacting with limestones of different isotopic starting compositions. All fluids have 10 000 ppm dissolved carbon and have HCO_3^- as the dominant carbonate species. Overall, depending on the starting composition of the undolomitized Waulsortian limestone that is encountered by the seawater, a minor portion of the analysed planar dolomite values can be produced by the interaction. However, the majority of the values are inconsistent with this particular fluid–rock interaction and require either more negative initial carbonate $\delta^{18}O$ values, or a different fluid (differing either in oxygen isotopic composition or higher temperature) (Fig. 5). Figure 6 illustrates the FIPs that are thought to be required to encompass all of the planar dolomites based on the previous discussion. The FIP 16 fluid has a temperature of 50 °C and $\delta^{18}O$ value of –3‰ SMOW, whereas FIP 17 has a temperature of 70 °C (Fig. 6). The fluid path for seawater with a $\delta^{18}O$ value of 0‰ SMOW at 70 °C (FIP 17) is essentially indistinguishable from FIP 16, except for the equilibrium $\delta^{18}O$ planar dolomite value. Using these parameters, the analysed planar dolomites appear compatible with formation from a fluid (seawater) with a $\delta^{18}O$ value between –3 and 0‰ SMOW at temperatures between 50 and 70 °C. Figure 7 illustrates the fact that, regardless of the fluid and its carbon species, if it is above 50 °C and has a relatively plausible chemistry it can produce the desired values for planar dolomite if modelled solely in

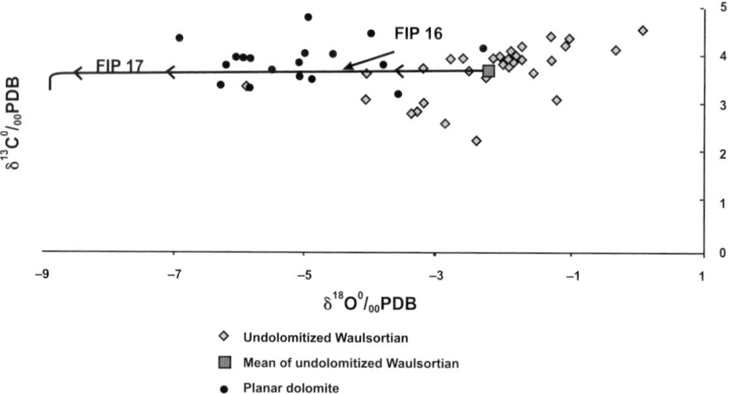

Fig. 6. $\delta^{18}O$ v. $\delta^{13}C$ PDB cross-plot of undolomitized Waulsortian and planar dolomite. Also illustrated are the FIPs 16 and 17 of the modified fluids from Figure 5 interacting with the mean value of the undolomitized Waulsortian. Note superimposition of FIP 16 on FIP 17.

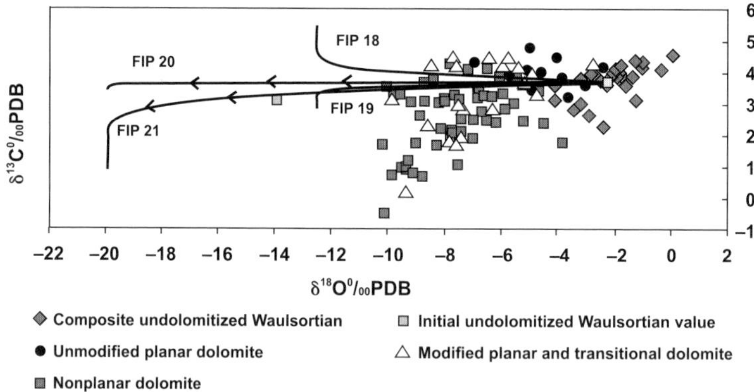

Fig. 7. $\delta^{18}O$ v. $\delta^{13}C$ PDB cross-plot of the undolomitized Waulsortian and all the dolomite data. Also illustrated are the FIPs 18–21 of higher temperature fluids interacting with the mean composition of the undolomitized Waulsortian. Note FIP 18 and 20 are modelled as HCO_3^--dominant fluids, whereas FIP 19 and 21 are CO_2-dominant.

the $\delta^{18}O$ v. $\delta^{13}C$ system. The FIPs (18–21) illustrated range in temperature from 100 to 200 °C and all have the same carbon concentration.

To test the sensitivity of the $\delta^{18}O$ v. $\delta^{13}C$ models to changing parameters, the concentrations of dissolved carbon in the fluid and the porosity were varied. Changing the fluid composition from 10 000 ppm to 150 ppm *does* change the shape of the FIP path. But because the majority of the shape of the FIP is controlled by $\delta^{18}O$ value changes, the variation between a FIP path with a fluid containing 150 ppm and one containing 10 000 ppm *is not visible at the scale used to represent the analysed data.* The most notable change resulting from decreasing the dissolved carbon concentration in the fluid will be an increase in the N values needed to reach a particular value. To reach a $\delta^{13}C$ PDB value of approximately 3.67‰ requires an N value of 3.26 at a concentration of 10 000 ppm, whereas at 150 ppm the N value required is 188, i.e. an approximately 50-fold increase in the fluid–rock ratio. Although, this large difference in N would be significant at early stages of the fluid–rock interaction, both fluids will reach equilibrium at relatively high N values (above 2000).

Changing the porosity of the host rock from 28 to 10% does not change the shape of the FIP, but changes the N value for which the $\delta^{18}O$

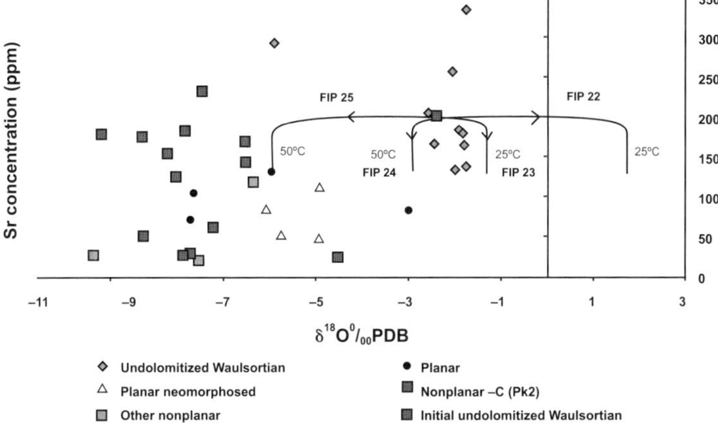

Fig. 8. Sr concentration (ppm) v. $\delta^{18}O$ PDB cross-plot of the dolomite data and the undolomitized Waulsortian (from Douthit *et al.* 1993). Also illustrated are the FIPs 22–25 of seawater interacting with the Waulsortian.

value equilibrates. At 28% porosity the N_e value would be approximately 135, whereas at 10% it is 39, roughly 3.5 times greater at the higher porosity.

$\delta^{18}O$ v. Sr concentration modelling

Fluid–rock interaction modelling of the genesis of the planar dolomites using $\delta^{18}O$ v. Sr concentration is similar in pattern to the previous models and uses many of the same parameters. The initial rock value used is the mean of the undolomitized Waulsortian ($\delta^{18}O$ PDB = –2.25‰, 200 ppm Sr). The Sr and Ca concentrations and $^{87}Sr/^{86}Sr$ ratio are the same for all four of the interacting fluids, 8 ppm and 415 ppm, and 0.7079, respectively, represented in Figure 8. FIP 23 is modelled with a –3‰ $\delta^{18}O$ SMOW fluid. Figure 8 illustrates that the fluid–rock interaction (FIPs 22 and 23) with a 25 °C fluid of a variable ($\delta^{18}O$ SMOW = 0 to –3‰) seawater composition is not compatible with any of the dolomite analyses. FIPs 24 and 25 illustrate that increasing the temperature of the model to 50 °C (Fig. 8) creates very differently shaped FIPs, as illustrated for other systems containing $\delta^{18}O$. The models at 50 °C appear compatible with the planar and planar neomorphosed dolomite analyses that have Sr concentrations of >125 ppm (Fig. 8). As some of the planar dolomites have Sr concentrations below 125 ppm, a different composition of the fluid must be modelled for their genesis. The changes in the fluid are not dramatic. An increase of 325 ppm Ca in the fluid (new value 740 ppm) will result in FIP models that can encompass all of the planar dolomites with respect to Sr concentration (Fig. 9). Changing the K_D^{Sr-Ca} from 0.03 to 0.01 (see Fig. 3) will also decrease the Sr content. FIP 26 has the same parameters as FIP 8 except that the Ca content of the fluid is 740 ppm. Because of the change in Ca concentration, the shape of the FIP is slightly different. On Figure 9, FIPs 27 and 28 are superimposed. These are the model pathways for the fluids with 0‰ $\delta^{18}O$ SMOW at 70 °C and –3‰ $\delta^{18}O$ SMOW at 50 °C with the higher Sr concentration. FIP 29 models the increase in temperature to 70 °C for the –3‰ $\delta^{18}O$ SMOW fluid (Fig. 9). It appears that seawater with 740 ppm Ca can intersect all the analysed planar dolomite values. The overall range of $\delta^{18}O$ PDB planar dolomite values appears compatible with either fluid changes between 0 and –3‰ $\delta^{18}O$ SMOW (with some temperature variation) or a temperature change between *c.* 62 and *c.* 33 °C for a –3‰ $\delta^{18}O$ SMOW fluid.

Discussion

Discussion of the fluid–rock interaction must address the genesis of the planar dolomite with respect to the fluids, temperatures and parameters of the models. Potentially viable dolomitizing fluids must be put in a robust geological framework for the Lower Carboniferous of Ireland.

The formation of the Lower Carboniferous planar dolomite is broadly encompassed by two hypotheses. First, if the dolomite 'crystallization' and classification model developed by Gregg & Sibley (1984) and Sibley & Gregg (1987) is

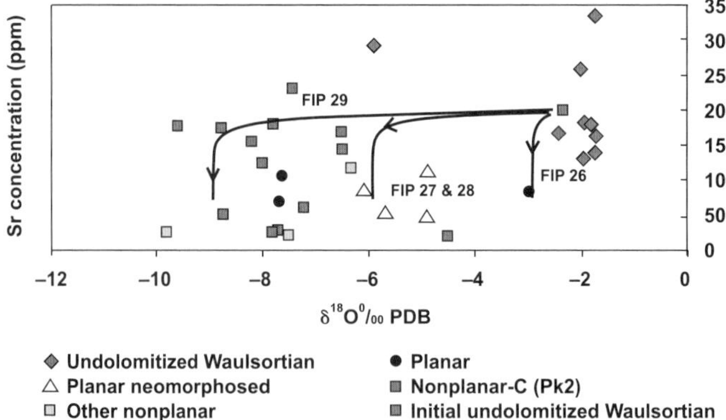

Fig. 9. Sr concentration (ppm) v. $\delta^{18}O$ PDB cross-plot of the dolomite data and the undolomitized Waulsortian (from Douthit *et al.* 1993). Also illustrated are the FIPs 26–29 of seawater interacting with the Waulsortian. Note the slight change in shape of the FIPs relative to those in Figure 8.

considered valid for planar dolomite, a general low formation temperature is inferred. However, this temperature is poorly constrained because of the difficulties in precipitating dolomite at low temperatures. Gregg & Sibley (1984) and Sibley & Gregg (1987) postulated that above a temperature of 50 °C and below 100 °C there is a critical roughening temperature (CRT), *beneath* which dolomites with a *planar* texture will form and *above* which *nonplanar* dolomite will dominate.

Secondly, a genetic link between the planar and nonplanar dolomite was proposed where both were thought to form from higher temperature hydrothermal fluids (Shearley *et al.* 1996; Braithwaite & Rizzi 1997; Eyre 1998; Hitzman *et al.* 1998). Modelling of the genesis of the planar dolomite must therefore be consistent with the fluid-inclusion analyses of nonplanar dolomite if it formed from the 'same' hydrothermal fluids.

Either hypothesis for the genesis of planar dolomite must account for: (1) the regional extent and implicit timing of its formation; (2) the replacive nature; (3) a source of Mg^{2+}; (4) more negative $\delta^{18}O$ PDB values than those expected from an estimated Early Carboniferous marine water; and (5) lower Sr concentrations and more radiogenic $^{87}Sr/^{86}Sr$ ratios than undolomitized limestones or marine dolomite.

An estimated marine dolomite (EMD) for the Lower Carboniferous has never been directly established with regard to its geochemical parameters, but is illustrated on the models as the low-temperature seawater FIP resultant dolomite. An Irish Lower Carboniferous EMD will have a $\delta^{18}O$ PDB value of 1.24 to –1.67‰, a $\delta^{13}C$ PDB value of 2–5‰, a Sr concentration (on average) of approximately 100 ppm and a $^{87}Sr/^{86}Sr$ ratio of 0.7079. The EMD is based on using a 25 °C equatorial seawater temperature for the Early Dinantian, minimal carbon isotope temperature fractionation, an equal dolomite to calcite K_D^{Sr-Ca} (Banner 1995) and a $^{87}Sr/^{86}Sr$ ratio of approximately 0.7079, if the late Ivorian–early Chadian is considered the time of precipitation. The EMD provides the point with which to compare the data generated in this study.

Low-temperature planar dolomite formation

Formation of the planar dolomite from unmodified (25–50 °C) seawater appears incompatible with the models generated for the Sr concentration v. $^{87}Sr/^{86}Sr$ ratio and $\delta^{18}O$ PDB v. $^{87}Sr/^{86}Sr$ ratio systems (Figs 1 and 2). In the $\delta^{18}O$ PDB v. $\delta^{13}C$ PDB system some planar dolomite analyses appear compatible with formation by 50 °C seawater with a $\delta^{18}O$ SMOW value of 0‰, but only by interacting with the most negative $\delta^{18}O$ value extremes of the undolomitized Waulsortian limestone (Fig. 2). If the seawater has $\delta^{18}O$ SMOW values of –3‰ at 50 °C, or 0‰ at an increased temperature of 70 °C, either fluid is capable of generating the planar dolomites by fluid–rock interaction (Fig. 2).

Formation of the planar dolomite by 25 °C seawater also is incompatible when modelled in the $\delta^{18}O$ PDB v. Sr concentration system (Fig. 3). Increasing the temperature of the seawater, whether it is originally a 0‰ or –3‰

fluid, can generate the planar dolomite with Sr concentrations >125 ppm (Fig. 4). Increasing the Ca concentration of the seawater to 740 ppm will make it compatible with all the planar dolomite analysed (Fig. 4).

Thus, the planar dolomite is compatible with formation from thermally and chemically modified seawater. The modifications to the fluid are minor for most of the systems modelled. A slight increase of 325 ppm in the Ca concentration of the seawater is required to make the Sr concentrations compatible.

A substantial modification to seawater is required to account for the $^{87}Sr/^{86}Sr$ ratio. From the work of Bruckschen *et al.* (1995, 1999) and Veizer *et al.* (1999), it is obvious that *at no time* during the Early Carboniferous was seawater as radiogenic as the planar dolomite analyses suggest. If unmodified seawater were to be used it would have to be either substantially *older* (Cambrian or Late Silurian–Early Devonian) or *younger* (Late Tertiary–Early Quaternary) (Veizer *et al.* 1999). The most geologically reasonable Early Devonian seawater has an $^{87}Sr/^{86}Sr$ ratio that is still only as radiogenic as the least radiogenic samples of planar dolomite. Therefore, the Early Carboniferous or possibly Late Devonian seawater must have interacted with sediments/or rocks with substantially higher and more variable $^{87}Sr/^{86}Sr$ ratios.

Thermal modification of the seawater to temperatures between 50 and 70 °C requires only shallow burial. Calculations of the Early Carboniferous geothermal gradient vary substantially. Hitzman (1995) noted the two palaeotemperature studies of Jones (1992) that give a possible range of geothermal gradient from 75 °C km^{-1} in the Dublin Basin to 40 °C km^{-1} in the central Irish Midlands (Fig. 10). If a high gradient such as 40 °C km^{-1} is used, the fluids could have been at a depth of only about 625 m. A more conservative estimate of the geothermal gradient is 25 °C km^{-1} and with a 25 °C fluid only 1 km of burial is required to reach the 50 °C temperature needed by the models. Sevastopulo & Redmond (1999) calculated, based on sediment thickness, that a 45–50 °C km^{-1} palaeogeothermal gradient would have existed at the base of the Waulsortian by the beginning of the Asbian. As the thickness of individual units, especially the Waulsortian, is so variable, calculating geothermal gradients by sediment thickness is difficult. In parts of the Dublin Basin and the southern Midlands (Shannon Trough) the thickness of the Waulsortian alone can exceed 900 m (Strogen *et al.* 1996) (Fig. 10). It is proposed that by the end of the Arundian sediment thicknesses in the Central and Southern Midlands were great enough at the base of the Waulsortian for dolomitizing fluids to reach 50 °C, whereas in the Dublin Basin a younger, probably Asbian, age is required.

Increasing the $^{87}Sr/^{86}Sr$ ratio of the seawater is not overly problematic. Several sources could contribute the requisite radiogenic Sr to the fluid. Intrabasinal interaction of the seawater with any one of a number of units underlying the Waulsortian would elevate the $^{87}Sr/^{86}Sr$ ratio. Directly beneath the Waulsortian, throughout most of the Irish Midlands, is the ABL (Argillaceous Bioclastic Limestone). Abundant but discontinuous argillaceous–shale horizons characterize this unit (Shearley *et al.* 1996; Strogen *et al.* 1996). Interaction of seawater with these Rb^{2+}-rich argillaceous seams would elevate the $^{87}Sr/^{86}Sr$ ratio of the fluid to the required values if the Rb^{2+} had decayed for roughly 20 Ma, coinciding with the Holkerian–Asbian stage, depending on the geological timescale used (data from Walshaw pers. comm.). Other sub-Waulsortian 'intrabasinal' units with lithologies suitable for increasing the $^{87}Sr/^{86}Sr$ ratio include the Lower Limestone Shales and the Devonian Old Red Sandstone (ORS). There are no Sr isotopic data for the Lower Limestone Shales, but the shale portions would have a similar $^{87}Sr/^{86}Sr$ ratio to that of the ABL.

Two other potential sources for Sr *more radiogenic* than Early Carboniferous seawater or Lower Carboniferous units are the variable compositions of the 'basement' rocks of the Lower Palaeozoic (greenschist – metasedimentary and metavolcanic rock) and the Caledonian granites (notably the Leinster Massif: see Fig. 10). Limited Sr isotopic data for the Old Red Sandstone and the Lower Palaeozoic rocks indicate that these units have dramatically higher $^{87}Sr/^{86}Sr$ ratios than the Lower Carboniferous seawater. Four samples of the ORS yield $^{87}Sr/^{86}Sr$ ratios of 0.70825–0.71699 at 350 Ma, with Sr contents ranging between 27 and 132 ppm (Walshaw pers. comm.). The Lower Palaeozoic samples (Silurian shales and graywackes, $n = 5$) yield $^{87}Sr/^{86}Sr$ ratios that range from 0.70967 to 0.71907, with Sr concentrations of 63–287 ppm at 350 Ma (Walshaw pers. comm.). Aureole rocks of the Leinster Granite, thought to represent equivalent Lower Palaeozoic graywackes SE of and distal to the granite, also have high $^{87}Sr/^{86}Sr$ ratios of 0.71983–0.72326 at 350 Ma.

The $^{87}Sr/^{86}Sr$ data for the Leinster Granite (Mohr 1991) indicate substantial variability, but the mean $^{87}Sr/^{86}Sr$ ratio of 0.7114 is decidedly

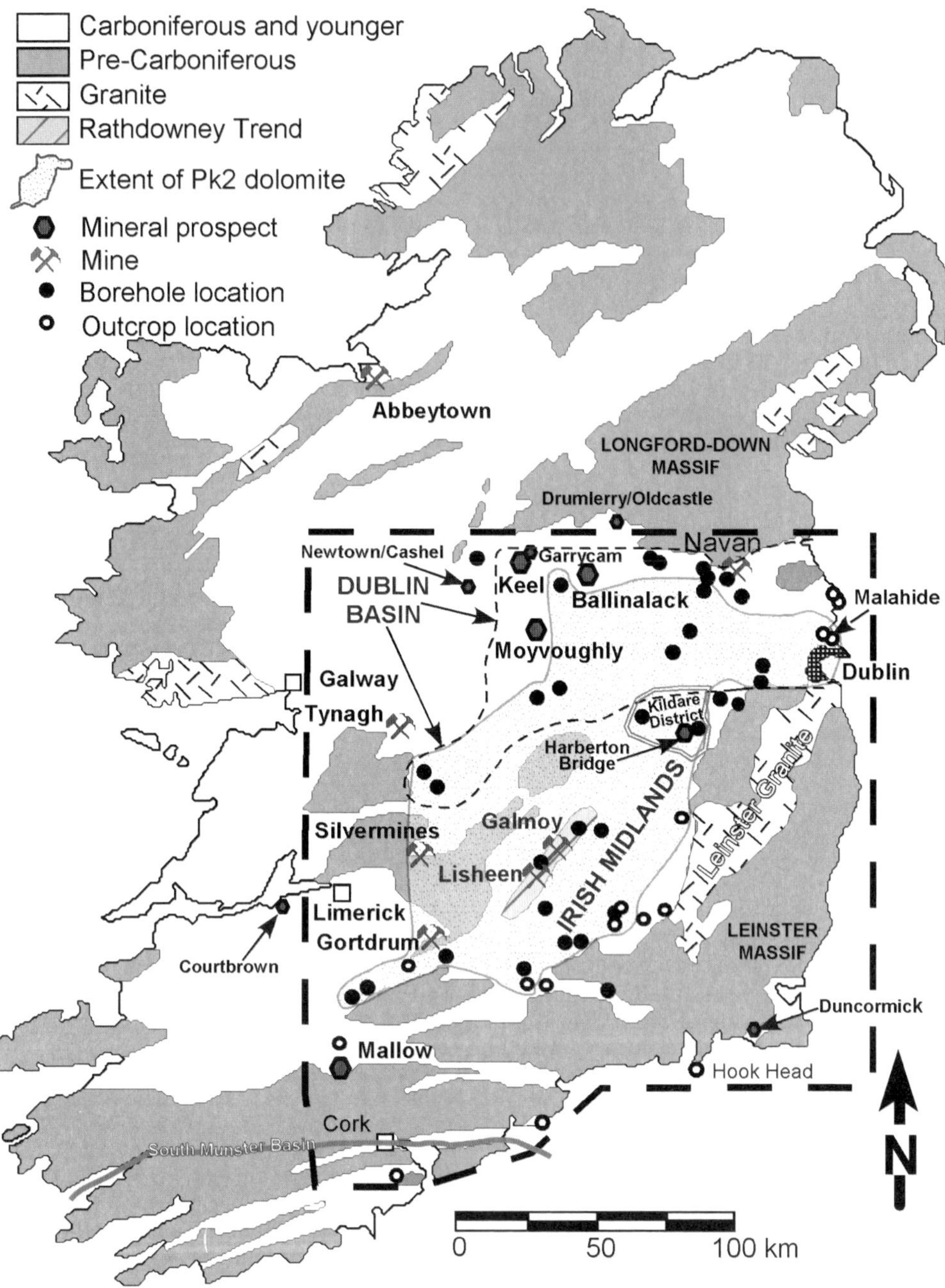

Fig. 10. Generalized geology map of Ireland with the study area, sampling points of Wright (2001) and major physiographical provinces.

more radiogenic than Lower Carboniferous seawater at 350 Ma, and has Sr concentrations ranging from 61 to 283 ppm. Note, however, that these values are not based on the entire data set of Mohr (1991). The least and most radiogenic analyses from each unit and rock type were summed, in an effort to reflect the overall variability.

The modelled Sr concentration (8 ppm) of seawater appears compatible with generation of the planar dolomite if only the $^{87}Sr/^{86}Sr$ ratio v. Sr concentration system is investigated (Fig. 8). Analysis of the models for the Sr concentration v. $\delta^{18}O$ system reveals that the Sr concentration (8 ppm) may in fact be applicable for Early Carboniferous seawater generating some of the planar dolomites. To generate all of the Sr concentration values via low Sr seawater, one of *two* processes needs to occur.

First, a lower K_D^{Sr-Ca} (FIP 3, 0.01 v. FIP 4, 0.03) may be applicable to the system (Fig. 1); or, secondly, the Ca concentration of the fluid must be elevated with respect to the original seawater (415 ppm). A third possibility is that the concentrations of Sr in the planar dolomites are a function of the diagenetic advancement of the precursor carbonate. Using the assumption that half of the Sr content of a precursor carbonate is incorporated into the resultant dolomite (Sr primarily substitutes for Ca, calcite with a lower $^{m}Sr/^{m}Ca$ ratio than seawater, where m = molar), it is theoretically plausible that the low Sr concentration Waulsortian samples (*c.* 133 ppm Sr) could produce the low Sr (*c.* 70 ppm Sr) planar dolomite analyses. However, when this is modelled using the parameters of the fluid discussed, the *shape* of the FIP and the *initial* value changes, but the *resultant* Sr value is the same as in the earlier models.

Increasing the Ca concentration of the seawater may be accomplished several ways. First, normal shallow-burial diagenesis and dissolution–precipitation reactions of calcite could result in pore fluids with increased Ca concentrations in the Lower Carboniferous carbonate rocks. Secondly, the transformation and alterations of clay minerals in shales of the ABL, Lower Limestone Shales and the Upper ORS could provide Ca to seawater circulating into these units. Thirdly, the ORS, Lower Palaeozoic rocks and the Leinster Granite could provide calcium to the fluid via alteration of feldspars and micas. Fourthly, evaporation of seawater, prior to its migration into the sedimentary pile, would also increase the Ca concentration. Without further analysis of the samples using halide and bromide systematics, the exact cause of the increase in Ca cannot be ascertained.

Alternative low-temperature dolomitization models

Low-temperature alternatives to producing planar dolomite by circulating slightly modified seawater include precipitation by meteoric or mixed waters. Keeping the temperature at approximately 25 °C, but making the fluid meteoric water or a mixture of marine and meteoric water, will produce the $\delta^{18}O$ values of the planar dolomite because the $\delta^{18}O$ value of the fluid is more negative than seawater. A –5‰ $\delta^{18}O$ SMOW fluid at 25 °C would produce a planar dolomite with a $\delta^{18}O$ PDB value of –3.61‰ and, at 35 °C, with a value of –5.54‰. Producing the appropriate $\delta^{18}O$ values of the planar dolomite is not problematic, *but* difficulties arise when the other trace-element and isotopic systems are considered.

One obstacle to producing planar dolomite from meteoric water is that the $\delta^{13}C$ PDB value of such water is also commonly very negative, reflecting incorporation of oxidized organic matter; $\delta^{13}C$ PDB values of –5‰ are relatively common and values as low as –14‰ have been reported (Cander 1994). The dolomites precipitated from a meteoric fluid would have progressively more negative $\delta^{13}C$ PDB values at relatively low N values and the most negative would be below 0‰. Figures 5 and 6 illustrate that none of the planar dolomites have $\delta^{13}C$ PDB values below 3‰. However, if the meteoric fluid contained low concentrations of dissolved carbon, the fluid $\delta^{13}C$ values could have been buffered by reaction with the host-rock limestone.

Although it is feasible that meteoric water of slightly elevated temperature could have produced planar dolomites there are other obstacles. If the water is to remain undiluted, it must come from a proximal source. There are scant sedimentological data supporting the existence of meteoric water sources proximal to planar dolomite occurrences during the Early Carboniferous. Shallow-water oolitic units occur both below and above the Waulsortian, but are not regionally extensive. The main oolitic unit is the Lisduff Member of the ABL beneath the Waulsortian. Meteoric water would have needed to enter the sedimentary pile during oolite formation, remain unmodified and be stored in the Sub-Waulsortian lithologies for later release upwards into the Waulsortian.

A distal source for meteoric water is even more problematic. Two possible source areas existed during earliest Carboniferous. The first is 'St George's Land' (the western end of the Wales–Brabant Massif), represented by the unroofed area of/and surrounding the Leinster Granite (Fig. 10). The second potential source would have been at the northern leading edge of the marine transgression, which by the Late Courceyan (during extensive development of the Waulsortian) would have been at the extreme north of present-day Ireland (see Hitzman 1995 for facies reconstruction). Meteoric water from either source would have had to travel long distances through an extensive thickness of carbonate rocks and contemporaneous seawater without being buffered or altered in $\delta^{18}O$ or $\delta^{13}C$ values. Regional flow of meteoric water during early diagenesis seems highly unlikely given the regional nature of the planar dolomite; however, it may have been viable on more local scales.

As the planar dolomites are more radiogenic than the Early Carboniferous seawater or the limestones, they also need a Sr source. Pure meteoric water could potentially interact with earliest Carboniferous, Devonian or Lower Palaeozoic rocks to gain the requisite radiogenic Sr. However, during this interaction the fluid would potentially lose its meteoric $\delta^{18}O$ or $\delta^{13}C$ signature by interactions with the carbonate portion of the stratigraphy. If the radiogenic Sr came from a distal source, it could be from either of the two previously outlined, 'St George's Land or the northern shoreline, but it faces the same problem of dilution and interaction. If radiogenic meteoric water were migrating regionally it would need a substantial recharge area and substantial topography to drive it into the basin. The recharge area appears to be adequate, but topography may be a problem. The siliciclastic source on the northern edge of the transgressive front is thought to have had low–moderate relief at most during the Early Carboniferous. To the south, the Leinster Granite was unroofed in the Late Devonian, but was covered relatively quickly by the advancing marine transgression depositing dominantly carbonate sediments during the Latest Devonian and Early Carboniferous (Mohr 1991). It is still considered a structural 'high' by many researchers, but does not appear capable of providing topographic head for meteoric fluid flow during the Early Carboniferous.

High-temperature planar dolomite formation

The formation of Lower Carboniferous planar dolomite at temperatures higher than approximately 70 °C can be addressed by comparison with the models generated for the low- to high-temperature brines used in modelling the nonplanar dolomites. If, as proposed by Eyre (1988) and Hitzman *et al.* (1998), the planar dolomites are hydrothermal in origin, they should share geochemical characteristics with the nonplanar dolomite that fluid-inclusion data indicate formed at higher temperatures.

A common misconception in understanding the formation of the planar dolomite is that nonplanar dolomite generations intimately associated are considered to be products of the same fluid and precipitation environment (regional dolomite of Eyre 1998 and Hitzman *et al.* 1998). To assess whether or not the two dolomites are genetically linked, the fluid models for the nonplanar generations are compared to the planar dolomite analyses. Johnson *et al.* (2001) described three chemically and thermally distinct fluid types (type 1, T_m –39.7 to –16.7 °C, T_h 72–136 °C; type 2, T_m –8.8 to –0.1 °C, T_h 103–136 °C; type 3, T_m –12.9 to –0.9 °C, T_h 169–271 °C – where T_m is the melting temperature and T_h is the homogenization temperature) forming the Irish nonplanar dolomites. These salinities and temperatures are consistent with previous observations on Irish sulphide and gangue minerals (e.g. Samson & Russell 1987; Eyre 1998). The three fluids and their associated models can be compared to the planar dolomite analyses. If the planar and nonplanar dolomites are genetically linked, the most useful interpretive geochemical information should come from the modelling and cross-plots of paired dolomite analyses from the same samples. The discussion of the generation of the planar dolomites by high-temperature fluids is simpler than that for low-temperature formation because high-temperature brines already contain the requisite Sr concentration, trace-element and Sr isotopic characteristics that had to be derived for the low-temperature seawater models.

In the Sr concentration v. $^{87}Sr/^{86}Sr$ system, a high-temperature radiogenic brine interacting with Waulsortian limestone will produce planar dolomite values along the interaction pathway at intermediate N values (FIP 5, Fig. 2). The properties of the nonplanar dolomites can be compared to those of the undolomitized Waulsortian and the planar dolomites in Figure 11.

The variable Sr concentrations of the planar dolomite and their relationship to the nonplanar dolomite on Figure 11 require either multiple fluids with different Sr concentrations or a single fluid with an evolving Sr concentration to produce planar followed by nonplanar dolomite.

Early dolomites usually contain higher Sr concentrations than their late counterparts (Veizer 1983; Tucker & Wright 1990, p. 381). Some of the dolomite analyses appear to follow this trend of decreasing Sr concentration, from

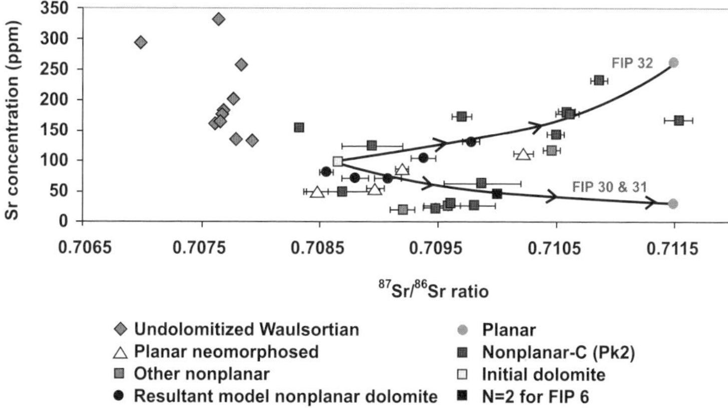

Fig. 11. Sr concentration v. $^{87}Sr/^{86}Sr$ ratio cross-plot for all dolomite data and undolomitized Waulsortian (from Douthit *et al.* 1993). Also illustrated are FIPs 30–32 of seawater interacting with the planar dolomite. Note FIP 32 follows the same pathway as FIP 31.

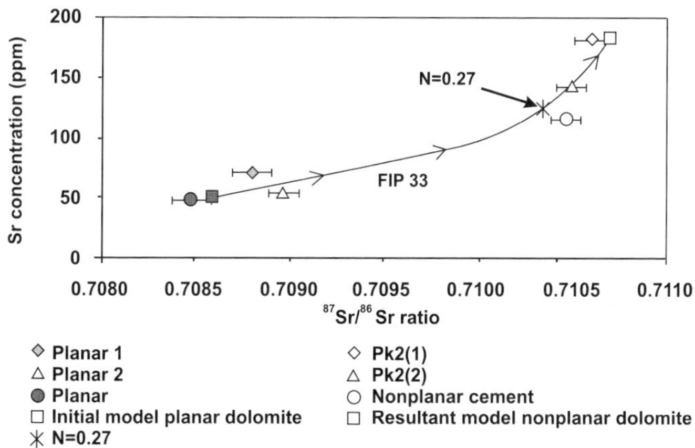

Fig. 12. Sr concentration v. $^{87}Sr/^{86}Sr$ ratio cross-plot of paired samples from the Waulsortian dolomite data. Also illustrated is FIP 33 of a brine interacting with the planar dolomite. Note $N = 0.27$ represents the ratio of fluid–rock interaction.

the undolomitized Waulsortian through the planar dolomite, ending with the nonplanar dolomite. They would fall along a FIP similar to FIP 4 in Figure 2 but continuing the pathway onto that of FIPs 30 and 31 in Figure 11. However, this trend is not consistent with analyses of the 'paired' samples with both a planar and nonplanar generation in the same sample (Fig. 12). Only one analysis pair of dolomite from the Supra-Waulsortian appears to follow this trend. It, therefore, appears unlikely that the regionally extensive early planar replacement was by the same high-temperature fluid as that forming the nonplanar dolomite cements.

The modelling of $\delta^{18}O$ values v. the $^{87}Sr/^{86}Sr$ ratio for the nonplanar dolomite by interaction with a radiogenic seawater is also consistent with formation of the associated planar dolomites at lower N values along the same pathways (Fig. 4). Although the FIPs on Figure 4 are consistent with both planar and nonplanar dolomite formation, they are modelled at a lower temperature than is required by the fluid-inclusion analyses (50–70 °C v. 100–270 °C). To have the fluid–rock

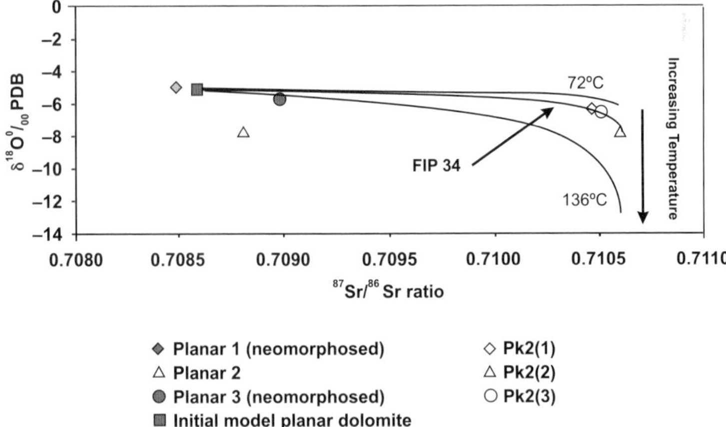

Fig. 13. $\delta^{18}O$ v. $^{87}Sr/^{86}Sr$ ratio cross-plot for Waulsortian planar and nonplanar (Pk2) dolomite data. Illustrated is FIP 34 the best-fit 'brine'. Also illustrated are the temperature-dependent changes on a 0‰ $\delta^{18}O$ SMOW fluid with the remaining parameters the same as FIP 34.

interaction pathways intersect both the planar and nonplanar dolomites at higher temperatures, the $\delta^{18}O$ value of the fluid must be different from that represented in Figure 4. If the $\delta^{18}O$ fluid values used are those calculated for equilibrium with the nonplanar dolomite isotope and fluid-inclusion analyses, the models appear compatible for planar dolomite precipitation followed by nonplanar dolomite from the same fluids.

The data from paired analyses provide a test of this possibility. Only samples from the Waulsortian are compatible with sequential dolomite formation from a single fluid (Fig. 13). The lower-temperature model (upper part of diagram) is for 72 °C and the higher-temperature model (lower part of diagram) is for 136 °C (Fig. 13). Comparison of these models indicates that the lower-temperature fluids more closely match the analysed data. The best-fit FIP would be at 82 °C and, as temperature increases, the FIP diverges more from the analysed data. At temperatures consistent with the highest temperatures provided by fluid-inclusion analyses the FIP bears little relationship to the analysed data.

Interpretation of models for the $\delta^{18}O$ v. $\delta^{13}C$ system for a high-temperature genesis of planar dolomites is indeterminate. In essence, all fluids at temperatures above 50 °C will produce a fluid–rock interaction pathway that is, first, dominated by changes in $\delta^{18}O$ and, subsequently, by changes in $\delta^{13}C$ (Fig. 7). If the starting composition is the undolomitized Waulsortian limestone, planar dolomites will lie along the initial portion of any of the high-temperature FIPs. All of the planar dolomite analyses could be accounted for on the modelled FIPs by changing the initial Waulsortian value used (Fig. 7). Regardless of the temperature of the fluid, all the planar dolomite would be produced with relatively low fluid–rock interaction (N).

Paired analyses of the dolomite also provide few data for unequivocal interpretations. In general, the Waulsortian nonplanar dolomites have more negative $\delta^{13}C$ and $\delta^{18}O$ values than their earlier counterpart dolomite generations (Fig. 14). Few of the paired analyses of the planar, planar neomorphosed and planar–nonplanar dolomite fall along the same FIP as their associated nonplanar dolomite (Fig. 14). Changing the initial limestone composition from undolomitized Waulsortian to planar dolomite with an equivalent $\delta^{13}C$ PDB value does not change the resultant dolomite or the shape of the FIP. Therefore, high-temperature fluid–rock interaction with the undolomitized Waulsortian would not produce models in which all of the dolomite generations would lie along the same FIP. Overall, the genesis of the planar dolomite by high-temperature fluids cannot be ruled out, but the paired dolomite analyses suggest that it is unlikely.

Modelling of Sr concentration v. $\delta^{18}O$ values indicates that there are additional problems in producing the planar dolomites via interaction of high-temperature fluids with the undolomitized Waulsortian. For interaction with an initial undolomitized Waulsortian, the FIPs are all dominated by changes in the $\delta^{18}O$ values before changes in Sr concentrations (Figs 8 and 9). This

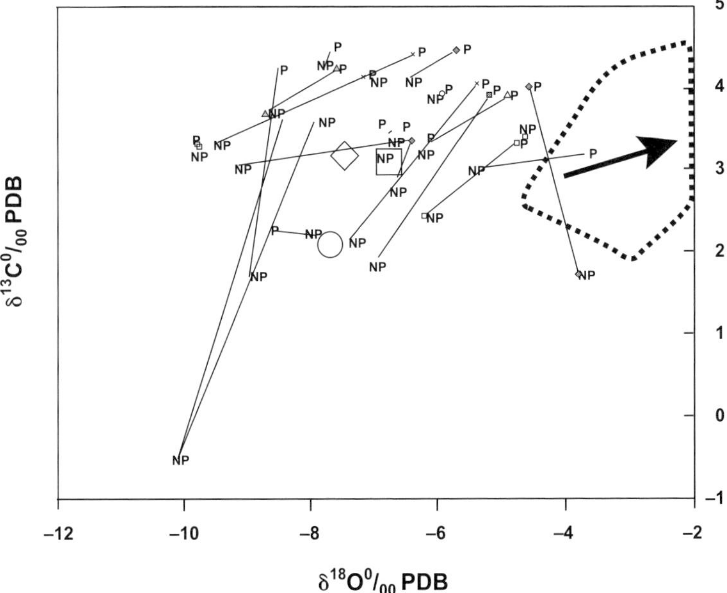

Fig. 14. ^{18}O v. $\delta^{13}C$ PDB cross-plot of the Waulsortian paired analyses of the different dolomite generations within single samples. 'Pairs' of analyses are connected by tie-lines, with P being either neomorphosed planar or planar–nonplanar dolomite, and the NP representing nonplanar, saddle cement or Pk2 dolomite. The dolomites that would form in equilibrium with the fluid-inclusion constrained fluids modelled are also indicated. The large open square, circle and diamond represent the low-temperature fluid of the Waulsortian, low-temperature fluid of the Supra-Waulsortian and high-temperature fluid of the Waulsortian, respectively. These compositions are derived from samples with both isotopic and fluid-inclusion analyses. The dashed polygon and large arrow indicate the region of the chart populated by the analyses for the undolomitized Waulsortian.

produces FIPs with an interaction path that will first migrate through dolomites with high Sr concentrations before reaching the lower Sr concentrations at higher N values. It is therefore difficult to produce planar and nonplanar dolomites from the same fluid(s) because most of the nonplanar dolomites have higher Sr contents. Theoretically, this results in a paragenetic sequence that is the reverse of that documented for the host rocks using petrography.

Genesis of the planar dolomite and summary of modelling

High-temperature models for planar dolomite formation require fluid interactions and paragenetic sequences of formation that are geologically implausible and contradict detailed petrography. The genesis of planar dolomite appears more compatible with a low-temperature environment. The generation of the planar dolomite by slightly modified Early Carboniferous seawater requires fewer modifications to the fluid and can be placed in a reasonable geological and paragenetic context. Numerical and geochemical simulations of dolomitization by Jones *et al.* (2002) and Whitaker *et al.* (2002) indicate that the temperatures, conditions and fluid-drive mechanism advocated in this paper for the Irish Lower Carboniferous planar replacement dolomite are viable on a global scale.

Although this discussion is generally restricted to trying to explain the genesis of the planar replacement dolomite in the Waulsortian, the data, models and conclusions can be extended to occurrences of planar replacement dolomite above and below. The genesis of the nonplanar dolomites was discussed and modelled by Wright (2001), and this exercise allowed realistic geochemical constraints to be applied to the modelling and analysis of the planar replacement dolomites. The genesis, interaction history (modelling data) and relationship to Zn–Pb mineralization of the nonplanar dolomites occurring in the Sub-Waulsortian, Waulsortian and Supra-Waulsortian are detailed in Wright

(2001). Although planar and nonplanar dolomites are commonly spatially associated, the nonplanar generations consistently occur as pore-filling cements, not as replacement phases. They mantle and otherwise fill porosity in planar dolomitized rocks. The fluids precipitating nonplanar dolomites therefore have a diagenetic interaction history that involves earlier generations of planar dolomite and not the undolomitized limestone present during replacive phases.

Geological parameters and possible mechanisms of dolomitization

Dolomitization of the Lower Carboniferous carbonate rocks of Ireland can be described as *massive* in both geographic and stratigraphic distribution. Simple volumetric calculations of the dolomite present and inferred testify to this. The area sampled in this study is approximately 43 000 km^2. If a conservative estimate of half of this contained carbonate units, and the range of thickness of dolomitized intervals is 0–300 m, an approximate thickness of 25 m is spread across the entire sample volume, giving >500 km^3 of dolomite present. A 25 m dolomite thickness is very conservative given that the combined stratigraphic thickness of the Lower Carboniferous could be 1–2 km. Based on calculations by Land (1985) and Lumsden & Caudle (2001), approximately 0.28 l cm^3 of seawater are required to precipitate 1 cm^3 of dolomite. If a conservative original porosity of the carbonate is used (28% assumed in the geochemical modelling in this study), and an end porosity after dolomitization of 5%, a 23% porosity reduction by dolomite is required. Approximately 6.0×10^4 l of seawater are needed to dolomitize 1 m^3 of limestone. If we reduce the amount of dolomite present to 5.0×10^5 m^3 of dolomite in the Lower Carboniferous of Ireland, we end up with 3.23×10^{11} l of seawater needed to form just this tiny amount of dolomite.

Although, these calculations are perhaps crude and oversimplified, they illustrate that the dolomitization of the Irish Lower Carboniferous limestones required immense amounts of fluid. Therefore, the mechanism used to transport this volume of fluid and the character of the fluids would appear to be relatively uniform, as the majority of planar dolomite identified above, below and within the Waulsortian appears petrographically and geochemically similar.

Evaporitic and meteoric mechanisms

The dominant depositional conditions inferred from the undolomitized and massively dolomitized units is inconsistent with an evaporative environment. Dolomitization mechanisms that require evaporation of seawater in an intertidal, supratidal or sabkha environment for the genesis of the planar dolomite are then *generally* ruled out. Nagy *et al.* (2004) has noted the evaporitic character of some of the Supra-Waulsortian lithologies and their associated dolomites. These observations have a potentially profound bearing on the origin of the associated nonplanar dolomites, and also the type, origin and composition of fluids responsible for Zn–Pb mineralization. However, an evaporitic origin for the planar dolomites cannot be extended to a regional scale and, no matter how important it may be locally, evaporation does not appear to be a dominant influence in the formation of replacement dolomite.

Similar conclusions can be reached for geological settings with a freshwater–meteoric recharge area proximal to the dolomitized units. Evidence for localized emergent highs and meteoric recharge is rare and often equivocal in interpretation. A role for meteoric water cannot be ruled out, but if this dolomitization mechanism was involved in the genesis of the planar replacement dolomites it would necessarily be restricted to areas surrounding the Leinster (Brabant) Massif and exposed areas on the north side of the Dublin Basin (i.e. the Longford Down Massif: Fig. 10).

Ruling out dolomitization models involving evaporative and meteoric components leaves only models involving the migration of seawater (or possibly brines) into the carbonate host rocks, subsurface. The mechanisms commonly proposed for these environments include Kohout convection, ocean current pumping, shallow seawater reflux, shale dewatering in adjacent basins or regional migrations of brines (Kohout 1967; Saller 1984).

Shale dewatering

If shale dewatering is proposed as the source of fluids for the planar replacement of dolomite, a timing constraint of the dolomitization is imposed. The Dublin Basin and the smaller interconnected basins in the area did not form until at least early Arundian and thick accumulations of the basinal carbonates would not have developed until at least the Holkerian (Fig. 10). The dominant obstacle for replacement dolomitization via this type of fluid, in Ireland and elsewhere, is the small volumes produced via this process. As indicated, large volumes of fluid are required to create even a small percentage of the dolomite present in the Lower Carboniferous rocks of Ireland. As well as the volumetric constraint, no petrographic or field observations support this model and it is expected that substantial dolomite would also have been created in the basinal rocks of the Dublin Basin and in the platform areas adjacent to it (Fig. 10). The proportion of dolomite would decrease across the platform away from the basin. No geometry of dolomite distribution fitting this pattern was identified, and thick massively dolomitized sections occur over a wide area and stratigraphic interval, proximal and distal to the Dublin Basin.

Petrographic and geochemical data must be evaluated and interpreted to assess the applicability of the remaining dolomitization mechanisms (Kohout convection, ocean current pumping, shallow seawater reflux and regional brine migration) to the Irish Lower Carboniferous dolomite. The crystal morphology of the replacement dolomite is dominantly either planar-s (subhedral) or planar-e (euhedral). Dolomites associated with burial and regional migration of brines are dominantly nonplanar, and microthermometric analyses of these confirm fluid temperatures commonly in excess of 100 °C (Radke & Mathis 1980).

Regional brine migration

No reliable fluid-inclusion data have yet been collected on the replacive planar dolomites in the Irish system. However, back-calculation of fluid temperatures from oxygen isotope data are an alternative and provide maximum fluid temperatures of 50 °C using $\delta^{18}O$ SMOW = –3‰ seawater. The planar replacive dolomite is consistent with formation during early–intermediate diagenesis and commonly appears to have formed before complete 'burial' cementation of the limestone host. Fluid models involving high temperatures such as burial or regional brine dolomitization require unrealistic fluid interaction and paragenetic sequences that contradict petrography and are geologically implausible.

Shallow seawater reflux

Of the three dolomitization models remaining, shallow seawater reflux might be expected to be most applicable to the shallow-water platform facies. However, the geochemistry of the planar replacement dolomites appears to preclude direct precipitation from *unmodified* seawater. The calculated temperatures require a depth of burial ranging from 330 m to 1 km. The burial depth is ambiguous because of two factors. First, the overburden for any given formation is highly variable, especially in the units overlying the Waulsortian where interfingering of facies is common along the platform margins. Second, no consistent and uncontested geothermal gradient has yet been established for the Lower Carboniferous of Ireland. If a conservative 25 °C km^{-1} gradient is used, the sediments would need 1 km of burial. However, calculated gradients for Ireland are much higher, at 75 °C (Athboy borehole near Navan) and 40 °C (Roselawn borehole near Kildare District), and result in burial depths of approximately 625 and 330 m, respectively (Fig. 10).

An alternative to linking the depth of burial to the temperature requirements is to have the $\delta^{18}O$ values of the replacement dolomite reflect the variable geothermal gradient. If a unit has a uniform thickness of 400 m in an area with a 40 °C km^{-1} geothermal gradient, fluids at the base would be approximately 41 °C, assuming the original fluid was equatorial 25 °C seawater. In the same depositional environment with a geothermal gradient of 75 °C km^{-1}, a 400 m-thick unit would be 55 °C at the base. Using calculations presented earlier (see Table 1), these temperature changes alone would result in a >2‰ shift in the $\delta^{18}O$ PDB value of the replacement dolomite, if the same fluid is used in both instances. Although highly speculative, a $\delta^{18}O$ ‰ PDB contour plot of the planar dolomite may actually reflect the dependence of oxygen isotope values on the palaeogeothermal gradient (Wright *et al.* 2000*a*; Wright 2001). In general, if the $\delta^{18}O$ values are a direct function of temperature of the fluids present in a region of elevated geothermal gradient, the prescribed burial depths will be relatively modest when compared to dolomites forming in areas with lower ambient temperatures.

Later alteration in the burial environment provides an alternative means of producing dolomites that have more negative $\delta^{18}O$ values than those expected in a low-temperature environment. Petrographic observation of 'unaltered' planar replacement dolomites reveals no alteration such as overgrowths of later cements, or corrosion within cathodoluminescence (CL) microstratigraphies. Mottling visible in CL in Pk1 (planar) dolomites has been deemed as evidence of alteration (Reed pers. comm.), but is interpreted here as reflecting inclusions and defects incorporated into the crystals during replacement of the precursor. In polymodal (*sensu stricto* Gregg & Sibley 1984) dolomite samples, the lack of comparable CL zonation (intensity and colour) between the nonplanar dolomite and the Pk1 'pristine' dolomite is taken as evidence supporting the lack of alteration by later fluids. Extensive investigation of the neomorphism and/or alteration of planar replacement dolomite is documented by Wright *et al.* (2003), and provides both petrographic and geochemical evidence of the distinction between these types.

The planar replacement dolomites require a fluid with a higher $^{87}Sr/^{86}Sr$ ratio than Early Carboniferous seawater and these must therefore have interacted with units suitable to this purpose. Several units below the Waulsortian could provide radiogenic Sr to the fluid, depending on the timing inferred for fluid migration. Because planar dolomitization occurred relatively early in the diagenetic history of the host rocks, only the Lower Palaeozoics, ORS and Lower Limestone Shales/Navan Group satisfy the requirements. Units such as the ABL could only provide the requisite Sr to planar dolomites forming in the Holkerian–Asbian. The Sr concentrations of the planar dolomite require a Ca concentration higher than that of seawater and the source for that could be either the same as that for the radiogenic Sr or host-rock pore fluids.

Kohout convection, ocean current pumping and a proposed model for Irish Lower Carboniferous planar replacement dolomitization

The theory that genesis of the planar dolomites occurred via replacement by a low-temperature, slightly modified seawater is supported by fluid–rock interaction modelling of the gathered geochemical data. Kohout convection and ocean current pumping can operate irrespective of depositional environment and both methods of fluid migration satisfy the current geochemical and geological constraints. It is envisaged that a modified dolomitization mechanism dominantly using a process similar to Kohout convection (Kohout 1967; Saller 1984), but possibly with a component of ocean current pumping, caused migration of fluids into the Lower Carboniferous carbonates. After the initiation of Waulsortian mudbank development, because of an increased thermal regime beneath the sedimentary pile, fluids were drawn into the carbonate sequence from deeper water. At the transition from the carbonate platform into deeper basinal areas (Munster Basin and the younger Dublin Basin) ocean currents probably operated concurrently with Kohout convection. As fluids migrated inward, they encountered units beneath the Waulsortian. Having entered these underlying units (Lower Limestone Shale, Old Red Sandstone and the contact with Lower Palaeozoic rocks) the seawater scavenged radiogenic Sr, possibly calcium and additional magnesium. The now warmer water migrated up-slope into the overlying Waulsortian and younger platform carbonates of the Supra-Waulsortian that were in the early–intermediate stage of diagenesis. Ca in pore fluids, provided from dissolution–precipitation reactions of calcite, was incorporated into the modified fluid, resulting in the variability noted in Sr concentrations of the planar replacement dolomite. Upward fluid migration would have been facilitated by structures such as faults (mainly synsedimentary), debris flows and pinch-outs in the units beneath the Waulsortian. The lack of suitable nucleation sites (carbonate poor), inhibition caused by incompatible Mg/Ca ratios or other cation suppression are possible reasons why dolomitization did not occur pervasively in the underlying units (Lower Palaeozoic, Devonian and Lower Carboniferous). Numerical simulations by Jones *et al.* (2000, 2002) and Whitaker *et al.* (2002) illustrate the ability of a Kohout convective system to massively dolomitize carbonate platforms at the temperatures and geochemcial conditions outlined in this paper.

The changes in porosity and permeability of the limestone initiated by replacive dolomitization directly affected the later migration of brines in the system and in essence created new reservoirs. Understanding the genesis of the planar replacive dolomite is important in its own right and vital to understanding the further diagenetic evolution of those rocks.

Understanding the timing and genesis of early dolomitization, as in Ireland, may hold the key to predicting and evaluating fluid migration associated with mineralization and petroleum in dolomitized reservoirs worldwide. The models and methodologies presented for understanding dolomite genesis in the Lower Carboniferous rocks of Ireland can be applied to any dolomitized reservoir. Establishing the physical and chemical conditions present during dolomitization permits a small level of predictability to be attributed to the distribution of the dolomite and its associated petrophysical characteristics. Predicting the occurrence and distribution of dolomite is vitally important in hydrocarbon prospectivity and field development. Within the Middle East, the Jurassic Arab Formation is a prolific hydrocarbon reservoir. The Arab Formation and the laterally equivalent Asab Formation are variably dolomitized. In many instances dolomitization has a positive effect on reservoir quality, there are, however, important exceptions. Ascertaining whether dolomitized mudstones of the Arab Formation were genetically linked to the dolomitized ooid grainstones of the Asab Formation could potentially provide information on the geometry of the dolomitized zones, fluid vectors and the timing of dolomitization relative to hydrocarbon emplacement. The Permian–Triassic Khuff Formation also has an extremely complicated diagenetic history involving dolomitization, but the nature of the dolomitizing fluids and the time when dolomitization occurred are ambiguous. Dolomitization models are sometimes applied to dolomitized intervals in hydrocarbon reservoirs without testing the ability of the fluids to migrate through the units and dolomitize them. The methodologies outlined provide a method for narrowing down feasible models in extremely complicated systems.

Conclusions

(1) Modification of Early Carboniferous seawater by an increase in temperature to a minimum of 50 °C and a maximum of 70 °C, and the addition of radiogenic Sr and a slight increase in Ca content, provided a suitable fluid for producing the planar replacement dolomite in the Lower Carboniferous limestones of Ireland.

(2) Fluid–rock interaction as dissolution–reprecipitation reactions during replacement of the limestones appears, from geochemical modelling, to be the dominant process for producing the planar dolomites.

(3) High-temperature models for replacement planar dolomite formation require fluid interactions and paragenetic sequences of dolomite formation incompatible with observations.

(4) Kohout convection is the most likely regionally important mechanism for circulating modified Early Carboniferous seawater through the various carbonate lithologies to dolomitize them via replacement.

(5) Early 'low'-temperature diagenetic replacive dolomitization created and/or redistributed porosity and permeability in the Waulsortian host rock. Retention of porosity and permeability during subsequent diagenesis was critical in creating aquifers for exploitation by regionally migrating brines.

Portions of this study were initiated under a grant from the J. William Fulbright Foreign Scholarship Board and the United States Information Agency to J. M. Gregg. We acknowledge support from Enterprise Ireland (Forbairt) to I. D. Somerville, the National Science Foundation (NSF-INT-9729653) to J. M. Gregg and K. L. Shelton, the donors of the Petroleum Research Fund, administered by the American Chemical Society (PRF 35893-AC8) to J. M. Gregg and K. L. Shelton, the University of Missouri Research Board (RB00-117) to K. L. Shelton, grants from SEG and AAPG to A. W. Johnson, Jefferson Smurfitt Corporation to W. R. Wright, and BHP Ltd to I. D. Somerville and W. R. Wright. We also like to thank all the mining and exploration companies, operating in Ireland, for access to their mines and or core for sampling, as well the excellent discussions and input from the many geologists at Anglo-American, Arcon, BHP, CSA, Minorco, Navan Resources, Noranda, Tara Mines/Outokumpu and Rio Algom. We would also like to thank the technical staff at University College Dublin, especially M. Murphy for his help and tutelage on mass spectrometry and T. Culligan for producing the thin sections. We would also like to especially thank S. Daly and R. Walshaw for the discussion of and the use of their unpublished data. Additonally, we would like to thank G. Rizzi and an anonymous referee for their constructive comments, as well as the editors G. Rizzi, C. Braithwaite and G. Darke in making this a better manuscript.

References

BANNER, J.L. 1995. Application of the trace element and isotope geochemistry of strontium to studies of carbonate diagenesis. *Sedimentology*, **42**, 805–824.

BANNER, J.L. & HANSON, G.N. 1990. Calculation of simultaneous isotopic and trace element variations during water–rock interaction with applications to carbonate diagenesis. *Geochimica et Cosmochimica Acta*, **54**, 3123–3137.

BANNER, J.L., HANSON, G.N. & MEYERS, W.J. 1988. Determination of initial Sr isotopic compositions of dolostones from the Burlington–Keokuk Formation (Mississippian); constraints from cathodoluminescence, glauconite paragenesis and analytical methods. *Journal of Sedimentary Petrology*, **58**, 673–687.

BRAITHWAITE, C.J.R. & RIZZI, G. 1997. The geometry and petrogenesis of hydrothermal dolomites at Navan, Ireland. *Sedimentology*, **44**, 421–440.

BRUCKSCHEN, P. & VEIZER, J. 1997. Oxygen and carbon composition of Dinantian brachiopods: Palaeoenvironmental implication for the Lower Carboniferous of Western Europe. *Palaeogeography, Palaeoclimatology, Palaeoecology*, **132**, 243–264.

BRUCKSCHEN, P., BRUHN, F., VEIZER, J. & BUHL, D. 1995. $^{87}Sr/^{86}Sr$ isotopic evolution of Lower Carboniferous seawater: Dinantian of western Europe. *Sedimentary Geology*, **100**, 63–81.

BRUCKSCHEN, P., OESMANN, S. & VEIZER, J. 1999. Isotope stratigraphy of the European Carboniferous: proxy signals for ocean chemistry, climate and tectonics. *Chemical Geology*, **161**, 127–163.

CANDER, H.S. 1994. An example of mixing-zone dolomite, Middle Eocene Avon Park Formation Floridian Aquifer System. *Journal of Sedimentary Research*, **A64**, 615–629.

DIXON, P.R. 1990. *The role of basement circulated fluids in the origin of sediment-hosted Zn–Pb–Ba mineralization in Ireland*. PhD thesis, Yale University, New Haven, CT.

DOUTHIT, T.L., MEYERS, W.J. & HANSON, G.N. 1993. Nonmonotonic variation of seawater $^{87}Sr/^{86}Sr$ across the Ivorian/Chadian boundary (Mississippian, Osagean); evidence from marine cements within the Irish Waulsortian Limestone. *Journal of Sedimentary Petrology*, **63**, 539–549.

EVERETT, C.E. 1999. *Tracing ancient fluid flow pathways: A study of the Lower Carboniferous base metal ore field in Ireland*. PhD thesis, Yale University, New Haven, CT.

EYRE, S.L. 1998. *Geochemistry of dolomitization and Zn–Pb mineralization in the Rathdowney Trend, Ireland*. PhD thesis, Imperial College, London.

GREGG, J.M. & SIBLEY, D.F. 1984. Epigenetic dolomitization and the origin of xenotopic dolomite texture. *Journal of Sedimentary Petrology*, **54**, 908–931.

GREGG, J.M, JOHNSON, A.W, SHELTON, K.L, SOMERVILLE, I.D. & WRIGHT, W.R. 2001. Dolomitization of the Waulsortian Limestone (Lower Carboniferous) Irish Midlands. *Sedimentology*, **45**, 745–766.

HITZMAN, M.W. 1995. Geological setting of the Irish Zn–Pb–(Ba–Ag) orefield. *In*: ANDERSON, K., ASHTON, J., EARLS, G., HITZMAN, M. & TEAR, S. (eds) *Irish Carbonate-hosted Zn–Pb Deposits*. Society of Economic Geology, Littleton, Guidebook Series, **21**, 3–24.

HITZMAN, M.W., ALLAN, J.R. & BEATY, D.W. 1998. Regional dolomitization of the Waulsortian Limestone in southeastern Ireland: evidence of large-scale fluid flow driven by the Hercynian orogeny. *Geology*, **26**, 547–550.

JOHNSON, A.W., SHELTON, K.L., GREGG, J.M., SOMERVILLE, I.D. & WRIGHT, W.R. 2001. Fluid inclusion evidence for the presence of multiple fluids in the Zn–Pb hosting Carboniferous carbonate rocks in the Irish Midlands: initial findings. *In*: HAGNI, D. (ed.) *Studies on Ore Deposits, Mineral Economics, and Applied Mineralogy: With Emphasis on Mississippi Valley-type Base Metal and Carbonatite-related Ore Deposits*. University of Missouri-Rolla Press, Rolla, MD, 18–30.

JONES, G., WHITAKER, F.F., SMART, P. & SANFORD, W. 2000. Numerical modelling of geothermal and reflux circulation in Enewetak Atoll: implications for dolomitization. *Journal of Geochemical Exploration*, **69–70**, 71–75.

JONES, G., SMART, P., WHITAKER, F.F., ROSTRON, B. & MACHEL, H.G. 2002. Numerical modelling of reflux dolomitisation in the Grosmont Platform Complex (Upper Devonian), Western Canada sedimentary basin. (Abs.) *In*: *The Geometry and Petrogenesis of Dolomite Hydrocarbon Reservoirs*. Petroleum Group, Geological Society, London.

JONES, G.LL. 1992. Irish Carboniferous conodonts record maturation levels and the influence of tectonism, igneous activity and mineralisation. *Terra Nova*, **4**, 238–244.

KOHOUT, F.A. 1967. Ground-water flow and the geothermal regime of the Floridian Plateau. *Transactions of the Gulf Coast Association of Geological Societies*, **17**, 339–354.

LAND, L.S. 1985. The origin of massive dolomite. *Journal of Geological Education*, **33**, 112–125.

LUMSDEN, D.N. & CAUDLE, G.C. 2001. Origin of massive dolostone: The upper Knox model. *Journal of Sedimentary Research*, **71**, 400–409.

MACHEL, H.G. & BURTON, E.A. 1994. Golden Grove Dolomite, Barbados: Origin from modified seawater. *Journal of Sedimentary Research*, **A64**, 741–751.

MOHR, P. 1991. *Origin and cooling history of the Leinster granite: an isotope study*. PhD thesis, University College Dublin, Dublin, Ireland.

NAGY, ZS.R, GREGG, J.M., SHELTON, K.L., BECKER, S.P., SOMERVILLE, I.D. & JOHNSON, A.W. 2004. Early dolomitization and fluid migration through the Lower Carboniferous carbonate platform in the SE Irish Midlands: implications for reservoir attributes. *In*: BRAITHWAITE, C.J.R., RIZZI, G. & DARKE, G. (eds) *The Geometry and Petrogenesis of Dolomite Hydrocarbon Reservoirs*. Geological Society, London, Special Publications, **235**, 367–392.

RADKE, B.M. & MATHIS, R.L. 1980. On the formation and occurrence of saddle dolomite. *Journal of Sedimentary Petrology*, **50**, 1149–1168.

SALLER, A.H. 1984. Petrologic and geochemical constraints on the origin of subsurface dolomite: An example of dolomitization by normal seawater, Enewetak Atoll. *Geology*, **12**, 217–220.

SAMSON, I.M. & RUSSELL, M.J. 1987. Genesis of the Silvermines zinc–lead–barite deposit, Ireland; fluid inclusion and stable isotope evidence. *Economic Geology*, **82**, 371–394.

SEVASTOPULO, G.D. & REDMOND, P. 1999. Age of mineralization of carbonate-hosted, base metal deposits in the Rathdowney Trend, Ireland. *In*: MCCAFFREY, K.J.W., LONEGRAN, L. & WILKINSON, J.J. (eds) *Fractures, Fluid Flow and Mineralization*. Geological Society, London, Special Publications, **155**, 303–311.

SHEARLEY, E., REDMOND, P., KING, M. & GOODMAN, R. 1996. Geological controls on the mineralization and dolomitization of the Lisheen Zn–Pb–Ag deposit, Co. Tipperary, Ireland. *In*: STROGEN, P., SOMERVILLE, I.D. & JONES, G.L.l. (eds) *Recent Advances in Lower Carboniferous Geology*. Geological Society, London, Special Publications, **107**, 23–33.

SIBLEY, D.F. & GREGG, J.M. 1987. Classification of dolomite rock textures. *Journal of Sedimentary Petrology*, **57**, 967–975.

STROGEN, P., SOMERVILLE, I.D., PICKARD, N.A.H., JONES, G.LL. & FLEMING, M. 1996. Controls on ramp, platform and basinal sedimentation in the Dinantian of the Dublin Basin and Shannon Trough, Ireland. *In*: STROGEN, P., SOMERVILLE, I.D. & JONES, G.LL. (eds) *Recent Advances in*

Lower Carboniferous Geology. Geological Society, London, Special Publications, **107**, 263–279.

TUCKER, M.E. & WRIGHT, V.P. 1990. *Carbonate Sedimentology*. Blackwell Scientific, Boston, MA.

VEIZER, J. 1983. Chemical diagenesis of carbonates: Theory and application of trace element technique. *In*: ARTHUR, M.A. *et al.* (eds) *Stable Isotopes in Sedimentary Geology*. Society of Economic Paleontologists and Mineralogists, Short Course, **10**, 3–100.

VEIZER, J., ALA, D. *ET AL.* 1999. $^{87}Sr/^{86}Sr$, $\delta^{13}C$ and $\delta^{18}O$ evolution of Phanerozoic seawater. *Chemical Geology*, **161**, 59–88.

WHITAKER, F., WILSON, A., SANFORD, W. & SMART, P. 2002. Spatial patterns and rates of dolomitisation during geothermal convection in carbonate platforms. (Abs.) *In*: *The Geometry and Petrogenesis of Dolomite Hydrocarbon Reservoirs*. Petroleum Group, Geological Society, London, 21.

WRIGHT, W.R. 2001. *Dolomitization, fluid-flow and mineralization of the Lower Carboniferous rocks of the Irish Midlands and Dublin Basin Regions*. PhD thesis, University College Dublin, Belfield, Ireland.

WRIGHT, W.R., SOMERVILLE, I.D., GREGG, J.M. & SHELTON, K.L. 1999. Dolomite CL and $\delta^{18}O$ $\delta^{13}C$ data from Irish midlands and Dublin Basin, Carboniferous. *In*: STANLEY, C.J. *et al.* (eds) *Mineral Deposits: Processes to Processing*. Proceedings of the 5th Biennial SGA Meeting, London, Volume 2. Balkema, Rotterdam, 913–916.

WRIGHT, W.R., JOHNSON, A.W., SHELTON, K.L., SOMERVILLE, I.D. & GREGG, J.M. 2000*a*. Fluid Migration and rock interactions during dolomitisation of the Dinantian Irish Midlands and Dublin Basin. *Journal of Geochemical Exploration*, **69–70**, 159–164.

WRIGHT, W.R., SOMERVILLE, I.D., GREGG, J.M., JOHNSON, A.W. & SHELTON, K.L. 2000*b*. Dolomitization and neomorphism of Irish Lower Carboniferous (Early Mississippian): evidence from petrographic and isotopic data. (Abs.) *In*: *Permo-Carboniferous Carbonate Platforms and Reefs*. Joint Society of Economic Paleontologists and Mineralogists and International Association of Sedimentologists, Research and Field Conference, El Paso, TX, 149.

WRIGHT, W.R., SOMERVILLE, I.D., GREGG, J.M., SHELTON, K.L. & JOHNSON, A.W. 2001. Application of regional dolomite cement CL (cathodoluminescence) microstratigraphy to the genesis of Zn–Pb mineralization in Lower Carboniferous rocks, Ireland: Similarities to the Southeast Missouri Pb–Zn District?. *In*: HAGNI, D. (ed.) *Studies on Ore Deposits, Mineral Economics, and Applied Mineralogy: With Emphasis on Mississippi Valley-type Base Metal and Carbonatite-related Ore Deposits*. University of Missouri-Rolla Press, Rolla, MO, 1–17.

WRIGHT, W.R., SOMERVILLE, I.D., GREGG, J.M., SHELTON, K.L. & JOHNSON, A.W. 2002. The petrogenesis of dolomite, regional patterns of dolomitisation and fluid flow in the Lower Carboniferous of the Irish Midlands and Dublin Basin. (Abs.) *In*: *The Geometry and Petrogenesis of Dolomite Hydrocarbon Reservoirs*. Petroleum Group, Geological Society, London, 10.

WRIGHT, W.R., SOMERVILLE, I.D., GREGG, J.M., JOHNSON, A.W. & SHELTON, K.L. 2003. Dolomitization and neomorphism of Irish Lower Carboniferous (Early Mississippian) limestones: Evidence from petrographic and isotopic data. *In*: AHR, W., HARRIS, A.P., MORGAN, W.A. & SOMERVILLE, I.D. (eds) *Permo-Carboniferous Carbonate Platforms and Reefs*. Society for Economic Paleontologists and Mineralogists, Special Publications, **78**, 395–408.

Dolomitization: from conceptual to numerical models

FIONA F. WHITAKER[1], PETER L. SMART[2] & GARETH D. JONES[3]
[1]*Department of Earth Sciences, University of Bristol, Bristol BS8 1RJ, UK (e-mail: Fiona.Whitaker@bristol.ac.uk)*
[2]*School of Geographical Sciences, University of Bristol, Bristol BS8 1SS, UK*
[3]*ExxonMobil Upstream Research Company, P.O. Box 2189, Houston, TX 77027, USA*

Abstract: Dolomitization requires not only favourable thermodynamic and kinetic conditions, but also a fluid-flow mechanism to transport reactants to and products from the site of dolomitization. This paper reviews work that seeks to provide a quantitative framework for conceptual models of dolomitization, using analytical and, particularly, numerical simulation models of fluid flow and rock–water interaction. This approach is starting to yield new insights into the major controls on the rate and pattern of fluid flux, and the resultant dolomitization.

Three sets of forces can drive the fluid flow required for dolomitization: elevation (topographic) head of meteoric water and/or seawater; gradients in fluid density due to variation in salinity and/or temperature; and pressure due to sedimentological and/or tectonic compaction. However, in many situations individual flow mechanisms may not operate in isolation. Rather fluid flow will commonly be a product of a number of different drives acting simultaneously. The balance between drives will change over time with variations in relative sea-level, climate, platform geometry and palaeogeography (which collectively comprise the critical boundary conditions). The simplistic prediction of dolomite body geometry from a single driving force may be misleading, as fluid flow will critically depend both on the boundary conditions and the distribution of permeability. Indeed, even for single driving forces, model predictions change significantly as simplistic assumptions are relaxed and these key parameters are specified with increasing realism.

The coupled modelling of dolomitization reactions within the flow field is less tractable than that of groundwater circulation because the kinetics of dolomitization are less well understood, particularly at lower temperatures. Dolomitization is likely to occur along a reaction front, where a favourable balance is struck between mass transport and reaction kinetics. For instance, in simulations of geothermal convection dolomitization focuses along the 50–60 °C isotherm. Dolomitization reactions are favoured by higher temperatures in deeper zones, but rates are limited by low flow because of lower permeability. Although flow rates are higher in shallow more permeable carbonates, lower temperatures limit reactions. High flow rates during reflux of platform-top brines give rapid dolomitization. This is associated with porosity occlusion in front of and behind the broad zone of replacement dolomitization driven by anhydrite cementation and overdolomitization, respectively. Lithological heterogeneities strongly affect the pattern of dolomitization, which is highly focused within more permeable beds and those with a higher reactive surface area.

While we focus here on dolomitization, models can also provide insights into diagenetic processes such as marine calcite cementation and aragonite, calcite and evaporite dissolution by refluxing brines, and by seawater circulation below the aragonite and calcite compensation depths. However, it is important to be aware of the assumptions and limitations of the numerical model(s) used. Particular attention must be paid to specification of boundary conditions, permeability and reactive surface area. The uncritical application of numerical techniques to particular cases of dolomitization is at best uninformative and at worst misleading. Careful application of these techniques offers great promise for well-constrained field problems, with greater inclusion of natural heterogeneity and time-variant boundary conditions. We also need to model feedbacks between diagenesis and porosity–permeability, and to include platform growth in simulations of slower diagenetic processes.

Studies of modern carbonate platforms suggest that relatively small quantities of primary synsedimentary dolomite can form in incompletely lithified supratidal, peritidal and salina sediments (Hardie 1977; Perkins *et al.* 1994; Juster *et al.* 1997). However, it is universally recognized that large-scale dolomite bodies common in the rock record must result from replacement reactions with fluids importing Mg and removing Ca (Land 1985; Machel &

From: BRAITHWAITE, C. J. R., RIZZI, G. & DARKE, G. (eds) 2004. *The Geometry and Petrogenesis of Dolomite Hydrocarbon Reservoirs*. Geological Society, London, Special Publications, **235**, 99–139. 0305-8719/$15.00

Mountjoy 1986; Hardie 1987). Fundamentally, any model for the formation of massive dolomite bodies must satisfy two basic criteria: first, thermodynamic and kinetic conditions must be favourable for dolomitization; and, secondly, there must be a fluid-flow mechanism by which reactants and products can be transported to and from the site of dolomitization.

There are numerous types of surface and subsurface fluids, many derived from seawater, that are thermodynamically supersaturated with respect to dolomite. However, laboratory studies using such fluids have failed to synthesize significant volumes of dolomite under near-surface conditions without organic mediation (Lippman 1973; Usdowski 1994; Arvidson & Mackenzie 1999). Attention has focused on the kinetic barriers to dolomitization. A range of factors has been identified that are kinetically favourable for dolomitization, including elevated Mg/Ca, CO_3^{2-} activity and temperature, low ionic strength, slow crystal growth, the presence of seed crystals and microbial activity (Machel & Mountjoy 1986; Tribble *et al.* 1995; Budd 1997; Arvidson & Mackenzie 1999; Machel 2004).

In natural systems the time available for dolomitization is commonly considerable and geochemical acceleration of the process may not be necessary. The absence of dolomite in many Pleistocene carbonates may reflect the lengthy 'induction period' that has been observed in hydrothermal experiments prior to the formation of detectable dolomite (Nordeng & Sibley 1994, 2003). Following this induction period, dolomite formed rapidly from the precursor phase, a very-high-magnesium calcite (VHMC with >36 mol% magnesium) that is highly unstable in meteoric waters. A number of modern platforms host groundwater apparently similar to normal seawater composition, but depleted in magnesium (e.g. Fanning *et al.* 1981; Whitaker *et al.* 1994; Smith *et al.* 2002). This suggests that dolomitization occurs due to circulation of seawater-derived fluids, either throughout the flow system or at depth where fluids contact carbonates in which the induction period has been exceeded without meteoric dissolution of precursor VHMC.

Possibly because of the general acceptance that special geochemical conditions may be necessary for dolomitization, traditional rock-based studies have focused largely on the geochemistry of the dolomitizing fluids as inferred from trace-element, stable-isotope and fluid-inclusion techniques. Having identified a dolomitizing fluid, such studies generally invoke a conceptual model for circulation of this fluid, often with little consideration of the dynamics of the proposed circulation. Unlike the thermodynamically stable dolomites formed at depth, most dolomites that originally form at shallow depth and/or from evaporitic brines appear to be poorly ordered calcian dolomites and tend to recrystallize with time and burial (Steefel & Van Cappellen 1990; Nordeng & Sibley 1994; Machel 2004). With increasing temperature, pressure and exposure to a range of fluid compositions the originally metastable dolomite phases become thermodynamically highly unstable, and recrystallization may alter their texture, structure and/or composition. It is has been suggested that several different fluids can be involved in this recrystallization process during burial (Malone *et al.* 1996). Machel (2004) considers that few other well-studied dolomites reveal unambiguous evidence of stepwise recrystallization at multiple times, highlighting the absence of stepwise recrystallization in the Devonian Rimbey–Meadowbrook reef trend of western Canada despite its considerable lateral extent and range of burial depths. However, for most dolomite bodies any temporal sequence of recrystallization would be difficult to detect and the geochemical signature of the most recent fluid will dominate, obscuring those of any earlier formative fluids (Land 1985).

Over the last 15 years there has been increasing recognition that replacement dolomitization not only requires propitious geochemical conditions, but also an efficient fluid-flow mechanism. Thus, most present-day models for the genesis of massive dolomites are essentially hydrological models (Machel 2004). Compared with the challenge of prescribing the geochemical conditions favourable to dolomitization, the mass-transfer requirement for dolomitization is a much more tractable problem. These models can be differentiated according to the three fundamental forces that drive fluid circulation, namely those derived from differences in elevation head, fluid density and applied stress. The nature of fluid flow driven by these mechanisms will differ in platforms surrounded by seawater ('active platforms' *sensu* Simms 1984) compared to those buried by other sediments, and also possibly at elevated temperature and pressure, or uplifted into the continental realm.

The exploitation of dolomite hydrocarbon reservoirs has historically provided an important motive for the development and application of predictive models of dolomitization. Dolomites provide porous and permeable hosts for many commercial reservoirs. For example, Wendte *et al.* (1998) estimated that in the

Devonian strata of the Western Canada Sedimentary Basin of Alberta and British Columbia, approximately 50% of all producible oil, and as much as 75% of all producible gas, occurs in dolomite reservoirs. Although at shallow depth dolomites tend to be less porous than limestones, they retain their porosity with burial (Schmoker & Halley 1982). Saller & Henderson (1998) point out that 90% of the porosity in dolomites is diagenetic, and demonstrate (as have many others) that there is a clear association between the distribution of this porosity and the hydrological system responsible for circulation of the dolomitizing fluids. For 78% of the economically important dolomite reservoirs surveyed by Sun (1995) the development of the dolomite was related to early eogenic processes, although porosity was enhanced in 45% of these by later processes such as karstification and fracturing. In addition to the prediction of dolomite porosity and permeability, an understanding of the spatial distribution and connectivity of dolomite bodies is essential for reservoir simulation. Dolomite bodies are not, however, necessarily stratally determined. Thus, an understanding of the flow system that generated them and its controls is therefore critical for satisfactory prediction of their distribution and extent (see for example the recent study of Thiry *et al.* 2003 for dolomite bodies in the Chalk of the Paris Basin).

An increasing body of work published over the last 10–15 years provides a quantitative framework for conceptual models of dolomitization. Mathematical models of dolomitization range from simple calculations of mass budgets, through analytical models of fluid flow assuming very simple boundary conditions, to sophisticated fully coupled fluid-flow and reaction-transport models. These studies include both generic simulations of flow and dolomitization in stylized platforms and specific case studies ranging from small-scale modern atolls to extensive (epieric) ramp and platform settings. The aim of this paper is to review this work, evaluating the degree to which we can answer two sets of important questions. First, what is the rate and pattern of fluid flow in carbonate platforms in response to different drives, what are the major controls on this circulation and how might different drives interact over time? Secondly, what is the rate and distribution of dolomitization driven by these flow systems? Early studies used analytical models to address these questions, and this approach continues to provide insights into flow mechanisms in simple settings. However, major advances in numerical techniques and computing power have permitted more sophisticated numerical simulations, which increasingly couple groundwater flow, heat and/or mass transport, and, more recently, geochemical reactions. We consider how the predictions from these simulations have changed as simplistic assumptions have been relaxed and increasingly realistic specifications of key parameters such as permeability have been adopted. Finally, we discuss the limitations of existing models and the challenges faced by current researchers, issuing a warning to geologists against uncritical application of numerical simulations to particular cases of dolomitization.

Conceptual models of fluid flow linked to dolomitization

Elevation- (topographic-) head-driven flow

Elevation- (topographic-) head-driven flow is generally considered in terms of flow of meteoric waters, ranging in spatial scale from shallow freshwater lenses developed below small islands on exposed parts of the platform top (Vacher 1988) to regional discharge through carbonate units within or attached to a continental hinterland (Garven & Freeze 1984). Flow results from the hydraulic gradient created by differences in water-table elevation and fluid fluxes are generally high (Fig. 1A). However, few workers consider such flow systems to be significant for dolomitization because of the low Mg concentration of meteoric waters (Hardie 1987). Meteoric flow can also cause displacement of more Mg-rich fluids, such as basinal brines in foreland basins (Garven 1995), or seawater in active platforms exposed by a fall in relative sea-level (Demming & Nunn 1991). The total volume of Mg-rich fluid displaced is limited to that in storage. Thus, extensive dolomitization appears to be unlikely unless the fluids involved have a high Mg content and/or considerable focusing of flow occurs, for instance along permeable fairways. In the case of the Cambrian Bonneterre Formation in SE Missouri, basinal fluids focused through the permeable Lamotte Sandstone have dolomitized a thin basal layer of the limestone and shale sequence over a regionally extensive area (Gregg 1985; Gregg & Shelton 1989). A topographic drive, caused by orogeny and uplift, was suggested by Yao & Demicco (1995, 1997) for dolomitization of the Cambro-Ordovician carbonates of the southern Canadian Rocky Mountains by a residual evaporitic brine that through time was diluted by meteoric water.

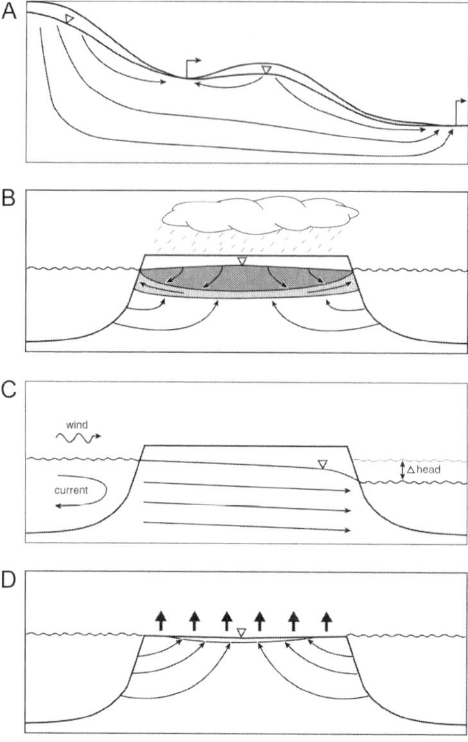

Fig. 1. Fluid-flow driven by elevation head. (**A**) Continental-style topographic-driven flow of meteoric waters. (**B**) Submixing-zone seawater circulation driven by meteoric recharge in coastal aquifers. (**C**) Trans-platform flow driven by differences in sea-surface elevation maintained by winds or current flow. (**D**) Evaporative pumping of seawaters to replace fluids lost by evaporation from salinas. Note large vertical exaggeration to clarify boundary conditions.

The discharge of meteoric water at continental and platform margins may generate a compensatory inflow of seawater at depth to sustain the discharge of salt water in the meteoric–marine mixing zone (Fig. 1B) (Bear 1972). The mixing zone or 'Dorag' model for dolomitization was widely invoked in the 1970s on the basis of favourable thermodynamic conditions (Hanshaw *et al.* 1971; Badiozamani 1973). However, it has been largely discredited as modern mixing zones reveal a paucity of dolomites, although extensive dissolution (Machel & Mountjoy 1986; Hardie 1987; Smart *et al.* 1988) and mineral stabilization (Budd 1988) can occur. More significant may be the compensatory inflow of salt water below the mixing zone (Fig. 1B), first suggested by Machel & Mountjoy (1990), and subsequently invoked in dolomitization of the Little Bahama Bank (Vahrenkamp & Swart 1994).

Less frequently considered is subsurface seawater flow driven by spatial variations in sea-surface elevation. This may be generated at a range of timescales by the interaction between platform topography and tides, currents, winds and waves (Fig. 1C) (Whitaker & Smart 1993). High-frequency exchange of seawater in response to tidal forcing will be limited to superficial sediment veneers and platform margins (Land *et al.* 1989). However, Carballo *et al.* (1987) proposed that tidal pumping is actively dolomitizing supratidal sediments in the Florida Keys. Wind set-up and ocean currents, particularly in zones of upwelling, can generate sustained gradients in sea-surface elevation (Patterson & Kinsman 1982; Marshall 1986). Montañez & Read (1992), for instance, suggested that wave swash, which led to accumulation of water behind coastal barriers, could have been a drive during dolomitization of the Ordovician Knox Group. Also Humphrey (1988) explained windward/leeward contrasts in the distribution of Pleistocene dolomites in Barbados in terms of trans-island head differences.

Finally, evaporation of water from a basin or area of platform top isolated from the surrounding ocean can lower hydraulic head and generate inflow from the ocean and/or continental groundwaters (Fig. 1D). This mechanism, termed evaporative pumping by Hsü & Siegenthaler (1969), relies on a completely closed basin and an arid climate to maintain a deficit in the water balance (Lucia 1972). Downward reflux of brines is prevented by the presence of a basal aquitard, with inflow through permeable reefs (Kendall 1989) and/or laterally through the sill (Logan 1987). Kendall (1989) accounts for the dolomitization of Lower Palaeozoic carbonates of the Williston Basin by evaporative pumping associated with the Mid Devonian Elk Point Basin. In addition to the dolomitization potential of inflowing seawater, this mechanism may drive dissolution of evaporites, as described by Logan (1987) for the MacLeod Salt Basin in Western Australia.

Density-driven flow

Density-driven flow can result from contrasts in groundwater temperature and/or salinity. Geothermal heating of fluids within a platform surrounded by cold ocean waters may result in warm platform waters rising to discharge at

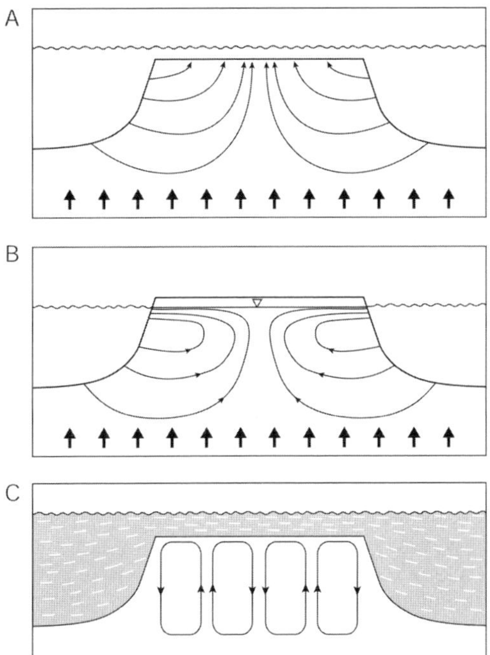

Fig. 2. Thermal convection in an isolated carbonate platform driven by geothermal heating. Forced convection in an active platform; (**A**) submerged and (**B**) emergent relative to sea-level; (**C**) free convection within a platform buried beneath low-permeability sediments.

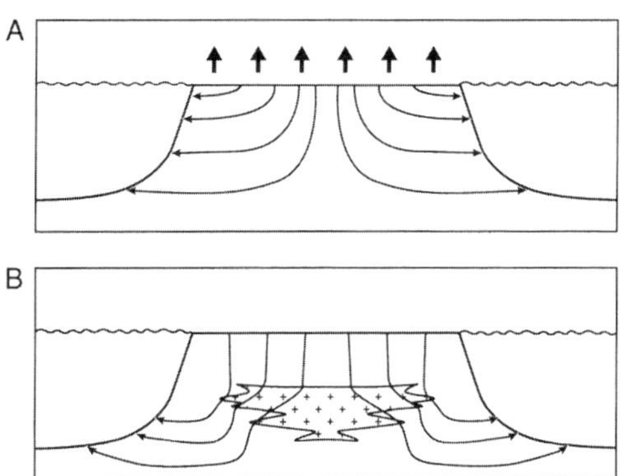

Fig. 3. Reflux of elevated salinity brines. (**A**) Generated across the platform top by evaporation and (**B**) generated in the subsurface by dissolution of buried evaporites.

shallow depth, with inflow of cold ocean waters at depth. This 'open-cycle' thermal convection (Fig. 2A & B) was first recognized in southern Florida by Kohout and co-workers (Kohout 1965; Henry & Kohout 1972; Kohout *et al.* 1977). The raised temperature of the platform-interior groundwaters may reduce kinetic barriers to dolomite nucleation and growth, and also lower the Mg/Ca ratio required to drive fluids into the dolomite stability field (Hardie 1987). A number of workers explain dolomitization of Late Tertiary and Eocene atolls in the Pacific by open-cycle thermal convection (e.g. Saller 1984; Samaden *et al.* 1985; Aharon *et al.* 1987). Given adequate permeability thermal circulation can continue to operate post-burial, but is often in the form of closed-cycle convection cells (Fig. 2C) (Wood & Hewlett 1982; Machel & Anderson 1989). Deep convection of hydrothermal brines has been suggested as the cause of subsurface hydrothermal dolomites in the Manatoe Formation (Morrow *et al.* 1986) and Michigan Basin (Conglio *et al.* 1994). Within such closed cells the restricted supply of magnesium remains a severe limitation to dolomitization potential unless the cells interact with forced convection, a style of flow termed 'mixed convection' (Raffensperger & Vlassopoulos 1999).

However, even adjacent to a heat source, such as a magmatic intrusion, the degree to which density can be modified by geothermal heating is limited compared with the effect of variations in salinity. Thus, for example, an increase in temperature from 30 to 40 °C is equivalent to a reduction in salinity of <3.2‰. Changes in salinity can be generated at the surface by evaporation provided that replenishment of fluid is limited, for instance when circulation of seawater becomes restricted on shallow banks or in lagoons. The gravity-driven downward flow of these denser fluids, termed reflux (Fig. 3A), not only occurs in hypersaline settings (Logan 1987), but also where seawater is only of slightly elevated salinity (40‰: Simms 1984; Whitaker & Smart 1990). Hypersaline brines that have precipitated evaporite minerals, such as gypsum, will also have an elevated Mg/Ca ratio, enhancing their potential for dolomitization. Refluxing waters may penetrate to considerable depth, but buried carbonates may also be affected by fluids that have dissolved evaporites in the subsurface (Fig. 3B). Dolomitization by hypersaline reflux was first proposed for the Permian reef complex of west Texas by Adams & Rhodes (1960) and recently has been revisited by Melim & Scholle (2002). However, it has also been suggested that brines formed from the common evaporites gypsum, anhydrite and halite will be impoverished in magnesium, restricting Mg flux (Allan & Wiggins 1993).

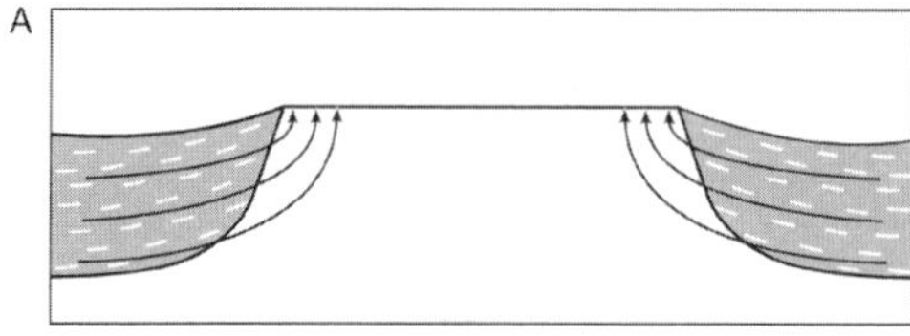

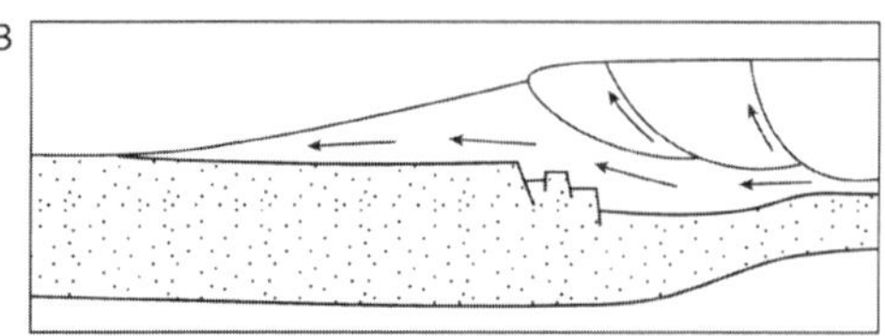

Fig. 4. Compaction-driven fluid flow. (**A**) Preferential compaction of basinal sediments (grey) surrounding a more resistant carbonate platform expels fluids along more permeable interlayers (high anisotropy) towards the platform at the edge of the basin. (**B**) Tectonic compaction in a zone of convergence expels brines from sediments beneath the thrust sheet and overlying low-permeability crust (stippled).

Compaction-driven flow

Compaction-driven flow is known to be significant during the consolidation of rapidly accumulating permeable siliclastic deposits such as are found in intracratonic basins (Bethke 1985) and deltas (Harrison & Summa 1991). As sediments are buried and compacted, increased stress raises pore pressures and drives flow towards the surface or laterally into less rapidly compacting sediments (Fig. 4A). Carbonate sediments are highly susceptible to early cementation in the shallow subsurface, particularly when exposed to meteoric waters (James & Choquette 1984). Thus, the compaction of most platform carbonates is small relative to that of unconsolidated and non-carbonate sediments in the adjacent basins, giving the potential for lateral flow of fluids from the compacting basin towards the platform. Fluids expelled by compaction of basinal shales are geochemically evolved, with potential for dolomitization (Illing 1959). Basinal brines expelled from compacting shales have been invoked by Mattes & Mountjoy (1980) to dolomitize the Upper Devonian Miette build-up.

Compaction may also be caused by tectonic loading and compression during the development of orogenic belts such as thrust zones (Fig. 4B), as proposed by Oliver (1986). Oliver likens this system to a giant 'squeegee', which, he states, could drive flow over tens to hundreds of kilometres. Because compaction expels only a single pore volume of fluid from the source rocks, many have questioned its ability to satisfy the mass-balance requirements of dolomitization (Demming *et al.* 1990; Garven 1995; Machel & Calvell 1999). Only if the compactional water from a large volume source were channelled through a small cross-section of permeable rocks or along permeable faults will there be sufficient magnesium flux for extensive dolomitization. Tectonic compaction has been cited as the dominant factor responsible for the dolomitization of some Devonian reefs in western Canada (Drivet & Mountjoy 1997) and as a possible explanation for the strontium isotope ratios of some Caribbean island dolomites (Machel 2000).

A considerable range of conceptual models is thus available to describe the circulation systems that may provide the magnesium flux necessary for formation of replacement dolomite. The literature is replete with examples where different flow models have been applied to account for the same dolomite body. For example, on the Pacific island of Niue, where the dolomites are yet to undergo burial, no less than four different circulation systems have been suggested (Wheeler & Aharon 1997). In part this may arise from the tendency for geological studies to invoke a single flow mechanism operating in isolation. In real platforms this may not always be the case. For example, all active platforms will be subject to geothermal heating and this may reinforce or act in opposition to other drives. Furthermore, most circulation systems are sensitive to changes in relative sea-level, climate and platform geometry. For example, generation of reflux brines is highly sensitive to sea-level change, being terminated either by a rise in relative sea-level that floods the platform top, or a fall that leads to the development of a meteoric lens. Thus, the steady-state circulation systems implicit in most conceptual models will be the exception rather than the rule.

Even for a single drive, the range of cartoons of fluid flow in the literature serves mainly to illustrate the degree of uncertainty about actual flow patterns and their dependence on controlling variables such as the distribution of permeability. The uncertainty is even greater for depictions of the form of dolomite bodies associated with different circulation systems (Simms 1984; Wilson *et al.* 1990; Whitaker & Smart 1993) (e.g. Fig. 6 later), as they are poorly constrained with respect to critical hydrogeological and geochemical parameters. There is thus a critical need to address these issues in quantitative rather than qualitative terms.

In the remainder of this paper we review analytical and hardware models, and then numerical simulations of different flow mechanisms and assess insights they provide into dolomitization potential. Note that all flow velocities quoted are Darcian velocities, i.e. flow rate per unit area. The pore or linear velocity of a fluid in a porous medium is the Darcian velocity per unit interconnected pore space (effective porosity). Permeability is intrinsic permeability and is quoted in m^2, where 1 m^2 equals 1.013×10^{12} darcies (D).

Analytical and hardware models

Whether or not fluid flow will occur in response to a particular driving mechanism in a given geological setting, together with specific characteristics of the flow, can be determined quantitatively using long-established equations describing the physics of flow in porous media (e.g. Nield 1968; Bear 1972; Phillips 1991). Analytical solutions are based on solving the differential equations governing flow in two dimensions within a simple flow domain (assuming, for example, an aquifer of uniform thickness and isotropic homogeneous permeability) and simple boundary conditions at steady state. Results from such calculations have been supported by hardware (also termed physical scale or analogue) model studies, suggesting that such problems do scale. However, this work has been limited mainly to geothermal convection, reflux circulation and evaporative pumping.

Elevation- (topographic-) head-driven flow

Elevation- (topographic-) head-driven flow in continental aquifers has been described analytically in the seminal study of Toth (1963) and subsequently by many workers (e.g. Phillips 1991). A summary of these numerous studies, which document the dependence of flow on aquifer configuration and properties, is beyond the scope of this paper. We do briefly note some significant findings relating to the effects of topographically driven flow on the distribution of groundwater temperatures, which are known to be of critical importance for the kinetics of dolomitization. When stratigraphy and flow are predominantly horizontal, the local temperature gradient is largely unperturbed as the Peclet number (the ratio of advective to conductive forces) is usually small compared with the aspect ratio of the basin. However, analytical models indicate that significant temperature perturbations occur in recharge and discharge zones, even with uniform geothermal heating. Temperatures are significantly elevated in the distal and discharge zones, making these favourable locations for dolomitization, particularly given the retrograde solubility of calcite (inversely dependent on temperature). Cathles (1993) used analytical models to examine the thermal consequences of topographically driven flow in the Mississippi Valley lead–zinc district. He demonstrated that, irrespective of groundwater flow rates, topographically driven flow could not account for associated dolomitization at the distal end of the flow system as specification of unrealistically high heat flux was necessary to generate the temperatures indicated by fluid inclusions in the dolomites. This problem was resolved by the numerical simulations of Nunn & Lin (1997) invoking an insulating blanket of low thermal conductivity sediments that was subsequently lost to erosion.

The development of groundwater resources in oceanic islands and coastal areas has provided an important impetus to studies of elevation-head-driven circulation in coastal aquifers (see review by Reilly & Goodman 1985). Analytical models generally assume steady-state equilibrium and no mixing between freshwater and salt water, giving a sharp interface. The geometry of the interface between freshwater and salt water is derived from simple hydrostatics according to the Ghyben–Herzberg relationship (based on relative density) and assuming horizontal flow (the Dupuit Assumption). This analytical approach has most usefully been applied to predict the effect of island size, recharge and permeability on the depth and morphology of the freshwater lens developed beneath carbonate islands (Vacher 1988; Budd & Vacher 1990). However, such models provide only approximate solutions because they are limited to strip islands (length greater than 4 × width). Furthermore, they do not account for solute advection and dispersion that generate a zone of mixing between freshwater and salt water. Analytical models, such as that of Volker *et al.* (1985), that attempt to describe this zone of dispersion generally fail to describe the thickening of the mixing zone towards the coast observed in many present-day islands (Whitaker & Smart 1997). However, this is successfully predicted in the model of Bear & Todd (1960) based on the assumption that the zone of mixing is largely generated by tidally induced oscillation rather than dispersion along the flow path. Bear & Todd (1960) also provide an equation that can be used to determine the total flux of salt discharging through the mixing zone and thus the magnitude of induced saline inflow. However, the distribution of fluid flux with depth and distance from the shore cannot be determined.

There has been very little attention to flow of seawater in response to gradients in sea-surface elevation. Using Darcy's law we can calculate a velocity of 10^{-8} m s^{-1} in response to a sustained head gradient of only 0.1 m across a 10 km-wide platform with a permeability of 10^{-10} m^2. This velocity scales in magnitude with permeability and head gradient. Thus, with high permeability, quite small gradients in sea-surface elevation can generate significant flow, especially within small platforms and reefs. However, the degree to which fluids move consistently from one side of a platform to the other, rather than shuttling back and forth, depends on the frequency of temporal variations in sea-surface elevation. For example, Phillips (1991) presents an analysis of groundwater oscillations in response to external forcing by tides or seasonal variations in water level. Groundwater oscillations scale in magnitude with the tidal (or seasonal) amplitude, and the penetration distance depends on permeability, porosity and frequency of the forcing oscillation. Diurnal tidal perturbations can penetrate 1–100 m in a material with a permeability of 10^{-10} m^2, with flow velocities of 10^{-4}–10^{-6} m s^{-1}. This effect could be suppressed somewhat by the effects of capillarity beneath exposed (island) carbonates and the density difference between any freshwater lens and underlying saline water.

Hsü & Siegenthaler (1969) investigated evaporative pumping with a series of column experiments using homogenous sediments (silt–sand size) saturated with freshwater that were subjected to different intensities of evaporative heating. The vertical flow rate induced by evaporative pumping ranged from 7×10^{-8} m s^{-1} for a relatively weak evaporation rate (<1 m $year^{-1}$) to 1×10^{-6} m s^{-1} for extremely high evaporative rates (15 m $year^{-1}$). Counter-intuitively, the flow rates through coarse sand and through very fine silt were approximately the same under the same rates of evaporation, because the hydraulic gradient generated in response to evaporation is inversely proportional to grain size. Hsü & Siegenthaler (1969) concluded that evaporative pumping could be an efficient mechanism for dolomitizing low-permeability sediments of arid peritidal environments.

Density-driven flow: thermal convection

Elder (1967*a*, *b*) presents a theoretical analysis of convection, supported by 'hot-plate' experiments heating a saturated porous medium to generate pairs of convection cells with upwards and outward-directed flow, that are comparable to patterns observed in natural geothermal systems (Elder 1965). A number of analytical solutions for free convection of this type have been developed, most notably in papers by Wood & Hewlett (1982, 1984) who suggested that free convection in both sloping and horizontal aquifers could result in significant diagenetic effects.

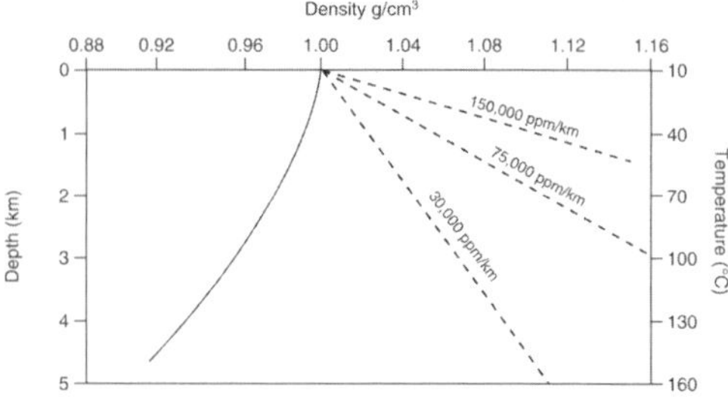

Fig. 5. Density of water as a function of temperature and salinity. Salinity depth gradients are representative of those commonly observed in sedimentary basins offset the thermal expansion of water. After Bjorlykke & Palm (1988).

However, more recently Bjorlykke *et al.* (1988) demonstrated that even for relatively permeable units (10^{-12} m^2), a substantial bed thickness (331 m for a one-layer case, 540 m for a three-layer case) free of even minor shale breaks is needed for free convection to occur. Even in carbonates this is unlikely, except in massive reef facies. Bjorlykke *et al.* (1988) also pointed out that for sloping beds Wood & Hewett's (1982) assumption that the isotherms are parallel to the slope of the bed is unrealistic and gives rise to an overestimation of the magnitude of the free convective flux. They showed that with more realistic assumptions typical of many clastic basins, including lower permeability (10^{-16} compared to 10^{-15} m^2) and thinner beds (10 compared to 100 m), much lower fluxes were obtained than in the original study (10^{-10} compared to 10^{-8} m s^{-1} for a 15°-sloping layer) and that this reduced even further when the isotherm slope was reduced to 1° (10^{-12} m s^{-1}). The latter is sufficiently low that for distances of up to 100 m diffusion would dominate over advection. They therefore concluded that thermally driven free convection was unlikely to be significant in diagenesis in most sedimentary basins, particularly when the opposing effect of increasing salinity with depth was included (Fig. 5).

Where mixed (free and forced convection)

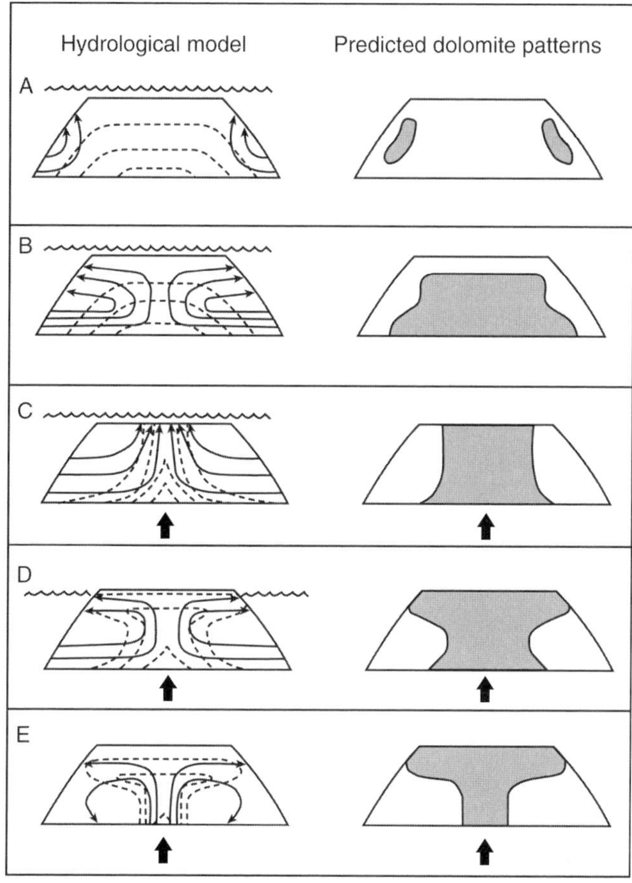

Fig. 6. Geothermal convection and predicted distribution of resultant dolomite. Arrows indicate flow direction; dashed lines are isotherms; dolomite is stippled. (**A**) and (**B**) show geothermal convection driven by normal geothermal heat flux; in (**A**) $K_V > K_H$, and in (**B**) $K_V < K_H$. (**C**) and (**D**) show geothermal convection with elevated heat flux beneath the platform; in (**D**) the platform top is emergent (submerged in all other simulations). Scale is 'km scale'. After Wilson *et al.* (1990).

systems occur, as is likely in many basins, Prats (1966) showed that forced convection becomes dominant when the magnitude of the forced convection flux approaches that of free convection. This conclusion has recently been supported by numerical modelling by Raffensperger & Vlassopoulos (1999). Note that in mixed systems, free convection cells migrate laterally in the direction of forced convection.

Henry & Hilleke (1972) were the first to apply Elder's (1967*a*, *b*) analysis to carbonate platforms and demonstrated that the density contrast between cold ocean waters and warm platform waters generated pressure differences sufficient to drive geothermal convection of seawater and alter the temperature field. Wilson *et al.* (1990) considered in more detail patterns of geothermal convection and temperature in small-scale (3 km-wide) isolated platforms. They identified two fundamental controls (Fig. 6): the Rayleigh number (the ratio of viscous to buoyant forces) and the horizontal:vertical anisotropy of permeability (K_V:K_H). In low Rayleigh number (cool) systems, conduction dominates heat transfer and isotherms parallel the platform top. Convection is best developed at the margins but can penetrate to the interior of the small platform at higher K_V:K_H. In contrast, in high Rayleigh number (hotter) systems, convective heat transport becomes significant and distorts isotherms. Thus, the margins are cooled and the platform is heated at shallow depth, in the

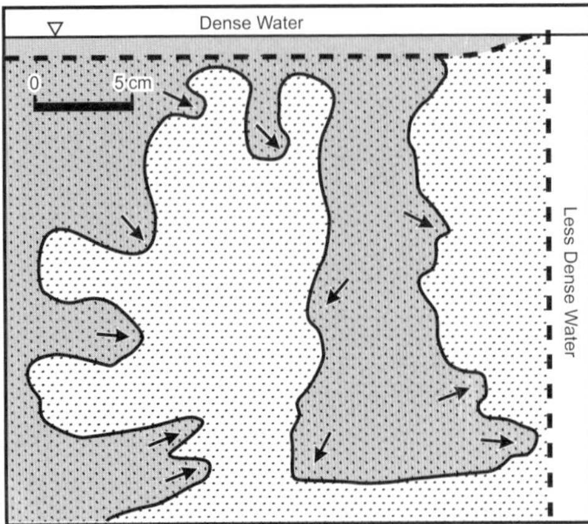

Fig. 7. Small-scale hardware model to simulate reflux of 42‰ mesohaline waters displacing normal seawater. Position of plumes of denser water is shown after 24 h. After Simms (1984).

interior for a submerged platform top, or across the entire emergent platform top.

More recently LeClerc *et al.* (2000) introduced a one-dimensional (1-D) analytical model of geothermal circulation within atolls, validated by 2-D numerical simulations. The 1-D model demonstrates that the temperature contrast between cold deep ocean water and the warm platform top can maintain convection even in the absence of a geothermal heat flux. Where geothermal heating does occur, the flow rate increases with heat flux but, counter-intuitively, it decreases with a reduction in deep ocean temperature. This is because the colder water drawn into the platform at depth reduces the contrast between the platform groundwaters and those in the adjacent ocean. Similarly, reduction in salinity with depth in the ocean water can promote circulation, an effect which if ignored can lead to underestimation of flow rates by more than a third. Jean-Baptiste & LeClerc (2000) suggest that the negative feedback between convective flow and heat transfer leads to an asymptotic flow regime, with an upper limit on geothermally driven groundwater fluxes of some 10^{-8} m s^{-1}, irrespective of the permeability of the carbonates.

Density-driven flow: saline reflux

A number of analytical studies have investigated fluid flow in a porous medium arising from increased salinity, including those of Nield (1968), Sarkar *et al.* (1995), Wooding (1997) and Sharp *et al.* (2001). However, only Simms (1984) and Phillips (1991) have focused on reflux circulation in carbonate platforms. The propensity for reflux depends on the Rayleigh number, here a function of the salinity contrast, permeability and the thickness of the flow domain. Using stability theory, Nield (1968) established that the criterion for the onset of reflux was a Rayleigh number greater than 0; that is to say, that any vertical inversion of density is unstable. Thus, although reflux is generally considered to result from platform-top brine generation, reflux of mesosaline brines with salinities only slightly greater than seawater would be expected. For instance, based on the Great Bahama Bank, Simms (1984) calculated a Rayleigh number of 80 for a flow domain 333 m thick with a permeability of *c.* 10^{-10} m^2 and platform-top fluids of only 42‰. Using a sandbox experiment, he then demonstrated that with this Rayleigh number lobes of dense fluid penetrated rapidly downwards, displacing underlying less dense pore fluid before spreading out at the base of the sandbox (Fig. 7). Increasing the permeability and the salinity contrast to the higher values observed in some platforms would further increase the Rayleigh number and enhance reflux.

Although the Rayleigh number provides a useful single value to characterize the stability of specific systems with respect to free convection (including those driven by the thermal density contrasts discussed above), Simmons *et al.* (2001) drew attention to a number of limitations when it is applied to a heterogeneous porous medium. The use of average hydraulic properties ignores the presence of heterogeneities, which may trigger instability. Furthermore, the dimensional analysis that leads to the Rayleigh number makes an assumption that the flow is steady state. This is not the case in many simulations and Oostrom *et al.* (1992) therefore suggested that plume thickness is a more appropriate length scale for use in calculation of the Rayleigh number than the aquifer thickness that is conventionally employed. Finally, Simmons *et al.* (2001) pointed out that in transient flow situations aquifer storage may have a significant impact on the onset of instabilities, but storativity is not included in the Rayleigh number.

The Darcy–Overbeck–Boussinesq equations can be solved to derive an approximate solution for vertical flow velocity under reflux conditions (Simms 1984). For material of 10^{-10} m^2 permeability and fluid with salinity of 42‰, reflux can occur at a velocity of 5×10^{-6} m s^{-1}. With increasing brine concentration velocities increase at a rate of approximately 8×10^{-7} m s^{-1} per 1‰ increase in salinity. This is equivalent to that for an increase in permeability of 1.2 times, as velocity scales directly with permeability. However, the permeability of platform-top sediments varies over more than five orders of magnitude (Enos & Sawatsky 1981), with even greater variation where evaporites are present (Bredehoeft 1988). In contrast, seawaters reach halite saturation within an order of magnitude increase in salinity. This suggests that permeability rather than brine concentration will be the major control on the pattern of reflux in real platforms. Given the close association between reflux and evaporite deposition, it is important to resolve the uncertainty that remains regarding the permeability of evaporite crystal mushes at shallow depth (Sonnenfeld 1984).

Compaction-driven flow

The earliest analytical model relevant to sediment compaction is the 1-D solution of Gibson (1958). Gibson's linear theory, which considers consolidation under a constant sedimentation rate, permeability, sediment density and small strains, became the benchmark against which numerical solutions were tested. However, a linear theory is clearly incompatible with conditions in sedimentary basins, and a number of authors have considered compaction for variable permeability and density, and finite strains (Gibson *et al.* 1967; Smith 1971; Sharp 1976; Bethke & Corbet 1988). Non-linear theory was further refined by Audet & McConnell (1992) who incorporated principles from soil mechanics. They demonstrated that the extent of abnormal fluid pressures, which reflects the rate of fluid flow, is governed by a dimensionless number that is the ratio of sediment hydraulic conductivity to sediment accumulation rate.

One-dimensional models assume that compactional fluid flow is directed upwards towards the sediment–water interface. However, at the scale of sedimentary basins, fluid flow is generally directed towards the edges as a result of different loading rates and inhomogeneous distribution of sediment hydraulic parameters. Magara (1976) developed an analytical model for the direction of flow and volume of fluids expelled by compaction from sediments of variable hydraulic properties. For a thick sequence of shales flow is vertically upwards, and total fluid flux increases with depth to a maximum at intermediate depth, then declines to the base of the sediment column. Interbedded sequences of sands and shales with contrasting permeability (K sand:K shale = 1000) resulted in predominantly lateral flow. Total fluid flux increases towards the edge of the basin and at a given distance, with depth. However, the total volume of fluid expelled from compacting sediments is limited. Even at a distance of 8 km from the centre of the basin, Magara (1976) estimated that total flux does not exceed 3.3×10^3 m^3 m^{-2}.

The larger scale of circulation driven by tectonic compaction alleviates somewhat the volumetric limitations. For instance, Knoop *et al.* (2002) considered syntectonic fluid migration across carbonate units in the western ranges of the Rocky Mountain foreland from deeper siliciclastic rocks during the Mesozoic contraction. Using a 1-D flow model they inferred a time-integrated fluid flux of some 2.4 $\times 10^8$ m^3 m^{-2}. More permeable conduits, including low-displacement faults and regional thrust faults, probably experienced considerably higher fluxes, with fluid temperatures reaching 300–400 °C.

Comparison of flow rates

Simms (1984) demonstrated the potential utility of analytical models when he compared representative velocities for reflux, geothermal and submixing-zone seawater circulations. His calculations suggested that flow rates for geothermal circulation are comparable with those for submixing-zone seawater circulation (although an order of magnitude less than circulation within the mixing zone). Flow rates for the reflux of mesosaline waters (42‰) are an order of magnitude lower, but no figures were presented for more saline brines that would generate higher velocities. However, such comparisons demand careful consideration, particularly where numbers are based on calculations that have different parameter values and boundary conditions. Furthermore, most analytical solutions fail to provide information on the distribution of fluid flows in space, which is critical for determination of the distribution of dolomitization. They are commonly also limited to platform geometries and property distributions that are grossly oversimplified compared to those observed in nature (Bethke 1989), for instance generally assuming constant permeability with depth.

Despite these inadequacies, analytical models can still provide valuable insights into the controls on and behaviour of groundwater flow systems in carbonate platforms, as exemplified by the recent studies of Cathles (1993) and LeClerc *et al.* (2000). They also provide practical guidelines for estimating which flow mechanisms are likely to be important in given hydrogeological situations. For instance, Kooi (1999) presents an analytical expression for the flow field arising from competition between topographic- and compaction-driven flow within a confined aquifer. This demonstrates elegantly that very unusual circumstances are required for compactional flow to reverse topographic flow, although temporal variation in sedimentation rates might lead to episodic flow-reversal and fluid-expulsion events. Finally, both hardware and analytical solutions provide important benchmarks for testing the performance of numerical codes using well-defined problems. For instance, the analytical solutions to the Henry (1964) salt-water intrusion and Elder (1967*b*) salt-convection problems both provide benchmarks for density-dependent groundwater codes (Simpson & Clement 2003). However, these authors detail pragmatic issues associated with this testing. Furthermore, Konikow & Bredeheoft (1992) argue on philosophical grounds that groundwater models cannot be verified or validated because numerical results are always non-unique and, unlike the model domain, natural flow systems are not closed.

Groundwater flow models

Numerical models are based on a mathematical approach fundamentally different from that of the analytical solutions discussed above. The differential equations governing flow are approximated by a difference form of the equation that is solved at nodes that comprise a 2-D or 3-D grid. This method permits specification of different sets of hydraulic parameters and recharge fluxes to individual nodes or sets of nodes, replicating the sorts of heterogeneities seen in the real world. With the advent of high-speed computers and the development of sophisticated numerical groundwater flow models, we are now able to investigate much more complex hydrogeological scenarios than were accessible using analytical models. However, the availability of exact analytical solutions has provided an important set of test cases that have proved useful for validation of the numerical modelling schemes (e.g. Voss 1984). Numerical models have been employed to investigate the role of groundwater flow in a wide range of geological situations that include ore emplacement (Garven & Freeze 1984), generation of overpressure in siliclastic basins (Harrison & Summa 1991), thrusting (Ge & Garven 1992) and continental rifting (Wieck *et al.* 1995).

Numerical models solve the interdependent partial differential equations governing fluid flow, heat and/or solute transport in porous media. Darcy's law is substituted into the mass-balance equation to obtain the groundwater flow direction and velocity. The heat transport equation is obtained in a similar manner by substituting Fourier's law into an energy balance equation, and both fluid pressures and heat are assumed to be at steady state (Sanford *et al.* 1998). Because the density and viscosity of fluids are dependent on pressure, groundwater temperature and salinity, it is critical in model studies that appropriate equations of state are employed (this is also, of course, the case in calculating dimensionless numbers such as the Rayleigh number). Many early (and some more recent, e.g. Bitzer 1996) numerical modelling studies relied on the Boussinesq assumptions (constant viscosity and linear dependence of density on temperature), but most recent studies incorporate non-linear temperature–density and viscosity relationships (see Raffensperger & Vlassopoulos 1999). The effect of pressure on viscosity and density is generally minor (<5%) to depths of 5 km (Sanford *et al.* 1998), but the effect of salinity is very significant and should be included.

More recent studies also solve for solute transport, treating this as a transient problem by combining mechanical dispersion and chemical diffusion with mass transport by advection. Although carbonate platforms commonly have significant secondary dissolutional and fracture porosity, modelling is generally at a spatial scale sufficiently large that it is possible to assume that fluid behaviour within the rock is equivalent to that in a porous medium (Renard & Marsily 1997). Fluid velocities are therefore assumed to be sufficiently slow that flow is laminar and Darcy's Law can be applied.

It is important to note that most simulations use linear (Cartesian) coordinates to simulate flow in a 2-D cross-section parallel to the flow direction. This implies that flow convergence or divergence is unimportant, as will be the case where platform length is large compared to platform width. This is a reasonable assumption for extensive rimmed shelves and ramps, but a linear coordinate system is inappropriate for the majority of small carbonate platforms such as circular atolls. Flow in these may, however, be simulated using a radial coordinate system. For instance, Sanford *et al.* (1998) demonstrate for geothermal convection that in circular geometries little flow reaches the very centre of the platform due to the reduction in the cross-sectional area for flow towards the interior

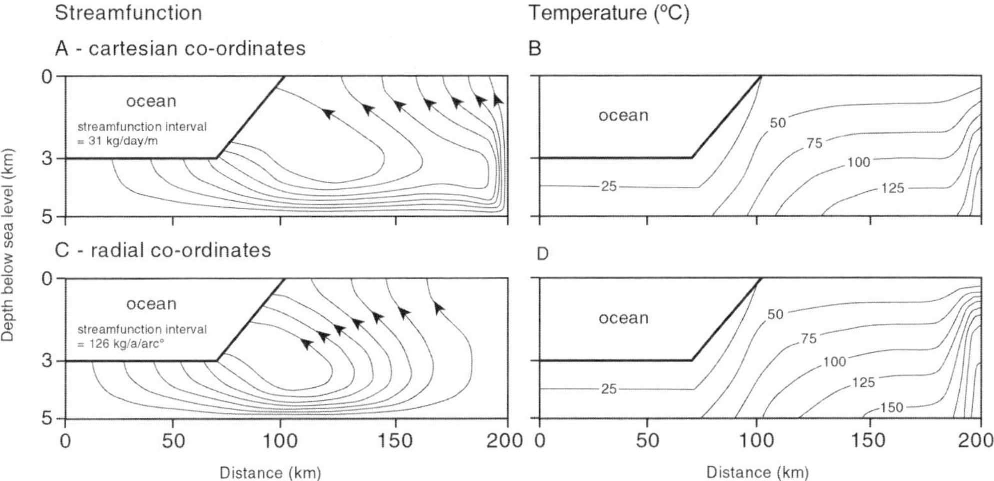

Fig. 8. Effect of coordinate systems on geothermal circulation and temperature for: (**A**) and (**B**) Cartesian coordinates; and (**C**) and (**D**) radial coordinates. After Sanford *et al.* (1998).

(Fig. 8). However, the convergence of flow and advected heat results in higher temperatures at the platform interior than for strip platforms of similar width. Conversely, for reflux of brines, flow divergence would significantly reduce groundwater fluxes towards the platform margins, as demonstrated for Enewetak Atoll by Jones *et al.* (2000).

With an analytical approach both the problem and the solution can be simply expressed in terms of parameters (such as dimensionless numbers). In contrast, each numerical simulation is quite specific and several cases must be solved to yield general information. Two approaches have been adopted to experimental simulation of circulation of saline groundwater in carbonate platforms: sensitivity analyses performed to test the role of individual variables using generic platforms and specific case studies to investigate possible circulation in specific platforms. The generic sensitivity studies commonly provide insight into the critical parameters that need to be parameterized for effective simulation of specific case studies, both in the modern (where limited hydrogeological data are available for model parameterization and comparison with model output) and in the ancient. In the latter, however, parameterization is often difficult because of changes in critical variables such as porosity and permeability that may be profound during both dolomitization and subsequent diagenesis. Indeed, Bethke (1989) notes that this is a major limitation in studies of palaeohydrological simulations.

Elevation- (topographic-) head-driven flow

The hydrodynamic potential of elevated topography to drive circulation of meteoric waters deep into foreland basins over distances of several hundred kilometres was first quantified by Garven & Freeze (1984). Permeability contrasts were shown to focus subsurface flow in transmissive carbonate units within the succession and fluids could attain velocities of up to 3×10^{-7} m s^{-1} (Garven 1995). At these high flow rates topography dominates over all other drives. Yao & Demicco (1995) suggest that a 10–30 km-wide zone along the 200 km-long margin of the Cambrian carbonate platform of the Western Canada Sedimentary Basin was dolomitized by topographically driven recharge (Fig. 9). The flow pattern is consistent with that inferred from the geometry of the dolomite body and flow velocities average 10^{-8} m s^{-1}, but questions remain regarding the source of magnesium in such systems. Free convection was also modelled but fluxes were much lower and did not accord with the geometry of the dolomite body (Yao & Demicco 1997).

Because of the high flow velocities, there is also a significant redistribution of heat towards the distal end of topographically driven regional flow systems. This generates elevated temperatures in the more transmissive limestones at depth and significant cooling as the groundwaters move towards discharge points at the surface (Garven & Freeze 1984). Under these conditions the retrograde solubility of

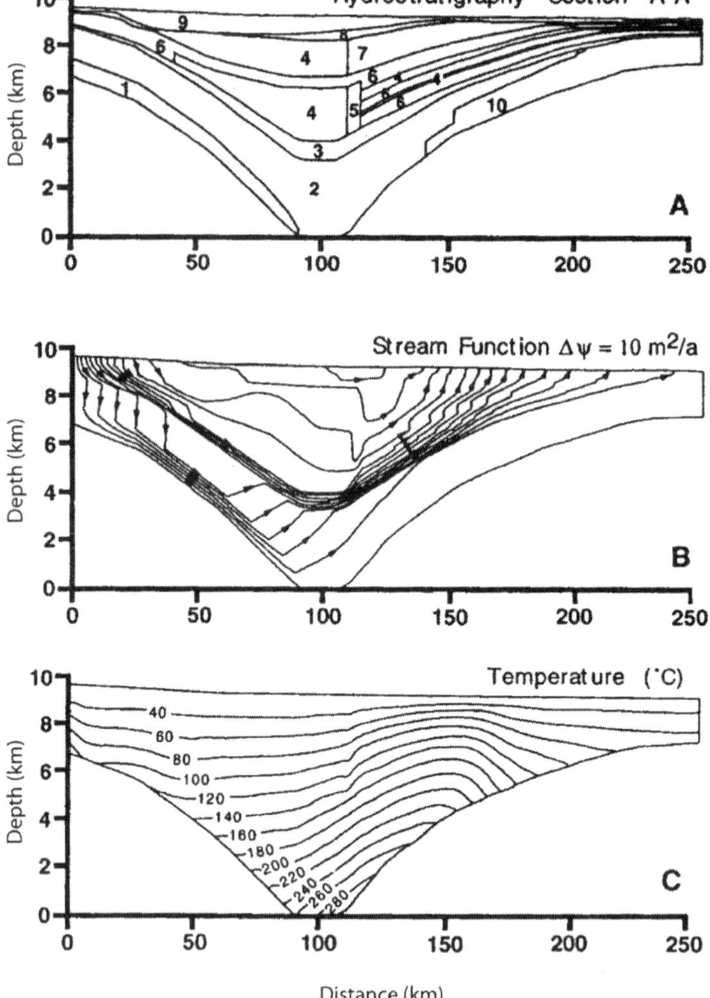

Fig. 9. Elevation-head driven circulation in the Cambrian Cathedral Formation, Southern Rocky Mountains. (**A**) Hydrostratigraphy (Unit 5 is the dolomitized platform-margin grainstones). (**B**) Temperature. (**C**) Streamfunction. After Yao & Demmico (1995).

limestones can lead to dissolution of calcite and precipitation of dolomite. For instance, a 2.3 molal Cl and 0.15 molal Ca basinal brine equilibrated with respect to calcite and dolomite will precipitate 8 cm^3 of dolomite as it cools from 200 to 20 °C (Bethke & Marshak 1990). This process is suggested to be important in the formation of the dolomitized dissolution breccias frequently associated with carbonate-hosted lead–zinc mineralization (Garven & Freeze 1984; Bethke *et al.* 1988; Garven *et al.* 1993).

Numerical models incorporating solute transport have also been utilized in water-resource investigations of coastal meteoric aquifers in order to simulate flow in both the meteoric and freshwater–salt water mixing zone (e.g. Herman *et al.* 1986; Voss & Souza 1987; Ghassemi *et al.* 1990; Wicks & Herman 1995). These provide an insight into the control exerted by relative sea-level (via its effect on exposed platform width), recharge and permeability on the size, shape and flow dynamics of the freshwater lens. However, such models have performed relatively poorly in the simulation of the distribution and thickness of mixing zones because dispersion (which determines these properties) is scale and fluid-density dependent (Herbert &

Lloyd 2000). Furthermore, as has been demonstrated by non-steady-state simulations, ocean tides exert a major control on the thickness of the freshwater–salt water mixing zone (Oberdorfer *et al.* 1990; Underwood *et al.* 1992). Thus, in a study of the configuration of the Home Island freshwater lens by Ghassemi *et al.* (2000) simulations that ignored tidal effects required dispersivities some two–three orders of magnitude higher than more sophisticated ones that included them. These authors concluded that the inability to incorporate the intrinsic heterogeneity of the karstic aquifer is the major limitation in improving the reliability of the simulations in this atoll island.

The focus in these studies has been primarily on various aspects of seawater intrusion and little attention has been paid explicitly to submixing zone seawater circulation. In many cases the boundary of the modelled flow domain is relatively shallow (see for example Nielsen 1999) and thus has a significant impact on the modelled circulation. The only previously published simulations specifically directed at quantifying submixing-zone flow are those of Kaufman (1994). Kaufman simulated flow beneath a 40 km-wide island (recharge unspecified) generating a rather thick (150 m) lens within which meteoric waters flow seaward at up to 1.4×10^{-1} m year^{-1}. However, the circulation pattern defined differs fundamentally from that known to occur in coastal aquifers. Simulated flow was down and out from the surface of the island, not only within the meteoric and mixing zones but also throughout the saline zone, to a maximum simulated depth of 1 km. Furthermore, statements that a thicker vadose zone and longer run times (higher magnitude and longer duration of sea-level fall) give a thicker lens and higher flow rates are incompatible with well-established hydrological theory and cast doubt on the reliability of this study.

We have recently undertaken simulations of submixing-zone circulation beneath a 4.5 km-wide island, with a permeability that reduces with depth from 10^{-10} m^2 at the surface and anisotropy of 100 (Fig. 10). Recharge of 0.5 m year^{-1} generates a freshwater lens up to 30 m thick, which discharges through the underlying mixing zone. Discharge is focused within the shallow (fresher) part of the mixing zone, with a maximum fluid flux of 5×10^{-7} m s^{-1} adjacent to the coast. It draws in seawater over a zone that extends to at least 1 km offshore, and this circulates to depths approaching 200 m. The magnitude of the fluid flux in the submixing-zone circulation is reduced for lower

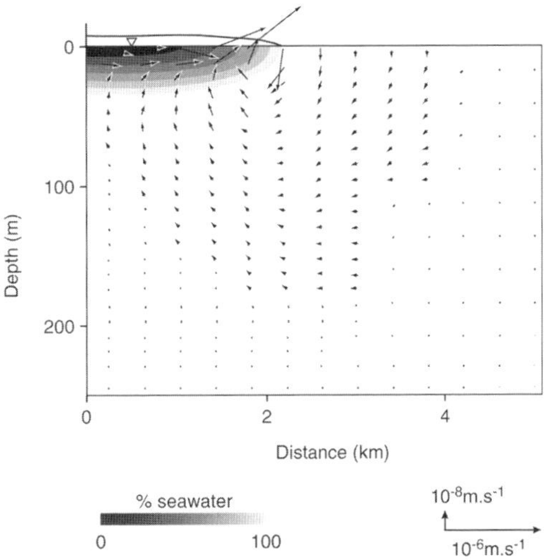

Fig. 10. Elevation-head across a 4.5 km-wide island drives discharge of freshwaters via the mixing zone, and compensatory submixing-zone circulation of seawater. Salinity is indicated by shading and contours (0.1 and 0.4 moles salinity). Arrows show flow field. Permeability reduces with depth from a maximum of 10^{-10} m^2, anisotropy is 100.

permeability simulations that also give a thicker lens. Increasing permeability anisotropy from 50 to 1000 gives greater lateral penetration of the submixing-zone circulation beneath a lens of reduced thickness and aerial extent. Thus, significant submixing-zone circulation appears to be focused close to the coast, where maximum fluid fluxes are comparable with the highest measured for geothermal circulation. Because the principle flows are in the same direction, it is probable that circulation due to the buoyant and geothermal drives will enhance submixing-zone seawater circulation, although the fluid fluxes will remain limited compared to those for topographically driven flow.

Few studies have focused on trans-platform flow driven by differences in sea-surface elevation maintained by winds or currents. Jones (2000) simulated flow in response to a head of 10 cm across a 100 km-wide platform, with permeabilities representative of packstone/wackestones reducing exponentially with depth. This head gradient generates flow velocities of up to 10^{-3} m year^{-1} that increase to 10^{0} m year^{-1} for grainstone permeabilities. Velocities scale directly with hydraulic gradient and thus increase by two orders of magnitude for the same head difference for a 1 km-wide atoll island. Interaction with head-driven circulation beneath the exposed island would be

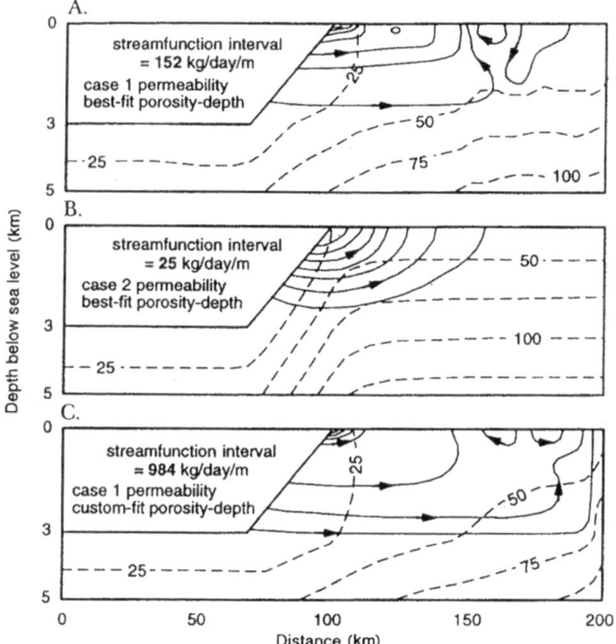

Fig. 11. Effect of the vertical distribution of permeability on geothermal circulation and temperature. (**A**) Homogenous permeability of 10^{-13} m^2. Exponential decline in permeability with depth for (**B**) permeable (grainstone) and (**C**) less permeable (packstone–wackestone) platforms. In all cases anisotropy is 1:1000. After Sanford *et al.* (1998).

expected to modify this circulation but was not simulated.

Density-driven flow: thermal convection

Kaufman (1994) reported a simulation for a steep-margined 500 m-thick platform, with an exponential decrease of permeability with depth from 10^{-14} m^2 at the surface and a low anisotropy (2.5). This indicated geothermal circulation restricted to the platform slope. The maximum flow rate was 7.0×10^{-10} m s^{-1} and the temperature gradients remained conductive throughout. Higher rates (up to 2.4×10^{-9} m s^{-1}) were predicted for a closed convection cell developed above an area of elevated heat flux ($2.5 \times$ background values) in the centre of a flat-topped platform.

Sanford *et al.* (1998) presented the first systematic analysis of controls on the rate and pattern of geothermal convection and the resulting temperature distributions. They simulated flow in a 100 km (half-width) platform surrounded by a 3 km-deep ocean. Permeability was 10^{-13} m^2 and uniform with a high anisotropy (1000), considered to represent the effect of horizontal layering of contrasting lithofacies typical of many shallow-water carbonate sequences. The specified permeability was sufficiently high to allow rather higher fluid fluxes (2×10^{-9} m s^{-1}) than those predicted by Kaufman (1994) with an almost identical heat flux, and significant advective cooling was apparent from the temperature field (as predicted by analytical models). With homogeneous permeability, thermal circulation extended to almost 2 km below the ocean floor, with a total fluid flux of 10^{-12} m^2. But a more realistic exponential reduction of porosity and permeability with depth (from 10^{-10} m^2 at the surface to 10^{-14} m^2 at 3 km representative of grainstones) focused flow largely in the upper part of the platform (Fig. 11) and reduced total fluid flux by three orders of magnitude. At depth there was reduced advective cooling compared to the homogeneous permeability simulation, resulting in temperatures up to 50 °C higher around the platform margin, but also less shallow warming in the platform interior.

Both fluid flux and platform temperatures are critically dependent on the magnitude and degree of anisotropy of permeability. Sensitivity analysis indicates three different modes of heat

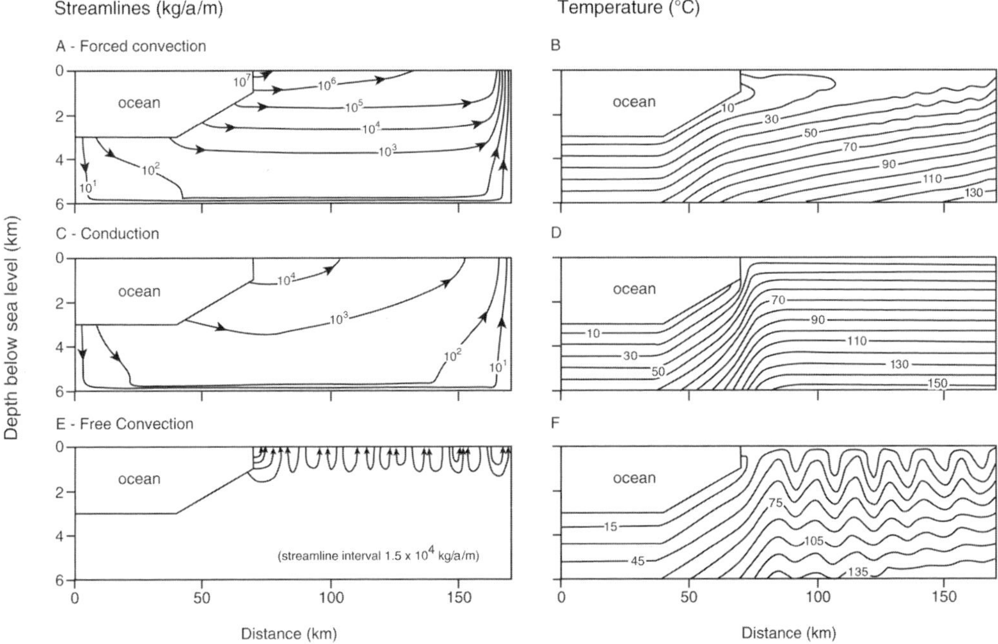

Fig. 12. Effect of permeability on geothermal circulation and temperature for different heat-transport mechanisms: (**A**), (**B**) forced convection in a permeable (grainstone) platform; (**C**), (**D**) conduction in a less permeable (packstone–wackestone) platform. For both simulations anisotropy is 1000. (**E**), (**F**) Free convection in a less permeable (packstones–wackestone) platform with anisotropy reduced to 10 by increasing vertical permeability. After Jones (2000).

transport can occur within the platform (Figs 12 and 13): forced convection (as described above) when permeability is high and $K_V < K_H$; free convection when permeability is high and $K_V \geq K_H$; and conduction when permeability is low. Reducing the depth-dependent permeability (for instance from values representing grainstones to those representing packstone–wackestones) reduces the fluid flux by an order of magnitude, without significant alteration of the pattern of flow. The whole platform is warmed by up to 45 °C as conduction rather than convection dominates heat transport. Simulations with low anisotropy, isolated permeable units (e.g. buried reefs) and anomalously high geothermal heat flux generate free convection cells. In real platforms these may be inhibited by small-scale heterogeneities and they are, however, only of significance for dolomitization when open to the platform surface as reactants will rapidly become exhausted in closed circulation.

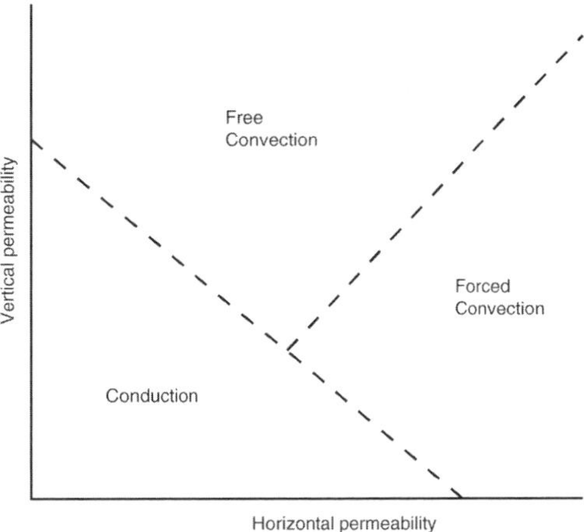

Fig. 13. Dominant heat-transport mechanism in geothermal systems as a function of permeability and permeability anisotropy. After Sanford *et al.* (1998) and Jones (2000).

These simulations also disprove some commonly held assumptions about thermal convection. Surprisingly, platform geometry and relative sea-level appear to be only second-order controls on the magnitude of convective fluid flux. Platform-top submergence does affect the pattern of discharge, as suggested by Wilson

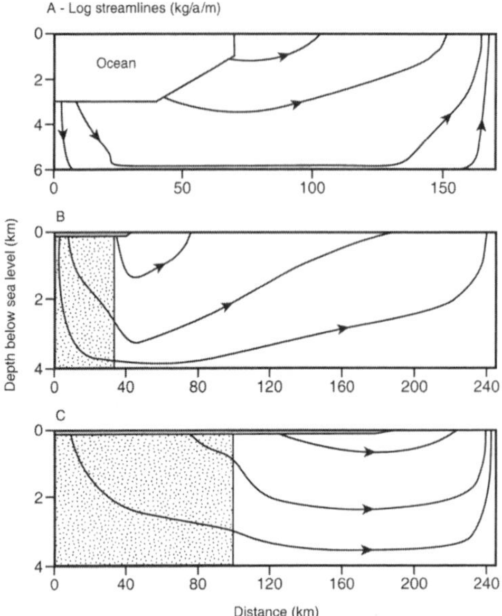

Fig. 14. Geothermal circulation in: (**A**) rimmed shelf, (**B**) distally-steepened ramp and (**C**) homoclinal ramp. Model domain is identical for (**B**) and (**C**). Stipple represents less permeable wackestone–mudstone for basin (B) and outer ramp (C). Other units are packstones. After Jones (2000).

et al. (1990). The magnitude of fluid flux is largely unaffected by the relative position of sea-level, even when the platform top is drowned to considerable depth. Sanford *et al.* (1998) demonstrated that convection is still active in much smaller platforms (20 km half-width compared to the 100 km described above), although the outflow face for discharge shifts from the interior for large platforms to the slopes on smaller ones. The conceptual models of Burchette & Wright (1992) and Whitaker & Smart (1993), which suggest that geothermal convection will be limited to the margins of extensive carbonate ramps, are questioned by the numerical simulations of Jones (2000) for homoclinal ramps. Whereas fluid flux in these was an order of magnitude lower than that for the rimmed platform, substantial geothermal convection did occur, particularly in the inner ramp where specified permeabilities were higher. In contrast, as suggested by the conceptual models, flow in distally steepened ramps was focused around the break of slope due to the less permeable inner ramp (Fig. 14). Note also that the reduced fluid flux implies that thermal convection in ramps might be relatively easily influenced or suppressed by other synchronous drives.

Concern over the potential for groundwater contamination following underground nuclear tests in Pacific atolls has prompted detailed simulations of present-day geothermal convection. Simulations of Enewetak by Samaden *et al.* (1985), using a five-layered permeability model, illustrate thermally driven advective cooling by radial flow, with a maximum flow rate of 1.2×10^{-7} m s^{-1} in the margin at the base of the carbonate sequence. Jones *et al.* (2000) obtained a similar flow distribution and a maximum flow rate of 7.9×10^{-7} m s^{-1} for Enewetak using a different code, and with an exponential decrease in permeability with depth. Henry *et al.* (1996) focused on geothermal circulation within the volcanic basement of Mururoa, but also predicted flow within the carbonates, with a maximum rate of 1.8×10^{-8} m s^{-1} at the base of the carbonate unit. A homogeneous permeability of 3×10^{-11} m^2 was specified for the carbonates in order to generate the negative temperature profiles observed in boreholes. Further simulations of Mururoa by LeClerc *et al.* (1999) demonstrated that the temperature and salinity profile of the ocean impacts on geothermal circulation (as discussed above). A high-permeability unit was introduced at the base of the carbonates. Flow rates increased with permeability of this unit until, at very high permeability, the transmissivity of the overlying carbonates became limiting, defining a maximum velocity of 4.3×10^{-7} m s^{-1}. Introduction of a second high-permeability layer at shallower depth further improved fit with actual temperature profiles but greatly reduced flow in the overlying near-surface carbonates.

Density-driven flow: reflux

Kaufman (1994) simulated reflux beneath a small-scale (10 km) silled lagoon. He found that the magnitude and lateral extent of reflux was proportional to the concentration of the brine and the permeability of underlying carbonates. Maximum flow rates increased from 2.1×10^{-10} m s^{-1} for mesosaline (41‰) brines to 3.0×10^{-9} m s^{-1} for brines concentrated to halite saturation (290‰). The control exerted by brine concentration was confirmed by Jones *et al.* (1997) for a much larger (100 km) simulated platform (Fig. 15). More rapid reduction of permeability with depth and high anisotropy focused flow laterally in the shallow (<1 km) subsurface. Maximum fluid-flux driven by reflux of brines concentrated up to halite saturation was up to two orders of magnitude greater than

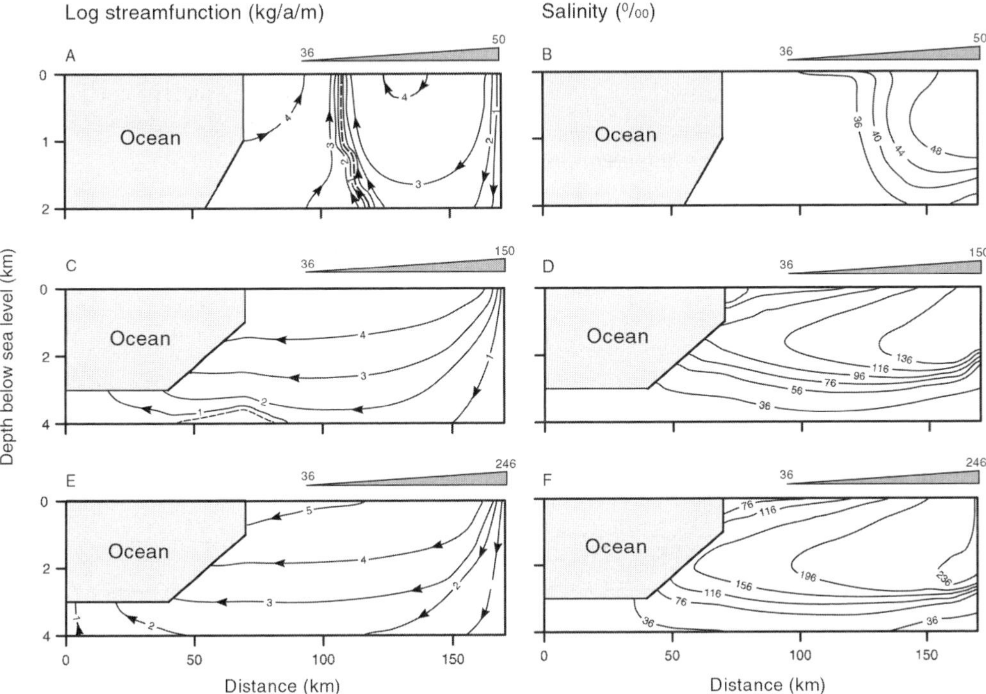

Fig. 15. Extent of reflux circulation after 1 Ma for: (**A**) mesohaline brines (up to 50‰), (**B**) brines concentrated up to gypsum saturation and (**C**) to halite saturation. Note (A) and (B) only display the upper 2 and 4 km of simulated area, respectively. After Jones (2000).

for thermal convection for the same platform flooded with seawater of normal salinity.

Jones *et al.* (2004) present a systematic and detailed analysis of the interaction between reflux and thermal circulation, and the effect and hierarchy of the parameters controlling reflux. They show progressive development of reflux circulation from the interior of the platform towards the margin through time, with a corresponding reduction in the magnitude and extent of geothermal influence (Fig. 16). The rate at which this occurs is controlled by the concentration of platform top brines, the duration of brine generation and permeability distribution. The extent and location of brines on the platform top are secondary controls. High-permeability facies focus refluxing brines and enhance flow, but precipitation of evaporites that are generally of low permeability can severely restrict reflux. Jones *et al.* (2004) suggest that reflux circulation will also be important in carbonate ramps. Even slightly elevated salinities at the top of the inner ramp are sufficient to offset the weak density gradients that drive thermal circulation within them (Fig. 17).

Jones *et al.* (2002) considered the role and fate of reflux brines after brine generation ceases. Following flooding of the platform by seawater, subsurface brines continue to flow through it and entrain seawater, a variant of reflux termed 'latent reflux'. Brines concentrated up to gypsum saturation (150‰) remain within the platform for more than 100 times the duration of the brine generation period (Fig. 18). Thus, brines with varying degrees of dilution are likely to be a normal component of the deep groundwater system in many carbonate platforms.

Case studies of reflux largely serve to illustrate these general controls. Jones *et al.* (2000) simulated reflux of brines concentrated up to 80‰ from the 40 km-wide lagoon of the near-circular Enewetak atoll. For the specified permeability, which decreased exponentially from 10^{-11} m^2 at the surface, the maximum flow velocity generated by reflux (4.8×10^{-9} m s^{-1}) was more than twice that for geothermal circulation (1.9×10^{-9} m s^{-1}). However, the platform margins were not invaded by brines until 500 ka after the onset of reflux, due to the persistence of thermal circulation at the margins and the requirement to flush out water of seawater

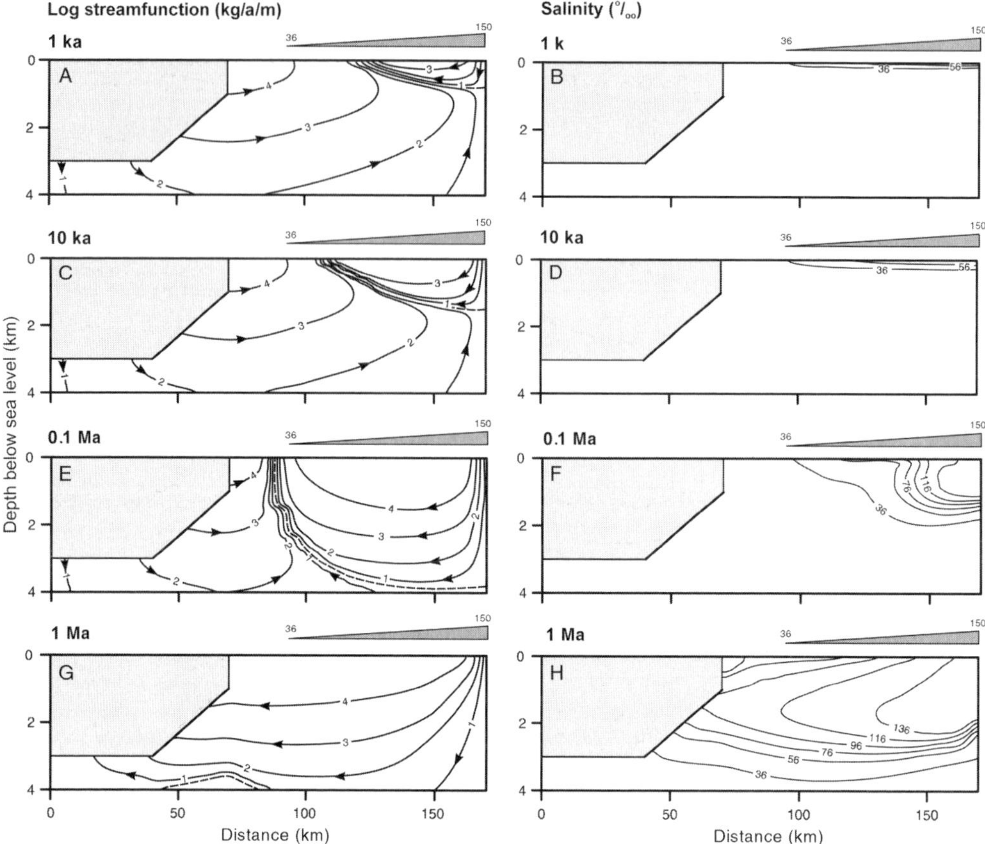

Fig. 16. Temporal evolution of geothermal and reflux circulation with platform-top brines concentrated up to gypsum saturation. After Jones *et al.* (2002).

salinity. Note, however, that due to numerical stability constraints, this permeability is some two orders of magnitude lower than that indicated by matching simulations of present-day geothermal circulation to measured temperature profiles.

The potential for reflux to dolomitize regional-scale Devonian platforms of the Western Canada Sedimentary Basin was investigated numerically by Jones & Rostron (2000) and Jones *et al.* (2003). Revisiting the analytical model of Shields & Brady (1995), Jones & Rostron (2000) demonstrated the effect of relaxing the unrealistic assumption of no flow across the platform top beyond the area of brine generation. Even with high-permeability anisotropy, most brines will discharge onto the platform top within 30 km of their source and cannot extend over the length scales of hundreds of kilometres suggested by the analytical model.

Jones *et al.* (2003) simulated dolomitization of the Grosmont platform of the Western Canada Sedimentary Basin. A significant aspect of this study was the incorporation of transient effects, including the evolving platform architecture, deposition of low permeability units and variation in the salinity of bank-top waters. The simulations showed, counter-intuitively, that latent reflux during cyclic platform drowning serves to increase circulation during succeeding periods of active reflux. This is due to the dilution of the brines beneath the platform top, increasing the density gradient driving subsequent reflux. As suggested by generic simulations, more permeable stratigraphic units serve to focus flow, although their effectiveness depends on their connectivity and position relative to brine generation (Fig. 19). Despite low-permeability shale breaks, there is also significant cross-formation flow towards these more connected flow paths. Free thermohaline

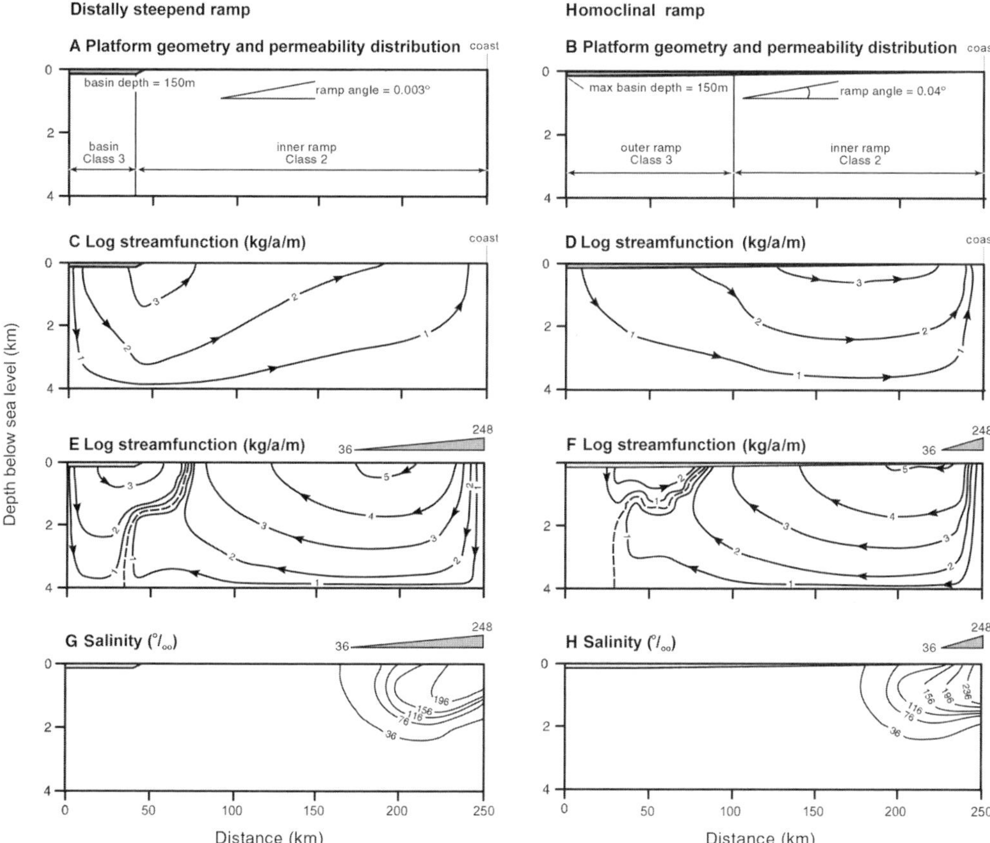

Fig. 17. Effect of platform geometry on circulation after 0.1 Ma with platform-top brines concentrated up to halite saturation, for a distally steepened ramp case (A, C, E and G) and a homoclinal ramp case (B, D, F and H). Note greater width of platform (250 km) and changes in the permeability and platform-top salinity distribution relative to our rimmed shelf case. Simulations use depth-dependent wackestone permeability (class 2 > class 3). After Jones *et al.* (2004).

convection cells were shown to develop within vertically extensive reef bodies (Fig. 19), distributing magnesium-bearing fluids that entered from the transmissive pathways.

Compaction-driven flow

Bethke (1985) modelled sedimentary-compactional flow in intracratonic basins such as the Michigan Basin using a Lagrangian frame of reference. He showed that during progressive deposition of 5 km of sediments over a period of 100 Ma, fluids tended to flow towards the basin margin because of the effective anisotropy of the 50% sandstone 50% shale sequence. Only in the upper 1 km of the sediments was flow upwards. He suggested that vertical flow was predominant when the ratio of the vertical to lateral permeability exceeded the ratio of the lengths of the vertical and horizontal pathways. Flow was cumulative towards the basin margin, but the maximum velocities were very small (6.3×10^{-11} m s^{-1}) and there was thus no deflection of the subsurface isotherms from the conductive case. Thermal expansion of pore fluids was shown to be unimportant compared to compaction in contributing to fluid flow. Interestingly, in the case of intracratonic basins where excess hydraulic potentials are not developed, fluid velocities scale with sedimentation rate and are independent of permeability. In contrast, in rapidly sedimenting basins, such as the Gulf Coast, substantial overpressure may develop in low-permeability sediments. For a sedimentation rate of 5 mm year^{-1}, fluid pressure is fixed at the lithostatic gradient for

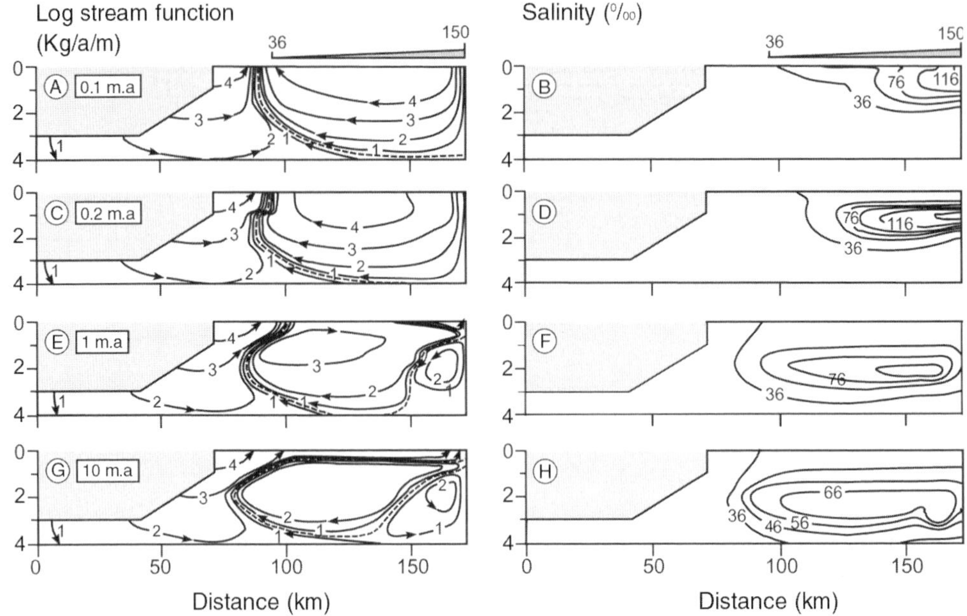

Fig. 18. Temporal evolution of groundwater circulation for up to 10 Ma after a single reflux event lasting 0.1 Ma with brines concentrated up to gypsum saturation. After Jones *et al.* (2002).

permeabilities of $<3 \times 10^{-18}$ m^2 and fluid velocity then scales with permeability and is independent of sedimentation rate.

Flow driven by the interaction within a basin between consolidation due to sediment deformation and topography was considered by Bitzer (1996) who identified two phases of consolidation fluid flow. Early during basin formation the entire basin is generally submarine and flow is driven only by sedimentary compaction. Simulated flow velocities from consolidation alone in a 3 km-deep basin were generally less than 6×10^{-13} m s^{-1}, with a maximum at shallow depth of 4×10^{-12} m s^{-1}, equivalent to almost half the sedimentation rate. Later, when subsidence and sedimentation slow or cease, the basin may become partially emergent and topographic flow dominates, giving flow rates several orders of magnitude higher. Christiansen & Garven (2003) modelled the interaction between compaction of sediments overlying a permeable crust within which geothermal heating drives free convection. They showed that free convection dominates with a high permeability crust, but compaction dominates where low permeability of the crust ($<10^{-13}$ m^{-2}) inhibits free convection. In mixed systems both flow systems operate and fluids move down from the sediment into the crust where, instead of closed cells developing, they flow laterally across the basin, albeit at low rates ($<10^{-13}$ m s^{-1}).

The study of Kaufman (1994) is unique in modelling the compaction-driven flow of pore fluids from basin-filling shales into a carbonate platform developed at the basin edge (Fig. 20). For homogeneous shales filling the 150 km-wide basin pore fluids move vertically towards the surface and fluid fluxes are very low (maximum 2×10^{-12} m s^{-1}). Vertical flow is maintained within the shales during deposition of a relatively permeable carbonate unit at the edge of the basin, as predicted by analytical models. However, on burial, fluids expelled from shales within 10 km of the forereef deposits move into the platform at a rate of up to 3×10^{-10} m s^{-1} (Fig. 20A). Shale permeability is an important control, with extension of the distribution and magnitude of lateral flow at lower permeabilities and higher permeability anisotropy. Further simulations investigated the effect of heterogeneous sedimentation in the basin down-dip of the carbonate platform, including mixed and interbedded carbonates and shales and a basin-wide permeable unit (Fig. 20B). The more permeable units focus compactional fluids from the shales over a distance of up to 200 km from the platform at rates up to 1.9×10^{-9} m s^{-1}. As shales continue to accumulate, burying the more permeable units, fluid flux through the

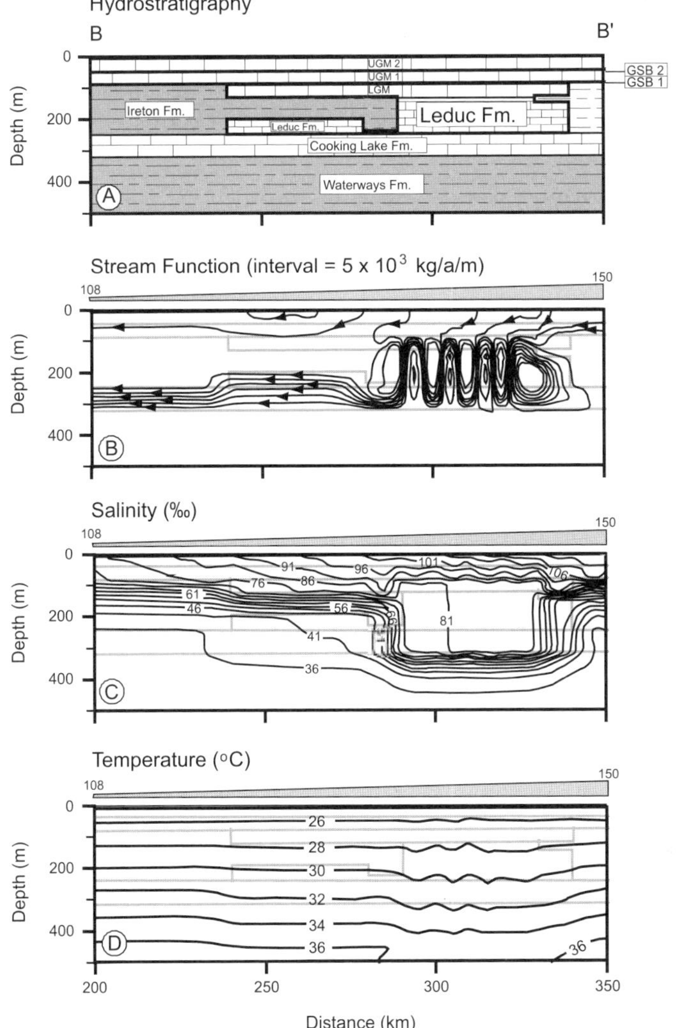

Fig. 19. Reflux from the Upper Devonian Grossmont platform of the Western Canada Sedimentary Basin. (**A**) Hydrostratigraphy of simulated section along a northern extension of the Rimbey–Meadowbrook reef trend (shales are grey, carbonates are white), (**B**) fluid flux, (**C**) salinity and (**D**) temperature after 100 000 years of reflux of brines up to gypsum saturation. Hydrostratigraphy is ghosted in (B) and (C). After Jones *et al.* (2003).

carbonates decreases as vertical flow dominates at shallower depths and consolidation of deeper shales approaches completion.

Tectonic compaction has been modelled by a number of workers, particularly with regard to its potential to form MVT deposits and influence migration of hydrocarbons. Ge & Garven (1994) simulated the effects of compression and vertical loading of thrust sheets up to 10 km thick. This demonstrated that tectonic compaction can drive flow at rates up to 3×10^{-7} m s^{-1} through the basinal aquifer system of a foreland basin. However, the resultant regional-scale flow remains subordinate to that resulting from the topographic relief of the orogen. Slightly lower flow rates of 10^{-9} and 10^{-8} m s^{-1} are predicted by Demming *et al.* (1990) and Garven *et al.* (1993) for compression-driven flow associated with thrusting, and simulations of accretionary prisms (Screaton *et al.* 1990) indicate flow rates of 10^{-9}–10^{-10} m s^{-1}.

However, flow rates decline very rapidly as stress dissipates (Garven *et al.* 1993) and extreme focusing of flow is required to generate

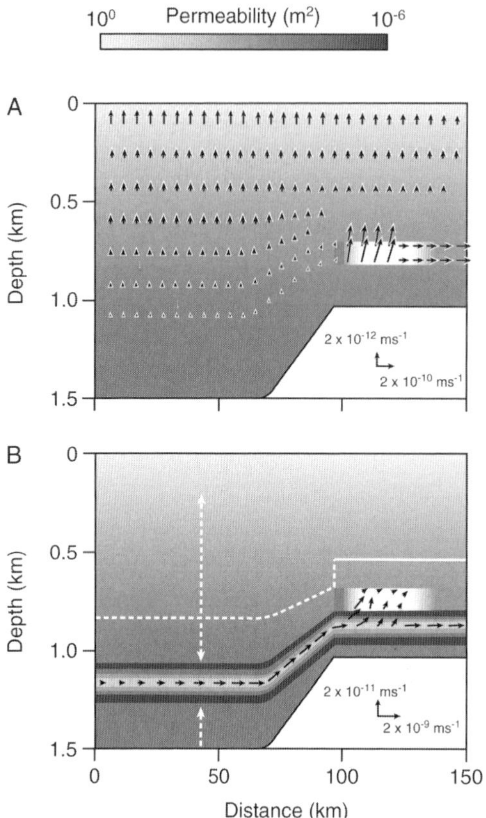

Fig. 20. Compaction-driven flow. (**A**) 150 m-thick platform–reef complex passing laterally into forereef and shales, within shale-filled basin. (**B**) 50 m-thick basin-wide sandstone unit below the platform–reef complex. Solid arrows show flow field; dashed white arrows show flow direction in areas of very low flow; shading shows distribution of horizontal permeability. Anisotropy is 2.5 in carbonates and sandstone, 100 in shales. After Kaufman (1994).

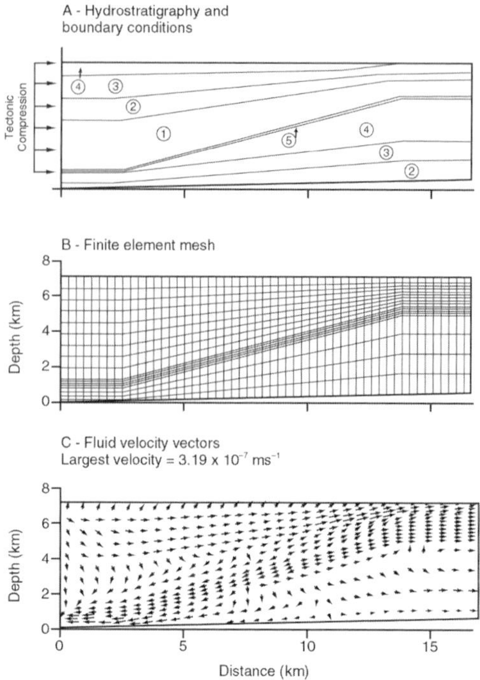

Fig. 21. Cross-section of the McConnel Thrust, Canadian Rockies. (**A**) Hydrostratigraphy and model boundary conditions; (**B**) finite-element mesh; and (**C**) fluid velocity vectors 5 years after an instantaneous horizontal stress of 300 MPa was applied to the upper part of the left (eastern) boundary. After Ge & Garven (1994).

the fluid inclusion temperatures observed in dolomite cements (Demming *et al.* 1990) in the absence of a 'thermal blanket' (Nunn & Lin 1997). Furthermore, even for such a large-scale system, the total volume of fluid flow is limited. For example, Ge & Garven (1994) estimate some 10^5 m^3 of fluids were expelled per metre width of section from the McConnell Thrust in the Canadian Rockies over tens to hundreds of years, with an average 100 m of thrust movement (Fig. 21).

Comparison of flow rates

Figure 22 compares rates of fluid flux predicted from numerical modelling of groundwater circulation associated with different drives. The values given are those for maximum flow rates and provide an order of magnitude indication only. Such comparative tables should be used with great care. In this case, rates are compiled from a range of sources, including simulations of generic carbonate platforms and individual case studies, and they are not exclusively from carbonates. The studies use a variety of geometries, boundary conditions and permeabilities that may give very different flow rates and distribution. However, several of the important flow systems have been modelled by our group at Bristol using standardized permeability–depth relationships and anisotropy that facilitate comparison.

As anticipated, permeability is the major control on flow. Thus the considerable range of fluxes for forced thermal convection reflects in part the wide range of permeability scenarios modelled. Boundary conditions are also important. For example, the maximum fluid flux ranges over two orders of magnitude from mesosaline

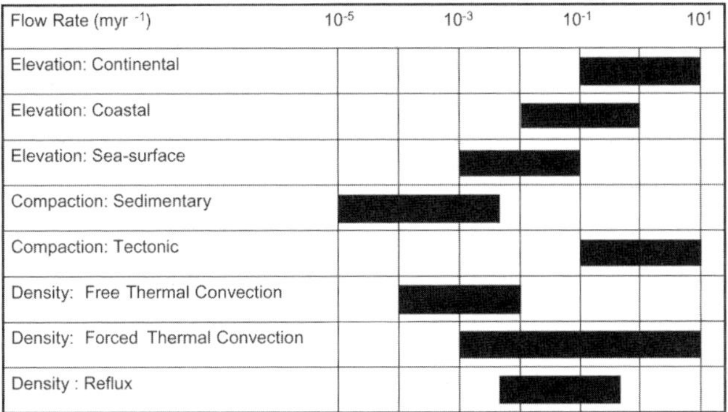

Fig. 22. Summary of maximum flow rates simulated for different hydrological drives.

to hypersaline reflux for a given permeability distribution. The maximum groundwater flux also gives no indication of the temporal persistence or spatial extent of the different flow systems. For instance, although topographically driven flow and tectonic compaction both give rise to similarly high maximum flow rates, the total flux from compaction may be limited by the finite volume of compactional fluids available and the brief duration of the excess pressure. Similarly, although reflux can give rise to high flow rates, this system is dependent on maintenance of very specific conditions on the platform top. Development of elevated salinity on the platform top is highly dependent on platform architecture and on amplitude and frequency of sea-level variations. In contrast, both submixing-zone saline circulation and particularly forced convection are likely to be active for much longer parts of the sea-level cycle.

However, groundwater flux is only one of the critical parameters controlling the supply of magnesium. The concentration and availability of magnesium in the circulating fluids is also important. Thus, the high Mg concentration of reflux brines increases their potential for dolomitization, while, although topographically driven circulation of meteoric waters generates high groundwater fluxes, the rates of magnesium supply are much lower because of the low magnesium concentration. We consider this further in the next section of the paper.

Modelling dolomitization

Mass-balance techniques

The volume of rock that can be dolomitized within a given time can be derived using mass-balance considerations from the volumetric fluid flux, the magnesium concentration of the fluid and the mass of magnesium required to dolomitize a unit volume of limestone (Land 1985; Hardie 1987). The latter is dependent on the porosity and the dolomitization exchange efficiency. Some 350 kg of magnesium is required to dolomitize 1 m^3 of calcite (Cooper & Tindall 1994) and this equates to 237 m^3 of seawater (Mg concentration 1.35 kg m^3) assuming an exchange efficiency of 100%. For a porosity of 40%, this is equivalent to almost 600 pore volumes of seawater. As the magnesium concentration in the fluid increases, this volume reduces significantly to only 52 m^3 of brine at gypsum saturation. Note, however, that many authors invoke substantially lower exchange efficiencies to account for geochemical equilibria. For example, Simms (1984) assumes only 10% of available magnesium is consumed by dolomitization, implying some 6000 pore volumes of seawater are needed (also compare Montañez & Read 1992 and Lumsden & Caudle 2001 for the Ordovician Upper Knox Group). Exchange efficiencies increase with salinity and have been estimated at 36% for anhydrite-saturated brines and 47% for halite-saturated brines (Shields & Brady 1995).

One of the first applications of mass-balance techniques was by Deffeyes *et al.* (1965) to estimate whether reflux was capable of providing sufficient magnesium to dolomitize Plio-Pleistocene subtidal sediments on Bonaire. More recently, a number of authors have undertaken mass-balance calculations to assess the dolomitizing potential of a variety of fluid flow systems (e.g. Sears & Lucia 1980; Simms 1984; Machel & Anderson 1989; Wilson *et al.* 1990; Montañez & Read 1992; Amthor *et al.* 1993; Kaufman 1994; Shields & Brady 1995; Wendte *et al.* 1998; Lumsden & Caudle 2001). However, there are significant uncertainties in magnesium mass-balance calculations, both regarding the amount of dolomite formed and the dolomitization potential of available fluids.

The starting point, the amount of a particular phase of dolomite present in the formation, can itself be problematic. This is particularly the case where dolomitization is incomplete or involves multiple dolomite generations, where there is uncertainty about the lateral extent of the dolomite body into adjacent formations with complex stratal geometries and shales (Wendte *et al.* 1998) or combinations of these. It is also important to include the volume of porosity-reducing dolomite cements in the calculation (compare Montañez & Read 1992 and Lumsden & Caudle 2001 for the Ordovician Upper Knox Group).

Of more interest in this context are uncertainties about the volumes of dolomitizing fluid available, and its composition. Thus, Wendte *et al.* (1998) calculated residual brine volumes from estimates of evaporite thickness, but could not be certain that there was significant loss of evaporites by dissolution or that evaporite precipitation was 100% efficient. The estimated volume of brines available was thus probably a minimum, whereas the concentration of magnesium would be a maximum. It is also important to remember that there have been quite substantial variations in the Mg concentration of seawater through time. For instance, Holland & Zimmermann (2000) estimate an 18 mmol kg^{-1} increase in average oceanic Mg since the Eocene. However, Stanley & Hardie (1998) do not believe that such variations are significant in explaining the variations in the amounts of global dolomite. In many situations the duration of the phase of active dolomitization, and thus total flux carried by any fluid circulation system, is also poorly constrained (see, for example, Montañez & Read 1992 where rates were compared with the duration of eustatic cycles).

In general, mass-balance studies ignore spatial variations in fluid flux, which as we have seen differ systematically with the drive(s) for fluid flow and also permeability. This can lead to inaccurate and misleading interpretations, as shown by Jones & Rostron (2000) for the Devonian of the Western Canada Sedimentary Basin. These authors demonstrated that the mass-balance calculations of Shields & Brady (1995), which assume spatially uniform fluid flux and no flow across the platform top beyond the area of brine generation, overestimate the potential for regional-scale dolomitization. Even with high permeability anisotropy most brines would have discharged onto the platform top within 30 km of the brine source and would not have extended over the length scales of hundreds of kilometres suggested by Shields & Brady (1995). Jones & Rostron (2000) computed an advancing reaction front between dolomite and unaltered limestone within a series of stream tubes between groundwater flow streamlines (Fig. 23). This suggested that an area within 15 km of the brine source could be dolomitized within the 16 Ma available, but that much greater times would be needed to dolomitize larger areas (in excess of 10^4 Ma for >30 km from the source). Thus, brine reflux cannot account for the Devonian dolomites in the Western Canada Sedimentary Basin unless higher permeabilities or brine concentrations are used. This disparity illustrates the difficulty of adequately parameterizing models of palaeoflow systems.

Jones (2000) used this stream tube approach to compare the distribution and magnitude of fluid flux and the resultant patterns of dolomitization for different flow systems. Simulations of an isolated steep-sided platform indicate that geothermal convection would generate an asymmetric dolomite body, thickest at the margin and thinning into the interior. In contrast, both reflux and flow driven by differential sea-surface elevation would form elongate tabular dolomite bodies at very shallow depth where the permeability is highest. However, with continuing reflux the dolomite body would eventually resemble that predicted for geothermal convection. Only submixing-zone circulation would generate a dolomite geometry sufficiently distinct to be diagnostic, thinning inland from a maximum seaward of the periphery of the island lens. This study suggests, in contradiction to the widely held paradigm (Wilson *et al.* 1990; Whitaker & Smart 1993), that in many cases the shape and distribution of a dolomite body may not be diagnostic of the drive for fluid flow.

This problem is exacerbated by several other factors: more than one drive may be active, giving hybrid circulation systems; boundary conditions will not remain constant, causing variation in the flow; and potential feedbacks between permeability and the extent of dolomitization will affect the distribution of fluid flux. For all circulation systems, flow is strongly focused at shallow depth because of the significant reduction of permeability with depth. More permeable layers will also focus flow, enhancing the rate of dolomitization, whereas lower permeability zones act as barriers, restricting dolomitization. Such patterns are widely recognized in the rock record, with preferential dolomitization within and around open fractures and conduits

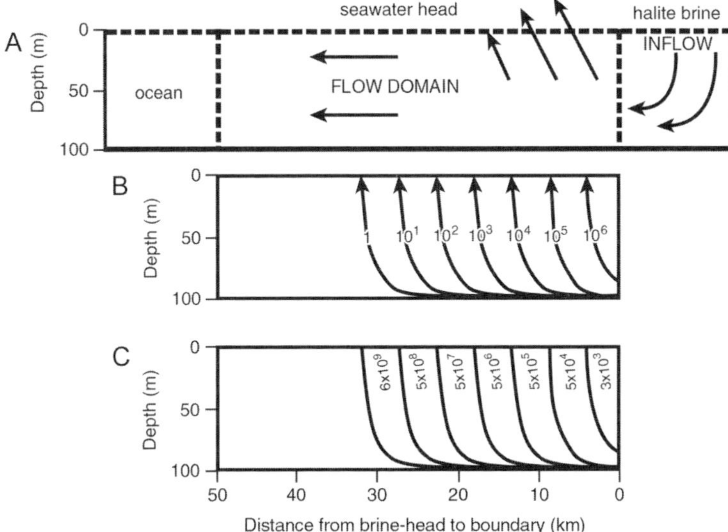

Fig. 23. (**A**) Hydrogeological model of reflux with halite-saturated brines generating a differential head of 17 m (based on the Western Canada Sedimentary Basin). (**B**) Streamlines (kg $year^{-1}$ m^{-1}) assuming permeability of 10^{-10} m^2 and anisotropy of 1000. (**C**) Corresponding distribution of calculated dolomitization time (years) for individual streamlines assuming 7% porosity and 47% exchange efficiency. After Jones & Rostron (2000).

(Wilson *et al.* 1990; Yao & Demicco 1997; Melim & Scholle 2002).

The major limitation of the mass-balance approach is that it ignores geochemical reaction and kinetic controls on dolomite nucleation and growth rate, simply assuming an advancing front at which all magnesium is consumed. The system is thus implicitly transport limited, whereas in some cases, and especially at low temperature, dolomitization may be reaction rate limited. The inclusion of a low exchange efficiency fails to address the critical issue that rates will vary spatially with temperature, salinity and other geochemical controls. In the absence of more sophisticated techniques, mass-balance calculations can provide a source of minimum estimates of fluid volume required for dolomitization, but the uncertainties mentioned above should always be considered.

Reaction path calculations

An indication of the basic behaviour of any geochemical system can be provided by reaction path calculations. These involve the progressive evolution of a closed geochemical system at a rate controlled by the reaction kinetics and evolving solution composition. Wilson *et al.* (2001) used reaction path calculations to demonstrate the controls on dolomitization of calcite by seawater in the absence of fluid flow and insulated against heat transfer. They assumed that the dolomitization process involved dissolution of calcite and precipitation of dolomite (Machel & Mountjoy 1986), thus the rate of dolomitization reflects the rate of dolomite precipitation, and calcite dissolution follows according to equilibrium thermodynamics. The rate of dolomite precipitation is described by the kinetic expression of Arvidson & Mackenzie (1999) extrapolated to temperatures less than 100 °C, and ignoring the effects of organic reactions and sulphate reduction (Wilson *et al.* 2001).

In baseline simulations at 50 °C (Fig. 24A) the rate of calcite dissolution exceeds that of dolomite precipitation. This generates a small increase in porosity consistent with the volumetric calculations of Weyl (1960). Dolomitization is thus accompanied by a slight calcium enrichment of pore fluids consistent with the calcium-rich brines observed in many sedimentary basins, and a decrease in pH due to removal of some carbonate from the fluid. Wilson *et al.* (2001) suggest that reactive surface area is an important control on dolomitization rate. For heterogeneous sediments, this is affected most significantly by the degree of preferential flow through more permeable pathways. There also appears to be a sharp increase in reaction rate with temperature, due almost entirely to the increase in the kinetic rate

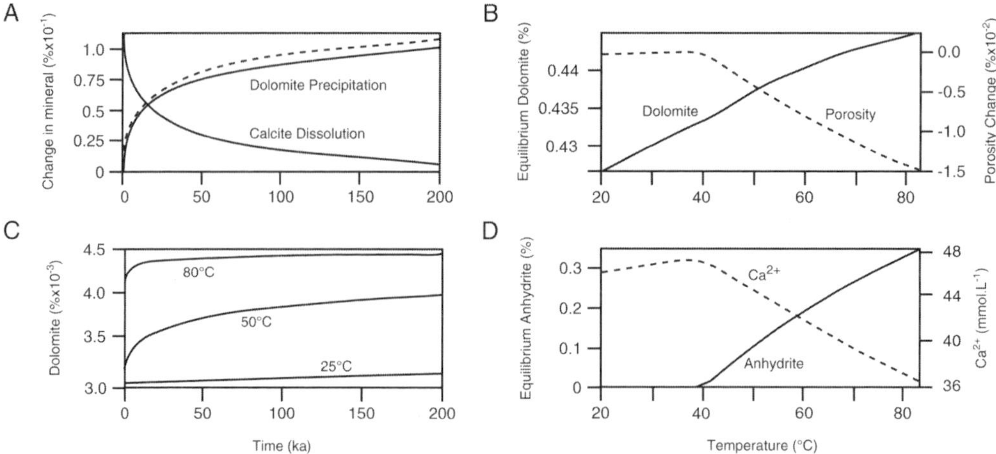

Fig. 24. Reaction path calculations (after Wilson *et al.* 2001). (**A**) Change in volume of calcite and dolomite over time at 50 °C and 104 cm^2 g^{-1} specific surface area (dashed line mirrors calcite volume to show porosity generation). (**B**) Volume of dolomite calculated at equilibrium and porosity change (dashed line) as a function of temperature. (**C**) Volume of dolomite over time at 25, 50 and 80 °C. (**D**) Volume of anhydrite calculated at equilibrium and equilibrium aqueous calcium concentration (dashed line) as a function of temperature. After Wilson *et al.* (2001).

constant (Fig. 24C). Equilibrium calculations indicate the final state towards which the kinetic system progresses at a given temperature (Fig. 24B & D). The equilibrium calcium concentration increases until the fluid becomes saturated with respect to anhydrite (at temperatures >39 °C), driving further calcite dissolution and countering the porosity increase associated with dolomitization. This model of anhydritization synchronous with dolomitization, sourcing sulphate from seawater, supports observations by Machel (1986) of anhydritization of the Upper Devonian Nisku build-ups in Alberta.

Phillips (1991) and Raffensperger (1996) broadly classify transport-controlled reactions occurring in sedimentary basins into three types: (1) *isothermal reactions*, which occur at fronts that separate mineralogically distinct zones and propagate in the direction of flow; (2) *gradient reactions*, which result from movement of fluid through temperature and pressure gradients, and give more progressive alteration; and (3) *mixing reactions*, which occur when two or more fluids of different compositions mix, commonly producing a zone of highly localized alteration. The 4% increase in dolomite volume between 20 and 80 °C shown in the equilibrium calculations (Fig. 24C) is the amount of dolomite that would precipitate along a flow path in a thermal gradient reaction. In contrast, for a reaction front process the kinetic reaction path calculations (Fig. 24A) indicate a potential 45% increase in dolomite volume. Thus, reaction path modelling indicates that more dolomite is likely to be precipitated by a reaction front than by gradient reactions.

Reactive-transport models

Coupled groundwater flow and reactive-transport models offer the potential to predict regional patterns and rates of mineral formation during circulation of groundwaters within carbonate platforms. This approach may use analytical solutions for simple systems, but, more generally, numerical simulations have been adopted. These are capable of describing more complex systems with realistic boundary conditions, and also enable prediction of the spatial distribution of secondary minerals such as dolomite. Reactive transport models have been used to investigate the role of groundwater flow for diagenesis in a range of settings including: cementation driven by geothermal convection in sandstones (Wood & Hewlett 1982); mineralization in rocks surrounding cooling plutons (Steefel and Lasaga 1994); unconformity-related uranium ore deposition (Raffensperger 1996); formation of Mississippi Valley type lead–zinc ores (Garven *et al.* 1999); and porosity evolution in coastal mixing zones in carbonates (Sanford & Konikow 1989).

The simplest form of reactive transport models are 1-D models that describe the

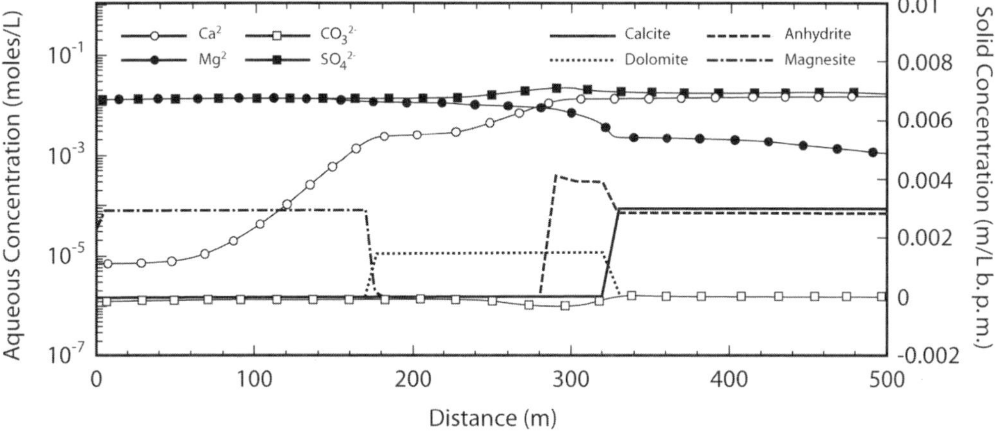

Fig. 25. One-dimensional simulation of reaction front evolution for an isothermal flow path and domain comprising the mineral phases calcite, dolomite, magnesite and anhydrite, with calcite and anhydrite originally present. The percolating fluid is in equilibrium with magnesite only. Longitudinal dispersivity 10 m. After Raffensperger (1996).

transmission of reaction fronts along an isothermal flow path. Such models normally make an assumption of local equilibrium (i.e. instant precipitation or dissolution depending on local fluid composition) that is only valid at a large length scale and timescale. For instance, Raffensperger (1996) demonstrates the potential complexity of 1-D reaction fronts. He defines a domain comprising the mineral phases calcite, dolomite, magnesite and anhydrite, with quartz (unreactive here), calcite and anhydrite originally present. The percolating fluid is in equilibrium with magnesite only. Rather than a simple dissolution front for calcite and associated precipitation front for magnesite (a result of the increased carbonate content of the fluid), dolomite is formed as an intermediate product and occupies a distinct reaction zone (Fig. 25). This zone increases in length with time as the two reaction fronts defining it advance at different rates. An additional complexity is that anhydrite is predicted to precipitate just upstream of the calcite dissolution front, but dissolves completely a few metres after the front (Fig. 25).

The earliest 2-D reactive transport model of dolomitization was that of Wood (1987) for gradient reactions. This was based on the seminal work of Wood & Hewlett (1982), who modelled quartz cementation and dissolution when saturated pore fluids migrate across a temperature field. Within a confined carbonate unit that pinches and swells, topographically driven flow was predicted to drive dolomitization in areas of upward flow (due to cooling) with calcite precipitation in areas of downward flow (due to warming). Reaction rates were assumed to be proportional to the angle at which flow lines crossed the isotherms, which were specified to be horizontal, with no advective heat transfer between the isothermal upper and lower boundaries. Wood & Hewlett (1982) also assumed that the mass-transfer coefficients for dolomite and calcite precipitation are similar in magnitude but of opposite sign, and that there was no limitation on the magnesium supply in their topographically driven system. More recent work suggests that these assumptions were rather unrealistic. Wood & Hewlett (1982) also introduced the concept of a rock alteration index (RAI) as a basis for prediction of the direction, distribution and magnitude of gradient reactions. This is determined from the product of modelled groundwater flux and temperature gradient (Phillips 1991). The RAI can be useful for comparison of different circulation systems. For instance, Raffensperger & Vlassopoulos (1999) show that, for mixed convection systems, the pattern and rate of rock alteration are relatively unaffected by the magnitude of the forced convective flux until the latter approaches that of the free convective circulation. Thereafter, both the overall rate and the degree of spatial differentiation of maxima and minima in the RAI are considerably reduced.

An early attempt to incorporate kinetics used the US Geological Survey (USGS) groundwater and solute transport model, SUTRA (Voss 1984), modified to incorporate heat and

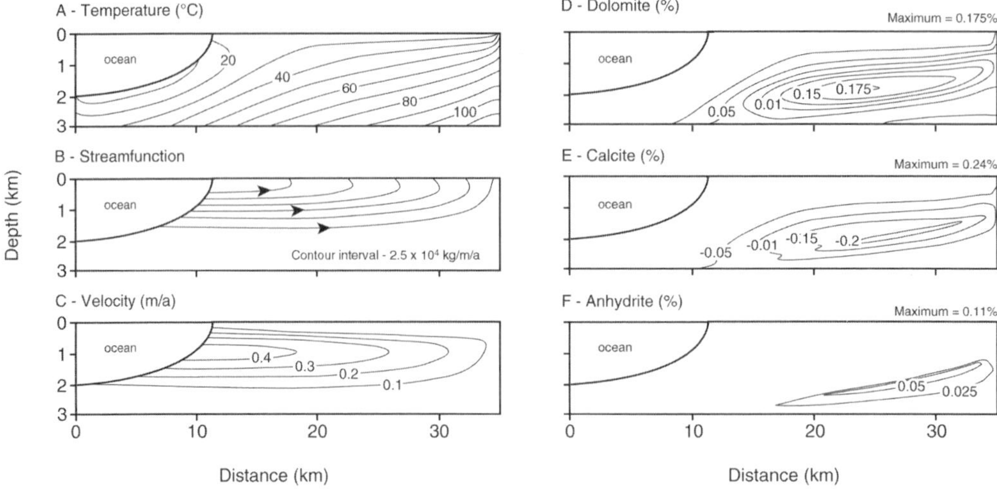

Fig. 26. Reactive-transport simulations of geothermal convection. (**A**) Temperature, (**B**) streamfunction, (**C**) fluid-flow rate (maximum rate is 1.3×10^{-8} m s^{-1}); (**D**), (**E**) and (**F**) volume of dolomite, calcite and anhydrite after 100 000 years. Simulations use depth-dependent permeability representative of packstones–wackestones (maximum 7×10^{-8} m^2), with anisotropy of 1000. After Wilson *et al.* (2001).

multispecies transport, to model dolomitization by geothermal circulation of seawater in an isolated rimmed platform (Whitaker *et al.* 1993). The rate equation for dolomitization incorporated the relative abundance of calcite and dolomite, and the concentration of 'excess magnesium', defined as the difference between the actual magnesium concentration and that at equilibrium with respect to both calcite and dolomite. The latter is highly temperature sensitive, thus predicted dolomitization was critically dependent on groundwater temperature. The simulations illustrated the sensitivity of reactive transport models to the Damkholer number, the ratio of the chemical reaction rate to advective transport (Jones *et al.* 1997). At high Damkholer number the distribution of dolomite is a function of magnesium mass flux, and the dolomite body developed is an asymmetric wedge that is thickest at the platform margin and thins into the interior. This form is similar to the conceptual models of Wilson *et al.* (1990) and Whitaker & Smart (1993). At low Damkholer number kinetic factors dominate and dolomitization reflects temperature and/or salinity. Thus, the advective cooling of the platform margin by geothermal convection serves to limit dolomitization that is greatest in the warm platform interior.

More recently, Wilson *et al.* (2001) have simulated dolomitization by geothermal convection within an isolated platform, some 40 km across and 2 km deep, in which porosity and permeability decline with depth. The pattern of advective cooling predicted accords with previous simulations under similar conditions (Sanford *et al.* 1998; Jones 2000) and the maximum flow rate is 1.3×10^{-8} m s^{-1}. The model predicts dolomitization in a broad, quasi-stationary zone extending across the platform and coincident with the 60–70 °C isotherm (Fig. 26). Lower temperatures at shallower depths limit dolomitization in the upper parts of the platform despite ample fluid flux because of the sluggish reaction rate, while at greater depths fluid flux is limiting. Dolomitization is accompanied by calcite dissolution, with only a minor change in porosity, and also anhydrite precipitation at temperatures >80 °C. The maximum rate of dolomitization is 1.35% Ma^{-1} after 50 000 years, suggesting that in excess of 60 000 years would be required to dolomitize a calcitic platform of the specified porosity and permeability by geothermal circulation.

The rate and distribution of dolomitization appears to be critically dependent on permeability, both through its effect on mass transport and through feedback to groundwater temperature via heat transport (Fig. 27). In less permeable platforms dolomitization is transport-controlled and occurs only near the margins, while for high-permeability platforms the lower temperatures cause dolomitization to be focused in the (warmer) centre. The reactive surface area is also important, with reduced surface area slowing dolomitization and shifting

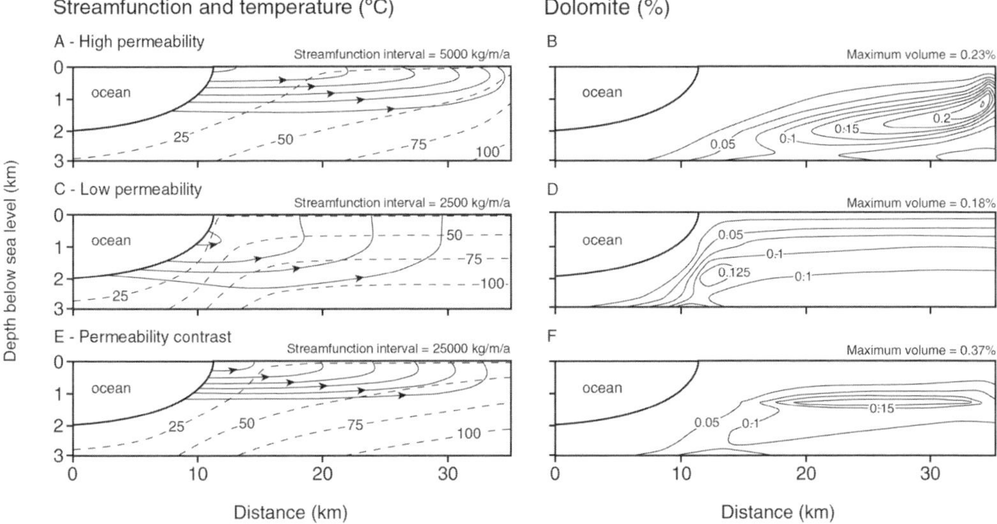

Fig. 27. Temperature and stream function (left) and volume of dolomite after 100 000 years (right) predicted by reactive-transport simulations of geothermal convection. In all cases permeability is depth-dependent, anisotropy is 1000, and flow is inward and upward from basin to platform top. (**A**) Low-permeability representative of mudstones (maximum 9×10^{-9} m^2). (**B**) High-permeability representative of grainstones (maximum 2×10^{-5} m^2). (**C**) 1.2 km of moderate-permeability packestones–wackestones. After Wilson *et al.* (2001).

the zone of reaction into deeper, warmer parts of the platform. Dolomitization is characteristically diffuse and slow in simulations with a uniform sediment type; complete dolomitization requiring at least 60 Ma. However, more realistic simulations incorporating beds with higher permeability and reactive surface area show focused dolomitization at rates twice those of the baseline simulation. The complex shape and location of dolomitized areas predicted by these simulations suggest that great caution is required when inferring fluid flow from field observations of dolomite geometry.

As predicted by reaction path calculations, geothermal dolomitization is accompanied by precipitation of anhydrite along the 80 °C contour and at greater temperatures (Wilson *et al.* 2001). Simulations predict rapid dolomitization, with up to 0.11% anhydrite by volume forming within 100 000 years, although rates may be lower where significant subsurface sulphate reduction occurs. This process of 'anhydritization' cogenetic with dolomitization is observed in the rock record and can significantly reduce porosity and permeability (Kendall & Walters 1978). In the Upper Devonian Nisku build-ups of Alberta void-filling anhydrite cements are present where dolomitization exceeds 70–80 vol%, but are absent in limestone areas (Machel 1986), although these anhydrites may have recrystallized from gypsum.

An important feedback, which has not been incorporated into these simulations (where the percentage dolomite formed is very low), is that between the volume of the changing mineral phases and porosity. This has been included in some simulations, such as that of Lee (1997) who simulated the diagenetic changes resulting from topographically driven circulation in the Illinois Basin during Permian times following uplift of the Pascola arch. Late dolomite cements precipitated in the deep aquifers, and on the northern flank of the basin where warm fluids ascended at rates of 10° m a^{-1} and cooled. The rate of precipitation was sufficiently rapid that the volumes of dolomite cements typically seen in the Ordovician carbonate strata (*c.* 1%) could have formed in as little as 1 Ma. Although the porosity–permeability changes here were probably not large enough to perturb the flow system, they suggest that the cementation was sufficient to inhibit subsequent migration of petroleum. More significant changes in the flow system were demonstrated by Raffensperger (1996) for free convective circulation in sandstones with quartz cementation. Here changes in porosity substantially disturbed the original pattern of groundwater

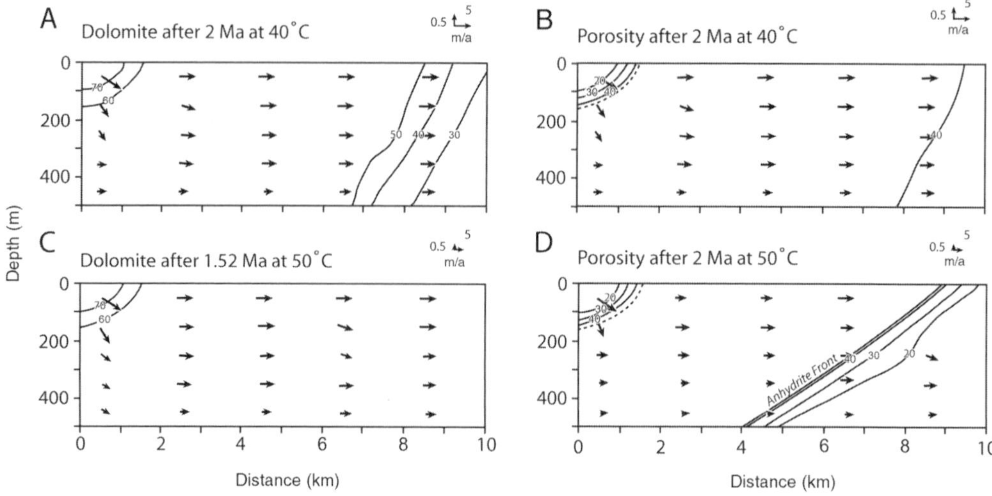

Fig. 28. Reactive-transport simulations of reflux. (**A**), (**B**) Distribution of dolomite and porosity at 40 °C after 2 Ma. (**C**), (**D**) Distribution of dolomite and anhydrite cement at 50 °C after 1.52 and 2 Ma, respectively.

convection, eventually causing the development of two separate cells where only one had been present initially.

Feedbacks between diagenesis, permeability and flow are also included in our recent reactive transport simulations of reflux dolomitization (Fig. 28) using the Xt2 code (Bethke 1997). We specify the same rate law and reactive surface area as the thermal simulations of Wilson *et al.* (2001). The specified carbonate flow domain is 10 km wide and 500 m thick, with an initial homogeneous porosity of 35%, horizontal permeability of 10^{-11} m^2 and a mineralogy that is 100% aragonite. The platform is filled initially with seawater, and gypsum-saturated brine (128‰ salinity and 93.1 mmol Mg^{2+}, based on salina water from West Caicos, G. D. Jones unpublished data) is injected in the upper left cell. The rate of reflux is initially specified as 6×10^{-8} m s^{-1} based on previous numerical simulations of reflux (Jones 2000) by adjusting both the hydraulic gradient and the rate of brine injection. A 1% change in porosity is assumed to result in a permeability change of 2.8×10^{-13} m^2. The upper and lower boundaries are specified as no flow.

These preliminary simulations show a broad (2–3 km-wide) and near-vertical dolomitization front that advances through the platform at the relatively fast rate of *c.* 3 km Ma^{-1} at 40 °C. After 2 Ma there is complete dolomitization over the full 500 m depth and 7 km of the flow domain extending from the source (Fig. 28A). Replacement dolomitization generates up to 8% porosity after 2 Ma. However, within 1.5 km of the brine source the precipitation of dolomite cements occludes up to 20% porosity, replicating a process described by Saller & Henderson (1998) as 'overdolomitization' (Fig. 28B).

The rate of advancement of the dolomitization front increases with temperature to approximately 6 km Ma^{-1} at 50 °C, such that the flow domain is completely dolomitized in 1.5 Ma (Fig. 28C). At temperatures greater than 40 °C porosity-occluding anhydrite cements precipitate ahead of the advancing dolomite front, but this anhydrite is completely dissolved during replacement dolomitization. The zone of anhydrite cements is broad (up to 5 km) and the porosity occlusion is moderate (0–20%) for approximately 1 km behind the anhydrite front and severe (predominantly 25%) thereafter (Fig. 28D). Thus, we might rarely expect to see large volumes of anhydrite cement associated with reflux dolomites from the geological record because of the high susceptibility of anhydrite to post-emplacement dissolution, particularly where the dolomitization front has swept through rocks cemented with anhydrite.

Although simple, these preliminary reactive transport simulations suggest that reflux is a relatively efficient dolomitization mechanism that can generate porosity during replacement. However, porosity is occluded in front of and behind the zone of replacement dolomitization by anhydrite cementation and 'overdolomitization', respectively. For instance, precipitation of anhydrite cements ahead of the dolomitization front deceases the flow rate from 6×10^{-8} to 3×10^{-9} m s^{-1} for the 50 °C simulation (Fig. 28C & D). There are, however, considerable challenges in describing the evolving relationship between porosity and permeability during precipitation of cements, as these may preferentially occlude pore throats having a much more significant impact on permeability than indicated by the porosity reduction. Similarly, where dolomitization is not fabric selective, whole-scale reorganization of the pore network will occur. The normal assumption of a simple and unchanging relationship between porosity and permeability (Steefel & Lasaga 1994; Lee 1997) is thus probably not warranted.

Discussion

Mathematical models provide an important complement to field and laboratory based investigations of dolomitization. Their particular utility lies in their ability to reproduce processes that occur at relatively slow rates and over large time and length scales (Person *et al.* 1996), and to provide quantitative estimates of rates of dolomite formation and distribution. Mathematical models of dolomitization range from simple but very useful calculations of mass budgets, through analytical models of fluid flow to sophisticated fully coupled reaction-transport models. The choice of model will depend on the nature of the questions posed and the degree to which the system can be constrained with respect to critical parameters. However, increasingly the general availability of sophisticated and well-tested numerical models of groundwater flow, solute transport and geochemical reaction mean that these are the models of choice.

Analytical solutions for simplified representations of carbonate platforms are often undervalued. They can provide both rapid estimates of the relative significance of fluid-flow mechanisms for a given hydrogeological situation (e.g. Kooi 1999), and also significant insights into the controls on, and behaviour of, fluid-flow systems (e.g. LeClerc *et al.* 2000). Dimensional analysis offers the ability to quantify in a generalized manner the response of the system to changes in platform geometry and boundary conditions, and to rock and fluid properties, and some sophisticated schemes have been developed (for instance that of Fan *et al.* 1997 for playa lake basin reflux). In particular, dimensionless numbers such as the Raleigh number can be used to characterize the nature of the specific system under investigation, and recent studies have significantly improved our understanding of the determination of such numbers (Sharp *et al.* 2001).

However, numerical techniques also offer the ability to investigate more complex hydrogeological scenarios that may not be amenable to analytical modelling, but which characterize many circulation systems resulting in dolomitization. Because of the variety of elevation-head, density and pressure drives for fluid flow in carbonate platforms, the ability to simulate several potential drives simultaneously is important. Thus, numerical schemes that couple mass, heat and solute transport must be employed, for instance, in describing the competitive interactions between thermal convection and reflux (Jones *et al.* 2002, 2004). Such systems are intrinsically non-steady state. Thus single snap shots in time as used in some early studies may be misleading in both the extent and the magnitude of circulation predicted. Furthermore, the circulation may continue to evolve once the active drive has ceased, as demonstrated by Jones *et al.* (2002) for latent reflux after cessation of brine generation on the platform top. Thus, consideration of time-varying boundary conditions is a difficult but important requirement of modelling studies. For example, a minor rise or fall in relative sea-level can transform a system from one with platform-top brines to complete marine inundation, or exposure and development of meteoric conditions. There are also simultaneous and subsequent changes in the architecture of both the carbonate platform and wider sedimentary basin, with further sedimentation and erosion. There is a particular requirement to model dolomitization in evolving platforms and basins where rates of dolomitization are relatively slow. To date, models of carbonate platforms that incorporate such changing model domains have done so in a rather simplistic manner (Jones *et al.* 2003, 2004), although work on other more dynamic systems is considerably more sophisticated, for instance Person & Garven's (1994) study of fluid flow in rifting basins.

It is generally thought that on burial carbonates are less affected by changes in boundary conditions and that fluid flow rates are considerably reduced. This suggestion is supported by the much lower total fluid fluxes modelled in response to sedimentary compaction and free convection. However, modelling has also demonstrated that for platforms that remain in connection with surrounding seawater, processes such as reflux and forced thermal convection can drive flow to many kilometres depth. This confirms geochemical indications of the prevalence of open-system conditions in platforms drilled by ODP (Ge *et al.* 2003). Furthermore, many studies of ore genesis demonstrate that large fluid fluxes can result from topographically driven circulation generated by uplift, especially where flow is focused in permeable fairways.

A major purpose of groundwater flow models is to provide spatially distributed numerical estimates of fluid flux that can be used in mass-balance calculations. As would be expected from first principles, permeability is a critical control on the determined flux, the exception being compaction-driven flow when conditions remain hydrostatic. Indeed, Budd (1997) suggested that the phases of dolomitization observed in many Plio-Pleistocene carbonates are a result of enhanced seawater circulation following meteoric permeability enhancement during the preceding sea-level lowstand(s). Parameterization of permeability is often difficult, particularly where later diagenetic changes (including dolomitization) have occurred (Bethke 1989). A particular problem for reliable estimation of fluid flux is the recognition and characterization of high permeability fairways that focus flow and give rise to high water–rock ratios. Such fairways may be of depositional origin and can be mapped by subsurface studies, but many carbonates also have significant secondary fracture porosity and permeability. This is much more difficult to characterize, as large-scale rather than core-sample tests are needed for measurement, and these are rarely available from the deep subsurface.

Finally, dolomitization and other diagenetic reactions cause changes in the porosity and permeability of the rock. There is therefore a need to incorporate the feedback between fluid flow and permeability in model studies, but to date such studies (e.g. Steefal & Lasaga 1994) have been very limited in number, and fail to incorporate the complex control of permeability by evolving pore network and geometry. In simulating fluid flow during Mississippi Valley-type (MVT) mineralization in SE Missouri, Appold & Garven (1999) parameterized initial porosity and permeability from the careful backstripping of cements and analysis of the petrophysical evolution of late burial dolomites of Gregg *et al.* (1993). The higher premineralization porosity of the dolomites was a significant control on the simulated flow, highlighting the importance of careful petrographic work to unravel the paragenesis of cementation, dissolution and recrystallization of dolomites (Gregg 2004).

Whilst dolomitization requires fluid flow to source reactants and remove products, the extent and rate of reaction and thus the distribution of dolomite is controlled by the reaction thermodynamics and kinetics. Given the strong temperature dependence of the dolomitization reaction rate, and the considerable potential for advective heat transport in the relatively permeable carbonates, it is clear that the coupling of geochemical and groundwater–heat and solute-transport models is needed to adequately predict the distribution of dolomite. Such models have been widely and successfully employed in the study of ore bodies (e.g. Garven *et al.*, 1999; Swenson & Person 2000), but the comparable work on dolomitization is more limited. One particular difficulty is that the kinetic rate constant for dolomitization has only been obtained at high temperatures (>100 °C), and there is thus considerable uncertainty in extrapolation to the lower temperatures dominant before deep burial. Further insight may be gained from the application of Nordeng & Sibley's recent (2003) experimental and petrographic work on dolomite kinetics.

Dolomitization within large-scale fluid flow systems is a product of a complex set of hydrological, mechanical, thermal and chemical mass-transfer and reaction processes. The application of coupled models of these processes can reveal qualitatively new effects and system behaviours (Chen *et al.* 1990), as well as quantifying known ones. However, two factors hamper our ability to conduct numerical experiments at the basin scale (Raffensperger 1996). First, computer processing time and memory limitations have proved a major barrier in the past (Garven *et al.*'s 1999 fully coupled heat-, solute- and reactive-transport models of ore emplacement in the Viburnum trend took 6 days to run). Second, parameterizing the increasingly complex models is challenging, and sufficiently detailed spatial data on physical parameters such as porosity and permeability and the distribution and chemical composition of mineral phases are generally not available. Whilst the former limitation is rapidly diminishing, it is likely the latter will persist. Indeed, Bethke (1989) considers the major limitations to modelling palaeohydrology are the difficulties in back-stripping porosity, allocating permeability and defining boundary conditions.

Although the primary focus of this review is dolomitization, the fluid-flow systems discussed also have wider implications for carbonate diagenesis. Related processes include marine calcite cementation (Harris *et al.* 1985), shallow aragonite and calcite dissolution by refluxing brines (Sun 1992), and deep dissolution below the aragonite–calcite compensation depth (Saller 1984). An understanding of patterns and rates of flow and their effects on temperatures is also central to issues of generation and migration of hydrocarbons (Bethke *et al.* 1991; Person *et al.* 1995) and, with the increasing utilization of deep groundwaters for waste disposal and cooling waters, has direct application to management of groundwater resources (Shinn *et al.* 1994). Fluid flow is of particular importance in platform carbonates because of their generally high permeability, but occurs in most subsurface systems. Even those with very low permeability are rarely hydrologically closed (Neuzil 1995). Thus, for example, brine-driven flow systems comparable to reflux occur in a range of geological settings, including around salt domes (Evans & Nunn 1989), beneath playa lakes (Fan *et al.* 1997) and in artificial brine disposal ponds (Simmons & Narayan 1997). Just as our understanding of fluid-flow systems driving dolomitization has been informed by studies of similar flow systems in other environments, insights gained from modelling carbonate platforms will have wider implications.

Summary

One of the purposes of this overview has been to quantitatively assess the dolomitizing potential of different circulation systems. To incorporate the geochemical controls on dolomitization requires coupled reactive transport modelling. However, this approach is still in its infancy, and we have to rely on estimates of the distribution and magnitudes of fluid flux derived from groundwater flow models. The results of such models are strongly dependent on the specified platform geometries, boundary conditions and permeabilities that vary widely between different studies (although the Bristol-based simulations do provide some basis for intercomparison). Thus, even the apparently straightforward comparisons of Figure 22 should be treated with caution, and are best viewed as order of magnitude approximations. Few flow systems should be considered as steady state, both because they evolve through time (e.g. the progressive occupation of platforms by refluxing brines) and because of changes in boundary conditions, such as short-term changes in climate and relative sea level and, at longer timescales, platform architecture. Finally, in natural systems fluid flow will normally be the product of interaction between a number of drives operating simultaneously, and thus simplistic prediction of fluid flux for a single driving force can be misleading.

Notwithstanding these issues, we have come a long way over the last 20 years, from simplistic conceptual cartoons towards the development of a quantitative understanding of replacement dolomitization. Numerical models will rarely provide absolute answers but they can show what is plausible (and equally important what is not!). In some cases this may not require sophisticated fully coupled models. Despite their impressive comprehensive nature, given the uncertainty in parameterizing such demanding models, the potential of analytical approaches should not be overlooked. Nevertheless, the dolomitization process remains one that includes significant fluid flow and geochemical limitations, and thus fully coupled reactive-transport models offer considerable promise for the future. For the non-expert seeking generalizations from this body of work *caveat emptor*! Uncritical and inappropriate application of results from modelling studies is at best uninformative and at worst misleading. Careful model application does, however, offer considerable promise for understanding the broad controls on dolomitization and interpreting well-constrained field examples. Particular attention must be paid to specification of permeability and time-variant boundary conditions. We also need better estimates of parameters in the rate equation for dolomitization at low temperature. Finally, where reactions rates are slow relative to those of sedimentary processes, we also need to model evolving geometries of carbonate platforms during accumulation and burial.

This paper has benefited significantly from discussion with many colleagues, in particular W. Sanford, A. Wilson, H. Machel and J. Gregg. We are also grateful for the careful editorial work of C. Braithwaite. We thank Y. Xiao at ExxonMobil Upstream Research Company for assisting with the reactive transport simulations of reflux. G. D. Jones acknowledges support from the University of Bristol Philip Mary Morris Scholarship. The views expressed in this paper by G. D. Jones are his own and not necessarily those of ExxonMobil.

References

ADAMS, J.E. & RHODES, M.L. 1960. Dolomitization by seepage reflux. *AAPG Bulletin*, **44**, 1912–1920.

AHARON, P., SOCKI, R.A. & CHAN, L. 1987. Dolomitisation of atolls by seawater convection flow; tests of a hypothesis at Niue, South Pacific. *Journal of Geology*, **95**, 187–203.

ALLAN, J.R. & WIGGINS, W.D. 1993. *Dolomite Reservoirs: Geochemical Techniques for Evaluating Origin and Distribution*. American Association of Petroleum Geologists, Continuing Education Course Notes, **36**.

AMTHOR, J.E., MOUNTJOY, E.W. & MACHEL, H.G. 1993. Subsurface dolomites in the Upper Devonian Leduc Formation build-ups, central part of Rimbey–Meadowbrook reef trend, Alberta, Canada. *Bulletin of Canadian Petroleum Geology*, **41**, 164–185.

APPOLD, M.S. & GARVEN, G. 1999. The hydrology of ore formation in the Southeast Missouri District: Numerical models of topography-driven fluid flow during the Ouachita Orogeny. *Economic Geology*, **94**, 913–936.

ARVIDSON, R.S. & MACKENZIE, F.T. 1999. The dolomite problem: control of precipitation kinetics by temperature and saturation state. *American Journal of Science*, **299**, 257–288.

AUDET, D.M. & MCCONNELL, J.D.C. 1992. Forward modelling of porosity and pore pressure evolution in sedimentary basins. *Basin Research*, **4**, 147–162.

BADIOZAMANI, K. 1973. The Dorag dolomitisation model – application to the Middle Ordovician of Wisconsin. *Journal of Sedimentary Petrology*, **43**, 965–984.

BEAR, J. 1972. *Dynamics of Fluids in Porous Media*. Elsevier, New York.

BEAR J. & TODD. D.K. 1960. *The Transition Zone Between Fresh and Salt Waters in Coastal Aquifers.* Hydraulic Laboratory, University of California, Berkeley. Water Resource Center Contribution, **29**.

BETHKE, C.M. 1985. A numerical model of compaction-driven groundwater flow and heat transfer and its application to the paleohydrology of intracratonic sedimentary basins. *Journal of Geophysical Research*, **90**, 6817–6828.

BETHKE, C.M. 1989. Modelling subsurface flow in sedimentary basins. *Geologische Rundschau*, **78**, 129–154.

BETHKE, C.M. 1997. *The Xt Model of transport in Reacting Geochemical Systems*. University of Illinois, 83 pp.

BETHKE, C.M. & CORBET, T.F. 1988. Linear and non-linear solutions for one-dimensional compaction flow in sedimentary basins. *Water Resources Research*, **24**, 461–467.

BETHKE, C.M. & MARSHAK, S. 1990. Groundwater migrations across North America – The plate tectonics of groundwater. *Annual Review of Earth and Planetary Science*, **18**, 287–315.

BETHKE, C.M., HARRISON, W.J., UPSON, C. & ALTANER, S.P. 1988. Supercomputer analysis of sedimentary basins. *Science*, **239**, 261–267.

BETHKE, C.M., REED, J.D. & OLTZ, D.F. 1991. Long-range petroleum migration in the Illinois Basin. *AAPG Bulletin*, **75**, 925–945.

BITZER, K. 1996. Modelling consolidation and fluid flow in sedimentary basins. *Computers & Geosciences*, **22**, 467–478.

BJORLYKKE, K., MO, A. & PALM, E. 1988. Modelling of thermal convection in sedimentary basins and its relevance to diagenetic reactions. *Marine and Petroleum Geology*, **5**, 338–351.

BREDEHOEFT, J.D. 1988. Will salt repositories be dry? *Transactions of the American Geophysical Union*, **69**, 121–131.

BUDD, D.A. 1988. Aragonite to calcite transformation during fresh water diagenesis of carbonates: insights from pore water chemistry. *Geological Society of America Bulletin*, **100**, 1260–1270.

BUDD, D.A. 1997. Cenozoic dolomites of carbonate islands: their attributes and origin. *Earth Science Reviews*, **42**, 1–47.

BUDD D.A. & VACHER H.L. 1990. Predicting fresh water lenses in carbonate paleo-islands. *Journal of Sedimentary Petrology*, **61**, 43–53

BURCHETTE, T.P. & WRIGHT, V.P. 1992. Carbonate ramp depositional systems. *Sedimentary Geology*, **79**, 3–57.

CARBALLO, J.D., LAND, L.S. & MISER, D.E. 1987. Holocene dolomitisation of supratidal sediments by active tidal pumping, Sugarloaf Key, Florida. *Journal of Sedimentary Petrology*, **57**, 153–156.

CATHLES, L.M. 1993. A discussion of flow mechanisms responsible for alteration and mineralization of the Cambrian aquifers of the Ouachita–Arkoma–Ozark system. *In*: HORBURY, A.D. & ROBINSON, A.G. (eds) *Diagenesis and Basin Development*. American Association of Petroleum Geologists, Studies in Geology, **36**, 99–112.

CHEN, W., GHAITH, A., PARK, A. & ORTOLEVA, P. 1990. Diagenesis through coupled processes: Modeling approach, self-organization, and implications for exploration. *American Association of Petroleum Geologists, Memoir*, **49**, 103–130.

CHRISTIANSEN, L.B. & GARVEN, G. 2003. A theoretical comparison of buoyancy-driven and compaction-driven fluid flow in oceanic sedimentary basins. *Journal of Geophysical Research*, **108**, 2130.

CONGLIO, M., SHERLOCK, R., WILLIAMS-JONES, A.E., MIDDLETON, K. & FRAPE, S.K. 1994. Burial and hydrothermal digenesis of Ordovician carbonates from the Michigan Basin, Ontario, Canada. *In*: PURSER, B.H., TUCKER, M.E. & ZENGER, D.H. (eds) *Dolomites: A Volume in Honour of Dolomieu*. International Association of Sedimentologists, Special Publications, **21**, 231–254.

COOPER, R.C. & TINDALL, J.A. 1994. Model for dolomite formation in northwest Florida. *Journal of Hydrology*, **157**, 367–391.

DEFFEYES, K., LUCIA, F.J. & WEYL, P.K. 1965. Dolomitisation of recent and Plio-Pleistocene sediments by marine waters on Bonaire, Netherlands Antilles. *In*: PRAY, L.C. & MURRAY, R.C. (eds) *Dolomitization and Limestone Diagenesis*. Society of Economic Paleontologists and Mineralogists, Special Publications, **13**, 71–89.

DEMMING, D. & NUNN, J.A. 1991. Numerical simulations of brine migration by topographically-driven recharge. *Journal of Geophysical Research*, **96**, 2485–2499.

DEMMING, D., NUNN, J.A. & EVANS, D.G. 1990. Thermal effects of compaction-driven groundwater flow from overthrust belts. *Journal of Geophysical Research*, **95**, 6669–6683.

DRIVET, E. & MOUNTJOY, E.W. 1997. Dolomitization of the Leduc Formation (Upper Devonian), southern Rimbey–Meadowbrook reef trends, Alberta. *Journal of Sedimentary Research*, **67**, 411–423.

ELDER, J.W. 1965. *Terrestrial Heat Flow*. American Geophysical Union, Geophysical Monograph, **8**.

ELDER, J.W. 1967*a*. Steady free convection in a porous medium heated from below. *Journal of Fluid Mechanics*, **27**, 29–48.

ELDER, J.W. 1967*b*. Transient convection in a porous medium heated from below. *Journal of Fluid Mechanics*, **27**, 609–623.

ENOS, P. & SAWATSKY, L.H. 1981. Pore networks in Holocene carbonate sediments. *Journal of Sedimentary Petrology*, **51**, 961–985.

EVANS, P.G. & NUNN, J.A. 1989. Free thermohaline convection in sediments surrounding a salt column. *Journal of Geophysical Research*, **94**, 12 413–12 422.

FAN, Y., DUFFY, C.J. & OLIVER, D.S., JR. 1997. Density-driven groundwater flow beneath a closed desert basins: field investigations and numerical experiments. *Journal of Hydrology*, **196**, 139–184.

FANNING, K.A., BYRNE, R.H., BRELAND, J.A. & BETZER, P.R. 1981. Geothermal springs of the West Florida continental shelf: evidence for dolomitization and radionuclide enrichment. *Earth and Planetary Science Letters*, **52**, 345–354.

GARVEN, G. 1995. Continental scale groundwater flow and geological processes. *Annual Review of Earth and Planetary Sciences*, **23**, 89–117.

GARVEN, G. & FREEZE, R. 1984. Theoretical analysis of the role of groundwater flow in the genesis of stratabound ore deposits. *American Journal of Science*, **284**, 1085–1174.

GARVEN G., APPOLD, M.S., TOPTYGINA, V.I. & HAZLETT, T.J. 1999. Hydrogeologic modelling of the genesis of carbonate-hosted lead–zinc ores. *Hydrogeology Journal*, **7**, 108–126.

GARVEN, G., GE., S., PERSON, M.A. & SVERJENSKY, D.A. 1993. Genesis of strata-bound ore deposits in the mid-continent basins of North America, 1. The role of regional groundwater flow. *American Journal of Science*, **293**, 497–568.

GE, S. & GARVEN, G. 1992. Hydrochemical modelling of tectonically-driven groundwater flow with application to the Canadian Rocky Mountains. *Journal of Geophysical Research*, **99**, 9119–9144.

GE, S. & GARVEN, G. 1994. A theoretical model for thrust-induced deep groundwater expulsion with application to the Canadian Rocky Mountains. *Journal of Geophysical Research*, **99**, 13 851–13 868.

GE, S., BEKINS, B. *ET AL*. 2003. Fluid flow in subseafloor processes and future ocean drilling. *Eos*, **84**(16), 145–152.

GHASSEMI, F.J., ALAM, K. & HOWARD, K.W.F. 2000. Freshwater lenses and practical limitations of their three-dimensional simulation. *Hydrogeology Journal*, **8**, 521–537.

GHASSEMI, F.J., JAKEMAN, A.J. & JACOBSEN, G. 1990. Mathematical modelling of seawater intrusion, Nauru Island. *Hydrological Processes*, **4**, 269–281.

GIBSON, R.E. 1958. The progress of consolidation in a clay layer increasing in thickness with time. *Geotechnique*, **8**, 171–182.

GIBSON, R.E., ENGLAND, G.L. & HUSSEY, M.J.L. 1967. The theory of one-dimensional consolidation of saturated clays, I. Finite non-linear consolidation of thick homogeneous layers. *Geotechnique*, **17**, 261–273.

GREGG, J.M. 1985. Regional epigenetic dolomitisation in the Bonneterre dolomite (Cambrian), southeastern Missouri. *Geology*, **13**, 503–506.

GREGG, J.M. 2004. Basin fluid flow, base-metal sulphide mineralization and the development of dolomite petroleum reservoirs. *In*: BRAITHWAITE, C.J.R., RIZZI, G. & DARKE, G. (eds) *The Geometry and Petrogenesis of Dolomite Hydrocarbon Reservoirs*. Geological Society, London, Special Publications, **235**, 157–175.

GREGG, J.M & SHELTON, K.L. 1989. Minor and trace element distributions in Bonneterre Dolomite (Cambrian), southeast Missouri: Evidence for possible multiple basin fluid sources and pathways during lead–zinc mineralization. *Geological Society of America Bulletin*, **101**, 221–230.

GREGG, J.M., LAUDON, P.R., WOODY, R.E., & SHELTON, K.L. 1993. Porosity evolution of the Bonneterre Dolomite (Cambrian) southeastern Missouri, U.S.A. *Sedimentology*, **40**, 1153–1169.

HANSHAW, B.B., BACK, W. & DEIKE, R. 1971. A geochemical hypothesis for dolomitisation by groundwater. *Economic Geology*, **66**, 710–724.

HARDIE, L.A. 1977. *Sedimentation on the Modern Carbonate Tidal Flats of Northwest Andros Island, Bahamas*. John Hopkins University Studies in Geology, **22**.

HARDIE, L.A. 1987. Dolomitisation: a critical view of some current views. *Journal of Sedimentary Petrology*, **57**, 166–183.

HARRIS P.M., KENDALL, C.G. ST. C. & LERCHE, I. 1985. Carbonate cementation – a brief review. *In*: SCHNEIDERMANN, N. & HARRIS, P.M. (eds) *Carbonate Cements*. Society of Economic Paleontologists and Mineralogists, Special Publications, **36**, 79–95.

HARRISON, W.J. & SUMMA, L.L. 1991. Paleohydrology of the Gulf of Mexico Basin. *American Journal of Science*, **291**, 109–176.

HENRY, H.R. 1964. Effects of dispersion on salt encroachment in coastal aquifers, seawater in coastal aquifers. *US Geological Survey, Water Supply Paper*, **1613C**, 70–84.

HENRY, H.R. & HILLEKE, J.B. 1972. *Exploration of multiphase fluid flow in a saline aquifer system affected by geothermal heating*. Bureau of Engineering Research Report, **150–118**.

HENRY, H.R. & KOHOUT F.A. 1972. Circulation patterns of saline groundwater affected by geothermal heating – as related to waste disposal. *American Association of Petroleum Geologists, Memoir*, **18**, 202–221.

HENRY, P., GUY, C., CATIN, R., DUDOIGNON, P., SORNEIN, J.F. & CARISTAN, Y. 1996. A convective model of groundwater flow in Mururoa basalts. *Geochimica et Cosmochimica Acta*, **60**, 2087–2109.

HERBERT, A.W. & LLOYD, J.W. 2000. Approaches to the modelling of saline intrusion for assessment of small island water resources. *Quarterly Journal of Engineering Geology and Hydrogeology*, **33**, 77–86.

HERMAN, M.E., BUDDEMEIER, K.W. & WHEATCRAFT, S.W. 1986. A layered aquifer model of atoll island hydrology: validation of a computer simulation. *Journal of Hydrology*, **84**, 303–322.

HOLLAND, H.D. & ZIMMERMANN, H. 2000. The dolomite problem revisited. *International Geology Review*, **42**, 481–490.

HSÜ, K.J. & SIEGENTHLER, C. 1969. Preliminary experiments on hydrodynamic movement induced by evaporation and their bearing on the dolomite problem. *Sedimentology*, **12**, 11–25.

HUMPHREY, J.D. 1988. Late Pleistocene mixing zone dolomitisation, West Indies. *Sedimentology*, **35**, 327–348.

ILLING, L.V. 1959. Deposition and diagenesis of some upper Paleozoic carbonate sediments in western Canada. *In*: *Proceedings of the 5th World Petroleum Congress, New York*, Section 1, 23–52. Portland Press, London.

JAMES, N.P. & CHOQUETTE, P.W. 1984. Diagenesis 9. Limestones, the meteoric diagenetic environment. *Geoscience Canada*, **11**, 161–194.

JEAN-BAPTISTE, P. & LECLERC, A.M. 2000. Self-limiting geothermal convection in marine carbonate platforms. *Geophysical Research Letters*, **27**, 743–746.

JONES, G.D. 2000. *Numerical modelling of saline groundwater circulation in carbonate platforms*. PhD thesis, University of Bristol.

JONES, G.D. & ROSTRON, B.J. 2000. Analysis of fluid flow constraints in regional-scale reflux dolomitization: constant versus variable-flux hydrogeological models. *Bulletin of Canadian Petroleum Geology*, **48**, 230–245.

JONES, G.D., WHITAKER, F.F., SMART P.L. & SANFORD, W.E. 1997. Dolomitization of carbonate platforms by saline groundwater: Coupled numerical modelling of thermal and reflux circulation mechanisms. *In*: HENDRY, J., CAREY, P., PARNELL, J., RUFFELL, A. & WORDEN, R. (eds) *Proceedings of Geofluids II Conference*, 378–381. Queen's University, Belfast.

JONES, G.D., WHITAKER, F.F., SMART, P.L. & SANFORD, W.E. 2000. Numerical modelling of geothermal and reflux circulation in Enewetak Atoll: Implications for dolomitization. *Journal of Geochemical Exploration*, **69–70**, 71–75.

JONES, G.D., WHITAKER, F.F., SMART, P.L. & SANFORD, W.E. 2002. Fate of reflux brines in carbonate platforms. *Geology*, **30**, 371–374.

JONES, G.D., SMART, P.L., WHITAKER, F.F., ROSTRON, B.J. & MACHEL, H.G. 2003. Numerical modelling of reflux dolomitization in the Grosmont platform complex (Upper Devonian), Western Canada Sedimentary Basin. *AAPG Bulletin*, **87**, 1273–1298.

JONES, G.D., WHITAKER, F.F., SMART, P.L. & SANFORD, W.E. 2004. Numerical analysis of seawater circulation in carbonate platforms: II. The dynamic interaction between geothermal and reflux circulation. *American Journal of Science*, **304**, 250–284.

JUSTER, T., KRAMER, P.A. & VACHER, H.L. 1997. Groundwater flow beneath a hypersaline pond, Cluett Key, Florida Bay, Florida. *Journal of Hydrology*, **197**, 339–369.

KAUFMAN, J.K. 1994. Numerical models of fluid flow in carbonate platforms: implications for dolomitization. *Journal of Sedimentary Research*, **A64**, 128–139.

KENDALL, A.C. 1989. Brine mixing in the Middle Devonian of western Canada and its possible significance to regional dolomitisation. *Sedimentary Geology*, **64**, 271–285.

KENDALL, A.C. & WALTERS, K.L. 1978. The age of metasomatic anhydrite in Mississippian reservoir carbonates, southeastern Saskatchewan. *Canadian Journal of Earth Sciences*, **15**, 424–430.

KNOOP, S.R., KENNEDY, L A. & DIPPLE, G.M. 2002. New evidence for syntectonic fluid migration across the hinterland-foreland transition of the Canadian Cordillera. *Journal of Geophysical Research*, **107(B4)**, article 2071 doi: 1029/2001JB000217.

KOHOUT, F.A. 1965. A hypothesis concerning cyclic flow of salt water related to geothermal heating in the Floridan aquifer. *Transactions of the New York Academy of Sciences, Series 2*, **28**, 249–271.

KOHOUT, F.A., HENRY, H.R. & BANKS, J.E. 1977. Hydrology related to geothermal conditions of the Floridan Plateau. *In*: SMITH, K.L. & GRIFFIN, G.M. (eds) *The Geothermal Nature of the Floridan Plateau*. Florida Department of Natural Resources Bureau of Geology Special Publications, **21**, 1–34.

KOOI, H. 1999. Competition between topography- and compaction-driven flow in a confined aquifer: some analytical results. *Hydrogeology Journal*, **7**, 245–250.

KONIKOW, L.F. & BREDEHOEFT, J.D. 1992. Groundwater-models cannot be validated. Validate or invalidate? *Advances in Water Resources*, **15**, 75–83.

LAND, L.S. 1985. The origin of massive dolomite. *Journal of Geological Education*, **33**, 112–125.

LAND, L.S., LUND, H.J. & MCCULLOUGH, M.L. 1989. Dynamic circulation of interstitial seawater in a Jamaican fringing reef. *Carbonates and Evaporites*, **4**, 1–7.

LECLERC, A.M., JEAN-BAPTISTE, P. & BROC, D. 2000. Physics of atoll groundwater flow. *Mathematical Geology*, **32**, 301–317.

LECLERC, A.M., JEAN-BAPTISTE, P. & TEXIER, D. 1999. Density-induced water circulation in atoll coral reefs: a numerical study. *Limnology and Oceanography*, **44**, 1268–1281.

LEE, M.-K. 1997. Predicting diagenetic effects of groundwater flow in sedimentary basins: a modeling approach with examples. *In*: MONTAÑEZ, I.P., GREGG, J.M. & SHELTON, K.L. *Basinwide Fluid Flow and Associated Diagenetic Patterns: Integrated Petrographic, Geochemical, and Hydrologic Consideration*. Society of Economic Paleontologist and Mineralogists, Special Publications, **57**, 3–14.

LIPPMAN, F. 1973. *Sedimentary Carbonate Minerals*. Springer, New York.

LOGAN, B.W. 1987. *The MacLeod Evaporite Basin, Western Australia*. American Association of Petroleum Geologists, Memoir, **44**.

LUCIA, F.J. 1972. Recognition of evaporite-carbonate shoreline sedimentation. *In*: RIGBY, J.K. & HAMBLIN, W.K. (eds) *Recognition of Ancient Sedimentary Environments*. Society of Economic Paleontologists and Mineralogists, Special Publications, **16**, 180–191.

LUMSDEN, D.N. & CAUDLE, G.C. 2001. Origin of massive dolostone: the Upper Knox model. *Journal of Sedimentary Research*, **71**, 400–409.

MACHEL, H.G. 1986. Early lithification, dolomitisation and anhydritization of Upper Devonian Nisku Buildups, subsurface Alberta, Canada. *In*: SCHROEDER, J.H. & PURSER, B.H. (eds) *Reef Diagenesis*. Springer, New York, 336–356.

MACHEL, H.G. 2000. Dolomite formation in Caribbean islands – driven by plate tectonics? *Journal of Sedimentary Research*, **70**, 977–984.

MACHEL, H.G. 2004. Concepts and models of dolomitization: a critical reappraisal. *In*: BRAITHWAITE, C.J.R., RIZZI, G. & DARKE, G. (eds) *The Geometry and Petrogenesis of Dolomite Hydrocarbon Reservoirs*. Geological Society, London, Special Publications, **235**, 7–63.

MACHEL, H.G. & ANDERSON, J.H. 1989. Pervasive subsurface dolomitization of the Nisku Formation in Central Alberta. *Journal of Sedimentary Petrology*, **59**, 891–911.

MACHEL, H.G. & CALVELL, P.A. 1999. Low-flux, tectonically-induced squeegee fluid flow ('hot flash') into the Rocky Mountain Foreland Basin. *Bulletin of Canadian Petroleum Geology*, **47**, 510–533.

MACHEL, H.G. & MOUNTJOY, E.W. 1986. Chemistry and environments of dolomitization – a reappraisal. *Earth Science Reviews*, **23**, 175–222.

MACHEL, H.G. & MOUNTJOY, E.W. 1990. Coastal mixing zone dolomite, forward modelling and massive dolomitisation of platform-margin carbonates – Discussion. *Journal of Sedimentary Petrology*, **60**, 1008–1012.

MAGARA, K. 1976. Water expulsion from clastic sediments during compaction – directions and volumes. *AAPG Bulletin*, **60**, 543–553.

MALONE, M.J., BAKER, P.A. & BURNS, S.J. 1996. Recrystallization of dolomite: An experimental study from 50–200°C. *Geochimica et Cosmochimica Acta*, **60**, 2189–2207.

MARSHALL J.F. 1986. Regional distribution of submarine cements within an epi-continental reef system: Central Great Barrier Reef, Australia. *In*: SCHROEDER, J.M. & PURSER, B.H. (eds) *Reef Diagenesis*. Springer, New York, 8–26.

MATTES, B.W. & MOUNTJOY, E.W. 1980. Burial dolomitisation of the Upper Devonian Miette buildup, Jasper National Partk, Alberta. *In*: ZENGER, D.H., DUNHAM, J.B. & ETHINGTON, R.L. (eds) *Concepts and Models of Dolomitization*. Society of Economic Paleontologists and Mineralogists, Special Publications, **28**, 259–297.

MELIM, L.M. & SCHOLLE, P.A. 2002. Dolomitisation of the Capitan Formation forereef facies (Permian, west Texas and New Mexico): seepage reflux revisited. *Sedimentology*, **49**, 1207–1227.

MONTAÑEZ, I.P & READ, J.F. 1992. Eustatic control on early dolomitization of cyclic peritidal carbonates: evidence form the Early Ordovician Upper Knox Group, Appalachians. *Bulletin of the Geological Society of America*, **104**, 872–886.

MORROW, D.W., CUMMING, G.L. & KOEPNICK, R.B. 1986. Manetoe facies – a gas bearing, megacrystalline Devonian dolomite, Yukon and Northwest Territories, Canada. *AAPG Bulletin*, **70**, 702–720.

NIELD, D. 1968. Onset of thermohaline convection in a porous medium. *Water Resources Research*, **4**, 553–560.

NEILSEN, P. 1999. Groundwater dynamics and salinity in coastal barriers. *Journal of Coastal Research*, **15**, 732–740.

NEUZIL, C.E. 1995. Abnormal pressures as hydrodynamic phenomena. *American Journal of Science*, **295**, 742–786.

NORDENG, S.H. & SIBLEY, D.F. 1994. Dolomite stoichiometry and Ostwald's step rule. *Geochimica et Cosmochimica Acta*, **58**, 191–196.

NORDENG, S.H. & SIBLEY, D.F. 2003. The induction period and dolomite kinetics in experimental and natural systems. *In*: *12th Bathurst Meeting of Carbonate Sedimentologists*, July 2003, Durham, England, 74, University of Durham.

NUNN, J.A. & LIN, G. 1997. Thermal insulation by Late Paleozoic carbonaceous sediments: Implications for formation of Mississippi Valley-Type ore deposits. *In*: HENDRY, J., CAREY, P., PARNELL, J., RUFFELL, A. & WORDEN, R. (eds) *Proceedings of Geofluids II Conference*, 201–204. Queen's University, Belfast.

OBERDORFER, J.A., HOGAN, P.J. & BUDDEMEIR, R.W. 1990. Atoll island hydrology; flow and fresh water occurrence in a tidally dominated system. *Journal of Hydrology*, **120**, 327–340.

OLIVER, J. 1986. Fluids expelled tectonically from orogenic belts: their role in hydrocarbon migration and other geologic phenomena. *Geology*, **14**, 99–102.

OOSTROM, M., HAYWORTH, J.S., DANE, J.H. & GUVEN, O. 1992. Behavior of dense aqueous leachate plumes in homogeneous porous media. *Water Resources Research*, **28**, 2123–2134.

PATTERSON, R.J. & KINSMAN, D.J.J. 1982. Formation of diagenetic dolomite in coastal sabkha along the Arabian (Persian) Gulf. *AAPG Bulletin*, **66**, 28–43.

PERKINS, R.D., DWYER, G.S., ROSOFF, D.B., FULLER, J., BAKER, P.A. & LLOYD, R.M. 1994. Salina sedimentation and diagenesis: West Caicos Island, British West Indies. *In*: PURSER, B.H., TUCKER, M.E. & ZENGER, D.H. (eds) *Dolomites: A Volume in Honour of Dolomieu*. International Association of Sedimentologists, Special Publications, **21**, 37–54.

PERSON, M. & GARVEN, G. 1994. A sensitivity study of the driving forces during continental-rift basin evolution. *Geological Society of America Bulletin*, **106**, 461–475.

PERSON, M., RAFFENSPERGER, J.P., GE, S. & GARVEN, G. 1996. Basin-scale hydrogeologic modeling. *Reviews of Geophysics*, **34**, 61–87.

PERSON, M., TOUPIN, D. & EADINGTON, P. 1995. Effects of convective heat transfer on the thermal history of sediments and petroleum generation within continental rift basins. *Basin Research*, **7**, 81–96.

PHILLIPS, O.M. 1991. *Flow and Reactions in Permeable Rocks*. Cambridge University Press, Cambridge, 285 pp.

PRATS, M. 1966. The effect of horizontal fluid flow on thermally induced convection currents in porous mediums. *Journal of Geophysical Research*, **71**, 4835–4838.

RAFFENSPERGER, J.P. 1996. Numerical simulation of sedimentary basin-scale hydrochemical processes. *In*: CORAPCIOGLU, M.Y. (ed.) *Advances in Porous Media*, Volume 3, Elsevier, New York, 185–305.
RAFFENSPERGER, J.P. & VLASSOPOULOS, D. 1999. The potential for free and mixed convection in sedimentary basins. *Hydrogeology Journal*, **7**, 505–520.
REILLY, T.E. & GOODMAN, A.S. 1985. Quantitative analysis of saltwater–freshwater relationships in groundwater systems: a historical perspective. *Journal of Hydrology*, **80**, 125–160.
RENARD, P.H. & MARSILY, G. 1997. Calculating equivalent permeability: a review. *Advances in Water Resources*, **20**, 253–278.
SALLER, A.H. 1984. Petrologic and geochemical constraints on the origin of subsurface dolomite, Enewetak Atoll: an example of dolomitization by normal seawater. *Geology*, **12**, 217–220.
SALLER, A.H. & HENDERSON, N. 1998. Distribution of porosity and permeability in platform dolomites: insight from the Permian of West Texas. *AAPG Bulletin*, **82**, 1528–1550.
SAMADEN, G., DALLOT, P. & ROCHE, R. 1985. L'atoll d'Enewetak: système géothermique insulaire a l'état naturel. *La Houille Blanche*, **2**, 143–151.
SANFORD, W.E. & KONIKOW, L.F. 1989. Simulation of calcite dissolution and porosity changes in salt water mixing zones in coastal aquifers. *Water Resources Research*, **25**, 655–667.
SANFORD, W.E., WHITAKER, F.F., SMART, P.L. & JONES, G.D. 1998. Numerical Analysis of seawater circulation in carbonate platforms: I geothermal circulation. *American Journal of Science*, **298**, 801–828.
SARKAR, J., NUNN, J.A. & HANOR, S. 1995. Free thermohaline convection beneath allochthonous salt sheets: an agent for salt dissolution and fluid flow in Gulf Coast sediments. *Journal of Geophysical Research*, **100**, 18 085–18 092.
SCHMOKER, J.W. & HALLEY, R.B. 1982. Carbonate porosity versus depth: a predictable relation for South Florida. *AAPG Bulletin*, **66**, 2561–2570.
SCREATON, E.J., WUTHRICH, D.R. & DREISS, S. 1990. Permeability, fluid pressures and flow rates in the Barbados ridge complex. *Journal of Geophysical Research*, **95**, 8997–9007.
SEARS, S.O. & LUCIA, F.J. 1980. Dolomitisation in northern Michigan Niagara reefs by brine refluxion and freshwater/seawater mixing. *In*: ZENGER, D.H., DUNHAM, J.B. & ETHINGTON, R.L. (eds) *Concepts and Models of Dolomitization*. Society of Economic Paleontologists and Mineralogists, Special Publications, **28**, 215–235.
SHARP, J.M. 1976. Momentum and energy balance equations for compacting sediments. *Mathematical Geology*, **8**, 305–322.
SHARP, J.M., FENSTEMAKER, T.R., SIMMONS, C.T., MCKENNA, T.E., & DICKINSON, J.K. 2001. Potential salinity driven free convection in a shale-rich sedimentary basin: Example from the Gulf of Mexico basin in south Texas. *AAPG Bulletin*, **85**, 2089–2110.
SHIELDS, M.J. & BRADY, P. 1995. Mass balance and fluid flow constraints on regional scale dolomitization, late Devonian, Western Canada Sedimentary Basin. *Bulletin of Canadian Petroleum Geology*, **43**, 371–392.
SHINN, E.A., REESE, R.S. & REICH, C.D. 1994. *Fate and Pathways of Injection-well Effluent in the Florida Keys*. US Geological Survey, Open File Report, **94–276**.
SIMMONS, C.T. & NARAYAN, K.A, 1997. Mixed convection processes beneath a saline disposal basin. *Journal of Hydrology*, **194**, 263–285.
SIMMONS, C.T., FENSTEMAKER, T.R. & SHARP, J.M. 2001. Variable-density groundwater flow and solute transport in heterogeneous porous media: approaches, resolutions and future challenges. *Journal of Contaminant Hydrology*, **52**, 245–275.
SIMMS, M.A. 1984. Dolomitization by groundwater flow systems in carbonate platforms. *Transactions of the Gulf Coast Association of Geological Sciences*, **24**, 411–420.
SIMPSON, M.J. & CLEMENT, T.P. 2003. Theoretical analysis of the Henry and Elder problems as benchmarks of density-dependent groundwater flow models. *Advances in Water Resources*, **26**, 17–31.
SMART, P.L., DAWANS, J.M & WHITAKER, F.F. 1988. Carbonate dissolution in a modern mixing zone, South Andros, Bahamas. *Nature*, **335**, 811–813.
SMITH, J.E. 1971. The dynamics of shale compaction and evolution of pore-fluid pressures. *Mathematical Geology*, **3**, 239–263.
SMITH, L.S., WHITAKER, F.F., PARKES, R.J., SMART, P.L. & BEDDOWS, P.A. 2002. Dolomitisation by saline groundwaters in the Yucatan Peninsula. *In*: RIZZI, G., DARKE, G. & BRAITHWAITE, C. (convenors) *The Geometry and Petrogenesis of Dolomite Hydrocarbon Reservoirs*. Final Programme and Abstracts. Petroleum Group Geology Society, London.
SONNENFELD, P. 1984. *Brines and Evaporites*. Academic Press, Orlando.
STANLEY, S.M. & HARDIE, L.A. 1998. Secular oscillations in the carbonate mineralogy of reef-building and sediment producing organisms driven by tectonically forced shifts in seawater chemistry. *Paleogeography, Paleoclimatology, Paleoecology*, **144**, 3–19.
STEEFEL, C.I. & LASAGA, A.C. 1994. A coupled model for transport of multiple chemical species and kinetic precipitation/dissolution reactions with application to reactive flow in single phase hydrothermal systems. *American Journal Science*, **294**, 529–592.
STEEFEL, C.I. & VAN CAPPELLEN, P. 1990. A new kinetic approach to modelling water–rock interaction. The role of nucleation, precursors and Ostwald ripening. *Geochimica et Cosmochimica Acta*, **54**, 2757–2677.
SUN, S.Q. 1992. Skeletal aragonite dissolution from hypersaline seawater a hypothesis. *Sedimentary Geology*, **77**, 249–257.
SUN, S.Q. 1995. Dolomite reservoirs: porosity evolution and reservoir characteristics. *AAPG Bulletin*, **79**, 186–204.
SWENSON, J.B & PERSON, M. 2000. The role of basin-scale transgression and sediment compaction in stratiform copper-mineralization: implications from White Pine, Michigan, USA. *Journal of Geochemical Exploration*, **69–70**, 239–243.
THIRY, M., HANOT, F. & PIERRE, C. 2003. Chalk

dolomitisation beneath localized subsiding Tertiary depressions in a marginal marine setting in the Paris Basin (France). *Journal of Sedimentary Research*, **73**, 157–170.

TÓTH, J. 1963. A theoretical analysis of groundwater flow in small drainage basins. *Journal of Geophysical Research*, **68**, 4795–4812.

TRIBBLE, J.S., ARVIDSON, R.S., LANE, I.M. & MACKENZIE, F.T. 1995. Crystal chemistry and thermodynamic and kinetic properties of calcite, apatite and biogenic silica: applications to petrologic problems. *Sedimentary Geology*, **95**, 11–37.

USDOWSKI, E. 1994. Synthesis of dolomite and geochemical implications. *In*: PURSER, B.H., TUCKER, M.E. & ZENGER, D.H. (eds) *Dolomites: A Volume in Honour of Dolomieu*. International Association of Sedimentologists, Special Publications, **21**, 345–360.

UNDERWOOD, M.R., PETERSON, F.L. & VOSS, C.I. 1992. Groundwater lens dynamics of atoll islands. *Water Resources Research*, **28**, 2889–2902.

VACHER, H.L. 1988. Dupuit–Ghyben–Herzberg analysis of strip-island lenses. *Geological Society of America, Bulletin*, **100**, 580–591.

VAHRENKAMP, V.C. & SWART, P.K. 1994. Late Cenozoic dolomites of the Bahamas: metastable analogues for the genesis of ancient platform dolomites. *In*: PURSER, B.H., TUCKER, M.E. & ZENGER, D.H. (eds) *Dolomites: A Volume in Honour of Dolomieu*. International Association of Sedimentologists, Special Publications, **21**, 133–153.

VOLKER, R.E., MARINO, M.A. & ROLSTON, D.E. 1985. Transition zone width in groundwater of ocean atolls. *Journal of Hydraulic Engineering*, **111**, 659–676.

VOSS, C.I. 1984. *SUTRA – A finite element simulation model for saturated–unsaturated fluid density dependent groundwater flow with energy transport and chemically-reactive single species transport*. US Geological Survey, Water Resources Investigative Report, **84–4369**.

VOSS, C.I. & SOUZA, W.R. 1987. Variable density flow and solute transport simulation of regional aquifers containing a narrow freshwater–saltwater transition zone. *Water Resources Research*, **23**, 1851–1866.

WENDTE, J., QING, H., DRAVIS, J.J., MOORE, S.L.O., STAIUK, L.D. & WARD, G. 1998. High-temperature saline (thermoflux) dolomitisation of Devonian Swan Hills platform and bank carbonates, Wild River area, west-central Alberta. *Bulletin of Canadian Petroleum Geology*, **46**, 210–265.

WEYL, P.K. 1960. Porosity through dolomitisation: conservation of mass requirements. *Journal of Sedimentary Petrology*, **30**, 85–90

WHEELER, C. & AHARON, P. 1997. *Geology and Hydrogeology of Niue*. Developments in Sedimentology, **54**, 537–564.

WHITAKER, F.F. & SMART, P.L. 1990. Circulation of saline groundwaters through carbonate platforms: evidence from the Great Bahama Bank. *Geology*, **18**, 200–204.

WHITAKER, F.F. & SMART, P.L. 1993. Circulation of saline groundwaters in carbonate platforms: a review and case study from the Bahamas. *In*: HORBURY, A.D. & ROBINSON, A.G. (eds) *Diagenesis and Basin Development*. American Association of Petroleum Geologists, Studies in Geology, **36**, 113–134.

WHITAKER, F.F. & SMART, P.L. 1997. *Hydrology and Hydrogeology of the Bahamian Archipelago*. Developments in Sedimentology, **54**, 183–216.

WHITAKER, F.F., SANFORD, W.E. & SMART, P.L. 1993. Numerical modelling of Geothermal convection in carbonate platforms: controls on circulation and implications for dolomitisation. *In*: MACQUAKER, J. (ed.) *Abstract Volume of the Annual Meeting British Sedimentological Research Group*, 155. University of Manchester.

WHITAKER, F.F., SMART, P.L., VAHRENKAMP, V.C., NICHOLSON, H. & WOGELIUS, R. 1994. Dolomitisation by near-normal seawater? Field evidence from the Bahamas. *In*: PURSER, B.H., TUCKER, M.E. & ZENGER, D.H. (eds) *Dolomites: A Volume in Honour of Dolomieu*. International Association of Sedimentologists, Special Publications, **21**, 111–132.

WICKS, C.M. & HERMAN, J.S. 1995. The effect of zones of high porosity and permeability on the configuration of the saline–freshwater mixing zone. *Groundwater*, **33**, 733–740.

WIECK, J., PERSON, M. & STRAYER, L. 1995. A finite element method for simulating fault block motion and hydrothermal fluid flow within rifting basins. *Water Resources Research*, **31**, 3241–3258.

WILSON, A.M., SANFORD, W.E., WHITAKER, F.F. & SMART, P.L. 2001. Spatial patterns of diagenesis during geothermal circulation in carbonate platforms. *American Journal of Science*, **301**, 727–752.

WILSON, E.N., HARDIE, L.A. & PHILLIPS, O.M. 1990. Dolomitization front geometry, fluid flow patterns, and the origin of dolomite: the Triassic Latemar buildup, northern Italy. *American Journal of Science*, **290**, 741–796.

WOOD, J.R. 1987. A model for dolomitisation by pore fluid flow. *In*: *Proceedings of the Conference on Physics and Chemistry of Porous Media II, American Institute of Physics*, 17–36. American Institute of Physics, New York.

WOOD, J.R., & HEWLETT, T.A. 1982. Fluid convection and mass transfer in porous sandstones – a theoretical model. *Geochimica et Cosmochimica Acta*, **47**, 1707–1713.

WOOD, J.R. & HEWLETT, T.A. 1984. Reservoir diagenesis and convective fluid flow. *In*: MCDONALD, D.A. & SURDAN, R.C. (eds) *Clastic Diagenesis*. American Association of Petroleum Geologists, Memoir, **37**, 99–111.

WOODING, R.A. 1997. Steady state free thermal convection of a liquid in a saturated permeable medium. *Journal of Fluid Mechanics*, **2**, 273–285.

YAO, Q. & DEMICCO, R.V. 1995. Paleoflow patterns of dolomitising fluids and paleohydrology of the Southern Rocky Mountains: evidence from dolomite geometry and numerical modeling. *Geology*, **23**, 791–794.

YAO, Q. & DEMICCO, R.V. 1997. Dolomitisation of the Cambrian carbonate platform, southern Canadian Rocky Mountains: dolomite front geometry, fluid inclusion geochemistry, isotopic signature and hydrogeological modeling studies. *American Journal of Science*, **297**, 892–938.

Origin and petrophysics of dolostone pore space

F. JERRY LUCIA

Bureau of Economic Geology, John A. and Katherine G. Jackson School of Geosciences, The University of Texas at Austin, Austin, TX 78713-8924, USA

Abstract: The common claim that dolomitization creates 12% porosity is based on the mole-for-mole replacement equation. However, in the past 50 years, data have been collected demonstrating that dolomitization does not create porosity. Instead, dolostones inherit the porosity and fabric of the precursor limestone, and porosity is reduced by overdolomitization. The porosity of the precursor limestone depends on the diagenetic history up to the time of dolomitization. Data show that: (1) carbonates are born with high porosity and lose porosity gradually over a long period; and (2) mud-dominated fabrics compact more readily than grain-dominated fabrics. The problem of estimating the time of dolomitization is minimized by confining observations to young limestones and associated dolostones. Limited data from Holocene dolomitic sediments suggest no change in porosity with dolomitization. Study of Plio-Pleistocene carbonates in Bonaire, Netherlands Antilles, demonstrates that precursor limestones are more porous than dolostones. Limestones average 25% porosity, whereas dolostones average 11% porosity. Data from the Neogene of the Great Bahama Bank show that dolostones and adjacent limestones both have 40% porosity. Porosity studies of the Jurassic Arab D Formation show that dolostones and associated grain-dominated limestones have similar porosity ranges and that the decrease in porosity with increasing dolomitization results from compaction of the mud-dominated fabrics. These data suggest that porosity in dolostone is not created by a mole-for-mole replacement mechanism. Instead, dolostone porosity is: (1) inherited from the precursor limestone; and (2) occluded by the process of overdolomitization. Palaeozoic dolostones, however, are commonly more porous than juxtaposed limestones. The explanation for this observation is that limestones lose porosity through compaction and cementation, whereas dolostones resist compaction and retain much of their porosity.

Permeability studies have demonstrated that dolomitization of grain-dominated limestones usually does not change porosity–permeability relationships. Instead, precursor limestone fabric controls pore-size distribution. The dolomite crystal size of a mud-dominated dolostone may, however, be larger than the carbonate mud size, improving the porosity–permeability relationship substantially. Hence, there is a predictable relationship between interparticle (grains or crystals) porosity, permeability, precursor grain size and dolomite crystal size.

The origin and petrophysics of dolostone have been linked since 1837, when Elie De Beaumont proposed that the conversion of limestone to dolostone creates 12% porosity because the molar volume of dolomite is smaller than the molar volume of calcite (De Beaumont 1837). This chemical model was used to explain the presence of vuggy pore space in dolostones of the Tyrolean Alps, and after 165 years it is still the one most commonly used to explain porous dolostones. Although challenged by Landes (1946), Schmoker & Halley (1982), Halley & Schmoker (1983) and, most recently, by Lucia & Major (1994), it is nonetheless the most common dolomitization model taught in colleges and universities (Tucker 2001).

The purpose of this paper is to demonstrate that dolostone porosity is not the result of dolomitization and that the concept that porous dolostone is formed from non-porous limestone is incorrect. Data will be presented to show that many young limestones are very porous and that this porosity is inherited by replacement dolostone. Furthermore, the observation that many Palaeozoic limestones are less porous than associated dolostones will be shown to result from differential compaction, limestones being more susceptible to compaction than dolostones. The excellent petrophysical quality of many dolostones will be related to: (1) an increase in crystal size in dolomitized mud-dominated fabrics; and (2) the ability of dolostone to retain porosity.

Terminology

Terminology is discussed here briefly because its improper use adds to the confusion that

From: BRAITHWAITE, C. J. R., RIZZI, G. & DARKE, G. (eds) 2004. *The Geometry and Petrogenesis of Dolomite Hydrocarbon Reservoirs*. Geological Society, London, Special Publications, **235**, 141–155. 0305-8719/$15.00

surrounds this issue. *Porosity* is defined as pore volume divided by bulk volume and in geological literature is commonly expressed as per cent porosity, i.e. porosity × 100. *Pore space* or *pore volume* is that portion of the bulk volume not occupied by minerals. When the term *porosity* is used, the *bulk volume* of the sample should be clearly understood. In core analysis, the bulk volume is the volume of the sample measured. In thin section, the bulk volume is the volume of the thin section. However, in many instances, the bulk volume is not clear and may not even be considered, as in such comments as 'intergrain porosity' or 'a description of porosity'. Intergrain or intercrystal porosity is the pore volume between grains or crystals divided by the bulk volume under consideration and is expressed as a number. It cannot be described except by a number or a reference to a value (high porosity, for example). It is correct to refer to intergrain porosity, such as in 'the intergrain porosity is 25%'. It is incorrect to use porosity with a descriptor, such as 'large and well-connected porosity', but it is correct to say 'large and well-connected pore space'. This distinction becomes important to the thesis of the paper, which is that dolomitization does not create porosity. However, it will be demonstrated that dolomitization can create pore space and modify pore size.

The problem

The principal culprit responsible for the confusion is the mole-for-mole replacement equation for dolomitization:

$$Mg^{+2} + 2CaCO_3 = MgCa(CO_3)^2 + Ca^{+2}. \quad (1)$$

Using this equation, De Beaumont (1837) demonstrated that the mole-for-mole conversion of calcite to dolomite results in a 12% decrease in volume because the molar volume of dolomite is smaller than the molar volume of calcite. He proposed, therefore, that epigenetic dolostones should have 12% porosity. He argued that this fact explains the presence of vugs observed in the dolostones in the Tyrol. The equation has been the model for dolomitization most used since that time.

The problem with this model is that it assumes that only magnesium, and no calcium or carbonate, is added from the dolomitizing water. Weyl (1960), using a conservation of mass argument, attempted to demonstrate that no carbonate could be added from the formation waters because of the small volume of carbonate found in groundwaters. However, it has recently been demonstrated that calcium and carbonate are extracted from the dolomitizing fluid, along with magnesium (Lucia & Major 1994). In addition, there are numerous observations of 'pore-filling dolomite' in limestones and dolostones. Gregg *et al.* (1993) demonstrated that a large volume of pore space was occluded by late dolomite. This pore-filling dolomite did not require the replacement of $CaCO_3$ to precipitate. Murray (1960) addressed this problem by introducing the concept of local and distant sources of carbonate, local source being equivalent to the volume of carbonate available from the dissolution of the local limestone and distant source being carbonate transported in with the dolomitizing fluid. Although Murray (1960) envisioned that distant-source dolomitization was a later phase, it is more reasonable to assume that it would be active to some degree during the replacement phase. Halley & Schmoker (1983) demonstrated that young limestones in south Florida are very porous and suggested that the associated dolostones inherited their porosity from the precursor limestone. They proposed a pore-filling phase to account for young dolostones having less porosity than associated limestones and introduced the term *overdolomitization* to describe the pore-filling phase. A more correct chemical model for dolomitization is therefore one that includes the addition of carbonate and magnesium during the replacement phase, and the addition of calcium, magnesium and carbonate during the pore-filling phase.

Morrow (1990) proposed the following equation (equation 2) for volume-for-volume replacement. The volume of carbonate needed to replace the difference in molar volumes is 0.1 for aragonite and 0.25 for calcite. It is reasonable to assume that this small amount of carbonate will continue to be available for dolomite precipitation after all the calcite or aragonite has been replaced until the water reaches equilibrium with dolomite and calcite/aragonite. This is the pore-filling phase, also referred to as 'overdolomitization' by Halley & Schmoker (1983) and Lucia & Major (1994) (equation 3).

Replacement phase:

$$(2 - x)CaCO_3 + Mg^{2+} + xCO_3^{-2} = CaMg(CO_3)_2 + (1 - x)Ca^{2+} \quad (2)$$

where x = moles of carbonate added from the dolomitizing fluid.

Pore-filling phase (overdolomitization):

$$Ca^{2+} + Mg^{2+} + 2CO_3 = CaMg(CO_3)_2. \quad (3)$$

Approach

The thesis of this paper is that dolostone porosity is related to the porosity of the precursor limestone and the amount of carbonate added from the dolomitizing fluid. The approach taken to investigate this thesis focuses on three basic questions: (1) What was the porosity of the precursor limestone? (2) What volume of carbonate has been added during and after the replacement phase? (3) How much porosity has been lost to compaction? By limiting the investigation to young carbonates from non-tectonic settings, the effects of compaction and late dolomitization are minimized, allowing work to focus on the nature of the precursor limestone and the volume of carbonate added.

Porosity of precursor limestones

The porosity of dolomitized limestone will depend on whether dolomitization occurred early or late in the diagenetic history of a limestone. It can occur syndepositionally or it can occur 160 Ma after deposition, as described by Kupecz & Land (1991) for the Ellenburger (Lower Ordovician) of west Texas. It is well known that carbonate sediments are very porous, having porosity values that range from 45 to 70%, depending on the volume of lime mud (Enos & Sawatsky 1981). Limestones do not have porosity values higher than this range and, with rare exception, have porosities much lower. Schmoker *et al.* (1985) reported an average porosity value of 12% for limestone reservoirs in the United States with a range of 9–17%. The basic question, therefore, is not how limestone porosity was created but how the porosity was destroyed and at what rate. The data presented here are from passive margins because the porosity history of tectonically active areas is complicated by the high compressive forces and temperatures typical of that environment. For example, Miller & Folk (1994) reported a porosity of 0.1 percent for the thrust and folded Triassic limestones of northern Italy.

In passive margins limestones tend to retain their porosity, losing it only gradually over geological time by cementation and compaction. Grainstones of Eocene age in south Florida located between depths of 92 and 490 m have an average porosity of 32%, and wackestone 36% (Budd 2002) (Fig. 1). Grain-dominated fabrics and wackestones from Pliocene–Miocene limestones at depths mostly between 200 and 600 m on the Great Bahama Bank both have an average porosity of 40% (Melim *et al.* 2001) (Fig. 1). Neither of these examples have evidence of meteoric diagenesis. In contrast, outcrops of Plio-Pleistocene limestone on Bonaire, Netherlands Antilles, have an average porosity of 25% (Lucia & Major 1994). Here there is ample evidence of extensive meteoric cementation and dissolution that resulted in the low porosity values.

Porosity in Cretaceous and Jurassic limestones commonly decreases as mud content increases, a phenomenon illustrated by Loucks (2002) using data from Cretaceous limestones in south Texas at depths ranging from 2000 to 2350 m. Grain-dominated fabrics have an average porosity of 13%, whereas mud-dominated fabrics average 7% porosity (Fig. 2). Cruz (1997) described Cretaceous reservoirs found at a depth of 4750 m in the Santos Basin, offshore Brazil, that have grain-dominated fabrics with 16% average porosity and mud-dominated fabrics with 2% average porosity.

The same cannot be said for the Cretaceous Shuaiba mud-dominated fabrics in the Middle East that typically have high porosity values. For example, in the Yibal Field, Oman, mudstones and wackestones have an average porosity of 35% at a depth of around 1450 m. In contrast, Rudist grain-dominated fabrics found in the Al Huwesiah Field have less porosity, averaging 18% (Fig. 2). High porosity in the mud-dominated fabrics is not unusual when compared with that of the Tertiary of Florida and the Great Bahama Bank, but it is unusual when compared with other Cretaceous carbonates. The Yibal Field is the shallowest of the three Cretaceous examples presented here, and the high porosity most probably results from a lack of deep burial or lack of cement, as suggested by Moshier (1989). Other investigators have suggested that the high porosity of the Shuaiba mud-dominated fabrics results from dissolution (Harris & Frost 1984). My observations, however, agree with those presented by Moshier (1989), who concluded that the high porosity did not result from dissolution but from incomplete cementation. The lower porosity value in rudist, grain-dominated fabrics most probably results from meteoric diagenesis of the rudist shoal.

An example from the Jurassic Arab D reservoir in the Ghawar Field, Haradh area, Saudi Arabia, from depths between 2100 and 2200 m, is illustrated in Figure 3. Although the data are highly scattered, average values show a trend of decreasing porosity with increasing lime mud (Lucia *et al.* 2001), a situation similar to that commonly observed in Cretaceous limestone reservoirs.

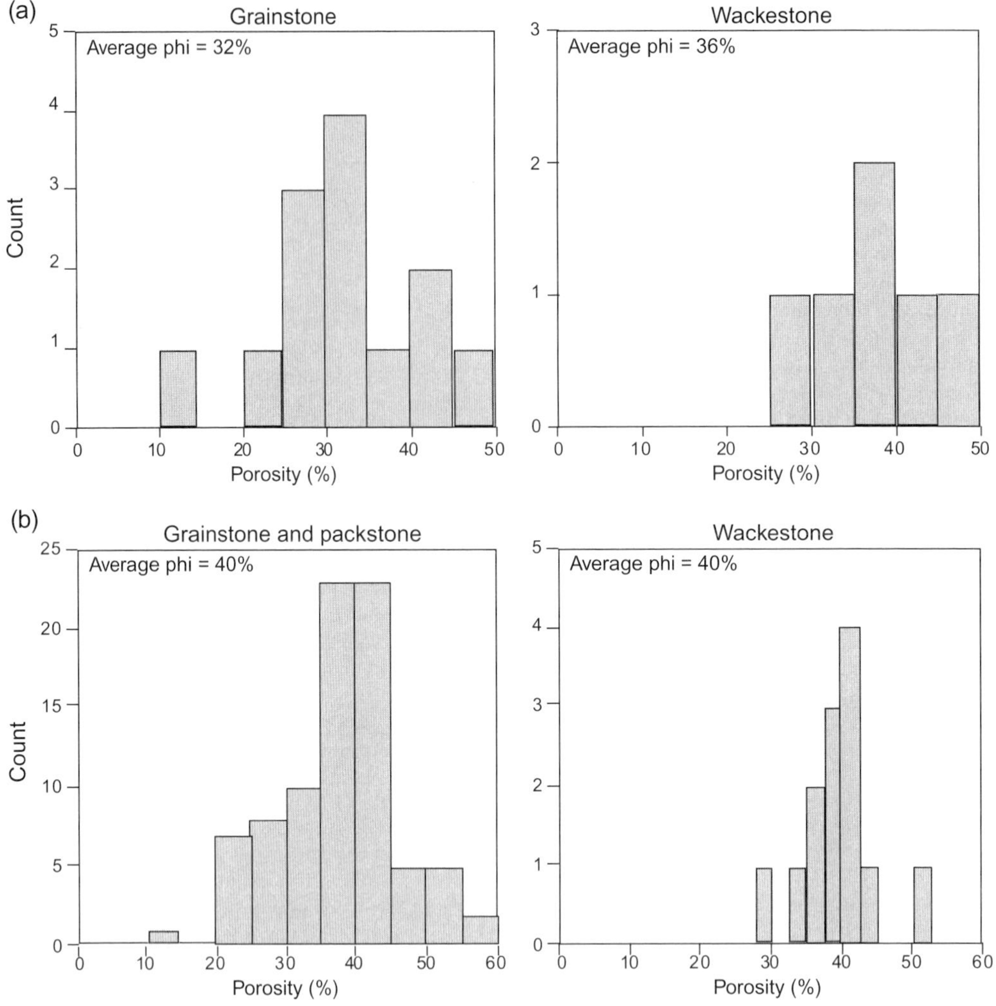

Fig. 1. Porosity histograms for grain-dominated and mud-dominated Tertiary limestones. (**a**) Palaeogene of west-central south Florida (Budd 2002). (**b**) Neogene of the Great Bahama Banks (Melim *et al.* 2001).

These data are consistent with the conclusion of Schmoker & Hally (1982) that, in a passive-margin environment, limestones lose their porosity gradually, probably related to their burial history (Fig. 4). Limestones less than 40 Ma old can have 30–40% porosity. The exception is limestone that has experienced extensive meteoric diagenesis, in which case porosity is reduced to the 20–30% range. There is no distinction between grain- and mud-dominated fabrics. Beginning with Cretaceous limestone, however, mud-dominated fabrics gradually become less porous than grain-dominated fabrics. In the author's experience, this trend continues into Palaeozoic rocks, where mud-dominated fabrics are commonly dense and grain-dominated fabrics are commonly porous.

Porosity of associated dolostones

Limestones lose their porosity gradually over tens of millions of years, and dolomitization can occur syndepositionally or millions of years after deposition. The porosity of the precursor limestone will, therefore, partly depend on the time of dolomitization. Determining the age of dolomitization relative to the host limestone is never easy. The dolomitization of peritidal sediments is assumed to be syndepositional. Dolomitized subtidal facies that are continuous

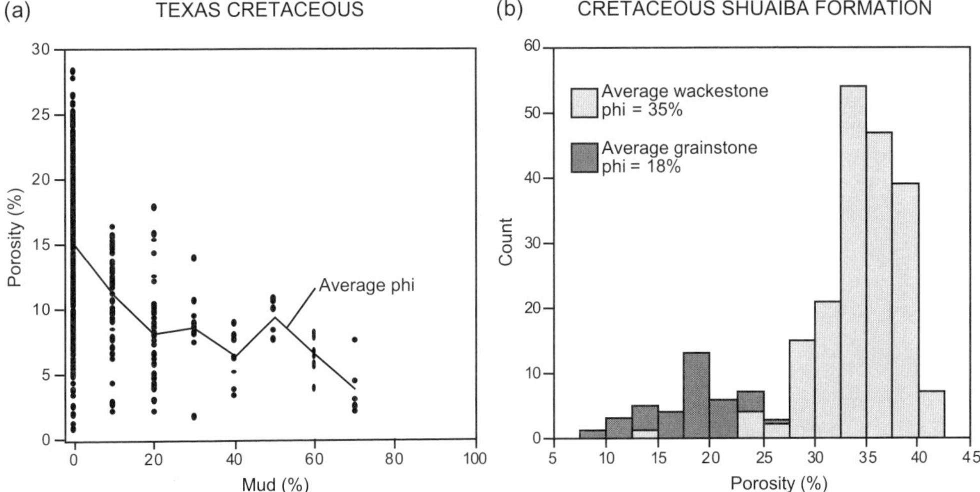

Fig. 2. Porosity data for Cretaceous limestones. (**a**) Plot of porosity v. lime mud content for south Texas Cretaceous samples, showing that average porosity decreases with increasing lime mud content (Loucks 2002). (**b**) Porosity histograms for Shuaiba (Cretaceous) samples from Oman. Mud-dominated samples from the Yibal Field have an average porosity of 35%. Rudist grainstone samples from the Al Huwiasiah Field have an average porosity of 18%.

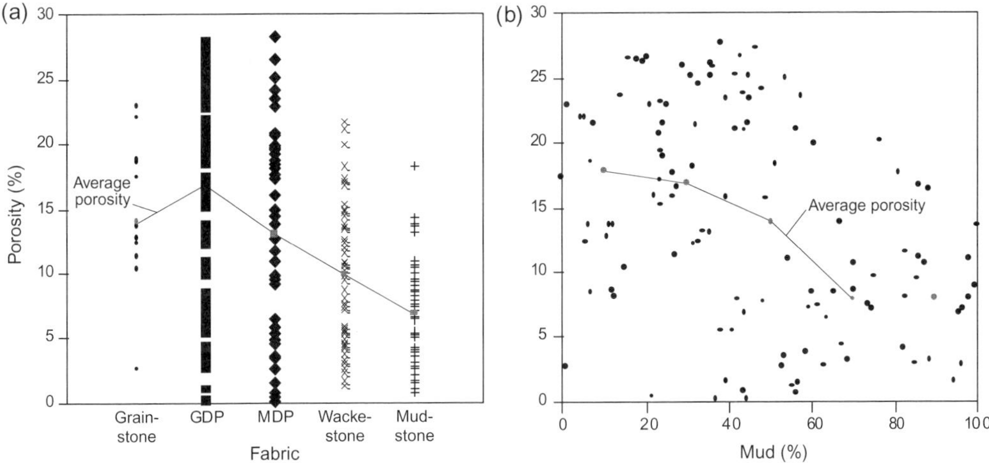

Fig. 3. Cross-plot from the Arab D reservoir in the Haradh area of the Ghawar Field, Saudi Arabia. (**a**) Porosity v. rock fabric. GDP: grain-dominated packstone; MDP: mud-dominated packstone. (**b**) Porosity v. per cent lime mud. Although there is considerable scatter in the data, the average trends show decreasing porosity with increasing lime mud content (Lucia *et al.* 2001).

with and lie beneath peritidal dolostone are generally thought to be the same age as the tidal-flat dolomite. The age of subtidal dolostone bodies not continuous with tidal-flat dolostones is difficult to determine, and the time of dolomitization is often determined based on geochemical data or relating dolostone bodies to some other datable diagenetic or structural event. Oxygen and strontium isotopes have been used to give a Permian age to the dolomitization of Permian carbonates in west Texas (Bein & Land 1982). Dolomitization of the El Paso Group occurred after collapse of an Ordovician cavern system, a collapse that

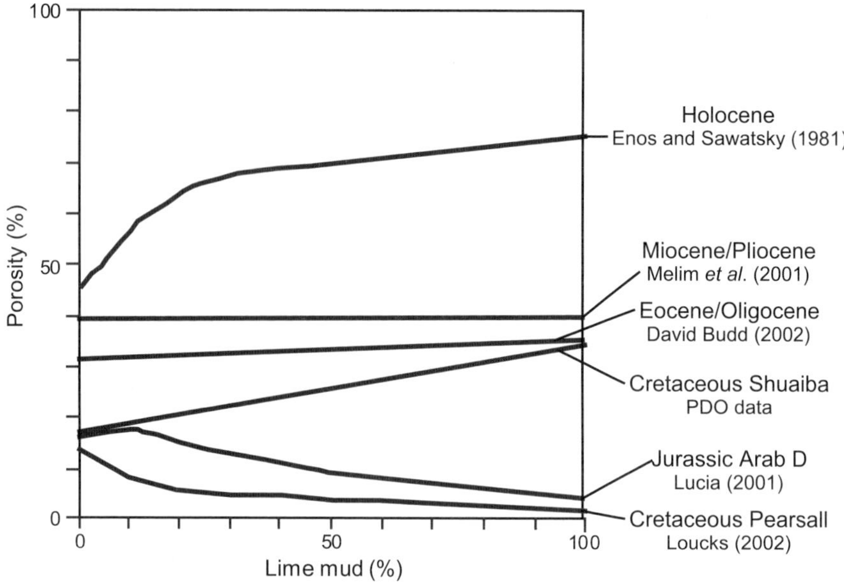

Fig. 4. Plot of per cent lime mud v. per cent porosity for Holocene, Neogene, Palaeogene, Cretaceous and Jurassic limestones. The plot shows: (1) the gradual loss of porosity with time (and presumably burial); and (2) greater porosity loss with increasing mud content starting in the Cretaceous, or 100 Ma after deposition.

continued through the Silurian (Lucia 1995*a*). Late dolomitization, therefore, occurred at least 100 Ma, and perhaps as much as 160 Ma, after deposition, assuming the Pennsylvanian age for the dolomitization proposed by Kupecz & Land (1991) for the time-equivalent Ellenburger Group. Wilson *et al.* (1990) dated late dolomitization in the Italian Alps by physical relationships to igneous intrusives, and Zempolich & Hardie (1997) dated late dolomitization by physical relationships to Alpine folding and thrusting.

To minimize the possible time gap between deposition and dolomitization, porosity information is collected from young carbonates where limestones and dolostones are juxtaposed. The youngest dolomite known is Holocene in age and is commonly found near the surface in evaporitic tidal-flat deposits. Unfortunately, there is little porosity information published about Holocene dolomitic sediments or dolostones. What few data are available have been published by Lucia (1999) and indicate that Holocene dolomitic sediments are very porous (Fig. 5), showing no trend of increasing or decreasing porosity through dolomitization. Clearly, porosity is not created by dolomitization but inherited from the sediment. Lasemi *et al.* (1989) concluded, on the basis of textural evidence, that Holocene dolomite from Andros Island is pore filling. However, no porosity data were presented to demonstrate porosity gain or loss.

The transition from dolostone to limestone found in Plio-Pleistocene carbonate outcrops on Bonaire, Netherlands Antilles, was described by Lucia & Major (1994). They attributed dolomitization to the evaporite reflux model and concluded that these carbonates have never been buried. The limestone is composed of grain-dominated fabrics with an average porosity of 25% and a porosity range of 10–40% (Fig. 5). Porosity values are lower than those found in carbonate sediments because of the extensive meteoric diagenesis during Pleistocene low stands (Lucia & Major 1994). The porosity values of dolomitic limestones range between 20 and 30%, a range similar to that of limestone porosity. The adjacent dolostones have an average porosity of 11% and a porosity range of 3–30%. If magnesium only and no carbonate were added during dolomitization, as is assumed by the mole-for-mole replacement model, the average dolostone porosity should be 35% instead of 11% (Fig. 6). The mole-for-mole dolomitization equation therefore does not explain porosity values. Some carbonate, as well as magnesium, must be added to keep the

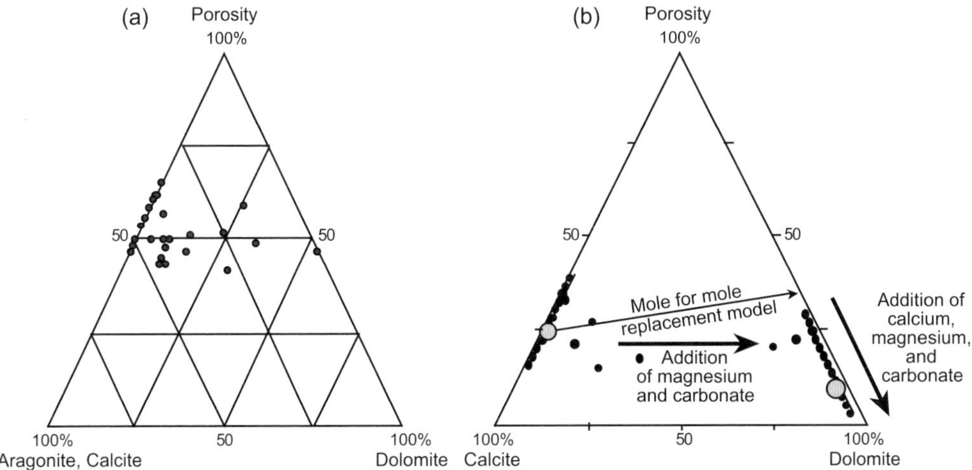

Fig. 5. Ternary diagrams of porosity, dolomite and $CaCO_3$ data from the Holocene and Plio-Pleistocene. (**a**) Holocene data from Qatar and Bonaire showing no recognizable change in porosity with increasing amounts of dolomite (Lucia 1999). (**b**) Plio-Pleistocene data from Bonaire showing no increase in porosity during calcite replacement and loss of porosity during overdolomitization (Lucia & Major 1994).

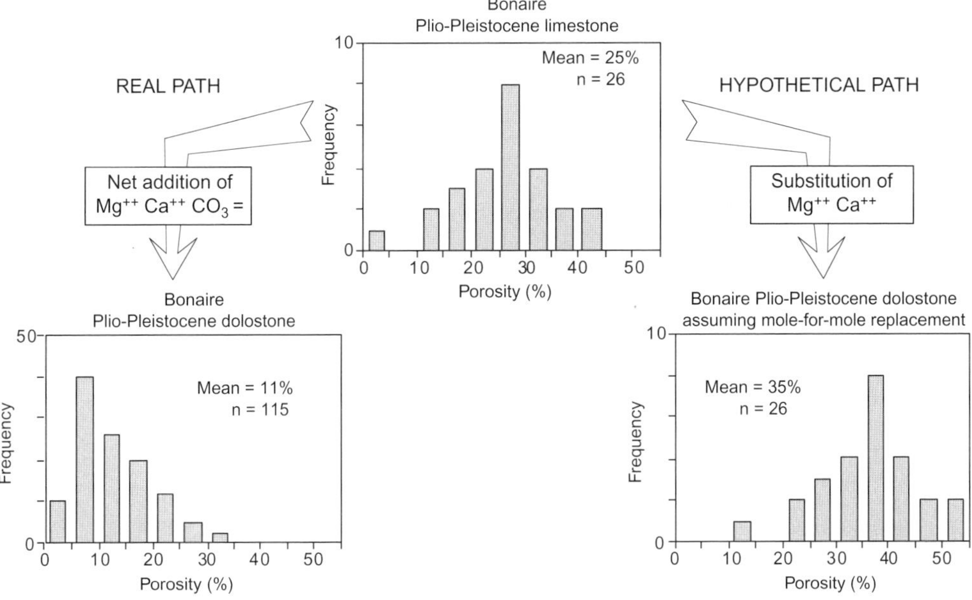

Fig. 6. Porosity–frequency plots for dolostone, the precursor limestone and a hypothetical dolostone calculated from the limestone frequency plot, assuming mole-for-mole replacement (Lucia & Major 1994), showing that the mole-for-mole model does not explain the dolostone porosity.

porosity from increasing during the replacement phase, and calcium, magnesium and carbonate must be added after the replacement phase to reduce average porosity to 11%.

Dolostone of Miocene age (McNeill *et al.* 2001) is found in the Unda well drilled on the Great Bahama Bank at a depth of 270–354 m. The dolostone has an average porosity of 40% (Melim *et al.* 2001), which is equal to the average porosity of limestones in the Unda and

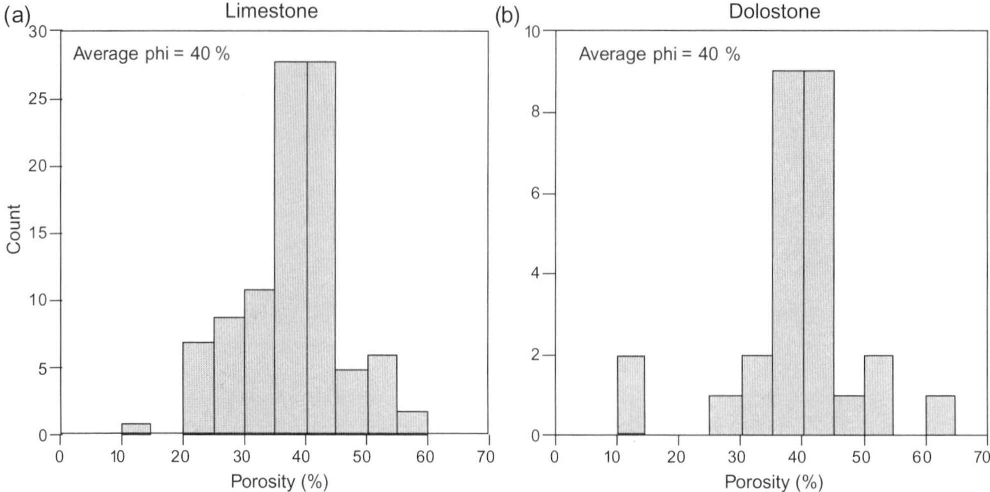

Fig. 7. Porosity histograms for (**a**) Neogene limestones from the Unda and Clino wells and (**b**) Miocene dolostones from the Unda well, Great Bahama Banks (Melim *et al.* 2001). Both limestone and dolostone have 40% porosity, showing that dolostone porosity is inherited from limestone.

Clino wells (Fig. 7). Little is understood about the origin of the dolomitizing water, except that it was most probably modified seawater, and little is known about the time of dolomitization (Swart & Melim 2000). There is no evidence that these carbonates were ever subjected to extensive meteoric diagenesis. The originally high porosity of about 70% in the wackestones was therefore reduced to the present value of 40% by simple burial cementation and compaction, whereas the grain-dominated fabrics retained their depositional porosity of 40%. It is likely that dolomitization occurred after the wackestones were compacted to 40% porosity. The high porosity in muddy sediments is lost in the first 100 m of burial, according to Goldhammer (1997), indicating that dolomitization occurred after 100 m of burial. The similarity between limestone and dolostone porosity values suggests that: (1) dolostone porosity was not created through the process of dolomitization but was simply inherited from the precursor limestone; and (2) no carbonate was added after the replacement phase of dolomitization.

Unfortunately, no limestone–dolostone pairs are available from Cretaceous reservoirs. A search of the literature indicates that there are but a few Cretaceous dolostone reservoirs. Data from the Arab D reservoir in the Ghawar Field, Saudi Arabia, of Jurassic age provide insight into the porosity history of interbedded limestones and dolostones. Rock fabrics from the Arab D reservoir were originally described by Powers (1962) and more recently by Lucia *et al.* (2001). There is no indication of extensive meteoric diagenesis. The age of dolomitization and the nature of the dolomitization water are poorly understood, although geochemical data suggest modified marine water (Cantrell *et al.* 2001). A plot of dolomite as a *per cent of bulk volume* (BV) v. porosity shows that porosity tends to decrease as per cent BV dolomite increases (Fig. 8). At about 80% BV dolomite, porosity increases to a point similar to that of the porosity of the limestone. From this point on, the samples are all dolostone, except for a few anomalous calcitic dolostones, and the porosity of the dolostone decreases. This relationship is similar to one described by Powers (1962). The decrease in porosity from 0 to 80% BV dolomite is attributed to an increase in mud-dominated fabrics and a corresponding increase in compaction. The compaction is thought to occur much later than dolomitization, and the porosity of the precursor limestone and dolomitic-limestone fabrics is suggested to be between 30 and 40%. Carbonates with more than 80% BV dolomite resisted compaction because the dolomite forms a supporting framework. The grain-dominated fabrics also resisted compaction. The similarity between the porosity of grain-dominated limestones and dolostones suggests that dolostone porosity is inherited from the precursor limestone. The loss of porosity in the dolostone is attributed to variations in the porosity of the precursor and to the addition of carbonate after the replacement phase (overdolomitization).

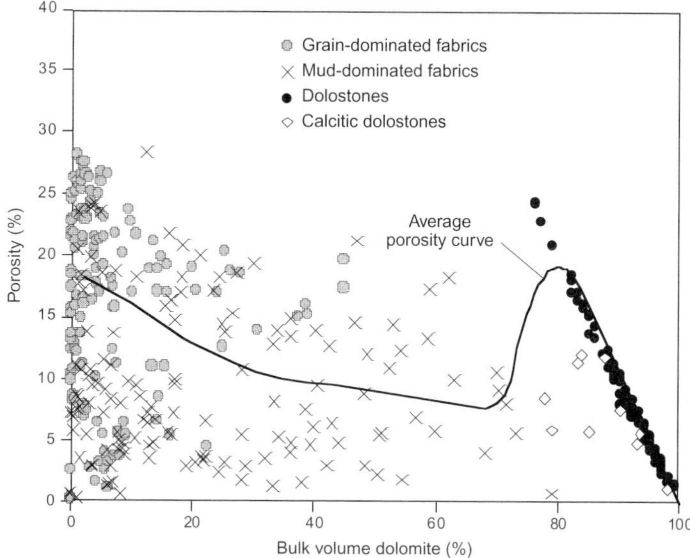

Fig. 8. A plot of porosity v. bulk-volume dolomite (dolomite as a per cent of mineral volume plus porosity) for various fabrics, showing that average porosity decreases with increasing dolomite until little or no calcite remains. This loss of porosity is attributed to differential compaction related to the increase in mud-dominated fabrics, which are most susceptible to compaction. Dolostone porosities can be as high as limestone porosity but decrease as dolomite cement is added (overdolomitization).

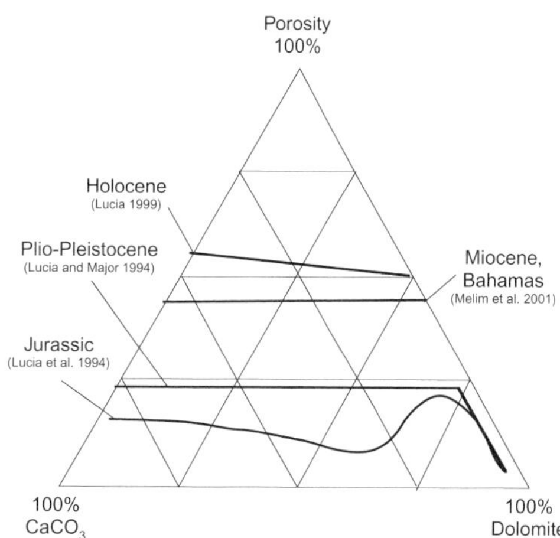

Fig. 9. Ternary diagram of porosity, $CaCO_3$, and dolomite from the Holocene, Plio-Pleistocene, Miocene and Jurassic. There is little difference between limestone and dolostone porosity in the Holocene and Miocene. The decrease in porosity with increasing dolomite in the Jurassic is attributed to differential compaction of limestone v. dolostone. Both Plio-Pleistocene and Jurassic data show loss of porosity in the dolostone through overdolomitization.

The data from Holocene and Tertiary limestone–dolostone pairs demonstrate that young dolostones inherit porosity from precursor limestones (Fig. 9). The Miocene example has less porosity than the Holocene example reflecting dolomitization after shallow burial. The Bonaire example demonstrates that if the limestone has undergone extensive meteoric diagenesis, with accompanying porosity loss, porosity in the dolostone reflects this loss. In addition, dolomite cementation continues after replacement dolomitization. The Jurassic example illustrates the effect of burial diagenesis and overdolomitization on porosity values. The dolostones resist compaction and preserve porosity. However, the addition of pore-filling dolomite after replacement reduces porosity below the porosity of the precursor limestone. The mud-dominated limestones, and to some degree grain-dominated limestones, lose porosity through compaction. Porosity reduction with increasing dolomite is therefore not related to the dolomitization process but to the susceptibility of muddy sediments to compaction.

A recent defence of the mole-for-mole model was offered by Wilson *et al.* (1990) and Zempolich & Hardie (1997). In both papers the precursor limestone is assumed to be dense, but no porosity values are reported for the current limestone and no discussion of the diagenetic history of the limestone is presented. Wilson *et al.* (1990) referred to 'porous dolomite' but offered no porosity measurements to substantiate or quantify this observation. Zempolich & Hardie (1997) presented a well-documented history of porosity evolution observed at limestone–dolomite contacts. Visible porosity in thin section and polished slabs was determined to be 10–15% using point-counting methods. Micropores are not visible in thin sections of slabs, so point-count porosity values are typically understated. The large crystal and grain sizes of these fabrics suggest, however, that the visible porosity values are close to total porosity. The age of the pore space relative to dolomitization is not clear. Most of the pores are oomolds that can be reasonably assumed to have formed early in the diagenetic history of the limestone and before dolomitization. It is nevertheless argued that oomolds are found only in association with the dolomitization front. Also, the limestone is a deep-water oolite shed off a shallow-water platform, minimizing the likelihood of oomolds being formed by meteoric dissolution. In addition, oomolds are commonly produced by dissolution of unstable aragonite, and aragonite is uncommon in the greenhouse environment of the Triassic. Interestingly, both these papers focus on diffusion of magnesium for dolomitizing the dense limestone, as did Miller & Folk (1994). This is a very different dolomitization model from the fluid-flow model envisioned in this paper and could result in no additional carbonate being available during dolomitization.

Petrophysical properties

If dolostone simply inherits porosity from the precursor limestone, and may actually reduce porosity through overdolomitization, why are dolostones often better reservoirs than limestones? One reason is that dolostones are less susceptible to compaction than limestones, maintaining their porosity more effectively. Another reason is the change in pore structure that occurs during dolomitization. Permeability and water saturation are two important petrophysical parameters that describe a carbonate reservoir. They are controlled fundamentally by pore-size distribution, and pore-size distribution has been shown to be controlled by particle size and sorting (Lucia 1995*b*). The pore-size distribution in a dolograinstone is similar to that in the precursor limestone and typically has little relationship to crystal size (Lucia 1995*b*). The pore space and porosity are inherited from the precursor limestone, not created through dolomitization. Similarly, the pore-size distribution in grain-dominated dolopackstones is controlled by the grain size and sorting of the precursor limestone, and the pore space and porosity are inherited from the precursor limestone.

In mud-dominated dolostones, however, crystal size pays a dominant role in pore-size distribution (Fig. 10). As the dolomite crystal size increases, the pore size also increases. In this case, the porosity of the dolostone is inherited from the precursor limestone, but the pore space is newly created. If a wackestone is converted to dolostone with a fine (less that 20 μm) crystal size, the dolostone will have pore sizes similar to those of the precursor limestone because lime mud is commonly composed of crystals and grains that are less than 20 μm in size. However, if the dolomite crystals are larger than 20 μm, the pore size will be larger than that of the precursor limestone, and permeability will be higher and water saturation lower for equivalent conditions. Assuming that interparticle porosity is constant, permeability of mud-dominated dolostones is a direct function of dolomite crystal size, and permeability increases by about a factor of 10 as crystal size increases from fine to medium to large (Fig. 10).

Figure 11 illustrates the process of converting a wackestone with micropores to a medium crystalline dolowackestone with large intercrystal pores. As dolomite is formed, it replaces not only the lime mud particles but also pore space between them. The adjacent lime mud is dissolved to supply most of the carbonate needed to fill this pore space. A small amount of carbonate comes from the dolomitizing water as well, keeping the porosity constant. As dolomitization progresses, the crystals impinge on one another, and intercrystal pore space is created. The size of this pore space is controlled by the size of the dolomite crystals and is larger than the micropores. When all or most of the lime mud has been dissolved, carbonate must be supplied to the growing rhombs by dissolution of the large allochems, resulting in the formation of grain moulds, a type of separate vug (Lucia 1995*b*, 1999). Porosity is not increased by this process because the dissolved carbonate is reprecipitated locally on the growing dolomite crystals, along with a small amount of carbonate from the water. At this stage, the wackestone and dolowackestone have the same porosity but different pore space; the pore space in the dolowackestone was created

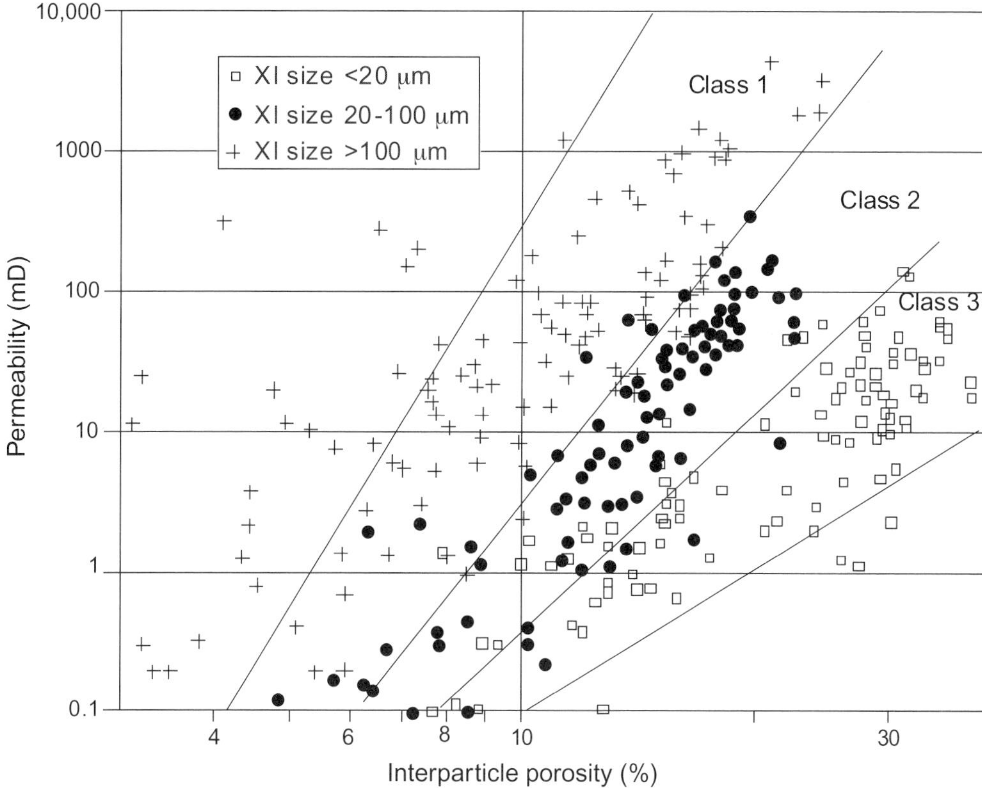

Fig. 10. Cross-plot of interparticle porosity (intercrystal porosity) and permeability showing the effect of crystal size in mud-dominated dolostones (Lucia 1995*b*). Petrophysical classes 1, 2 and 3 are shown for comparison purposes.

through dolomitization. After all the limestone has been dissolved, the dolomite crystals continue to grow using magnesium, calcium and carbonate from the dolomitizing water until the water reaches equilibrium. This overdolomitization can reduce porosity to 5% or less. With burial, the partially dolomitized wackestones lose porosity by compaction, whereas the dolostones resist compaction and retain their porosity.

Figure 12 is a cross-plot of interparticle porosity and permeability for all dolostones. Large crystalline dolostones with any precursor fabric, as well as dolograinstones, plot in the petrophysical class 1 field (see Lucia 1995*b*). Medium crystalline, mud-dominated dolostones and grain-dominated dolopackstones plot in the petrophysical class 2 field, and fine crystalline, mud-dominated dolostones plot in the petrophysical class 3 field. According to Lucia (1995*b*), grainstone fabrics plot in the class 1 field, grain-dominated packstones plot in the class 2 field and mud-dominated fabrics plot in the class 3 field. The effect of dolomitization on petrophysical properties is to increase the petrophysical characteristics of mud-dominated fabrics from class 3 to classes 2 and 1. In a dolostone reservoir the mud-dominated fabrics are therefore commonly excellent reservoir rocks, along with grain-dominated dolostones, whereas in a limestone reservoir only grain-dominated fabrics have excellent reservoir quality.

Conclusions

The data presented here suggest the following model for porosity development in dolostones in a passive-margin environment. Assuming that dolomitization occurred within the first 40 Ma after deposition, it can be reasonably assumed that the limestone porosity was in the 30–40% range. During the replacement phase, the dolostone will inherit this porosity and any change in molar volume will be offset by the precipitation of pore-filling dolomite using

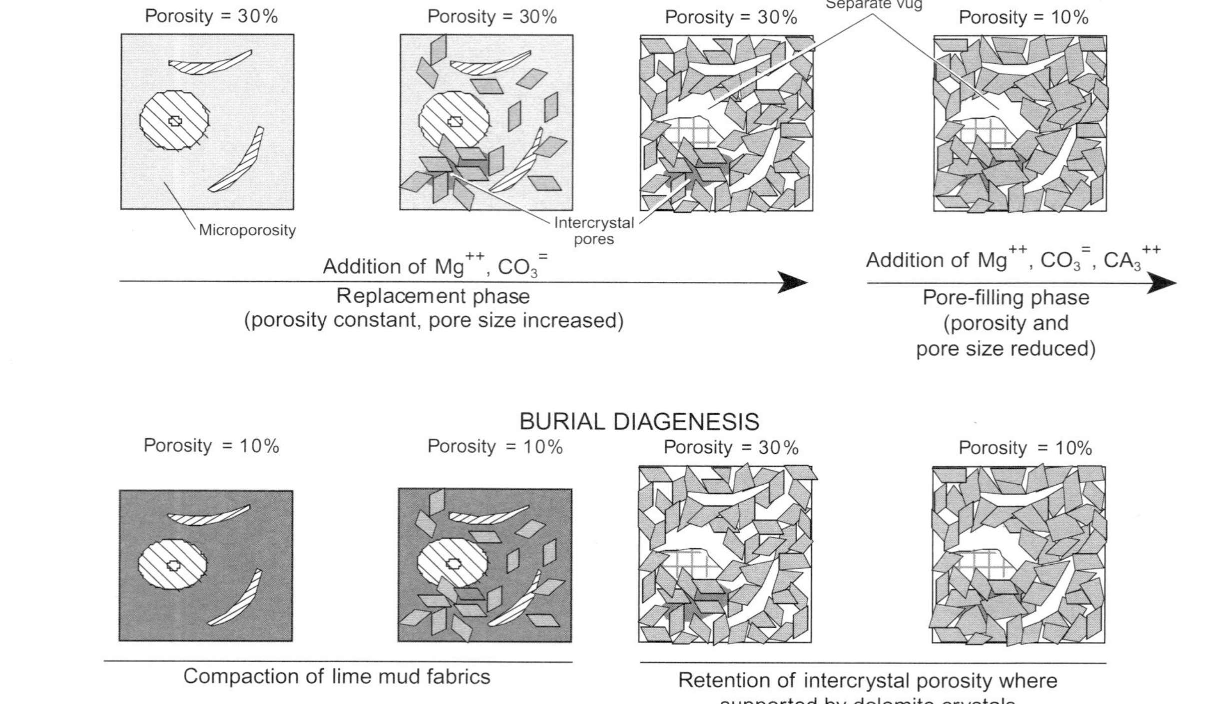

Fig. 11. Cartoon illustrating the development of intercrystal porosity in a medium crystalline dolowackestone. The wackestone has 30% micropores. As dolomite is formed, it replaces not only the lime mud particles but also pore space between the mud particles. Lime mud is dissolved from the adjacent mud to supply most of the carbonate needed to fill this pore space. As dolomitization progresses, the crystals impinge upon each other and intercrystal pore space is created, which is larger than the micropores. When all or most of the lime mud has been dissolved, carbonate must be supplied to the growing rhombs by the large allochems dissolving to form separate vugs. During the replacement process, a small amount of carbonate comes from the dolomitizing water, keeping the porosity constant at 30%. At this stage, wackestone and dolowackestone have the same porosity but different pore sizes. Pore space in the dolowackestone was created through dolomitization. After all the limestone has been dissolved, the dolomite crystals continue to grow using magnesium, calcium and carbonate from the dolomitizing water until the water reaches equilibrium, reducing porosity to as low as 5%. With burial, partially dolomitized wackestones lose porosity by compaction to perhaps 10%, whereas dolostones resist compaction and retain their porosity.

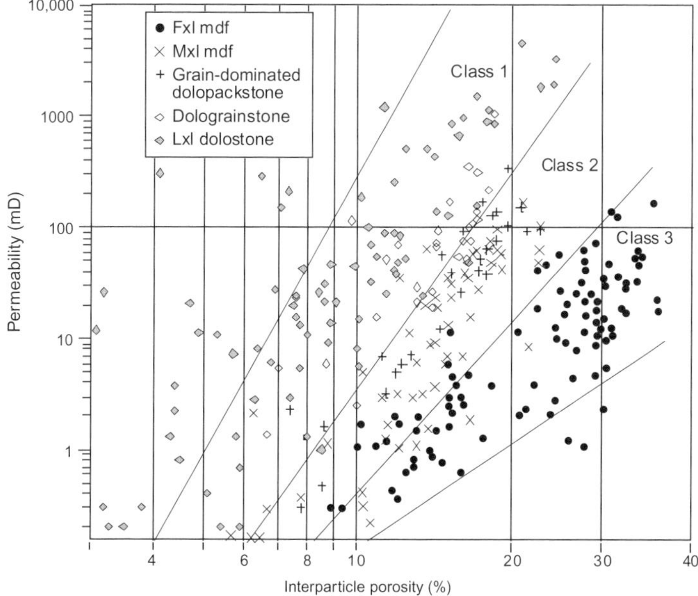

Fig. 12. Cross-plot of interparticle porosity (intercrystal and intergrain porosity) and permeability comparing distribution of the five basic dolostone rock fabrics with the petrophysical class fields. Dolograinstone and large crystalline (Lxl) dolostone plot in the class 1 field, grain-dominated dolopackstone and medium crystalline mud-dominated dolostone (Mxl mdf) plot in the class 2 field, and fine crystalline mud-dominated dolostone (Fxl mdf) plot in the class 3 field (Lucia 1995*b*).

carbonate transported in the dolomitizing water. After the replacement phase, dolomite continues to precipitate in the pore space and overdolomitization occurs reducing porosity to as low as 5%.

The dolomitization model must account for the addition of magnesium as well as carbonate during the replacement and the overdolomitization phase.

Replacement phase (Morrow 1990)

$$(2-x)CaCO_3 + Mg^{2+} + xCO_3^{2-} = CaMg(CO_3)_2 + (1-x)Ca^{2+} \quad (2)$$

where x = moles of carbonate added from the dolomitizing fluid.

Pore-filling phase (overdolomitization)

$$Ca^{2+} + Mg^{2+} + 2CO_3 = CaMg(CO_3)_2. \quad (3)$$

Limestone porosity is lost to compaction and cementation very gradually over tens of millions of years, with mud-dominated fabrics losing porosity faster than grain-dominated fabrics. Dolostones resist compaction more than limestones, although initially dolostones will have the same or less porosity than adjacent limestones. The limestone will lose porosity preferentially with burial owing to compaction, resulting eventually in porous dolostones and dense limestones. This process, however, will take hundreds of millions of years in a passive-margin setting.

This model is based on an open-circulation system in which flowing dolomitizing water brings in magnesium, calcium and carbonate hydrodynamically. In a diffusion-dominant dolomitization model, it is possible that only magnesium and calcium are transported in and out of the limestone along the dolomitization front and that carbonate is not available to form additional dolomite to compensate for the molar volume changes. As long as there is fluid flow, however, the availability of carbonate from the water to add dolomite to offset the volume change must be considered.

Dolostones commonly make excellent reservoirs because they retain porosity and because mud-dominated dolostones can have larger pore sizes than mud-dominated limestones. Grain-dominated dolostones tend to have the same pore structure as the precursor limestone. Dolomitized mud-dominated fabrics, however, can have significantly larger pores than the precursor limestone because dolomite crystal sizes are commonly larger than the precursor mud size. Permeability can be predicted if interparticle (grains or crystals) porosity, the precursor rock fabric and dolomite crystal size are known.

I would like to acknowledge the many thoughtful discussions with my colleagues C. Kerans, S. Ruppel and R. Loucks here at the Bureau of Economic Geology. The manuscript was edited by L. Dietrich to whom I am deeply indebted. Financial support came from the many sponsors of the Reservoir Characterization Research Laboratory project, an Industrial Associates project at the Bureau of Economic Geology, John A. and Katherine G. Jackson School of Geosciences, The University of Texas at Austin.

References

BEIN, A. & LAND, L.S. 1982. *San Andres Carbonates in the Texas Panhandle: Sedimentation and Diagenesis Associated with Magnesium–Calcium–Chloride Brines*. The University of Texas at Austin, Bureau of Economic Geology Report of Investigations, **121**.

BUDD, D.A. 2002. The relative roles of compaction and early cementation in the destruction of permeability in carbonate grainstones: a case study of the Paleogene of west-central Florida, U.S.A. *Journal of Sedimentary Research*, **72**, 116–128.

CANTRELL, D.L., SWART P.K., HANDFORD, R.C., KENDALL, C.G. & WESTPHAL, H. 2001. Geology and production significance of dolomite, Arab-D reservoir, Ghawar field, Saudi Arabia. *GeoArabia*, **6**, 45–60.

CRUZ, W.M. 1997. *Study of Albian carbonate analogs: Cedar Park Quarry, Texas, USA, and Santo Basin reservoirs, southeast offshore Brazil*. PhD dissertation, The University of Texas at Austin, Austin, TX.

DE BEAUMONT, E. 1837. Application du calcul a l'hypothèse de la formation par epigenie des anhydrites, des gypses et des dolomies. *Bulletin de la Société Géologique de France*, **8**, 174–177.

ENOS, P. & SAWATSKY, L.H. 1981. Pore networks in Holocene carbonate sediments. *Journal of Sedimentary Petrology*, **51**, 961–985.

GOLDHAMMER, R.K. 1997. Compaction and decompaction algorithms for sedimentary carbonates. *Journal of Sedimentary Research*, **67**, 26–56.

GREGG, J.M., LAUDON, P.R., WOODY, R.E. & SHELTON, K.L. 1993. Porosity evolution of the Cambrian Bonneterre Dolomite, Southeastern Missouri, USA. *Sedimentology*, **40**, 1153–1169.

HALLEY, R.B. & SCHMOKER, J.W. 1983. High-porosity Cenozoic carbonate rocks of south Florida: progressive loss of porosity with depth. *AAPG Bulletin*, **67**, 191–200.

HARRIS, P.M. & FROST, S.H. 1984. Middle Cretaceous carbonate reservoir, Fahud Field and Northwestern Oman. *AAPG Bulletin*, **68**, 649–658.

KUPECZ, J.A. & LAND, L.S. 1991. Late-stage dolomitization of the Lower Ordovician Ellenburger Group, West Texas. *Journal of Sedimentary Petrology*, **61**, 551–574.

LANDES, K.K. 1946. Porosity trough dolomitization. *AAPG Bulletin*, **30**, 305–318.

LASEMI, Z., BOARDMAN, M.R. & SANDBERG, P.A. 1989. Cement origin of supratidal dolomite, Andros Island, Bahamas. *Journal of Sedimentary Petrology*, **59**, 249–257.

LOUCKS, R.G. 2002. Controls on reservoir quality in platform-interior limestone through the Gulf of Mexico: example form the Lower Cretaceous Pearsall Formation in South Texas. *Gulf Coast Association of Geologic Societies Transactions*, **52**, 659–672.

LUCIA, F.J. 1995*a*. Lower Paleozoic development, collapse, and dolomitization, Franklin Mountains, El Paso, Texas. *In*: BUDD, D.A., SALLER, A.H. and HARRIS, P.A. (eds) *Unconformities and Porosity in Carbonate Strata*. American Association of Petroleum Geologists, Memoir, **63**, 279–300.

LUCIA, F.J. 1995*b*. Rock-fabric/petrophysical classification of carbonate pore space for reservoir characterization. *AAPG Bulletin*, **79**, 1275–1300.

LUCIA, F.J. 1999. *Carbonate Reservoir Characterization*. Springer, Berlin.

LUCIA, F.J. & MAJOR, R.P. 1994. Porosity evolution through hypersaline reflux dolomitization. *In*: PURSER, B.H., TUCKER, M.E. & ZENGER, D.H. (eds) *Dolomites: A Volume in Honour of Dolomieu*. International Association of Sedimentologists, Special Publications, **21**, 325–341.

LUCIA, F.J., JENNINGS, J.W., JR, RAHNIS, M. & MEYER, F.O. 2001. Permeability and rock fabric from wireline logs, Arab-D Reservoir, Ghawar Field, Saudi Arabia. *GeoArabia*, **6**, 619–646.

MCNEILL, D.F., EBERLI, G.P., LIDZ, B.H., SWART, P.K. & KENTER, J.A.M. 2001. Chronostratigraphy of a prograded carbonate platform margin: a record of dynamic slope sedimentation, western Great Bahama Bank. *In*: GINSBURG, R. (ed.) *Subsurface Geology of a Prograding Carbonate Platform Margin, Great Bahama Bank: Results of the Bahamas Drilling Project*. Society of Economic Paleontologists and Mineralogists, Special Publications, **70**, 61–100.

MELIM, A.L., ANSELMETTI, R.S. & EBERLI, G.P. 2001. The importance of pore type on permeability of Neogene carbonates, Great Bahama Bank. *In*: GINSBURG, R. (ed.) *Subsurface Geology of a Prograding Carbonate Platform Margin, Great Bahama Bank: Results of the Bahamas Drilling Project*. Society of Economic Paleontologists and Mineralogists, Special Publications, **70**, 217–240.

MILLER, J.K. & FOLK, R.L. 1994. Petrographic, geochemical and structural constraints on the timing and distribution of postlithification dolomite in the Rhaetian Portoro ('Calcare Nero') of the Portovenere Area, La Spezia, Italy. *In*: PURSER, B.H., TUCKER, M.E. & ZENGER, D.H. (eds) *Dolomites: A Volume in Honour of Dolomieu*. International Association of Sedimentologists, Special Publications, **21**, 187–202.

MORROW, D.W. 1990. Dolomite – part 1: The chemistry of dolomitization and dolomite precipitation. *In*: MCILREATH, I.A. & MORROW, D.W. (eds) *Diagenesis*. Geoscience Canada, Reprint Series **4**, 113–123.

MOSHIER, S.O. 1989. Development of microporosity in a micritic limestone reservoir, Lower Cretaceous, Middle East. *Sedimentary Geology*, **63**, 217–240.

MURRAY, R.C. 1960. Origin of porosity in carbonate rocks. *Journal of Sedimentary Petrology*, **30**, 59–84.

POWERS, R.W. 1962. Arabian Upper Jurassic carbonate reservoir rocks. *In*: HAM, W.E. (ed.) *Classification of Carbonate Rocks: A Symposium*. American Association of Petroleum Geologists, Memoir, **1**, 122–192.

SCHMOKER, J.W. & HALLEY, R.B. 1982. Carbonate porosity versus depth: a predictable relation for South Florida. *AAPG Bulletin*, **66**, 2561–2570.

SCHMOKER, J.W., KRYSTINIK, K.B. & HALLEY, R.B. 1985. Selected characteristics of limestone and dolomite reservoirs in the United States. *AAPG Bulletin*, **69**, 733–741.

SWART, P.K. & MELIM, L.A. 2000. The origin of dolomites in Tertiary sediments from the margin of Great Bahama Bank. *Journal of Sedimentary Petrology*, **70**, 738–748.

TUCKER, M.E. 2001. *Sedimentary Petrology; An introduction to the Origin of Sedimentary Rocks*. Blackwell Scientific, Oxford.

WEYL, P.K. 1960. Porosity through dolomitization: conservation-of-mass requirements. *Journal of Sedimentary Petrology*, **30**, 59–84.

WILSON, E.N., HARDIE, L.A. & PHILLIPS, O.M. 1990. Dolomitization front geometry, fluid flow patterns, and the origin of massive dolomite: the Triassic Latemar buildup, Northern Italy. *American Journal of Science*, **290**, 741–796.

ZEMPOLICH, W.G. & HARDIE, L.A. 1997. Geometry of dolomite bodies within deep-water resedimented oolite of Middle Jurassic Vajont limestone, Venetian Alps, Italy: analogs for hydrocarbon reservoirs created through burial dolomitization. *In*: KUPECZ, J.A., GLUYAS, J.G. & BOLCH, S. (eds) *Reservoir Quality Prediction in Sandstones and Carbonates*. American Association of Petroleum Geologists Memoirs, **69**, 127–162.

Basin fluid flow, base-metal sulphide mineralization and the development of dolomite petroleum reservoirs

JAY M. GREGG

Department of Geology & Geophysics, School of Mines and Metallurgy, 125 McNutt Hall, University of Missouri-Rolla, Rolla, MO 65409-0410, USA (e-mail: greggjay@umr.edu)

Abstract: Saline basinal fluids, at temperatures from 60 to 250 °C, have affected almost every sedimentary basin in the world including rocks from Palaeoproterozoic to Cenozoic age. These fluids commonly precipitate base-metal sulphides (pyrite, sphalerite, galena, etc.) and associated minerals (barite, fluorite, calcite, dolomite, etc.) ranging in volume from trace amounts to large economic ore deposits. Such deposits are commonly referred to as Mississippi Valley-type (MVT) after the large Palaeozoic deposits of this kind found in the Mississippi Valley of North America. They are primarily hosted by platform carbonates, typically dolomite, and are usually associated with hydrocarbons.

Dolomites not affected by mineralizing fluids commonly display micron- to decimicron-size planar textures, and have well-developed micro- and mesoporosity networks dominated by intercrystal and vug porosity. However, these and other carbonate rocks affected by basinal fluids may undergo massive geochemical and textural alteration. This occurs even when the affected rocks are distal from the main loci of sulphide mineralization. Alteration includes: dolomitization of limestone; neomorphic recrystallization of existing dolomite; and precipitation at intervals of large volumes of open-space-filling dolomite, calcite and quartz cements alternating with dissolution. Dolomitization of limestone and/or neomorphic recrystallization of dolomite, at elevated temperatures, commonly results in centimicron and larger size crystals, and development of nonplanar textures that increase pore-throat tortuosity. Open-space-filling dolomite, calcite and quartz cementation causes a dramatic reduction of porosity and blockage of pore throats. Periods of carbonate dissolution, proximal to intense sulphide mineralization, result in the development of large-scale macroporosity such as breccias that are commonly superimposed on karst and tectonic fractures.

Exposure to mineralizing basinal fluids substantially alters porosity and permeability distribution, and thus the potential reservoir properties of the dolomite. The resulting reservoir may have little resemblance to its precursor. Understanding the epigenetic history of a dolomite is critical, therefore, as this will ultimately affect its development strategy and production history.

Base-metal sulphide mineralization, ranging in scale from trace occurrences of pyrite, sphalerite, galena and other minerals to major ore districts, is ubiquitous in sedimentary basins throughout the world (Sangster 1990, 1996). Mineralization occurs in rocks ranging in age from Palaeoproterozoic (Wheatley *et al.* 1986; Baugaard *et al.* 2001) to Cenozoic (Land & Prezbindowski 1985; Kyle & Saunders 1996; Leach *et al.* 2001), although known large (economic ore) deposits are apparently restricted to pre-Cenozoic rocks (Anderson & Macqueen 1982). Sulphides are most commonly hosted in carbonate rocks, usually dolomite, although sandstone-hosted deposits are known (Bjørlykke & Sangster 1981). Carbonate-hosted ore deposits are generally referred to as Mississippi Valley-type (MVT), after the large deposits of this kind found on the Mid-continent of North America (Fig. 1). MVT metal sulphides are believed to have been deposited at relatively shallow depths (mostly ≤1 km) by warm (60–250 °C), saline, basinal fluids. Fluid-inclusion studies indicate complex, highly saline Na–Ca–Cl brines, and probable mixing of a number of end-member components (Anderson & Macqueen 1982; Shelton *et al.* 1992). Recent work indicates a component of highly evaporated seawater and/or disolved evaporites in many of these fluids (Gleeson *et al.* 1999; Banks *et al.* 2002; Kendrick *et al.* 2002).

Mineralization was accompanied by precipitation of large volumes of open-space-filling gangue minerals such as calcite, dolomite and quartz, and less commonly by fluorite, barite, celestite and anhydrite. MVT deposits are almost always associated with massive dolomitization (Anderson & Macqueen 1982; Sangster 1996). Both MVT and related metal sulphide deposits are also almost always associated with

From: BRAITHWAITE, C. J. R., RIZZI, G. & DARKE, G. (eds) 2004. *The Geometry and Petrogenesis of Dolomite Hydrocarbon Reservoirs*. Geological Society, London, Special Publications, **235**, 157–175. 0305-8719/$15.00

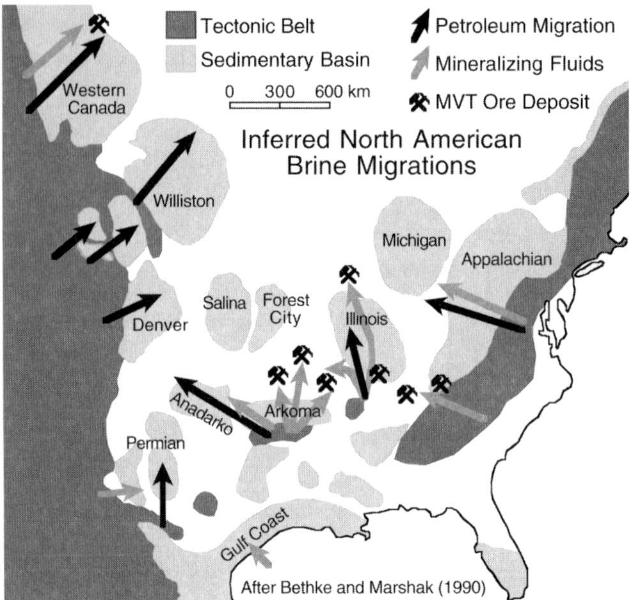

Fig. 1. North America showing location of major tectonic belts, sedimentary basins, Mississippi Valley-type ore deposits, and inferred mineralizing fluid and petroleum migration pathways. Modified after Bethke & Marshak (1990).

organic matter and evidence of petroleum generation and migration or both (Fowler 1933; Jackson & Beals 1967; Marikos *et al.* 1986; Mazzullo *et al.* 1986; Voss *et al.* 1989; Anderson 1991; Kyle & Saunders 1996; Gregg *et al.* 2001; Baugaard 2002). A genetic link between MVT sulphide mineralization and petroleum is well-established (Connan 1979; Gize & Hoering 1980; Macqueen & Powell 1983; Sverjensky 1984; Olson 1984; Powell & Macqueen 1984; Gize & Barnes 1987; Montacer *et al.* 1988; Leventhal 1990; Kesler *et al.* 1994).

The purpose of this paper is to review the effect of basinal brines, associated with sulphide mineralization, on dolomite rock texture. Specifically, it considers the effects that these fluids have on the derived properties of porosity and permeability. An understanding of these processes will help in predicting the properties of potential carbonate reservoirs in regions where such mineralization has occurred.

Origin and distribution of sulphide mineralization

The apparent wide variety of processes and geological settings leading to carbonate-hosted sulphide mineralization has resulted in a lack of a single unifying descriptive model for the origin of these deposits (Sangster 1983). They probably represent a continuum of hydrological processes that extends from deep convection and sea-floor exhalation of warm brines, as suggested for many of the deposits in Ireland (Russell 1986; Boyce *et al.* 1999; but see Hitzman & Beaty 1996), to large-scale cross-basin flow of brines in deep aquifers (Bethke & Marshak 1990; Cathles 1993; Garven & Raffensperger 1997). The former processes result in what are commonly referred to as sedimentary exhalative (SEDEX) deposits and the latter in 'true' MVT deposits (Sangster 1990). In reality, many carbonate-hosted sulphide deposits, whether called MVT or not, probably result from a hybrid of these end-member processes (e.g. Shelton *et al.* 1995; Garven *et al.* 1999).

Figure 1 shows the distribution of major carbonate-hosted sulphide ore deposits on the North American continent. Inferred fluid-flow pathways and petroleum-migration pathways (Bethke & Marshak 1990) indicate that large volumes of rock over very wide geographical areas are affected by the mineralizing fluids. For instance, the fluid-flow system responsible for the vast MVT ore deposits of SE Missouri, hosted in the Bonneterre Dolomite (Cambrian),

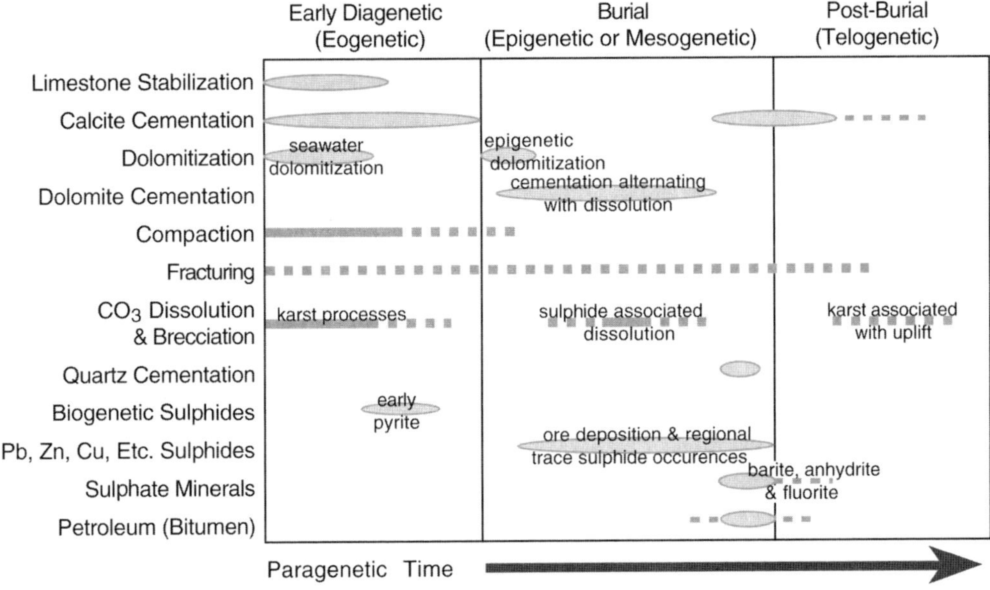

Fig. 2. An idealized paragenesis for Mississippi Valley-type and other carbonate-hosted mineral deposits.

has been shown to extend from northern Arkansas, into western Illinois, and westward to eastern Kansas, and are likely to include links to the systems responsible for the northern Arkansas and Tri-State MVT districts and the Missouri barite districts (Gregg 1985; Leach & Rowan 1986; Gregg & Shelton 1989; Voss *et al.* 1989; Farr 1992; He *et al.* 1997; Coveney *et al.* 2000). Although the exact mechanism(s) of ore precipitation are not yet fully understood, large-scale (ore-grade) sulphide mineralization is usually associated with porous structures such as reef or bioherm trends, with palaeokarst, or with tectonically fractured carbonate rocks (Anderson & Macqueen 1982; Sangster 1983; Ohle 1985) suggesting an association with development of large-scale porosity networks. Minor sulphide mineralization commonly is scattered over larger regions (e.g. Coveny & Goebel 1983) and does not appear as dependant on large-scale porosity development.

Paragenesis of sulphide metal deposits

Every carbonate-hosted base-metal sulphide occurence has a unique paragenesis (Sangster 1983). However, a 'standard' paragenesis (Fig. 2), incorporating diagenetic and epigenetic processes can be constructed based on work in the SE Missouri districts (Hagni 1986; Voss *et al.* 1989) and numerous observations in other regions.

Host-rock preparation for sulphide mineralization typically involves eogenetic processes that occur at relatively low temperatures. These include stabilization of high-Mg calcite and aragonite to low-Mg calcite, by neomorphic processes, early limestone cementation (Bathurst 1975), early dolomitization (see discussion below), initial compaction and fracturing, and karst dissolution during periodic sea-level low stands. 'Burial' processes involve warm (>60 °C) basinal fluids and may occur at relatively shallow depths (<<1 km) rendering the term burial less discriptive than 'epigenetic' or 'mesogenetic'. Undolomitized limestones typically become dolomitized by these fluids and early formed dolomites may undergo neomorphic recrystallization (Gregg & Shelton 1990). These processes are both accompanied and followed by precipitation of open-space-filling dolomite cement. Dolomite cementation alternates with periods of dissolution (accompanied by brecciation in many areas) and precipitation of metal sulphide minerals. It is commonly followed by quartz cementation and less commonly by silicification of some carbonate rocks. Rarely, quartz cementation occurs concurrently with dolomite cementation. Calcite cementation follows dolomite cementation in every area investigated by the author and always occurs towards the end of the epigenetic process. Sulphide mineralization may range in volume from trace amounts to large-scale economic ore deposits. Sulphide minerals, in the usual order of abundance, may include pyrite, marcasite, sphalerite, galena, chalcopyrite and numerous other metal sulphides. Late-stage mineralization may include precipitation of sulphate minerals (barite, celestite and anhydrite) and fluorite. Evidence of hydrocarbon emplacement, in the form of bitumen, pyrobitumen and, occasionally, liquid petroleum, encountered in fluid inclusions is most commonly associated with the period of calcite precipitation. Telogenetic processes include continued calcite cementation, karst development and mineral alteration associated with uplift into the zone of oxidation.

Early diagenetic dolomitization

Nearly every occurrence of carbonate-hosted sulphide mineralization is associated with large-scale dolomitization in one form or another (Anderson & Macqueen 1982). Large-scale replacement of limestone by dolomite has presented a major problem in sedimentary petrology. Dolomitization of limestones requires a source of large volumes of Mg^{2+}, a hydrological mechanism to move the magnesium into the proximity of the limestone to be dolomitized and a means of overcoming the kinetic barriers to the dolomitization reaction (Fairbridge 1957; Morrow 1990*a*, *b*). A number of models have been suggested for massive dolomitization (see reviews by Warren 2000; Machel 2004). These can generally be divided into two major categories: (1) those that occur relatively early in the diagenetic process, usually involving seawater at relatively low temperatures (<60 °C); and (2) those that occur later, in the (deep?) subsurface, usually involving basinal or basement fluids, at relatively high temperature (≥60 °C).

Land (1985) points out that basinal fluids typically are depleted in magnesium due to chloritization of clays. Further, early diagenetic cementation commonly occludes porosity and permeability in limestones, limiting access by dolomitizing fluids (Gregg *et al.* 2001). Although dolomitization by basinal fluids is known (Mountjoy & Amthor 1994) seawater was probably the original source of Mg^{2+} for many, if not most, massive dolomites that act as petroleum reservoirs and hosts for sulphide mineralization. Dolomitization models involving seawater fall into three general categories: (1) evaporated seawater (e.g. Adams & Rhodes 1960; McKenzie 1981); (2) diluted seawater (e.g. Badiozamani 1973; Folk & Land 1975); and (3) unmodified seawater (e.g. Saller 1984; Land 1991).

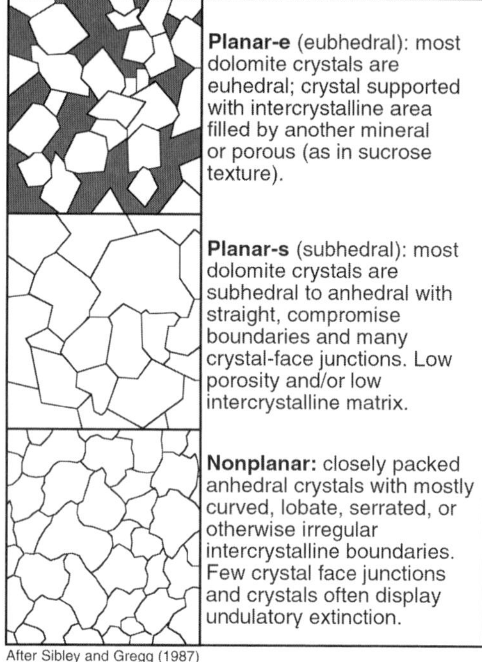

Fig. 3. Classification system for dolomite crystal textures from Sibley & Gregg (1987).

Eogenetic dolomite textures

Holocene dolomites are typically characterized by micron-sized (<1.0 μm–0.01 mm) (terminology of Friedman 1965) mosaics of planar crystals (Fig. 3) and relatively high porosity (Fig. 4A). They are usually calcium rich, ranging from approximately 35 to 47 mole% $MgCO_3$ and display attenuated ordering reflections in X-ray diffraction (Illing *et al.* 1965; McKenzie 1981; Mazzullo *et al.* 1987; Budd 1997). Cenozoic dolomites, that have not been exposed to basinal fluids, typically display somewhat larger mean crystal size (micron–decimicron sized, <1 μm–0.1 mm) (Fig. 4B & C), but share the properties of planar texture, high calcium content and poor ordering with Holocene dolomites (Sibley 1982; Halley & Schmoker 1983; Vahrenkamp *et al.* 1991; Budd 1997).

Chemical and accompanying textural change in dolomite occurs through the process of neomorphic recrystallization (Land 1980; Gregg & Sibley 1984; Sibley & Gregg 1987). In the

Fig. 4. Early diagenetic dolomite fabrics. (**A**) Micron-sized planar-e dolomite from peritidal flats on Belize (Mazzullo *et al.* 1986). Scale = 2 μm. (**B**) Decimicron-sized planar-e dolomite (sucrose texture) from the Seroe Domi Formation (Pliocene) of Aruba. Scale = 0.25 mm. (**C**) Mimetic and non-mimetic replacement of skeletal grains by micron–decimicron-sized planar dolomite preserving original inter- and intraparticle porosity of the original grainstone. Seroe Domi Formation (Pliocene) of Bonaire. Scale = 0.5 mm. (**D**) Oolitic grainstone replaced by decimicron-sized planar-e to planar-s dolomite. Smackover Limestone (Jurassic) of the Texas Gulf Coast. Scale = 0.25 mm.

diagenetic environment, neomorphism proceeds by a dissolution–reprecipitation mechanism in the presence of an aqueous fluid (Bathurst 1975). Neomorphism does not happen by magic, but must be driven by processes that lower the total free energy of the system. Probable driving mechanisms for dolomite that will achieve this include: (1) increasing cation ordering and reducing lattice defects; (2) increasing stoichiometry (by decreasing the calcium to magnesium ratio to near 50 mole% $MgCO_3$); and (3) decreasing surface free energy by increasing mean crystal size (Ostwald ripening). Dolomite neomorphism, resulting in textural and chemical change, may begin very early in diagenesis (Gregg *et al.* 1992) and continue into the epigenetic environment (Gregg & Sibley 1984; Gregg & Shelton 1990). Mesozoic and Palaeozoic dolomites, regardless of their texture, typically display better cation ordering and stoichiometry than their Cenozoic counterparts, indicating that dolomite undergoes chemical stabilization as a normal part of the diagenetic process (Land 1980; Gregg & Shelton 1990).

Dolomite reservoirs, prior to the influx of basinal fluids associated with sulphide mineralization, texturally resemble the Pliocene dolomites of the Netherlands Antilles (Sibley 1982) (Fig. 4B & C), Cenozoic dolomite aquifers of southern Florida (Halley & Schmoker 1983) or the Jurassic Smackover Formation of the Gulf Coast (Ottman *et al.* 1976; Moore 1984) (Fig. 4D). These are typified by micron–decimicron-sized (<0.001–0.1 mm) planar textures (Fig. 3). Dolomite crystal size distribution is largely controlled by the nucleation density of the aragonite or Mg calcite substrate (Sibley 1982; Randazzo & Zachos 1984; Braithwaite 1991) and size distributions may be polymodal (Sibley *et al.* 1993). Such dolomites are characterized by high intercrystal and vug porosity with relatively uniform pore-throat size and distribution (Woody *et al.* 1996).

As with crystal size, the distribution of porosity in these rocks is largely controlled by the original porosity of the substrate. For instance, a skeletal grainstone with both interparticle and intraparticle porosity will probably retain this property after dolomitization, especially if dolomitization is non-destructive (Fig. 4C). Porosity and permeability in dolomite are commonly further modified, during premineralization diagenesis, by compaction, fracturing, karst dissolution, and early diagenetic cementation and neomorphism (Gregg *et al.* 1992, 1993); however, these processes do not greatly reduce the initial porosity (Halley & Schmoker 1983).

Eogenetic porosity development

Porosity in dolomite has been explained in terms of a volume reduction during dolomitization of limestone resulting in textures with high intercrystal porosity (Murray 1960; Weyl 1960). This model assumes a local source for CO_3^{2-}. If an outside source for carbonate exists, dolomitization of limestone may actually result in a total volume increase and a net loss of porosity (Lucia & Major 1994). Choquette & Pray (1970) categorize porosity into two types: (1) fabric-selective, including interparticle, intraparticle, intercrystal, mouldic, fenestral, shelter and growth framework; and (2) non-fabric-selective, including fracture, channel, vug and cavern. The geometry and sizes of fabric-selective pore types are dependent on the depositional and diagenetic textural relationships of the carbonate rock. Non-fabric-selective pore types develop post-depositionally.

The relationships between dolomite porosity and petrophysical properties have been addressed in a number of studies. Wardlaw (1976) divided the total porosity of a carbonate rock into two different regions: relatively large pores, representing the larger volume of porosity; and pore-throats, which comprise a relatively smaller volume. Permeability in dolomite is dependent on the connectivity of the pores via pore-throats and is not directly related to total porosity or crystal size. Wardlaw used resin pore casts to describe carbonate pore systems as polyhedral or tetrahedral pores connected by sheet-like pore-throats occurring at compromise boundaries. He found that pore-throat networks may be disrupted by increasing crystal size as pores are reduced from polyhedral to tetrahedral, and finally to interboundary sheet pores. The geometry of intercrystal pore space is related to the size and shape of the crystals, and to the amount and distribution of compaction, open-space-filling cement, mineralization and crystal dissolution (Lucia 1983). Empirical models predicting reservoir performance, based on physical observations of fluid-flow parameters such as capillary-pressure data, were developed by Ghosh & Friedman (1989). They characterize pore systems in dolomite using the sizes and shapes of pores, roughness of the pore surfaces, and length and number of pore throats.

Effect of basinal fluids on dolomite

Epigenetic dolomite textures

Exposure of carbonate rocks to warm basinal fluids will result in dolomitization of limestone (Mountjoy & Amthor 1994) and cause significant textural changes in existing dolomite (Gregg & Sibley 1984; Gregg & Shelton 1990). As discussed above, the frequency of large-scale dolomitization of limestones in the epigenetic environment remains controversial. Nevertheless, the process has been well-documented in a number of studies (e.g. Zenger 1983; Machel & Mountjoy 1987; Mountjoy & Amthor 1994). This is especially true in instances where dolomite has apparently replaced residual or previously undolomitized limestone strata in regions hosting sulphide mineralization (Gregg 1985; Gregg & Shelton 1990; Braithwaite & Rizzi 1997). The source of Mg^{2+} for this remains controversial but may involve complex geochemical processes during basin fluid flow (Lee 1997) or dissolution and remobilization of Mg^{2+} from previously formed dolomite (Gao *et al.* 1992).

Early diagenetic dolomitization typically results in a rock with micron–decimicron crystal size (<0.1 mm), commonly a polymodal crystal size distribution, depending on the limestone substrate, and relatively high intercrystal porosity (see the discussion above). Dolomite replacing limestone at temperatures in excess of 60 °C typically displays centimicron–centimetre-sized nonplanar or planar texture, or both (Gregg & Sibley 1984; Sibley & Gregg 1987) with relatively unimodal crystal size distributions (Sibley *et al.* 1993) (Figs 3 and 5A & D). Similar coarse crystalline planar–nonplanar crystal textures result from neomorphic recrystallization of early diagenetic dolomite (Sibley & Gregg 1987) (Fig. 5B). At this stage recrystallization is most probably driven by reduction in surface energy (Ostwald ripening). Alternately, reduction of residual non-stoichiometry or lattice disorder in the dolomite could also provide the drive for neomorphism (Gregg & Shelton 1990). Neomorphic recrystallization is

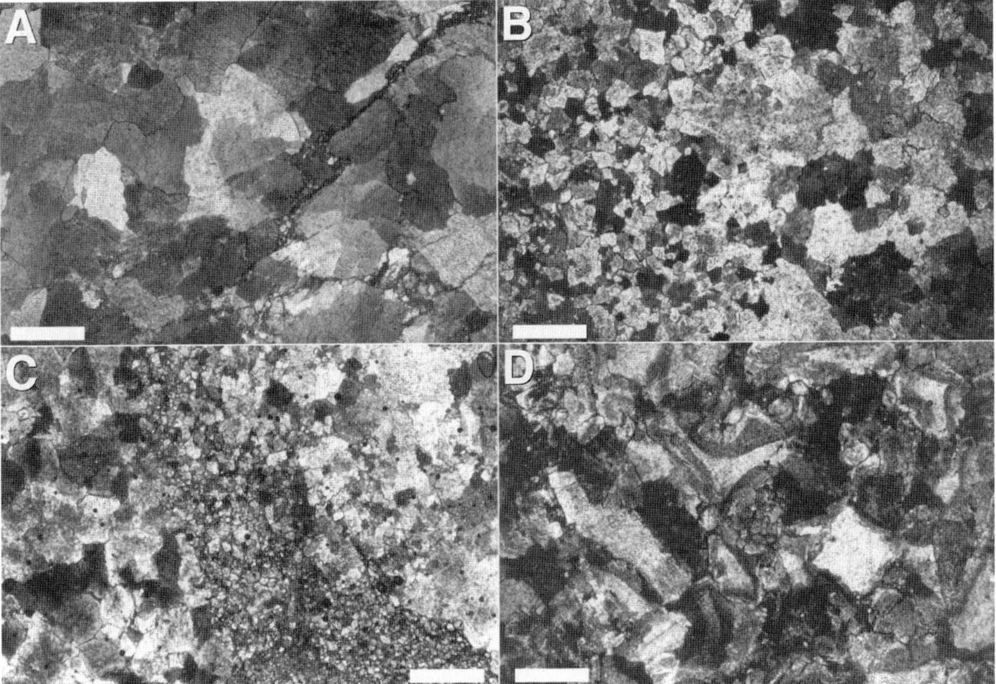

Fig. 5. Dolomite fabrics resulting from exposure to basinal fluids. Scale bars on all = 0.5 mm. (**A**) Oolitic grainstone replaced by centimicron-sized nonplanar dolomite. This dolomite is associated with minor pyrite and trace sphalerite mineralization. Smackover Limestone (Jurassic) of the Texas Gulf Coast. Compare with Fig. 4D. (**B**) Decimicron-sized planar-s dolomite (left field) grading in to centimicron-sized nonplanar dolomite (right field). This complex fabric may be the result of partial neomorphic recrystallization of the nonplanar dolomite (Gregg & Shelton 1990). Bonneterre Dolomite (Cambrian) SE Missouri. (**C**) Dolomitized breccia in the Trenton Limestone (Ordovician), Albian Scipio Field, Michigan. Clast (centre to lower right) is replaced by decimicron-sized dolomite and the matrix by centimicron-sized nonplanar dolomite. (**D**) Crinoidal grainstone of the Malahide formation (Lower Carboniferous), Ireland, replaced by centimicron-sized nonplanar dolomite. Dolomitization here occurred adjacent to a fault and is associated with minor sulphide mineralization.

most likely to occur in areas of high fluid flux due to initial high porosity, fracturing or brecciation. It is important to understand that not all of the dolomite necessarily undergoes neomorphic recrystallization with the accompanying textural changes. Therefore, after early diagenetic dolomitization and overprinting by epigenetic dolomitization and neomorphism during exposure to mineralizing fluids, the resulting carbonate may be a hybrid fabric displaying a complex mixture of planar and nonplanar crystal textures (Fig. 5B & C) (Gregg & Shelton 1990; He *et al.* 1997; Yoo *et al.* 2000; Gregg *et al.* 2001).

Woody *et al.* (1996) studied the contrasting petrophysical properties of planar and nonplanar dolomite (Fig. 3). They found that both planar-e (euhedral) and planar-s (subhedral) dolomite display a linear relationship between porosity and log-permeability (Fig. 6A). However, the increase in permeability with porosity is lower in planar-s dolomite than in planar-e dolomite. Mercury injection capillary pressure curves for planar-e dolomites have intermediate–concave shapes (Fig. 7A) (see Kopaska-Merkel & Friedman 1989 for a discussion of capillary pressure curve shapes) reflecting their well-connected pore space and relatively uniform pore-throat sizes (Fig. 8A). Planar-s textures, by contrast, display intermediate–convex-shaped curves (Fig. 7B) reflecting their lower pore connectivity (Fig. 8B). This is due to continued pore-filling crystal growth during diagenesis (Wardlaw 1976). Patchy cementation, which tends to block pore throats rather than causing a significant loss in

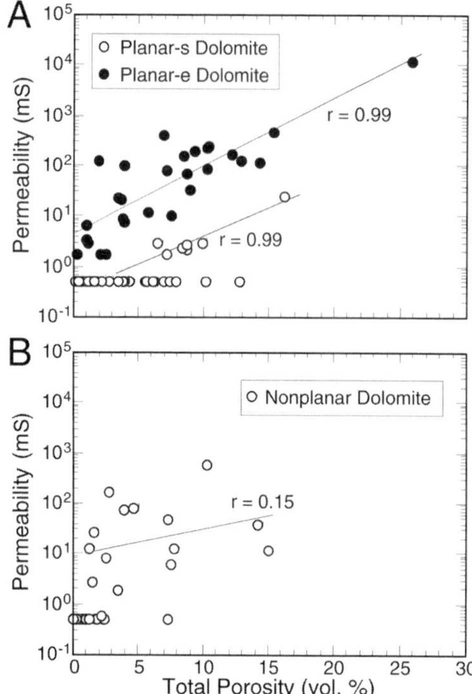

Fig. 6. Porosity v. log-permeability of: (**A**) planar-e and planar-s dolomite showing strong correlation between porosity and permeability ($r = 0.99$); (**B**) the same plot for nonplanar dolomite indicates a very low correlation between porosity and permeability. It should be noted that values for planar-s and nonplanar dolomite that lie below the determination limit of the permeameter (0.5 mD) were not used in calculating the correlation coefficient. From Woody *et al.* (1996).

total porosity (Lucia 1983), can result in similar capillary pressure curves.

There is no linear relationship observed between porosity and log-permeability in nonplanar dolomite (Woody *et al.* 1996) (Fig. 6B). Capillary pressure curves tend to be bimodal due to the existence of poorly connected–nonconnected pore systems and polymodal or nonuniform pore-throat size distribution. In general, nonplanar dolomites have lower porosity than planar dolomites, smaller pore-throat sizes and pore-throat geometries that are tortuous, restricted or closed (Fig. 8C). Secondary porosity such as fracture and channel porosity is commonly a factor in nonplanar dolomite permeability (Woody *et al.* 1996).

The study of Woody *et al.* (1996) indicates that dolomite texture can have an overriding influence on pore-throat geometries, on porosity–permeability relationships and, ultimately, on the characteristics of a dolomite as a reservoir rock. Dolomitized strata that have been affected by basinal fluids can be expected to contain a range of fabrics comprising crystal textures formed under both early diagenetic and epigenetic conditions. Such reservoir rocks cannot be expected to have uniform petrophysical properties, and will require careful petrological and petrophysical study in order to maximize petroleum production.

Epigenetic cementation

One of the prominent characteristics of carbonate-hosted sulphide mineralization is the precipitation of large volumes of open-space-filling dolomite cements, commonly accompanied or followed by precipitation of open-space-filling calcite, quartz, sulphides, sulphates and other minerals (Fig. 2) (e.g. Ebers & Kopp 1979; Voss & Hagni 1985; Misra *et al.* 1989; McManus & Wallace 1992; Hollis & Walkden 1996). Typically, dolomite (and associated calcite) cements display characteristic compositional microstratigraphies that are visible using cathodoluminescence petrography, staining and other techniques (Fig. 9). Compositional microstratigraphies in cements associated with sulphide mineralization have made it possible to trace mineralizing fluid migration over large regions, demonstrating that the effects of these fluids are far ranging and not just restricted to areas where there is heavy sulphide mineralization (e.g. Wallace *et al.* 1991; Muchez *et al.* 1994; Montañez 1994, 1996, 1997; Zeeh *et al.* 1997; Yoo *et al.* 2000; Wright *et al.* 2003). For instance, in southern Missouri, a number of researchers have demonstrated that fluids associated with the SE Missouri mineral districts precipitated open-space-filling cements in Cambrian carbonate rocks over a region of more than 54 000 km^2 (Gregg 1985; Rowan 1986; Farr 1989; Voss *et al.* 1989). Individual compositional zones of cement have been correlated with specific stages of sulphide mineral precipitation, as well as other cements, and periods of carbonate dissolution (Fig. 10) and hydrocarbon emplacement (Voss *et al.* 1989).

Gregg *et al.* (1993) studied the effect of cementation, over time, on the porosity of the Bonneterre Dolomite (Cambrian), the principal host for sulphide mineralization in the SE Missouri mineral districts. Using the distinctive cathodoluminescence (CL) microstratigraphy of epigenetic dolomite cements in southern Missouri (Fig. 9A), they demonstrated a progressive loss of micro- and mesoporosity

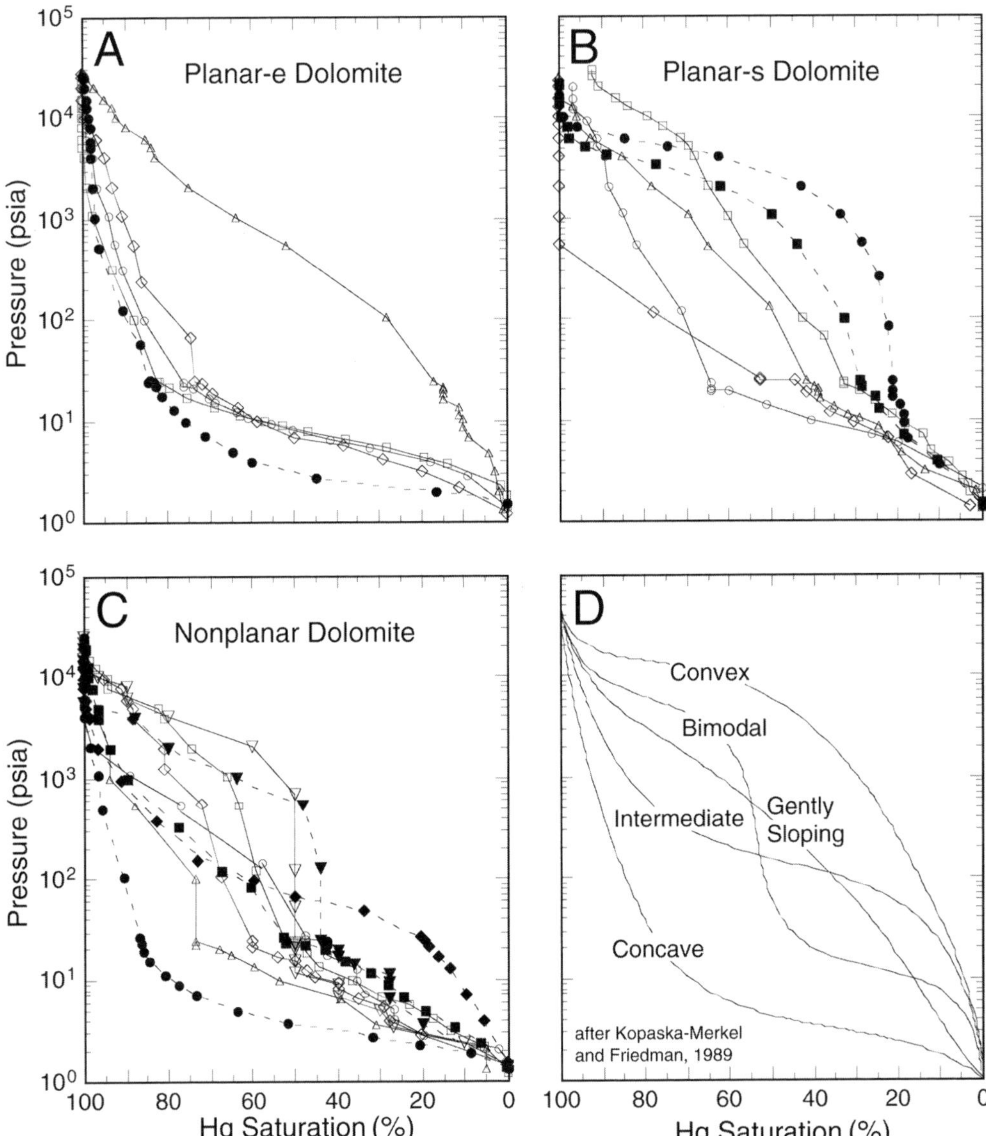

Fig. 7. High-pressure mercury capillary pressure curves (from Woody *et al.* 1996). (**A**) Planar-e dolomite displaying curves that are largely concave, indicating well-connected pore systems with low pore to throat size ratios. (**B**) Planar-s dolomite displaying curves that range from intermediate to convex shapes, indicating pore systems that are not as well connected as in planar-e dolomite and having a relatively large pore to throat size ratio. (**C**) Nonplanar dolomite displaying mostly bimodal–intermediate curves with steep gradients, indicating pore systems with high tortuosity and large pore to throat size ratios. (**D**) The terminology used to describe the shape and gradient of capillary pressure data (from Kopaska-Merkel & Friedman 1989).

(intercrystal and vug) during sulphide mineralization. Samples were selected from mineral exploration cores and were analysed in thin section and, on selected samples, by helium porosimetry to determine porosity. Volumes of open-space-filling dolomite and quartz cement were estimated by point-counting successive compositional zones in dolomite and unzoned quartz using CL petrography (Figs 9A and 10). These zones were then 'backstripped' to

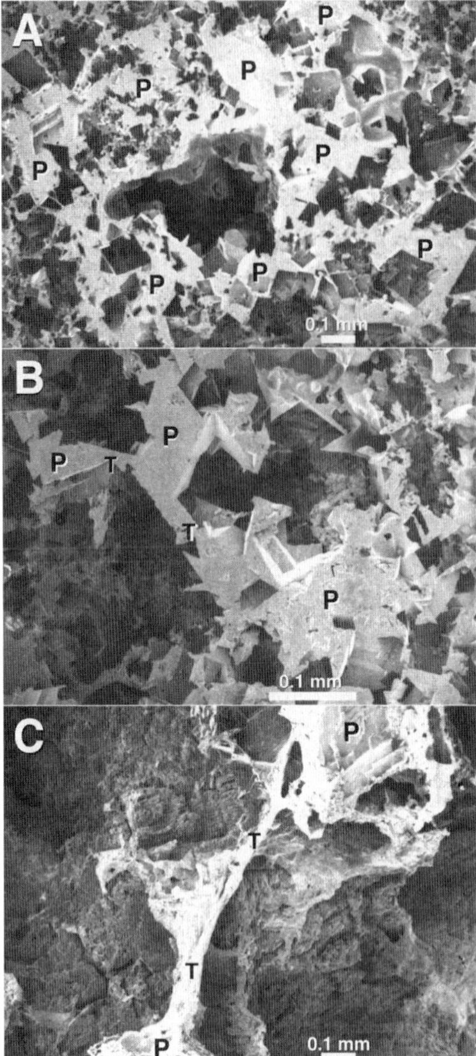

Fig. 8. Scanning electron microscope (SEM) micrographs of pore casts from the Bonneterre Dolomite. (**A**) Planar-e dolomite with well-interconnected intercrystal porosity (P) and a very low pore to throat size ratio. (**B**) Planar-s dolomite with several vugs (P) and minor intercrystal porosity connected by relatively restricted pore throats (T). (**C**) Nonplanar dolomite displaying small pores (P) connected by a narrow, curved pore throat (T).

determine the volumes of cements and successive porosity loss. Initial and final porosities were found to vary, depending on the particular facies of the Bonneterre Dolomite measured (Fig. 11). It was determined that epigenetic cementation was responsible for a dramatic reduction in porosity throughout the Bonneterre, from an average initial porosity of nearly 19% to an average final porosity of less than 4% (Figs 10 and 11).

Variations in permeabilities were not determined by Gregg *et al.* (1993), but can be inferred from the later work of Woody *et al.* (1996). Successive precipitations of dolomite as syntaxial cements on planar dolomite crystals resulted in the development of a planar-s texture in what was previously a planar-e crystal texture. This resulted in disrupted pore-throat networks as discussed by Wardlaw (1976) and indicated by intermediate–gently sloping capillary pressure curves (Fig. 7B). Even a single episode of cementation in a dolomite is likely to have a serious adverse effect on pore-throat geometry (Wardlaw *et al.* 1988). Cementation in nonplanar dolomite will have the effect of closing already restricted pore throats, as well as filling remaining vug, fracture and channel porosity.

Dissolution and brecciation

Major carbonate-hosted ore deposits commonly concentrate in rocks affected by karst dissolution or tectonic fracturing and faulting (Anderson & Macqueen 1982; Ohle 1985). Such megaporosity is then typically further modified by alternating periods of carbonate precipitation and dissolution (Voss & Hagni 1985, Montañez 1992). Chemical reactions involving petroleum, and occurring during sulphide precipitation, have been cited as a possible source of sulphur for the sulphide minerals and for the acidity that causes carbonate dissolution during periods of intense sulphide mineralization (Connan 1979; Sverjensky 1984; Spirakis & Heyl 1988; Leventhal 1990; Kesler *et al.* 1994). Large-scale dissolution can, therefore, be expected proximal to major sulphide mineral deposits, but it does not appear to affect rocks distal from such deposits as extensively as open-space-filling cementation.

Examples of MVT deposits associated with palaeokarst include the Pine Point deposit in the Northwest Territories, located in a Middle Devonian, dolomitized bioherm complex. Here, ore is concentrated in karst that developed on palaeohighs, prior to mineralization, and was later modified by basinal fluids (Rhodes *et al.* 1984). Similarly, breccias related to zinc–lead ore bodies in Proterozoic dolomites on Baffin Island are attributed to several stages of karst development. In this case petroleum, filling the caverns, is suggested as the reducing agent leading to sulphide mineralization (Olson 1984). Karst development and brecciation in the

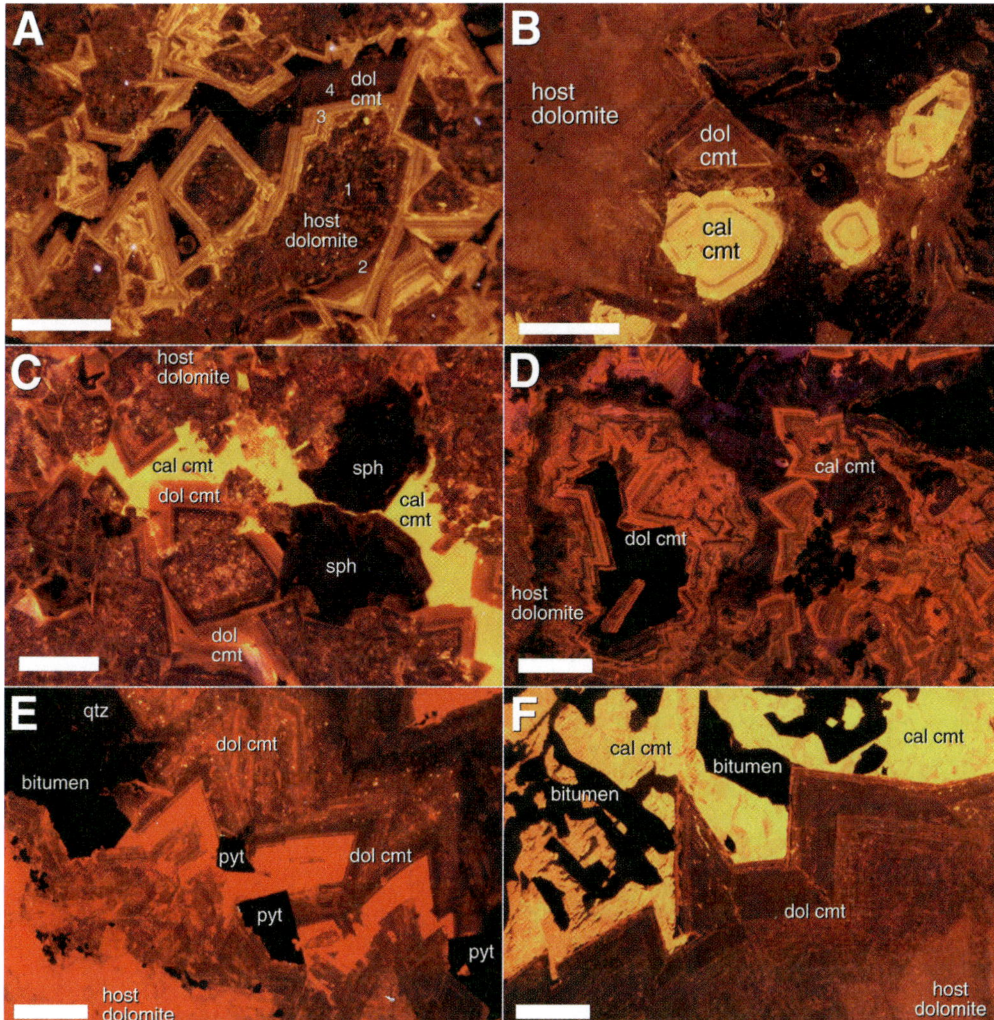

Fig. 9. Cathodoluminescence photomicrographs of epigenetic open-space filling cements associated with basinal fluids and sulphide mineralization. (**A**) Bonneterre Dolomite (Cambrian), Fletcher Mine, Viburnum Trend, Missouri showing the four-zoned stratigraphy of Voss *et al.* (1989). The host dolomite (zone 1), which replaced an oolitic grainstone in this case is a substrate upon which later dolomite cement generations (zones 2–4) grew as syntaxial overgrowths occluding porosity. Scale = 0.25 mm. (**B**) Dolomite and calcite cement (dol cmt and cal cmt, respectively) growing into a vug in a dolomitized section of the Trenton Limestone (Ordovician), NW Ohio. These cements are associated with pyrite, marcasite, sphalerite, galena and petroleum, and formed in an 'Albion-Scipio-type' fracture system along the Findley Arch (Haefner *et al.* 1988). Photomicrograph by R. Haefner. Scale = 0.25 mm. (**C**) Vug filling epigenetic dolomite and calcite cement associated with pyrite, sphalerite (sph) and galena in Lower Permian carbonates on the southern Central Basin Platform adjacent to the Delaware Basin, west Texas (Mazzullo *et al.* 1986). Scale = 0.10 mm. (**D**) Dolomite and calcite cement displaying zonation typical of the East Tennessee zinc–lead district and surrounding region within the Appalachian Basin (Ebers & Kopp 1979; Montañez 1994). This example is from the Douglas Lake Member (Ordovician) that overlies the top of the Knox Group. Photomicrograph by F. Furman. Scale = 0.05 mm. (**E**) Open-space-filling dolomite and quartz cement with associated pyrite and bitumen at Bushy Park. These rocks are in the Palaeoproterozoic Transvaal Sequence, South Africa and associated with economic sphalerite and galena deposits. Photomicrograph by W. Baugaard. Scale = 0.5 mm. (**F**) Epigenetic dolomite and calcite cement and bitumen filling a vug in a dolomitized section of the Waulsortian Limestone (Viséan), Ireland. These rocks are associated with the large sphalerite and galena deposits of the Rathdowney Trend (Gregg *et al.* 2001). Scale = 0.5 mm.

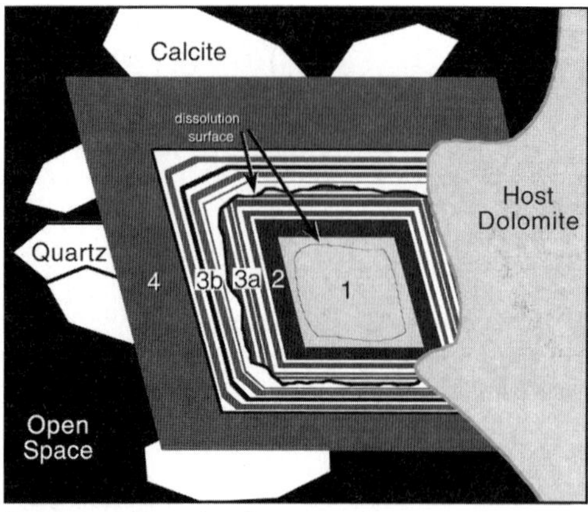

Stages of Porosity Loss:	
Initial porosity (after host and zone 1)	18.7%
Intermediate B porosity (after zones 2 & 3)	7.3%
Intermediate A porosity (after zone 4)	4.7%
Final porosity (after quartz cement)	3.7%

Fig. 10. Cartoon showing SE Missouri cement stratigraphy after Voss *et al.* (1989) and average porosity loss during successive stages of cementation in the Bonneterre dolomite (data from Gregg *et al.* 1993). Note the period of dissolution after dolomite cement zone 3A followed by cement zones 3B and 4.

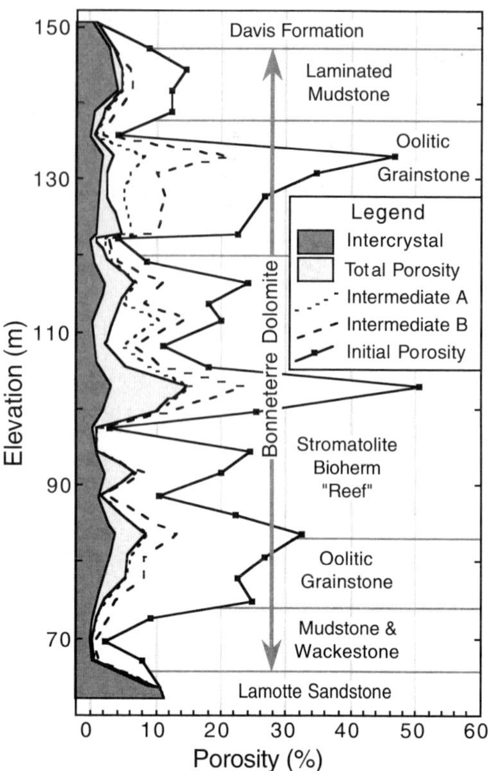

Fig. 11. Porosity log of a core through the Bonneterre Dolomite in the Viburnum Trend of SE Missouri. Relative abundances of intercrystal and other porosity types are shown along with estimates of initial porosity at the onset of mineralization and at successive stages during mineralization and cementation. Intermediate A represents porosity after precipitation of dolomite cement zone 4 and prior to quartz cement. Intermediate B is porosity after dolomite cement zones 2 and 3 but prior to zone 4. (From Gregg *et al.* 1993.)

Upper Knox Group (Ordovician) has a controlling effect on the concentration of sulphide mineralization in the east and central Tennessee mineral districts (Fig. 12A) (McCormick *et al.* 1971; Haynes & Kesir 1994). However, Montañez (1992) determined that the regional distribution of MVT mineralization is concentrated in porosity resulting from eustatic sea-level rises, rather than from karst, suggesting that these rocks were the principal pathways for fluids. In the Lower Carboniferous of Ireland, large zinc and lead sulphide deposits are hosted by fault breccias (Fig. 12B) that have undergone large-scale dissolution during mineralization (Hitzman & Beaty 1996). The periods of dissolution alternated with periods of cementation by dolomite and calcite (Fig. 9F). Breccias resulting from faulting, carbonate dissolution and cementation during sulphide mineralization opened megaporosity in fracture-related oil fields in the Trenton Limestone (Ordovician) in Michigan and Ohio (Fig. 5C) (Shaw 1975; Gregg & Sibley 1984; Haefner *et al.* 1988). At the Bushy Park zinc–lead deposit in South Africa, epigenetic brecciation (Fig. 12C) and carbonate and quartz cementation (Fig. 9E) in platform dolomites of the Transvaal Group were superimposed on both eogenetic karst and tectonic fracturing (Baugaard 2002).

In the Viburnum Trend lead–zinc district of SE Missouri, dissolution of the Bonneterre Dolomite (Cambrian) during sulphide mineralization modified earlier formed karst features and enhanced megaporosity, but had a very limited effect on meso- and microporosity. A major period of carbonate dissolution can be observed in epigenetic dolomite cement within cement zone 3 of Voss & Hagni (1985) (Fig. 13) and occasionally beneath zone 2 (Fig. 10). Voss & Hagni (1985; also Voss *et al.* 1989) observed that epigenetic dolomite cement growing on breccia clasts in the Bonneterre commonly displays only those zones following the zone 3

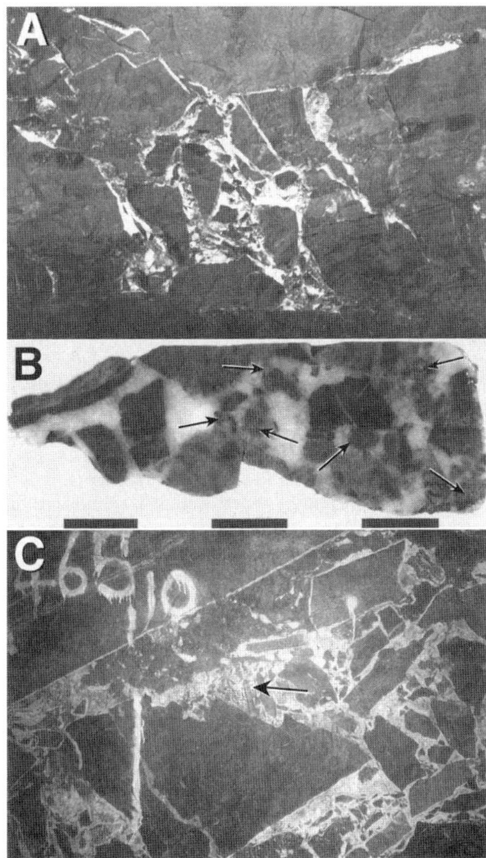

Fig. 12. (**A**) Large-scale brecciation and mineralization in the Upper Knox Group (Lower Ordovician), Young Mine, east Tennessee. This breccia was developed on pre-existing karst features. The breccia is cemented with epigenetic dolomite and calcite (white) that displays a 'classic' snow-on-the-roof fabric overlain by sphalerite (dark). Vertical-scale of photograph is approximately 2 m. Photograph by F. Furman. (**B**) Core showing brecciated Waulsortian (Lower Carboniferous) dolomite from the Rathdowney Trend zinc district, Ireland. Breccia clasts are cemented by dolomite, calcite and sphalerite (arrows). Scale is in cm. (**C**) Large-scale brecciation superimposed on eogenetic karst and tectonic fracture features at the Bushy Park mine in South Africa. Breccia porosity is almost completely filled by dolomite, calcite, pyrite and sphalerite. Note hammer (arrow) for scale. Photograph by W. Baugaard.

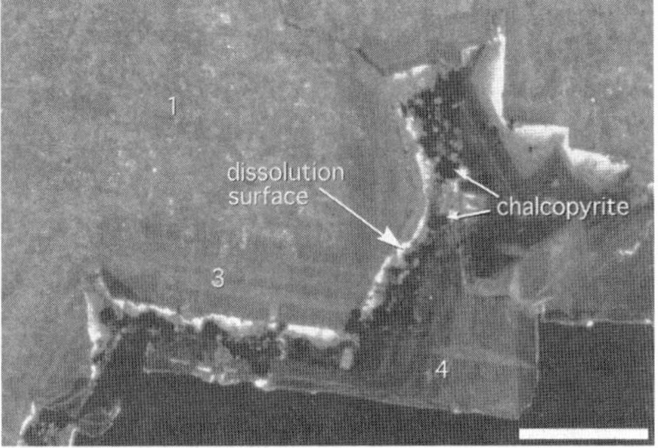

Fig. 13. A dissolution surface (large arrow) is clearly visible in cathodoluminescence Zone 3 (multiple banded zone of Voss & Hagni 1985) of epigenetic dolomite cement in the Bonneterre Dolomite, Fletcher Mine, Missouri. This dissolution surface is paragenetically related to the major period of dissolution that resulted in large-scale brecciation in the Viburnum Trend mineral district. Opaque chalcopyrite is visible on the dissolution surface (small arrows). Note that cathodoluminescence Zone 2 is missing in this example (see Voss *et al.* 1989 for explanation). Scale = 0.5 mm.

period of dissolution, indicating that the major period of brecciation occurred at that time. These periods of dissolution had less effect on meso- and microporosity than on megaporosity in the Bonneterre Dolomite because the fluid flux through the smaller pores was less than in the larger pores. In fact, much of the meso- and microporosity was already closed before the major dissolution period or was subsequently closed by cementation (Gregg *et al.* 1993).

As a result of neomorphic recrystallization and epigenetic cementation during sulphide mineralization, the Bonneterre Dolomite underwent significant textural change and dramatic porosity loss due to epigenetic cementation. At the same time karst-related megaporosity was enhanced in the proximity of sulphide precipitation by carbonate dissolution and brecciation. This changed the Bonneterre from a rock with porosity and permeability, characteristics similar to those of Smackover dolomite petroleum reservoirs (Fig. 4D) to a rock with very poor porosity and permeability, except for those areas where existing megaporosity was enhanced by dissolution and brecciation (Gregg *et al.* 1993). Similar changes in texture and in porosity and permeability networks are evident in other dolomites affected by mineralizing basinal fluids. It can be expected, therefore, that a dolomite that has been exposed to mineralizing basinal fluids will have little resemblance to the original rock. Micro- and mesoporosity (intercrystal and small vug) will be reduced and, in some cases, eliminated by recrystallization and cementation. Remaining porosity and permeability will probably be dominated by fracturing, and, in some cases, by large-scale enhancement of megaporosity during periods of mineralization.

Modelling fluid flow in dolomites

A basic problem in numerical modelling of fluid flow through carbonate sections is setting the physical parameters used for the modelling. Among these parameters is hydraulic conductivity, which is dependent on porosity and permeability. This is particularly difficult to model if porosity and permeability are changing during the period that is being examined (Whitaker *et al.* 2004). In modelling fluid flow during MVT mineralization in SE Missouri Appold & Garven (1999) used the data of Gregg *et al.* (1993) to set the initial conditions for their models. They noted that the porosity of the Bonneterre dolomite was considerably higher at the onset of mineralization than at present (as shown by Imes 1990). However, they did not vary hydraulic conductivity while running their models. Using the petrographic techniques discussed in this paper it should be possible to model variation in conductivity in instances where the paragenesis of cementation, dissolution and recrystallization are well understood.

Summary and conclusions

Basinal fluid flow and associated base-metal sulphide mineralization have affected carbonate rocks of all ages in most sedimentary basins throughout the world. These fluids affect the reservoir properties of dolomite in a number of ways.

Dolomitization by warm (>60 °C) basinal fluids, and neomorphic recrystallization of existing dolomite by the same fluids, results in nonplanar dolomite textures. Nonplanar dolomite has a lower permeability to porosity relationship because of its higher pore-throat tortuosity. Precipitation of massive amounts of carbonates, quartz, sulphides and other minerals as open-space-filling cements occlude much of the intercrystal and vug (micro- and meso-) porosity in dolomites. Those that have been altered by basinal fluids commonly display complex fabrics made up of larger than or approximately centimicron-sized planar and nonplanar crystals, formed during mineralization, and micron–decimicron-sized planar textures, remnants of eogenetic dolomitization. Much of their porosity is attenuated by epigenetic, open-space-filling cementation.

Large-scale dissolution and brecciation alternating with periods of carbonate cementation occurs in proximity to sulphide precipitation. Large open-space cavities, fractures and karst features (macroporosity) may be enhanced due to the much larger fluid flow compared to micro- and mesoporosity that may have been attenuated by cementation and neomorphism resulting in textural change. In this way, a dolomite that was originally dominated by interconnected micro- and mesoporosity (intercrystal and vug) can become a rock where porosity and permeability are dominated by features such as fractures and breccias.

This review largely relies on previously published work conducted by myself with colleagues and students on carbonate-hosted base-metal deposits all over the world. I would like to thank the St. Joe Minerals Corporation and the Doe Run Company for access to samples and to their mines. I would like to thank W. Baugaard, S. Becker, F. Furman, Z. He, A. Johnson, T. Keller, P. Laudon, Z. Nagy, R. Woody and W. Wright for their graduate research in carbonate petrology and sediment-hosted deposits that provided much of the material for this paper. I would also like to thank my long-time colleague, K. L. Shelton, for his many contributions to the work cited herein. This paper profited from peer reviews of colleagues and especially from the careful editorial work of C. Braithwaite. The author wishes to acknowledge support from the donors of the Petroleum Research Fund, administered by the American Chemical Society (PRF 35893-AC8), and the National Science Foundation (ERA-0106388) for continuing work on carbonate-hosted base-metal sulphide mineralization.

References

ADAMS, J.E. & RHODES, M.L. 1960. Dolomitization by seepage refluxion. *AAPG Bulletin*, **44**, 1912–1920.

ANDERSON, G.M. 1991. Organic maturation and ore precipitation in southeast Missouri. *Economic Geology*, **86**, 909–926.

ANDERSON, G.M. & MACQUEEN, R.W. 1982. Ore deposits models – 6. Mississippi Valley-type lead zinc deposits. *Geoscience Canada*, **9**, 108–117.

APPOLD, M.S. & GARVEN, G. 1999. The hydrology of ore formation in the Southeast Missouri District: Numerical models of topography-driven fluid flow during the Ouachita Orogeny. *Economic Geology*, **94**, 913–936.

BADIOZAMANI, K. 1973. The Dorag dolomitization model – Application to the Middle Ordovician of Wisconsin. *Journal of Sedimentary Petrology*, **43**, 965–984.

BANKS, D.A., BOYCE, A.J. & SAMSON, I.M. 2002. Constraints on the origins of fluids forming Irish Zn–Pb–Ba deposits. Evidence from the composition of fluid inclusions. *Economic Geology*, **97**, 471–480.

BATHURST, R.G.C. 1975. *Carbonate Sediments and Their Diagenesis*. Elsevier Scientific, New York.

BAUGAARD, W.D. 2002. *Stratigraphy and diagenesis of the Paleoproterozoic Bushy Park zinc–lead deposit, South Africa*. PhD dissertation, University of Missouri-Rolla, Rolla, MO.

BAUGAARD, W.D., GREGG, J.M. & AHLER, B. 2001. Stratigraphy and diagenesis of the Paleoproterozoic Bushy Park zinc–lead deposit, South Africa. *Geological Society of America, Abstracts with Programs*, **33**, A-337.

BETHKE, C.M. & MARSHAK, S. 1990. Brine migrations across North America – the plate tectonics of groundwater. *Annual Reviews of Earth and Planetary Science*, **18**, 287–315.

BJØRLYKKE, A. & SANGSTER, D.F. 1981. An overview of sandstone lead deposits and their relation to red-bed copper and carbonate hosted lead–zinc deposits. *In*: SKINNER, B.J. (ed.) *Economic Geology Seventy-fifth Anniversary Volume*. The Economic Geology Publishing Co., El Paso. 179–213.

BOYCE, A.J., FALLICK, A.E., LITTLE, C.T.S., WILKINSON, J.J. & EVERETT, C.E. 1999. A hydrothermal vent tube worm in the Ballynoe baryte deposit, Silvermines Ireland. Implications for timing and ore genesis. *In*: STANLEY, C.J. ET AL. (eds) *Mineral Deposits: Processes to Processing*. Proceedings of the Fifth Biennial SGA Meeting and the Tenth Quadrennial IAGOG Symposium. Vol. 2. A.A. Balkema, Rotterdam, 825–827.

BRAITHWAITE, C.J.R. 1991. Dolomites, a review of origins, geometry and textures. *Transactions of the Royal Society of Edinburgh: Earth Sciences*, **82**, 99–112.

BRAITHWAITE, C.J.R. & RIZZI, G. 1997. The geometry and petrogenesis of hydrothermal dolomites at Navan, Ireland. *Sedimentology*, **44**, 421–440.

BUDD, D.A. 1997. Cenozoic dolomites of carbonate islands: their attributes and origin. *Earth Science Reviews*, **42**, 1–47.

CATHLES, L.M. 1993. A discussion of flow mechanisms responsible for alteration and mineralization in the Cambrian aquifers of the Ouachita–Arkoma Basin–Ozark system. *In*: ROBINSON, A.G. & HORBURY, A.D. (eds) *Diagenesis and Basin Development*. American Association of Petroleum Geologists, Studies in Geology, **36**, 99–112.

CHOQUETTE, P.W. & PRAY, L.C. 1970. Geologic nomenclature and classification of porosity in sedimentary carbonates. *AAPG Bulletin*, **54**, 207–250.

CONNAN, J. 1979. Genetic relationship between oil and ore in some Pb–Zn–Ba ore deposits. *In*: ANHAEUSSER, C.R., FOSTER, R.P. & STRATTON, T. (eds) *A symposium on mineral deposits and the transportation and deposition of metals*. Geological Society of South Africa, Special Publications, **5**, 263–274.

COVENEY, R.M., JR & GOEBEL, E.D. 1983. New fluid inclusion homogenization temperatures for sphalerite from minor occurrences in the midcontinent area. *In*: KISVARSANYI, G., GRANT, S.K., PRATT, W.P. & KOENIG, J.W. (eds) *International Conference on Mississippi Valley-type Lead Zinc Deposits. Proceedings Volume*. University of Missouri-Rolla, Rolla, MO, 234–242.

COVENEY, R.M., JR, RAGAN, V.M. & BRANNON, J.C. 2000. Temporal benchmarks for modeling Phanerozoic flow of basinal brines and hydrocarbons in the southern Midcontinent based on radiometrically dated calcite. *Geology*, **28**, 795–798.

EBERS, M.L. & KOPP, O.C. 1979. Cathodoluminescence microstratigraphy in gangue dolomite, the Mascot–Jefferson City District, Tennessee. *Economic Geology*, **74**, 908–918.

FAIRBRIDGE, R.W. 1957. The dolomite question. *In*: LEBLANC, R.J. & BREEDING, J.G. (eds) *Regional Aspects of Carbonate Deposition*. Society of Economic Paleontologists and Mineralogists, Special Publications, **5**, 164–170.

FARR, M.R. 1989. Compositional zoning characteristics of late dolomite cement in the Cambrian Bonneterre Formation, Missouri. *Carbonates and Evaporites*, **4**, 177–194.

FARR, M.R. 1992. Geochemical variation of dolomite within the Cambrian Bonneterre Formation, Missouri: Evidence for fluid mixing. *Journal of Sedimentary Petrology*, **62**, 636–651.

FOLK, R.L. & LAND, L.S. 1975. Mg/Ca ratio and salinity: two controls over crystallization of dolomite. *AAPG Bulletin*, **56**, 434–453.

FOWLER, G.M. 1933. Oil and oil structures in Oklahoma-Kansas zinc–lead mining field. *AAPG Bulletin*, **17**, 1436–1445.

FRIEDMAN, G.M. 1965. Terminology of crystallization textures and fabrics in sedimentary rocks. *Journal of Sedimentary Petrology*, **35**, 643–655.

GAO, G., LAND, L.S. & FOLK, R.L. 1992. Meteoric modification of early dolomite and late dolomitization by basinal fluids, Upper Arbuckle Group, Slick Hills, southwestern Oklahoma. *AAPG Bulletin*, **76**, 1649–1664.

GARVEN, G., APPOLD, M.S., TOPTYGINA, V.I. & HAZLETT, T.J. 1999. Hydrogeologic modeling of the genesis of carbonate-hosted lead–zinc ores. *Hydrogeology Journal*, **7**, 108–126.

GARVEN, G. & RAFFENSPERGER, J.P. 1997. Hydrogeology and geochemistry of ore genesis in sedimentary basins. *In*: BARNES, H.L. (ed.) *Geochemistry of Hydrothermal Ore Deposits*, 3rd edn. Wiley, New York, 125–189.

GHOSH, S.K. & FRIEDMAN, G.M. 1989. Petrophysics of a dolostone reservoir: San Andres Formation (Permian), west Texas. *Carbonates and Evaporites*, **4**, 45–119.

GIZE, A.P. & BARNES, H.L. 1987. The organic geochemistry of two Mississippi Valley-type lead–zinc deposits. *Economic Geology*, **82**, 457–470.

GIZE, A.P. & HOERING, T.C. 1980. The organic matter in Mississippi Valley-type deposits. *Carnegie Institution Geophysical Laboratory Yearbook*, **79**, 384–388.

GLEESON, S.A., BANKS, D.A., EVERETT, C.E., WILKINSON, J.J., SAMSON, I.M. & BOYCE, A.J. 1999. Origin of mineralising fluids in Irish-type deposits: constraints from halogens. *In*: STANLEY, C.J. ET AL. (eds) *Mineral Deposits: Processes to Processing. Proceedings of the Fifth Biennial SGA Meeting and the Tenth Quadrennial IAGOG*

Symposium. Vol. 2. A.A. Balkema, Rotterdam, 857–860.

GREGG, J.M. 1985. Regional epigenetic dolomitization in the Bonneterre Dolomite (Cambrian), southeastern Missouri. *Geology*, **13**, 503–506.

GREGG, J.M. & SHELTON, K.L. 1989. Minor and trace element distributions in the Bonneterre Dolomite (Cambrian), southeast Missouri: Evidence for possible multiple basin fluid sources and pathways during lead–zinc mineralization. *Geological Society of America Bulletin*, **101**, 221–230.

GREGG, J.M. & SHELTON, K.L. 1990. Dolomitization and dolomite neomorphism in the back reef facies of the Bonneterre and Davis Formations (Cambrian), southeast Missouri. *Journal of Sedimentary Petrology*, **60**, 549–562.

GREGG, J.M. & SIBLEY, D.F. 1984. Epigenetic dolomitization and the origin of xenotopic dolomite texture. *Journal of Sedimentary Petrology*, **54**, 908–931.

GREGG, J.M., HOWARD, S.A. & MAZZULLO, S.J. 1992. Early diagenetic recrystallization of Holocene (<3000 years old) peritidal dolomites, Ambergris Cay, Belize. *Sedimentology*, **39**, 143–160.

GREGG, J.M., LAUDON, P.R., WOODY, R.E. & SHELTON, K.L. 1993. Porosity evolution of the Bonneterre Dolomite (Cambrian) southeastern Missouri, U.S.A. *Sedimentology*, **40**, 1153–1169.

GREGG, J.M., SHELTON, K.L., JOHNSON, A.W., SOMERVILLE, I.D. & WRIGHT, W.R. 2001. Dolomitization of the Waulsortian Limestone (Lower Carboniferous) Irish Midlands. *Sedimentology*, **48**, 745–766.

HAEFNER, R.J., MANCUSO, J.J., FRIZADO, J.P., SHELTON, K.L. & GREGG, J.M. 1988. Crystallization temperatures and carbon and oxygen isotopes of Mississippi Valley-type carbonates and sulfides of the Trenton Limestone, Wyandot Co., Ohio. *Economic Geology*, **83**, 1061–1069.

HAGNI, R.D. 1986. Mineral paragenetic sequence of the lead–zinc–copper–cobalt–nickel ores of the Southeast Missouri Lead District. *In*: CRAIG, J.R., HAGNI, R.D. ET AL. (eds) *Mineral Paragenesis*. Theophrastus Publications, Athens, Greece, 93–132.

HALLEY, R.B. & SCHMOKER, J.W. 1983. High-porosity Cenozoic carbonate rocks of South Florida: progressive loss of porosity with depth. *AAPG Bulletin*, **67**, 191–200.

HAYNES, F.M. & KESIR, S.E. 1994. Relation of mineralization to wall-rock alteration and brecciation, Mascot–Jefferson City Mississippi-Valley-type district, Tennessee. *Economic Geology*, **89**, 51–66.

HE, Z., GREGG, J.M., SHELTON, K.L. & PALMER, J.R. 1997. Sedimentary facies control of fluid flow and mineralization in Cambro-Ordovician strata, southern Missouri. *In*: MONTAÑEZ, I.P., GREGG, J.M. & SHELTON, K.L. (eds) *Basinwide Fluid Flow and Associated Diagenetic Patterns: Integrated Petrologic, Geochemical and Hydrologic Considerations*. Society of Economic Paleontologists and Mineralogists, Special Publications, **57**, 81–99.

HITZMAN, M.W. & BEATY, D.W. 1996. The Irish Zn–Pb–(Ba) orefield. *In*: SANGSTER, D.F. (ed.) *Carbonate Hosted Lead–Zinc Deposits*. Society of Economic Geologists, Special Publications, **4**, 112–143.

HOLLIS, C. & WALKDEN, G.M. 1996. The use of burial diagenetic calcite cements to determine the controls upon hydrocarbon emplacement and mineralization on a carbonate platform, Derbyshire, England. *In*: STROGEN, P., SOMERVILLE, I.D. & JONES, G.L. (eds) *Recent Advances in Lower Carboniferous Geology*. Geological Society, London, Special Publications, **107**, 35–49.

ILLING, L.V., WELLS, A.J. & TAYLOR, J.C.M. 1965. Penecontemporary dolomite in the Persian Gulf. *In*: PRAY, L.C. & MURRAY, R.C. (eds) *Dolomitization and Limestone Diagenesis*. Society of Economic Paleontologists and Mineralogists, Special Publications, **13**, 89–111.

IMES, J.L. 1990. *Major Geohydrologic Units in and Adjacent to the Ozark Plateaus Province, Missouri, Arkansas, Kansas, and Oklahoma – St. Francois Aquifer*. US Geological Survey, Hydrologic Investigations Atlas, **HA-117-C**, scale 1:750 000.

JACKSON, S.A. & BEALES, F.W. 1967. An aspect of sedimentary basin evolution: the concentration of Mississippi Valley-type ores during late stages of diagenesis. *Bulletin of Canadian Petroleum Geology*, **15**, 383–433.

KENDRICK, M.A., BURGESS, R., LEACH, D. & PATTRICK, R.A.D. 2002. Hydrothermal fluid origins in Mississippi valley-type ore districts: Combined noble gas (He, Ar, Kr) and halogen (Cl, Br, I) analysis of fluid inclusions from the Illinois-Kentucky fluorspar district, Viburnum Trend, and Tri-State districts, midcontinent United States. *Economic Geology*, **97**, 453–469.

KESLER, S.E., JONES, H.D., FURMAN, F.C., SASSEN, R., ANDERSON, W.H. & KYLE, J.R. 1994. Role of crude oil in the genesis of Mississippi Valley-type deposits: Evidence from the Cincinnati arch. *Geology*, **22**, 609–612.

KOPASKA-MERKEL, D.C. & FRIEDMAN, G.M. 1989. Petrofacies analysis of carbonate rocks: example from lower Paleozoic Hunton Group of Oklahoma and Texas. *AAPG Bulletin*, **73**, 1289–1306.

KYLE, J.R. & SAUNDERS, J.A. 1996. Metallic deposits of the Gulf Coast basin: Diverse mineralization styles in a young sedimentary basin. *In*: SANGSTER, D.F. (ed.) *Carbonate-Hosted Lead–Zinc Deposits*. Society of Economic Geologists, 75th Anniversary Volume Special Publications, **4**, 218–229.

LAND, L.S. 1980. The isotopic and trace element geochemistry of dolomite: the state of the art. *In*: ZENGER, D.H., DUNHAM, J.B. & ETHINGTON, R.L. (eds) *Concepts and Models of Dolomitization*. Society of Economic Paleontologists and Mineralogists, Special Publications, **28**, 87–110.

LAND, L.S. 1985. The origin of massive dolomite. *Journal of Geological Education*, **33**, 112–125.

LAND, L.S. 1991. Dolomitization of the Hope Gate Formation (north Jamaica) by seawater:

Reassessment of mixing-zone dolomite. *In*: Taylor, H.P., Jr, O'Neil, J.R. & Kaplan, I.R. (eds) *Stable Isotope Geochemistry: A Tribute to Samuel Epstein*. Geochemical Society, Special Publications, **3**, 121–133.

Land, L.S. & Prezbindowski, D.R. 1985. Chemical constraints and origins of four groups of Gulf Coast reservoir fluids. *AAPG Bulletin*, **69**, 119–121.

Leach, D.L. & Rowen, E.L. 1986. Genetic links between Ouachita foldbelt and the Mississippi Valley-type deposits of the Ozarks. *Geology*, **14**, 931–934.

Leach, D.L., Bradley, D., Lewchuk, M.T., Symons, D.T.A., Marsily, G.D. & Brannon, J. 2001. Mississippi Valley-type lead–zinc deposits through geological time: implications from recent age-dating research. *Mineralium Deposita*, **36**, 711–740.

Lee, M.K. 1997. Predicting diagenetic effects of groundwater flow in sedimentary basins: A modeling approach with examples. *In*: Montañez, I.P., Gregg, J.M. & Shelton, K.L. (eds) *Basinwide Fluid Flow and Associated Diagenetic Patterns: Integrated Petrologic, Geochemical and Hydrologic Considerations*. Society of Economic Paleontologists and Mineralogists, Special Publications, **57**, 3–14.

Leventhal, J.S. 1990. Organic matter and thermochemical sulphate reduction in the Viburnum Trend, southeast Missouri. *Economic Geology*, **85**, 622–631.

Lucia, F.J. 1983. Petrophysical parameters estimated from visual descriptions of carbonate rocks: A field classification of carbonate pore space. *Journal of Petroleum Technology*, **35**, 629–637.

Lucia, F.J. & Major, R.P. 1994. Porosity evolution through reflux dolomitization. *In*: Purser, B.H., Tucker, M.E. & Zenger, D.H. (eds) *Dolomites: A Volume in Honour of Dolomieu*. International Association of Sedimentologists, Special Publications, **21**, 325–341.

Machel, H.G. 2004. Concepts and models of dolomitization: a critical reappraisal. *In*: Braithwaite, C.J.R., Rizzi, G. & Darke, G. (eds) *The Geometry and Petrogenesis of Dolomite Hydrocarbon Reservoirs*. Geological Society, London, Special Publications, **235**, 7–63.

Machel, H.G. & Mountjoy, E.W. 1987. General constraints on extensive pervasive dolomitization – and their application to the Devonian carbonates of western Canada. *Bulletin of Canadian Petroleum Geology*, **35**, 143–158.

Macqueen, R.W. & Powell, T.G. 1983. Organic geochemistry of the Pine Point lead–zinc ore field and region, Northwest Territories, Canada. *Economic Geology*, **78**, 1–25.

Marikos, M.A., Laudon, R.C. & Leventhal, J.S. 1986. Solid insoluble bitumen in the Magmont West Orebody, southeast Missouri. *Economic Geology*, **81**, 1983–1988.

Mazzullo, S.J., Reid, A.M. & Gregg, J.M. 1986. Extensive dolomitization of high-Mg calcite supratidal deposits, Ambergris Cay, Belize. *In*: *Society of Economic Paleontologists and Mineralogists, Annual Midyear Meeting, Abstracts*, Volume **III**, 72 Tulsa, Oklahoma.

Mazzullo, S.J., Reid, A.M. & Gregg, J.M. 1987. Dolomitization of Holocene supratidal deposits, Ambergris Cay, Belize. *Geological Society of America Bulletin*, **98**, 224–231.

McCormick, J.E., Evans, L.L., Palmer, R.A. & Rasnick, F.D. 1971. Environment of the zinc deposits of the Mascot–Jefferson City district, Tennessee. *Economic Geology*, **66**, 757–762.

McKenzie, J.A. 1981. Holocene dolomitization of calcium carbonate sediments from coastal sabkhas of Abu Dhabi, U.A.E.: A stable isotope study. *Journal of Geology*, **89**, 185–198.

McManus, A. & Wallace, M.W. 1992. Age of Mississippi Valley-type sulfides determined using cathodoluminescence cement stratigraphy, Leonard Shelf, Canning Basin, Western Australia. *Economic Geology*, **87**, 189–193.

Misra, K.C., Kopp, O.C., Paris, T.A. & Linkous, M.A. 1989. Mississippi Valley-type fluorite-barite-sphalerite mineralization in the Sweetwater District, Tennessee. *Carbonates and Evaporites*, **4**, 211–230.

Montacer, M., Disnar, J.R., Orgeval, J.J. & Trichet, J. 1988. Relationship between Zn–Pb ore and oil accumulation processes: Example of the Bou Grine deposit (Tunisia). *Organic Geochemistry*, **13**, 423–432.

Montañez, I.P. 1992. Controls of eustasy and associated diagenesis on reservoir heterogeneity in Lower Ordovician, Upper Knox carbonates, Appalachians. *In*: Candelaria, M.P. & Reed, C.L. (eds) *Paleokarst, Karst-related Diagenesis, and Reservoir Development: Examples from Ordovician–Devonian Age Strata of West Texas and Mid-Continent*. Society of Paleontologists and Mineralogists, Permian Basin Section Publications, **92–33**, 165–181.

Montañez, I.P. 1994. Late diagenetic dolomitization of Lower Ordovician, Upper Knox carbonates: A record of the hydrodynamic evolution of the southern Appalachian Basin. *AAPG Bulletin*, **78**, 1210–1239.

Montañez, I.P. 1996. Regional diagenetic events and fluid flow conduits associated with southern Appalachian Mississippi Valley-type mineralization delineated by cathodoluminescent cement stratigraphy. *In*: Sangster, D.W. (ed.) *International Field Conference on Carbonate-hosted Pb-Zn Deposits*. Society of Economic Geologists, 75th Anniversary Volume, Special Publications, **4**, 432–447.

Montañez, I.P. 1997. Secondary porosity and late diagenetic cements of the Upper Knox Group, Central Tennessee Region: A temporal and spatial history of fluid flow conduit development within the Knox regional aquifer. *In*: Montañez, I.P., Gregg, J.M. & Shelton, K.L. (eds) *Basinwide Diagenetic Patterns: Integrated Petrological, Geochemical, and Hydrologic Considerations*. Society of Economic Paleontologists and Mineralogists, Special Publications, **57**, 100–117.

MOORE, C.H. 1984. The upper Smackover of the Gulf rim: depositional systems, diagenesis, porosity evolution and hydrocarbon production. *In*: VENTRESS, W.P.S., BEBOUT, D.G., PERKINS, B.F. & MOORE, C.H. (eds) *The Jurassic of the Gulf Rim*. Gulf Coast Section, Society of Economic Paleontologists and Mineralogists Foundation, 283–308.

MORROW, D.W. 1990*a*. Dolomite – Part 1: The chemistry of dolomitization and dolomite Precipitation. *In*: MCILREATH, I.A. & MORROW, D.W. (eds) *Diagenesis*. Geoscience Canada Reprint Series, **4**, 113–124.

MORROW, D.W. 1990*b*. Dolomite – Part 2: Dolomitization models and ancient dolostones. *In*: MCILREATH, I.A. & MORROW, D.W. (eds) *Diagenesis*. Geoscience Canada Reprint Series, **4**, 125–139.

MOUNTJOY, E.W. & AMTHOR, J.E. 1994. Has burial dolomitization come of age? Some answers from the Western Canada Sedimentary Basin. *In*: PURSER, B.H., TUCKER, M.E. & ZENGER, D.H. (eds) *Dolomites: A Volume in Honour of Dolomieu*. International Association of Sedimentologists, Special Publications, **21**, 203–229.

MUCHEZ, P., MARSHALL, J.D., TOURET, J.L.R. & VIAENE, W.A. 1994. Origin and migration of paleofluids in the Upper Visean of the Campine Basin of northern Belgium. *Sedimentology*, **41**, 133–145.

MURRAY, R.C. 1960. Origin of porosity in carbonate rocks. *Journal of Sedimentary Petrology*, **30**, 59–84.

OHLE, E.L. 1985. Breccias in Mississippi Valley-type deposits. *Economic Geology*, **80**, 1736–1752.

OLSON, R.A. 1984. Genesis of paleokarst and strata-bound zinc–lead sulphide deposits in a Proterozoic dolostone, northern Baffin Island, Canada. *Economic Geology*, **79**, 1056–1103.

OTTMANN, R.D., KEYES, P.L. & ZIEGLER, M.A. 1976. Jay Field, Florida – A Jurassic stratigraphic trap. *In*: BRAUNSTEIN, J. (ed.) *North American Oil and Gas Fields*. American Association of Petroleum Geologists, Memoir, **24**, 276–286.

POWELL, T.G. & MACQUEEN, R.W. 1984. Precipitation of sulfide ores and organic matter: Sulfide reactions at Pine Point, Canada. *Science*, **224**, 63–66.

RANDAZZO, A.F. & ZACHOS, L.G. 1984. Classification and description of dolomitic fabrics of rocks from the Floridan aquifer, USA. *Sedimentary Geology*, **37**, 151–162.

RHODES, D., LANTOS, E.A., LANTON, J.A., WEBB, R.J. & OWENS, D.C. 1984. Pine Point Orebodies and their relationship to stratigraphy, structure, dolomitization, and karstification of the Middle Devonian barrier complex. *Economic Geology*, **79**, 991–1055.

ROWAN, E.L. 1986. Cathodoluminescent zonation in hydrothermal dolomite cements: Relationship to Mississippi Valley-type Pb–Zn mineralization in southeastern Missouri and northern Arkansas. *In*: HAGNI, R.D. (ed.) *Process Mineralogy* **VI**. The Metallurgical Society, Warrendale, PA, 69–87.

RUSSELL, M.J. 1986. Extension and convection: a genetic model for the Irish Carboniferous base metal and barite deposits. *In*: ANDREW, C.J., CROWE, R.W.A., FINLAY, S., PENNELL, W.M. & PYNE, J.F. (eds) *Geology and Genesis of Mineral Deposits in Ireland*. Irish Association for Economic Geology, Dublin, 545–554.

SALLER, A.H. 1984. Petrologic and geochemical constraints on the origin of subsurface dolomite, Enewetak Atoll: An example of dolomitization by normal seawater. *Geology*, **12**, 217–220.

SANGSTER, D.F. 1983. Mississippi Valley-type deposits: A geological mélange. *In*: KISVARSANYI, G., GRANT, S.K., PRATT, W.P. & KOENIG, J.W. (eds) *International Conference on Mississippi Valley-type Lead Zinc Deposits*. Proceedings Volume. University of Missouri-Rolla, Rolla, MO, 7–19.

SANGSTER, D.F. 1990. Mississippi Valley-type and Sedex lead-zinc deposits: a comparitive Examination. *Transactions of the Institution of Mining and Metallurgy, Section B: Applied Earth Sciences*, **99**, B21–B34.

SANGSTER, D.F. (ed.). 1996 *Carbonate-hosted Lead–Zinc Deposits*. Society of Economic Geologists, 75th Anniversary Volume, Special Publications, **4**.

SHAW, B. 1975. *Geology of the Albion–Scipio trend, southern Michigan*. MS thesis, University of Michigan, Ann Arbor.

SHELTON, K.L., BAUER, R.M. & GREGG, J.M. 1992. Fluid inclusion studies of regionally extensive epigenetic dolomites, Bonneterre Dolomite (Cambrian), southeast Missouri: Evidence of multiple fluids during dolomitization and lead–zinc mineralization. *Geological Society of America Bulletin*, **104**, 675–683.

SHELTON, K.L., BURSTEIN, I.B., HAGNI, R.D., VIERRETHER, C.B., GRANT, S.K., HENNIGH, Q.T., BRADLEY, M.F. & BRANDOM, R.T. 1995. Sulfur isotope evidence for penetration of MVT fluids into igneous basement rocks, southeast Missouri, U.S.A. *Mineralium Deposita*, **30**, 339–350.

SIBLEY, D.F. 1982. The origin of common dolomite fabrics, clues from the Pliocene. *Journal of Sedimentary Petrology*, **52**, 1087–1100.

SIBLEY, D.F. & GREGG, J.M. 1987. Classification of dolomite rock textures. *Journal of Sedimentary Petrology*, **57**, 967–975.

SIBLEY, D.F., GREGG, J.M., BROWN, R.G. & LAUDON, P.R. 1993. Dolomite crystal size distribution. *In*: REZAK, R. & LAVOIE, D. (eds) *Carbonate Microfabrics*. Springer, New York, 195–204.

SPIRAKIS, C.S. & HEYL, A.V. 1988. Possible effects of thermal degradation of organic matter on carbonate paragenesis and fluorite precipitation in Mississippi Valley-type deposits. *Geology*, **16**, 1117–1120.

SVERJENSKY, D.A. 1984. Oil field brines as ore-forming solutions. *Economic Geology*, **79**, 23–37.

VAHRENKAMP, V.C., SWART, P.K. & RUIZ, J. 1991. Episodic dolomitization of late Cenozoic carbonates in the Bahamas: Evidence from strontium isotopes. *Journal of Sedimentary Petrology*, **61**, 1002–1014.

VOSS, R.L. & HAGNI, R.D. 1985. The application of cathodoluminescence microscopy to the study of sparry dolomite from the Viburnum Trend, southeast Missouri. *In*: HAUSEN, D.M. & KOPP, O.C. (eds) *Process Mineralogy V, Mineralogy – Applications to the Minerals Industry*. AIME, New York, 51–68.

VOSS, R.L., HAGNI, R.D. & GREGG, J.M. 1989. Sequential deposition of zoned dolomite and its relationship to sulfide mineral paragenetic sequence in the Viburnum Trend, southeast Missouri. *Carbonates and Evaporites*, **4**, 195–209.

WALLACE, M.W., KERANS, C., PLAYFORD, P.E. & MCMANUS, M.W. 1991. Burial diagenesis in the Upper Devonian reef complexes of the Geikie Gorge Region, Canning Basin, Western Australia. *AAPG Bulletin*, **75**, 1018–1038.

WARDLAW, N.C. 1976. Pore geometry of carbonate rocks as revealed by pore casts and capillary pressures. *AAPG Bulletin*, **60**, 245–257.

WARDLAW, N.C., MCKELLAR, M. & YU, L. 1988. Pore throat and size distribution determined by mercury porosimetry and by direct observation. *Carbonates and Evaporites*, **3**, 1–15.

WARREN, J. 2000. Dolomite: occurrence, evolution and economically important associations. *Earth Science Reviews*, **52**, 1–81.

WEYL, P.K. 1960. Porosity through dolomitization. Conservation-of-mass requirements. *Journal of Sedimentary Petrology*, **30**, 85–90.

WHEATLEY, C.J.V., FRIGGENS, P.J. & DOOGE, F. 1986. The Bushy Park carbonate-hosted zinc–lead deposit, Griqualand West. *In*: ANHAEUSSER, C.R. & MASKE, S. (eds) *Mineral Deposits of Southern Africa*. Geological Society of South Africa, Johannesburg, 891–900.

WHITAKER, F.F., SMART, P.L. & JONES, G.D. 2004. Dolomitization: from conceptual to numerical models. *In*: BRAITHWAITE, C.J.R., RIZZI, G. & DARKE, G. (eds) *The Geometry and Petrogenesis of Dolomite Hydrocarbon Reservoirs*. Geological Society, London, Special Publications, **235**, 99–139.

WOODY, R.E., GREGG, J.M. & KOEDERITZ, L.F. 1996. Effect of texture on petrophysical properties of dolomite: Evidence from the Cambrian–Ordovician of southeastern Missouri. *AAPG Bulletin*, **80**, 119–132.

WRIGHT, W.R., SOMERVILLE, I.D., GREGG, J.M., JOHNSON, A.W. & SHELTON, K.L. 2003. Dolomitization and neomorphism of Irish Lower Carboniferous (Early Mississippian) limestones: evidence from petrographic and isotopic data. *In*: AHR, W.M., HARRIS, P.M., MORGAN, W.A. & SOMERVILLE, I.D. (eds) *Permo-Carboniferous Carbonate Platforms and Reefs*. SEPM/AAPG, Special Publications, **78**, 395–408.

YOO, C.M., GREGG, J.M., & SHELTON, K.L. 2000. Dolomitization and dolomite neomorphism: Trenton and Black River Limestones (Middle Ordovician) northern Indiana, U.S.A. *Journal of Sedimentary Research*, **70**, 265–274.

ZEEH, S., WALTER, U., KUHLEMANN, J., HERLEC, U., KEPPENS, E. & BECHSTÄDT, T. 1997. Carbonate cements as a tool for fluid flow reconstruction: A study in parts of the eastern Alps. *In*: MONTAÑEZ, I.P., GREGG, J.M. & SHELTON, K.L. (eds) *Basin-wide Fluid Flow and Associated Diagenetic Patterns: Integrated Petrologic, Geochemical and Hydrologic Considerations*. Society of Economic Paleontologists and Mineralogists, Special Publications, **57**, 167–206.

ZENGER, D.H. 1983. Burial dolomitization in the Lost Burrow Formation (Devonian), east-central California, and the significance of late diagenetic dolomitization. *Geology*, **11**, 519–522.

Predicting and characterizing fractures in dolostone reservoirs: using the link between diagenesis and fracturing

JULIA F. W. GALE[1], STEPHEN E. LAUBACH[1], RANDALL A. MARRETT[2], JON E. OLSON[3], JON HOLDER[3] & ROBERT M. REED[1]

[1]*Bureau of Economic Geology, Jackson School of Geosciences, The University of Texas at Austin, Austin, TX 78713-8924, USA (e-mail: julia.gale@beg.utexas.edu)*

[2]*Department of Geological Sciences, Jackson School of Geosciences, The University of Texas at Austin, Austin TX 78712-0254, USA*

[3]*Department of Petroleum & Geosystems Engineering, The University of Texas at Austin, Austin, TX 78712-0228, USA*

Abstract: Fracture geometries and fracture-sealing characteristics in dolostones reflect interactions among mechanical and chemical processes integrated over geological timescales. The mechanics of subcritical fracture growth results in fracture sets having power-law size distributions where the attributes of large, open fractures that affect reservoir flow behaviour can be accurately inferred from observations of cement-sealed microfractures and other microscopic diagenetic features, which are widespread in dolostones. Fracture porosity is governed by the competing rates of fracture opening and cement precipitation during fracture growth and by cements that post-date fracture opening.

Combined analysis of structural and diagenetic features provides the best approach for understanding how fracture systems influence fluid flow. We review previous work and integrate new data on fractures and diagenetic features in cores from the Lower Ordovician Ellenburger and Permian Clear Fork formations in West Texas, and the Lower Ordovician Knox Group in Mississippi, together with outcrop samples of Lower Cretaceous Cupido Formation dolostones from the Sierra Madre Oriental, Mexico, in order to illustrate our approach.

Dolostones form important hydrocarbon reservoirs, and fluid flow within them is commonly influenced by opening-mode fractures. For example, many of the Ordovician dolostone reservoirs in North America, including the Ellenburger, Knox and Trenton–Black River groups, are fractured. An understanding of fundamental processes that govern the fluid-flow properties of such reservoirs, namely diagenesis and fracturing, has remained elusive, making characterization challenging and fracture prediction difficult.

Opening-mode fractures can form throughout burial and exhumation under a wide spectrum of loading conditions, partly because rocks have low tensile strength. Although fractures may develop at any time during the history of a rock, individual fractures may look superficially alike, and the timing of any given fracture and its cause are difficult to pin down. On the basis of fracture-mechanics principles, it has been concluded that some differences in fracture shape are expected for fractures developed at different stages of burial and exhumation (Nelson 2001; Marrett & Laubach 2001), but these are rarely sufficient for determining the timing, cause and pattern of fracturing. The role of fractures in dolostone reservoir behaviour is difficult to assess because predictions of the fracture patterns that govern fluid flow are frequently made on the basis of sparse observations of individual fractures, and inferences about the timing and loading conditions that produced them.

Better indicators of fracture timing may be found by first recognizing that diagenesis and fracturing are intimately linked processes. A complex diagenetic sequence, if it can be unravelled, has the potential to indicate the timing of fractures in relation to other events because a series of precipitation, dissolution and mineral reaction events are captured, relative to growth of fractures. In essence this approach is similar to the way in which metamorphic and structural studies are combined to determine tectonic histories of rocks at higher pressure–temperature conditions than are typical for sedimentary basins (e.g. papers in Knipe & Rutter 1990). Because diagenetic products are far more readily sampled in the subsurface than large

From: Braithwaite, C. J. R., Rizzi, G. & Darke, G. (eds) 2004. *The Geometry and Petrogenesis of Dolomite Hydrocarbon Reservoirs*. Geological Society, London, Special Publications, **235**, 177–192. 0305-8719/$15.00

fractures, diagenetic observations can throw light on fracture attributes, even if the diagenetic processes are not fully understood. Laubach *et al.* (2000) used the term *structural diagenesis* to encompass the combination of structural and diagenetic processes associated with mechanical property changes during fracture growth and sealing, and we use this term here.

Attempts to model fracture systems on a reservoir or field scale have been made using a variety of approaches. Commonly, the focus of these studies is on the modelling method itself rather than on the input data. The reason for this is that the complexity of a natural fracture system cannot easily be replicated by a model and what is needed is an approximation. One approach is to use empirically based or deterministic methods. For example, Hennings *et al.* (2000) collected outcrop data to create an analogue model for the subsurface, whereas Trice (1999) used borehole image logs and the relationship of fractures to the stress field to construct a three-dimensional (3-D) static model of productive fracture systems. Ericsson *et al.* (1998) created a 3-D deterministic model in which fracture density was related to facies distribution and structural curvature, and then the model was verified with empirical borehole image data. This fracture model was then used to improve transmissibility prediction for reservoir simulation. Rawnsley & Wei (2001) constructed a model using seismic-scale reservoir structures, geological layering and observed intersections with flowing fractures in wells, and then calibrated the model against well test data. As an alternative, Gauthier *et al.* (2002) used a geostatistical method that integrated all available data, including seismic, well and production data, and then tested the model using a new well. Several workers have attempted to use neural networks to achieve a similarly stochastic model (e.g. Thomas & LaPointe 1995; Ouenes 2000). Those favouring a geostatistical approach regard deterministic models as unsuitable because they do not address the uncertainty in fracture architecture. Here, we emphasize the importance of high-quality input data and note that a successful modelling approach must utilize data that relate to the growth of fractures during diagenesis. Our contention is that these key data have not yet been sufficiently incorporated into any type of reservoir fracture model.

The purpose of this contribution is to highlight critical topics in structural diagenesis that must be addressed if we are to progress in our understanding and ability to make predictions about fractures in dolostone reservoirs. We review previous work and integrate new data on fractures and diagenetic features in cores from the Lower Ordovician Ellenburger and Permian Clear Fork formations in West Texas, and the Lower Ordovician Knox Group in Mississippi, together with outcrop samples of Lower Cretaceous Cupido Formation dolostones from the Sierra Madre Oriental, Mexico, in order to illustrate our approach.

Fracture characterization in dolostones

Previous studies of fractures in dolostone

Previous studies of fractures in dolostones have focused on a range of attributes. Fracture architecture at outcrop and in the subsurface, and the relationship of fractures to major structures or stratigraphy, are common topics (e.g. Joubert & Rice 1997; Antonellini & Mollema 2000; Muldoon *et al.* 2001). Mechanical properties of dolostones have been measured (Williams & McNamara 1992; Palchik & Hatzor 2002), and their tendency to develop fractures more readily than limestones is well documented (Nelson 2001; Ortega & Marrett 2001). Marquez & Mountjoy (1996) determined the timing of fracture events in a paragenetic sequence to gain insight into how fractures were generated, and Montañez (1997) considered how they might have affected fluid flow and cementation.

Extending fracture attribute predictions into areas where fractures have not been sampled requires an understanding of the conditions that prevailed during fracture growth. Some empirical studies have focused on palaeostress and palaeopore-fluid pressure conditions (e.g. Mollema & Antonellini 1999; Cozzi 2000; Gillespie *et al.* 2001). Other studies, however, show that the interaction of diagenesis with fracture growth also has a fundamental control on fracture patterns (Montañez 1997; Troudi *et al.* 2000; Marrett & Laubach 2001; Monroy Santiago *et al.* 2001). All of these studies provide useful information, but each focuses on different specific aspects of fracture development. There is no one systematic approach to predicting key attributes of a fracture system, namely orientation, intensity, population size distribution, spatial architecture, openness and connectivity.

Structural diagenesis: a new approach

Structural diagenesis has the potential to provide a unifying concept for understanding fracture development in dolostones. Key rock properties and fracture attributes can be measured on a site-specific basis and placed in

Fig. 1. Micrographs of fractures in Lower Ordovician Ellenburger Formation comparing SEM-based imagery with conventional petrography. (**a**) SEM CL image of pale-grey luminescent dolomite (L) forming the rock matrix and dark-grey luminescent dolomite (D) partly filling the large fracture, sealing the small fracture, and filling some pores in the matrix. Fracture boundaries are indicated by the solid line. A very dark, almost non-luminescent, iron-rich dolomite may be seen in the large fracture, forming the outer zone of some dark-grey crystals. Fracture porosity (P) is present in the large fracture. (**b**) Plane-polarized-light photomicrograph of the same area as in (a). Note that the small fracture is not visible and there is no distinction between the different dolomite cements. (After Gomez *et al.* 2001.)

the context of a paragenetic sequence. Predictive fracture models can then be built by determining how mechanical rock properties evolved with time and how they controlled developing fracture patterns.

This approach has an important practical advantage in that we may not need to understand the origin and cause of diagenetic events completely in order to model their effects on fracture initiation, growth and sealing. This point is especially important in the case of fractured dolostones where causes of diagenetic events are notoriously difficult to determine (Tucker & Wright 1990 chapter 8; Purser *et al.* 1994).

Imaging microstructure

A structural–diagenetic sequence may be established by analysing the timing of diagenetic events relative to fracturing. A tool that has proven powerful for this purpose is scanning-electron-microscope-based cathodoluminescence (SEM CL) (Milliken & Laubach 2000; Reed & Milliken 2003). Structures not visible with conventional petrographic techniques, but that provide important timing information, were picked out using SEM CL (Fig. 1). Problems of persistent fluorescence in carbonate samples were overcome by using a blue filter (Reed & Milliken 2003). Secondary electron imagery (SEI), energy-dispersive spectral analysis and element mapping were conducted on the same areas of polished sections as CL detection, aiding in identifying minerals and in recognizing compositional variation (Fig. 2). By using a combination of secondary electron and CL images, details of composition, microstructure and relict porosity were discerned that are otherwise not easily seen.

Structural diagenesis: key topics

Diagenetic microstructure

The dolostones we studied contain arrays of cement-filled microfractures that do not contribute to present-day fluid flow. Yet, the

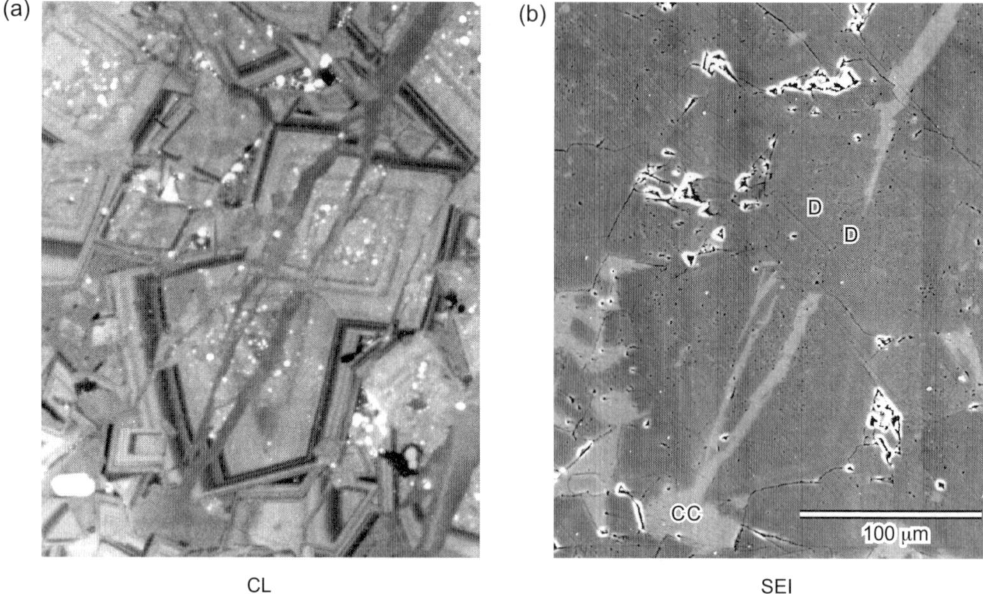

Fig. 2. Use of both SEM CL (**a**) and SEI (**b**) of the same area of thin section to determine cement composition and fill pattern. Both dolomite (D) and calcite (CC) cements are present in the fractures. Dolomite is compositionally indistinguishable from the host-rock dolomite in the SEI but has different luminescence. Calcite also fills pores in the host rock. Sample is of Lower Ordovician Knox Group dolostone from Mississippi. Vertical lines in (b) are a scanning artifact.

composition and sequence of cement fill, orientation and aspect ratios of these microfractures are consistent with macrofractures in the same rocks, indicating that the microfractures are part of the same population as the larger fractures (Laubach *et al.* 2000). Core samples may contain a large population of microfractures with apertures at the submillimetre scale that are orders of magnitude more abundant than those at the centimetre scale. Problems related to scale of observation in well bores and cores may be overcome by using microstructural observations to predict macrofracture attributes such as orientation, aperture and openness.

Scaling of fracture intensity

Outcrops provide access to fractures at a range of scales that may not be seen in core, allowing calibration of attribute scaling relationships. Marrett *et al.* (1999) have shown examples of power-law distributions of opening-mode fracture apertures in sandstone and limestone, and shear displacements in tuff. Here we demonstrate that similar relationships apply to fracture populations in dolostones and that microfracture data may be used to predict intensity of macrofractures at any size by function extrapolation.

Cretaceous Cupido Formation dolostones. Cretaceous Cupido Formation dolostones are well exposed in Canyon Boquilla Corral de Palmas, Monterrey Salient, Sierra Madre Oriental, NE Mexico. Laramide-age folds with wavelengths and amplitudes on the kilometre scale post-date many opening-mode fractures in the shallow part of the platform sequence (Marrett & Laubach 2001). Fractures predating early solution-collapse breccias are commonly associated with early, soft-sediment deformation, such as pinch and swell, and faults with growth strata (figs 7 and 8 in Hooker *et al.* 2002).

Kinematic apertures of pre-Laramide opening-mode fractures were measured along scanlines normal to fracture strike (Fig. 3). The kinematic aperture is the wall-to-wall distance normal to the fracture at the point crossed by the scanline irrespective of whether the fracture is open or sealed. Fractures at this location are largely confined to dolomitized layers at the tops of parasequences (Goldhammer *et al.* 1991; Ortega & Marrett 2001). The mechanical layer in which measurements were made consists of

(a)

(b)

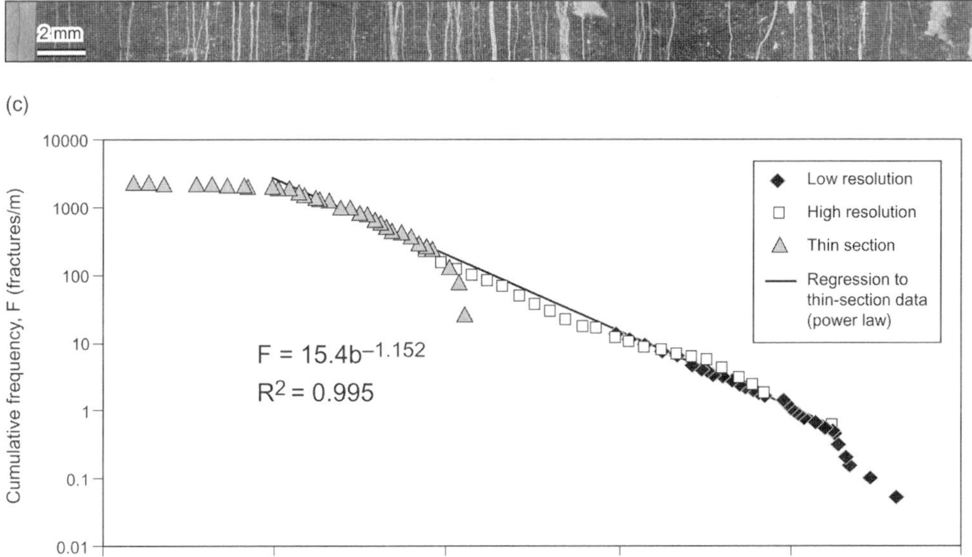

Fig. 3. Aperture-size scaling in Lower Cretaceous Cupido Formation dolostones, Sierra Madre Oriental, Mexico. (**a**) Field photograph of sealed fractures in a 30 cm-thick dolostone cycle top. The underlying unit is limestone with evaporite nodules that have been replaced by calcite. Note how fractures taper and terminate in the limestone layer. (**b**) Fractures in thin section of sample from the dolostone outcrop in (a). (**c**) Aperture-size distributions for fractures in outcrop (high- and low-resolution data) and thin section. The power-law regression is to thin-section data only.

two beds, which taken together are approximately 30 cm thick (Fig. 3a).

Data were measured at two resolutions. In the high-resolution set, all fractures wider than 0.05 mm (575 fractures) were measured along a scanline approximately 2.5 m long. In the low-resolution data set, all fractures wider than 0.95 mm (262 fractures) were measured along a scanline approximately 21 m long. The high-resolution data were taken from the first 2.5 m of the low-resolution scanline, so that fractures wider than 0.95 mm are common to both data sets. In addition, 88 fracture apertures were measured along a 38 mm-scanline normal to fracture strike in a standard petrographic thin section prepared from a sample of the fractured layer (Fig. 3b). A cumulative frequency plot of fracture apertures, normalized to scanline length, shows both high- and low-resolution data and thin-section data on the same plot (Fig. 3c). The thin-section data form a straight-line segment with truncation and censoring artifacts (Marrett *et al.* 1999). A best fit to the thin-section data yields a function with a power-law segment of F (fractures m^{-1}) = 15.4 aperture $(mm)^{-1.152}$, with a least-squares regression correlation coefficient of 0.995. Significantly, this function is a good fit for the data sets at the other two resolutions, and the straight-line sections of the three curves approximate a continuous straight line for more than three orders of magnitude. These aperture data confirm that opening-mode fracture sets in the Cupido Formation dolostone follow power-law distributions and support the contention that the micro- and macrofractures are different size fractions of the same fracture set. Microfractures may therefore be used to predict macrofracture intensities at a given aperture size.

Ordovician Knox Group dolostone. An Ordovician Knox Group dolostone core sample from Oketibbeha County, Mississippi, from a depth of approximately 4439 m (14 578 ft) was imaged using SEM CL and SEI, and two opening-mode fracture sets were identified (Fig. 4). Earlier fractures (F_1) have variable orientations, from NW to NNW, are curved or branching with hooking terminations and irregular walls, and are completely sealed. The later fracture set (F_2) comprises abundant, subparallel ENE-striking fractures with sharp, planar walls and some preserved fracture porosity. F_1 fractures are shorter and wider than the F_2 fractures, which have length-to-width ratios of greater than 100:1. There were insufficient F_1 fractures for scaling analysis, but the F_2 fracture apertures were measured along a scanline 5.955 mm long, constructed normal to the fracture orientation. The scanline yielded 31 microfractures with apertures ranging from 0.3 μm to 0.01 mm. The aperture-size distribution is described by a power law (Fig. 4b), where F (fractures mm^{-1}) = 3.3×10^{-4} aperture $(mm)^{-1.402}$. Extrapolation of the power law to larger aperture sizes than those measured indicates that there would be, on average, 0.33 fractures 1 mm wide or greater in 1 m of scanline.

Fracture morphology

Fracture morphology in dolostones is variable. Some fractures are straight sided, and the fracture walls are easily distinguished from the host rock (Fig. 4c & d). In other cases fractures have irregular traces, and the walls are not clearly defined (Fig. 1). Fracture walls may appear irregular if they have been modified by dissolution or by subsequent precipitation of dolomite or other cements. If dolomitization post-dates a fracture event, then fracture walls may be modified substantially. Irregular traces may also result if fractures propagate along rhombohedral dolomite grain boundaries or intragrain zone boundaries (Fig. 5). Fractures also commonly propagate around allochems in carbonates.

Fracture fill and diagenesis

There are two patterns in how fractures seal. First, cement that precipitates on the walls of fractures during fracture growth is termed synkinematic cement (Laubach 2003). Synkinematic cement is present in all fractures in a given set, tending to fill microfractures, and to line and bridge large fractures. Secondly, cements that post-date fracture opening, termed post-kinematic cements (Laubach 2003), fill remaining porosity in some large fractures but are not necessarily uniformly distributed throughout a fracture set.

Microfractures in the dolostones we studied are typically lined or filled with dolomite. Where polymineralic fill is present in microfractures, textural relations show that dolomite precipitated first. Initial dolomite precipitation occurs in dolostones containing relict limestone, as well as later anhydrite and calcite cements, and also holds where more than one fracture set is present. For example, in the Knox Group dolostone sample described above (Fig. 4) two sets of fractures are present, each with synkinematic dolomite and post-kinematic calcite. Dolomite precipitated on fracture walls in the later fracture set (F_2) despite the fact that post-kinematic calcite previously occluded remaining fracture porosity in the earlier fracture set (F_1). These relationships suggest that rock composition influences the composition of the initial cement precipitated in new fractures. This influence is not surprising if fracturing is a relatively rare event and geochemical reactions are

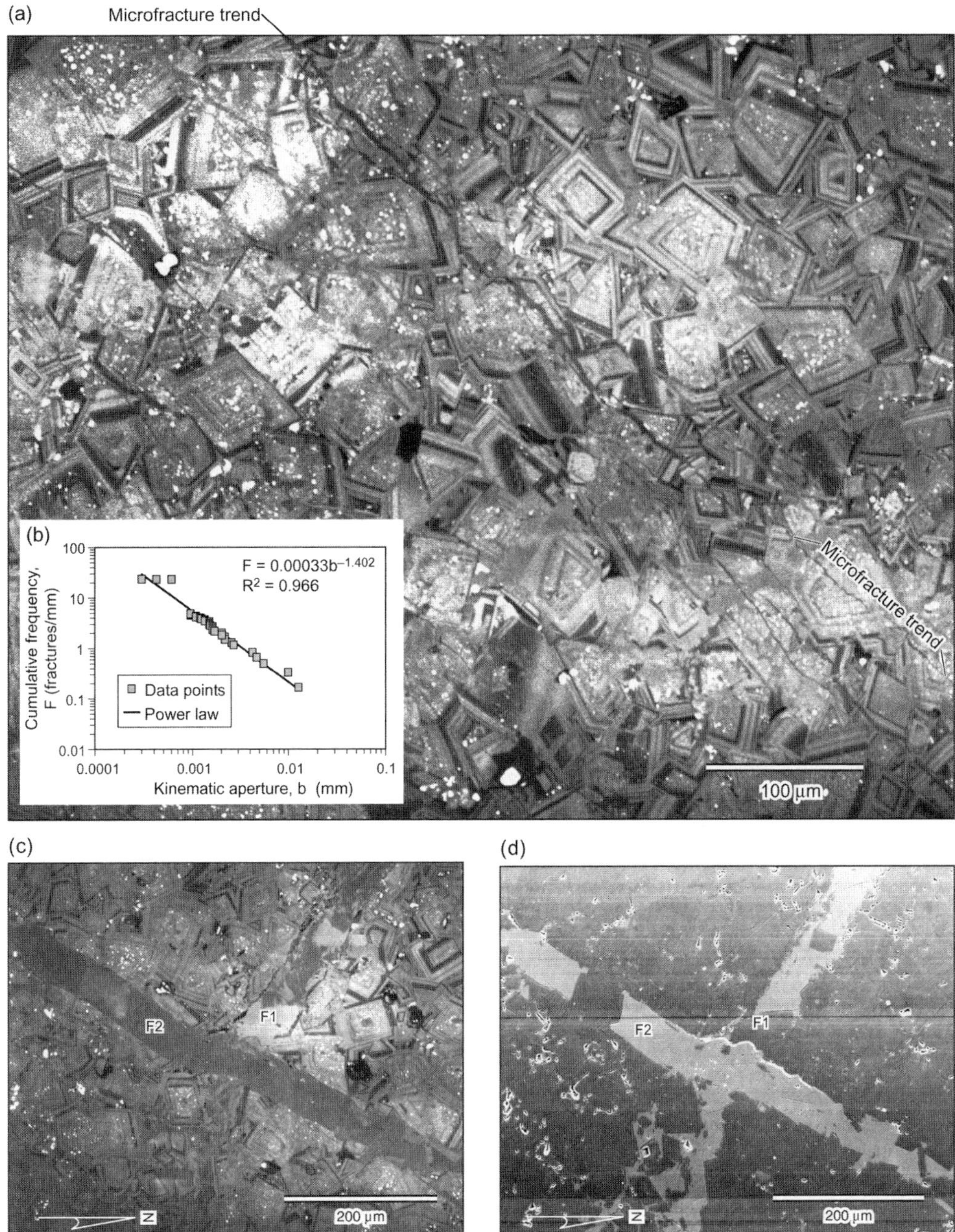

Fig. 4. Ordovician Knox Group dolostone from a depth of approximately 4440 m (14 580 ft) contains multiple microfractures. (**a**) SEM CL of microfractures with consistent orientation, top left to bottom right, cutting across zoned dolomite crystals. (**b**) Power-law aperture-size distribution of microfractures from this sample. (**c**) SEM CL and (**d**) SEI of cross-cutting fractures in Knox Group dolostone. The older fracture (F_1) has a thin dolomite lining and bridges and is otherwise filled with twinned and zoned calcite. The younger fracture (F_2) also has dolomite lining and bridges and is filled with later calcite.

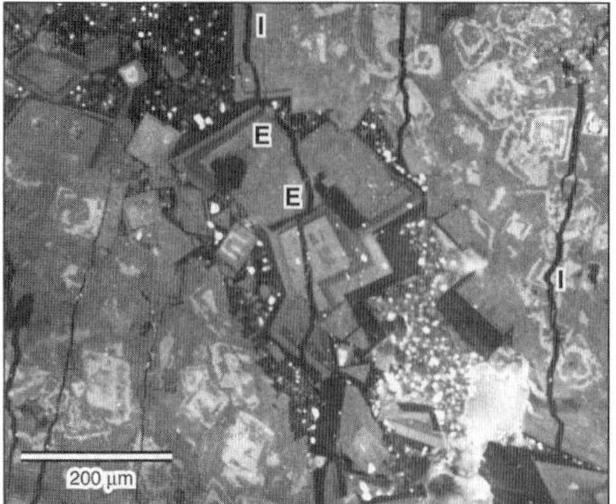

Fig. 5. Irregular fracture pathways around dolomite rhombs (E) in Ellenburger dolostone and along internal zones (I) within rhombs indicate that fracturing post-dated dolomitization and an earlier phase of fracturing.

Fig. 6. Crack–seal texture in Ellenburger dolostone. Slivers of host rock are surrounded by fracture-filling dolomite cement. The dark cement is synkinematic dolomite.

(a)

(b)

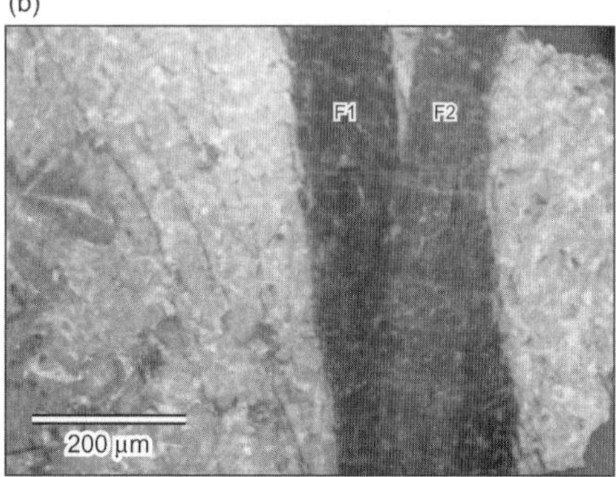

Fig. 7. (**a**) Plane-polarized-light photomicrograph and (**b**) light-microscope-mounted CL of synkinematic fibrous dolomite fill in the same fractures in Ellenburger dolostone. The fibrous nature of the fill is only seen in plane polarized light, but the fact that there are two fractures is only revealed by the CL image. Fibres in the younger fracture (F_2) grow in continuity with those in the older fracture (F_1).

dominated by the composition of the host dolostone.

Within any given fracture set there may be a transition from completely sealed to partly open fractures, termed the *emergent threshold* (Laubach 2003). The fracture aperture size at which this transition occurs is governed by fracture-opening rate, cement-precipitation rates and the time that the rock has been exposed to cementation after fracturing. Microfractures and larger fractures also commonly contain textural evidence of crack–seal processes. Crack–seal structure develops in synkinematic cement as fractures repeatedly open and refill with cement. Laubach (1988, 2003) drew attention to the value of crack–seal structures for deciphering fracture histories. An example of crack–seal structure (Fig. 6) is indicated by multiple slivers of host grains that have become entrained within the fracture fill as the fractures repeatedly opened and sealed. Subtle variation in CL intensity between one stage of fracture sealing and the next may also reveal internal crack–seal structure within cement. The presence of crack–seal structure in dolomite-filled fractures in dolostones, as well as in quartz-filled fractures in sandstones (Laubach *et al.* 2002), suggests that similar mechanical and cementation processes operate in the two rock types. Fibrous crystal morphologies (Fig. 7a & b) may develop

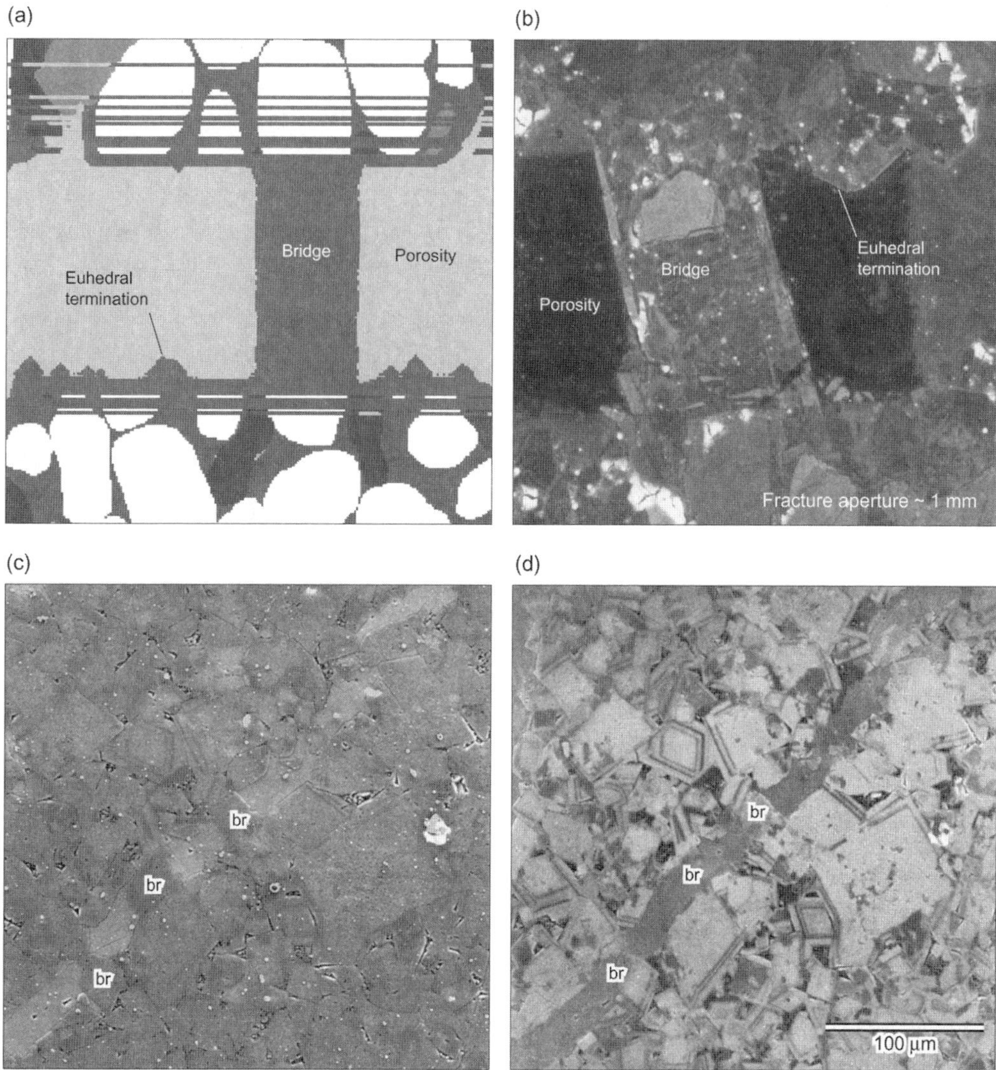

Fig. 8. (**a**) Model of a large, open fracture with a quartz bridge and fracture lining, and smaller, parallel sealed microfractures in the wall rock (Lander *et al.* 2002). (**b**) SEM CL image of quartz bridge with crack–seal texture and fracture wall lining in sandstone. (**c**) and (**d**) Composite SEM CL–SEI of dolomite bridges (br) in fracture in Knox Group dolostone. The two images are constructed using different image-strength ratios of CL and SE images from the same location. Dolomite bridges surrounded by calcite cement are seen in (c), whereas the growth zoning in dolomite in the host rock is clearer in (d).

in synkinematic cement during fracture opening. In the example from the Ellenburger Formation, CL reveals two fractures (Fig. 7b). Fibres in the older fracture continue in optical continuity into the younger fracture (Fig. 7a), demonstrating the importance of nucleation kinetics in controlling cement microstructure.

In many fractures dolomite forms cement bridges from one wall to the other. Dolomite bridges and fracture-wall linings bear some resemblance to structures seen in fractured quartz-cemented sandstones and to models of quartz cementation in fractures (Lander *et al.* 2002). An example from Lander's model (Fig. 8a) and a fracture in sandstone (Fig. 8b) are compared with a fracture in Knox Group dolostone (Fig. 8c & d). These structures may appear similar because similar structural diagenetic processes are at work. Similar processes include repeated reopening of the fractures, formation of cement bridges, rapid initial cement growth on the fracture surface followed by slower growth into the fracture void space (Lander *et al.* 2002), and the tendency for both quartz and dolomite to grow on a mineralogically similar substrate (Arvidson & Mackenzie 1997; Lander & Walderhaug 1999). By analogy with models of quartz precipitation in fractures, the relative rates of growth and fracture opening may govern the emergent threshold from sealed to open fractures and determine whether fractures above that size are bridged or merely lined with a thin veneer of euhedral crystals (Lander *et al.* 2002). In fractured dolostones the synkinematic cement is commonly dolomite, and emergent thresholds of less than 0.1 mm have been observed, meaning that appreciable porosity can be preserved in submillimetre fractures. At this aperture size, however, cement linings and substantial numbers of bridges are likely to limit pore connectivity. At larger aperture sizes, linings are thin relative to the open part of the fracture and there are fewer bridges, so pores are more likely to be connected.

Porosity-reducing cements in large opening-mode fractures mostly post-date fracture opening. The mainly post-kinematic character of these cements suggests that fluid-transport-limited processes operate mostly after fractures formed under conditions where loading and pore-fluid pressure do not promote fracture growth. Heterogeneous cement patterns could reflect a spectrum of flow pathways. Post-kinematic cements fill remaining pore space in the fractures and host rock after fractures have ceased opening in some cases (Figs 2 & 4) but are absent in other cases (Fig. 1). Post-kinematic cements in dolostones include calcite, ferroan dolomite, anhydrite, barite and quartz. They have the greatest effect in sealing large fractures and commonly have highly heterogeneous distributions (Laubach 2003). Mechanisms controlling post-kinematic fracture fill are probably different from those that control the synkinematic cement fill because fresh surfaces are no longer being generated by fracturing. Fluid flow through the fracture system may govern how these fractures seal (Noh & Lake 2002).

Mechanical rock properties

Many aspects of natural opening-mode fracture patterns can be explained by considering the mechanics of subcritical fracture growth (Atkinson 1984; Olson 1993). Chemical reactions near the fracture tip allow fractures to grow without reaching critical stress intensity. A rock property, the subcritical crack index (Atkinson 1984; Olson 1993), governs subcritical fracture growth. The value of the subcritical crack index (n) is related to the velocity of crack growth (v) by the equation

$$v = A\ (K_I/K_{IC})^n \quad (1)$$

where A is the critical fracture propagation velocity constant, K_I is the stress intensity at the crack tip and K_{IC} is the critical stress intensity required for failure (the fracture toughness). The growth of a fracture modifies the surrounding stress field, and the interaction of neighbouring fractures controls the developing fracture pattern. Olson (1993) and Olson *et al.* (2001) used boundary-element modelling to show that for constant strain, mechanical-layer thickness and other elastic rock properties, fracture patterns are dependent on the subcritical crack index. In particular, they showed that fracture clustering depends on the value of the subcritical crack index (Fig. 9).

We can now consider what governs subcritical crack index and mechanical-layer thickness. Initial depositional composition, grain size and porosity affect the subcritical index (Holder *et al.* 2001). Mechanical-layer thickness depends only partly on bed thickness and stacking patterns of beds that are determined by what position they have in a depositional system and whether that system is greenhouse or icehouse (Read 1995). Superimposed on these effects, diagenesis will control which combinations of beds acted as mechanical layers at different times. Having established that fracture architecture is largely controlled by the subcritical crack index, we must emphasize that it is this value *at the time of fracturing* that governs the fracture pattern. During the early stages of carbonate sequence lithification, profound mechanical changes occur through diagenetic processes. Moreover, these changes are not likely to be uniform throughout the sequence. For example, dolomitization of cycle tops in a shallow carbonate platform sequence may render the topmost layers more brittle than the intervening limestones and, hence, more susceptible to early fracturing (Ortega & Marrett 2001). The stage at which a fracture set grows, from earliest lithification to present day, is crucial to its development because mechanical rock properties, including the subcritical crack index, may change with time as diagenesis proceeds. The details of these changes will depend on the burial history of the rock and the materials that make up the rock. Wood & Holm (1980) demonstrated on the coarsest of scales that deformation styles are influenced by the age of the carbonate sequence. The timing of deformation may thus affect structural

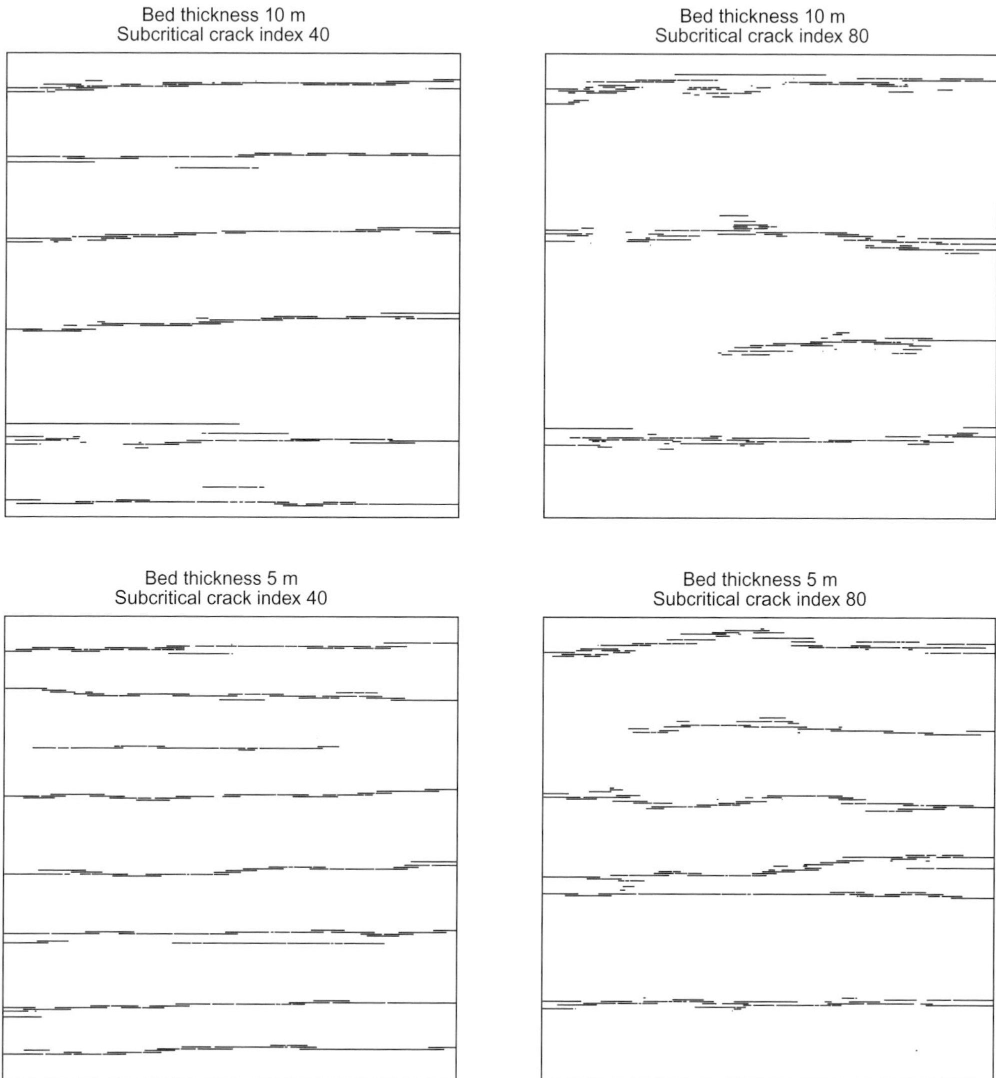

Fig. 9. Map views of geomechanical models of fractures in Permian Clear Fork dolostone (from Philip *et al.* 2002). Subcritical crack index and bed thickness are the only input parameters varied for the four realizations.

behaviour to as great an extent as the initial sedimentary architecture and composition of the starting materials.

Here we draw attention to the potential of linking diagenetic history with the evolution of mechanical behaviour through time and using this correlation as a tool for understanding how opening-mode fractures grow and seal. If a structural–diagenetic sequence is known for a given rock, and if subcritical crack indices for different stages of diagenesis of that rock were available, then rock properties at the time of fracturing could be estimated. These properties, in particular the subcritical crack index, could then be used to condition geomechanical models of fracture growth that would replicate fracture patterns growing under those conditions.

Holder *et al.* (2001) developed an apparatus for measuring subcritical crack index in sedimentary rocks. Rijken *et al.* (2002) tested the idea that there may be a systematic variation in subcritical crack index with petrological features in sandstones. Rijken's study showed that subcritical fracture index increases with decreasing grain size and subcritical index decreases with increasing carbonate cement. Tests were also run adding artificial cement to previously tested samples, resulting in increased strength but decreased subcritical index. A similar approach in dolostones may yield relationships that could be used for estimating subcritical crack indices through different stages of diagenesis.

Integration of diagenesis and fracture mechanics in fluid-flow modelling: an example from the Clear Fork Formation, west Texas

Fracture characterization of Permian Clear Fork Formation dolostones was carried out in outcrops and core (Moros Otero 1999; Gale *et al.* 2002). These results were used by Philip *et al.* (2002), in conjunction with subcritical crack index measurements, to produce geomechanical models of fracture systems and then to simulate coupled fracture–matrix fluid flow in those systems.

Samples yielded a range of subcritical indices, from 38 to 81, with good reproducibility. Philip *et al.* (2002) modelled fracture development for two n values (40 and 80) and considered two bed thicknesses, which were thought to be representative of Clear Fork Formation mechanical bed thicknesses (5 and 10 m). Input strain values for the model were determined through measurements of fracture apertures in scanlines. The fracture models are displayed as maps of fracture traces (Fig. 9). For $n = 40$, en echelon arrays of fractures form linked arrays that are spaced proportionally to bed thickness (Fig. 9a & c). With $n = 80$, the fractures are more clustered and have wider spaces between the clusters (Fig. 9b & d). Philip *et al.* (2002) studied the effect of variation in the amount of synkinematic cement on permeability for a bed thickness of 10 m and subcritical crack index of 40. They reduced fracture apertures by factors ranging from zero (no reduction) to half to simulate synkinematic cementation and found that permeability decreased with increasing synkinematic cement thickness. They concluded that permeability reduction was caused mostly by reduction in open fracture length because fractures have narrow tips, and the tips and narrow segments of fractures in en echelon arrays seal with small amounts of cement, effectively reducing the connectivity of the system. Philip *et al.* (2002) also modelled occlusion of fracture porosity by post-kinematic cement. They filled fractures according to the amount of post-kinematic cement found in thin-section analysis (Gale *et al.* 2002) but applied this randomly to reflect its heterogeneous distribution. Permeability reduction in this case was also attributed to reduction of open fracture length.

Discussion

Characteristics of fractured dolostones

The essential character of opening-mode fracture systems in all the dolostones we studied is similar: namely, that fracture apertures follow power-law distributions, that the dominant synkinematic cement is dolomite and that aperture size exerts some control on potential fracture porosity, although this porosity may be occluded by post-kinematic cements. We postulate that these attributes are widespread in other dolostones, although variation in the power-law function, emergent threshold and amount of post-kinematic cement would be expected. Variation in fracture spatial organization is determined largely by the subcritical crack index at the time of fracturing.

Intensity and size range of opening-mode fractures

The intensities of fractures 1 mm wide or greater that we have measured in carbonates range from 0.1 (Austin Chalk) to 15.4 fractures m^{-1} (Cupido Formation). Dolostones tend to fall at the high-intensity end of this spectrum because of their tendency to accommodate strain by brittle failure rather than by other deformation mechanisms.

Fracture apertures in dolostones in this study range over four orders of magnitude between less than 1 μm and 42 mm wide. The lower size limit of fractures in these sets is not known because we are limited by our method of observation in recording the narrowest fractures. An upper size limit for fracture apertures might be obtained through direct observation of outcrop. For subsurface fractures, where direct observation is restricted to a small sample and we use a power-law model to predict intensity of large fractures, an alternative approach is needed. Geomechanical relationships between fracture height and aperture, where height is governed by mechanical layer thickness, might be used to infer the maximum likely aperture (Olson 2003). If fractures were widened by dissolution, however, then the power-law function describing aperture-size distribution might break down. Observation of fracture walls and fill is necessary to indicate whether dissolution was important.

Subcritical crack indices and fluid composition

A subcritical crack index measured in the laboratory applies to the sample in its present state and might not apply to fractures formed during an earlier stage in its history. If we could reconstruct subcritical index values for rocks throughout their history, then fractures developed at any stage could be modelled. Suites of samples from different settings, ranging from modern, semi-lithified, carbonate sediments through to ancient, deeply buried carbonates, would allow us to explore the possible relationships between dolostone petrography and subcritical crack index. Clearly, in dolostone reservoirs, the processes and timing of dolomitization would be key to understanding how the subcritical crack index changes with time.

Fluid composition is known to influence subcritical crack propagation and cement precipitation or dissolution. Consideration of saturation indices of fluids in fractures, and of kinetic inhibitors of dolomite precipitation, would be necessary to better understand fracture growth and sealing mechanisms.

Use of outcrops in fractured dolostone characterization

Although outcrops may be analogues for the subsurface in certain circumstances, they should be considered in the context of their burial history (e.g. Gray *et al.* 2001, burial history for the Sierra Madre Oriental). Outcrops may provide information from rocks that have substantially different burial histories from those in the subsurface. Additionally, modern, semi-lithified sediment may provide a window into the early history of much older rocks, allowing us to observe processes operating during early stages of structural diagenesis. Burial history will not only affect deformation response in a rock package, but will also control diagenetic history. It is important to establish burial history so that maximum depths of burial and the time at each depth can be incorporated into the deformation and diagenetic history.

Fracture intensity and major structures

Opening-mode fractures commonly develop in association with larger structures, such as faults and folds, and methods that predict where fractures are most likely to develop in relation to larger structures, for example curvature analysis, have been applied with success in several cases (e.g. Hennings *et al.* 2000). However, not all fractures are related to major structures, and a genetic relationship should not be assumed. It is especially dangerous to rely on fracture orientation as an indicator of association. The range of fracture orientations predicted to develop synchronously as a result of folding through tangential longitudinal strain is large (Price 1966; Hancock 1985), and, allowing for local variation, most fracture orientations could be attributed to one category or another. In the absence of careful documentation of fracture timing, some fractures may be identified as being associated with a larger structure when, in fact, they pre- or post-date it.

A further potential pitfall in placing fracture development into a wider tectonic picture is the commonly-held assumption that fractures in the subsurface must parallel present-day maximum horizontal stress $S_{H_{max}}$ in order to be open (Queen & Rizer 1990; Crampin 1994; Heffer *et al.* 1997). The reasoning behind this idea is that open fractures having any orientation other than parallel to $S_{H_{max}}$ will be closed by the present-day stress field, particularly if pore fluid pressures in a hydrocarbon reservoir are reduced by drawdown. Contrary to this idea, Stowell *et al.* (2001) showed many examples where $S_{H_{max}}$ is at a high angle to open fractures in the subsurface. Fractures may be open in apparently unfavourable orientations if the rock in which they formed has been diagenetically altered by cementation subsequent to their development and thereby strengthened. The formation of mineral bridges and growth of minerals on fracture walls further stabilizes fractures against closure (Dyke 1995; Stowell *et al.* 2001). Mineral growth on walls and concomitant cementation in the host rock are ubiquitous features of all but the most recent of fractures. Thus, fractures developing in active tectonic environments are likely to be open parallel to $S_{H_{max}}$ both because they could currently grow in that direction and because they and their host rocks have had no post-fracture cement precipitation. For other natural fracture sets there is no reason to expect that present-day $S_{H_{max}}$ and open fractures should be parallel unless stress field orientations have remained constant from the time of fracturing to the present.

Conclusions

Natural opening-mode fracture systems in the dolostones described in this paper show many characteristics that are similar to those in sandstones. Aperture-size distributions follow power-law distributions, and the degree of fracture clustering depends on the subcritical crack index. In dolostones the synkinematic cement is typically dolomite, just as in sandstones quartz is the synkinematic cement, meaning that synkinematic cementation is a rock-dominated process. Synkinematic dolomite lines and bridges or seals fractures according to whether they are above or below a certain fracture size, the emergent threshold. Synkinematic cements are also present in pores in the host rock. Crack–seal structures and bridges in large fractures are similar in morphology to those in quartz cements in fractures in sandstones. Models that explain these structures in quartz are likely to be relevant for dolostones. Post-kinematic cements, if present, seal the remaining fracture porosity, but their distribution is heterogeneous.

In the case of dolostones, the complexity of diagenetic processes might seem to be a hindrance to fracture prediction, but the reverse is true. A complex diagenetic signal can help resolve fracture events that are otherwise marked by nearly indistinguishable opening-mode fractures. Moreover, fracture events punctuate and may help us unravel diagenetic history. The evolution of key mechanical rock properties with diagenesis can then be incorporated into geomechanical models of fracture growth, giving site-specific predictions of fracture architecture. Improving the understanding of fractured dolostone reservoirs is most likely to be achieved through a combination of structural and diagenetic work.

Work on fracture characterization was funded through the Fracture Research and Application Consortium, The University of Texas at Austin, whose members include ChevronTexaco, Devon Energy Corporation, Ecopetrol, EOG Resources, Instituto Mexicano del Petroleo, Marathon Oil, Pemex Exploración y Producción, Petroleos de Venezuela, Petrobras, Repsol-YPF-Maxus, Saudi Aramco, Schlumberger, TotalFinaElf and Williams Exploration & Production. Additional funding was provided by The University of Texas System through the University Lands Joint Advanced Recovery Initiative. Cores were obtained from Goldrus Production Company. This paper is published with permission of the Director of the Bureau of Economic Geology, The University of Texas at Austin.

References

ANTONELLINI, M. & MOLLEMA, P.N. 2000. A natural analog for a fractured and faulted reservoir in dolomite: Triassic Sella Group, Northern Italy. *AAPG Bulletin*, **84**, 314–344.

ARVIDSON, R.S. & MACKENZIE, F.T. 1997. Tentative kinetic model for dolomite precipitation rate and its application to dolomite distribution. *Aquatic Geochemistry*, **2**, 273–298.

ATKINSON, B.K. 1984. Subcritical crack growth in geological materials. *Journal of Geophysical Research*, **89**, 4077–4114.

COZZI, A. 2000. Synsedimentary tensional features in Upper Triassic shallow-water platform carbonates of the Carnian Prealps (northern Italy) and their importance as palaeostress indicators. *Basin Research*, **12**, 133–146.

CRAMPIN, S. 1994. The fracture criticality of crustal rocks. *Geophysics Journal International*, **118**, 428–438.

DYKE, C.C. 1995. How sensitive is natural fracture permeability at depth to variation in effective stress? *In*: MYER, L.R., COOK, N.G.W., GOODMAN, R.E. & TSANG, C.-F. (eds) *Fractured and Jointed Rock Masses*. A.A. Balkema, Rotterdam, 81–88.

ERICSSON, J.B., MCKEAN, H.C. & HOOPER, R.J. 1998. Facies and curvature controlled 3D fracture models. *In*: JONES, G., FISHER, Q.J. & KNIPE, R.J. (eds) *Faulting, Fault Scaling and Fluid Flow in Hydrocarbon Reservoirs*. Geological Society, London, Special Publications, **147**, 299–312.

GALE, J.F.W., LAUBACH, S.E., REED, R.M., MOROS OTERO, J.G. & GOMEZ, L. 2002. Fracture analysis of Clear Fork outcrops in Apache Canyon and cores from South Wasson Clear Fork field. *In*: LUCIA, F.J. (ed.) *Integrated outcrop and subsurface studies of the interwell environment of carbonate reservoirs: Clear Fork (Leonardian-Age) Reservoirs, West Texas and New Mexico*. Final report, Contract No. DE-AC26-98BC15105, US DOE (2002), 191–226.

GAUTHIER, B.D.M., GARCIA, M. & DANIEL J.-M. 2002. Integrated fractured reservoir characterization: a case study in a north Africa field. *Society of Petroleum Engineers Reservoir Evaluation and Engineering*, **5**, 284–294.

GILLESPIE, P.A., WALSH, J.J., WATTERSON, J., BONSON, C.G. & MANZOCCHI, T. 2001. Scaling relationships of joint and vein arrays from The Burren, Co. Clare, Ireland. *Journal of Structural Geology*, **23**, 183–201.

GOLDHAMMER, R.G., LEHMAN, P.J., TODD, R.G., WILSON, J.L., WARD, W.C. & JOHNSON, C.R. 1991. *Sequence Stratigraphy and Cyclostratigraphy of the Mesozoic of the Sierra Madre Oriental, Northeast Mexico*. Gulf Coast Section, Society of Economic Paleontologists and Mineralogists.

GOMEZ, L.A., GALE, J.F.W., RUPPEL, S.C. & LAUBACH, S.E. 2001. Fracture characterization using rotary-drilled sidewall cores: an example from the Ellenburger Formation. *In*: *Proceedings, West Texas Geological Society Fall Meeting*. West Texas Geological Society, Publication, **01–110**, 81–89.

GRAY, G.G., POTTORF, R.J., YUREWICZ, D.A., MAHON, K.I., PEVEAR, D.R. & CHUCLA, R.J.

2001. Thermal and chronological record of syn- to post-Laramide burial and exhumation, Sierra Madre Oriental, Mexico. *In*: BARTOLINI, C., BUFFLER, R.T. & CANTU-CHAPA, A. (eds) *The Western Gulf of Mexico Basin: Tectonics, Sedimentary Basins and Petroleum Systems*. American Association of Petroleum Geologists, Memoir, **75**, 159–181.

HANCOCK, P.L. 1985. Brittle microtectonics: principles and practice. *Journal of Structural Geology*, **7**, 437–457.

HEFFER, K.J., FOX, R.J., MCGILL, C.A. & KOUTSABELOULIS, N.C. 1997. Novel techniques show links between reservoir flow directionality, earth stress, fault structure and geomechanical changes in mature waterfloods. *Society of Petroleum Engineers Journal*, **2**, 91–98.

HENNINGS, P., OLSON, J.E. & THOMPSON, L.B. 2000. Combined outcrop data and three-dimensional structure models to characterize fractured reservoirs: An example from Wyoming. *AAPG Bulletin*, **84**, 830–849.

HOLDER, J., OLSON, J.E. & PHILIP, Z. 2001. Experimental determination of subcritical crack growth parameters in sedimentary rock. *Geophysical Research Letters*, **28**, 599–602.

HOOKER, J.N., MARRETT, R.A. & LAUBACH, S.E. 2002. Origin of faults and extension fractures in Cretaceous carbonates, Northeast Mexico. *Gulf Coast Association of Geological Societies, Transactions*, **52**, 421–428.

JOUBERT, T. & RICE, C. 1997. Fractures of a Triassic carbonate reservoir in northeastern British Columbia, Canada. (Abs.) *AAPG Bulletin*, **81**, 1386.

KNIPE, R.J. & RUTTER, E.H. (eds). 1990. *Deformation Mechanisms, Rheology and Tectonics*. Geological Society, London, Special Publications, **54**.

LANDER, R.H. & WALDERHAUG, O. 1999. Porosity prediction through simulation of sandstone compaction and quartz cementation. *AAPG Bulletin*, **83**, 433–449.

LANDER, R.H., GALE, J.F.W., LAUBACH, S.E. & BONNELL, L.M. 2002. Interaction between quartz cementation and fracturing in sandstone. (Abs.) *AAPG Annual Convention Program*, **11**, A98–A99.

LAUBACH, S.E. 1988. Subsurface fractures and their relationship to stress history in East Texas Basin sandstone. *Tectonophysics*, **156**, 37–49.

LAUBACH, S.E. 2003. Practical approaches to identifying sealed and open fractures. *AAPG Bulletin*, **87**, 561–579.

LAUBACH, S E., MARRETT, R.A. & OLSON, J.E. 2000. New directions in fracture characterization. *The Leading Edge*, **19**, 704–711.

LAUBACH, S.E., REED, R.M., OLSON, J.E., LANDER, R.H. & BONNELL, L.M. 2002. Interaction between diagenetic reactions and mechanics of opening-mode fractures in sandstones: the message from crack–seal texture. (Abs.) *Geological Society of America, Abstracts with Programs*, **34**, A10–A11.

MARQUEZ, X.M. & MOUNTJOY, E.W. 1996. Microfractures due to overpressures caused by thermal cracking in well-sealed Upper Devonian Reservoirs, Deep Alberta Basin. *AAPG Bulletin*, **80**, 570–588.

MARRETT, R.A. & LAUBACH, S.E. 2001. Fracturing during burial diagenesis. *In*: MARRETT, R.A. (ed.) *Genesis and Controls of Reservoir-scale Carbonate Deformation, Monterrey Salient, Mexico*. Bureau of Economic Geology, The University of Texas at Austin, Guidebook, **28**, 83–107.

MARRETT, R.A., ORTEGA, O. & KELSEY, C.M. 1999. Extent of power-law scaling for natural fractures in rock. *Geology*, **27**, 799–802.

MILLIKEN, K.L. & LAUBACH, S.E. 2000. Brittle deformation in sandstone diagenesis as revealed by cathodoluminescence imaging with application to characterization of fractured reservoirs. Chapter 9. *In*: PAGEL, M., BARBIN, V., BLANC, P. & OHNENSTETTER, D. (eds) *Cathodoluminescence in Geosciences*. Springer, Berlin, 225–244.

MOLLEMA, P.N. & ANTONELLINI, M. 1999. Development of strike slip faults in the dolomites of the Sella Group, Northern Italy. *Journal of Structural Geology*, **21**, 273–292.

MONROY-SANTIAGO, F., LAUBACH, S.E. & MARRETT, R.A. 2001. Preliminary diagenetic and stable isotope analyses of fractures in the Cupido Formation, Sierra Madre Oriental. *In*: MARRETT, R.A. (ed.) *Genesis and Controls of Reservoir-scale Carbonate Deformation, Monterrey Salient, Mexico*. Bureau of Economic Geology, The University of Texas at Austin, Guidebook, **28**, 83–107.

MONTAÑEZ, I.P. 1997. Secondary porosity and late diagenetic cements of the Upper Knox Group, Central Tennessee region: a temporal and spatial history of fluid flow conduit development within the Knox regional aquifer. *In*: MONTAÑEZ, I.P., GREGG, J.M. & SHELTON, K.L. (eds) *Basin-wide Diagenetic Patterns: Integrated Petrologic, Geochemical, and Hydrologic Considerations*. Society of Economic Paleontologists and Mineralgists, Special Publications, **57**, 101–117.

MOROS OTERO, J.G. 1999. Relationship between fracture aperture and length in sedimentary rocks. MS thesis, The University of Texas at Austin.

MULDOON, M.A., UNDERWOOD, C.A., COOKE, M.L. & SIMO, T. 2001 Stratigraphic controls on fracture patterns within the Silurian dolomite of northeastern Wisconsin. (Abs.) *Geological Society of America, Abstracts with Programs*, **33**, 45.

NELSON, R.A. 2001. *Geologic Analysis of Naturally Fractured Reservoirs*, 2nd edn. Gulf Publishing, Houston, TX.

NOH, M.H. & LAKE, L.W. 2002. Geochemical modeling of fracture filling. *In*: *Proceedings of the 13th Symposium on Improved Oil Recovery*. Society of Petroleum Engineers, Tulsa, OK, SPE 75245.

OLSON, J.E. 1993. Joint pattern development: effects of subcritical crack-growth and mechanical crack interaction. *Journal of Geophysical Research*, **98**, 12 251–12 265.

OLSON, J.E. 2003. Sublinear scaling of fracture

aperture versus length: an exception or the rule? *Journal of Geophysical Research*, **108**(B9), article no. 2413.

OLSON J.E., QIU, J., HOLDER, J. & RIJKEN, P. 2001. Constraining the spatial distribution of fracture networks in naturally fractured reservoirs using fracture mechanics and core measurements. *In*: *Proceedings of the Society of Petroleum Engineers Annual Technical Conference*. Society of Petroleum Engineers, New Orleans, LA, SPE 74312.

ORTEGA, O.J. & MARRETT, R.A. 2001. Stratigraphic controls on fracture intensity in Barremian–Aptian carbonates, northeastern Mexico. *In*: MARRETT, R.A. (ed.) *Genesis and Controls of Reservoir-scale Carbonate Deformation, Monterrey Salient, Mexico*. Bureau of Economic Geology, The University of Texas at Austin, Guidebook, **28**, 57–82.

OUENES, A. 2000. Practical application of fuzzy logic and neural networks to fractured reservoir characterization. *In*: MOHAGHEGH, S. (ed.) *Applications of Virtual Intelligence to Petroleum Engineering*. *Computers & Geosciences*, **26**, 953–962.

PALCHIK, V. & HATZOR, Y.H. 2002. Crack damage stress as a composite function of porosity and elastic matrix stiffness in dolomites and limestones. *Engineering Geology*, **63**, 233–245.

PARKS, W.L. & GALE, M.S. 1999. Exploring overpressured, naturally fractured reservoirs in the Powder River Basin, Wyoming: A multi-disciplinary effort. *In*: *AAPG Regional Meeting, Powder River Basin Symposium*, 2–11.

PHILIP, Z.G., JENNINGS, J.W. JR, OLSON, J.E. & HOLDER, J. 2002. Modeling coupled fracture–matrix fluid flow in geomechanically simulated fracture networks. *In*: *Proceedings of the Society of Petroleum Engineers Annual Technical Conference and Exhibition*. Society of Petroleum Engineers, San Antonio, TX, SPE 77340.

PRICE, N.J. 1966. *Fault and Joint Development in Brittle and Semi-brittle Rock*. Pergamon Press, London.

PURSER, B.H., TUCKER, M.E. & ZENGER, D. (eds). 1994. *Dolomites: A Volume in Honour of Dolomieu*. International Association of Sedimentologists, Special Publications, **21**.

QUEEN, J.H. & RIZER, W.D. 1990. An integrated study of seismic anisotropy and the natural fracture system at the Conoco Borehole Test facility, Kay County, Oklahoma. *Journal of Geophysical Research*, **95**, 11 255–11 273.

RAWNSLEY, K. & WEI, L. 2001. Evaluation of a new method to build geological models of fractured reservoirs calibrated to production data. *Petroleum Geoscience*, **7**, 23–33.

READ, J.F. 1995. Overview of carbonate platform sequences, cycle stratigraphy and reservoirs in greenhouse and icehouse worlds. *In*: READ, J.F., KERANS, C., WEBER, L.J., SARG, J.F. & WRIGHT, F.M. (eds) *Milankovitch Sea Level Changes, Cycles and Reservoirs on Carbonate Platforms in Greenhouse and Ice-house Worlds*. Society of Economic Paleontologists and Mineralogists, Short Course Notes, **35**, 102.

REED, R.M. & MILLIKEN, K.L. 2003. How to overcome imaging problems associated with carbonate minerals on SEM-based cathodoluminescence systems. *Journal of Sedimentary Research*, **73**, 326–330.

RIJKEN, P., HOLDER, J. & OLSON, J.E. 2002. Natural fracture characterization of the Travis Peak Formation, East Texas: application of new rock testing and modeling methods. *Gulf Coast Association of Geological Societies, Transactions*, **52**, 837–847.

STOWELL, J.F.W., LAUBACH, S.E. & OLSON, J.E. 2001. Effect of modern state of stress on flow-controlling fractures: a misleading paradigm in need of revision. *In*: ELSWORTH, D., TINUCCI, J.P. & HEASLEY, K.A. (eds) *Rock Mechanics in the National Interest*, Volume 1. *Proceedings of the 38th Annual US Rock Mechanics Symposium*. A.A. Balkema, Rotterdam, 691–698.

THOMAS, A.L. & LA POINTE, P.R. 1995. Conductive fracture identification using neural networks. *In*: DAEMEN, J.J.K. & SCHULTZ, R.A. (eds) *Rock Mechanics; Proceedings of the 35th US Symposium on Rock Mechanics*, Volume 35. A.A. Balkema, Rotterdam, 627–632.

TRICE, R. 1999. Application of borehole image logs in constructing 3D static models of productive fracture networks in the Apulian Platform, Southern Apennines. *In*: LOVELL, M., WILLIAMSON, G. & HARVEY, P. (eds) *Borehole Imaging; Applications and Case Histories*. Geological Society, London, Special Publications, **159**, 155–176.

TROUDI, H., M'RABET, A., ACHECHE, M.H., GHARIANI, H. & GHARIENI, N. 2000. Diagenetic (lithification and dolomitization) and tectonic (fracturing) controls on the Senonian Abiod carbonate reservoir in Tunisia, North Africa. *AAPG 2000 Annual Meeting Expanded Abstracts*, **9**, 149.

TUCKER, M.E. & WRIGHT, V.P. 1990. *Carbonate Sedimentology*. Blackwell, Oxford, 365–400.

WILLIAMS, D.M. & MCNAMARA, K. 1992. Limestone to dolomite to dedolomite conversion and its effect on rock strength; a case study. *The Quarterly Journal of Engineering Geology*, **25**, 131–135.

WOOD, D.S. & HOLM, P.E. 1980. Quantitative analysis of strain heterogeneity as a function of temperature and strain rate. *In*: SCHWERDTNER, W.M., HUDLESTON, P.J. & DIXON, J.M. (eds) *Analytical Studies in Structural Geology*. *Tectonophysics*, **66**, 1–14.

Reservoir properties of Arab carbonates, Al Rayyan Field, offshore Qatar

DAVID CLARK[1], JOHN HEAVISIDE[2] & KASSIM HABIB[3]
[1]*Clark Research Ltd, High Wycombe, UK (e-mail: dnclark@clarkresearch.ltd.uk)*
[2]*BP Exploration Operating Co. Ltd, Sunbury-on-Thames, UK*
[3]*Qatar Petroleum, Doha, Qatar*

Abstract: Oil is present in the Upper Jurassic Arab A and Arab C carbonates of the Al Rayyan Field. The Arab A consists of porous limestone with thin units of tight dolomite at the top and bottom. Two reservoir intervals can be distinguished, separated by a slightly less permeable unit. The upper reservoir consists of stromatolites and cross-bedded peloidal-grainstones, whereas the lower reservoir consists predominantly of oolitic-grainstones. The less permeable unit comprises moderately cemented, cross-bedded peloidal-grainstones. Higher porosity and permeability generally occurs in the coarser grained, cross-bedded grainstones, where pores are mostly primary and intergranular; lower permeability is characteristic of grainstones with moderately developed early cement. Porosity is enhanced at some levels by the presence of leached intragranular pores, but these have little effect on permeability. The Arab C is strongly stratified and completely dolomitized. Three reservoir intervals can be recognized, separated by less permeable layers of poorer quality rock. The upper reservoir consists of partially leached and moderately cemented, cross-bedded bioclastic-peloidal-grainstones. Porosity is high because of the presence of leached intragranular pores in addition to primary intergranular pores but permeability is relatively low because of the presence of early cement. The main reservoir comprises loosely compacted and lightly cemented, cross-bedded peloidal-grainstones. Porosity is lower than that in the upper reservoir because leached intragranular pores are generally absent, but permeability is higher because cementation is relatively light. The lower reservoir is made up of peloidal-grainstones and coarse-grained bioclastic-grainstones. The upper and main reservoirs are separated by a thick interval of relatively tight, thinly bedded, strongly cemented peloidal-grainstones, whereas the main and lower reservoirs are separated only by a thin bed of tight, nodular anhydrite. The pore types and cements observed in the Arab C are similar to those of the Arab A, but dolomitization has resulted in an enhancement of permeability and capillary pressure properties compared to those of the Arab A. Thus, the overall water saturation of the Arab C is lower than that of the Arab A, and the displacement pressures and irreducible water saturations are also significantly lower than those of Arab A carbonates.

The Al Rayyan Field is located above the Qatar Arch, about 45 miles off the northern coast of Qatar (Fig. 1). The field was discovered in 1976 when the exploration well QMB-1 was drilled. The initial test rates from the field were uneconomic and development did not take place until 1996 when it was demonstrated that high productivities could be obtained from horizontal wells. (A typical vertical well produces 1400–1500 barrels per day from each reservoir, whereas a horizontal well can produce up to 10 000 barrels per day, depending on the length of the horizontal section, the zone in which it is placed and the capacity of the downhole pump (ESP) that is installed in the well.) Two thin, oil-bearing reservoirs are present, namely the Arab A and Arab C. Both contain relatively heavy, low gas–oil ratio (GOR) oil with a gravity of 23.9° API in the Arab A and 25.5° API in the Arab C (where API is the Americal Petroleum standard for expressing the specific gravity of oil). The oil is trapped in a low relief anticline with simple four-way dip closure (Fig. 2). This paper presents a study of the sedimentology, stratigraphic zonation and reservoir properties of the carbonates. The results have been input to a reservoir model to simulate production behaviour. An understanding of the properties of the different zones of the reservoirs also has been used to help design and drill horizontal wells in order to optimize oil recovery and maximize productivity.

The Arab A and Arab C form part of the Upper Jurassic Qatar Formation that comprises an alternating succession of carbonate and anhydrite units, starting with the Arab D

From: BRAITHWAITE, C. J. R., RIZZI, G. & DARKE, G. (eds) 2004. *The Geometry and Petrogenesis of Dolomite Hydrocarbon Reservoirs*. Geological Society, London, Special Publications, **235**, 193–232. 0305-8719/$15.00

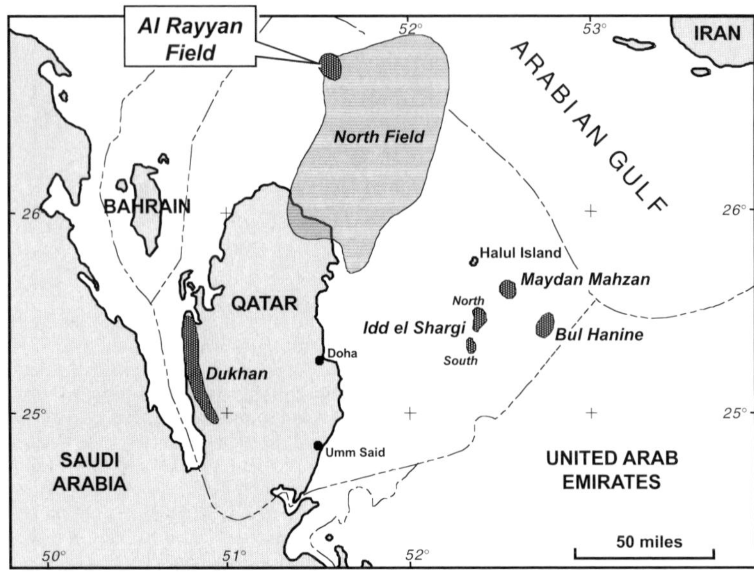

Fig. 1. Location of the Al Rayyan Field.

Carbonate at the base, and followed successively by the Lower Anhydrite, Arab C Carbonate, Middle Anhydrite, Arab B Carbonate, Upper Anhydrite and Arab A Carbonate at the top (Fig. 3; after Alsharhan & Nairn 1997). The Qatar Formation is succeeded by the Hith Anhydrite, which forms a well-developed regional seal above the Arab carbonate reservoirs. Thinner, but locally important, intraformational seals are also provided by the Upper, Middle and Lower Anhydrites. The carbonate units were deposited mainly in subtidal lagoonal, intertidal and supratidal environments. Deposition was interrupted from time to time by episodes of subaerial exposure and karst formation. The sulphates accumulated mainly as subaqueous gypsum in hypersaline lagoons. The overall environment of deposition of the carbonates appears to have become more restricted as time progressed, with relatively normal marine conditions with corals prevailing in the Arab D and cyanobacterial mats, representing more restricted (higher salinity) subtidal and intertidal conditions, becoming increasingly common in the Arab C, Arab B and Arab A. Episodes of hypersaline deposition also became more common, with successive sulphate units becoming progressively thicker and culminating in the deposition of about 265 ft of gypsum in the Hith.

Available data

Seven vertical exploration and appraisal wells have been drilled to date on the Al Rayyan Field, namely QMB-1, QMB-2, QMB-3, ALR-4, ALR-11, ALR-12 and ALR-13. Wireline logs are available from all of these wells, including gamma ray, resistivity and porosity logs. A correlation of these logs can be found in Figure 4. An extensive coring programme was undertaken in most of the wells, with Arab A cores available from five wells and Arab C cores from six wells. No cores are available from the Arab A at QMB-1 and ALR-13, and only the lower part of the reservoir is represented at QMB-2. No cores are available from the Arab C at ALR-13, and the lowermost part of the reservoir is not represented at QMB-1 and QMB-3. Measurements of porosity and permeability were made using core plugs taken at 1 ft-intervals throughout the carbonate units, and detailed sedimentological descriptions have been made of all of the cores. Depth plots of the core plug measurements can be found in Figure 5a & b.

Arab A

The Arab A has an average thickness of 56 ft, ranging from 53 ft at ALR-12 to 60 ft at ALR-13. It consists of porous, weakly bedded limestones with thin units of dense dolomite at the top and bottom. The average core porosity of the Arab A for the whole field is 22.7%, and

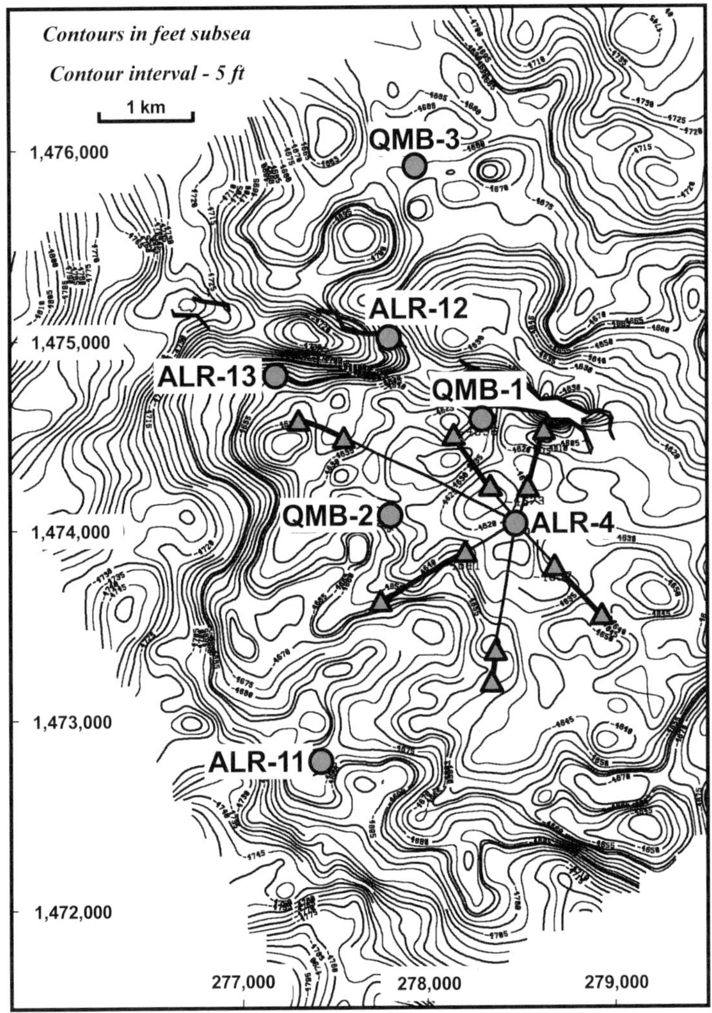

Fig. 2. Depth structure map for the top Arab C (contours in ft ss).

the geometric mean permeability is 115.6 mD. Primary intergranular pores contribute most of the porosity but secondary intragranular pores are also important at some levels. The crest of the Arab A structure is at 4505 ft ss (ft subsea, i.e. the depth below mean sea level), with a vertical dip closure of at least 60 ft. The free water level (FWL) is at 4566 ft ss, as determined from the Modular Formation Dynamics Tester (MDT) pressure measurements. The FWL is the level of 100% water saturation (S_w) in the reservoir.

Parasequences

Nine lithofacies can be identified in the Arab A: stromatolite; millimetre-bedded peloidal-grainstone; centimetre-bedded peloidal-grainstone; flaser-bedded peloidal-grainstone; cross-bedded peloidal-grainstone; peloidal-oolitic-grainstone; peloidal-pisolitic-grainstone; intraclastic-oolitic-rudstone; and dolomite. Most lithofacies described in this study are classified according to depositional texture after Dunham (1962), with predominant grain types being used as prefixes, following Folk (1959). The only exceptions to this are the terms stromatolite, after Riding (1999), which is preferred to Dunham's boundstone term, and rudstone which is used for mud-free grain-supported sediments of gravel grade, following Embry & Klovan (1971). These lithofacies can be grouped into three lithological units, each of which represents a separate parasequence or depositional cycle (Figs 6 and 7).

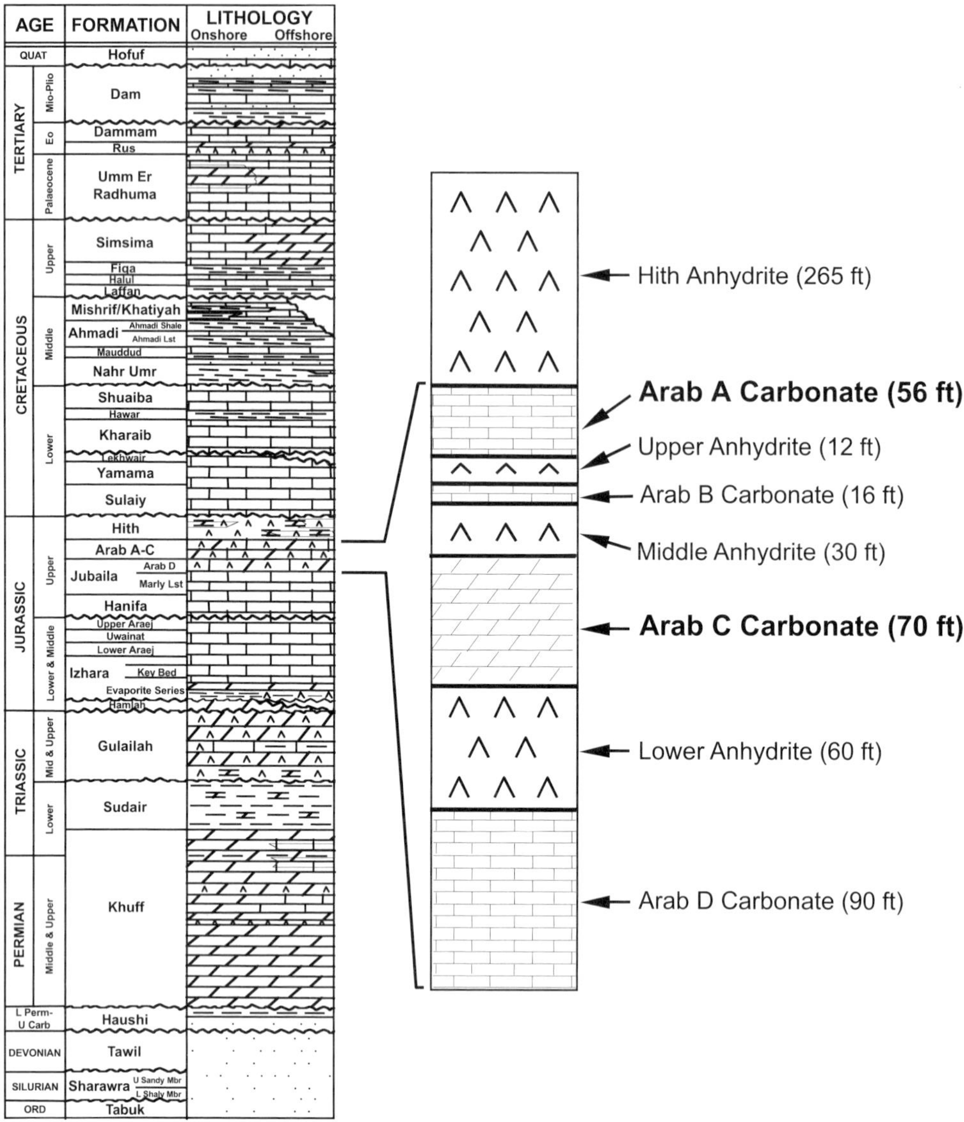

Fig. 3. Stratigraphy of the Upper Jurassic of Qatar.

The *Lower Arab A* (*A13–A10*) consists of cross-bedded oolitic-peloidal-grainstones, pisolitic-peloidal-grainstones, flaser-bedded peloidal-grainstones, stromatolites and intra-clastic-oolitic-rudstones (Figs 6 and 7). Collectively, these rocks probably accumulated in coastal barrier–lagoon complexes, similar to those found today in the Persian Gulf (Evans *et al.* 1973; Purser & Evans 1973). The cross-bedded oolitic-peloidal-grainstones were deposited on high-energy, subtidal sand bars and intertidal beaches, whereas the flaser-bedded peloidal-grainstones, stromatolites and intraclastic-oolitic-rudstones are thought to represent upper intertidal and supratidal flats or beaches. The pisolitic-peloidal-grainstones are interpreted to be the result of early diagenetic modification of lagoonal sediments that came into contact with hypersaline groundwater under vadose conditions (Loreau & Purser 1973; Shinn 1973). As pisoliths only occur at the top of the unit, it is possible that they formed in a sabkha environment that became established at the end of the first cycle of deposition. A

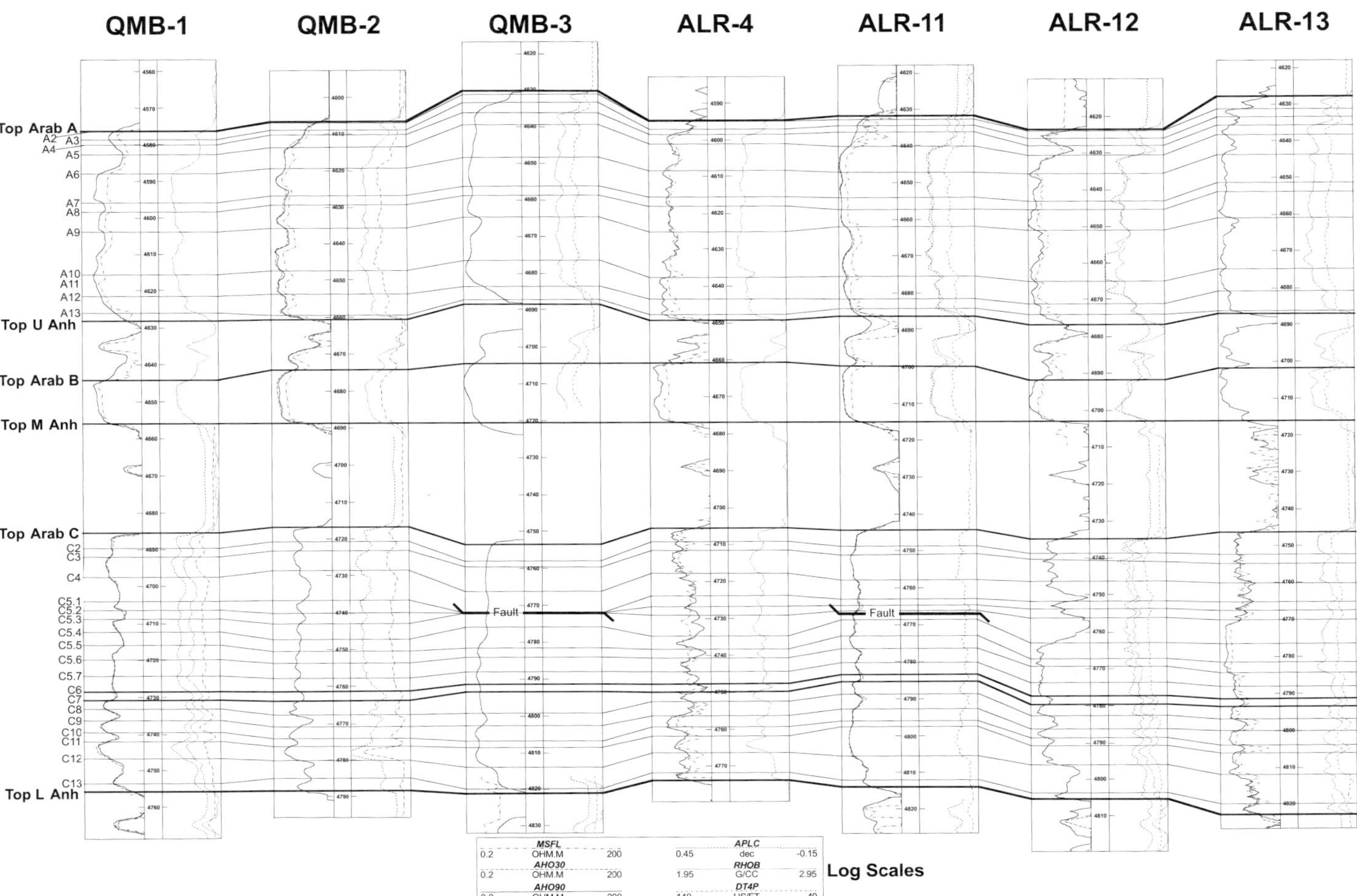

Fig. 4. Correlation of wireline logs of the Arab A and Arab C reservoirs.

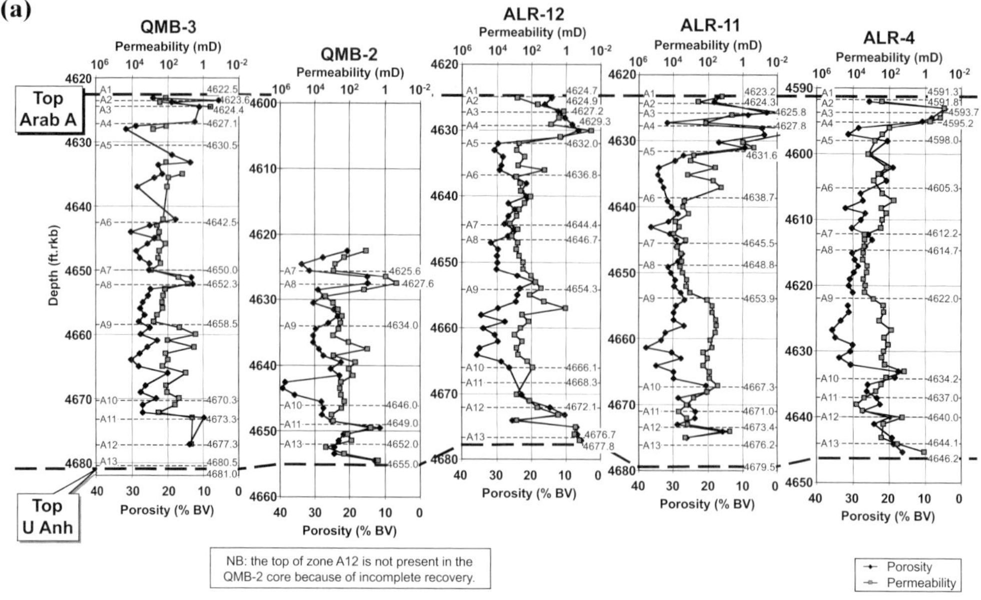

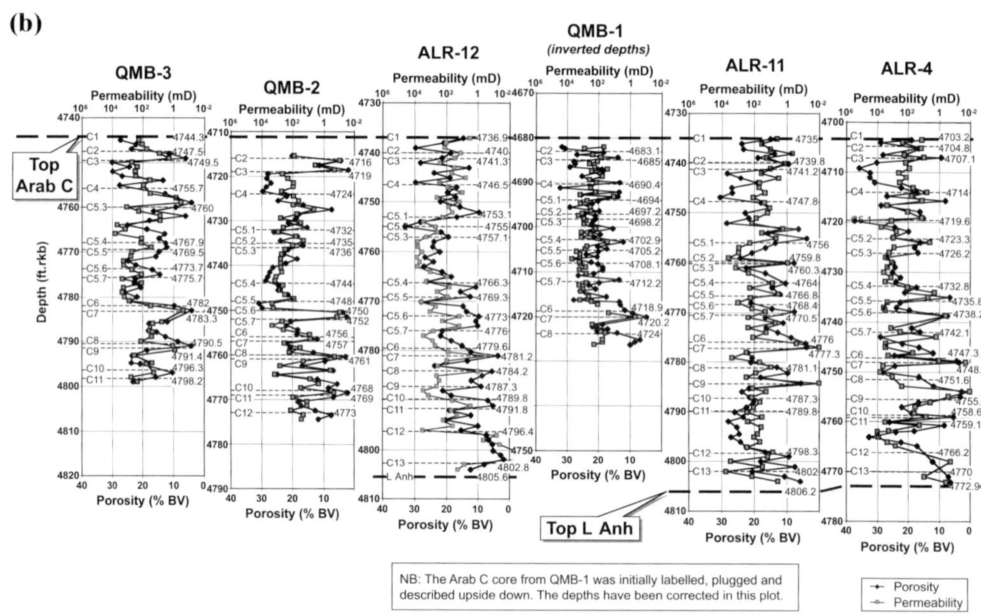

Fig. 5. (**a**) Depth plots of core analysis data for the Arab A carbonate. (**b**) Depth plots of core analysis data for the Arab C carbonate.

number of irregular bedding surfaces have been observed in the lower unit. These are interpreted as palaeokarst (Walkden 1979; Esteban & Klappa 1983), although some might have originated as hardgrounds (Bathurst 1971, p. 395). The basal beds of the unit are completely dolomitized at some localities, particularly at ALR-11.

The *Middle Arab A* (*A9–A7*) comprises cross-bedded or massive peloidal-bioclastic-grainstones, together with stromatolites at the base and pisolitic sediments at the top (Figs 6 and 7).

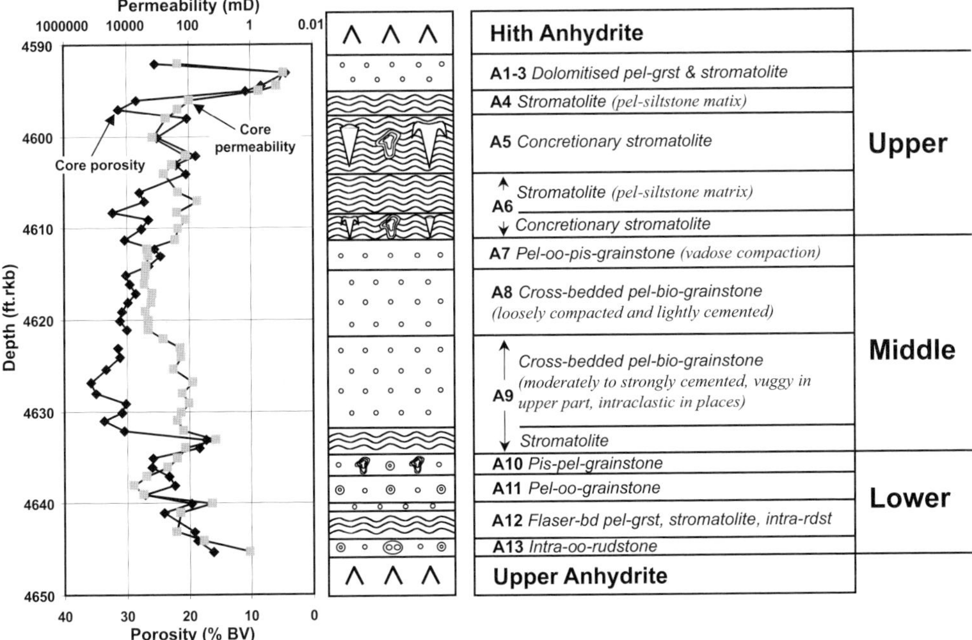

Fig. 6. Sedimentology and reservoir characteristics of the Arab A, ALR-4.

At some locations, such as ALR-4, ALR-11 and QMB-3, the Middle Arab A starts with a transgressive bed of upper intertidal stromatolite. In these areas, the stromatolite rests directly on an irregular exposure surface developed at the top the A10, and passes upwards into cross-bedded grainstones. At other locations, such as ALR-12 and QMB-2, the stromatolite is absent and the exposure surface is directly overlaid by cross-bedded grainstones. These form an amalgamated sequence of subtidal lagoonal and intertidal sediments more than 25 ft thick. The pisolitic sediments that occur at the top of the middle unit are interpreted as sabkha deposits, formed at the end of the second cycle of deposition, similar to those described from the top of the lower unit.

The *Upper Arab A* (*A1–A6*) is made up of a complex succession of stromatolites, concretionary stromatolites and cross-bedded, centimetre-bedded and millimetre-bedded peloidal-grainstones, capped by a variably developed unit of dolomite (Figs 6 and 7). The stromatolites were probably deposited in shallow lagoonal and intertidal environments (Logan *et al.* 1974), and subsequently modified in sabkhas in many cases (Shearman 1966; Butler 1969; Kendall & Skipwith 1969). Cross-bedded, centimetre-bedded and millimetre-bedded peloidal-grainstones are only present in the upper Arab A at ALR-11 and ALR-12. These grainstones probably reflect deposition in shallow lagoonal, lower intertidal and upper intertidal environments (Evans *et al.* 1973; Purser & Evans 1973). The dolomites appear to have formed by the replacement of peloidal-bioclastic-grainstones and stromatolites that occur immediately beneath the Hith Anhydrite. The depth to which dolomitization took place is variable, ranging from 4 ft below the Hith at ALR-4 to 7 ft below at ALR-12, reaching down to the upper part of the A5 at ALR-12.

Whereas the lithofacies in the Lower and Middle Arab A appear to be fairly uniformly developed across the field, the Upper Arab A is characterized by abrupt lateral changes between wells. Stromatolites and concretionary stromatolites are exclusively developed at ALR-4 and QMB-3 but are absent at ALR-11, where only cross-bedded, centimetre-bedded and millimetre-bedded peloidal-grainstones are present. The succession is different again at ALR-12 where all four lithofacies occur together in the Upper Arab A. It has not been possible to construct detailed maps of the lithofacies variations in the Upper Arab A because no cores are available from QMB-1, QMB-2 and ALR-13.

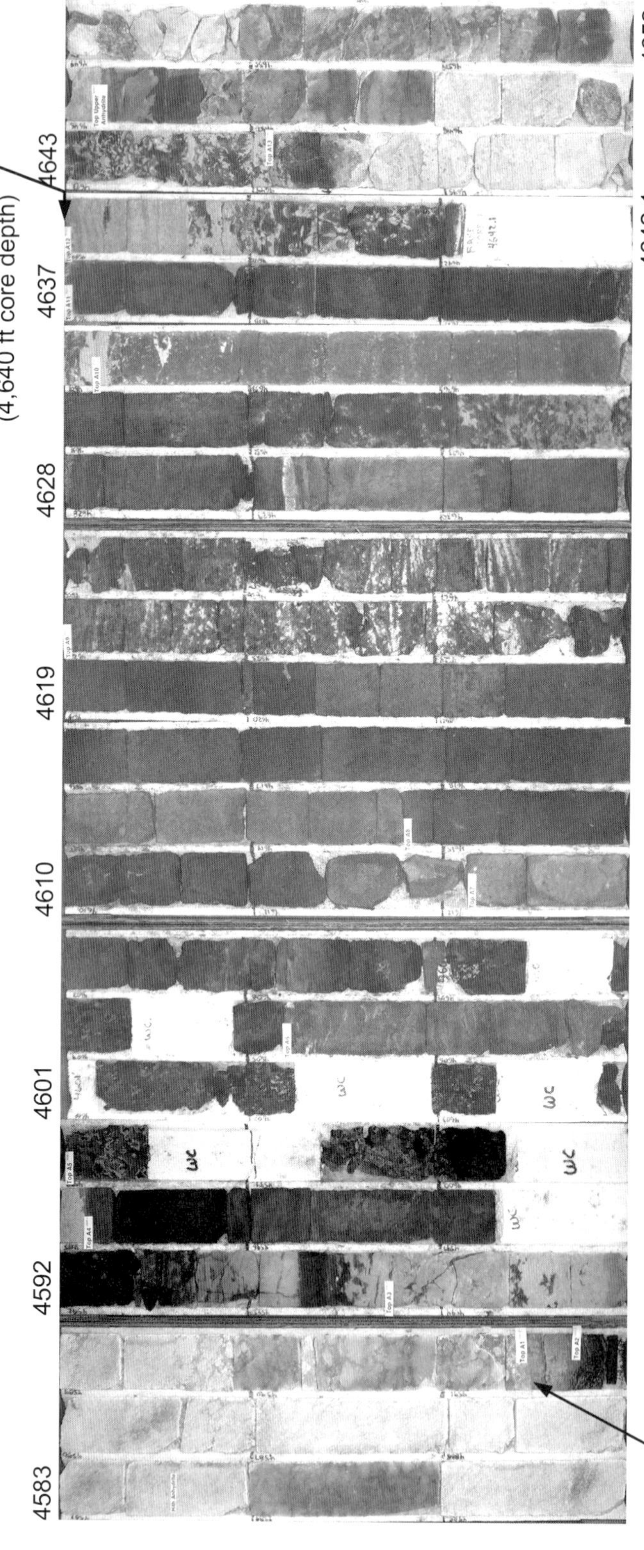

Fig. 7. Core slabs from the Arab A, ALR-4 (4583–4652 ft rkb). Tray length is 3 ft.

Stromatolite. This lithofacies comprises decimetre- or metre-bedded stromatolites with a peloidal-siltstone or very-fine-sand-grade peloidal-grainstone matrix (Fig. 8b). Three morphologies can be distinguished, wavy-bedded sheets, domes (laterally linked hemispheroids) and columns (vertically stacked hemispheroids) (Logan *et al.* 1964). Most of the stromatolites are wavy-bedded but large domes can be discerned in places (Fig. 8a). The domes generally exhibit faint, irregular lamination, whereas the spaces between are filled with mottled grainstone (Fig. 8b). They are probably quite widely developed but are difficult to recognize because they are much larger than the width of the core slabs. The columnar stromatolites are characterized by narrow, vertical columns with convex-up lamination. An example of this can be seen at a depth of 4641 ft rkb (relative to Kelly Bushing) in the ALR-4 core (Fig. 8c). Here, the columns are surrounded by wavy-bedded and cross-laminated peloidal-grainstone, and succeeded by flaser-bedded grainstone. A few thin layers of wavy-bedded 'cryptalgal' (Aitken 1967) stromatolite, with lenticular vugs or desiccation birds-eyes (Shinn 1968) and anhydrite nodules, occur in the uppermost part of the Arab A.

The stromatolites are interpreted as microbial mats composed of peloidal grains that were originally bound together by filamentous cyanobacteria (previously known as blue-green algae) (Ginsburg 1991; Riding 1999, 2003). The wavy-bedded stromatolites probably formed as soft sheet-like microbial mats, whereas the domal and columnar varieties are more likely to have grown as lithified mats (Ginsburg 1991; Riding 1999). The mottled grainstones can be interpreted either as thrombolites (Aitken 1967; Riding 1999), produced as cyanobacterial precipitates between domes, or simply as bioturbated sediment deposited around the domes. The absence of desiccation features such as mud cracks and birds-eyes in the domal and columnar varieties suggests that they formed in lower intertidal or possibly subtidal lagoonal environments, similar to modern stromatolites in Shark Bay (Logan *et al.* 1964). The vuggy 'cryptalgal' stromatolites, in contrast, were probably deposited in more exposed, upper intertidal–supratidal environments (Kendall & Skipwith 1968; Shinn *et al.* 1969), with the vugs originating as fenestrae or desiccation birds-eyes (Shinn 1968.).

Many of the stromatolites contain hard, conical, elongate, irregular, branching or pisolitic structures (Fig. 8d–g). These have a clotted peloidal texture in thin section with weak, wavy internal lamination evident in core slabs. They are interpreted as concretions and vadose pisoliths that formed when the stromatolites were subjected to early diagenesis in hypersaline sabkhas (Dunham 1969*a*; Shinn 1973; Scholle & Kinsman 1974; Clark 1980*a*). Stromatolites with concretions normally occur in close association with unaffected stromatolites, either forming alternating successions or passing laterally from one lithofacies into the other. This is particularly noticeable when drilling horizontal boreholes in the Upper Arab A, where lateral transitions from soft stromatolites to hard concretion-bearing stromatolites have been observed within the same bed (Fig. 9). These lateral changes are also evident on 'Logging While Drilling' (LWD) Image Logs, where the resistivity character of the rocks may vary markedly within a few tens of feet (Fig. 9).

Most of the stromatolites are loosely compacted and lightly cemented, with a predominantly intergranular porosity (Fig. 10a). Cementation probably took place during or soon after deposition, either as a result of bacterial activity (Ginsburg 1991; Riding 1999) or by inorganic precipitation from seawater in subtidal and intertidal environments (Harris *et al.* 1985). This prevented compaction and preserved most of the primary porosity of the sediments. Early freshwater leaching (Friedman 1964) is not widely developed but some intragranular pores are present in places, enhancing the overall porosity. More strongly cemented and dolomitized stromatolites occur at the top and bottom of the Arab A, adjacent to the Hith and Upper Anhydrites. Porosity is significantly lower in stromatolites where concretions predominate. This is because very little porosity is preserved within the concretions, although leached vugs are present in places. Permeability is not significantly affected because a loosely compacted peloidal matrix normally surrounds the concretions.

Millimetre-bedded peloidal-grainstone. This lithofacies has only been observed so far in the Upper Arab A at ALR-11 and ALR-12. It comprises millimetre-bedded peloidal-grainstones and siltstones with thin fragments of bivalves in places. (Millimetre-bedding has layers up to 1 cm thick, i.e. laminae, centimetre-bedding has layers 1–10 cm thick, decimetre-bedding has layers 10–100 cm thick and metre-bedding has layers greater thean 100 cm thick.) Most of the grainstones display flat lamination (Fig. 8h), but some ripple cross-lamination is evident in places. These rocks were probably deposited in an upper intertidal mudflat environment (Shinn *et al.* 1969). The thickest development of laminated grainstone occurs at ALR-11, where it rests directly on a wavy-laminated crust at the top of the A6 (4643.25 ft rkb). This crust is interpreted as a soilstone crust or calcrete associated with a possible exposure surface (Multer & Hoffmeister 1968; Shinn 1983), and the flat-laminated grainstones probably represent a transgressive unit at the base of the A5. The flat-laminated grainstones also occur in close association with centimetre-bedded peloidal-grainstones at ARL-11, and pass laterally into stromatolites towards ALR-4 and QMB-3. Together, these rocks probably represent an inner lagoonal complex, with environments varying from shallow subtidal to lower intertidal and upper intertidal (Evans *et al.* 1973; Purser & Evans 1973). Porosity is largely intergranular. Core permeability is relatively low (<50 mD), suggesting that significant early cementation (Friedman 1964; Harris *et al.* 1985) occurred relative to other peloidal-grainstones in the Arab A.

Centimetre-bedded peloidal-grainstone. As with the previous example, this lithofacies has only been observed in the Upper Arab A at ALR-11 and ALR-12. It comprises faintly centimetre-bedded peloidal-grainstones. Elongate intraclasts occur at some levels that probably originated as carbonate mud lumps. Some thin, millimetre-bedded intervals also occur. The centimetre-bedded grainstones are interpreted as lower–upper intertidal deposits (Shinn *et al.* 1969; Evans 1970; Purser & Evans 1973), with

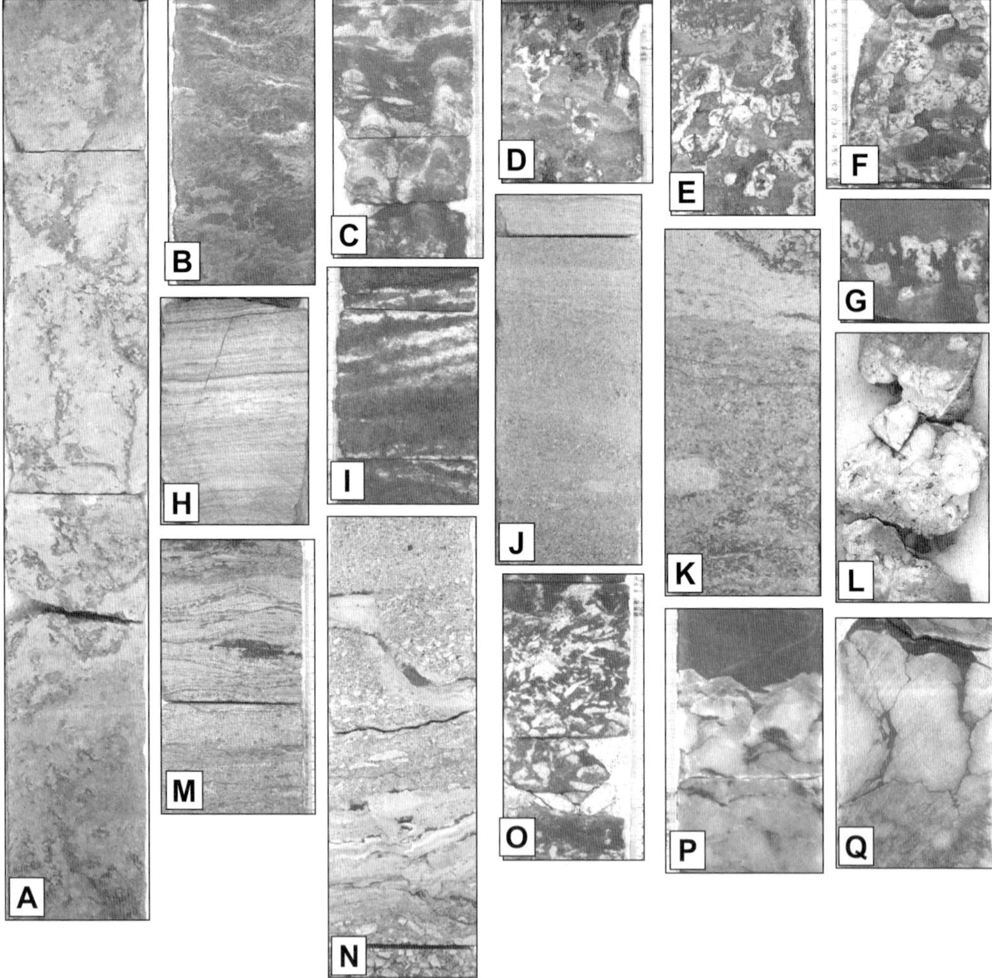

mud lumps derived from adjacent upper intertidal flats (Shinn 1983). Similar grainstones extend to the top of the A5 at ALR-11 (4631.6 ft rkb), where a layer of elongate grainstone intraclasts and a thin wavy-laminated crust caps them. This crust is interpreted as an exposure surface with a soilstone crust (Multer & Hoffmeister 1968; Shinn 1983) at the top of the A5, whilst the intraclasts are possible beachrock lumps (Evamy 1973; Purser & Loreau 1973; Clark 1980*a*). Porosity is largely primary and intergranular. Core permeability is significantly higher than in the millimetre-bedded grainstones (>600 mD), suggesting that early cementation was less important in this lithofacies. Early leaching of grains is well developed at some levels, resulting in honeycomb textures. Cementation and partial replacement by anhydrite took place during deeper burial to form patches and bands of secondary anhydrite (Clark 1980*b*; Clark & Shearman 1980). These are very extensive in the upper part of the A5 at ALR-11.

Flaser-bedded peloidal-grainstone. This lithofacies consists of peloidal-bioclastic-grainstones with well-developed ripples and cross-lamination (Fig. 8m) that are interpreted as lower intertidal flat deposits (Reineck & Singh 1975, pp. 358–359). Such grainstones are only

Fig. 8. Examples of typical lithofacies comprising the Arab A (core slabs, normal light, scale in centimetres). (**A**) *Stromatolite* with a loosely compacted, mottled or wavy-bedded peloidal-siltstone–grainstone matrix. This slab shows a cross-section of large, irregularly stacked, convex-upwards domes and columns. Zone A9, 4665 ft rkb, ALR-11. (**B**) *Stromatolite* with a loosely compacted, oil-stained, mottled or wavy-bedded peloidal-siltstone matrix. This lithofacies is interpreted as a soft cyanobacterial mat that bound together the component peloidal grains. Zone A6, 4605.5 ft rkb, ALR-4. (**C**) *Columnar stromatolite*, with partially oil-stained convex-up laminations and a loosely compacted, oil-stained, wavy-bedded peloidal-siltstone–grainstone matrix. This lithofacies is interpreted as a subtidal or lower-intertidal cyanobacterial mat. It is capped by flaser-bedded peloidal-grainstones of probable intertidal origin. Zone A12, 4641.0 ft rkb, ALR-4. (**D**) *Concretionary stromatolite* with a loosely compacted, oil-stained, peloidal-siltstone matrix. Typically, this lithofacies contains numerous hard, conical features that are interpreted as concretions that formed from hypersaline groundwaters in a sabkha environment. These have little or no porosity but they are embedded in a porous, fine-grained peloidal matrix that is identical to that of the stromatolites shown in sample D. Zone A5, 4604.0 ft rkb, ALR-4. (**E**) *Concretionary stromatolite* similar to sample A but with upward-branching concretionss rather than conical ones. Zone A5, 4598.0 ft rkb, ALR-4. (**F**) *Concretionary stromatolite* similar to sample A but displaying irregular, branching concretions. The concretions predominate over the matrix in this example, resulting in a significant lowering of porosity compared to the stromatolite lithofacies. Permeability, in contrast, does not appear to have been reduced, presumably because of the presence of the loosely compacted peloidal matrix. Zone A5, 4599.5 ft rkb, ALR-4. (**G**) Detail of a *concretionary stromatolite* displaying irregular, branching concretions with downward growing edges. Zone A5, 4638.7 ft rkb, ALR-12. (**H**) *Laminated peloidal-grainstone*. Zone A5, 4637.8 ft rkb, ALR-11. (**I**) *Cross-bedded peloidal-bioclastic-grainstone*, loosely compacted, moderately cemented, Zone A9, 4627.0 ft rkb, ALR-4. (**J**) *Cross-bedded peloidal-bioclastic-grainstone*, loosely compacted, moderately cemented with low-angle, unidirectional, planar cross-bedded. Zone A6, 4643.2 ft rkb, ALR-11. (**K**) *Pisolitic-peloidal-grainstone* comprising peloids, rare bioclasts and irregularly laminated concretions, ranging from medium sand to granule grade. The concretions are interpreted as vadose pisoliths that formed by precipitation from hypersaline groundwater in a sabkha environment. Zone A10, 4667.5 ft rkb, ARL-11. (**L**) Detail of *pisolitic-peloidal-grainstone* showing irregularly laminated or downward-elongated, granule- and pebble-grade vadose pisoliths. Basal Zone A6, 4694.5 ft rkb, QMB-3. (**M**) *Flaser-bedded peloidal-bioclastic-grainstone*, moderately cemented with well-developed ripple cross-lamination. This lithofacies is interpreted as an intertidal flat deposit. A pore lining, early marine cement is also well developed resulting in a low permeability (*c.* 18 mD) and a very weak oil stain. Zone A12, 4640.0 ft rkb, ALR-4. (**N**) *Intraclastic-rudstone* and wavy-laminated *stromatolite* with a well-developed irregular exposure surface or palaeokarst. Zone A12, 4674.3 ft rkb, ALR-12. (**O**) *Intraclastic-oolitic-rudstone*, loosely compacted, moderately cemented, containing coated beachrock lumps (Fig. 10F). Note that the matrix is only lightly cemented and oil stained, whereas the intraclasts are well-cemented and unstained. This lithofacies is interpreted as an intertidal beach deposit based on the presence of coated intraclasts together with early marine cements. Zone A13, 4643.0 ft rkb, ALR-4. (**P**) *Anhydrite pseudomorphs* of large, upward-growing gypsum crystals displaying well-preserved growth zones towards the top. These crystals are interpreted as free-standing gypsum 'trees' that grew subaqueously on the floor of a hypersaline lagoon. A thin bed of dolomitized, ripple cross-laminated peloidal-grainstone overlies the pseudomorphs. Upper Anhydrite (just below the base Arab A Carbonate), 4646.7 ft rkb, ALR-4. (**Q**) *Anhydrite pseudomorphs* of large, upward-growing, twinned gypsum crystals in a peloidal-grainstone matrix. These crystals are also interpreted subaqueous 'trees' growing on the floor of a hypersaline lagoon. The three-dimensional structure of the twins is evident on the broken surface at the top of the core slab. Upper Anhydrite (just below the base Arab A) 4679.6 ft rkb, ALR-11.

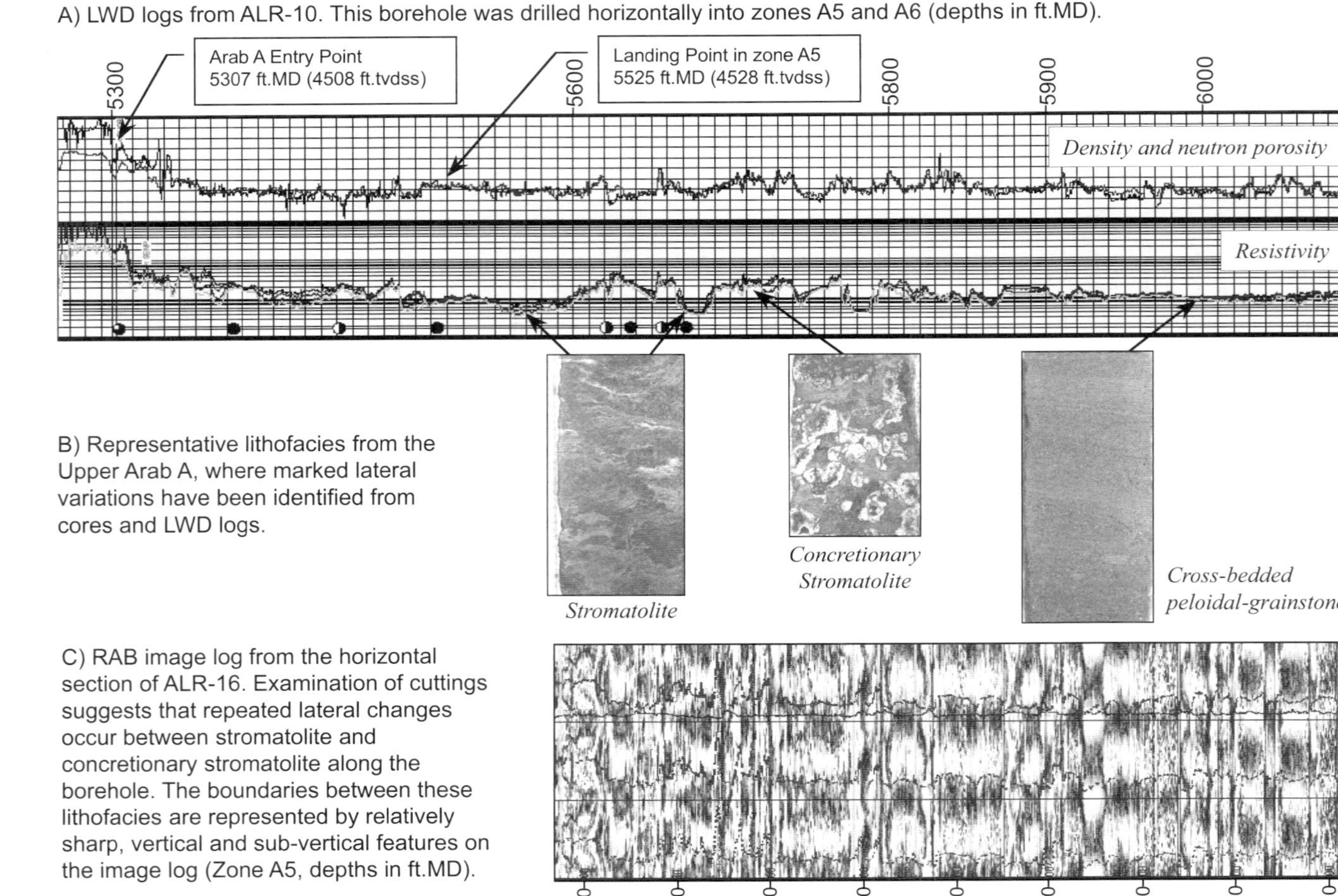

Fig. 9. LWD logs from the Upper Arab A of ALR-10 and ALR-16. All depths are measured depths (MD).

developed at the top of the A12 at QMB-3, ALR-4 and ALR-11, where they succeed wavy-bedded and columnar stromatolites, and at the base of Zone A8 at ALR-11, where they are followed by cross-bedded peloidal-bioclastic-grainstones. Patches of secondary anhydrite are present in places. These formed by partial cementation and replacement of the grainstone host during deeper burial (Fig. 10i) (Clark & Shearman 1980). Porosity is mostly intergranular (Fig. 10i). Typically, the pores are partially occluded by early cement (Fig. 10i), reducing permeability to less than 15 mD at some levels. Honeycomb textures with well-developed intragranular pores surrounded by cement rims are present in some laminae. Such textures probably resulted from the leaching of grains after early cementation (Friedman 1964), but in some cases the pore shapes are more akin to desiccation birds-eye structures (Shinn 1968).

Cross-bedded peloidal-grainstone. This lithofacies is widely developed in the Middle Arab A where it forms most of the rocks in Zones A8 and A9. It also occurs in Zone A6 at ALR-11. From a depositional viewpoint the rocks are very similar to the thickly bedded peloidal-grainstone lithofacies of the Arab C, but in this case no dolomitization has taken place. Typically, they comprise planar cross-bedded peloidal-grainstones and peloidal-bioclastic-grainstones (Fig. 8i & j), with partially micritized foraminifera, bivalves, gastropods, dascyclad algae and thinly coated, elongate grainstone intraclasts at some levels. The assemblage of grain types and the sedimentary structures suggests deposition in a predominantly subtidal, inner lagoonal setting, with periods of intertidal flat and beach sedimentation (Shinn *et al.* 1969; Evans 1970; Purser & Evans 1973). The coated intraclasts are interpreted as beachrock lumps (Evamy 1973; Purser & Loreau 1973; Clark 1980*a*). Evidence of intertidal and supratidal deposition appears to have been largely removed during periods of exposure and erosion, leaving an amalgamated succession of cross-bedded grainstones.

The porosity of the cross-bedded grainstones is predominantly primary and intergranular (Fig. 10c). In most cases the grainstones are loosely compacted with keystone vugs (Dunham 1971; Inden & Moore 1983) in evidence in thin section (Fig. 10d & e). In Zone A8, thin rims of early cement have preserved the primary pores and prevented compaction without affecting permeability (Fig. 10c). Here, permeability may reach 5000 mD or more. In Zone A9, in contrast, thicker rim cements have slightly reduced primary porosity but drastically reduced permeability to less than 200 mD and less than 50 mD at some levels (Fig. 10g). In some examples, the primary intergranular pores have been partially filled with a slightly later blocky calcite cement of possible freshwater phreatic origin (Fig. 10h) (Harris *et al.* 1985). This has further reduced the total porosity of the grainstones in Zone A9. At some levels early freshwater leaching of bioclasts (Friedman 1964) has enhanced porosity to produce honeycomb textures (Figs. 10d & e). A few layers of irregular vugs appear to have been generated as a result of early leaching. Locally, patches of secondary anhydrite replaced and cemented the grainstone host (Fig. 10g) (Clark & Shearman 1980).

Peloidal-oolitic-grainstone. This lithofacies only occurs in the Lower Arab A. It consists of metre-bedded, loosely compacted, medium–coarse-sand-grade peloidal-oolitic-grainstones, with partially micritized and coated bioclasts. The mixture of grain types suggests deposition in a relatively high-energy, outer lagoonal or coastal barrier environment with sand shoals and tidal deltas (Evans *et al.* 1973; Purser & Evans 1973). The level of compaction varies and early rim cements are only weakly developed. Consequently, primary porosity is moderately well preserved and permeability is very high (>1300 mD). Porosity is enhanced in places by secondary intragranular pores that developed as a result of early leaching (Friedman 1964). Elsewhere, vadose compaction (Dunham 1969*a*; Clark 1979) has reduced the total porosity without affecting permeability. Thin, subvertical dissolution pipes that may be related to early leaching are also evident in core slabs.

Peloidal-pisolitic-grainstone. This lithofacies consists of lightly cemented peloidal-pisolitic-grainstone (Fig. 8k & l) with a predominantly primary, intergranular porosity (Fig. 10b & f). Some intergranular pores have been reduced by vadose compaction (Dunham 1969*a*; Clark 1979) and grains have been preferentially dissolved at points of contact (Fig. 10f). The larger, irregularly shaped and thinly coated composite grains are interpreted as vadose pisoliths (Dunham 1969*b*). These probably formed from hypersaline brines in a sabkha environment where high-Mg calcite was precipitated around groups of grains (Loreau & Purser 1973; Shinn 1973). Pisolitic grainstones occur at only two levels in the Arab A, namely in Zone A10 at the top of the Lower Arab A, where they succeed peloidal-oolitic-grainstones, and in

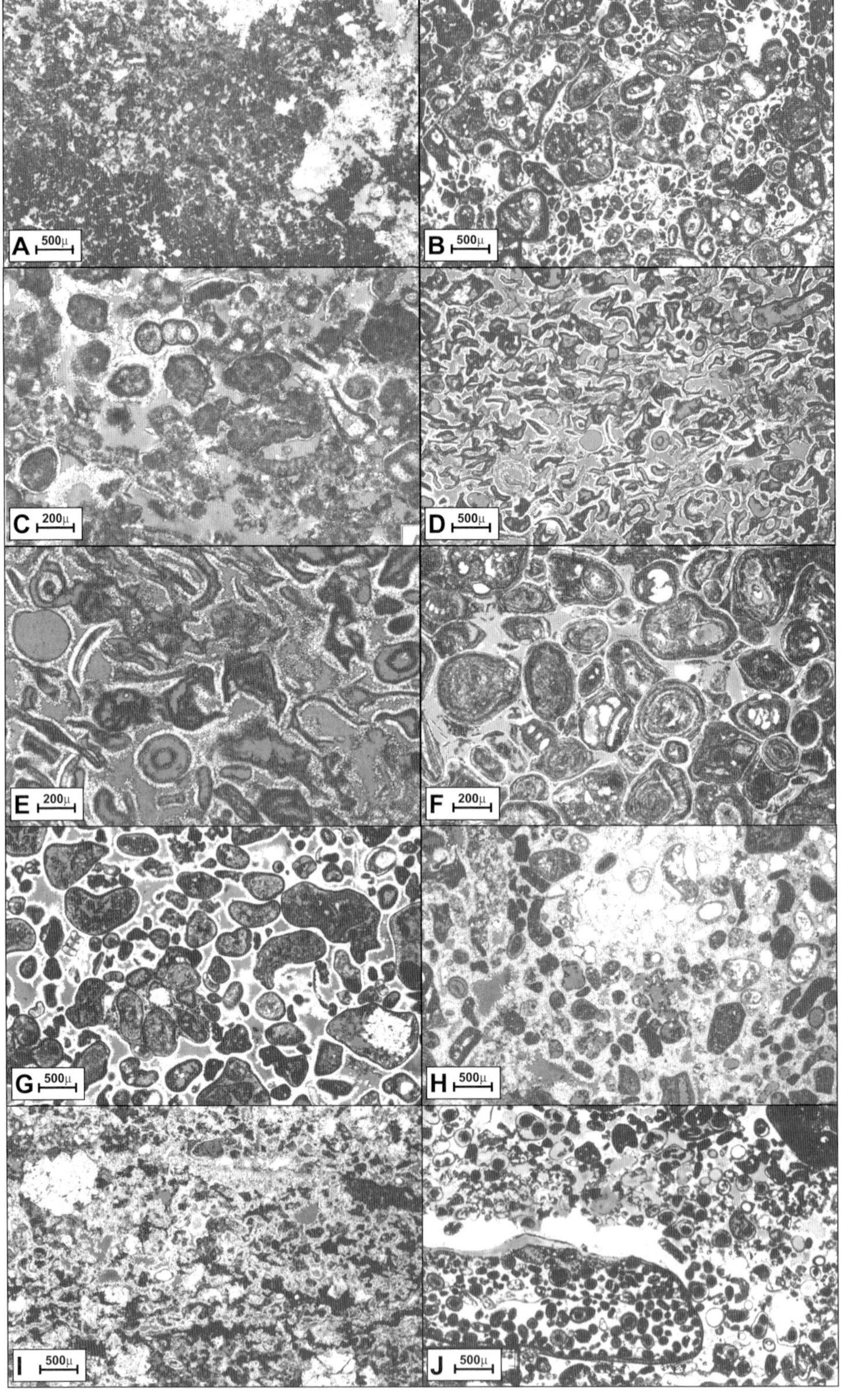
A 500μ
B 500μ
C 200μ
D 500μ
E 200μ
F 200μ
G 500μ
H 500μ
I 500μ
J 500μ

Zone A7 at the top of the Middle Arab A, above a thick sequence of cross-bedded peloidal-grainstones. The pisoliths probably reflect periods when sabkhas became established for prolonged periods at the end of each depositional sequence.

Intraclastic-oolitic-rudstone. This lithofacies only occurs beneath the stromatolites in the A13, at the base of the Arab A (Fig. 8n & o). It comprises loosely compacted and moderately cemented intraclastic-oolitic-rudstones. The intraclasts are characteristically elongate, flattened and coated with superficial lamination (Fig. 10j). Most are either flat lying or randomly arranged, but some layers are stacked to produce an 'edgewise' conglomerate with an imbricate fabric (Shinn 1983). They rest on an

Fig. 10. Photomicrographs of typical lithofacies comprising the Arab A. (**A**) Loosely compacted and lightly cemented *stromatolite* with a peloidal-siltstone matrix. This lithofacies is characterized by a predominantly primary intergranular porosity. Porosity 30.2%, permeability 85.3 mD. Zone A4, 4596.7 ft rkb, ALR-4. (**B**) Lightly cemented *pisolitic-oolitic-bioclastic-peloidal-grainstone* with a predominantly primary intergranular porosity locally together with intragranular pores. Some intergranular pores were reduced in size by vadose compaction and the grains have been preferentially dissolved at points of contact. The composite grains range from medium sand to pebble grade, and comprise irregularly coated aggregates of ooids, peloids and bioclasts. These aggregates are interpreted as vadose pisoliths that formed from hypersaline brines in a sabkha environment when high-Mg calcite was precipitated around groups grains. Porosity 25%, permeability 187 mD. Zone A10, 4635.3 ft rkb, ALR-4. (**C**) Loosely compacted and lightly cemented *cross-bedded peloidal-bioclastic-grainstone*. This lithofacies is made up mostly of peloids and partially micritized smaller benthonic foraminifera. The grains appear to have been cemented soon after deposition, and this has preserved and excellent primary intergranular porosity with no evidence of compaction. Porosity 32.5%, permeability 1140 mD. Zone A8, 4615.85 ft rkb, ALR-4. (**D**) Loosely compacted and lightly cemented *cross-bedded bioclastic-peloidal-grainstone* with a honeycomb texture. This example comprises mostly bioclasts, including smaller benthonic foraminifera and dascyclad algal fragments, together with some peloids. The grains appear to have been cemented soon after deposition and this has preserved an excellent intergranular porosity. In addition, many bioclasts appear to have been dissolved during an early phase of freshwater leaching to produce abundant secondary intragranular pores. In spite of the leaching, permeability is relatively low compared to porosity because of the presence of early cement that lines the intergranular pores and partially obstructs the pore throats. Porosity 37%, permeability 495 mD. Zone A9, 4628.75 ft rkb, ALR-4. (**E**) Detail of (D) showing early pore-lining cement, loose compaction and leached stems of dascyclad algae. Note that the early cement is not present in intragranular pores. A typical keystone vug is present in the bottom left corner of the photomicrograph. Porosity 37%, permeability 495 mD. Zone A9, 4628.75 ft rkb, ALR-4. (**F**) Detail of (B) showing residual intergranular and some intragranular pores and a distinctive 'fitted' or polygonal texture resulting from vadose compaction. This occurs when grains are flushed with freshwater prior to cementation and preferential leaching takes place at points of contact of the grains. It also results in a significant reduction of intergranular porosity. Porosity 25%, permeability 187 mD. Zone A10, 4635.3 ft rkb, ALR-4. (**G**) Loosely compacted and moderately cemented *cross-bedded bioclastic-peloidal-grainstone* with residual intergranular and intragranular pores. It comprises mostly rounded bioclasts, including some dascyclad algal stems, together with smaller peloids and grapestone lumps (composite grains). Cementation probably occurred in a marine environment soon after deposition. The cement typically forms an isopachous layer of acicular crystals that coat the outside surfaces of the grains. This prevented compaction and partially occluded primary intergranular pores. Most of the component bioclasts appear to have been partially dissolved by early freshwater leaching to produce secondary intragranular pores. As in the previous example, permeability is relatively low because of the presence of early cement. One small, castellated patch of anhydrite is evident in the lower right-hand corner that has partially replaced and cemented the host grainstone. Porosity 28.9%, permeability 492 mD. Zone A9, 621.85 ft rkb, ALR-4. (**H**) Loosely compacted and extensively cemented *cross-beddded bioclastic-peloidal-oolitic-grainstone* with residual intergranular and intragranular pores. As in the preceding examples, cementation probably occurred in a marine environment soon after deposition. In this case, however, it was much more pervasive and in places the intergranular pores were completely occluded. Most of the component bioclasts subsequently have been partially dissolved by early freshwater leaching to form secondary intragranular pores. Zone A9, 4623.0 ft rkb, ALR-4. (**I**) Moderately cemented *flaser-bedded peloidal-bioclastic-grainstone* with residual intergranular and some intragranular pores. Cementation probably occurred in a intertidal environment soon after deposition, partially occluding the intergranular pores. Preferential cementation of cross-lamination also occurred, severely reducing permeability. Some leaching of bioclasts is evident in places. Porosity 15%, permeability 12 mD. Zone A12, 4640.3 ft rkb, ALR-4. (**J**) Loosely compacted and moderately cemented *intraclastic-oolitic-bioclastic-peloidal-rudstone*. The intraclasts are typically elongate and superficially coated, and probably originated as beachrock lumps. The porosity of the intraclasts has been occluded by early cementation, whereas the primary intergranular porosity of the host sediment is only partially cemented. Leached intragranular pores are also developed in the host sediment and the cores of some grains have dropped down to form geopetal structures. Porosity 18.7%, permeability 34 mD. Zone A13, 4644.1 ft rkb, ALR-4.

irregular, exposure surface coated with a thin layer of black organic matter. The rudstone is interpreted as an intertidal beach deposit with the intraclasts originating as beachrock lumps (Evamy 1973; Purser & Loreau 1973; Clark 1980*a*; Inden & Moore 1983). The upper surface of the unit was colonized by a columnar stromatolite that also formed in an intertidal environment (Logan *et al.* 1974). The porosity of the intraclasts has been occluded by early cement, whereas the primary intergranular pores of the host sediment are only lightly cemented (Fig. 10j). Leached intragranular pores are also developed in the host sediment and the centres of some grains have dropped downwards to form geopetal structures. Patches of anhydrite cement that probably formed during deeper burial are present in some larger pores.

Dolomite. Dolomite is only found at the top and bottom of the Arab A where it forms thin zones of varying thickness that cut across stratigraphic boundaries (Fig. 6). The dolomites originated by replacement of limestones adjacent to the Hith and Upper Anhydrites. A number of lithofacies have been dolomitized, including stromatolite, intraclastic-rudstone, flaser-bedded grainstone and peloidal-grainstone. Ghosts of the original depositional textures are still recognizable in many of the dolomites. In some cases, the initial porosity and permeability have been severely reduced by the replacement process, but in others some intragranular, fenestral and vuggy porosity is preserved (Choquette & Pray 1970).

Capillary pressure curves

Five groups of capillary pressure curves can be distinguished for the Arab A (Fig. 11). These represent the main lithofacies present in the ALR-4 core and can be used to characterize similar lithofacies elsewhere in the field. This is particularly important when considering zones A5 and A6, which display marked lateral variations in lithofacies.

The stromatolites are characterized by three curves from the A4, A5, A6 and the lowest part of the A10 (Fig. 11). The most representative curve is from the A6 (4605.65 ft). This lithofacies is characterized by a wide range of pore-throat diameters, a low displacement pressure and high irreducible water saturation (Jennings 1987; Vavra *et al.* 1992). Porosity is generally high, consisting mainly of intergranular pores, and permeability is moderately good because well-developed rim cements are absent in intergranular pores. Water saturation tends to be higher because the grain size is relatively fine and hence the pore throats tend to be much narrower (Vavra *et al.* 1992). This has a direct impact on oil staining, as seen in cores, because the stromatolites generally exhibit a lighter yellowish brown stain, compared to the dark brown or black stain seen in coarser grainstones. No capillary pressure curves are available for the concretionary stromatolites, but they can be grouped together with the stromatolites because the matrix textures of the two lithofacies are very similar.

The peloidal-bioclastic-grainstones are represented by six curves from the A8 and A9, including examples of both lightly cemented, loosely compacted, and tighter, moderately cemented grainstones (Fig. 11). The lightly cemented grainstones are representative of the A8 zone (e.g. 4615.85 ft), whereas the A9 is made up of both lightly and moderately cemented examples, with the lower A9 being lightly cemented and the upper A9 moderately cemented (e.g. 4621.85 ft). The lightly cemented grainstones are characterized by a narrow range of pore-throat diameters, low displacement pressures and lower irreducible water saturations, whereas the moderately cemented ones have a larger range of pore-throat diameters, slightly higher displacement pressures and relatively high irreducible water saturations (Robinson 1967). Porosity is generally high, consisting mainly of intergranular pores. Permeability is higher and water saturation generally lower in the lightly cemented grainstones, and these display strong dark brown or black oil staining in cores. In contrast, permeability is significantly lower in cemented examples, with higher water saturations and lighter, patchy staining in cores.

The pisolitic-peloidal-grainstones are represented by three curves from the A7 and A10. The most typical curve is from 4635.3 ft in the A10 (Fig. 11). These rocks have similar characteristics to the peloidal-bioclastic-grainstones, but the overall porosity has been reduced by vadose compaction and the irreducible water saturations are generally higher. Consequently, individual pores are slightly smaller and permeability slightly lower. Water saturation is also higher than for the peloidal-bioclastic-grainstones, and the pisolitic-grainstones tend to have a lighter brown oil stain in cores.

Only one curve is available for the flaser-bedded peloidal-grainstones, from 4640.3 ft (Fig. 11). This represents a very thin layer within the A12 and therefore is relatively unimportant

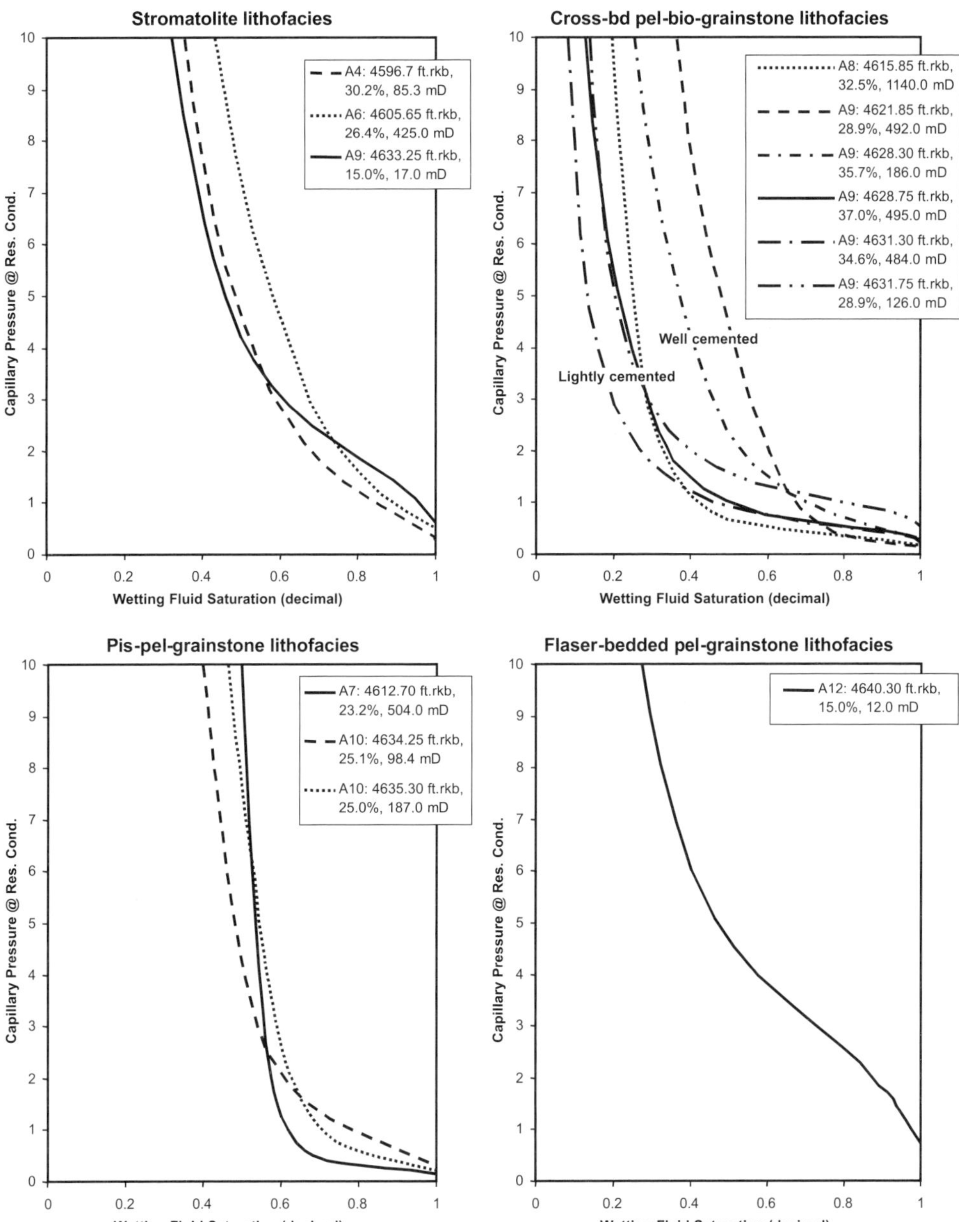

Fig. 11. Typical families of air–mercury injection capillary pressure curves for the Arab Aof ALR-4.

compared to the stromatolite and peloidal-bioclastic-grainstone lithofacies. Nevertheless, it is characteristic of tight, more cemented grainstones and can be used to represent other tight zones such as the A1, A2 and A3. The intraclast-oolitic-rudstones are also represented by a single capillary pressure curve from 4643.2 ft. As with the previous example, the rudstones are relatively unimportant because they only comprise a single bed, 3 ft in thickness, in the lower part of the A12. No capillary pressure curves are available for the millimetre- and

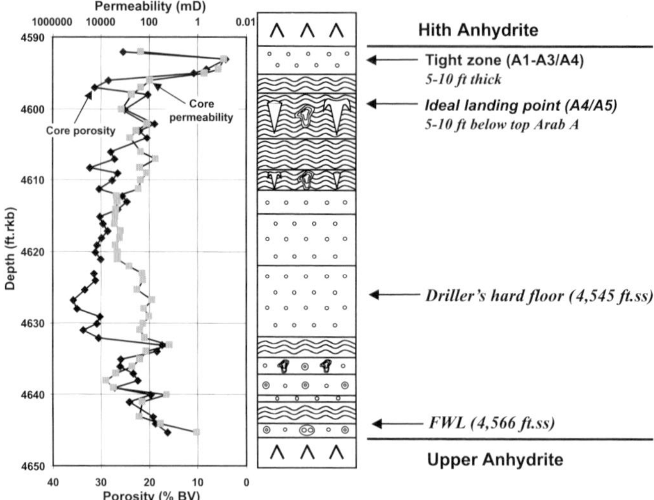

Fig. 12. Ideal landing point for Arab A horizontal wells drilled in the crestal area of the field.

centimetre-bedded peloidal-grainstones, nor for the dolomitized zones at the top and bottom of the Arab A.

Reservoir characteristics

The Arab A can be divided into 13 petrophysical zones (A1–A13), based on core analysis and wireline log data (Figs 4 and 5a). These zones can be grouped into two reservoir units, the upper (A5–A8) and lower (A10 and A11) reservoirs, and three less permeable intervals, the upper dolomite (A1–A4), the cemented unit (A9) and the basal unit (A12 and A13) (Fig. 6). Horizontal layering and hence barriers to vertical fluid flow are only weakly developed in the Arab A, although the partially cemented grainstones with lower permeability that occur in Zone A9 might form a baffle that restricts vertical flow to some extent. These grainstones are easy to identify in cores because they form an interval with patchy or banded oil staining between more permeable, strongly stained zones (4621–4624 ft in Figs 7 and 8e). Horizontal development wells are normally drilled into zones A4 and A5 (Fig. 12), in the uppermost part of the porous Arab A, immediately beneath the upper dolomite unit, in order to achieve high productivity and maximum oil recovery.

Upper dolomite (A1–A4). A thin unit of dense dolomite is present at the top of the Arab A that is assigned to the A1–A3 (Figs 5a and 6). This has low porosity and permeability, with average porosity ranging from 8 to 13%, dependent on the zone, and average permeability varying from 7 to 121 mD. Some residual intergranular porosity is preserved locally, although it rarely exceeds 15%, and a few bands of leached vugs or birds-eyes are present in dolomitized stromatolites. The A4 consists of tight dolomite at ALR-11 and ALR-12, where the average porosity and permeability are only 9% and 0.2 mD, respectively, and dolomitization extends down to the top part the A5 at ALR-12. The A4 has not been dolomitized at ALR-4 (Fig. 6), QMB-3 or possibly QMB-2, where it is composed of stromatolites with excellent intergranular porosity. The average porosity and permeability of the A4 are 30% and 200 mD, respectively, at these locations, with a maximum recorded permeability of 600 mD.

Upper reservoir (A5–A8). The upper reservoir consists of stromatolites, cross-bedded peloidal-grainstones and laminated peloidal-grainstones (Fig. 6). The A5 and A6 are characterized by strong lateral variations in lithofacies that have a significant impact on reservoir quality (Figs 9 and 13). There are also considerable vertical and lateral variations in the occurrence of concretions in the stromatolites, which further affect reservoir quality.

Zone A5 is made up exclusively of stromatolites with concretionary structures at ALR-4 and QMB-3. These typically have a loosely compacted peloidal-grainstone matrix with good intergranular porosity with an average porosity and permeability of 21% and 186 mD,

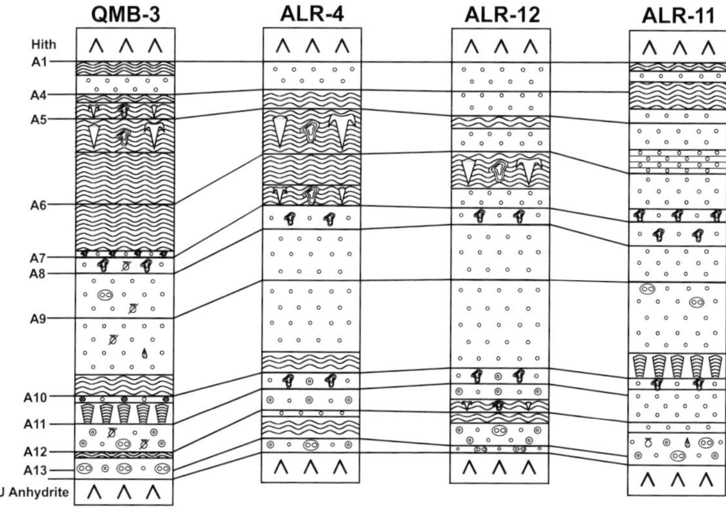

Fig. 13. Correlation of Arab A cores showing characteristic lateral variations in lithofacies particularly in the upper zones of the reservoir.

respectively. At ALR-12, the A5 comprises mainly unmodified stromatolites and centimetre- and millimetre-bedded peloidal-grainstones, with an average porosity and permeability of 30% and 234 mD. Here, the sediments have a texture similar to that of the matrix of the concretionary stromatolites but the average porosity is higher as there are no concretions. The top of Zone 5 is dolomitized at ALR-12, resulting in low porosity and permeability. No stromatolites are developed at ALR-11, and only centimetre- and millimetre-bedded peloidal-grainstones are present. Most of the porosity is intergranular but leached intragranular pores are also developed, giving a slightly higher average porosity of 32%. The average permeability at ALR-11 is 171 mD, similar to that at other localities, but permeability is generally lower in millimetre-bedded grainstones (<50 mD) and higher in centimetre-bedded units (>600 mD).

Zone A6 has similar lithological and petrophysical characteristics to those of the A5. Stromatolites and concretionary stromatolites are exclusively developed at ALR-4 and QMB-3, with an average porosity and permeability of 28% and 216 mD, respectively. The concretionary stromatolites tend to have slightly lower porosity than unmodified stromatolites because of the presence of concretions but permeability is unaffected. Concretionary stromatolites are also well developed at ALR-12, together with thin beds of cross-bedded peloidal-grainstone. Here, the average porosity and permeability are 25% and 454 mD, respectively. The average permeability is higher than at previous locations because of the occurrence of a permeable peloidal-grainstone (1683 mD) at the base of the zone. No stromatolites are developed at ALR-11, where a thick unit of cross-bedded peloidal-grainstone is present instead. These grainstones are very loosely compacted and lightly cemented, and hence porosity and permeability are high, with average values of 31% and 3178 mD, respectively, and a maximum permeability of 9112 mD. Most porosity is intergranular, but leached intragranular pores are also present at some horizons, resulting in total porosities up to 36%.

Zone A7 consists of pisolitic-peloidal-oolitic-grainstones with an average porosity and permeability of 24% and 1727 mD, respectively. Early cementation (Friedman 1964; Harris *et al.* 1985) was relatively light in this zone, allowing permeability to remain high. The dominant pores are intergranular but these have been widely affected by vadose compaction (Dunham 1969*a*; Clark 1979). This has resulted in a significant reduction in total porosity compared to adjacent zones, but permeability appears to be unaffected. Zone A8 is made up largely of cross-bedded peloidal-bioclastic-grainstones with an average porosity and permeability of 28% and 1118 mD, respectively. Early cementation is only lightly developed and permeability therefore is correspondingly high.

Cemented unit (A9). Zone A9 is made up largely of cross-bedded peloidal-bioclastic-grainstones. A relatively tight stromatolite is present at the base of the A9 at ALR-4 (Fig. 6), ALR-11 and QMB-3 but is not developed at ALR-12. The porosity of this lithofacies increases from 17% at the base to 30% at the top, with permeability ranging from 14 to 135 mD. The remainder of the A9 consists of loosely compacted and partially cemented, cross-bedded grainstones, with an average porosity and permeability of 29% and 97 mD, respectively. Residual intergranular pores that are partially occluded by carbonate cement contribute most of the porosity. Early fringing cements are widely developed at all locations, causing a significant reduction in permeability compared to the A7 Zone above. Leaching of bioclasts at some levels produced intragranular pores that have increased total porosity to nearly 38% without affecting permeability.

Lower reservoir (A10 and A11). The lower reservoir predominantly comprises oolitic-grainstones together with pisoliths at the top (Fig. 6). Zone A10 consists of pisolitic-peloidal-oolitic-grainstones with an average porosity and permeability of 18% and 217 mD, respectively. Early cementation is relatively light in this zone, allowing permeability to remain high. The A10 has similar characteristics to the A7, but reservoir quality is not quite as good because early cements are slightly more pervasive. The dominant pores are intergranular but these are widely affected by vadose compaction. This has resulted in a significant reduction in total porosity compared to adjacent zones, without affecting permeability. Zone A11 is composed mostly of peloidal-oolitic-grainstones, with an average porosity of 25% and average permeability of 2766 mD at ALR-4 and ALR-11. A concretionary stromatolite is present in the lower part of the zone at ALR-12, succeeded by peloidal-oolitic-grainstones. Here, the average porosity is reduced to 19% and average permeability is only 154 mD because of the influence of concretions. No thin sections are available from this zone so the exact nature of the pores is unknown, but examination of core slabs suggests that it is predominantly intergranular.

Basal unit (A12 and A13). Zone A12 comprises a thick bed of intraclastic-peloidal-oolitic-rudstone, followed by thinner beds of stromatolite and flaser-bedded peloidal-grainstone (Fig. 6). The overall reservoir quality is relatively poor, with an average porosity and permeability of 13% and 45 mD, respectively. Porosity is mostly intergranular, but leached intragranular pores are also present in the flaser-bedded unit and to a lesser extent in the rudstone. Reservoir quality has been adversely affected by cementation but large uncemented pores are locally preserved between intraclasts, resulting in a thin zone with up to 26% porosity and 1942 mD permeability. Zone A13 at the base of the Arab A consists of dolomitized intraclastic-rudstone with poor reservoir quality (<5% porosity).

Transition zone. As noted above, the FWL for the Arab A is interpreted to be at a depth of 4563 ft ss, based on MDT pressure data, but the distribution of oil shows in cores suggests that it might actually be significantly deeper. A plot of oil staining in relation to structural elevation shows that the base of strong continuous shows is only 3–4 ft above the FWL at ALR-12 and QMB-2, and coincides exactly with the FWL at ALR-4 (Fig. 14). Furthermore, patchy shows belonging to the transition zone are present up to 5 ft below the FWL at ALR-4, ALR-12 and QMB-2. This implies that the FWL should be placed at least 5 ft deeper than the present level, and might be up to 10 ft deeper after allowing for the capillary pressure properties of the rocks. In addition to the uncertainty about the depth of the FWL, the thickness of the capillary transition zone above it is strongly dependent on lithofacies. Thus, not only does the thickness of the transition zone vary according to stratigraphic zone, and hence reservoir quality, it also varies laterally in those zones where lithofacies changes occur. This makes the relationship between water saturation and structural elevation more difficult to understand in the Upper Arab A than in the Middle and Lower Arab A.

The transition zone in the tight, dolomitized unit at the top of the Arab A is at least 50–60 ft thick, depending on where the revised FWL is eventually placed. These dolomites are characterized by patchy oil shows consisting of alternations of strongly stained and unstained layers. Such staining occurs in all cores, irrespective of whether they are located on the flanks or the crest of the field. Consequently, most of the rocks in the A1–A3 are probably still in the transition zone even at the crest of the field. It also explains the absence of significant shows at ALR-11, where the top of the Arab A is 11 ft above the current FWL and poor quality rocks make up the top 10 ft of the succession.

The transition zone in the stromatolites and concretionary stromatolites of the A4, A5 and A6 is at least 3 ft thick, and possibly as much as 13 ft thick, dependent on the depth of the FWL.

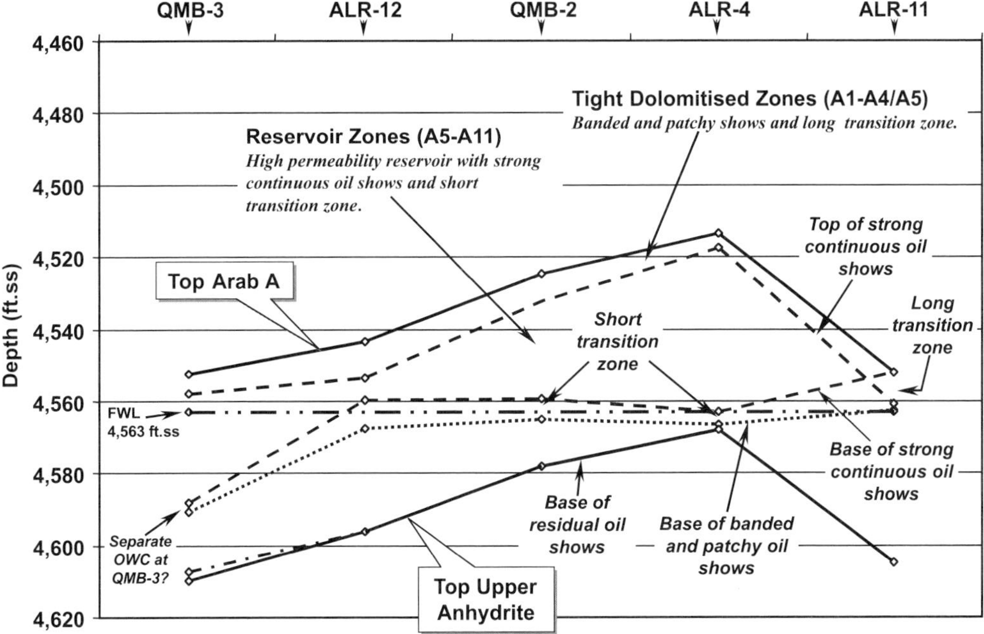

Fig. 14. Distribution of oil shows in Arab A cores.

This is based on the deepest occurrence of strongly stained stromatolites at 4560 ft ss in the ALR-12 core, below which patchy staining is present. The top of the transition zone in the poorly cemented, cross-bedded peloidal-grainstones is not seen in any of the cores. The deepest strongly stained example is found at 4563 ft ss in the ALR-4 core, and the highest patchy staining is at 4567 ft ss in the ALR-12 core. This implies that the top of the transition zone is up to 4 ft below the current FWL, whereas by analogy with similar peloidal-grainstones in the Arab C it should be about 5 ft above this level. The strongly cemented cross-bedded peloidal-grainstones of the upper A9 have a much thicker transition zone because permeability is generally much lower than that of the lightly cemented equivalents. This transition zone is at least 19 ft thick, based on the highest occurrence of patchy shows in the ALR-4 core, and could be as much as 29 ft thick if the FWL is lowered by 10 ft.

Arab C

The Arab C carbonate has an average thickness of 70 ft, ranging from 64 ft at QMB-3 to 77 ft at ALR-13. It is composed almost exclusively of well-bedded (i.e. strongly layered), peloidal and bioclastic grainstones, most of which have been completely dolomitized. The average core porosity of the Arab C for the whole field is 18.1%, and the geometric mean permeability is 85.7 mD. Primary intergranular pores contribute most of the porosity, but secondary intragranular pores are significant at some levels, together with some intercrystalline pores and vugs. The crest of the Arab C structure is at 4615 ft ss, with a vertical dip closure of about 75 ft (Fig. 2). The FWL is at 4688 ft ss.

Parasequences

Six lithofacies can be recognized in the Arab C: thin-bedded peloidal-grainstone; cross-bedded peloidal-grainstone; nodular anhydrite; bioclastic-peloidal-intraclastic-grainstone; coarse bioclastic-peloidal-grainstone and stromatolite. These lithofacies can be grouped into three lithological units, each representing a separate parasequence or depositional cycle (Fig. 15).

The *Lower Arab C (C13–C8)* commences with a thin bed of stromatolite deposited in an upper intertidal–supratidal environment (Figs 15 and 16) (Logan *et al.* 1974). This bed is transgressive across the top of the Lower Anhydrite. It is followed by a thick series of amalgamated coarse-grained bioclastic and intraclastic grainstones. These were deposited in an intertidal and supratidal beach setting, probably on the

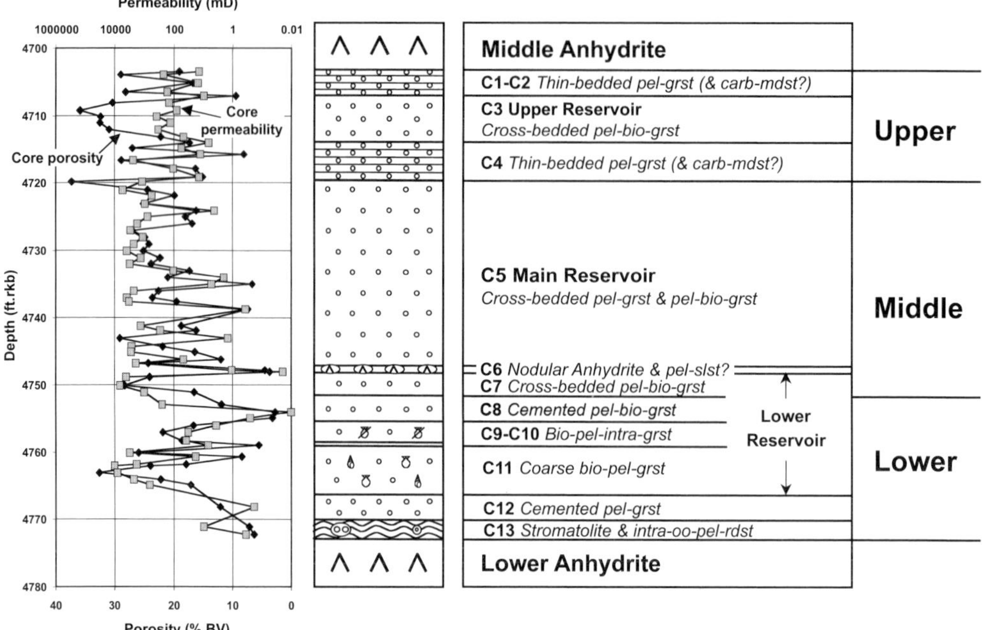

Fig. 15. Sedimentology and reservoir characteristics of the Arab C, ALR-4.

margins of a coastal lagoon complex (Evans *et al.* 1973; Purser & Evans 1973; Inden & Moore 1983). The individual beds of grainstone are separated by irregular surfaces interpreted as palaeokarst (Walkden 1979; Esteban & Klappa 1983).

The *Middle Arab C* (*C7–C5*) is made up predominantly of decimetre-bedded peloidal-grainstones, interbedded with thin, laminated peloidal-grainstones and carbonate-mudstones (Figs 15 and 16). Ideally, each bed consists of a first-order depositional cycle or parasequence that starts with a thick, cross-bedded peloidal-grainstone of subtidal lagoonal origin, and is followed by a thin unit of lower intertidal and laminated upper intertidal peloidal-grainstones and carbonate-mudstones (Fig. 17) (Shinn *et al.* 1969; Evans *et al.* 1973; Purser & Evans 1973; Shinn 1983). The cycle finishes with the emplacement of bands of nodular anhydrite deposited in a supratidal sabkha environment (Shearman 1966; Butler 1969; Kendall & Skipwith 1969). In practice, many sequences are incomplete and some are truncated by subaerial erosion surfaces. The laminated parts of parasequences are commonly missing and the thick-bedded grainstones form amalgamated stacks.

The *Upper Arab C* (*C4–C1*) is made up of two intervals of thin-bedded peloidal-grainstone (C4 and C2), alternating with two thicker bedded intervals (C3 and C1) (Figs 15 and 16). The latter are similar in character to those described above from the C7–C5. The thin-bedded intervals, in contrast, comprise cemented and relatively tight, centimetre- and millimetre-bedded peloidal-grainstones and carbonate-mudstones (Fig. 15). Several depositional cycles can be recognized. Some start with thick, cross-bedded lagoonal peloidal-bioclastic-grainstones (e.g. C3), and finish with thin units of lower intertidal and laminated upper intertidal peloidal-grainstones and carbonate-mudstones (Shinn *et al.* 1969; Evans 1970). Others consist merely of relatively thin alternations of centimetre- and millimetre-bedded grainstones, reflecting changes from lower to upper intertidal conditions.

Thin-bedded peloidal-grainstone. This lithofacies typically comprises cemented and dolomitized, centimetre- or millimetre-bedded peloidal-grainstones and siltstones, together with a few thin layers of carbonate-mudstone (Fig. 18a–c & e). In most cases, the component grains are almost exclusively structureless

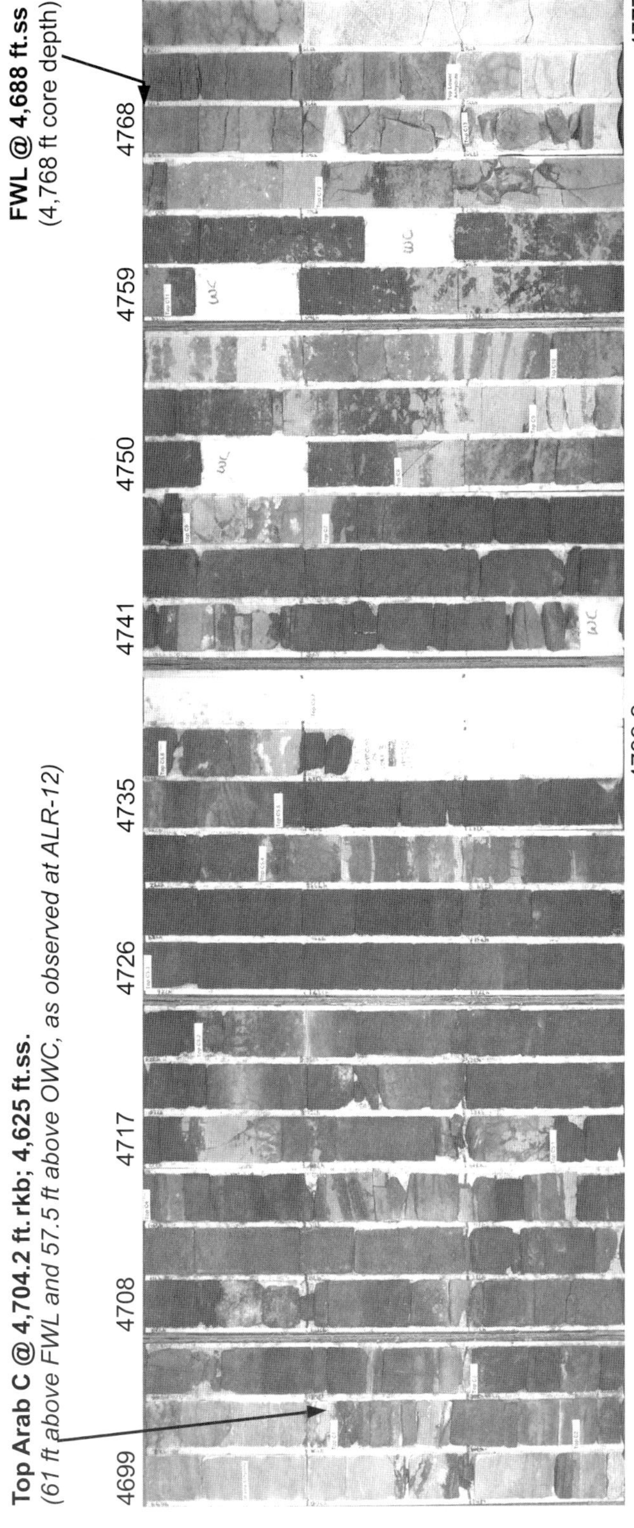

Fig. 16. Core slabs from the Arab C, ALR-4 (4699–4777 ft rkb). Tray length is 3 ft.

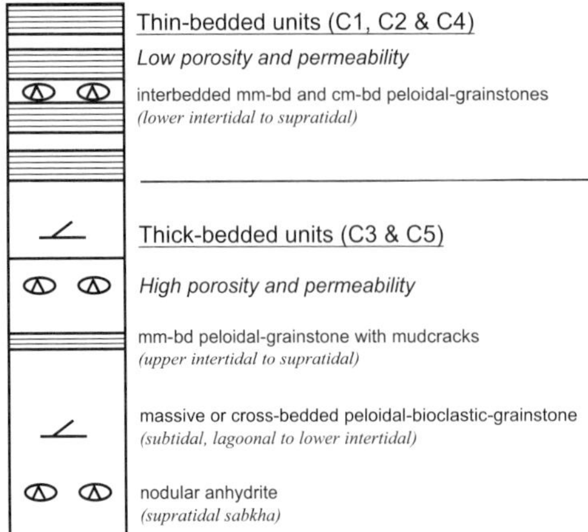

Fig. 17. Characteristics and interpretation of Middle and Upper Arab C lithofacies.

peloids, but in some examples admixtures of peloids and variably micritized and rounded bioclasts are present. Rare mud cracks and starved ripples (lenticular bedding) are present in the laminated units, and ripple cross-lamination is evident in some of the thicker beds. Thin bands of nodular anhydrite are closely associated with this lithofacies at some levels (Fig. 18b). These rocks are interpreted as upper intertidal and supratidal mudflat deposits that formed in a coastal lagoonal setting (Shinn *et al.* 1969; Evans 1970; Shinn 1983). The nodular anhydrite was probably emplaced during early diagenesis when supratidal sabkhas were established (Shearman 1966; Butler 1969; Kendall & Skipwith 1969).

The thin-bedded grainstones have been subjected to extensive early cementation (Friedman 1964; Harris *et al.* 1985) and recrystallization (Fig. 19d) (Folk 1965). This has destroyed most of the primary intergranular porosity and resulted in poor reservoir quality, with low porosity and permeability, and a general lack of oil staining (Fig. 16). Some of the laminated units may have formed as dolomitic crusts in a supratidal environment and this might have been another factor in destroying reservoir quality (Illing *et al.* 1965; Shinn *et al.* 1965). A few of the thicker units have retained some residual intergranular porosity and display a weak oil stain. In some cases, they have also been subjected to early freshwater leaching during periods of exposure, creating secondary intragranular pores (Friedman 1964). Another feature of this lithofacies is the presence of narrow, vertical or

Fig. 18. Examples of typical lithofacies comprising the Arab C (core slabs, normal light, scale in centimetres). (**A**) Centimetre-bedded *peloidal-bioclastic-grainstone* with an irregular exposure surface near the top with a lighter-coloured wavy-bedded calcrete crust immediately beneath it. Zone C5.4, 4766.0 ft rkb, ALR-11. (**B**) Partially oil stained, millimetre–centimetre-bedded *peloidal-grainstone* with a possible mud crack in the laminated unit in the middle of the slab and a layer of displacement anhydrite nodules immediately above. Together, these features suggest deposition in an upper intertidal to supratidal sabkha environment. Zone C5.6, 4741.1 ft rkb, ALR-11. (**C**) Partially oil-stained, millimetre–centimetre-bedded *peloidal-grainstone* with ripple cross-lamination in places. The tight, unstained layer is made up either of cemented *peloidal-siltstone* or *carbonate-mudstone*. Zone C4, 4716.0 ft rkb, ALR-4. (**D**) Partially oil-stained, centimetre-bedded *peloidal-grainstone* with an unstained cemented layer broken by oil-stained fractures. Zone C5, 4717.0 ft rkb, ALR-4. (**E**) Partially oil-stained, centimetre-bedded *peloidal-grainstone* with an unstained layer of cemented *peloidal-siltstone* or *carbonate-mudstone*. As in the previous example, the cemented layer is cut by oil-stained fractures. Zone C4, 4.0 ft rkb, ALR-4. (**F**) Cross-bedded *peloidal-grainstone* with a layer of anhydrite nodules near the base. This lithofacies was probably deposited in a lagoonal or intertidal beach environment. The anhydrite nodules, in contrast, probably grew displacively within the host when sabkha conditions were superimposed on the lagoonal sediments at a slightly later stage. Zone C5.6, 4770.0 ft rkb, ALR-11. (**G**) Loosely compacted, coarse-grained, vuggy *bioclastic-peloidal-grainstone* with large gastropods, bivalves and some coral fragments. This lithofacies is interpreted as an intertidal beach deposit. Zone C9, 4792.3 ft rkb, QMB-3. (**H**) Decimetre-bedded *peloidal-grainstone* with scattered nodules of anhydrite. As in the previous example, this lithofacies was probably deposited in a lagoonal environment with the anhydrite nodules emplaced later when sabkha conditions became established. Zone C5.7, 4774.0 ft rkb, ALR-11. (**I**) Cemented, decimetre-bedded *bioclastic-peloidal-intraclastic-grainstone* with bivalve fragments and large, thinly coated intraclasts. This lithofacies is tentatively interpreted as an intertidal beach deposit, with the intraclasts representing beachrock lumps. Zone C7, 4786.9 ft rkb, QMB-3. (**J**) Dense layer of *nodular anhydrite* with chicken-wire texture, in a host of tight *peloidal-grainstone* or *carbonate-mudstone*. The nodules in this layer are thought to have formed by displacive growth in a sabkha environment. A similar layer of nodules has been identified in Zone C6 in all of the vertical boreholes drilled into the Arab C Carbonate. Zone C6, 4776.0 ft rkb, ALR-11. (**K**) Loosely compacted, coarse-grained *bioclastic-peloidal-grainstone* with large gastropods, bivalves and two pebble-sized coral fragments. Zone C11, 4798.8 ft rkb, QMB-3.

inclined conjugate pairs of fractures in the tightly cemented grainstones and mudstone layers (Fig. 18b & c). These fractures are typically stained with oil confirming that they are natural features and not induced by coring. This is also suggested by the preferential development of patches and veins of replacement anhydrite along some fractures (Clark 1980*b*; Clark & Shearman 1980), particularly immediately beneath the Middle Anhydrite.

Cross-bedded peloidal-grainstone. This lithofacies forms the best quality reservoir rocks in the Arab C (Figs 15 and 16). It is made up of decimetre-bedded, loosely compacted and dolomitized peloidal-bioclastic-grainstones and bioclastic-grainstones (Fig. 18f). Most of the grainstones consist predominantly of fine-sand- or silt-grade peloids, but bioclasts dominate at some levels, particularly in the C3 and C5.1. The bioclasts are generally of fine- or medium-sand grade and comprise partially micritized and rounded fragments of bivalves, dascyclad algae and smaller benthonic foraminifera. The silt-sized peloids are thought to have originated as faecal pellets, whereas the sand-sized ones

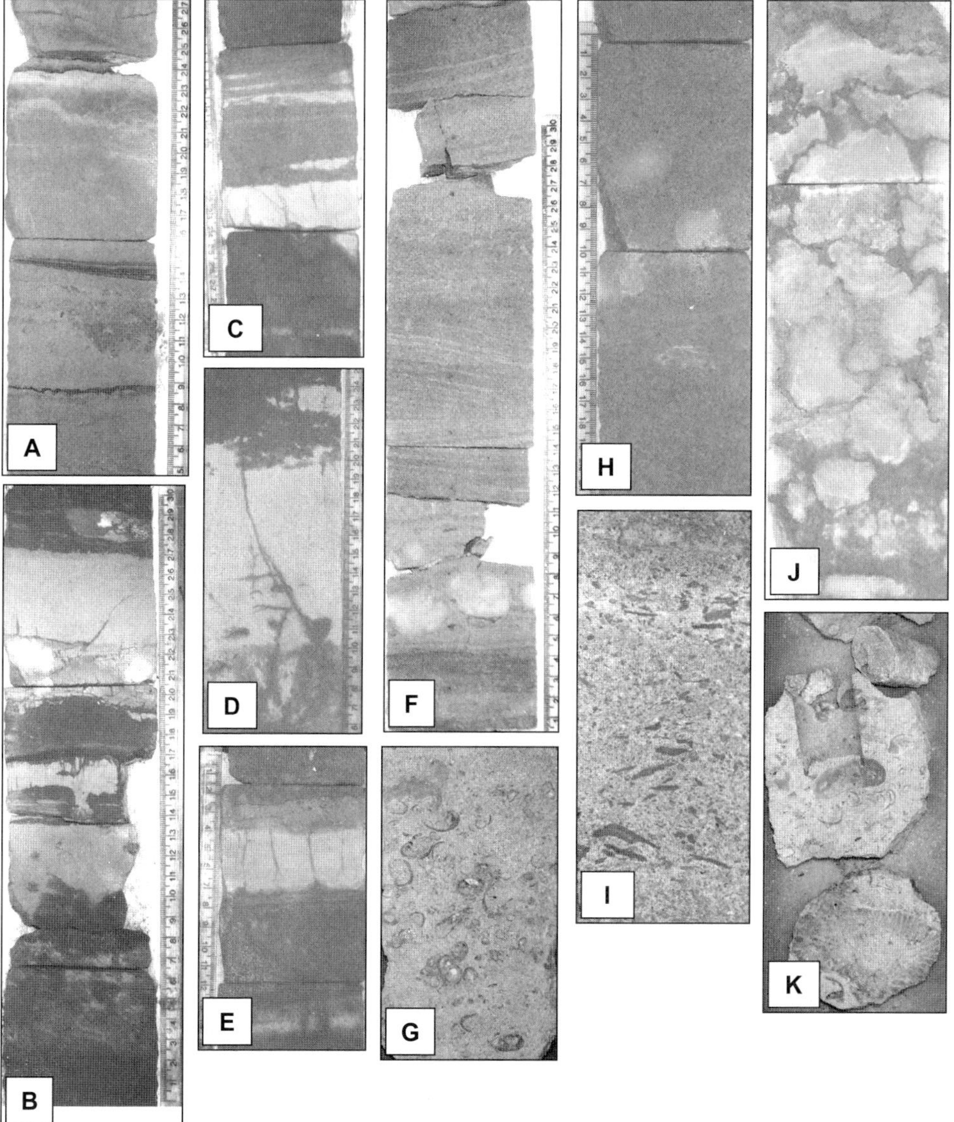

probably formed by micritization of bioclasts (Kendall & Skipwith 1969; Bathurst 1971). Low-angle, planar cross-bedding (Fig. 18f) is characteristically developed in some beds whilst others are either faintly mottled, possibly as a result of bioturbation, or lack sedimentary structures. Scattered nodules of anhydrite are developed in some beds (Fig. 18f & h). The presence of peloids together with dascyclads and smaller benthonic foraminifera suggests that the rocks were deposited in restricted subtidal lagoons and lower intertidal environments (Evans *et al.* 1973; Purser & Evans 1973). The anhydrite nodules were probably emplaced during early diagenesis when sabkha conditions were superimposed on the lagoonal sediments (Shearman 1966; Butler 1969; Kendall & Skipwith 1969), although some might be of later replacement origin (Clark 1980*b*).

Most of the grainstones in this lithofacies have been subjected to early marine cementation (Friedman 1964; Harris *et al.* 1985) and this has prevented any significant compaction from taking place. In most cases, the cements are thinly developed and an excellent intergranular porosity is preserved (Fig. 19e). Permeability is also very high and commonly exceeds several Darcies. Thin but laterally extensive bands of much tighter, weakly oil-stained grainstones are developed at some levels (Fig. 18d). Thicker fringing cements are present in their intergranular pores and, although porosity is only slightly lower, permeability is reduced to less than a few hundred millidarcies (Fig. 19f). In a few examples almost all of the primary porosity is occluded by blocky pore-filling cement, and permeability is further reduced to only a few millidarcies. These more pervasive cements probably formed in a phreatic environment close to the palaeowater table during periods of subaerial exposure (Friedman 1964; Harris *et al.* 1985).

Early freshwater leaching of grains (Friedman 1964) is not widespread in this lithofacies except where bioclasts are common, as in Zone C3 (Fig. 19b). Early leaching probably took place during phases of subaerial exposure and is closely associated with irregular palaeokarst surfaces (Fig. 18a) (Walkden 1979; Esteban & Klappa 1983). Where leaching occurred, secondary intragranular pores were created in addition to primary intergranular pores resulting in well developed honeycomb textures (cf. Fig. 10d). Although leaching has enhanced the overall porosity of the grainstones, it appears to have caused little or no change in permeability, which is largely controlled by the degree of early cementation. Cementation and early leaching of the peloidal-grainstones was followed by complete dolomitization. In some cases, the replacing dolomite crystals mimic depositional textures (Fig. 19b), but in others replacement has masked any pre-existing textures (Fig. 19f). In either case, permeability appears to have been enhanced by the replacement process. The reason for this is unclear, but it may be that calcite cement crystals around pore throats were replaced by larger, smoother dolomite rhombs. Partial leaching of dolomite crystals has resulted in a slight enhancement of the overall porosity of some grainstones by the creation of intercrystalline pores (Fig. 19b) (Choquette & Pray 1970). In places, the centres of some dolomite rhombs have been leached to produce intracrystalline pores (Fig. 19h). This leaching process is thought to have occurred at a relatively late stage in diagenesis, possibly in response to the decarboxylation of nearby source rocks during deeper burial (Clark 1980*b*).

Nodular anhydrite. This lithofacies consists of dense, impermeable layers of nodular anhydrite embedded in peloidal-grainstone or carbonate-mudstone (Fig. 18j). The nodules typically have an enterolithic or chicken-wire texture (Shearman 1966). The most prominent layer is 1–2 ft in thickness and occurs in the C6 (Fig. 15). It is present across the whole of the Al Rayyan field and is clearly recognizable in cores and on wireline logs (Figs 4 and 5b). Thinner, less dense layers and scattered nodules are present at other levels in the Middle and Upper Arab C, but are generally less extensive than the C6 layer (Fig. 18h). The thicker, denser layers are interpreted as nodules that precipitated from hypersaline groundwater in a supratidal sabkha environment (Shearman 1966; Butler 1969; Kendall & Skipwith 1969). The nodules grew displacively within the host sediment, eventually coalescing to form continuous bands of anhydrite. A similar interpretation can be applied to the scattered nodules, although some of these may have originated by replacement of the host during deeper burial rather than by displacive growth in sabkhas (Clark 1980*b*).

Bioclastic-peloidal-intraclastic-grainstone. This lithofacies consists of decimetre-bedded, bioclastic-peloidal-intraclastic-grainstones. The bioclasts consist mainly of large bivalve fragments together with high-spired cerithiid gastropods with superficial coatings, and smaller benthonic foraminifera in various stages of micritization. A few ooids are present in places. Flattened or elongate intraclasts are abundant at some locations (e.g. QMB-3, Fig. 18i) and these, like the gastropods, are superficially coated. Faint low-angle cross-bedding is visible in places with gastropods aligned along the cross-sets. These grainstones are tentatively interpreted as intertidal to supratidal beach deposits, possibly representing a series of beach ridges within a lagoonal environment (Evans *et al.* 1973; Purser & Loreau 1973). The coated intraclasts are thought to have formed as beachrock lumps, while coated gastropods have been described from modern supratidal beach ridges (Evamy 1973; Purser & Loreau 1973; Clark 1980*a*; Inden & Moore 1983).

Cementation of the beach ridges probably occurred in an intertidal environment soon after deposition and an isopachous layer of acicular cement crystals was precipitated on the surfaces of grains (Fig. 19i) (Inden & Moore 1983). Some of the bioclasts and most of the ooids were leached by freshwater soon after deposition to produce secondary intragranular pores (Friedman 1964). Vadose compaction (Dunham 1969*a*; Clark 1979) of uncemented ooids also occurred at this time, reducing the intergranular porosity at some levels but not significantly affecting permeability. Some residual intergranular pores were subsequently completely occluded by blocky calcite cement. Most leached intragranular pores were later partially filled with rhombs of dolomite cement. These probably formed when other sediments in the Arab C were being dolomitized, but the host grains escaped any significant replacement. The porosity of this lithofacies is relatively low because of the effects of early cementation and vadose compaction. Permeability is also relatively poor (<100 mD) because of the presence of thick rims of early cement that restrict pore throats.

Coarse bioclastic-peloidal-grainstone. This lithofacies is made up of dolomitized, vuggy, loosely compacted, coarse-grained bioclastic-peloidal-grainstones (Fig. 18g & k). Large gastropods, whole and fragmented bivalves (possibly oysters) and coral fragments (Fig. 18k) are visible in core slabs, and stems of dascyclad algae are evident in thin section together with small and loosely compacted peloids. Sedimentary structures are generally lacking, apart from irregular erosion surfaces and possible keystone vugs (Dunham 1971; Inden & Moore 1983) in the peloidal matrix. As in the previous example, this lithofacies is interpreted as an intertidal–supratidal beach deposit, possibly representing beach ridges (Evamy 1973; Purser & Loreau 1973; Clark 1980*a*; Inden & Moore 1983). The irregular surfaces are interpreted as palaeokarst (Walkden 1979; Esteban & Klappa 1983).

Compaction appears to have been prevented by early cementation and an excellent intergranular porosity is preserved together with keystone vugs (Fig. 19j). The latter were probably formed as a result of wave activity in an intertidal beach environment and preserved by rapid cementation (Inden & Moore 1983). Cementation was generally very light and hence the permeability of this lithofacies may be up to 10 D. More heavily cemented layers have lower permeability. In some places, large primary intraskeletal pores are preserved inside bivalve shells, but elsewhere early leaching of the shells has produced secondary intraskeletal vugs (Choquette & Pray 1970). Dolomitization occurred after cementation and many smaller peloidal grains were replaced and overgrown to form dolomite rhombs with cloudy centres (Fig. 19j). Superficially, this lithofacies appears to have intercrystalline (Choquette & Pray 1970) porosity, but closer examination reveals that the pores are predominantly intergranular. Porosity remained relatively constant as dolomitization progressed, but the shapes of the original pores were modified and pore throats were streamlined. This may explain the exceptionally high permeability of this lithofacies compared to similar undolomitized rocks. Some replacement and cementation by anhydrite (Clark & Shearman 1980; Clark 1980*b*) occurred after dolomitization, but this was not extensive and only had a minimal impact on reservoir quality (Fig. 19j).

Stromatolite. This lithofacies is composed of dolomitized stromatolites with a cemented or recrystallized peloidal-siltstone matrix. The stromatolites are typically mottled or wavy-bedded, locally with contortion structures, and small domes are present at some levels. Some organic laminae are preserved in places and a possible mud crack has been identified in mottled siltstones above an irregular exposure surface in the ALR-4 core (4722.6 ft rkb). The stromatolites are interpreted as cyanobacterial mats that grew on upper intertidal–supratidal salt marshes (Kendall & Skipwith 1968; Shinn *et al.* 1969). The mottled layers possibly reflect brief episodes of sedimentation in freshwater swamps or ponded environments in an otherwise upper intertidal or supratidal succession (Clark 1999*a*, slide 95). Stromatolites occur only at the base of the Arab C and are generally either tight or of very poor reservoir quality. Porosity is severely reduced by cementation and dolomitization, and, consequently, the stromatolites are only weakly stained with oil even at

ALR-4, which is located on the crest of the structure.

Capillary pressure curves

Four groups of capillary pressure curves can be distinguished for the Arab C, representing the main lithofacies identified in the ALR-4 core (Fig. 20). The bioclastic-grainstones are represented by three curves from the C3, the most typical of which is from 4708.15 ft. This lithofacies is characterized by a wide range of pore-throat diameters and relatively high displacement pressures and irreducible water saturations (Jennings 1987; Vavra *et al.* 1992). Porosity is generally high, consisting of a combination of intergranular, intragranular and intercrystalline pores, but permeability is only moderate because of the presence of well-developed rim cements in the intergranular pores. The irreducible water saturation is higher than for peloidal-grainstones partly because of the influence of rim cements and partly because of the presence of unconnected intragranular pores.

The fine-grained peloidal-grainstones are represented by three curves from the C5. Porosity is relatively high, comprising intergranular pores together with some leached intracrystalline pores. Generally, porosity is not as high as in the previous group because of the absence of intragranular pores, but permeability is much higher because of the scarcity of well-developed rim cements within the intergranular pores. Capillary pressure properties are largely dependent on the degree of compaction of the component grains, and on the presence or absence of intergranular cements. The most representative curve of the lightly cemented grainstones is from 4729.9 ft, where the pore-throat size distribution is very narrow, and the displacement pressure and irreducible water saturation are very low. An example of a poor quality, heavily cemented grainstone is found at 4712.95 ft, where the distribution of pore-throat diameters is much wider, and the displacement pressure and irreducible water saturation are much higher. The higher irreducible water saturation indicates that a large volume of unconnected pores is present and that effective porosity (Jennings 1987) is much lower than in lightly cemented grainstones.

The peloidal-bioclastic-grainstones are represented by one capillary pressure curve from 4720.9 ft in the C5.1. This group is texturally intermediate between the peloidal-grainstones and bioclastic-grainstones. Porosity is moderately high and consists predominantly of intergranular pores, but permeability tends to be lower than the fine-grained peloidal-bioclastic-grainstones because of the occurrence of rim cements. The displacement pressure is very similar to the lightly cemented grainstones, but the range of pore-throat diameters is slightly larger. The irreducible water saturation is also slightly higher suggesting that a larger proportion of unconnected pores are present than in the peloidal-grainstones (Jennings 1987). This is probably because of the presence of some isolated intragranular pores resulting from leaching of bioclasts.

The coarse-grained bioclastic-peloidal-grainstones are represented by three curves from the C9 and C11. Texturally, the rocks differ from previous groups in that the bioclasts are much larger. Porosity is predominantly intergranular, although it may be largely occluded by cementation and many of the larger bioclasts may have been leached to produce vugs. The capillary pressure curves of lightly cemented examples (e.g. 4760.7 ft) typically indicate a very narrow range of pore-throat diameters with low displacement pressures, similar to the fine-grained peloidal-bioclastic-grainstones, whereas heavily cemented examples (e.g. 4755.7 ft) show a much wider range of pore-throat diameters with higher displacement pressures. The irreducible water saturation is also much higher in the heavily cemented examples, indicating a large volume of unconnected pores.

Reservoir characteristics

The Arab C can be divided into 13 petrophysical zones (C1–C13), based on core analysis and wireline log data (Figs 4 and 5a). These zones can be grouped into three units with good reservoir quality, the upper (C3), main (C5) and lower (C7–C11) reservoirs, and four less permeable intervals of poorer quality rock, the upper (C1–C2) and lower (C4) thin-bedded units, the nodular anhydrite layer (C6) and the basal unit (C12 and C13) (Fig. 17).

Upper thin-bedded unit (C1 and C2). The upper part of this unit consists of a relatively thin porous bed of dolomitized peloidal-grainstone immediately beneath the Middle Anhydrite (Figs 5b and 16). This is assigned to the C1 zone. The lower part is made up of tighter, dolomitized, thin-bedded grainstones that are assigned to the C2. The average porosity of the C1 is 23.4%, consisting of a combination of primary intergranular and secondary intragranular pores. Early cementation is well developed in this zone and hence the average permeability is only 77 mD. The porosity and permeability of the C2 are much lower because most of the intergranular pores are occluded by cement. The overall reservoir quality and capillary pressure properties of this unit are very poor, and oil staining occurs in thin bands and discontinuous patches even at the crest of the field (as seen in the ALR-4 core). The distribution of oil shows in cores suggests that the capillary transition zone above the FWL is at least 50–60 ft thick (Fig. 21). As the vertical dip closure of the Arab C is only about 70 ft, this unit remains within the transition zone over most of the field and therefore water saturation is generally very high.

Upper reservoir (C3). The upper reservoir consists of porous and moderately permeable bioclastic-peloidal-grainstones assigned to the C3 (Figs 5b and 16). As in the C1, both primary intergranular and secondary intragranular pores are developed, with an average porosity of 25.8%. The overall porosity is slightly higher than in the C1 because early leaching was more extensive and intragranular pores are better developed. In some cases, the grains are partially leached, leaving secondary intercrystalline rather than simple intragranular pores. Early cementation is also well developed in this zone and hence the average permeability is only 121 mD. This is relatively low compared to the high porosity of the zone. The productivity of horizontal wells drilled into the C3 is relatively low compared to the main reservoir unit (C5). This is because the average permeability of the rocks is lower and therefore the *Kh* values of the wells are significantly smaller than for C5 wells. The water-cut of these wells also is much higher than observed from C5 wells. This is because the range of pore-throat diameters is relatively large and hence the overall water saturation of the reservoir is higher than for the C5 (Fig. 20). The transition zone above the FWL is about 18 ft thick (Fig. 21). The reservoir rocks exhibit a strong, continuous oil stain above the transition zone but this is lighter in colour than the C5, reflecting higher water saturations (Fig. 21).

Lower thin-bedded unit (C4). This unit comprises a relatively thick succession of tight, thin-bedded grainstones placed in the C4 (Figs 5b and 16). The diagenesis and pore types observed in the C4 are very similar to those described from the C2. The average porosity and permeability of the unit are 15.7% and 55 mD, respectively, but some thinner beds are heavily cemented and have much lower permeabilities. As with the thin-bedded unit at the top of the Arab C, the overall reservoir quality and capillary pressure properties are very poor and oil staining is discontinuous even at the crest of the field (Fig. 21). The transition zone above the FWL is probably about 50–60 ft thick. Some of the more porous beds may have thinner transition zones, but the heavily cemented layers never become oil stained and the transition zone for these is much thicker than the vertical closure of the Arab C. Water saturation is therefore generally quite high and the bulk of the rocks comprising the C4 are situated within the transition zone across the whole field (Fig. 20). The C4 is expected to act as a significant baffle to vertical fluid flow because of its thinly bedded nature and low permeability. This will have a strong impact on any communication between the C3 and C5 reservoirs, and it is likely that wells drilled into the C3 will produce very little oil from the underlaying C5 reservoir. Equally, wells completed in the C5 are unlikely to drain much oil from the C3.

Main reservoir (C5). The main reservoir of the Arab C comprises a thick succession of porous and very permeable peloidal-grainstones assigned to the C5 (Figs 5b and 16). Most of the grainstones are loosely compacted, with very thin layers of early cement lining the pores. Almost all of the porosity is comprised of primary intergranular pores preserved by early cementation. Leached intragranular pores are relatively rare and hence average porosities are significantly lower than in the C3, ranging from 19.1 to 23.9%, dependent on subzone. Permeability, however, is much higher than in the C3 because cementation is very light, with average values varying from 500 to over 1500 mD. Although the overall reservoir quality of the C5 is very good, several tighter layers are widely developed across the field. These are characterized by significantly lower average porosities and permeabilities, ranging from 12.3 to 17.4% and 121–450 mD, respectively. The lower porosity of these layers results from the presence of more strongly developed early cements and the partial occlusion of intergranular porosity. The cements also severely reduce permeability, even though the depositional texture of the grainstones is essentially the same as for the more permeable layers. The presence of the tighter layers has made it necessary to divide the C5 into seven subzones (C5.1–C5.7) for the purposes of reservoir simulation (Fig. 5b). However, the detailed correlation of these subzones has proved difficult because cemented bands do not necessarily occur at the same stratigraphic levels in all wells.

The majority of Arab C horizontal wells have been drilled into the upper part of the C5, just beneath the tight C4 (Figs 22 and 23). These are

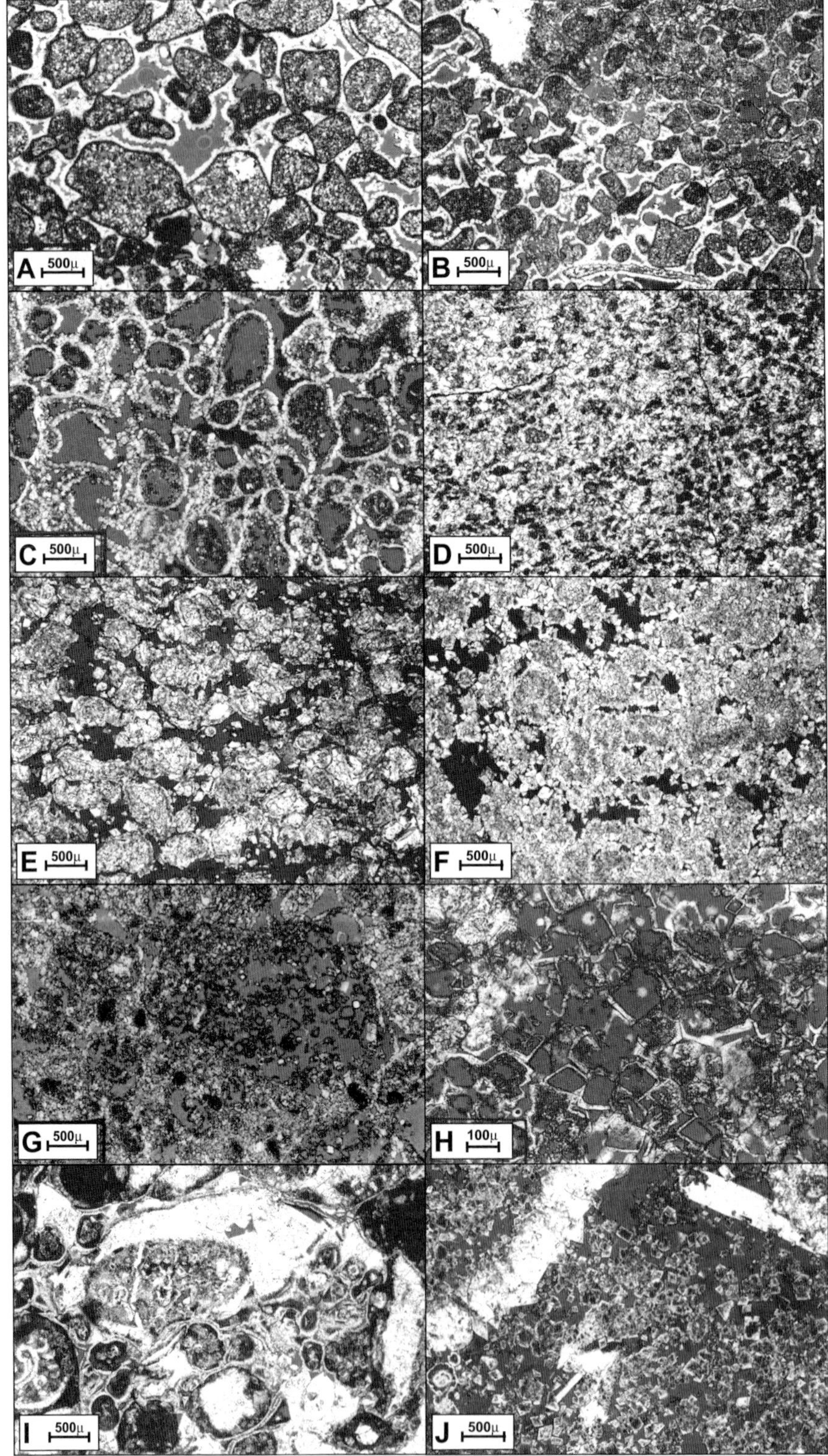
A 500μ
B 500μ
C 500μ
D 500μ
E 500μ
F 500μ
G 500μ
H 100μ
I 500μ
J 500μ

Fig. 19. Photomicrographs of typical lithofacies comprising the Arab C. (**A**) Loosely compacted, moderately cemented, partially leached and dolomitized *bioclastic-grainstone*, with intergranular and intragranular porosity. The cement typically forms an isopachous layer of acicular crystals coating the surfaces of the grains. The component bioclasts have been partially dissolved by early freshwater leaching to produce secondary intragranular porosity. Later dolomitization appears to have partially reorganized the intergranular pores into finer intercrystalline pores. Zone C2, 4682.5 ft rkb, QMB-1. (**B**) Loosely compacted, moderately cemented, and partially leached and dolomitized *bioclastic-grainstone*, with intergranular and intragranular porosity. This lithofacies consists almost entirely of leached bioclasts together with a few ooids. Cementation probably occurred in a marine environment soon after deposition. The cement typically forms an isopachous layer of acicular crystals coating the surfaces of the grains. This prevented compaction and partially occluded intergranular pores. Most of the component bioclasts were either completely dissolved by early freshwater leaching to produce secondary intragranular pores or, in most cases, partially dissolved to form intercrystalline porosity within the grains. Dolomitization occurred after early leaching and was mimetic, replicating the depositional and diagenetic textures of the sediment. Permeability is relatively low in relation to porosity because of the presence of early cement. Porosity 33%, permeability 195 mD. Zone C3, 4712.4 ft rkb, ALR-4. (**C**) Loosely compacted, moderately cemented, strongly leached and dolomitized *bioclastic-peloidal-grainstone* with intergranular and intragranular porosity. Marine cementation occurred soon after deposition to form an isopachous layer of crystals coating grains. The grains themselves were then completely dissolved by early freshwater leaching to produce a honeycomb texture and the residual framework of cement was finally completely dolomitized. Zone C3, 4689.7 ft rkb, QMB-1. (**D**) Heavily cemented and dolomitized *peloidal-grainstone*. This example is made up exclusively of fine sand or silt-sized peloidal grains and intergranular pores are almost completely occluded by cement. Dolomitization occurred after cementation, and the replacing dolomite crystals have destroyed the earlier textures so that only ghosts of the peloids remain. Porosity and permeability are very low because of the pervasive development of early cement. Porosity 7%, permeability 1 mD. Zone C4, 4717.35 ft rkb, ALR-4. (**E**) Loosely compacted, lightly cemented and dolomitized *peloidal-bioclastic-grainstone* with a predominantly intergranular porosity. This lithofacies comprises mainly peloids together with some small, rounded bioclasts. Cementation probably occurred in a marine environment soon after deposition, preventing compaction and preserving an excellent intergranular porosity. The cement typically forms a thin layer around the grains and hence the permeability is much higher than in previous examples. Dolomitization occurred after cementation, resulting in the replacement of peloids and masking of the earlier cement. Some dolomite cement was precipitated in intergranular pores as the host was being dolomitized. Locally, some dolomite crystals have been partially dissolved during a late phase of leaching, resulting in a slight enhancement of porosity. Porosity 30%, permeability 5446 mD. Zone C5.6, 4739.0 ft rkb, ALR-4. (**F**) Loosely compacted, moderately cemented and dolomitized *peloidal-bioclastic-grainstone* with residual intergranular porosity. As in the previous example, this lithofacies comprises mainly peloids together with small bioclasts. In this case, however, early cementation was more extensive, and hence porosity and permeability are much lower. Dolomitization occurred after cementation, resulting in the replacement of both the peloids and earlier cement. The residual intergranular porosity was largely unaffected, although a few small dolomite rhombs were precipitated in places. Porosity 9.25%, permeability 150 mD. Zone C5.7, 4744.45 ft rkb, ALR-4. (**G**) Heavily cemented, leached and dolomitized *peloidal-grainstone*. The primary intergranular pores have been largely occluded by early cement and the component grains were removed by leaching after cementation. Subsequent dolomitization has destroyed earlier depositional and diagenetic textures so that only ghosts of grains remain. Another phase of leaching, probably during deeper burial, dissolved the centres of some replacement dolomite rhombs. Porosity 25.1%, permeability 930 mD. Zone C5.1, 4723.1 ft rkb, ALR-4. (**H**) Leached and dolomitized *peloidal-grainstone* showing well-developed intracrystalline pores. These pores occur at the centres of replacement dolomite crystals and were probably formed by a late phase of leaching during deeper burial. Zone C4, 4693.3 ft rkb, QMB-1. (**I**) Coarse-grained, moderately cemented and slightly dolomitized *bioclastic-peloidal-oolitic-grainstone* with leached intragranular and residual intergranular porosity. This lithofacies comprises mainly bioclasts together with subordinate peloids and ooids. Cementation probably occurred in a marine environment soon after deposition and a typical isopachous layer of acicular cement crystals coats the surfaces of grains. Some bioclasts and most ooids were dissolved during an early phase of freshwater leaching to produce secondary intragranular pores. Some of the residual intergranular pores were subsequently occluded by blocky calcite cement and the leached intragranular pores were partially filled with rhombs of dolomite cement. Porosity 15%, permeability 6 mD. Zone C9, 755.7 ft rkb, ALR-4. (**J**) Coarse-grained, loosely compacted and dolomitized *bioclastic-peloidal-grainstone*. This lithofacies comprises large fragments of bivalves and gastropods together with dascyclad algae stems and small peloids. Compaction was prevented by early cementation and an excellent intergranular porosity is preserved in places. Dolomitization occurred after cementation, and many smaller peloidal grains were replaced and overgrown to form rhombs. Some replacement and cementation by secondary anhydrite also took place, as in the top right-hand corner of the photograph. Porosity 24%, permeability 2722 mD. Zone C11, 4760.3 ft rkb, ALR-4.

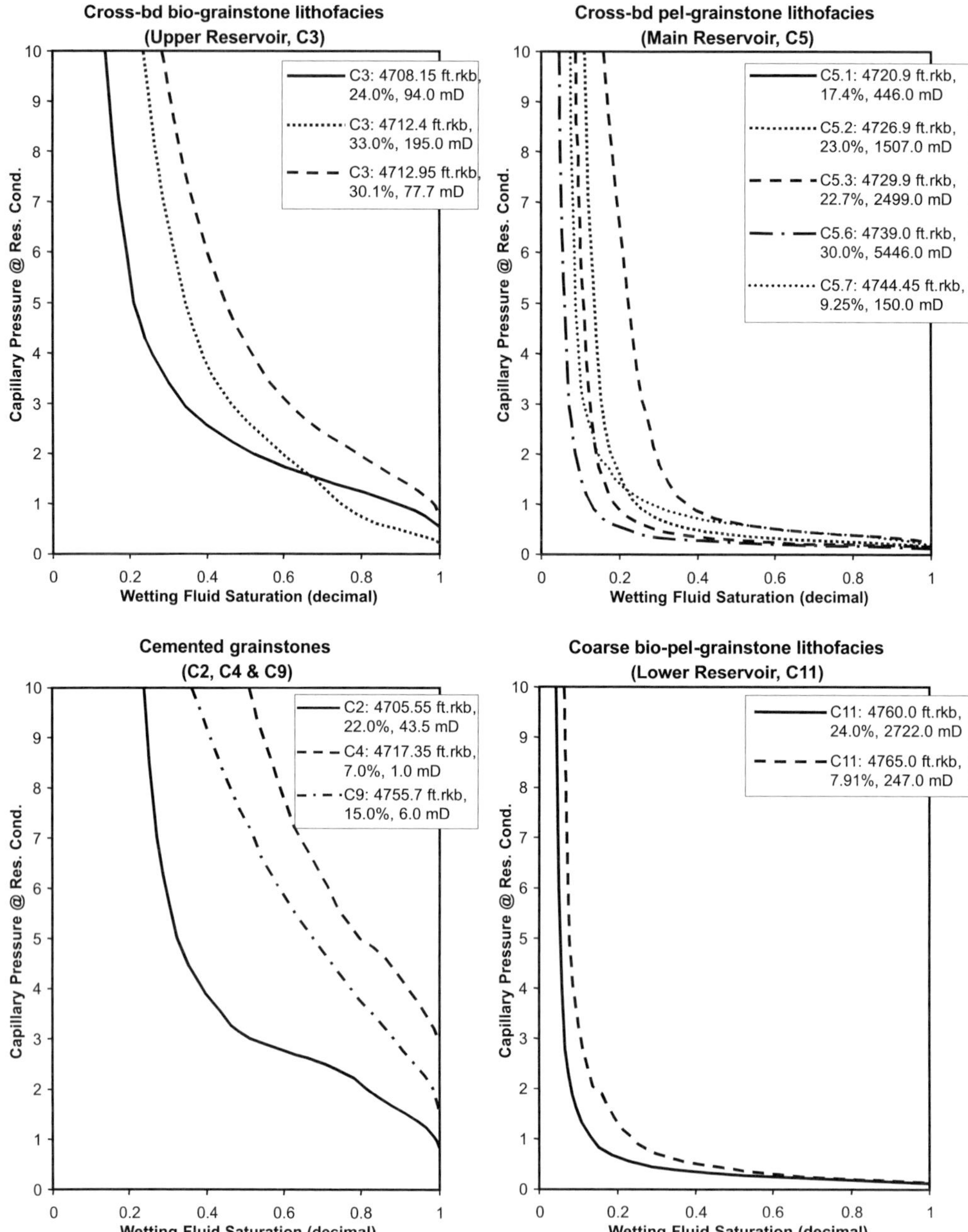

Fig. 20. Typical families of air–mercury injection capillary pressure curves for the Arab C of ALR-4.

able to access a much higher proportion of the oil reserves in the Arab C than wells in the C3. There are three reasons for this. First, the thickness of permeable oil-bearing rocks in the C5 is about 24 ft, whereas the C3 is only 6 ft thick. Secondly, the C5 wells are more productive because the average permeability is much higher and hence the *Kh* values for the C5 are larger those of the C3. Thirdly, the average water saturation is much lower than in the C3 (Fig. 20) and the water-cuts from C5 wells are generally lower than those from C3 wells. This

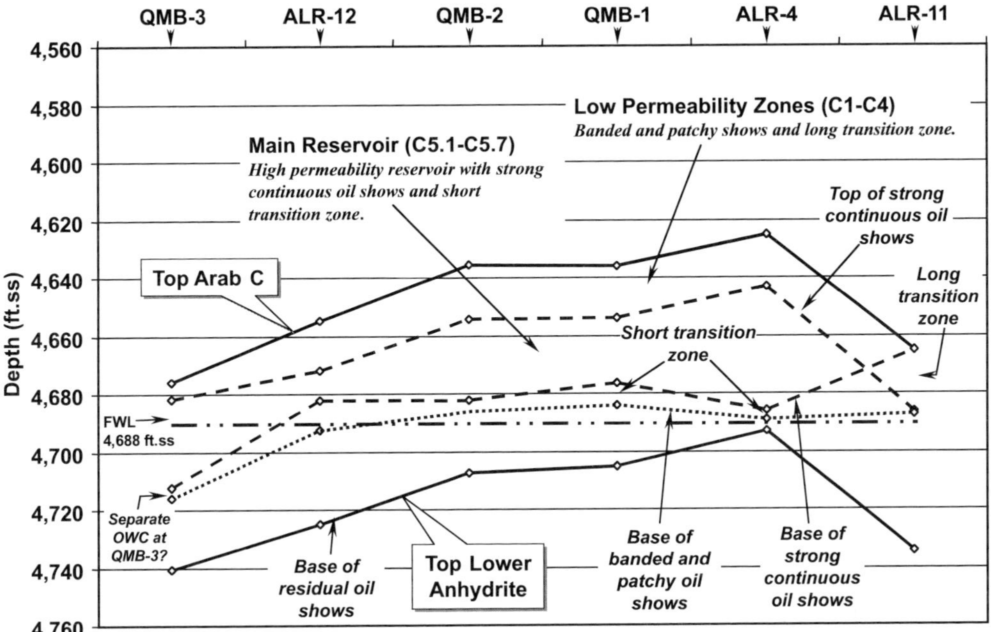

Fig. 21. Distribution of oil shows in Arab C cores.

is because the range of pore-throat diameters in C5 grainstones is much smaller, and the displacement pressure and irreducible water saturation are lower. Thus, the thickness of the transition zone above the FWL is only about 5 ft and oil shows are much stronger, with continuous dark brown or black staining occurring throughout the C5, except for the more cemented layers that have a lighter brown stain (Fig. 21).

Nodular anhydrite layer (C6). A persistent layer of nodular anhydrite, 1–2 ft thick, is present beneath the C5 (Figs 5b and 16). This is assigned to the C6. The anhydrite nodules occur in a tight, heavily cemented dolomite host, with an average porosity and permeability of 6.9% and 3 mD, respectively. These dolomites have very poor capillary pressure properties and are completely unstained by oil even at the crest of the field (Fig. 21). Although this zone is very thin, it is expected to form a baffle to vertical fluid flow because of its extremely low permeability and wide distribution. It is anticipated that it will provide some protection against the early breakthrough of bottom water in C5 horizontal wells.

Lower reservoir (C7–C11). The upper part of the lower reservoir is made up of porous and cemented peloidal-bioclastic-grainstones that are assigned to the C7 and C8 (Figs 5b and 16). The C7 is lithologically very similar to the more porous units within the C5, with a well-developed intergranular porosity, averaging 18.6%, and excellent average permeability of 586 mD. The C8, on the other hand, is more akin to the tighter layers in the C5 because porosity has been largely occluded by cementation. The average porosity and permeability for this zone are only 6.8% and 29 mD, respectively. The middle part of the lower reservoir is lithologically quite different from the upper part in that it consists of bioclastic and intraclastic grainstones. These are placed in the C9, with a cemented band at the base assigned to the C10. The average porosity of the C9 is 19.4%, consisting mainly of residual intergranular pores together with subordinate leached intragranular pores. Permeability is locally severely affected by early cementation but the average value for the zone is 426 mD. The cemented grainstones of the C10, in contrast, have an average porosity and permeability of only 9.5% and 36 mD, respectively. The lowest part of the reservoir consists of very porous and extremely permeable, coarse-grained bioclastic-grainstones that are attributed to the C11. The average porosity of this zone is 19.4%, consisting mostly of primary intergranular pores that have been preserved by early cementation. Some leached bioclasts are present giving rise to large vugs in places. Permeability is generally very high, averaging at 368 mD, but individual samples have values of up to 10 D. Porosity and permeability increase progressively towards the top of the zone as a result of decreasing cementation.

The lower reservoir is only oil-bearing in the central part of the field in the vicinity of ALR-4 (Fig. 21). Away from the crest it approaches and then drops below the FWL, as seen at ALR-12. Good quality reservoir rocks are present at several levels and these exhibit strong, continuous, dark brown oil staining (Fig. 21). Zone C7 is identical to the C5 in terms of reservoir quality, pore types and capillary pressure properties, and the transition zone is about 5 ft thick. The C11 is possibly the best quality reservoir zone in the whole of the Arab C, with high porosity and permeability, narrow distribution of pore-throat diameters, low displacement pressure and a thin transition zone of only 2 ft (Fig. 21). The more cemented beds, as represented by C8, C9 and C10, have a much wider distribution of pore-throat diameters, higher displacement pressures and thick transition zones, probably in excess of 50 ft. None of the horizontal wells drilled so far have been placed in the lower reservoir because of limited oil reserves and proximity to the FWL.

Basal unit (C12–C13). The basal part of the Arab C is subdivided simply on lithological grounds, with cemented peloidal-grainstones being placed in the C12 and stromatolites into the C13 (Figs 5b and 16). These zones have very poor reservoir quality as a result of extensive early cementation, recrystallization, and replacement by dolomite and anhydrite. Most of the rocks in the basal unit are situated below the FWL and patchy oil staining is only weakly developed in the uppermost part of the C12 in the ALR-4 core (Figs 16 and 21).

Controls on reservoir quality

The distribution of porosity, permeability and water saturation in the Arab A and Arab C carbonates is controlled by a combination of factors including depositional environment, grain size, early cementation and dolomitization. These factors determine how the reservoirs perform and how horizontal wells are drilled in order to obtain maximum oil production.

The Arab A is made up predominantly of stromatolites, cross-bedded peloidal-bioclastic-grainstones and peloidal-oolitic-grainstones, whereas the Arab C comprises laminated peloidal-grainstones, cross-bedded peloidal-bioclastic-grainstones and coarse bioclastic-peloidal-grainstones (Figs 6 and 17). The textures of these rocks are similar in that they are all grainstones made up of varying admixtures of peloids and bioclasts. Grain size tends to be fine- to medium-sand grade, but the stromatolites are almost exclusively silt or very-fine-sand grade. Porosity is predominantly primary and intergranular, but intragranular pores are common at some levels. Although the environments of deposition were broadly similar, varying from subtidal lagoonal, through intertidal to supratidal, there are significant differences between the intertidal rocks of the Arab A and Arab C. Strongly cemented and dolomitized, millimetre- and centimetre-bedded peloidal-grainstones represent intertidal environments in the Arab C, whereas poorly-bedded, porous and finer-grained stromatolites feature in the Arab A. The reasons for these differences are unclear but it may simply be that the salinity of the lagoons was higher in the Arab A, favouring the extensive development of cyanobacterial mats. As a result, the Arab C is characterized by lower water saturations and well-developed layering, particularly in the upper part, whereas the Arab A is characterized by higher water saturations and an absence of significant layering.

Vertical permeability barriers are well developed in the Arab C and these are important in influencing oil recovery and controlling bottom water influx. Communication between the Upper (C3) and Middle (C5) reservoirs is restricted by the tight, thinly bedded carbonates of the C4 (Figs 5b and 16). Thus, horizontal wells drilled into the C3 probably do not produce much oil from the C5, and C5 producers probably receive little contribution from the C3. Most of the Arab C producers are drilled into the C5 where higher permeability is encountered and larger oil reserves are present (Fig. 22). Additional wells or sidetracks will probably be required in the future to recover oil from the C3. The C6 anhydrite layer is an important barrier in the Arab C because it is thought to delay the breakthrough of bottom water into C5 wells. Although it is very thin, this anhydrite is present across the whole field and it is situated well above the FWL along the entire lengths of most C5 wells. Such bottom water protection is lacking in Arab A horizontal wells.

The average water saturation in the Arab A is much higher than in the Arab C (Table 1). This is a direct consequence of the capillary pressure properties of the carbonates. The Arab

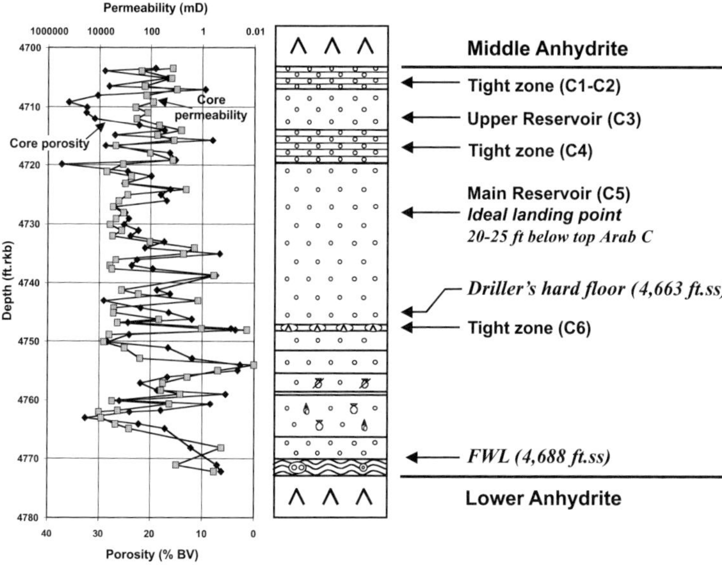

Fig. 22. Ideal landing point for horizontal wells drilled in the Arab C carbonate.

Table 1. *Comparison of reservoir properties of Arab A and Arab C carbonates at ALR-4, showing the relative influences of lithofacies, early cementation and dolomitization on porosity and permeability*

Feature	Arab A	Arab C
Average thickness	56 ft	70 ft
Lithology	Limestone	Dolomite
Layering	Weakly stratified	Strongly stratified
Lateral lithofacies variations	Marked lateral variations in Upper Reservoir	Insignificant lateral variations in Main Reservoir
Main lithofacies	Stromatolite Cross-bd pel-bio-grainstone Cross-bd pel-bio-grainstone	Mm-bd pel-grainstone Cross-bd pel-oo-grainstone Coarse bio-pel-grainstone
Environments of deposition	Supratidal Intertidal Shallow lagoonal	Supratidal Intertidal Shallow lagoonal
Diagenesis	Early leaching Early cementation	Early leaching Early cementation Dolomitization (Late leaching)
Predominant pore types	Intergranular Intragranular (Vugs)	Intergranular Intragranular (Intercrystalline)
Average porosity	23%	18%
Geometric average permeability	116 mD	86 mD
Average water saturation	48%	26%
Productivity	Lower *Kh* (Stromatolite, A4–A6)	Higher *Kh* (cross-bd pel-grst, C5)

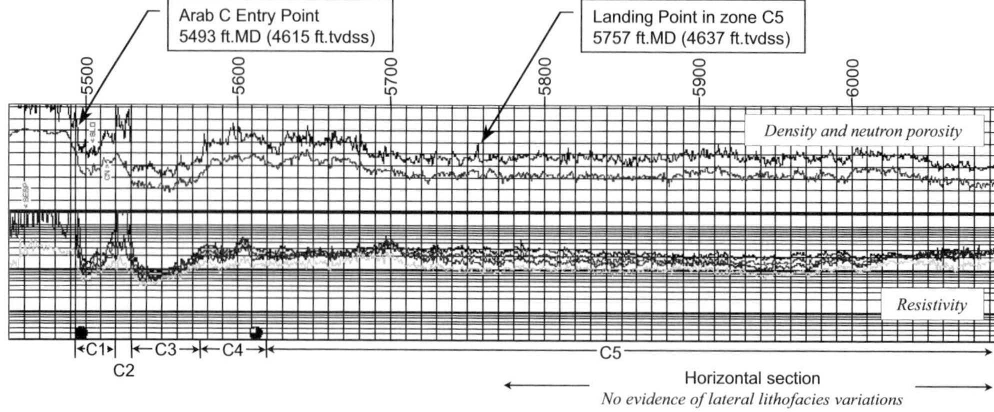

Fig. 23. LWD logs from the Upper Arab C of ALR-9. This borehole was drilled horizontally into zone C5. All depths are measured depths (MD).

A is generally characterized by a wide range of pore-throat diameters, and higher entry pressures and irreducible water saturations, whereas the Arab C exhibits a much narrower range of pore-throat diameters with lower displacement pressures and irreducible water saturations (Fig. 24). Thus, the transition zones tend to be thicker in the Arab A and water saturation is significantly higher for a given height above the FWL, compared to the Arab C. Arab A horizontal wells are drilled immediately beneath the upper dolomite unit, as high as possible above the FWL, in order to minimize water influx and maximize oil recovery (Fig. 9). Arab C wells, on the other hand, are drilled closer to the FWL in order to take advantage of highly productive grainstones in the C5 (Fig. 22).

Arab C horizontal wells tend to be drilled into the more permeable peloidal-grainstones of the C5, whereas most Arab A horizontal wells are targeted in the higher porosity but lower permeability stromatolites of the A5 and A6 (Table 2). Thus, the productivity of horizontal wells is probably higher in the Arab C than in the Arab A, although it is difficult to quantify this because of the variable lengths of the wells. Lithofacies variations in the Upper Arab A also appear to influence productivity and water-cut because of lateral changes in porosity, permeability and water saturation along well bores (Fig. 9).

The peloidal-bioclastic-grainstones of the Arab C generally possess better reservoir quality than comparable grainstones in the Arab A. This is not immediately evident from core data because field-wide average porosity and permeability values for the Arab C are lower than those for the Arab A (Table 1). These averages are misleading, however, because they include values for strongly cemented grainstones as well as other lithofacies. Early cements, in particular, have a strong impact on reservoir quality by blocking pore throats and reducing permeability (Table 2). In more extreme cases, the intergranular pores may be almost completely filled with blocky cement causing a drastic reduction in porosity and permeability. In fact, weakly cemented grainstones in the Arab C are characterized by higher permeability than comparable lithofacies in the Arab A (Table 2).

The differences are even more striking when comparing examples with approximately the same grain size and porosity. For instance, a typical peloidal-bioclastic-grainstone from the Arab C (Fig. 18e) has a permeability of 5446 mD, whereas a comparable example from the Arab A (Fig. 10c) has permeability of only 1140 mD. Both examples possess excellent intergranular porosity and both have a thin layer of pore-lining cement. The Arab C grainstone, however, has been dolomitized and this appears to have enhanced permeability without increasing porosity. The exact mechanism by which permeability is enhanced is not clear, but replacement has increased the crystal size of the host sediment while preserving the depositional texture. The pore-lining cement was also replaced and this may have resulted in the enlargement of pore throats by the growth of simpler dolomite crystals with fewer faces compared to the calcite precursors. As well as increasing permeability, replacement also appears to have improved the capillary pressure properties of the Arab C by decreasing the range of pore-throat diameters, and lowering the displacement pressures and irreducible

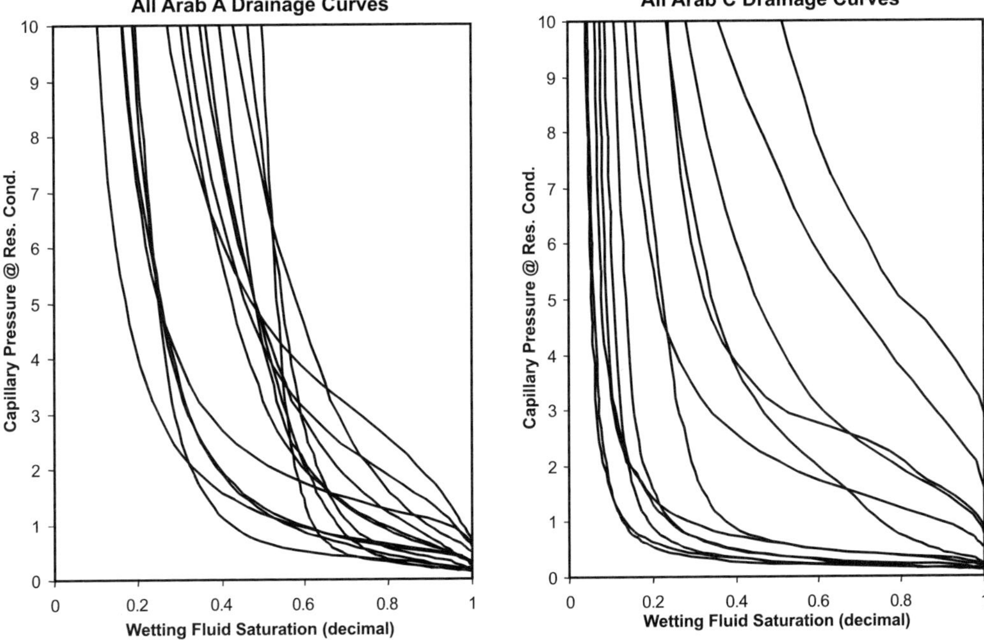

Fig. 24. Comparison of air–mercury injection capillary pressure curves for Arab A limestones and Arab C dolomites of ALR-4. Note the higher displacement pressures for Arab A, and the higher water saturations in the Arab A for a given height above the FWL.

Table 2. *Comparison of the main characteristics of Arab A and Arab C carbonates*

Lithofacies	Average core porosity (%)	Geometric average core permeability (mD)
Arab A – Upper Reservoir (A4–A8) – limestone		
All data (A4–A8)	27	619
Stromatolite (A4–A6)	26	241
Weakly cemented, cross-bd pel-grst (A8)	30	2,093
Cemented, cross-bd pel-grst (A9)	30	149
Arab C – Main Reservoir (C5) – dolomite		
All data (C5)	21	389
Weakly cemented, cross-bd pel-grst (C5)	23	2,294
Cemented cross-bd pel-grst (C5)	17	30

water saturations relative to undolomitized Arab A carbonates. This is probably the main reason why the overall water saturation of the Arab C reservoir is lower than that of the Arab A (Fig. 24).

Dolomitization

As noted previously, the Arab C carbonates have been almost completely replaced by dolomite, whereas the Arab A carbonates have only been dolomitized immediately adjacent to the Hith and Upper Anhydrites. In general, the depositional and early diagenetic textures of the host sediments are preserved in the dolomites. Pre-existing grains and cements and are still recognizable as ghosts but the textures are more coarsely crystalline than would be expected in the limestone precursors. Rhombic crystals of dolomite are present in places in the Arab A as a pore-filling cement and these appear to have been precipitated after an earlier phase of calcite (or aragonite?) cementation. A similar relationship has been described from Arab D carbonates in the Dukhan Field of Qatar (Clark 1999*b*, slides 700–705). Pore types and overall porosity were inherited largely unchanged from the limestone precursors, but permeability appears to have been enhanced by dolomite replacement.

The processes responsible for dolomitization have not been investigated in this study because no isotope or trace-element data are available for the Al Rayyan carbonates. Nevertheless, at least three possible replacement mechanisms can be postulated: penecontemporaneous dolomitization; seepage reflux dolomitization; and deeper burial dolomitization. Penecontemporaneous dolomitization may have occurred in some thinly bedded units of upper intertidal and supratidal sediments in the Arab C. This process has been described from modern supratidal flats of the Bahamas where laminated dolomitic crusts can form (Shinn *et al.* 1965), and from sabkhas in the Persian Gulf where intertidal and supratidal sediments are known to react with hypersaline brines to produce dolomite (Illing *et al.* 1965).

The bulk of the Arab C rocks, however, were probably dolomitized by seepage reflux during shallow burial, when residual evaporite brines percolated down through the sediment pile (Adams & Rhodes 1960). Such brines may have originated in the hypersaline lagoons and sabkhas that were responsible for the precipitation of subaqueous gypsum (Fig. 8p & q) and nodular anhydrite in the Hith, Upper and Middle Anhydrites (Fig. 3). A similar mechanism has been invoked for the dolomitization of Arab D carbonates in Abu Dhabi (Azer & Peebles 1998) and for Zechstein carbonates in NW Europe (Clark 1980*b*). In both examples, the replacement dolomites are closely associated with thick evaporites and they are characterized by the positive $\delta^{18}O$ ratios indicative of evaporitic dolomitization. Replacement during deeper burial was possibly responsible for the dolomitization of Arab A rocks adjacent to the Hith and Upper Anhydrites. In these cases, the proximity of dolomite to the anhydrite–carbonate boundaries suggests that replacement was related in some way to a reaction between the sulphates and carbonates. This may have been driven by higher temperatures and pressures related to increasing burial, but no isotope data are available to support such an interpretation.

Conclusions

The Arab A is made up predominantly of stromatolites and cross-bedded grainstones, whereas the Arab C comprises laminated and cross-bedded grainstones. These rocks were deposited in similar environments, ranging from lagoonal to supratidal, but there are important differences between the intertidal deposits of the Arab A, and those of the Arab C. Thick beds of porous stromatolite are present in the Arab A whereas thinly-bedded units of strongly cemented grainstone are developed in the Arab C that act as vertical permeability barriers.

Porosity is predominantly intergranular, although it is enhanced at some levels by early leaching. Primary porosity is widely preserved by early cementation, while permeability is reduced at some levels by more intense cementation. Dolomitization has significantly enhanced the permeability of cross-bedded grainstones in the Arab C, compared to undolomitized equivalents in the Arab A, but porosity is relatively unaffected. Dolomitization has also significantly improved the capillary pressure properties of the Arab C. Consequently, the overall water saturation of Arab C carbonates is lower than that of the Arab A, and displacement pressures and irreducible water saturations are significantly lower than those of Arab A carbonates.

Horizontal wells are drilled at the top of the porous Arab A in order to achieve maximum stand-off from the FWL. This minimizes water production and maximizes oil recovery in a reservoir with high water saturation and no bottom water protection. Arab C wells, in contrast, are drilled closer to the FWL in order to exploit the porous and permeable C5 zone, where water saturation is relatively low, productivity is higher and bottom water protection is afforded by the C6 anhydrite.

The authors would like to thank S. Willis of ARCO Qatar/BP and R. Al-Sulaiti of Qatar Petroleum for facilitating the study of the Al Rayyan Field and for granting permission to publish this paper. They would also like to acknowledge the help provided by colleagues in the Al Rayyan project team based initially at ARCO's offices in Plano, Texas, and later at BP's Sunbury office. Special thanks should go to T. Burchette of BP Sunbury for his help and advice during the study, and to M. Burstow of Johnston Controls and A. Trabelsi of Qatar Petroleum for their enthusiastic assistance with the examination of the cores.

References

ADAMS, J.E. & RHODES, M.L. 1960. Dolomitization by seepage reflux. *AAPG Bulletin*, **44**, 1912–1920.

AITKEN, J.D. 1967. Classification and environmental significance of cryptalgal limestones and dolomites, with illustrations from the Cambrian and Ordovician of southwestern Alberta. *Journal of Sedimentary Petrology*, **37**, 1163–1178.

ALSHARHAN, A.S. & NAIRN, A.E.M. 1997. *Sedimentary Basins and Petroleum Geology of the Middle East*. Elsevier, Amsterdam.

AZER, S.R. & PEEBLES, R.G. 1998. Sequence stratigraphy of the Arab A to C Members and Hith Formation, offshore Abu Dhabi. *GeoArabia*, **3**, 251–268.

BATHURST, R.G.C. 1971. *Carbonate Sediments and Their Diagenesis*. Developments In Sedimentology, **12**. Elsevier, Amsterdam.

BUTLER, G.P. 1969. Modern evaporite deposition and geochemistry of coexisting brines, the sabkha, Trucial Coast, Arabian Gulf. *Journal of Sedimentary Petrology*, **39**, 70–89.

CHOQUETTE, P.W. & PRAY, L.C. 1970. Geologic nomenclature and classification of porosity in sedimentary carbonates. *AAPG Bulletin*, **54**, 207–250.

CLARK, D.N. 1979. Patterns of porosity and cement in ooid reservoirs in Dogger (Middle Jurassic) of France: discussion. *AAPG Bulletin*, **63**, 676–679.

CLARK, D.N., 1980*a*. The sedimentology of the Zechstein 2 Carbonate Formation of Eastern Drenthe, the Netherlands. *Contributions to Sedimentology*, **9**, 131–165. Schweizerbart, Stuttgart.

CLARK, D.N. 1980*b*. The diagenesis of Zechstein carbonate sediments. *Contributions to Sedimentology*, **9**, 167–203. Schweizerbart, Stuttgart.

CLARK, D.N. 1999*a*. Description of slides from Middle Jurassic carbonates, Aquitaine Basin, SW France. *In*: SCHOLLE, P.A. & JAMES, N.P. (eds) *Triassic and Jurassic Carbonates, Evaporites and Associated Rocks*. Society of Economic Paleontologists and Mineralogists, Photo CD-18.

CLARK, D.N. 1999*b*. Description of slides from Arab D carbonates, Dukhan Oil Field, Qatar. *In*: SCHOLLE, P.A. & JAMES, N.P. (eds) *Triassic and Jurassic Carbonates, Evaporites and Associated Rocks*. Society of Economic Petrologists and Mineralogists, Photo CD-18.

CLARK, D.N. & SHEARMAN, D.J. 1980. Replacement anhydrite in limestones and the recognition of moulds and pseudomorphs: a review. *In*: ORTÍ CABO, F. (ed.) *Conferencias y Comunicaciones del i Symposium Sobre Diagénesis de Sedimentos y Rocas Sedimentarias*. Revista del Instituto de Investigaciones Geológicas, **34**, 161–186.

DUNHAM, R.J. 1962. *Classification of carbonate rocks according to depositional texture*. American Association of Petroleum Geologists, Memoirs, **1**, 108–121.

DUNHAM, R.J. 1969*a*. *Vadose compaction on Cayo Arenas, Campeche Banks*. Bellair Research Center, Shell Oil Company, Houston. (Unpublished report summarized by Clark, 1979.)

DUNHAM, R.J. 1969*b*. *Vadose Pisolite in the Capitan Reef (Permian), New Mexico and Texas*. Society of Economic Paleontologists and Mineralogists, Special Publications, **14**, 182–191.

DUNHAM, R.J. 1971. Meniscus cement. *In*: BRICKER, O. (ed.) *Carbonate Cements*. Johns Hopkins University Studies in Geology, **19**, 297–300.

EMBRY, A.F. & KLOVAN, J.E. 1971. A Late Devonian reef tract on northeastern Banks island, N.W.T. *Bulletin of the Canadian Petroleum Geologists*, **19**, 730–781.

ESTEBAN, M. & KLAPPA, C.F. 1983. Subaerial Exposure. *In*: SCHOLLE, P.A., BEBOUT, D.G. & MOORE, C.H. (eds) *Carbonate Depositional Environments*. American Association of Petroleum Geologists, Memoirs, **33**, 1–54.

EVAMY, B.D. 1973. The precipitation of aragonite and its alteration to calcite on the Trucial Coast of the Persian Gulf. *In*: PURSER, B.H. (ed.) *The Persian Gulf*. Springer, Berlin, 329–342.

EVANS, G. 1970. Coastal and nearshore sedimentation: a comparison of clastic and carbonate deposition. *Proceedings of the Geological Association, London*, **81**, 493–508.

EVANS, G., MURRAY, J.W., BIGGS, H.E.J., BATE, R. & BUSH, P.R. 1973. The oceanography, ecology, sedimentology and geomorphology of parts of the Trucial Coast barrier island complex, Persian Gulf. *In*: PURSER, B.H. (ed.) *The Persian Gulf*. Springer, Berlin, 233–277.

FOLK, R.L. 1959. Practical petrographic classification of limestones. *AAPG Bulletin*, **43**, 1–38.

FOLK, R.L. 1965. Some aspects of recrystallization in ancient limestones. *In*: PRAY, L.C. & MURRAY, R.C. (eds) *Dolomitization and Limestone Diagenesis; A Symposium*. Society of Economic Paleontologists and Mineralogists, Special Publications, **13**, 14–48.

FRIEDMAN, G.M. 1964. Early diagenesis and lithification in carbonate sediments. *Journal of Sedimentary Petrology*, **34**, 777–813.

GINSBURG, R.N. 1991. Controversies about stromatolites: vices and virtues. *In*: MULLER, D.W. & MCKENZIE, J. (eds) *Controversies in Modern Geology*. Academic Press, New York, 25–36.

HARRIS, P.M., KENDALL, C.G. & LERCHE, I. 1985. Carbonate cementation – a brief review. *In*: SCHNEIDERMANN, N. & LERCHE, I. (eds) *Carbonate Cements*. Society of Economic Paleontologists and Mineralogists, Special Publications, **36**, 79–95.

ILLING, L.V., WELLS, A.J. & TAYLOR, J.C.M. 1965. Penecontemporary dolomite in the Persian Gulf. *In*: PRAY, L.C. & MURRAY, R.C. (eds) *Dolomitization and Limestone Diagenesis; A Symposium*. Society of Economic Paleontologists and Mineralogists, Special Publications, **13**, 89–111.

INDEN, R.F. & MOORE, C.H. 1983. Beach Environment. *In*: SCHOLLE, P.A., BEBOUT, D.G. & MOORE, C.H. (eds) *Carbonate Depositional Environments*. American Association of Petroleum Geologists, Memoirs, **33**, 211–265.

JENNINGS, J.B. 1987. Capillary pressure techniques; application to exploration and development geology. *AAPG Bulletin*, **71**, 1196–1209.

KENDALL, C.G. St C. & SKIPWITH, P.A.D'E. 1968. Recent algal mats of a Persian Gulf lagoon. *Journal of Sedimentary Petrology*, **38**, 1040–1058.

KENDALL, C.G. St C. & SKIPWITH, P.A.D'E. 1969. Holocene shallow water carbonate and evaporite sediments of the Khor al Bazam, Abu Dhabi, SW Persian Gulf. *AAPG Bulletin*, **53**, 841–869.

LOGAN, B.W., REZAK, R. & GINSBURG, R.N. 1964. Classification and environmental significance of algal stromatolites. *Journal of Geology*, **72**, 68–83.

LOGAN, B.W., HOFFMAN, P. & GEBELEIN, C.D. 1974. *Algal mats, cryptalgal fabrics, and structures, Hamelin Pool, Western Australia: evolution and diagenesis of Quaternary carbonate sequences, Shark Bay, Western Australia*. American Association of Petroleum Geologists, Memoirs, **22**, 140–194.

LOREAU, J.-P. & PURSER, B.H. 1973. Distribution and ultrastructure of Holocene ooids in the Persian Gulf. *In*: PURSER, B.H. (ed.) *The Persian Gulf*. Springer, Berlin, 279–328.

MULTER, H.G. & HOFFMEISTER, J.E. 1968. Subaerial laminated crusts in the Florida Keys. *Geological Society of America Bulletin*, **79**, 183–192.

PURSER, B.H. & EVANS, G. 1973. Regional sedimentation along the Trucial Coast, SE Persian Gulf. *In*: PURSER, B.H. (ed.) *The Persian Gulf*. Springer, Berlin, 211–231.

PURSER, B.H. & LOREAU, J.-P. 1973. Aragonitic, supratidal encrustations on the Trucial Coast of the Persian Gulf. *In*: PURSER, B.H. (ed.) *The Persian Gulf*. Springer, Berlin, 233–277.

REINECK, H.-E. & SINGH, I.B. 1975. *Depositional Sedimentary Environments*. Springer, Berlin.

RIDING, R. 1999. The term stromatolite: towards an essential definition. *Lethaia*, **32**, 321–330.

RIDING, R. 2003. Bacterial carbonates: process and product in time and space. Unpublished extended abstract. 2003 British Sedimentological Research Group Annual General Meeting, Leeds, 20–22 December.

ROBINSON, R.B. 1967. Diagenesis and porosity development in Recent and Pleistocene oolites from southern Florida and the Bahamas. *Journal of Sedimentary Petrology*, **37**, 355–364.

SCHOLLE, P.A. & KINSMAN, D.J.J. 1974. Aragonite and high-Mg calcite from the Persian Gulf – a modern analogue for the Permian of Texas and New Mexico. *Journal of Sedimentary Petrology*, **44**, 904–916.

SHEARMAN, D.J. 1966. Origin of marine evaporites by diagenesis. *Transactions of the Institute of Mining and Metallurgy (B)*, **75**, 208–215.

SHINN, E.A. 1968. Practical significance of birdseye structures in carbonate rocks. *Journal of Sedimentary Petrology*, **38**, 215–223.

SHINN, E.A. 1973. Recent intertidal and nearshore carbonate sedimentation around rock highs, E Qatar, Persian Gulf. *In*: PURSER, B.H. (ed.) *The Persian Gulf*. Springer, Berlin, 193–209.

SHINN, E.A. 1983. Tidal flat environment. *In*: SCHOLLE, P.A., BEBOUT, D.G. & MOORE, C.H. (eds) *Carbonate Depositional Environments*. American Association of Petroleum Geologists, Memoirs, **33**, 172–210.

SHINN, E.A., GINSBURG, R.N. & LLOYD, R.M. 1965. Recent supratidal dolomite, Bahamas. *In*: PRAY, L.C. & MURRAY, R.C. (eds) *Dolomitization and Limestone Diagenesis; A Symposium*. Society of Economic Paleontologists and Mineralogists, Special Publications, **13**, 112–123.

SHINN, E.A., LLOYD, R.M. & GINSBURG, R.N. 1969. Anatomy of a modern carbonate tidal-flat, Andros Island, Bahamas. *Journal of Sedimentary Petrology*, **39**, 1202–1228.

VAVRA, C.L., KALDI, J.G. & SNEIDER, R.M. 1992. Geological application of capillary pressure: a review. *AAPG Bulletin*, **76**, 840–850.

WALKDEN, G.M. 1979. Palaeokarst surfaces in Upper Visean (Carboniferous) limestones of the Derbyshire Block, England. *Journal of Sedimentary Petrology*, **44**, 1232–1247.

Porosity and permeability in Miocene carbonate platforms of the Marion Plateau, offshore NE Australia: relationships to stratigraphy, facies and dolomitization

S. N. EHRENBERG
Statoil, N-4035 Stavanger, Norway

Abstract: Analyses of porosity and permeability are examined from cores drilled in two Miocene carbonate platforms cored by ODP Leg 194, seaward from the Great Barrier Reef, including one site in the Northern Marion Platform (NMP; mostly preserved as limestone) and two sites in the Southern Marion Platform (SMP; mostly dolomitized). The majority of plug samples are from coarse bioclastic facies and their dolomitized equivalents. Dolomitization probably occurred by circulation of normal to slightly modified seawater. Increasing fabric destruction at greater depth in the SMP may reflect overprinting of multiple dolomitization episodes in older strata, perhaps related to successive cycles of sea-level fluctuation. Both limestones and dolostones show similarly wide variation in porosity and permeability, reflecting extreme metre-scale heterogeneity of lithologies and diagenetic responses. The limestones show poorer porosity–permeability correlation, with generally lower permeability for given porosity compared with the dolostones, but with similar maximum permeabilities. Despite wide textural diversity, the dolostones cluster along a single trend that parallels the ideal relationship described by the Kozeny equation and characteristic of well-sorted sandstones. The low permeability-for-given-porosity of many limestones is explained by the fine grain size of some samples and, in other cases, by isolation of macropores behind mud matrix. Pore systems in the dolostones, however, tend to be dominated by interparticle macroporosiity, consisting of either relict intergranular pores (grainstones with fabric-preserving dolomitization) or intercrystalline pores (fabric-destructive dolostones and packstones with fabric-preserving dolomitization). Although many dolostones have vuggy pore systems, the vugs appear to be effective for fluid flow because they are connected by the enclosing system of interparticle macropores.

The scientific objectives for Leg 194 were developed based on results of ODP Leg 133, which drilled nine sites on the Queensland and Marion plateaus in 1990 (Fig. 1) (Davies *et al.* 1991; Pigram *et al.* 1992). The principal objective was to measure the magnitude of the major eustatic sea-level fall in latest Serravallian–base-Tortonian time (12.5–11.0 Ma), as recognized on the global sea-level curve of Haq *et al.* (1987). Other objectives were the study of carbonate depositional and palaeoenvironmental controls, dating and description of the main stratigraphic units recognized on seismic lines, and study of fluid flow and diagenesis (Isern *et al.* 2001, 2002).

This paper examines the patterns of dolomitization and porosity–permeability variation in cores of shallow-water carbonate lithologies from the two adjacent carbonate platforms cored by Ocean Drilling Program Leg 194 on the Marion Plateau (Fig. 2). These platforms were cored in only three of the Leg 194 sites (1193, 1196 and 1199; Figs 2 and 3). The other five 'off-platform' sites are of interest for constraining the ages and stratigraphic significance of the platform sites, but are not part of the present porosity–permeability data set. The lithologies examined consist entirely of high-energy bioclastic limestones and their dolomitized equivalents.

This data set is of significance as an example of how textural characteristics determine porosity–permeability distributions in shallowly buried (0–640 m below sea floor) and relatively young bioclastic carbonates lacking evaporite cements. As such, the Leg 194 data may be useful for comparison with porosity–permeability distributions of sedimentologically analogous, but more deeply buried, ancient or evaporite-cemented successions in order to visualize an early stage in the evolution of reservoir properties. Such insight could be valuable for interpreting the statistical and geometrical distributions of subsurface petrophysical properties. A particular focus in the context of the present volume is the degree to which the Leg 194 data set is relevant for understanding how carbonate porosity–permeability distributions change during dolomitization. Thus, the limestones are of interest as possibly representing lithologies from which the dolostones formed.

From: BRAITHWAITE, C. J. R., RIZZI, G. & DARKE, G. (eds) 2004. *The Geometry and Petrogenesis of Dolomite Hydrocarbon Reservoirs*. Geological Society, London, Special Publications, **235**, 233–253. 0305-8719/$15.00

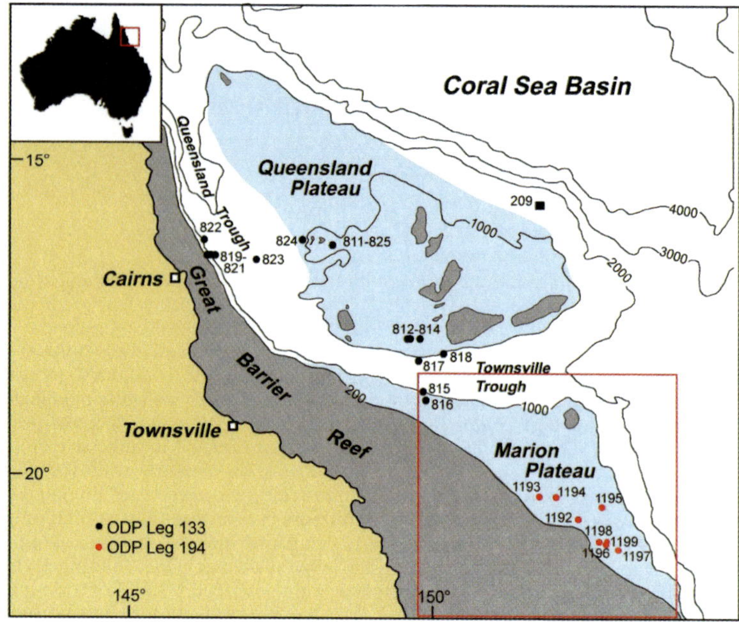

Fig. 1. Location of study area (modified from Isern *et al.* 2002).

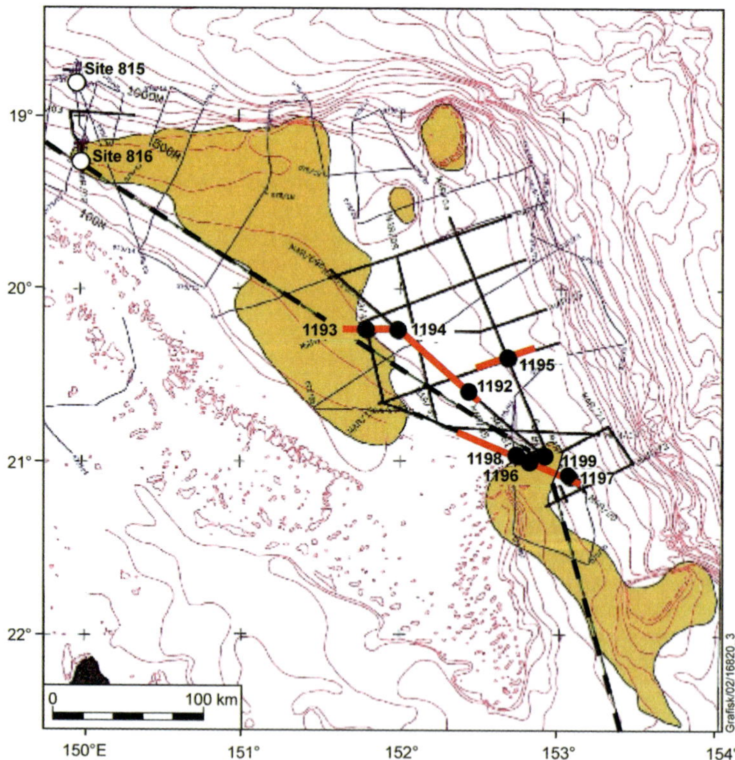

Fig. 2. Locations of ODP sites and seismic lines on the Marion Plateau (modified from Isern *et al.* 2002). Figure location is outlined on Figure 1. Carbonate platforms are coloured yellow. Seismic lines in Figure 3 are highlighted in red.

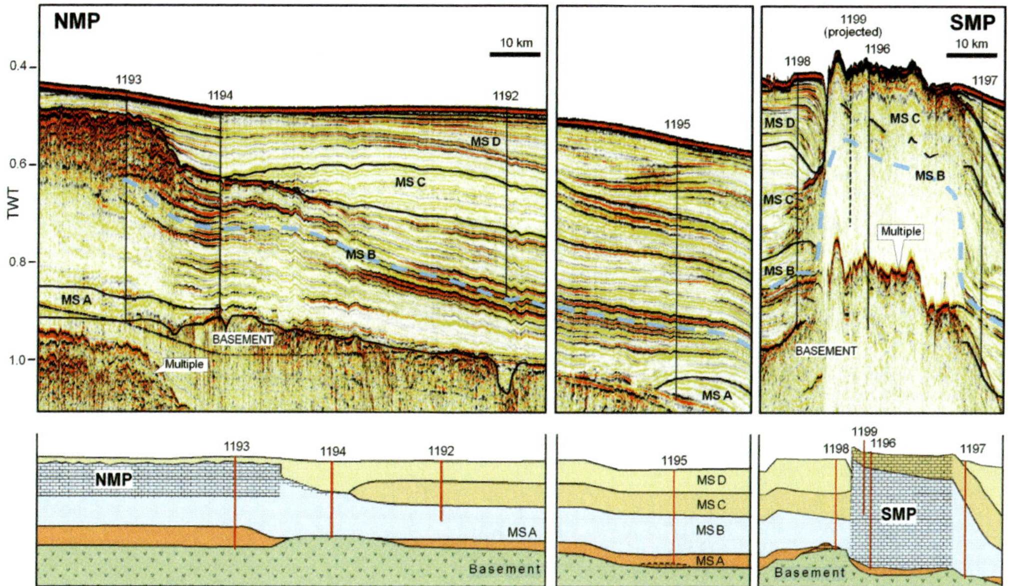

Fig. 3. Seismic lines and corresponding schematic cross-section through ODP Leg 194 sites, showing geometry of seismic megasequences (modified from Isern *et al.* 2002). Section locations are highlighted in red on Figure 2. Vertical scale on the seismic lines is two-way travel time (s). The brick pattern on the lower panel represents in-place neritic carbonate platforms. Site 1199 is located off the seismic line, 5 km NE of Site 1196.

The data set consists of porosity and permeability analyses of 407 horizontal 1 inch-plugs and 23 whole-core samples. A selection of 140 plugs and all 23 whole cores was examined petrographically to characterize facies, diagenesis and porosity types. These data are considered in the context of core descriptions and mineralogical analyses provided by the onboard laboratories of the ODP drillship *Joides Resolution* (Isern *et al.* 2002), as well as seismic images provided by ODP. Owing to technical problems, wireline logs were recorded only for sites 1196 and 1199, and some of the 1196 logging results were adversely affected by large borehole diameter ('washout').

A limitation of the present study is the widely varying sample spacing over the depth range of the platform sections from each site. Plug sampling was only possible within intervals with coherent core pieces of suitable size, and core recovery was poor over large intervals of the platform strata.

Analytical techniques

Porosity was measured by helium injection using a Boyle's law porosimeter, with bulk volume determined by caliper. For a small proportion of the samples, where porosity was suspected to have been overestimated because of irregularities in plug shape, porosity values were corrected based on measurement of plug volume by mercury displacement. Permeability values reported are Klinkenberg-corrected nitrogen permeabilities measured with a confining pressure of 20 bar. Permeability values reported as >50 D have been plotted as 50 D in the figures of this paper. Analyses were performed by Reservoir Laboratories AS, Stavanger. Full details of the analytical method are provided in Ehrenberg *et al.* (2003), together with complete tables of porosity, permeability and grain density data, and representative photomicrographs of thin sections from horizontal plugs and whole-core samples. Cathodoluminescence microscopy was conducted on a cold-cathode instrument constructed at Cambridge University, using accelerating potential of 26 kV, gun current of 600 mA, focused beam diameter of 1–10 mm and air chamber pressure of 0.01–0.05 torr.

Geological setting

The Marion Plateau is located off the coast of NE Australia (Figs 1 and 2), partly within the Great Barrier Reef Marine Park. Eight sites were drilled (1192–1199), each penetrating 500–700 m of sediment from water depths of 304 to 419 m. The sites were located along seismic lines

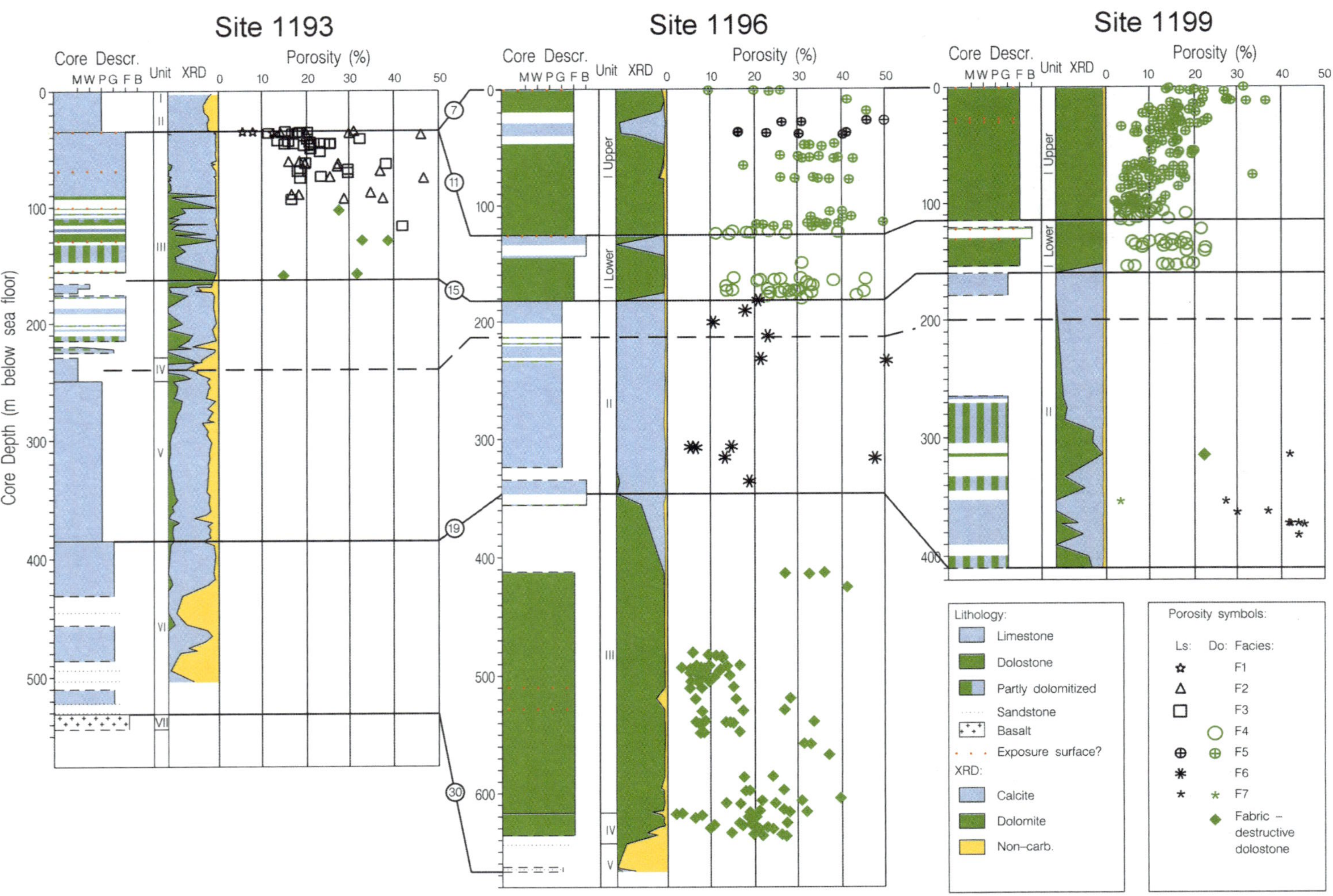
Site 1193
Site 1196
Site 1199
Core Descr.
M W P G F B
Unit
XRD
Porosity (%)
Core Depth (m below sea floor)
I Upper
I Lower
Lithology:
Limestone
Dolostone
Partly dolomitized
Sandstone
Basalt
Exposure surface?
XRD:
Calcite
Dolomite
Non-carb.
Porosity symbols:
Ls: Do: Facies:
F1
F2
F3
F4
F5
F6
F7
Fabric – destructive dolostone

(Figs 2 and 3) crossing the margins of two carbonate platforms, the Northern Marion Platform (NMP) and Southern Marion Platform (SMP).

Prior to drilling, seismic analysis was used to divide the Marion Plateau sediments into four stratal units, termed Megasequences A–D (Isern *et al.* 2002) (Fig. 3). These strata overlie an irregular basement topography, which Leg 194 drilling found to consist of variably altered basaltic volcanic rocks, probably formed during Late Jurassic–Early Cretaceous rifting in the Coral Sea region. Following rifting, the Marion Plateau gradually subsided on a passive continental margin, with initial marine transgression occurring in early Oligocene time (Ehrenberg *et al.* 2004). Carbonate platform growth took place throughout Miocene time, as palaeolatitude decreased from roughly 32° to 24°S (presently *c.* 20°S; Davies *et al.* 1989).

The steep-sided asymmetric platforms of the Marion Plateau have geometries similar to rimmed tropical platforms built by 'photozoan' assemblages (James 1997) dominated by coral reefs, green algae and ooids, such as in the Bahamas. However, the biotic assemblages of the Marion platforms were found to consist mainly of red algae, bryozoans, molluscs and benthic foraminifera, with corals and green algae being subordinate in nearly all cored intervals and commonly minor or absent. Ooids have not been observed. These assemblages thus constitute a dominantly 'heterozoan' association suggestive of cool subtropical water temperatures. Rather than reefal facies, early cemented coarse bioclastic debris form the main substance and support for the flat-topped Marion platforms. Strong influence of oceanic currents is inferred during Miocene platform growth and continuing to the present day, based on asymmetric slope geometries, sculptured drift deposits and current-swept phosphatic hardgrounds (Glenn & Kronen 1993; Isern *et al.* 2002).

Stratigraphy

The stratigraphy of the Marion platforms is described here in terms of the seismic 'megasequences' noted above and the 'lithological units' defined during shipboard core description. The megasequences are designated A–D, from oldest to youngest (Fig. 3), whereas the lithological units are numbered in the opposite sense, top downwards, using Roman numerals, according to standard ODP convention (Fig. 4). The lithological unit designations are specific to each site, such that identical unit numbers are not intended to indicate stratigraphic correlation of units between sites. An exception to this, however, is that the unit numbers do correspond to correlative units in sites 1196 and 1199, due to their proximity (5 km).

Seismic geometries indicate that the Marion platforms correspond with Megasequences B and C (Fig. 3), which were defined by seismic stratigraphy in the off-platform slope and drift sediments. Biostratigraphic study of Leg 194 cores from the off-platform sites showed that Megasequences B and C are of early–late Miocene age (*c.* 20–7 Ma). However, correlation from the seismically distinct off-platform sites into the more seismically transparent platform sites is associated with major uncertainties because the platform internal structure is poorly imaged seismically (Fig. 3) and biostratigraphic controls for the neritic biota of the platform sites tend to be much poorer than for the deeper-water biota of the off-platform sites.

Northern Marion Platform

The NMP at Site 1193 consists of 194 m of predominantly coarse bioclastic limestone and subordinate dolostone (lithological Unit III). This platform interval is overlain by 35 m of

Fig. 4. Summary of core descriptions, bulk-rock XRD analyses and plug-porosity profiles from the three Leg 194 sites studied. Core descriptions (left column) show depositional texture indicated by distance to right of depth scale (M, mudstone; W, wackestone; P, packstone; G, grainstone; F, floatstone and rudstone; B, boundstone). Lithology (limestone, dolostone, etc.) is indicated by colour and pattern fill. Owing to variable core recovery, there are many gaps in the information. Intervals containing cores with the same depositional texture have been joined by continuous vertical lines indicating constant texture, while changes in texture coinciding with core gaps are delimited by horizontal dashed lines. Similarly, intervals containing cores with the same lithology have been joined by continuous colour fill indicating constant lithology, while changes in lithology coinciding with core gaps are delimited by gaps in colour fill. The 'Unit' column indicates the vertical extent of lithostratigraphic units defined during shipboard core description. Bulk-rock XRD analyses (centre column; shipboard data reported in Isern *et al.* 2002) show percentages of calcite, dolomite and non-carbonate minerals. Plug-porosity profiles (right column) use plotting-symbols to identify facies and lithology (black, limestone; green, dolostone). Arrows on the upper right of sites 1196 and 1199 indicate shoaling cycles comprising Upper Unit I and the Lower Unit I of the SMP. Correlation lines connect isochronous surfaces based on dating constraints described in Ehrenberg *et al.* (2004). Circled numbers give the approximate age of each correlation horizon in Ma.

unlithified hemipelagic ooze (dated as <6 Ma and corresponding with Megasequence D). The NMP is underlain by 302 m of finer-grained lithologies, the upper portion of which corresponds with seismically imaged clinoforms, possibly representing the slope over which the NMP prograded eastwards (Fig. 3). Plug samples could be drilled only from the upper part of Unit III, and their frequency decreases markedly downwards (Fig. 4A).

Shipboard biostratigraphy (Isern *et al.* 2002) and Sr-isotope analyses of bioclasts (Ehrenberg *et al.* 2004) date the NMP (Unit III) as having been deposited from 16 to 11 Ma. This is consistent with seismic correlation between the rugged top of the NMP and the top of Megasequence B in the off-platform sites, biostratigraphically dated as 10.6–11.5 Ma (Fig. 3). The top surface of Unit III is capped by a thin layer (probably 10–50 cm thick; recovered only as a few core pieces <10 cm long) of tightly cemented, phosphatic, planktonic-and-benthic-foraminifer wackestone–packstone (facies F1 described below). Sr-isotope analysis gives a date of 9.6 Ma for this mud-rich cap, indicating deposition shortly after NMP demise, apparently during transgression of the karsted platform top (Ehrenberg *et al.* 2004).

Interpretation of the top of Unit III as a karst surface is based on the irregular seismic topography of the NMP top, the hiatus between the 11 Ma age of the platform-top and the 5.6 Ma date for the overlying hemipelagic ooze, and the identification of contemporaneous shallow-water deposits (lowstand wedge) downslope in Site 1194 (Isern *et al.* 2002).

Based on sedimentological core description, the 302 m-interval of finer-grained lithologies underlying the NMP (1193 Unit III) was divided into three lithological units (Isern *et al.* 2002):

- Unit IV – an upper 20 m of argillaceous mudstone, interpreted as distal periplatform deposits with palaeowater depth >120 m (biostratigraphically dated as 16 Ma).
- Unit V – 136 m of fine-grained packstone, interpreted as proximal periplatform (slope) deposits with palaeowater depth >100 m (biostratigraphically dated as 16–18 Ma).
- Unit VI – a basal 146 m of phosphatic grainstone with subordinate sandstone beds, interpreted as representing a high-energy, shallow-shelf setting. Biostratigraphy gives ages of 18–23 Ma for the lower part of this unit, whereas Sr-isotope dating gives 24–29 Ma.

Below this, the deepest 1193 cores (531–544 m below sea floor (BSF)) recovered 5.6 m of dark red, highly altered basaltic breccia (top of acoustic basement).

Southern Marion Platform

Stratigraphic correlations between the NMP and SMP based on available dating constraints (Ehrenberg *et al.* 2004) are shown in Figure 4. Based on sedimentological core description and wireline-log interpretation, the SMP section was divided into six units (Isern *et al.* 2002), numbered from the top downwards:

- Upper Unit I – 126–114 m (at sites 1196 and 1199, respectively) of coarse grainstone to floatstone with a grainstone–packstone matrix. Most of this interval is replaced by partly fabric-preserving dolomite, except for two limestone intervals in Site 1196: the capping 20–40 cm and a zone at least 12 m thick (possibly as thick as 28 m, however, due to uncertainty arising from core gaps; Fig. 4). No limestone is present in Site 1199. The biota is dominantly red algae, including both branched fragments and large rhodoliths, and larger benthic foraminifera. Upper Unit I is interpreted as a shoaling-upwards cycle, based on upward increase in rhodolith abundance and greater occurrence of coral and *Halimeda* in the upper 20 m. Upward-decreasing gamma ray activity through Upper Unit I (reflecting uranium content) may also be an indication of this shoaling trend. Sr-isotope dating of bioclasts gives ages of 6.5–6.9 Ma for the thin limestone cap at the sea floor and 6.8–11.2 Ma for the second limestone bed in Upper Unit I.
- Lower Unit I – 56–46 m (at sites 1196 and 1199, respectively) of coarse bioclastic carbonate dominated by red algae and larger foraminifera. The top 8–10 m of this unit contains beds at least a few tens of centimetres thick of coral-rich boundstone–floatstone that were poorly recovered in coring and are not represented by plug analyses. Like Upper Unit I, Lower Unit I is interpreted as a single shoaling-upward cycle, based on an upward increase in rhodoliths, coral and *Halimeda.* Furthermore, a karst surface is inferred capping Lower Unit I based on textures (reddish silt filling dissolution vugs) and from the presence of cavities reported from wireline responses in Site 1199. Most of Lower Unit I is replaced by partially fabric-preserving dolomite, except for the top

18 m in hole 1196A. Sr-isotope dating of bioclasts gives ages of 11.1–11.8 Ma for this limestone. These dates support interpretation of the inferred karst surface capping Lower Unit I as correlative with the top of the NMP and seismic Megasequence B. Sr-isotope dating constraints further indicate correlation of SMP Lower Unit I with the upper *c.* 160 m of the NMP (Fig. 4).

- Unit II – 164–250 m (at sites 1196 and 1199 respectively) of mostly undolomitized, dominantly fine–medium-grained grainstone rich in miliolid and Lepidocyclinid foraminifers. These facies are interpreted as representing a sand-shoal setting with water depths much shallower (<30 m and probably mostly <10 m) than the overlying and underlying units (mostly 60–120 m). Sr-isotope dating of bioclasts gives ages of 15–19 Ma for Unit II, indicating age-equivalence with the lower 60 m of the NMP, together with underlying NMP Units IV and V (Fig. 4).
- Unit III – 271 m of coarse dolostone with little preservation of primary texture except that ghosts of rhodoliths and moulds of large foraminifers and shells indicate a coarse bioclastic facies similar to Unit I.
- Unit IV – at least 19 m (maximum 26 m; actual thickness uncertain because of core gap) of strongly layered dolostone variably rich in quartz and phosphatic grains. Dating constraints from surrounding beds indicate correlation of SMP units III and IV with NMP Unit VI (Fig. 4).
- Unit V – the deepest 1196 cores (636–672 m BSF) recovered phosphatic sandstone and shale.

Depositional facies

The NMP and SMP plug samples may be grouped into six broad depositional facies categories, except where original textures have been obscured by dolomitization. The distribution of these facies (Fig. 4) illustrates the striking contrast in biota between the bryozoan-rich NMP and the red-algae-rich SMP. Several additional facies are present in the cores, but are not listed here because they are not represented by porosity–permeability analyses.

F1 – Planktonic-and-large-benthic-foraminifer wackestone–packstone

This very special lithology occurs only as a thin layer (probably 10–50 cm thick; recovered only as a few core pieces <10 cm long) capping the NMP at Site 1193. Both planktonic and benthic foraminifera commonly have dark brown material filling chambers, suggesting oxidation or phosphatization. Planktonic tests typically have geopetal mud fill followed by fibrous and then blocky calcite cements. The three plugs assigned to F1 have some of the lowest porosity and permeability values in the data set, due to abundant low-porosity mud and subordinate blocky calcite cement filling most intergranular space (Fig. 5E).

F2 – Bryozoan-large-benthic-foraminifer grainstone–packstone

Facies F2 (Fig. 5A) and F3 (Fig. 5C–D) together compose most of the NMP section of Site 1193. Both are dominated by ramose and bifoliate bryozoans and large benthic foraminifers (*Lepidocyclina*, *Amphistegina*, *Operculina*, *Miogypsina*, *Cycloclypeus*). Subordinate components are molluscs, echinoderms, red algae and *Halimeda*. Coral is very rare (one sample). Packstones assigned to F2 have subequal amounts of mud-free and mud-filled intergranular volume, whereas packstones assigned to F3 have either subordinate or zero mud-free shelter porosity. Facies F2 and F3 also include rudstone and floatstone (Embry & Klovan 1971), with many samples straddling the defining criteria of more than 10% of grains larger than 2 mm.

F3 – Bryozoan-large-benthic-foraminifer wackestone–packstone

The F3 biota is apparently similar to that of F2. The essential difference is the higher carbonate mud content of F3. Throughout much of the NMP core, mud-rich (F3) and mud-poor (F2) lithologies are interlayered on a scale of several tens of centimetres to a few metres, and both total (plug) porosity and macroporosity visible on the core surface are strongly affected by this layered variation in mud content. Facies F2 and F3 are interpreted as representing water depths <100 m (Isern *et al.* 2002).

F4 – Red-algae-large-benthic-foraminifer grainstone–packstone

This facies also includes rudstone and floatstone, with many samples straddling the defining criteria of more than 10% of grains larger than 2 mm. The distinction between facies F4 and F5 is based on whether or not rhodoliths (Martín *et al.* 1993) are a conspicuous fabric element visible on the slabbed core surface; thus, all samples in SMP Unit I have

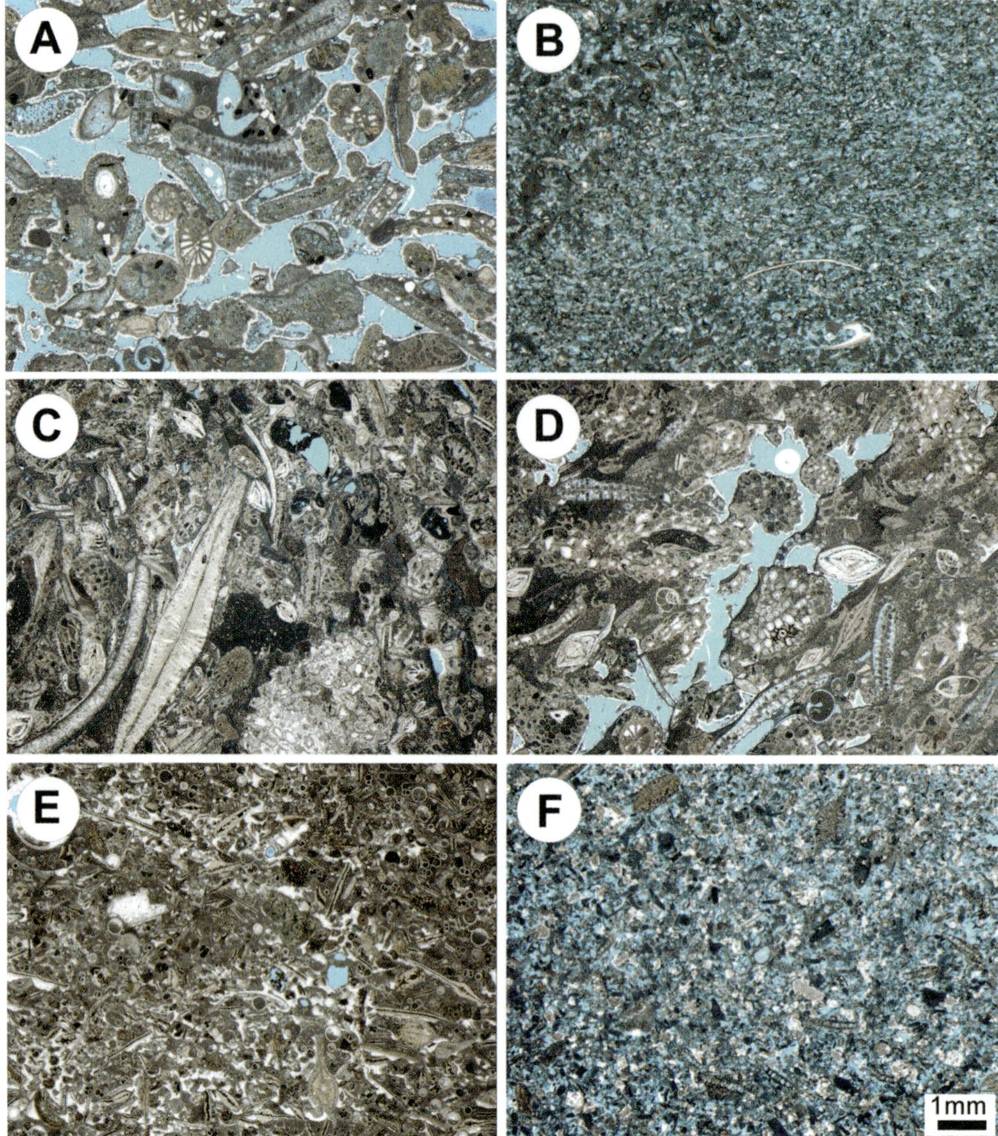

Fig. 5. Photomicrographs of limestones showing examples of facies and texture categories defined in the text. Porosity has been impregnated by blue-dyed epoxy. All views have the same magnification. (**A**) Coarse-grained grainstone; facies F2. Plug EHRE024 from 68.26 m BSF in Site 1193 (37% porosity, 28 D permeability). (**B**) Fine-grained grainstone; facies F6. Plug EHRE115 from 317.43 m BSF in Site 1196 (48% porosity, 224 mD permeability). (**C**) Packstone–wackestone without connecting vugs; facies F3. Plug EHRE025 from 68.41 m BSF in Site 1193 (29% porosity, 7 mD permeability). (**D**) Packstone with connecting vugs; facies F3. Plug EHRE019 from 46.48 m BSF in Site 1193 (23% porosity, 17 D permeability). (**E**) Packstone–wackestone without connecting vugs; facies F1. Plug EHRE008 from 35.27 m BSF in Site 1193 (8% porosity, 0.021 mD permeability). (**F**) Fine-grained grainstone; facies F7. Plug EHRE412 from 372.01 m BSF in Site 1199 (46% porosity, 526 mD permeability).

been assigned to facies F4 and F5 regardless of whether thin sections are available. Most F4 and F5 samples have been replaced by partly-fabric-preserving dolomite (Fig. 4), but original mud content can commonly be interpreted, based on the texture of dolomite crystals between relict grains (Fig. 6A–D), indicating that original textures vary from grainstone to packstone. The biota consists mainly of red algae (both solitary branches and less abundant encrusting forms) and larger benthic foraminifera. Subordinate components are molluscs, echinoderms, coral and *Halimeda*, although bioclast identities are commonly obscured by dolomitization.

F5 – Rhodolith-rich red-algae-large-benthic-foraminifer rudstone–floatstone

The biota is apparently the same as F4, except that rhodoliths are a major fabric element, with consequent importance of encrusting red algal forms. The rhodoliths vary widely in size (<1–6 cm), sorting, shape, abundance, and composition of both laminations and core. The matrix is poorly sorted, coarse-grained grainstone–packstone, ostensibly the same as F4. Rhodolith internal cavities commonly have geopetal mud fillings oriented parallel with bedding. Facies F4 and F5 are thought to represent water depths of 60–120 m (Isern *et al.* 2002).

F6 – Small-benthic-foraminifer-red-algae grainstone

Large parts of SMP Unit II of Site 1196 consist of fine–medium-grained grainstone rich in miliolid and soritid foraminifers, red algae, molluscs and micritized bioclasts, together with alveolinids and other larger foraminifera (Fig. 5B). Large (1–3 cm) whole gastropods and bivalves are common, while similarly large bryozoans, solitary corals and rhodoliths are locally present. In portions of the 1196 section, this facies contains centimetre-size fragments of organic matter suspected to be remnants of sea grass. Facies F6 is thought to represent a shallow-water (<30 m; probably <10 m) sand-shoal setting.

F7 – Lepidocyclina*-red-algae grainstone*

In Site 1199, the lower part of Unit II contains a distinctive assemblage of *Lepidocyclina*, branching red algae fragments, mud peloids and minor miliolids, and occurs as very-fine–coarse-grained grainstone (Fig. 5F). Facies F7 probably represents a setting similar to F6, but at somewhat greater water depths.

Diagenesis

The main diagenetic processes of likely importance for porosity and permeability are dissolution of bioclasts, calcite cementation and dolomitization. A fourth process, that can arguably be listed as diagenetic, is the post-depositional emplacement of carbonate mud into a coarse bioclastic framework, probably by downward infiltration of mud-laden water. Late mud emplacement, which is recognizable by geopetal textural relationships, is common in samples from the upper part of the NMP section and in rhodoliths of SMP Unit I.

Dissolution

Bioclasts of original aragonitic composition, including corals, *Halimeda* and most mollusc shells, have been completely dissolved throughout the intervals examined. Bifoliate bryozoans also show partial to complete dissolution. Other bioclasts, including foraminifers, ramose bryozoans, red algae and some bivalves, tend to have pristine appearance in the limestones, but are commonly represented by moulds in the dolostones, especially those showing greater degrees of fabric destruction.

Calcite cementation

Three sequential stages of calcite cement, each having characteristic morphology, are present in the limestones (Fig. 7A & C): (1) a thin (generally <10–20 μm; rarely up to 100 μm) isopachous coating of fibrous, inclusion-rich calcite; (2) a thicker (generally 30–100 μm) coating of acicular–prismatic ('dog-tooth') spar, which varies from isopachous to highly variable thickness; and (3) partial to complete filling of remaining space by fine to coarse mosaic spar. Different samples show widely varying proportions of these three cement types, as well as wide variations in total cement abundance.

Point counts of 43 limestone thin sections show that contents of mud and cement have reciprocal maximum values (Fig. 8). However, poor overall correlation between mud and cement reflects widely varying degrees to which intergranular and intragranular space has been filled, as well as variability in the total space available for filling.

A survey of cathodoluminescence (CL) properties of the calcite cements reveals that cements throughout the sampled (upper) portion of the NMP are characterized by strong bright–dark CL zoning (Fig. 7A & B), whereas morphologically similar cements from Upper Unit I of the SMP are entirely non-luminescent (Fig. 7C & D). Calcite cements from Unit II of the SMP (with dominantly fine–coarse blocky textures) are also non-luminescent or show only faint CL zoning (Fig. 7E & F). A possible

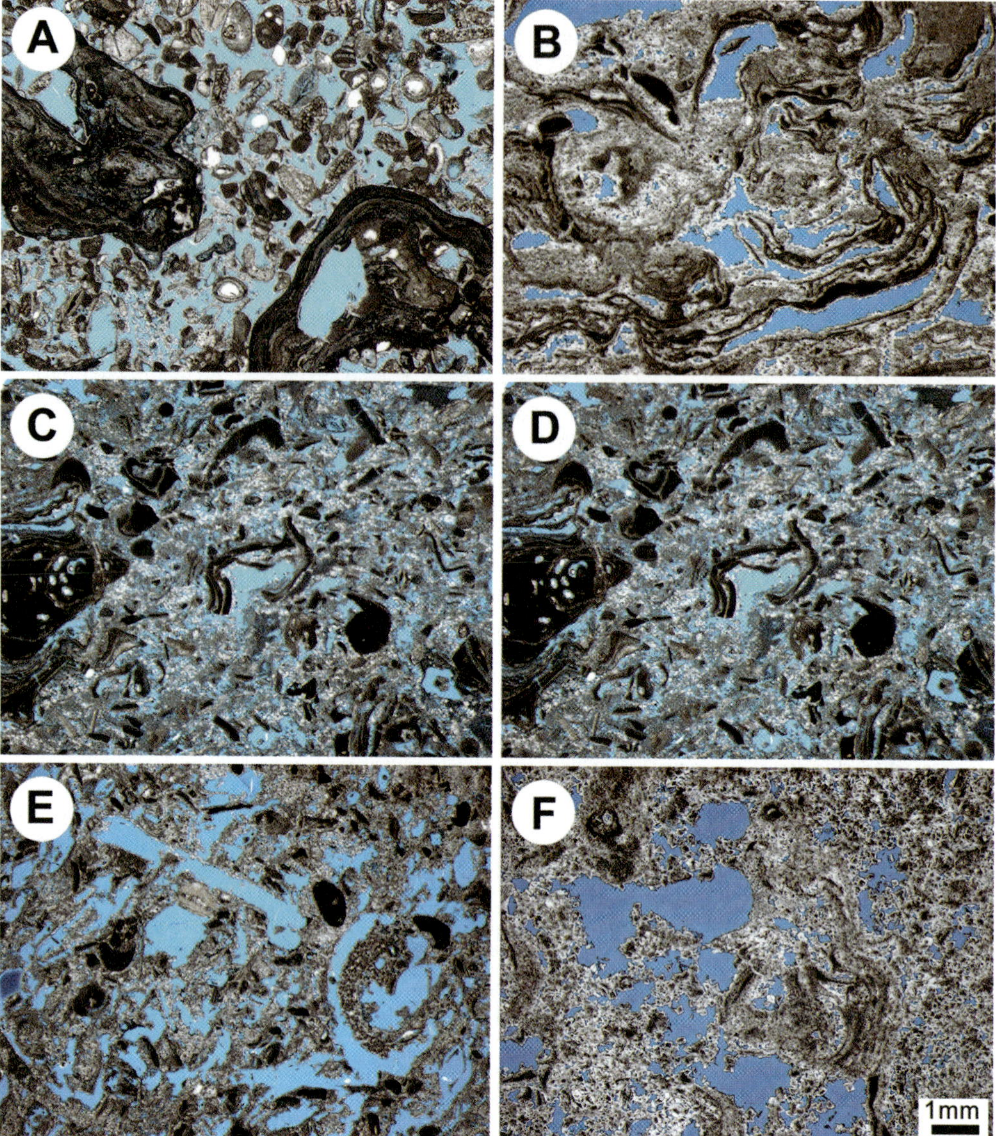

Fig. 6. Photomicrographs of dolostones showing examples of facies and texture categories defined in text. Porosity has been impregnated by blue-dyed epoxy. All views have the same magnification. (**A**) Intergranular-dominated texture; fabric-preserving dolostone of facies F5. Whole-core sample EHWR17 from 11.76 m BSF in Site 1199 (29% porosity, 3.6 D permeability). (**B**) Rhodolith-dominated texture; partially fabric-preserving dolostone of facies F5. Plug EHRE352 from 93.56 m BSF in Site 1199 (14% porosity, 3468 mD permeability). (**C**) Intercrystalline-dominated texture; partially fabric-preserving dolostone of facies F5. Plug EHRE072 from 58.38 m BSF in Site 1196 (33% porosity, 1.4 D permeability). (**D**) Intercrystalline-dominated texture in a fabric-destructive dolostone. Whole-core sample EHWR10 from 529.03 m BSF in Site 1196 (18% porosity, 3.6 D permeability). (**E**) Vug-dominated texture; partially fabric-preserving dolostone of facies F5. Plug EHRE245 from 0.15 m BSF in Site 1199 (32% porosity, 31 D permeability). (**F**) Vug-dominated texture in a fabric-destructive dolostone. Whole-core sample EHWR8 from 482.49 m BSF in Site 1196 (10% porosity, 5.9 mD permeability).

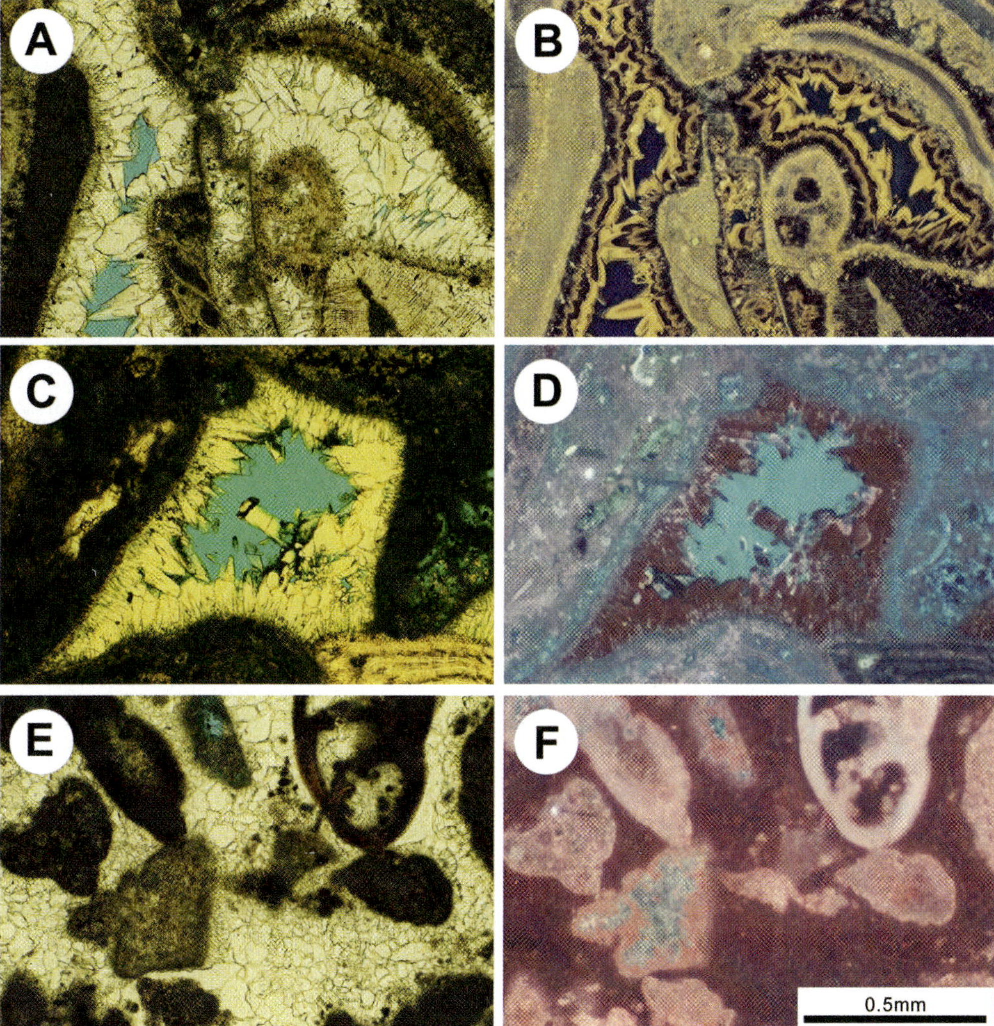

Fig. 7. Photomicrographs of three limestones in plane light (left) and cathodoluminescence (right). Luminescence intensities shown are proportional to both intrinsic luminescence intensity and camera exposure time, which is set automatically to acquire a standard overall light intensity. (**A**) and (**B**) Plug EHRE045 from 36.22 m BSF in Site 1193. CL exposure time: 20 s. An earliest isopachous coating of fibrous–very-finely-granular cement (10–20 µm thick) has bright luminescence. This probable neomorphosed high-Mg calcite marine-cement zone is absent within moulds of bioclasts (X). The isopachous marine cement is followed by oscillatory zoned acicular–blocky calcite cement having two main non-luminescent zones, each followed by a bright-luminescent zone. Interestingly, the boundary between fine-acicular and blocky mosaic cement-crystal morphology does not coincide in detail with the boundaries of the CL zones. Thus, the outer boundary of the earliest non-luminescent zone occurs within the fine-acicular cement in most places, but occurs elsewhere within the early part of the blocky mosaic cement, indicating that the outward change in cement morphology was not synchronous. (**C**) and (**D**) Plug EHRE061 from 29.49 m BSF in Site 1196. CL exposure time: 75 s. This sample displays a sequential development of fibrous-isopachous, acicular and blocky calcite-cement morphologies similar to (A), but all cement stages are non-luminescent. (**E**) and (**F**) Plug EHRE112 from 308.31 m BSF in Site 1196. CL exposure time: 100 s. Calcite cements in this sample are also entirely non-luminescent.

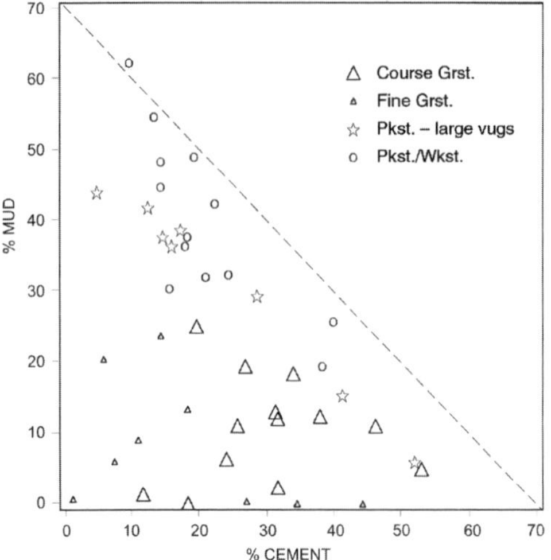

Fig. 8. Mud v. cement content of limestones (determined by point-counting thin sections; 300 points per slide). Maximum cement content has negative correlation with maximum mud content, suggesting that maximum total intergranular plus intragranular volume available for mud + cement infilling is around 60–70% of bulk volume in these strata. Many samples, however, have less than this apparent upper limit.

explanation for this contrast is that the greater isolation of the SMP from terrestrial detritus resulted in cement growth in an environment that was relatively depleted in the trace elements (mainly Mn) necessary to activate luminescence. The NMP, being more closely connected to the mainland, may have had a greater supply of Mn-rich terrestrial detritus. This hypothesis is supported by profiles of bulk chemical analyses (Ehrenberg *et al.* 2004), showing systematically higher contents of siliciclastic-associated components (alumina, silica, Ti, etc.) in the NMP samples.

Dolomitization

As is commonly the case elsewhere (Braithwaite 1991; Budd 1997) compelling evidence for a particular model of dolomitization is elusive for the Marion platforms. Nevertheless, various considerations are at least strongly suggestive that dolomitization occurred by circulation of normal or slightly modified seawater:

- The continent-margin setting of the two platforms and the semi-isolated setting of the more dolomitized SMP makes significant involvement of landward-derived waters unlikely.
- The absence of calcium sulphate minerals makes models of dolomitization by evaporative concentration and seepage reflux inappropriate.
- Strontium-isotope analyses of dolostones from the Marion and Queensland Plateaus show values matching the trend of global seawater composition for Miocene time (McKenzie *et al.* 1993; Ehrenberg *et al.* 2004), supporting the direct involvement of seawater as the dolomitizing agent.
- Trends of pore-water geochemistry v. depth immediately adjacent to the SMP indicate large-scale circulation of present-day seawater to depths of 200–300 m BSF, probably involving flow through the SMP edifice (Isern *et al.* 2002). Thus, the large-scale seawater flow necessary for extensive platform dolomitization is not only possible, but actually exists today.

A striking feature of the present data set is that most samples are either limestone or dolostone. Paucity of partly dolomitized rock is indicated by both petrographic estimation of dolomite content in thin sections (Fig. 9A) and shipboard bulk-rock X-ray diffraction (XRD) data (Fig. 9B). Apparently dolomitization, of the Marion platforms, tended to run to completion once it had started. According to Sibley *et al.* (1994; pers. comm. 2003), this characteristic follows from the kinetics of dolomite crystal growth, whereby a long 'induction period' of reaction with few observable products precedes a period of rapid reaction. A minor exception to the above generality is the common occurrence of brownish (Fe-rich?) fine-mosaic dolomite filling bryozoan zooecia in NMP facies F2 and F3. This chemical microenvironment apparently promoted early dolomitization of mud infillings, but reaction commonly did not continue into the enclosing skeletal material or intergranular matrix.

The relative abundance and patterns of occurrence of dolostones and limestones are very different in the NMP and SMP. In the NMP section (Site 1193), dolostones are subordinate to limestones and occur within several distinct layers (Fig. 4A). In contrast, the SMP is extensively dolomitized in the locations drilled, with limestone preserved only within limited intervals (Fig. 4B & C). The different styles of dolomite distribution in the two platforms could reflect the more seaward, isolated setting of the SMP, resulting in greater susceptiblity to large-scale seawater circulation driven by oceanic

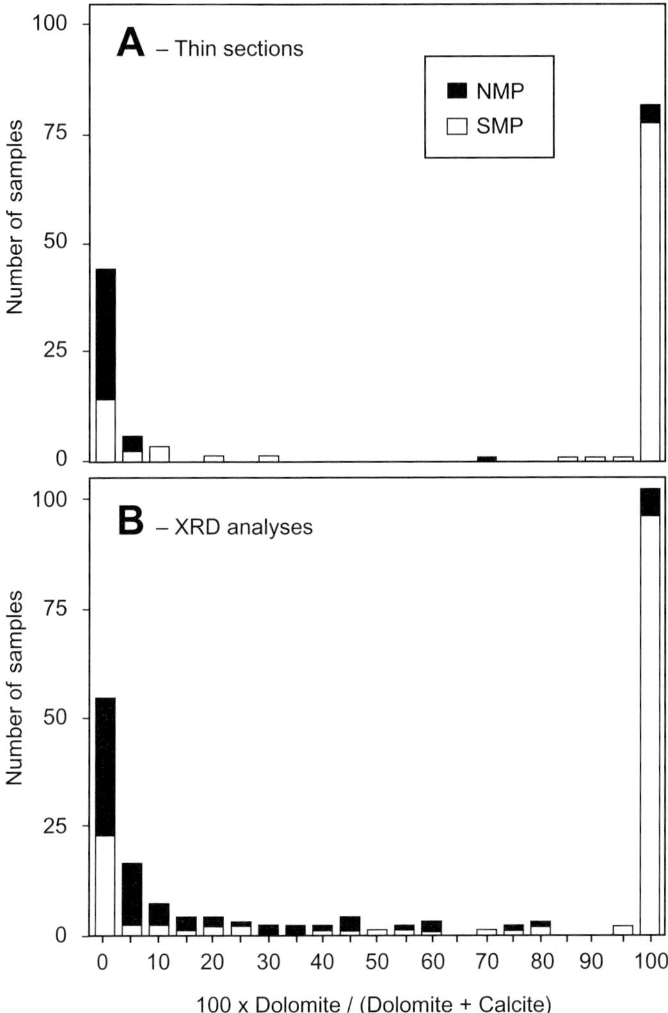

Fig. 9. Frequency distribution of limestones and dolostones in the northern and southern Marion platforms, as illustrated by 'dolomitization index' (DI) = 100 × % dolomite/(dolomite + % calcite). DI has been determined by (**A**) visual estimation from thin-sections and (**B**) shipboard bulk-rock X-ray diffraction analyses (Isern *et al.* 2002).

currents and cycles of relative sea-level fluctuation. In the NMP, concentration of dolomitization within a few distinct layers only below the top 50 m of platform section could reflect displacement of marine water by a meteoric-water lens at the top of the NMP soon after its deposition, combined with focused flow of modified seawater along higher-permeability, aquitard-bounded layers beneath this lens.

The dolostones are broadly divisible into two main groups depending on the degree of fabric preservation. The few dolostone plug samples from the NMP all have fabric-destructive texture, with relatively fine average crystal size (50–100 μm; Fig. 6E). In addition, the NMP contains several partially dolomitized intervals (represented by only one sample in the present data set), where undolomitized bioclasts occur in a dolomitized mud matrix. In the SMP, the degree of fabric destruction and also average dolomite crystal size increases with depth. Within Upper Unit I primary fabric tends to be well preserved, with bioclasts (red algae, foraminifera, *Halimeda* and echinoderm fragments) commonly mimetically replaced (Fig. 6A–D). Nevertheless, primary fabric is typically obscured to varying degrees. For example, it is commonly difficult to judge the degree to which intergranular dolomite matrix represents replacement of mud as opposed to cement growth. Within Lower Unit I, the primary fabric of the dolostones is recognizable, but is more obscured by recrystallization than in the upper cycle. Average crystal size of the intergranular matrix is generally 50–100 μm in Unit I. Deeper levels of the SMP (units III and IV) mostly have almost completely fabric-destructive dolomitization and coarse dolomite crystal size (200–400 μm on average; Fig. 6F). As commonly noted elsewhere (Budd 1997), red algae are the bioclasts most resistant to fabric destruction during dolomitization, but even red algae and rhodoliths have been partly to almost completely obliterated in most samples of units III and IV.

Similar relationships of increasing fabric destruction with depth and age have been observed in cores from the Little Bahamas Bank, where the boundary between middle and late Miocene also marks a major change in the intensity of dolomitization (Vahrenkamp & Swart 1994). This pattern may indicate that deeper, older levels have experienced more dolomitizing events, each possibly associated with a main cycle of sea-level rise and fall, and each increasing the extent of recrystallization.

Before being dolomitized, the dolostones presumably experienced the same processes of bioclast dissolution and early calcite cementation as the limestones. Many dolostones, however, appear to have experienced dissolution of calcitic bioclasts and mud matrix, to a more extensive degree than is observed in the limestones. This additional dissolution must have accompanied or followed dolomitization. Evidence for this additional dissolution includes the presence of moulds after larger benthic foraminifers (which do not show signs of dissolution in the limestones) and other unidentified bioclasts with abundances distinctly greater than are observed in the limestone thin sections, as well as partial dissolution of red algal fragments (which also do not show signs of dissolution in the limestones). Another indication of dissolution is the highly porous sucrosic dolomite texture of many dolostones that may reflect dissolution of remaining calcitic matrix and bioclasts following partial dolomitization (Murray 1960; Dawans & Swart 1988; Swart & Melim 2000).

A survey of the CL properties of representative dolostones reveals a systematic overall difference between the two platforms similar to that observed for the calcite cements. Dolomite of the NMP displays complex bright–dark oscillatory CL zoning, in which much of the total crystal volume is brightly luminescent, except for a thin outermost non-luminescent zone (Fig. 10A & B). Dolostones of the SMP also display bright–dark oscillatory CL zoning, but the bright zones are mainly restricted to the earliest growth stages, and there is a strong volumetric predominance of the non-luminescent zones (Fig. 10C–F). As proposed above to explain the contrast in calcite-cement luminescence between the NMP and SMP, the dominance of non-luminescent dolomite cements could reflect the lower content of land-derived fine siliciclastic detritus compared with the NMP.

Porosity and permeability

Limestones

Figure 11A shows the limestone data with plotting symbols identifying the main stratigraphic units and facies. Overall, the limestones show weak porosity–permeability correlation. Certain general differences are apparent between stratigraphic units, notably that the NMP limestones tend to have higher permeabilities for given porosity compared with the SMP limestones.

Because biotic assemblages, which are a main basis for defining depositional facies, may not be directly relevant for understanding porosity–permeability relationships, a separate limestone textural classification was devised where the main consideration is the nature of the pore system. The dolostones are covered by a different set of textural categories, described in the following section. After textural classification of the samples with thin sections, all remaining limestone plugs were assigned a texture based on macroscopic examination. Although some samples are transitional, most are readily assignable to one of the four textural categories described below.

Coarse-grained grainstone. These samples include both mud-free grainstones and mud-lean packstones. The defining characteristic is that the pore system is dominated by intergranular pores. Such rocks can be divided into two distinct grain size groups. Those with grain size >1 mm tend to have high permeability for given porosity (Fig. 5A).

Fine-grained grainstone. Grainstones with average grain size <1 mm are mostly from Unit II of the SMP, corresponding to facies F6 and

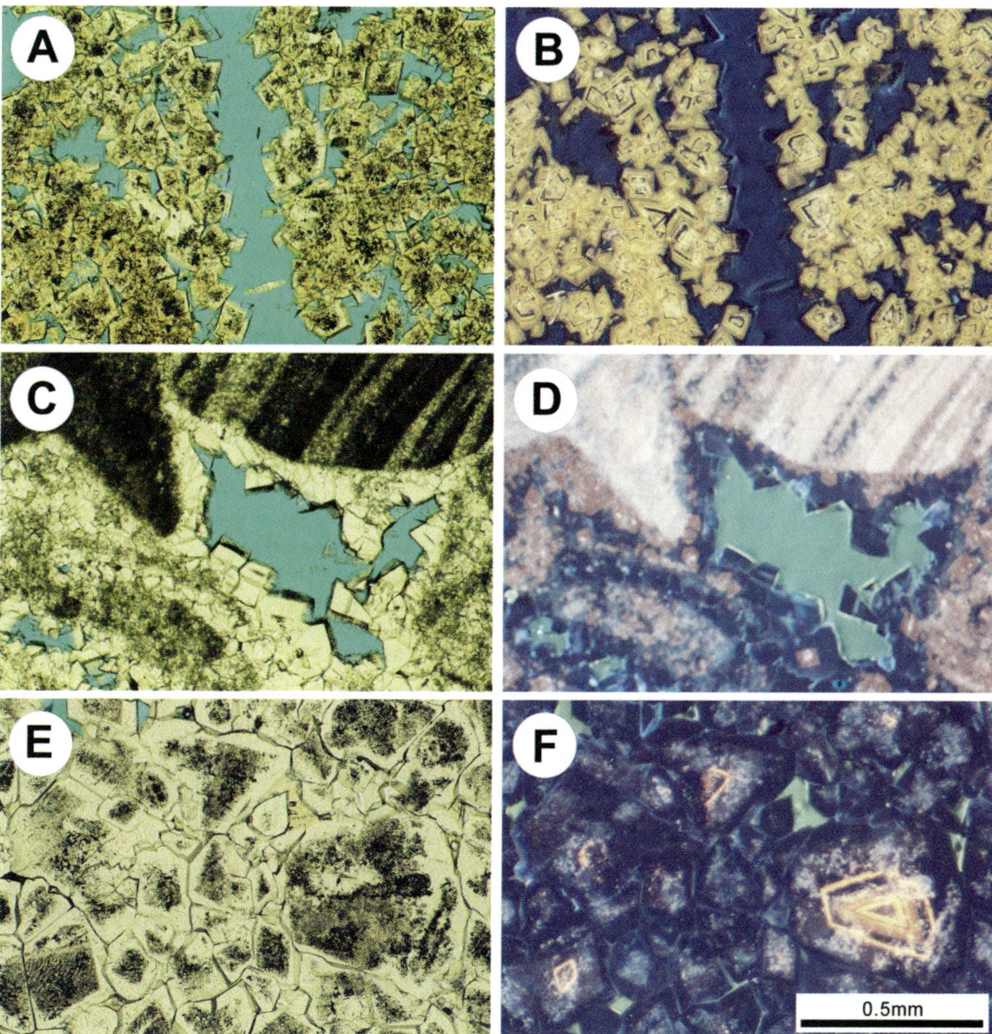

Fig. 10. Photomicrographs of three dolostones in plane light (left) and cathodoluminescence (right). Luminescence intensities shown are proportional to both intrinsic luminescence intensity and camera exposure time, which is set automatically to acquire a standard overall light intensity. (**A**) and (**B**) Plug EHRE041 from 129.03 m BSF in Site 1193. CL exposure time: 19 s. (**C**) and (**D**) Plug EHRE307 from 47.09 m BSF in Site 1196. CL exposure time: 90 s. Dolomite-replaced bioclasts show varying degrees of preservation of relict luminescence apparently inherited from precursor calcite compositions. Dolomite cement consists of an early bright-luminescent zone followed by volumetrically dominant non-luminescent cement. (**E**) and (**F**) Plug EHRE120 from 480.19 m BSF in Site 1196. CL exposure time: 100 s. Note that inclusion-rich areas display luminescence zoning. If these areas represent replaced bioclasts, then the replacement process has involved the growth of individual large dolomite crystal faces through former bioclast volumes and is mainly unrelated to original textural domains.

F7 (Fig. 5B & F), but a few F2 samples from the NMP also fall within this group.

Packstone with connecting vugs. The thin sections of these coarse grained, but mud-rich samples display one or more pores considerably larger than the largest grains present (Fig. 5D). Although it is unknown to what degree these vugs are actually connected or isolated on a reservoir scale, it seems likely that the vugs in

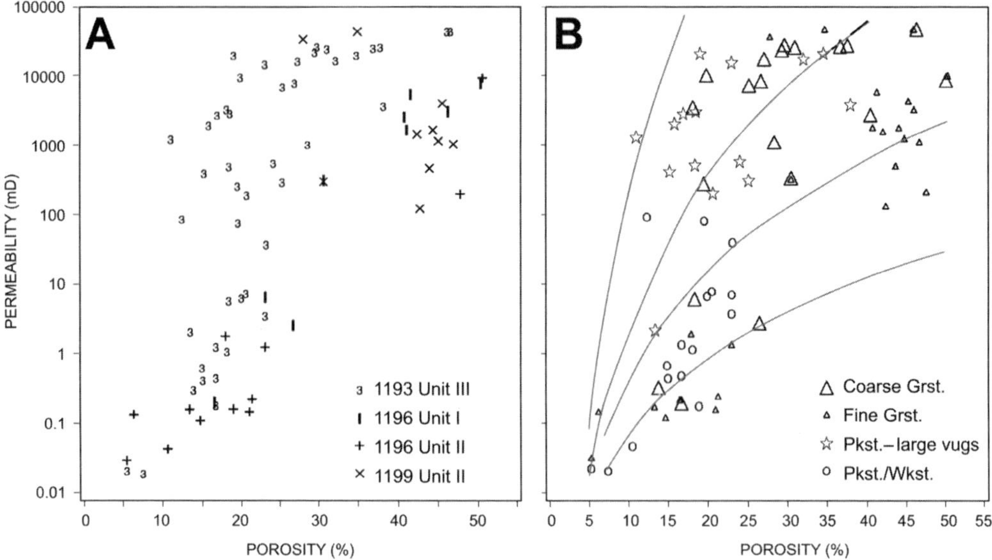

Fig. 11. Permeability v. porosity for limestones and partly dolomitized limestones. Plotting symbols indicate (**A**) stratigraphic units and drill sites; and (**B**) texture categories. Grey lines are boundaries of petrophysical rock fabric classes of Lucia (1999), calculated from the equation given in Jennings & Lucia (2001).

many cases have greatly increased the permeability measurable across the scale of a 1-inch plug.

Packstone–wackestone without connecting vugs. These samples have intergranular volumes mostly filled by mud. Vugs are common, but are smaller than the largest grains (Fig. 5C & E).

These textural categories at least partially sort the limestones into distinct porosity–permeability fields consistent with expected relationships (Fig. 11B). Although there are individual exceptions, higher permeability for given porosity tends to be associated with coarser grain size in the grainstones and presence of connected vugs in the packstones.

In Figure 11B, the limestone data are compared with the petrophysical rock fabric classes of Lucia (1995). A problem with this comparison is that the limestone porosities have not been corrected to remove 'separate-vug' porosity. Such correction was attempted by subtracting point-counted intragrain porosity for the samples represented by thin sections. This correction moves individual points 0–5 porosity units to the left, but does not change overall relationships much. On both uncorrected (Fig. 11B) and corrected plots, the limestones span the range of rock fabric classes 1–3 of Lucia (1995). Consistent with this system, the coarse grainstones mostly plot within class 1, while fine grainstones and mud-dominated fabrics tend to plot within classes 2 and 3. The high permeabilities of some packstones, plotting within the class 1 field, may reflect connectivity through vug systems that are larger than the 1-inch plug scale.

Dolostones

Figure 12A shows the dolostone data with plotting symbols identifying the main stratigraphic units. There are no obvious, systematic differences in porosity–permeability distribution between the main SMP dolostone units (units I, III and IV), or between SMP Upper Unit I (late Miocene) and Lower Unit I (middle Miocene). There is, however, a striking lateral difference within Unit I between sites 1196 and 1199. Both Upper and Lower Unit I dolostones from 1196 have porosities mostly >18% and permeabilities >800 mD, whereas Unit I dolostones from 1199 follow a much steeper permeability–porosity correlation and have lower porosity (mostly <23%) and permeability (<800 mD). This important pattern of lateral variation in reservoir quality is the subject of ongoing study.

The pore systems of the Leg 194 dolostones consist of three main components (Fig. 6): (1)

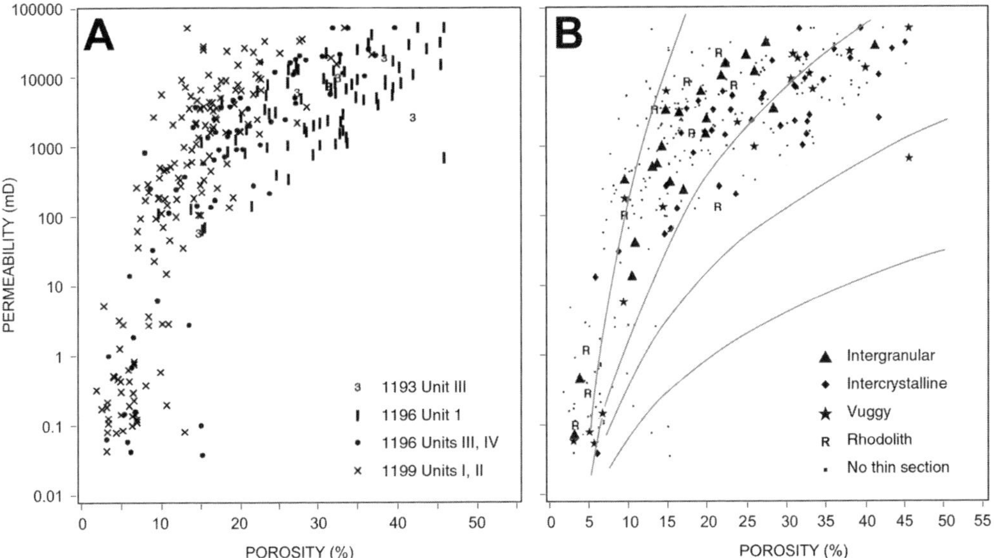

Fig. 12. Permeability v. porosity for dolostones. Plotting symbols indicate (**A**) stratigraphic units and drill sites; and (**B**) texture categories. Grey lines are boundaries of petrophysical rock fabric classes of Lucia (1999), calculated from the equation given in Jennings & Lucia (2001).

preserved intergranular pores; (2) vugs; and (3) intercrystalline pores. The relative dominance of these pore types, together with rhodolith dominance, is the basis for defining the following four dolostone textural categories. Average dolomite crystal size has not been used as a primary classification criterion because it is commonly highly variable in individual samples and because crystal size is mainly important for the samples with intercrystalline-dominated pore systems.

Intergranular dominated. The pore system is dominated by the preserved intergranular pores of an original grainstone texture (Fig. 6A). All samples in this group are from SMP Upper Unit I.

Intercrystalline dominated. The pore system is dominated by spaces between dolomite crystals that appear to replace mud matrix and bioclasts (Fig. 6C & D).

Vug dominated. The pore system is dominated by vugs, which typically appear to be bioclast moulds (Fig. 6E & F).

Rhodolith dominated. The pore system is dominated by the internal structure of a large rhodolith occupying much of the plug volume (Fig. 6B). All samples in this group are from SMP Unit I.

Despite their wide diversity, the dolostone textural categories do not occupy distinctly separate porosity–permeability fields (Fig. 12B). The data plot mainly within the field of petrophysical rock fabric class 1 of Lucia (1995), with some samples falling within the upper part of class 2, especially at the highest-porosity end of the data set (Figs 12B and 13B). As with the limestones, the porosity values in Figure 12 have not been corrected for the amount of separate vug porosity present. Application of this correction (using values estimated by eye from scanned images of the thin sections) moves the individual points 0–15 porosity units to the left on the plots, with many points falling to the left of the class 1 field, but overall relationships between textural groups do not appear to be significantly changed.

Distinction between partly fabric-preserving textures and fabric-destructive textures with regard to porosity–permeability characteristics can be evaluated in Figure 12A by comparing the samples from Unit I of sites 1196 and 1199 (fabric-preserving) with samples from units III and IV of site 1196 and Unit III of Site 1193 (fabric-destructive). Again, no systematic differences are apparent.

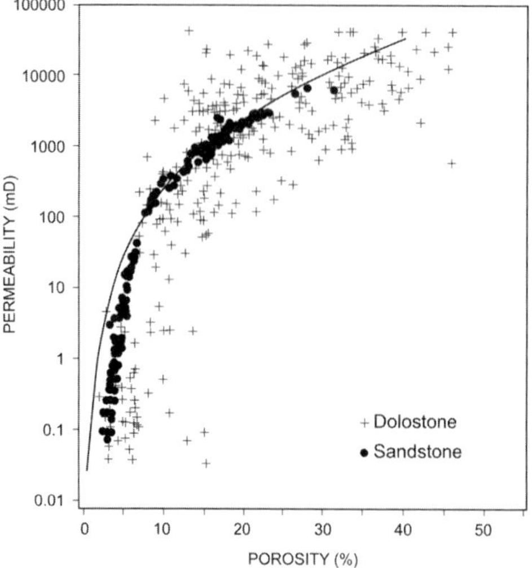

Fig. 13. Permeability v. porosity for dolostones compared with Fontainebleau Sandstone (data from Bourbie & Zinszner 1985). The line represents the Kozeny equation (Kozeny 1927): permeability (units of cm^2) = $CP^3/(TS^2)$, where C (grain shape constant) is assumed to be 1; P is fractional porosity, and T (tortuosity) is assumed to be 5. S (surface area per unit volume) is assumed = $6(1 - P)/D$, where D is the diameter in centimetres of particles comprising the sand pack (0.2 mm used here). Samples plotting to the right of this line could be explained as containing microporosity not involved in the interparticle macropore network (or particle size much finer than 0.2 mm for permeabilities >10 m), whereas samples plotting significantly above the line may have connected vugs bypassing the interparticle macropore network.

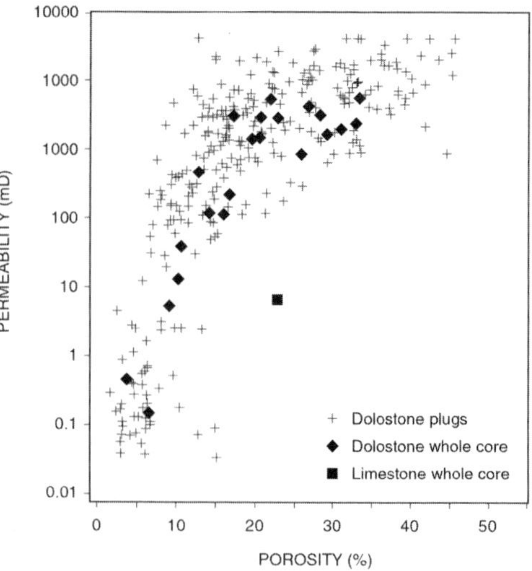

Fig. 14. Permeability–porosity plot comparing analyses of dolostone plugs and whole-core samples.

Taken together, the dolostones define a strongly curved trend on log-permeability v. linear-porosity coordinates, which roughly parallels the relationship for an ideal sand pack described by the Kozeny equation (Kozeny 1927) (Fig. 13). The increasing slope of the Kozeny curve with decreasing porosity reflects the semi-log coordinates used; the curve is linear on log–log coordinates. Similar trends are typical of clean, moderately-well-sorted sandstones, as illustrated by examples in Füchtbauer (1967) and Ehrenberg (1990). The most ideal and probably most widely known such trend is that of the Fontainebleau Sandstone (Fig. 13), which reflects porosity variation mainly due to increasing thickness of quartz-cement overgrowths around well-sorted grains of approximately 0.2 mm in diameter (Bourbie & Zinszner 1985). Fontainebleau Sandstone samples with porosity below about 8% porosity plot slightly below the Kozeny curve, possibly reflecting an additional factor besides quartz-cement abundance that distinguishes the pore systems of these from higher-porosity samples. In theory, the sandstone trend should attain zero permeability at about 3% porosity (Bryant *et al.* 1993).

The fact that all dolostone textural groups follow the same general trend, which parallels the trend of the Fontainebleau Sandstone, is interpreted as reflecting dominant control of dolostone permeability by the interparticle pore network. In the 'intergranular-dominated' dolostone texture group, this interparticle pore network corresponds with the preserved intergranular porosity. In the other dolostone texture groups, the interparticle pore network corresponds with the intercrystalline porosity of the dolomite matrix. Many dolostones have abundant vugs formed as moulds of bioclasts (Fig. 6E & F) or, in some cases, as primary cavities within rhodoliths (Fig. 6A & B). However, vug-dominated and rhodolith-dominated dolostone plugs do not plot in a systematically different field or trend compared with the dolostones with dominantly preserved-intergranular or matrix-intercrystalline pore systems (Fig. 12B). This similarity suggests that the vug porosity in the Marion Plateau dolostones functions as a part of the interparticle pore network.

Qualitative comparisons between individual thin sections and their positions with respect to the ideal Kozeny curve in Figure 13 suggest that the most important textural variable associated with displacement to the right of the Kozeny curve is not the presence of vugs, but rather the abundance of 'microporosity' (pores with cross-sections less than roughly 0.05 mm). Dolostones with higher apparent microporosity tend to plot relatively below or to the right of the Kozeny curve in Figure 13.

In some dolostones with larger vugs, for example plugs dominated by large rhodoliths, high permeabilities may reflect flow through a vug system that is larger than the scale of the 1-inch plug. The presence of large vugs connecting opposite ends of a plug could effectively 'short circuit' flow through the enclosing macropore system. On first consideration, this hypothesis appears to be supported by comparison between plug v. whole-core data (Fig. 14), because the plug permeabilities extend to much higher values (many from 10 to >50 D) than the whole cores (no values >7 D). The larger size of the whole-core samples would reduce the possibility for large vugs providing short-cuts for gas flow. However, examination of the actual samples involved suggests that 'flow short-cutting' is not a significant factor in the present data set. Of the 23 dolostone plug samples that have permeability >10 D and are also represented by thin sections, only three appear to contain large vug systems that might short-cut matrix flow. The other 20 samples all have relatively homogeneous-appearing coarse interparticle pore systems consisting of either relict intergranular pores (six samples) or intercrystalline pores (14 samples). Furthermore, examination of the 14 limestone plug samples with permeability >10 D and also represented by thin sections shows that only one of these is dominated by large vugs. The other 13 are characterized by relatively homogeneous-appearing intergranular pore systems that appear consistent with very high permeability.

Implications for dolostone reservoirs

This study demonstrates that extreme heterogeneity in porosity and permeability is an early formed and intrinsic characteristic of these bioclastic platform strata, despite their overall very good reservoir quality. Similarly heterogeneous porosity and permeability ranges characterize both limestones and dolostones, but the dolostones show better correlation between porosity and permeability, and follow a porosity–permeability trend more similar to that of sandstone. Plots of porosity v. depth, together with core observations, show that porosity–permeability heterogeneity mainly reflects metre-scale layering of varying lithologies and diagenetic responses. Nevertheless, a few large-scale stratigraphic trends are also apparent, including an upward porosity increase through the top 120 m of Site 1199 and upward porosity decrease through the top 60 m of Site 1193 (Fig. 4).

Clearly, most of the limestones of the present data set do not represent the same facies from which the dolostones formed. Most limestone plugs are from bryozoan-rich NMP facies F2 and F3, whereas most dolostone plugs are from red-algae-rich SMP facies F4 and F5. Facies F3 of the NMP may be more mud-rich and therefore less permeable than the SMP facies from which most of the sampled dolostones were derived. Also, the fine-grained grainstone textures of facies F6 and F7 are an important component of the limestone porosity–permeability distribution, but are probably represented by only a few dolomitized equivalents.

Despite these differences, comparison between the limestone and dolostone porosity–permeability distributions can be useful for illuminating some general principles, especially if comparison is restricted to the coarse bioclastic facies F2, F3, F4 and F5 (Fig. 15). The fabric-destructive dolostones are included because relict textures indicate that most of the core intervals from which these plugs are derived originally had coarse bioclastic texture. Dolomitized F5 plugs dominated by large rhodoliths are also excluded from this comparison.

As noted above, most dolostones plot within a single broad trend in Figure 15, whereas the limestones show relatively poor porosity–permeability correlation. High-permeability limestones (>100 mD) are grainstones and packstones with large and apparently connecting vugs. The low-permeability limestones are mainly packstones–wackestones with smaller, poorly connected vugs and heavily cemented (therefore vuggy) grainstones.

The present data set may be useful as an example of the petrophysical properties of a variably dolomitized, coarse bioclastic carbonate platform during early stages of burial (before the onset of stylolite-driven burial cementation). Both the limestone and dolostone layers comprising such a platform should be expected to display extreme short-range heterogeneity in reservoir quality. The dolostone layers should not necessarily have higher porosity or higher maximum permeabilities that their undolomitized lithological equivalents, but will have more regular and predictable trends of porosity–permeability correlation and higher average permeability for given porosity. These

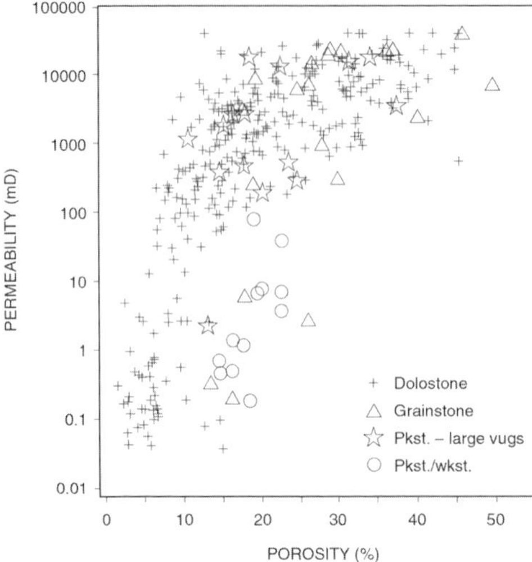

Fig. 15. Permeability–porosity plot comparing dolostones (excluding rhodolith-dominated plugs) and coarse bioclastic limestones. Open symbols divide the limestones into three texture groups: coarse-grained grainstone; packstone with large connecting vugs; and packstone/wackestone without apparent connecting vugs.

differences do not reflect the absence of vuggy porosity in the dolostones, but rather better connection of vugs through an enclosing network of intercrystalline macroporosity. This improvement arises through recrystallization of microporous lime mud to a macroporous dolomite matrix, resulting in a porosity–permeability trend resembling that of clean, well sorted sandstone.

My shipmates aboard the *Joides Resolution* are thanked for their hard work in developing the scientific foundation for this study (Isern *et al.* 2002). J. A. D. Dickson (Cambridge University) generously provided the cathodoluminescence photomicrographs in Figures 7 and 10. F. J. Lucia, G. P. Eberli and O. Walderhaug are thanked for contributing their ideas toward understanding the controlling processes involved with this data set. Reviews by J. Garnham and an anonymous referee were very helpful. This research used samples and data provided by the Ocean Drilling Program which is sponsored by the US National Science Foundation and participating countries under management of Joint Oceanographic Institutions, Inc. Travel costs for my participation in Leg 194 were paid by the Norwegian Research Council. Costs for my time and sample analyses were paid by Statoil. I especially thank K. H. Jakobsson for getting this project approved to begin with.

References

BOURBIE, T. & ZINSZNER, B. 1985. Hydraulic and acoustic properties as a function of porosity in Fontainebleau Sandstone. *Journal of Geophysical Research*, **90**, 11 542–11 532.

BRAITHWAITE, C.J.R. 1991. Dolomites, a review of origins, geometry and textures. *Transactions of the Royal Society of Edinburgh, Earth Sciences*, **82**, 99–112.

BRYANT, S., CADE, C. & MELLOR, D. 1993. Permeability prediction from geological models. *AAPG Bulletin*, **77**, 1338–1350.

BUDD, D.A. 1997. Cenozoic dolomites of carbonate islands: their attributes and origin. *Earth Science Reviews*, **42**, 1–47.

DAVIES, P.J., SYMONDS, P.A., FEARY, D.A. & PIGRAM, C.J. 1989. The evolution of the carbonate platforms of northeast Australia. *In*: CREVELLO, P.D., WILSON, J.L., SARG, J.F. & READ, J.F. (eds) *Controls on Carbonate Platform and Basin Development*. Society of Economic Paleontologists and Mineralogists, Special Publications, **44**, 234–258.

DAVIES, P.J., MCKENZIE, J.A., PALMER-JULSON, A. *ET AL.* 1991. *Proceedings of the Ocean Drilling Program, Initial Reports*, **133**. Ocean Drilling Program, College Station, TX.

DAWANS, J.M. & SWART, P.K. 1988. Textural and geochemical alternations in Late Cenozoic Bahamian dolomites. *Sedimentology*, **35**, 385–403.

EHRENBERG, S.N. 1990. Relationship between diagenesis and reservoir quality in sandstones of the Garn Formation, Haltenbanken, mid-Norwegian continental shelf. *AAPG Bulletin*, **74**, 1538–1558.

EHRENBERG, S.N., EBERIL, G.P. & BRACCO GARTNER, G.L. (eds). 2003. *Data Report: Porosity and Permeability of Miocene Carbonate Platforms on the Marion Plateau, ODP Leg 194. In*: *Proceedings of the Ocean Drilling Program, Scientific Results*, **194**. Ocean Drilling Program, College Station, TX.

EHRENBERG, S.N., MCARTHUR, J.M. & THIRLWALL, M.F. 2004. Growth, demise, and dolomitization of Miocene carbonate platforms, Marion Plateau, Australia: results from Sr-isotope stratigraphy (abstract). 66th European Association of Geoscientists & Engineers Conference & Exhibition, Paris.

EMBRY, A.F. & KLOVAN, J.E. 1971. A late Devonian reef tract on northeastern Banks Island, Northwest Territory. *Bulletin of Canadian Petroleum Geology*, **19**, 730–781.

FÜCHTBAUER, H. 1967. Influence of different types of diagenesis on sandstone porosity. *In*: *Proceedings of the 7th World Petroleum Congress*, Volume 2, 353–369. Elsevier, Barking, Essex.

GLENN, C.R. & KRONEN, J.D. 1993. Origin and significance of late Pliocene phosphatic hardgrounds on the Queensland Plateau, northeastern Australian margin. *In*: MCKENZIE, J.A., DAVIES, P.J., PALMER-JULSON, A. *ET AL.* (eds) *Proceedings of the Ocean Drilling Program, Scientific Results*, **133**. Ocean Drilling Program, College Station, TX, 525–534.

HAQ, B.U., HARDENBOL, J. & VAIL, P.R. 1987. Chronology of fluctuating sea levels since the Triassic. *Science*, **235**, 1156–1167.

ISERN, A.R., ANSELMETTI, F.S. & BLUM, P. 2001. Ocean drilling constrains carbonate platform formation and Miocene sea level on the Australian margin. *Eos*, **82**, 469–476.

ISERN, A.R., ANSELMETTI, F.S., BLUM, P. *ET AL.* 2002. *Proceedings of the Ocean Drilling Program, Initial Reports*, **194**. Ocean Drilling Program, College Station, TX.

JAMES, N.P. 1997. The cool-water carbonate depositional realm. *In*: JAMES, N.P. & CLARKE, J.A.D. (eds) *Cool-water Carbonates*. Society of Economic Paleontologists and Mineralogists, Special Publications, **56**, 1–20.

JENNINGS, J.W. & LUCIA, F.J. 2001. *Predicting Permeability From Well Logs in Carbonates with a Link to Geology for Interwell Permeability Mapping*. Society of Petroleum Engineers, SPE 71336.

KOZENY, J. 1927. Über kapillare leitung des wassers im boden (aufsteig, versickerung und andwendung auf die bewässerung). *Sitzungsberichte der Akademie der Wissenschaften in Wien, Abt. IIa*, **136**, 271–306.

LUCIA, F.J. 1995. Rock-fabric/petrophysical classification of carbonate pore space for reservoir characterization. *AAPG Bulletin*, **79**, 1275–1300.

MARTÍN, J.M., BRAGA, J.C., KONISHI, K. & PIGRAM, C.J. 1993. A model for the development of rhodoliths on platforms influenced by storms: middle Miocene carbonates of the Marion Plateau (northeastern Australia). *In*: MCKENZIE, J.A., DAVIES, P.J. *ET AL.* (eds) *Proceedings of the Ocean Drilling Program, Scientific Results*, **133**. Ocean Drilling Program, College Station, TX, 455–460.

MCKENZIE, J.A., ISERN, A., ELDERFIELD, H., WILLIAMS, A. & SWART, P.K. 1993. Strontium isotope dating of paleoceanographic, lithologic, and dolomitization events on the northeastern Australian margin, Leg 133. *In*: MCKENZIE, J.A., DAVIES, P.J. *ET AL.* (eds) *Proceedings of the Ocean Drilling Program, Scientific Results*, **133**, 489–497.

MURRAY, R.C. 1960. Origin of porosity in carbonate rocks. *Journal of Sedimentary Petrology*, **30**, 59–84.

PIGRAM, C.J., DAVIES, P.J., FEARY, D.A. & SYMONDS, P.A. 1992. Absolute magnitude of the second-order middle to late Miocene sea-level fall, Marion Plateau, Northeast Australia. *Geology*, **20**, 858–862.

SIBLEY, D.F., NORDENG, S.H. & BOROWSKI, M.L. 1994. Dolomitization kinetics in hydrothermal bombs and natural settings. *Journal of Sedimentary Research*, **A64**, 630–637.

SWART, P.K. & MELIM, L.A. 2000. The origin of dolomites in Tertiary sediments from the margin of the Great Bahama Bank. *Journal of Sedimentary Research*, **70**, 738–748.

VAHRENKAMP, V.C. & SWART, P.K. 1994. Late Cenozoic dolomites of the Bahamas: metastable analogues for the genesis of ancient platform dolomites. *In*: *Dolomites: A Volume in Honour of Dolomieu*. International Association of Sedimentologists, Special Publications, **21**, 133–153.

Dolomites in SE Asia – varied origins and implications for hydrocarbon exploration

ANDREW J. H. CARNELL[1,2] & MOYRA E. J. WILSON[3]

[1]*Robertson Research International, Llanrhos, Llandudno LL30 1SA, UK*

[2]*Present address: Shell Exploration and Production, EP Solutions, Seafield House, Wing 4C, Hill of Rubislaw, Anderson Drive, Aberdeen AB15 6BL, UK*

[3]*Department of Geological Sciences, University of Durham, South Road, Durham DH1 3LE, UK*

Abstract: Carbonates in SE Asia range in age from Palaeozoic to Recent, but are most important as reservoirs in the Neogene where they comprise a major target for hydrocarbon exploration (e.g. Batu Raja Formation, South Sumatra, Sunda and Northwest Java basins). Carbonates of pre-Tertiary, Palaeogene and Neogene age all show a strong diagenetic overprint in which dolomite occurs as both cementing and replacive phases associated with variable reservoir quality. This paper reviews published data on the occurrence and types of dolomites in SE Asian carbonates, and considers the models that have been used to explain the distribution and origin of dolomite within these rocks.

Pre-Tertiary carbonates form part of the economic basement, and are little studied and poorly understood. Although some, such as in the Manusela Formation of Seram, may form possible hydrocarbon reservoirs, most are not considered to form economic prospects. They are best known from the platform carbonates of the Ratburi and Saraburi groups. in Thailand, and the oolitic grainstones of the Manusela Formation of Seram. The Ratburi Group shows extensive dolomitization with dolomite developed as an early replacive phase and as a late-stage cement.

Palaeogene carbonates are widely developed in the region and are most commonly developed as extensive foraminifera-dominated carbonate shelfal systems around the margins of Sundaland (e.g. Tampur Formation, North Sumatra Basin and Tonasa Formation, Sulawesi) and the northern margins of Australia and the Birds Head microcontinent (e.g. Faumai Formation, Salawati Basin). Locally, carbonates of this age may form hydrocarbon reservoirs. Dolomite is variably recorded in these carbonates and the Tampur Formation, for example, contains extensive xenotopic dolomite.

Neogene carbonates (e.g. Peutu Formation, North Sumatra) are commonly areally restricted, reef-dominated and developed in mixed carbonate–siliciclastic systems. They most typically show a strong diagenetic overprint with leaching, recrystallization, cementation and dolomitization all widespread. Hydrocarbon reservoirs are highly productive and common in carbonates of this age. Dolomite is variably distributed and its occurrence has been related to facies, karstification, proximity to carbonate margins and faults. The distribution and origin of the dolomite has been attributed to mixing-zone dolomitization (commonly in association with karstic processes), sulphate reduction via organic matter oxidation, and dewatering from the marine mudstones that commonly envelop the carbonate build-up.

Dolomite has a variable association with reservoir quality in the region, and when developed as a replacive phase tends to be associated with improved porosity and permeability characteristics. This is particularly the case where it is developed as an early fabric-retentive phase. Cementing dolomite is detrimental to reservoir quality, although the extent of this degradation generally reflects the abundance and distribution of this dolomite.

Dolomitization is also inferred to have influenced the distribution of non-hydrocarbon gases. This is best documented in North Sumatra where carbon dioxide occurs in quantities ranging from 0 to 85%. There are a number of possible mechanisms for generating this CO_2 (e.g. mantle degassing), although the most likely source is considered to be the widely dolomitized Eocene Tampur Formation that forms effective basement for much of the basin. High heat flows are suggested to have resulted in the thermogenic decomposition of dolomite with CO_2 produced as a by-product.

Carbonates are geographically and temporally widespread in SE Asia, and comprise major reservoir units in many basins. They range in age from Precambrian (e.g. Mogok Group, Myanmar) to Recent (e.g. Spermonde Platform, Sulawesi and Pulau Seribu, Java), but are best

From: BRAITHWAITE, C. J. R., RIZZI, G. & DARKE, G. (eds) 2004. *The Geometry and Petrogenesis of Dolomite Hydrocarbon Reservoirs*. Geological Society, London, Special Publications, **235**, 255–300. 0305-8719/$15.00

known and most extensively studied in the Neogene, where they have major economic significance (e.g. Arun, Krisna and Ramba fields). Dolomite is widespread in the region (Tables 1–3) with its distribution reflecting the overall broad distribution of the carbonates. However, despite wide occurrence in space and time, it is volumetrically minor and patchily distributed. Many carbonates contain little or no dolomite (e.g. Tonasa Formation, Sulawesi and Wonosari Formation, Java) whilst others (e.g. Ratburi Group, Thailand) are extensively dolomitized. Where present, dolomite not only affects the physical and chemical characteristics of limestones but can also have a variable association with reservoir quality, with both constructive and destructive effects.

On the basis of their depositional characteristics and economic significance, SE Asian carbonates can be conveniently subdivided into groups of pre-Tertiary and Tertiary age. The latter can be subdivided into Palaeogene and Neogene age subgroups. Pre-Tertiary, Palaeogene and Neogene carbonates typically show a strong and varied diagenetic overprint, with dolomitization sometimes extensive. Pre-Tertiary carbonates show much compositional and diagenetic variation, but are commonly poorly differentiated as they form economic basement over most of the area. Maps and tables showing the known occurrences, and published information on dolomites in Pre-Tertiary, Palaeogene and Neogene carbonates, are presented as Figures 1–3 and Tables 1–3, respectively. Table 4 summarizes the published data on dolomites of different ages.

In Tertiary successions, carbonates are widespread, with most basins having at least temporary development of carbonate deposystems. Studies of carbonate biota suggest that Tertiary carbonates are broadly divisible into Palaeogene and Neogene age groups (Wilson & Rosen, 1998; Wilson 2000, 2002). In SE Asia, carbonate deposits often contain a significant proportion of volcaniclastic and/or terrestrially derived siliciclastic material (Wilson & Lokier 2002). Pure carbonate systems are most common in the Palaeogene where they tend to be dominated by coralline algae and larger benthic foraminifera, forming shoals and extensive carbonate platforms (e.g. Tonasa Limestone, South Sulawesi and Tampur Formation, North Sumatra). Corals are generally a minor component, but are locally important (e.g. Rajamandala Formation, West Java). Palaeogene carbonates typically show limited primary porosity and commonly experienced minimal diagenetic overprinting. Dolomite distribution in them has been little studied but is known to be locally extensive. Where dolomite is present it appears to significantly influence reservoir quality.

Neogene carbonates are better known than their Palaeogene precursors, largely as a result of their economic importance. Most were deposited in mixed carbonate–siliciclastic systems where they are best developed as reefal build-ups over palaeotopographic highs and in areas of limited siliciclastic input. Neogene carbonates commonly show high primary porosities and have commonly been affected by strong diagenetic overprinting. Dolomite is widely encountered, as both replacement and cement phases, and has a highly varied association with reservoir quality.

The origins of the dolomites in SE Asia are poorly understood and a number of models have been invoked (Table 4). The most popular and persistent of these models has been dolomitization in the mixing zone. More recently, an association of dolomitization and karstification has been noted. Many authors have suggested dolomitization by a fluid derived from dewatering of clays and/or Mg-rich fluids liberated by smectite–illite transformations. The origins invoked for dolomite formation in SE Asia differ from many current interpretations of dolomitization from other parts of the globe, where marine, reflux and evaporitic dolomites are now commonly reported. The question then is whether dolomites in equatorial regions have different origins from those in other areas, or whether a reassessment of dolomite origins is needed. As will be seen from the discussion below there is currently insufficient data available to adequately test the validity of the origins inferred for many dolomites in SE Asia. The aims of this paper are therefore to document the data available on dolomites, using case studies from carbonate formations of different ages. Although the origins and reservoir quality of the dolomites inferred by other workers are reported, these are topics that are flagged as in need of considerable further research.

Availability and quality of data

Although there are numerous publications on SE Asian carbonates, few discuss their diagenesis in detail. Even in extensively studied carbonates there is a tendency to concentrate on depositional and stratigraphic aspects, together with their relationship to basin evolution and the petroleum system with discussions of diagenesis being of a more general nature. For example, much has been published on the Peutu

Table 1. *Known dolomite occurences in Pre-Tertiary carbonate formations from SE Asia*

Formation name	Location	Age	Tectonic and depositional setting	Dolomite type, size and occurrence	% of unit	$\delta^{18}O/\delta^{13}C$	$^{86}Sr/^{87}Sr$	Associated with:							Reservoir quality	Notes	Principal references
								Mixed siliciclastics	Adjacent siliciclastics	Subaerial exposure	Dissolution	Fracturing	Coralline algae	Restricted facies			
Saraburi Group	Khorat Plateau Basin and environs	Permian	Platform carbonates deposited on a stable platform														Booth & Sattayarak 2000
Ratburi Group	Peninsular Thailand and Thai Basin	Permian	Platform carbonates deposited over Sibumasu Terrane	(1) Fabric retentative microcrystalline dolomite. (2) Fabric-destructive hypidiotopic finely crystalline dolomite. (3) Coarse euhedral vein cement	Extensive	$\delta^{18}O$ (1) 0 to –5; $\delta^{13}C$ (1) 0–5				*	*	*			Hydrocarbons discovered in Gulf of Thailand where Ratburi Group occurs as buried tower karst	(1) and (2) derived from marine–meteoric mixing or marine-water circulation. (3) Precipitated from fluids associated with granite emplacement	Baird & Bosence 1993; Chinoroje 1993
Chaiburi Formation	Southern Peninsular Thailand	Early–Late Triassic	Low- to high-energy platform carbonates deposited in a tectonically quiescent setting	Phukaothong Dolomite Member. Comprises light-grey dolomite with some chert nodules. Amount of dolomite decreases upwards into Chiak Limestone Member	0–100%												Ampornmaha 1995
Teunom Limestone Formation	NE Sumatra	Late Jurassic–Early Cretaceous	Open marine, shallow water	In part dolomitized	?	–	–								Not described		de Smet 1991
Kaloi Formation	Langsa, NE Sumatra	Probably Late Permian–Late Triassic	Open marine shallow water?	Massive limestones and dolomites	?	–	–								Not described		de Smet 1991
Ghegan Formation	Buru	Triassic	Littoral–neritic environment	More common towards base of formation. Replacive secondary dolomite	?	–	–								Formation includes bituminous shales	Origin of dolomite unclear. Dolomites replace reefal deposits	Tjokrosapoetra *et al.* 1993

Table continued overleaf

Table 1. *continued*

Formation name	Location	Age	Tectonic and depositional setting	Dolomite type, size and occurrence	% of unit	$\delta^{18}O/\delta^{13}C$	$^{86}Sr/^{87}Sr$	Associated with: Mixed siliciclastics	Adjacent siliciclastics	Subaerial exposure	Dissolution	Fracturing	Coralline algae	Restricted facies	Reservoir quality	Notes	Principal references
Manusela Formation	Seram Basin	Early–Middle Jurasic	Oolitic shoals on palaeotopo-graphic high (?relict from Triassic rifting) separated from Birds Head by palaeotopo-graphic low	Late-stage medium crystalline (*c.* 400 μm) tight equant/sucrosic dolomite replacing oolitic grainstones. Post- and predate compaction	Locally 0–100%	–	–		*			*			Dolomite associated with loss of porosity and reduced fracture intensity in potential reservoir rocks	Dolomite derived from Mg-rich fluids dewatering from shales at toe of thrust front? Fractures commonly developed in the dolomite lithofacies	Kemp 1992; Nilandaroe *et al.* 2002
Paglugaban Formation	Paglugaben Island, offshore NW Palawan, Philippines	Middle Carboniferous		Brecciated partly dolomitized limestones		–	–					*					Kiessling & Flugel 2000
Minilog Formation	NW Palawan, Philippines	Permian	Carbonate ramp and build-ups	Occasional dolomite. (1) Synsedimentary dolomite in oncoid gastropod peloid facies, dolomitization increases with increase in abundance of peloids (sabkha origin?). (2) Related to faulting	?	–	–					*		?	Tight and often recrystallized limestone, may be bituminous	Dolomites seen on Minilog Island, Dilumacad Island, Lagen Island, Matinloc Island	Kiessling & Flugel 2000
Coron Formation	Busuanga, Philippines	Late Triassic	Carbonate platform and reefs	Patches of euhedral dolomite associated with stylolites	?	–	–									One locality dolomite described	Kiessling & Flugel 2000
Maubisse Formation	Timor	Permian	Storm-dominated carbonate ramp	(1) Euhedral rhombs (150 μm) found in pressure dissolution seams. (2) Idiotopic and xenotopic mosaics (125–600 μm), sometimes stratiform	Very variable locally 0–95%	–	–								Interparticle porosity in the dolostones poorly developed	Stratiform dolomites replace silicicfied limestones	Barkham 1993

Facet Limestone	Fak Fak, Irian Jaya	Late Jurassic–Late Cretaceous	Shallow marine at base to deep marine at top	Dolomitic limestone in lower part	?	–	–	*		No information given	Mostly microcrystalline siliceous–chalky limestone interbedded with shale	Robinson *et al.* 1990*a*
Modio Dolomite	Waghete, Irian Jaya	?Silurian–Devonian	Probably marine	Well-bedded dark-grey–cream dolostone and light-grey dolomitic limestone	?	–	–			No information given	Equivalent to dolomite at base of well ASM-1, Mangguar and Brug Formations	Pigram & Panggabean 1989
Mogok Series	Shan Plateau, Myanmar	Precambrian		Local dolomite reported in crystalline (essentially regionally and thermally metamorphosed) carbonates				*	*			Bender 1983
Chaung Magyi Group	Shan Plateau, Myanmar	Precambrian–Cambrian	Shallow–deep marine in rapidly subsiding basin	Dolomite in beds up to 200 m thick				*	*		Outcrops show relic cross-bedding	Bender 1983
Molohein Group	Northern Shan State, Myanmar	Late Cambrian–Early Ordovician	?Shelfal carbonates and shelfal–deep-marine clastics	Thickly bedded finely crystalline dolomite interbedded with sandstone				*	*			Bender 1983
Pangyun Formation	Southern Shan State, Myanmar	Late Cambrian–Early Ordovician	Shelfal carbonates with shelfal–fluvial clastics	Dolomite interbedded with limestone and sandstone				*	*		Dolomites reported to be cross-bedded	Bender 1983
Shan Dolomite Group	Shan Plateau, Myanmar	Devonian–Triasic	Shelfal carbonates with local reef. Carbonate bioherms over clastics	Massive, but locally laminated					*		Comprises Maymo Dolomite Formation (Dev to ?Carb), Nawabangyi Dolomite Formation (Permo-Trias) and Kondeik Limestone (Trias)	Grobbett 1973; Bender 1983
Setul Limestone	Onshore Malaysia and adjacent areas of Straits of Malacca	Ordovician–Silurian		Locally extensively dolomitized								Fontaine 1992

Table 2. *Known dolomite occurences in Palaeogene carbonate formations from SE Asia (for more information on the carbonate deposits see Wilson 2002)*

Formation name	Location	Age	Tectonic and depositional setting	Dolomite type, size and occurrence	% of unit	$\delta^{18}O/\delta^{13}C$	$^{86}Sr/^{87}Sr$	Associated with: Mixed siliciclastics	Adjacent	Subaerial exposure	Dissolution	Fracturing	Coralline algae	Restricted facies	Reservoir quality	Notes	Principal references
Tampur Limestone Formation (TM)	Northern Sumatra, Langsa and Medan (North Sumatra Basin)	Probably Eocene–Early Oligocene	Margins of Sunda craton, prior to backarc basin formation. Shallow shelf, open marine, adjacent to land area	Massive, microcrystalline dolostone with fracture and vuggy macroporosity. Dolomite occurrences are common, and commonly highly brecciated and fractured	?	–	–					*			Alur Siwah-8 discovered gas in uppermost karstified dolomite of Tampur Formation	Some of the lithologies are metamorphosed. At Alur Siwah Miocene Peutu deposited on karstified Tampur Dolomites	Bennett *et al.* 1981; de Smet 1992; Collins *et al.* 1996; Barliana *et al.* 2000
Rajamandala Formation (RF)	W Java, Cianjur, near Bandung (South Java Basin)	Late Oligocene–Early Miocene	Block-faulted highs to south of Bogor Trough. Not clear if relationship with volcanic arc at this time. Open marine–restricted platform, reef margin, forereef slope	(1) Microcystalline (<50 μm) anhedral–rhombic replacement of argillaceous limestone. (2) Finely crystalline (>50 μm) subhederal–euhedral replacement associated with faults. (3) Fracture-filling saddle dolomite (50–250 μm)	–	–			*	*	*	*	*		No production or reservoir potential, but saddle dolomite is locally coated with residual hydrocarbon		Carnell 1996, 2000
Ngimbang Carbonates (NC)	Offshore Kangean, Sepangan area (East Java Basin)	Late Eocene mostly, sometimes middle Eocene–Early Oligocene	Backarc setting. >20 carbonate shoals/build-ups developed from broad, low-relief platform sequence, bounded by major fault	(1) Micrite replacing dolomite (<10 μm) present in minor amounts throughout. Most abundant in units with up to 10% siliciclastics. (2) Dolomite and crystalline calcite geopetal fill of breccia fractures (<100 μm).	up to 2%	–	–	*		*		*			Non-economic oil discovery JS53A-1 carbonates	(2) Probable carbonate exposure surface and associated karst(?) collapse breccia. (3) Fractures in late-stage dolomites overprinted; in early dolomites, affected by subsequent fracturing, fracture frequency is is high	Siemers *et al.* 1992

				(3) Later dolomite cements in tectonic fractures, fractures associated with exposure surfaces, or associated with stylolites (*c.* 100 μm)													
Berai (BR)	SE Kalimantan (Barito Basin and southern interior margin of Kutai Basin)	Mostly Oligocene, but also Late Eocene–recent	Large-scale block-faulted high, basin margin. Extensive carbonate platform, marginal deposits and build-ups	(1) Replacive planar-e (euhedral) microcystalline rhombs (<50 μm), seen in outcrop. (2) Dolomite cement (>1 mm) including non-planar saddle, seen in upper part of Berai in subsurface, associated with vuggy dissolution	low, but not given	$\delta^{18}O$ (1) –5 to –8. (2) –7.5 to –8.5. $\delta^{13}C$ (1) 0–3. (2) –0.5 to 0.5	–	*	*		*				Platform rim limestones where dolomite most common – average porosity 5.4% and permeability 44 mD	Dissolution is related to acidic waters compacting out of adjacent shales during moderate–deep burial (200–3000 m). Saddle dolomite ?87–145 °C	Saller & Vijaya 2002
Melinau Limestone (ML)	Sarawak (Sarawak–East Natuna Basin)	Late Eocene–Early Miocene	Isolated carbonate platform, basin margin	Planar-e dolomite rhombs (replacive and cement?). Pervasive at base of section and associated with calcilutitic calcarenite. *c.* 200 μm (plate 22b, Adams 1965)	<5%, locally pervasive	–	–		?			?	*		?	At Bukit Berar and base of Melinau Gorge section rock is almost pure dolomite	Adams 1965
Tonasa Formation (TN)	Western South Sulawesi (Sengkang Bone Basin)	Early/Middle Eocene–Middle Miocene	To west of volcanic arc on faulted high (backarc) large-scale syntectonic carbonate platform, segmented by faulting	(1) Microcystalline euhedral dolomite (<50 μm) associated with coralline algae. (2) Rare planar-e euhedral–sucrosic dolomite (50–100 μm) associated rarely with top of karst and fractures	<<0.01%	–	–			*		*	*	??	Possible although unlikely reservoir in subsurface	Dolomitization is post-depositional, may be associated with restricted facies	Crotty & Engelhardt 1993; Wilson 1995; Wilson pers. obs.
Tanpakura Formation (TP)	SE Sulawesi	Late Eocene–Early Oligocene	Rimmed shelf bordering microcontinental block	Planar-e rhombs (10–450 μm) (1) Dolomite replacing oncoids and matrix of oolitic-pisolitic facies. (2) Dolomite and birds-eye present in some wacke/packstone facies	?	–	–			?					No information given	Most common in lime mudstones, wackestones and a few in oolite. Most occur together with birds-eyes – prior to compaction upper intertidal–supratidal	Surono 1994

Table continued overleaf

Table 2. *continued*

Formation name	Location	Age	Tectonic and depositional setting	Dolomite type, size and occurrence	% of unit	$\delta^{18}O/\delta^{13}C$	$^{86}Sr/^{87}Sr$	Associated with: Mixed siliciclastics	Adjacent	Subaerial exposure	Dissolution	Fracturing	Coralline algae	Restricted facies	Reservoir quality	Notes	Principal references
Zaag Limestone (Z)	Misool, Irian Jaya (Salawati Basin)	Middle Eocene–Oligocene	Shallow-marine carbonates conformable on sandstone	Locally dolomitic	?	–	–								No information given	Recrystallized. Oolitic limestone locally dolomitic	Rusmana *et al.* 1989
Warpi Formation (WA)	Kaimana, Omba Waghete, Irian Jaya (Bintuni and Armeugah Nauka Basins)	Latest Cretaceous?–earliest Eocene?	Shallow-marine carbonate–siliciclastic shelf and carbonate platform on continental basement	Finely crystalline–sucrosic dolomites confined to upper part of Waripi. Common in Kamakawala anticline and in ASF-1X well	Minor–common	–	–	*	*						May form reservoir if suitable seal	Contains miliolids – suggests shallow deposits dolomitized. Locally glauconitic. Late gypsum and anhydrite in fractures	Pieters *et al.* 1983; Dow *et al.* 1990; Panggabean 1990; Robinson *et al.* 1990*b*; Brash *et al.* 1991
Faumi Limestone (F)	E. Birds Head, Taminabuan, Ransiki, Irian Jaya (Salawati and Bintuni Basins)	Middle Eocene–Oligocene	Shoal, carbonate bank or shallow-water shelf on continental basement high	Sporadic dolomitization	?	–	–	*							No information given	Locally may be dolomitized. Basal section of some wells dolomites	Lunt & Djaarfar 1991; Akhmad 2002
Puragi Formation	Taminabuan, Ransiki, W Irian Jaya	Late Cretaceous?–Middle Eocene	Probably sabkha or evaporitic backreef on continental basement	Dolomitic shale–clayey dolostone and anhydrite beds interbedded with shale and sandstone	?	–	–	*	*					*	No information given	In dolomite, anhydrite occurs as small solitary crystals	Pigram & Sukanta 1989; Pieters *et al.* 1983, 1990

Table 3. *Known dolomite occurences in Neogene carbonate formations from SE Asia (for more information on the carbonate deposits see Wilson 2002)*

Formation name	Location	Age	Tectonic and depositional setting	Dolomite type, size and occurrence	% of unit	$\delta^{18}O/\delta^{13}C$	$^{86}Sr/^{87}Sr$	Associated with: Mixed siliciclastics	Adjacent siliciclastics	Subaerial exposure	Dissolution	Fracturing	Coralline algae	Restricted facies	Reservoir quality	Notes	Principal references
Malacca Limestone Member – Belumai Formation (BF)	Malacca Straits, offshore N. Sumatra (North Sumatra Basin)	Early–Middle Miocene	Backarc. Over 70 carbonate build-ups developed on antecedent highs	Post-date lithification. (1) Replacive dolomite, initially of matrix, some sucrosic textures in NSB. (2) Hypidiotopic–xenotopic texture from crystal growth in J fields	?	–	–					*			Commercial gas and oil reservoirs. Porosity variations in the North Sumatra Basin dolomites ranges from very tight (1–2%) to excellent (30%)	Severe fracturing has affected the dolomites (tens of mD permeability)	McArthur & Helm 1982; Mundt 1982, 1983; Sulitra 1991
Arun Limestone (AL). Equivalent Limestone Member of Peutu Formation, etc.	Onshore N. Sumatra, on Arun or Lho Sukon high (North Sumatra Basin)	Early Miocene–Middle Miocene	Backarc carbonate platform developed on antecedent high	(1) Basal cryptocrystalline–microcrystalline tight dolomite replacing lime mud (2) Porous and permeable units of 315 m thickness	?	–	–								Arun Field	(2) Low-temperature mixing zone origin	Graves & Weegar 1973; Houpt & Kersting 1976
Peusangan Limestone/Sigili Limestone (PS)	Offshore N. Sumatra on Peusangan/ Western High and Sigili Highs (North Sumatra Basin)	Early–Middle Miocence	Backarc. Carbonate build-ups on N–S antecedent structural highs	Peusangan-C1–Reef limestone, the basal 40 m consist of tight and dense dolomitic limestone	?	–	–								Reservoir in subsurface, but basal dolomite <5% porosity		Hinton *et al.* 1987
Peutu Formation Limestone Member (LP)	Northern Sumatra, Takengon, Langsa (North Sumatra Basin)	Early–Middle Miocence	Backarc. Shallow marine, some build-ups on faulted highs	(1) Dolomite in the lower portion of build-ups. (2) Exposure-related features (oxides, phosphate replacement, dolomites and dissolution	?	–	–								Gas in subsurface S. Lho Sukon	Include thin shale stringers	Rory 1990; Collins *et al.* 1996

Table continued overleaf

Table 3. *continued*

Formation name	Location	Age	Tectonic and depositional setting	Dolomite type, size and occurrence	% of unit	$\delta^{18}O/\delta^{13}C$	$^{86}Sr/^{87}Sr$	Associated with: Mixed siliciclastics	Adjacent siliciclastics	Subaerial exposure	Dissolution	Fracturing	Coralline algae	Restricted facies	Reservoir quality	Notes	Principal references
Batu Raja Formation (BR)	S. Sumatra and offshore NW Java (South Sumatra, Sunda and Northwest Java Basins)	Late Early–Middle Miocene	Backarc, platform limestone, and/or reefal build-ups commonly developed on local fault-controlled highs	(1) Small dolomite rhombs (5–20/50 μm) in shaly beds. Dolomitization is significant only in carbonates with at least 7–10% clay. (2) Large euhedral rhombs (<50 μm) in karstic pores. (3) Ferroan dolomite rhombs and overgrowths (>200 μm). (4) Saddle dolomite as patchy fracture fill. (5) Dolomitic intraclasts in channel-fill deposit (Ramba Field)	Up to 27% (for clay-rich units)	$\delta^{18}O$ (2) –2 to –4. (3) –8 to –10 $\delta^{13}C$ (2) –1 to –4. (3) 5–6	–	*		*		*			Oil and gas reservoirs in build-ups. Porosities up to 30-40%. Dolomites may locally enhance permeability by increasing pore throat sizes in the finer-grained lithologies. However, porosity may be decreased by late-stage dolomite cements	(1) Dissolution associated with the clay seams probably contributed ions for these cements. (2) Progressive re-emplacement of marine fluids (i.e. mixing zone).(3) Methanogenic origin–meteoric origin, ?mixing of expelled basinal fluids with downward–percolating meteoric waters	Wight & Hardian 1982; Djuanda 1985; Longman *et al.* 1987, 1992; Crumb 1989; Tonkin *et al.* 1992; Park *et al.* 1995; Wicaksono *et al.* 1995
Mid-Main Limestone Member, Upper–Middle (MM) Cibulukan Formation (Main Carb. B)	Offshore NW Java & SE shelf edge and Seribu Platform (Northwest Java Basin)	Early Middle Miocene	Backarc setting. Laterally restricted carbonates and build-ups, grade laterally into deeper marine muds and silts	Dolomite is common, occurring as dolomicrite, dolomicrospar and a later ferroan dolomite	?	–	–	*							Low porosity	The limestone is grey, micritic and sometimes dolomitic	Arpandi & Patmosukismo 1975; Suherman & Syahbuddin 1986
Pre-Parigi Limestone Member (PP)–upper Cibulukan Formation	Offshore and onshore (Kromong) NW Java: SE shelf edge, Seribu Platform and W. Ardjuna (Northwest Java Basin)	Early Middle–Late Miocene	Backarc. Open, shallow-marine shelf, near clastic source (onshore). Offshore–stacked biostromal units and build-ups, pass laterally into deeper marine deposits	(1) Finely crystalline dolomite replaces of echinoderm plates with associated development of syntaxial overgrowths, and (2) finely crystalline matrix replacement (dolomicrite) with xenotopic and hypidiotopic crystal forms	Common–abundant in some sections	–	–	*							Gas in pre-Parigi	Glauconitic shale, some marl and dolomitic limestone intercalations are found in the lower part. Onshore thin intercalations of dolomitic limestone in argillaceous limestones	Arpandi & Patmosukismo 1975; Pringgo *et al.* 1977; Yaman *et al.* 1991

Parigi Limestone (PL)	Onshore and offshore NW Java, Sunda Straits (Northwest Java and Sunda Basins)	Late Middle Miocene–Pliocene, mostly late Miocene	Backarc. Carbonate build-ups from extensive stable marine platform. Biostromal–upper biohermal part	(1) Non-ferroan microcrystalline (5–40 μm) dolomite. (2) Finely crystalline idiotopic and locally hypidiotopic dolomite. Both affect the range of marine facies and a cryptalgal laminite	? Affects significant proportion	–	–	*		*	?	Porosity in the dolomite/calcareous dolomite ranges from 16 to 30% (mainly intercrystalline micropores and less moldic micropores	Most commonly dolomite has replaced carbonate mud and locally echinoid debris, red algae and calcite cement	Arpandi & Patmosukismo 1975; Bukhari *et al.* 1992
Madura Formation (MF) and Karren Limestone	Madura and offshore NE Java and Madura (Java Sea and East Java Basins)	(Late Miocene)–Pliocene. Offshore Oligocene–Pliocene	Backarc shallow-marine shelf (sometimes with siliciclastics)/open-marine shelf margin. Marls and limestone (GL Formation) in deeper water	Dolomitization, but not described	?	–	–	*				No information given		Situmorang *et al.* 1992
Paciran Formation (PC)	NE Java: Mojokerto and Jatirojo and Madura (Java Sea and East Java Basins)	Late Miocene–Pliocene ?	Backarc Marine shelf. Reefal limestone in upper part	Dolomitic limestone, but not described	?	–	–					No information given		Situmorang *et al.* 1992
Prupuh/Rancak Limestone (PU) Member of Kujung Formation (Kujung Unit 1)	NE Java – Tuban and offshore Kangean and Madura (Java Sea and East Java Basins)	Mostly Miocene (offshore Early–Middle Miocene), also Oligocene	Backarc on faulted highs, Miocene shelf and shelf margin/slope. Offshore, open marine, platforms and high-relief pinnacle build-ups	(1) BD field – Early sucrosic dolomite, Dolomite content decreases downwards from up to 50% to zero. (2) Poleng field – scattered rhombs of dolomite are common in the lower 300–400 ft of section	?	–	–	*				BD field. Sucrosic dolomite gives increases in porosity and permeability	Poleng Field dolomites in deep-water foraminiferal packstone/wackestone and laminated argillaceous limestones often with shale streaks	Kenyon 1977; Cucci & Clark 1993, 1995; Carnell pers. obs.
Batu Putih Limestone (BB)	East Kalimantan and in offshore area (Kutai Basin)	Oligocene–Late Miocene	Basin margin, now in compressional setting. Delta-front or shelf-edge shallow-marine carbonates	Partial or near-complete replacement of clay-rich matrix by planar-e microcrystalline (5–50 μm) rhombs	<3%	–	–	*		*		Low porosity <3% and low permeabilities	Present in clay-rich facies, more common at base of succession	Alam *et al.* 1999; Wilson pers. obs.

Table continued overleaf

Table 3. *continued*

Formation name	Location	Age	Tectonic and depositional setting	Dolomite type, size and occurrence	% of unit	$\delta^{18}O/\delta^{13}C$	$^{86}Sr/^{87}Sr$	Associated with: Mixed siliciclastics	Adjacent siliciclastics	Subaerial exposure	Dissolution	Fracturing	Coralline algae	Restricted facies	Reservoir quality	Notes	Principal references
Taballar Formation (TB)	Mangkalihat Peninsula and Maratua ridge	Late Eocene–Mio-Pliocene	Large-scale block-faulted high, basin margin. Extensive carbonate platform and some isolated build-ups	(1) Replacive planar-e microcystalline rhombs to idiotopic/sucrosic mosaic (40–400 μm). (2) Planar-e dolomite cement (200–800 μm). Both dolomite types restricted to 4 km-wide strip adjacent to fault juxtaposing limestones against Eocene mudstones	<1%, next to fault 95% dolomite	$\delta^{18}O$ (1) –6.3 t0 –10.3. (2) –5.9 to –9.0 $\delta^{13}C$ (1) –6.0 to 3.8. (2) –1.2 to 0.6			*			*			Variable, up to 20% porosity and 10s to 100s mD permeability. Best quality in sucrosic replacive dolomites with open fractures	Fracturing and degree of dolomitization increases towards fault	Wilson & Evans 2002
Vanda Limestone (V)	Offshore Tarakan basin, NE Kalimantan (Tarakan Basin)	Early Pliocene	Basin-margin, delta-front, shelf-edge setting. Reefs grown on compressional rollover anticlines	Replacive planar-e microcrystalline rhombs (<10 μm). Only replaces clayey matrix of argillaceous coral float/rudstone (at base of reef cycle	? <<20%	–	–	*	*	?					Porosity of facies 1.1–11%	May have taken place at same time as meteoric diagenesis, only affects units rich in clay	Netherwood & Wright 1992
Tigapapan Limestone (BT)	Offshore west Sabah (Sabah Basin)	Middle–Upper Miocene	Shelf, basin margin. Shallow-marine and some redeposited carbonate on dominantly clastic shelf	(1) Anhedral–euhedral sucrosic replacive and pore-filling rhombs (10–50 μm) in lower sandstone only. (2) Small (<50 μm) sucrosic rhombs scattered in organic- and clay-rich layers. (3) Fe-rich sucrosic–subhedral with unclear crystal boundaries (10–60 μm) scattered in organic- and clay-rich layers	Not given, locally up to 60%	$\delta^{18}O$ (1) 1.2–3.6. (2) –2.5 to –1.6. (3) –3.4 to –4.7 $\delta^{13}C$ (1) –32.2 to –37.6. (2) 3.0–7.7. (3) 1.7–4.6	(1) 0.708855. (2) 0.708886. (3) Not given	*	*	*					Low but not given	(1) Early methane derived during uplift and faulting, most likely during subaerial exposure. (2) Related to transformation of smectite–illite during burial compaction and shale dewatering. (3) Burial depths 1.4–1.8 km at 45–55 °C	Ali 1992, 1995

Luconia (LS) Cycles IV, V, VI	Central Luconia, (Sarawak–East Natuna Basin)	Middle (mostly)–Late Miocene. Some recent	Carbonate build-ups on faulted highs, basin margin	(1) Sucrosic dolomites developed during build-out and build-up and affecting muddy backreef and lagoonal deposits. Associated with meteoric leaching. (2) Rhombic dolomite cement derived from bacterial degradation of organic matter in muds	(1) ? Only during build-out (Cycle IV) and build-up	$\delta^{18}O$ (1) average –6 $\delta^{13}C$ (1) average –0.5. (2) –12	Most similar to near-by low-Mg calcite samples	*	*	*		*	15–30% porosity and permeabilities of 10-500 mD	Suggested from Sr data that most dolomitization occurred soon after deposition, some several Ma after deposition. Refluxing brines or mixing-zone dolomites ??	Epting 1980; Ali & Abolins 1999; Vahrenkamp 2000
Terumbu Limestone (TL)	Offshore NE Natuna (Sarawak–East Natuna Basin)	Mostly Middle–Late Miocene (also Early Miocene–Early Pliocene	Syntectonic isolated carbonate platforms and build-ups on faulted highs. Extensive shelf to east on continental basin margin	In coralline algal, echinoderm molluscan packstones from open to inner platform. Occur below sequence boundary related to exposure then flooding	0–60% for facies, overall <10%	–	–		*	*	*	*	Porosity *c.* 15–25% where dolomites common. Average porosity 14% ±5%. Average permeability 11 ± 32.3 mD	Associated with episodic emergence. High CO_2 results from thermal breakdown of deeply buried carbonates	May & Eyles 1985; Dunn *et al.* 1996
Nido Limestone (NL)	Offshore NW Palawan, Philippines (Palawan Basin)	Early Oligocene/ Early Miocene	Initial carbonate shelf on faulted continental basement developed into pinnacles/ build-ups	Minor dolomite cement in well MA-3, precipitated in fractures or vugs	? Very minor	–	–				*		Oil and gas reservoirs in reefal build-ups & forereef talus	Some early dolomites reported from dredge samples	Longman 1980; Wiedicke 1987; Grötsch & Mercadier 1999
Equivalent to Thong, Mang Cau and Nan Con Song Formation	Bac Bo–Yinggehai Basin (Vietnam)	Early–Middle Miocene–Pliocene	Carbonate shelves, platforms and build-ups in extensional to strike-slip basin	Extensive dolomitization reported in large build-ups	?	Limited data; not given	–		*				Gas with high CO_2 contents reported in build-ups	Mixing-zone dolomite associated with build-ups affected by repeated subaerial exposure	Mayall *et al.* 1977
Tacipi Formation (TA)	South Sulawesi (Sengkang Bone Basin)	Late Miocene	Carbonate shelf and coral patch reefs in forearc/intrarc setting	(1) Patchy microcrystalline dolomite replacing micrite (5–10 μm) associated with more porous zones. (2) Minor fissure fill and internal sediment associated with exposure (40–60 μm)		–	–		*			*	Patch reef build-ups form gas reservoir in subsurface	(1) Mixing-zone dolomite. (2) Dolomite sediment related to exposure. Dolomites described from Kapung Baru Field. No dolomite observed in outcrop	Grainge & Davies 1983, 1985; Mayall & Cox 1988; Ascaria *et al.* 1997

Table continued overleaf

Table 3. *continued*

Formation name	Location	Age	Tectonic and depositional setting	Dolomite type, size and occurrence	% of unit	$\delta^{18}O/\delta^{13}C$	$^{86}Sr/^{87}Sr$	Associated with: Mixed siliciclastics	Adjacent siliciclastics	Subaerial exposure	Dissolution	Fracturing	Coralline algae	Restricted facies	Reservoir quality	Notes	Principal references
Makale Formation (MB)	Kalosi area, western Central Sulawesi	Mostly Early–Middle Miocene (Makale may extend down to Late Eocene)	Platform carbonates on faulted high, pass laterally into slope and bathyal deposits. Formed in backarc, extensional or post-rift basin	Organic-rich muddy dolomitic intervals in Lower Makale Formation	?	–	–	*						?	No information given	Deposits formed in restricted (ponded) environment	Coffield *et al.* 1993
Tomori Formation	Tomori, East Arm of Sulawesi (Tomori Basin)	Eocene–Early Miocene	Compressional margin. Initially shallow shelf. Later, much shallower, platform limestones interfinger with reefal build-ups	(1) Microcrystalline, poorly formed dolomite replacing micrite matrix of Lower Platform Limestone. (2) Very fine–finely crystalline dolomite selectively replacing intraclasts and bioclasts in upper reefal limestones. Rare dolomite cement in molds after corals.	Generally not dolomitized	–	–					*			Oil reservoir in lower platformal limestone and upper platformal and reefal limestones gas reservoir	Formation of dolomite has only locally enhanced porosity. (1) Most notable in overthrust sections	Davies 1990
Ruta Formation	Bacan, Moluccas	Middle Miocene	Shallow-marine platform in backarc with local volcaniclastic input	Locally replacive dolomite rhombs occur in recrystallized limestones	?	–	–								No information given		Malaihollo 1993
Cablac/ Aliambata Formation (CF)	Timor	Early Miocene	Microcontinental block. Shallow shelf/platform and bathyal deposits	Dolomitization and dedolomitization are widespread	?	–	–								No information given		Audley Charles 1968
Kais Limestone (K)	Taminabuan, Mar, Irian Jaya (Salawati and Bintuni Basins)	Early–Late Miocene	Continental basement high. Barrier reef complex, shelf and build-ups	(1) Microcystalline (5–10 μm) euhedral–subhedral dolomite replaces mud/micrite matrix.	? Variable but may be pervasive	Whole rock	–	*	*	*	*	*	*		Oil/gas reservoirs in patch reefs/ build-ups along shelf margin,	Dolomitization attributed to: (1) mixing zone, hypersaline waters; (2)	Vincelette 1973; Hendarjo & Netherwood 1986; Dolan & Hermany 1988;

			comprising platform and reefal facies	(2) Euhedral cement partially infills porosity. (3) Late fracture filling. Dolomitization of matrix in backreef cyclic lagoonal facies. Dolomitization more advanced around periphery of reefal build-ups								e.g. Wiriagar (oil). Porosity 1.3–45%, permeabilities 10s to 100s mD	pervasive dolomitization at flanks: expulsion of connate waters from clay compaction	Gibson-Robinson *et al.* 1990; Livingstone *et al.* 1992; Nurzaman & Pujianto 1994
Ogar Limestone (OG) (NGLG)	Fak Fak, Palau Karas/Adi, Irian Jaya (Bintuni Basin)	Eocene–Late Miocene	On continental basement. Patch reefs and platform on open-marine shelf	Dolomite in places	?	–	–		*		*	No information given	Reef limestone, dolomitic in places, vuggy porosity; some chalky and thin shale interbeds	Robinson *et al.* 1990*b*
Onin Limestone (ON) (NGLG)	Fak Fak, Palau Karas/Adi, Irian Jaya (Bintuni Basin)	Early Eocene–Late Miocene	Mostly open marine, rarely shallow marine, shelf on continental basement	Rarely dolomitic	Rare	–	–		*			No information given	Rarely microcrystalline, granular and dolomitic	Robinson *et al.* 1990*a*
Rumbati Limestone Member (RT)	Fak Fak, Irian Jaya (Bintuni Basin)	Middle–Late Miocene	Backreef facies interfingering with reefal Onin and marly Tawar on continental basement	Fine-grained dolomitic limestone	?	–	–		*			No information given	Locally pyritic or scattered pyrite	Pieters *et al.* 1983; Robinson *et al.* 1991
Yawee Limestone (Y)	Omba, Waghete, Irian Jaya (Bintuni Basin)	Eocene–Middle Miocene	Open shallow-marine shelf on continental basement	Scattered dolostone	Minor dolostone	–	–	*	*			Potential reservoir lithology	–	Panggabean 1990; Brash *et al.* 1991
Lengguru Limestone (LN)	Steenkool, Kaimana, Omba, W Irian Jaya (Bintuni and Armeugah Nuaka Basins)	Eocene–Middle Miocene	Shallow carbonate platform on continental basement	Not described	Sparse	–	–	*				No information given	Predominantly micritic	Brash *et al.* 1991
Darai/Puri Limestone (DA)	Kubor Anticline, C. Highlands and Gulf of Papua, PNG (Papuan Basin)	Late Oligocene–Middle Miocene. Eocene in offshore areas	Shallow-water carbonates, shelf margin, build-ups and deeper water deposits in area of continental crust, foreland setting in Noegene	Associated with subaerial exposure surface, karstification and dissolution. Also during transgression	?	–	Measured for dating of dolomite, but not given: at least three phases of dolomitization: 21 Ma, 10 Ma and 55 Ma		*	*	*	Gulf of Papua, gas condensate discoveries in PASCA and Pandora Fields	21 Ma dolomites associated with subaerial exposure, 10 and 5.5 Ma associated with transgression, youngest also with siliciclastics	Sarg *et al.* 1995

Table continued overleaf

Table 3. *continued*

Formation name	Location	Age	Tectonic and depositional setting	Dolomite type, size and occurrence	% of unit	$\delta^{18}O/\delta^{13}C$	$^{86}Sr/^{87}Sr$	Associated with: Mixed siliciclastics	Adjacent siliciclastics	Subaerial exposure	Dissolution	Fracturing	Coralline algae	Restricted facies	Reservoir quality	Notes	Principal references
Miocene Limestones near Daru (LD)	Daru, W PNG (Papuan Basin)	Miocene	Barrier reef at margin of continental shelf	Some dolomite	?	–	–								No information given	Some dolomite	Willmott 1972
Naringel Limestone (NR)	Manus Island, NE PNG (Manus Basin)	Early Pliocene	Shallow-water reefs or banks around volcanic or land area	Argillaceous biogenic dolomite	?	–	–	*							Not described	Dolomites only mentioned in argillaceous facies	Francis 1988
Yalam Limestone (YL)	New Britain NE of PNG (Buna-Trobriand Basin)	Middle Miocene (sometimes Early Miocene–Early Pliocene	Shallow-water carbonates associated with (inactive?) volcanic arc	Patchy dolomite occurrence	?	–	–	*	*						No information given	–	Davies 1973

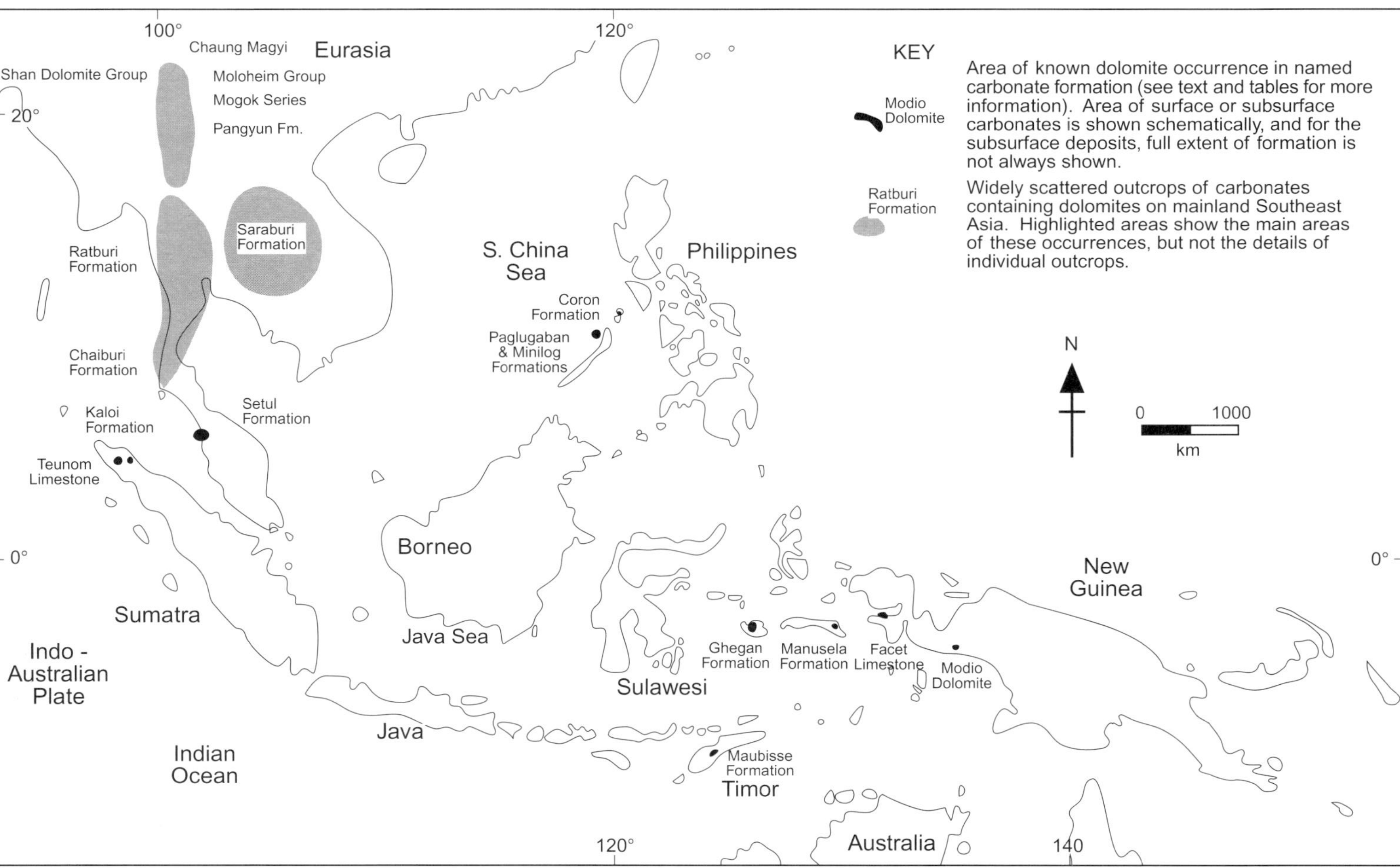

Fig. 1. Known dolomite occurrence in Pre-Tertiary carbonates.

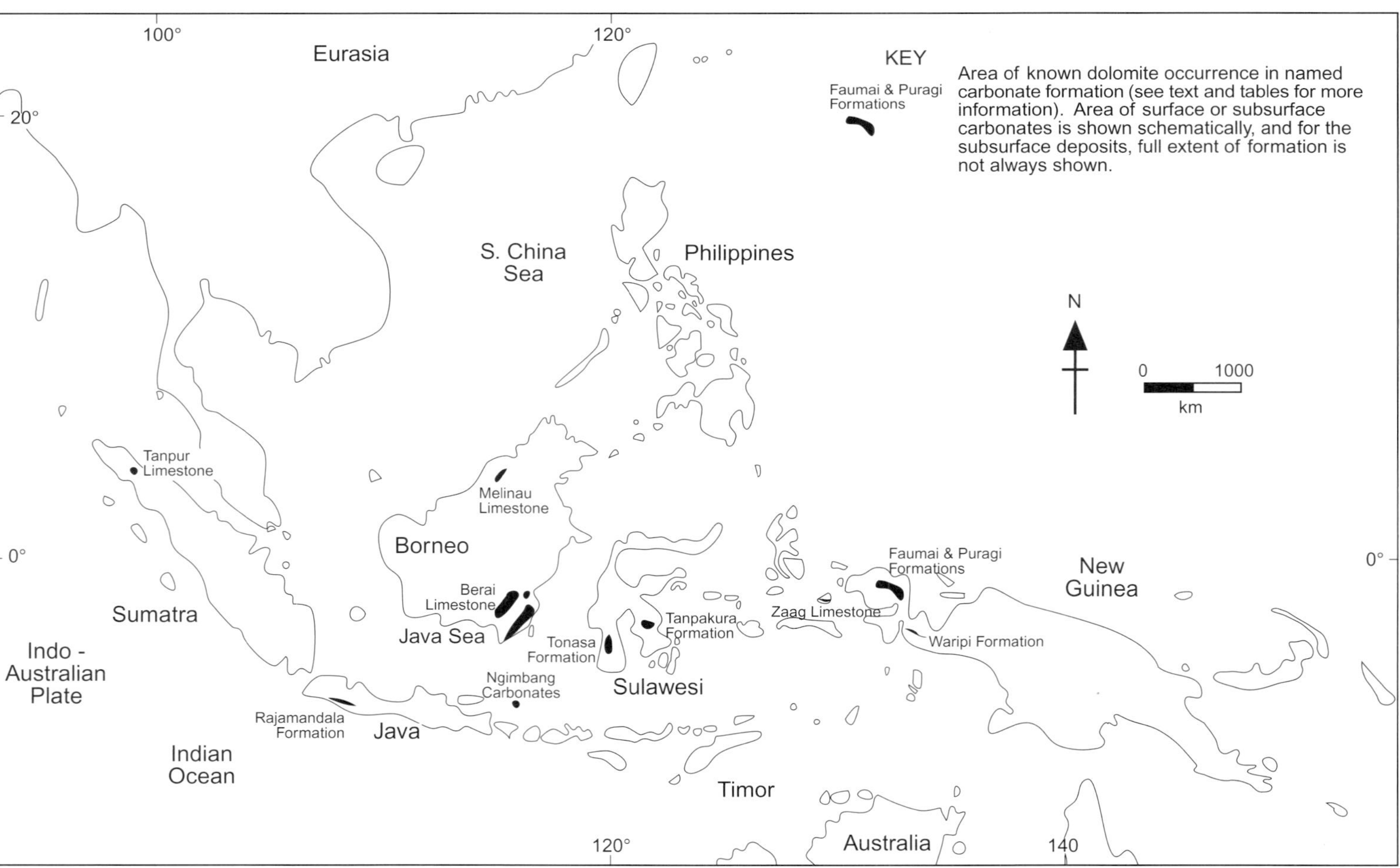

Fig. 2. Known dolomite occurrence in Palaeogene carbonates.

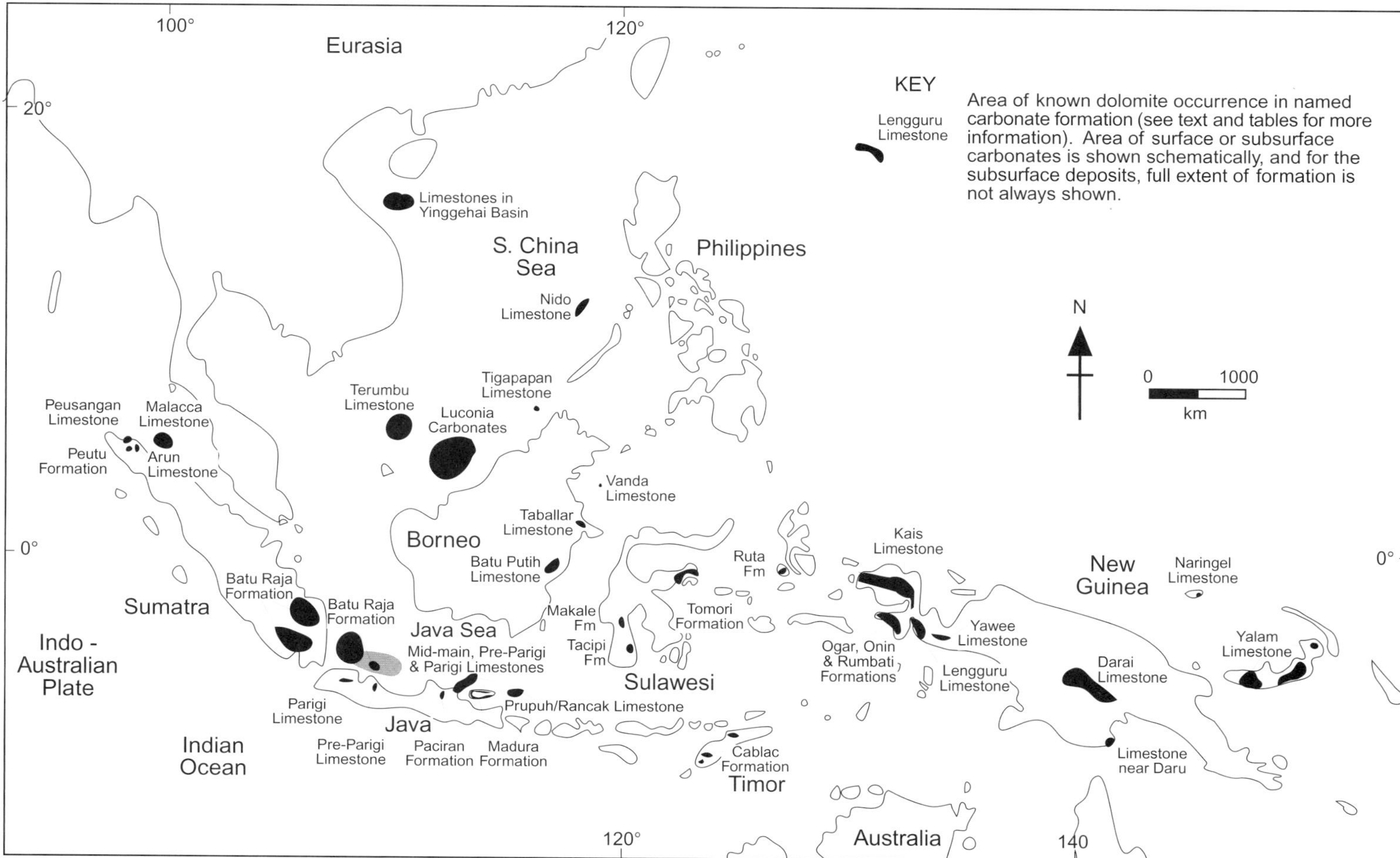

Fig. 3. Known dolomite occurrence in Neogene carbonates.

Table 4. *Reported dolomite occurrences in carbonate formations from SE Asia. Table also shows if isotopic data has been recorded, whether dolomites are associated with any specific features and if a dolomitization model has been inferred*

		Dolomite type				Isotopic data			Associated with								Inferred origin related to						
Age	Number of carbonate formations with reported occurrence of dolomite	Microcrystalline replacive	Replacive	Cement	Not reported	Delta ^{18}O	Delta ^{13}C	$^{86/87}Sr$	Mixed siliciclastics	Adjacent siliciclastics	Subaerial exposure	Dissolution	Fracturing	Coralline algae	Restricted facies	Not recorded	Marine–meteoric mixing	Marine water	Igneous emplacement	Shale compaction/dewatering	Subaerial exposure	Burial	Not reported
Pre-Tertiary	19	1	4	1	15	1	1	0	5	6	1	1	4	0	?1	9	1	1	1	1	0	0	17
Palaeogene	11	6	6	4	3	1	1	0	5	4, ?1	3, ?1	2	4, ?1	3	1	1	0	0	0	1	0	1	10
Neogene	34	14	10	10	16	4	4	3	17	10	9, ?1	5	7	6	?2	6	6, ?1	0	0	4	3, ?2	1	24
Total	64	21	20	15	34	6	6	3	27	20, ?1	13, ?2	8	15, ?1	9	1, ?3	16	7, ?1	1	1	6	3, ?2	2	51

Formation, North Sumatra, but only very limited information is available concerning diagenesis. The lack of published material concerning diagenesis is greater for dolomite, which, although commonly recorded, has received only limited discussion regarding its distribution, timing, origins or association with reservoir quality. It is beyond the scope of this paper to fully document the stratigraphic and sedimentological context of each formation discussed and shown in Tables 1–3. For these data the reader is referred to the references listed and given in Wilson (2002).

Much of the available material is directly or indirectly sponsored by the petroleum industry and tends to concentrate on those carbonates that are economically important (i.e. of Neogene age). As a result, there is a general paucity of publications on pre-Neogene carbonates. For example, the highly productive Batu Raja Formation, with numerous oil and gas fields, is the subject of many publications, whereas the non-prospective Setul Formation is only mentioned briefly in regional studies. The literature on pre-Neogene carbonates, although often of good quality, tends to concentrate on small geographic or stratigraphic ranges, or specific aspects of their depositional or diagenetic character.

Many older publications contain good observational information but carry interpretations that have been superseded by later data acquisition or research that may not always be in the public domain. They nevertheless remain good sources of information and in some instances make observations that have yet to be satisfactorily explained. For example, many carbonate formations containing admixed clays or adjacent to clay-rich formations are dolomitized, with dolomitizing fluids inferred to have been sourced from the compacting muds (see below). In other areas this mechanism is thought unlikely due to insufficient release of magnesium as a result of diagenetic changes in the clays. However, an alternative explanation, together with supporting data, has yet to be given for SE Asian dolomites associated with clays.

This paper considers the spatial and temporal distribution of dolomite within carbonates in SE Asia. Where appropriate, examples and case studies are used to illustrate and emphasise key points. Models to explain the origins and distribution of these dolomites are considered. The association of dolomites with reservoir quality is discussed, not only from the standpoint of porosity and permeability characteristics, but also with regard to generation of non-hydrocarbon gases. However, given the broad range of the study and paucity of published data, it is stressed that the examples are by no means exhaustive, and that there is considerable scope for further study of dolomitization in the region.

Dolomite in carbonates of Pre-Tertiary age

Pre-Tertiary carbonates are widely distributed across SE Asia where they range in age from Pre-Cambrian (e.g. Mogok Group, Myanmar) to Cretaceous (e.g. Nief Beds, Seram). They typically have restricted surface outcrop but are encountered in the subsurface where they are generally considered to form economic basement. However, hydrocarbons have been discovered in the Ratburi and Saraburi groups of Thailand, and the Manusela Formation of Eastern Indonesia. Owing to their perceived lack of economic importance Pre-Tertiary carbonates remain little studied and generally poorly understood. Published reports, although often of a high quality, typically deal with small geographic or stratigraphic ranges, or specific aspects of depositional or diagenetic character. Dolomitization is widespread but, although recognized in many studies, is usually only accorded a passing mention. For example, pre-Tertiary carbonates penetrated by wells in the Straits of Malacca (e.g. NSB-B1 and MSS-XA, Fig. 4) are recorded as being variably dolomitized (Fontaine 1992), but are not described in any detail. Thus, there is no consideration as to the timing or origin of this dolomite, or its influence on reservoir quality.

Pre-Tertiary carbonates are perhaps best known from the Ratburi and Saraburi groups of Thailand and the Manusela Formation of Seram, Indonesia. Of these, the Ratburi Group is the most widely studied and best known (e.g. Chinoroje 1993; Ampornmaha 1995; Heward *et al.* 2000). In the Thai Basin buried Ratburi Group tower karst has been successfully pursued as a basement play (Nang Nuan Field).

Dolomites of the Ratburi Group, Thailand

The Permian Ratburi Group is amongst the best known and most extensive of Pre-Tertiary carbonates in the region. These rocks are known from surface exposures and drilling occurrences across much of onshore and offshore Thailand (Fig. 5), and extend over most of the Sibumasu Terrane (as defined by Metcalfe 1990). However, there is some confusion over distribution as different names (e.g. Chuping Limestone of Malaysia) have been applied locally

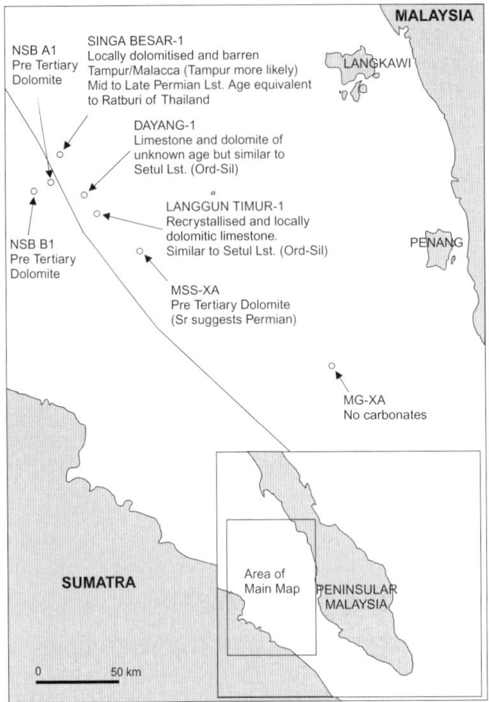

Fig. 4. Pre-Tertiary penetrations in the Malacca Straits. Although dolomites are noted to be present dolomite distribution, crystal habit, origin or impact on reservoir quality are not reported. (Adapted from Fontane 1992).

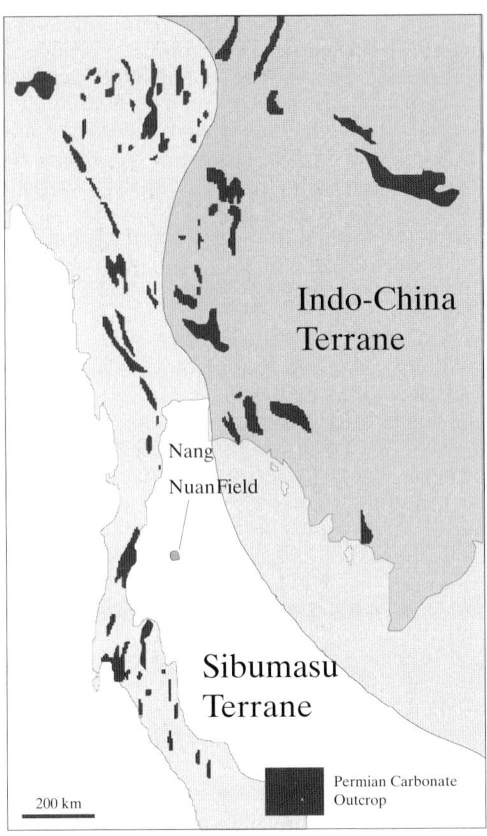

Fig. 5. Outcrops of Permian carbonates in Indo-China. The Ratburi Group and its equivalents crop out over much of the peninsula and western Thailand areas. Exploration activity has shown that Permian carbonates subcrop over much of Sibumasu Terrane.

(Shi & Archbold 1998). Thicknesses of 200–300 m are present in onshore areas of Peninsular Thailand where the Ratburi Group crops out as prominent upstanding tower karst. The Group is probably less extensive than is commonly perceived as it has been widely used as a 'stratigraphic bucket' for all pre-Tertiary carbonates in the Thailand area. For example, the Saraburi Group has commonly been misnamed as Ratburi despite accumulating on a different terrane separated from Sibumasu by the Tethys Ocean (Booth & Sattayarak 2000). The Triassic Chaiburi Group has commonly also been incorporated into the Ratburi Group (Ampornmaha 1995).

The Ratburi Group comprises a series of extensively dolomitized platform carbonates, which developed over the Sibumasu Terrane as it was translocated across Tethys. These carbonates are arranged into an overall upwards-shallowing succession in which five facies associations are recognized (Baird & Bosence 1993): (1) high-energy open platform; (2) low-energy open platform; (3) restricted platform; (4) peritidal and (5) carbonate slope.

Primary depositional fabrics have been extensively modified by diagenesis (Baird & Bosence 1993; Heward *et al.* 2000). An initial phase of syndepositional micritization and precipitation of fringing marine carbonate cements was followed by cementation by shallow burial calcite and extensive dolomitization (Fig. 6). This initial phase of dolomite includes both finely crystalline, texture-preserving, and more coarsely crystalline, hypidiotopic, fabric-destructive, forms. Isotope analysis yields low negative $\delta^{18}O$ and positive $\delta^{13}C$ values, which are close to those for Permian marine waters. Consequently, Baird & Bosence (1993) favour an interpretation in which dolomite precipitation resulted either from marine–meteoric mixing or marine-water circulation.

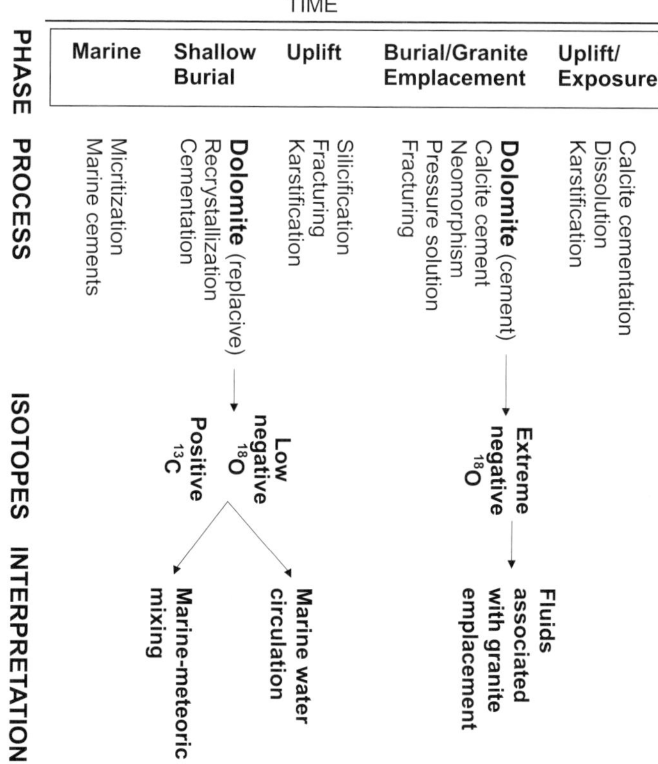

Fig. 6. Diagenesis of the Ratburi Group, showing two phases of dolomite formation. Initial dolomitization is pervasive and masks much of the primary fabric. Later dolomite (and calcite) is associated with granite emplacement. (Modified from Baird & Bosence 1993.)

Early dolomite formation was followed by uplift and karstification. Uplift was followed by a second phase of burial and granite emplacement during the Cretaceous. Some dissolution of early calcite and dolomite has been related to the passage of corrosive fluids associated with granite emplacement. A second phase of dolomite and calcite precipitation followed. This dolomite is coarsely crystalline and euhedral. Isotope studies (extreme negative $\delta^{18}O$) suggest precipitation from fluids associated with the granites (Baird & Bosence 1993). Hydrocarbons are present in carbonates of the Ratburi Group in the Gulf of Thailand. In these subsurface deposits hydrothermal processes rather than exposure and karstification are inferred to have generated secondary porosity (Heward *et al.* 2000). Overall, dolomitization has had a strong destructive influence on porosity in the Ratburi Group (Chinoroje 1993). It has been suggested that reservoir potential is limited away from areas directly influenced by hydrothermal fluids, or by rift-associated exposure and karstification.

Dolomites of the Manusela Formation, Seram

The Manusela Formation of Seram in eastern Indonesia comprises a series of variably dolomitized oolitic grainstones of mid–late Jurassic age (Kemp, 1992). These were deposited on a palaeotopographic high trending E–W, with a broadly similar orientation to modern-day Seram. The high was separated from the emergent Birds Head Microcontinent to the north by a topographic low, which acted as a siliciclastic by-pass. Diagenesis strongly affected the carbonates, and locally the primary microfabric has been completely overprinted. However, relic macrostructures such as cross-bedding remain, and it is suggested that the dolomitized rocks

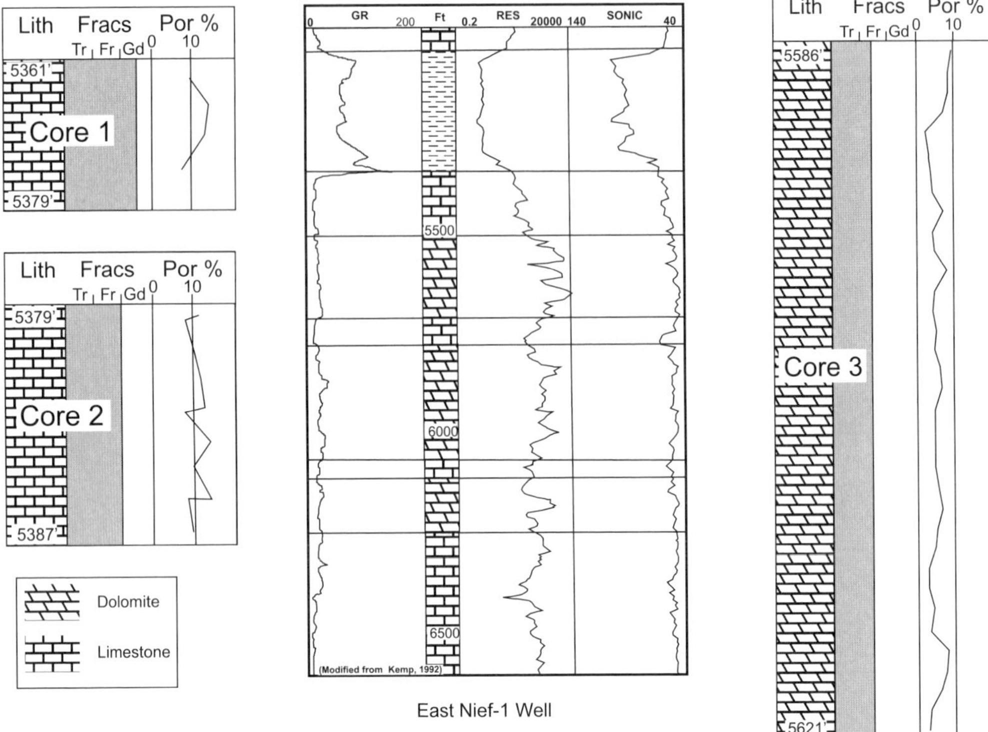

Fig. 7. Type log and core summary logs from East Nief-1 well. In limestone cores porosity generally ranges from 10 to 15%. This contrasts with dolomitic cores where porosity is typically about 5%. Similarly the intensity of open fractures is greatly reduced in dolomite cores. (Modified from Kemp 1992.)

were originally also oolitic grainstones (Kemp 1992). Seeps in the Nief gorge reveal the presence of hydrocarbons in the Manusela Formation. The occurrence of hydrocarbons at depth was confirmed during the drilling of East Nief-1 well in 1988.

Diagenesis caused widespread cementation (Fig. 7), leaching, dolomitization and compaction of the Manusela Formation. There is no evidence for early dolomitization, and all dolomites post-date a phase of burial and compaction. The extent of dolomitisation varies from minor amounts to complete replacement of the primary fabric by fine–medium equant, crystalline mosaics (Fig. 8). Kemp (1992) has suggested that the source of Mg-rich dolomitizing fluids was in dewatering clays in the Salas Formation. These fluids are inferred to have moved along an adjacent thrust fault before entering, and dolomitizing the Manusela Formation (Fig. 9).

Dolomites are associated with a net porosity reduction compared with undolomitized limestones (Fig. 7). In non-dolomitized intervals porosities of >10% are present as evenly distributed intergranular pores with lesser moulds and vugs. Dolomitized intervals have a greatly reduced porosity of about 5%, with patchily distributed intercrystalline and vuggy pores (Fig. 7). Reservoir quality and hydrocarbon recovery from the Manusela Formation is reliant on fracturing. Kemp (1992) stated that fracture density is lower in the dolomites than in the limestones. Recently, Nilandaroe *et al.* (2002) have revealed that fractures and brecciation are common in the dolomites, but are prone to cataclasis in which rock flour may act as an impermeable barrier. Economic reserves are only likely in dolomitized horizons where late fractures remain open (Kemp 1992).

Dolomite in carbonates of Palaeogene age

Carbonates of Palaeogene age are well developed in SE Asia where they occur as extensive platform carbonates developed around the margins of Sundaland (western Indonesia), and the northern margins of Australia and New

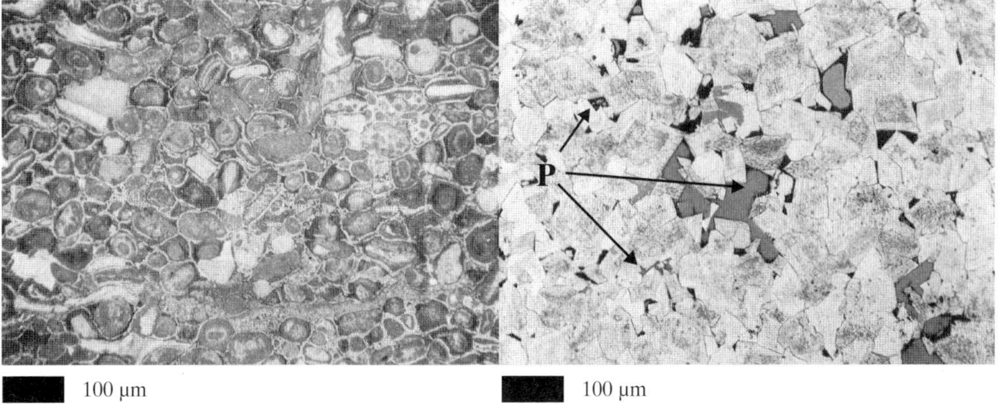

Fig. 8. In non-dolomitized Manusela Formation limestone the pore system is seen to comprise evenly distributed intergranular pores supplemented by lesser vugs and moulds. Where dolomitization has occurred there is no microscopic evidence of the primary fabric, and visible porosity (P) is patchily developed. (Modified from Kemp 1992.)

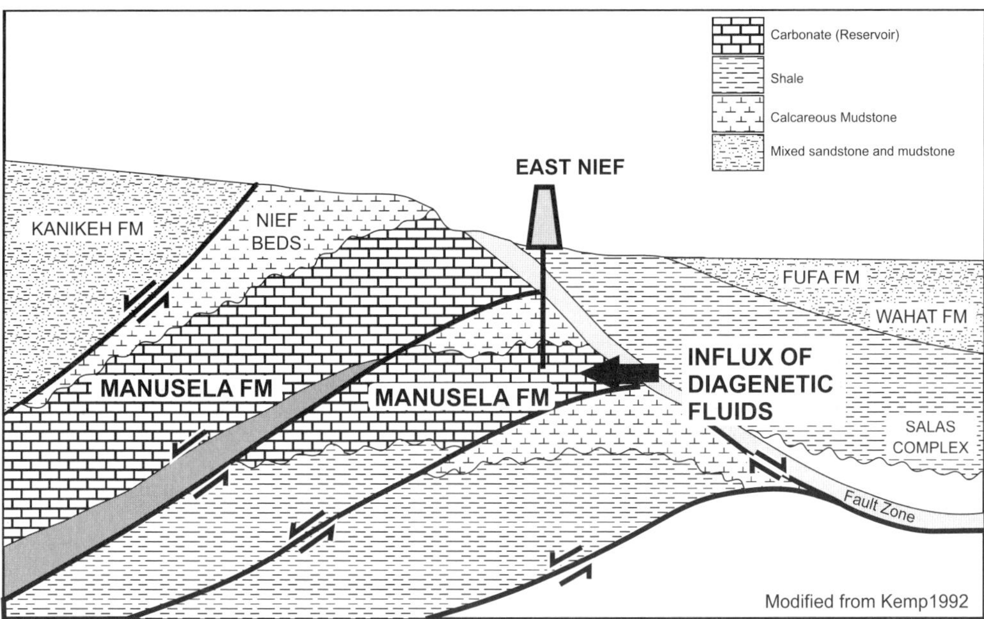

Fig. 9. Model for dolomitization of the Manusela Formation Mg-rich fluids expelled from compacting mudstones (Salas Formation) entered the toe end of a thrust fault, migrated along the fault and entered the Manusela Formation. (Modified from Kemp 1992.)

Guinea. They are typically dominated by larger benthic foraminifera and coralline algae, with corals of lesser importance (Wilson & Rosen 1998; Wilson 2002). Deposition during times of limited eustatic fluctuation in sea level, together with an original stable calcite mineralogy, results in the carbonates being less affected by meteoric diagenesis than Neogene carbonates that are more commonly aragonitic. As a consequence, primary porosity has a significant influence on reservoir quality with secondary porosity often of lesser importance. To date there has been minimal success in exploring for hydrocarbons in Palaeogene carbonates (e.g.

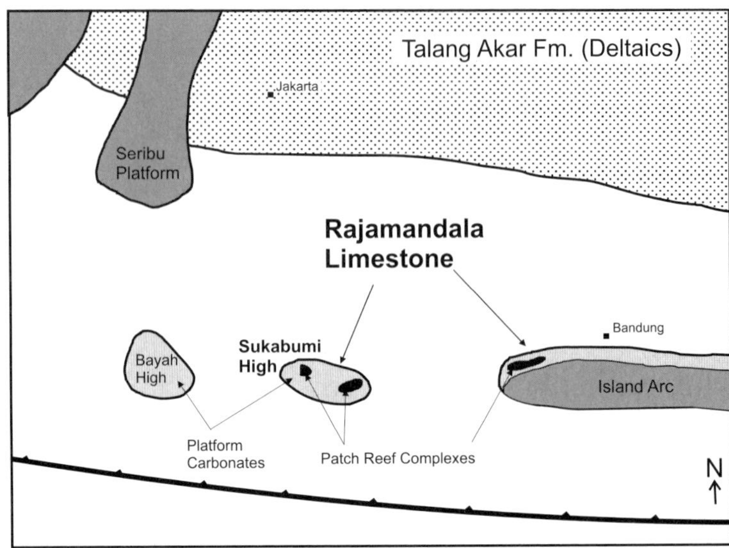

Fig. 10. Palaeogeographical reconstruction of the Rajamandala Limestone. Patch reefs and benthic foraminifera-dominated platforms are developed on palaeogeographic highs with clastics sourced from the Talang Akar Deltaics to the north channelled into the intervening lows (not to scale).

Kerendan Field, Kutei Basin: Saller & Vijaya 2002). Consequently, few outcrop or subsurface studies have been undertaken and much of this research does not focus on hydrocarbon aspects (Wilson & Boscence 1996).

Rajamandala Limestone, South Java Basin

The Rajamandala Limestone of West Java is late Oligocene in age and crops out as a series of topographic highs between Sukabumi in the west and Bandung in the east (Carnell 1996). It comprises a series of reefal build-ups surrounded by carbonate platform rocks developed on palaeotopographic highs (Fig. 10). Any siliciclastic material derived from the deltaic Talang Akar Formation was channelled into intervening lows. Seven depositional facies were recognized based on field and petrographic study: (1) reef; (2) back-barrier; (3) fore-reef debris; (4) foraminiferal/algal shelf; (5) open marine shelf; (6) *Eulepidina* facies; and (7) beach facies (Carnell 1996).

Diagenesis strongly affected the original lithologies and masked many of the primary fabrics. For the most part the diagenetic sequence is similar to that observed in other carbonates in the region, with extensive leaching, cementation, recrystallization and dolomitization. Tectonic activity in the current forearc setting of these carbonates has imparted a strong tectonic overprint to outcrops with well-developed E–W- and N–S-trending faults.

Within the Rajamandala Limestone dolomite occurs in three forms. It is most widespread as isolated rhombs in argillaceous, open marine shelf deposits (Fig. 11, left). It is also common as a coarsely crystalline fabric-replacive phase with an idiotopic–hypidiotopic fabric (Fig. 11, right). This dolomite occurs in several facies, and its distribution is apparently controlled by proximity to N–S-trending faults that cut the outcrops. Isotopic study has not been undertaken, and origins of these dolomites cannot, as yet, be inferred.

The least common form of dolomite is a late-phase cement consisting of well-formed saddle or baroque crystals confined to fractures (Fig. 12). Crystals are commonly coated by residual hydrocarbons, most probably derived from the underlying lacustrine mudstones of the Gunung Walat Quartzite Formation. Fracture-related dolomitization has also been reported in the Taballar Limestone of Kalimantan (Wilson & Evans 2002), where faults have juxtaposed carbonates against deep marine shales. More research is needed on the Rajamandala and Taballar limestones to ascertain whether the magnesium-bearing fluids forming the fault-associated dolomites were derived from adjacent clay-rich formations or another source.

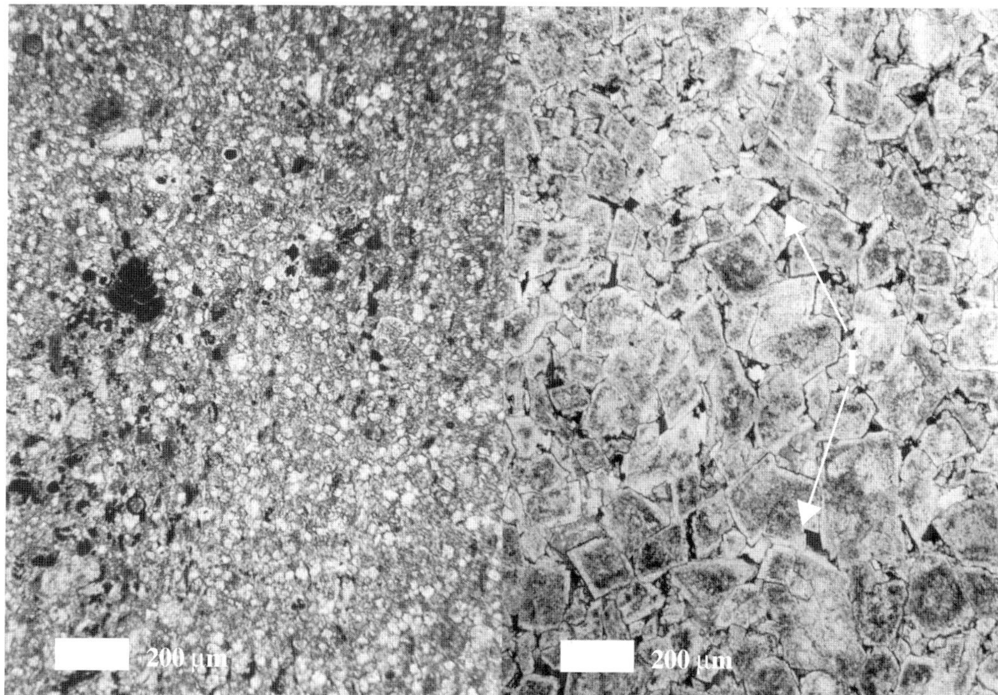

Fig. 11. Photomicrographs of dolomites from the Rajamandala Formation under plane-polarized light (PPL). Dolomite is widespread in the Rajamandala Limestone, most commonly as a microcrystalline replacive phase in more argillaceous limestones (left). It is less common but more complete where associated with faulting (right). Here dolomite consists of well-developed rhombs with minor intercrystalline porosity (I).

Tampur Formation, North Sumatra Basin

The Tampur Limestone developed as a carbonate platform in the North Sumatra Basin. Although extensive dolomitization has been reported from this formation (Collins *et al.* 1996; Barliana *et al.* 2000), there is no published research on its origin and distribution. Despite the limited understanding of deposition and diagenesis of the Tampur Formation, it is economically significant as a likely source for non-hydrocarbon gases, which are a major consideration in hydrocarbon exploration and production in the North Sumatra Basin (see below).

Ngimbang Carbonate Formation, East Java Basin

The middle–late Eocene Ngimbang Carbonate Formation is known from a number of wells in the East Java Basin. This formation is typical of many Palaeogene carbonates in that it formed on a platform dominated by larger benthic foraminifera with localized shales and carbonaceous units. Gas has been discovered at West Kangean in low porosity (typically <2%) carbonates (Siemers *et al.* 1992). Limited matrix porosity is supplemented by an extensive fracture network, which forms much of the total porosity (35–40% of total porosity) and is the main influence on permeability (about 70% of the permeability). At West Kangean microcrystalline dolomite is reported as a minor phase replacing matrix in argillaceous carbonates. Dolomite is locally associated with exposure surfaces where short fractures are filled with geopetal dolomite silt, and later dolomite and calcite cements. Overall, dolomite formation appears to be related to reduced reservoir quality. Replacement of the matrix has resulted in some creation or retention of porosity, but this has been offset by dolomite cementing fractures that are crucial to hydrocarbon deliverability in this field.

Dolomite in carbonates of Neogene age

Neogene carbonates are very extensive in SE Asia and in many basins comprise major

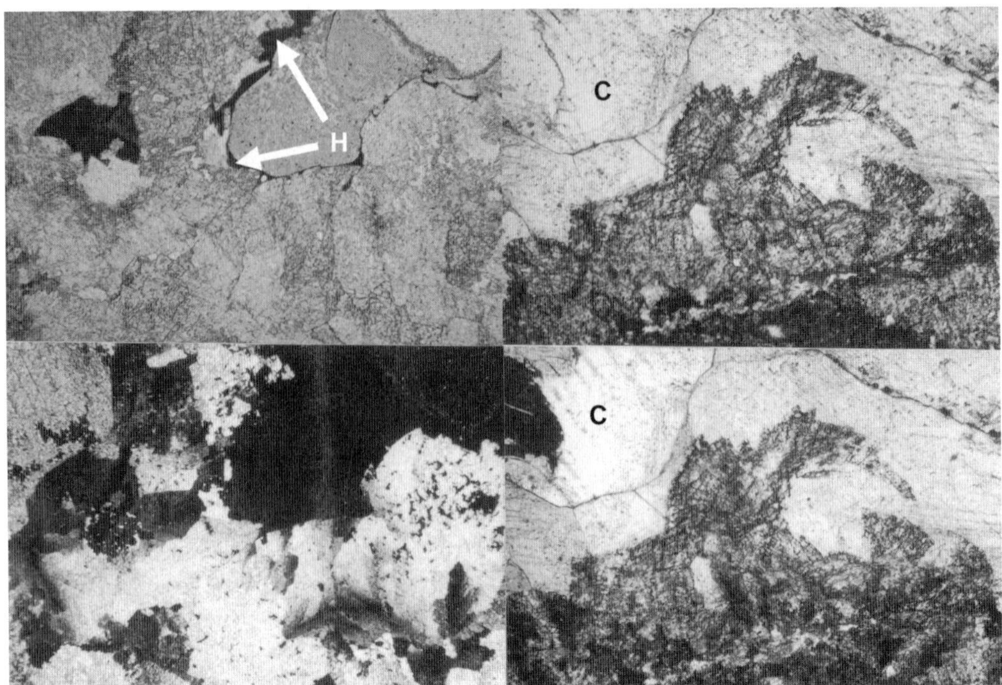

Fig. 12. PPL (top) and cross-polarized light (XPL) (lower) photomicrograph pairs of saddle dolomite (S) from the Rajamandala Formation. In the Rajamandala Limestone, saddle dolomite occurs as a relatively late fracture-filling cement. Locally pores are lined by residual hydrocarbons (H), although saddle dolomite precipitation and hydrocarbon emplacement are separated by a phase of calcite (C) cementation (right). Scale: 1 cm = 100 μm.

hydrocarbon reservoirs (e.g. Batu Raja Formation, South Sumatra, Sunda and North-west Java basins). Owing to their economic importance, there has been considerable study of carbonates of this age, particularly from the subsurface, with a resultant increase in the knowledge base and understanding.

In contrast with the larger foraminifera and coralline alga-dominated Palaeogene carbonates, Neogene carbonates commonly contain an abundant reefal component (e.g. coral and *Halimeda*). The resultant increase in aragonitic material makes them highly susceptible to diagenetic modification. In particular, meteoric alteration is common, due in part to glacio-eustatic fluctuations in sea level during the Neogene. Neogene carbonates are commonly areally restricted and developed in mixed carbonate–clastic systems where carbonates are confined to palaeotopographic highs and areas of low–moderate siliciclastic input.

The dolomite content of these carbonates ranges from absent to abundant. For example, Longman (1980) reported very minor dolomite in the Nido Limestone (offshore Palawan, Philippines), whereas dolomitization has been a major process in oil fields of the Salawati Basin, Irian Jaya (e.g. Gibson-Robinson *et al.* 1990). Despite its widespread occurrence and frequent association with reservoir rocks, the origins and distribution of the dolomite remain little studied.

Central Luconia, Sarawak – East Natuna Basin cycles IV, V and VI

The Central Luconia Platform of Sarawak, East Malaysia is a broad stable platform (Fig. 13), characterized by the extensive development of mid–late Miocene carbonates (Epting 1980). To date, some 200 build-ups have been seismically mapped with 65 of these drilled.

As with many Neogene carbonates in the region those of the Central Luconia Platform show a strong diagenetic overprint that has masked much of the primary depositional fabric. Extensive phases of leaching, cementation,

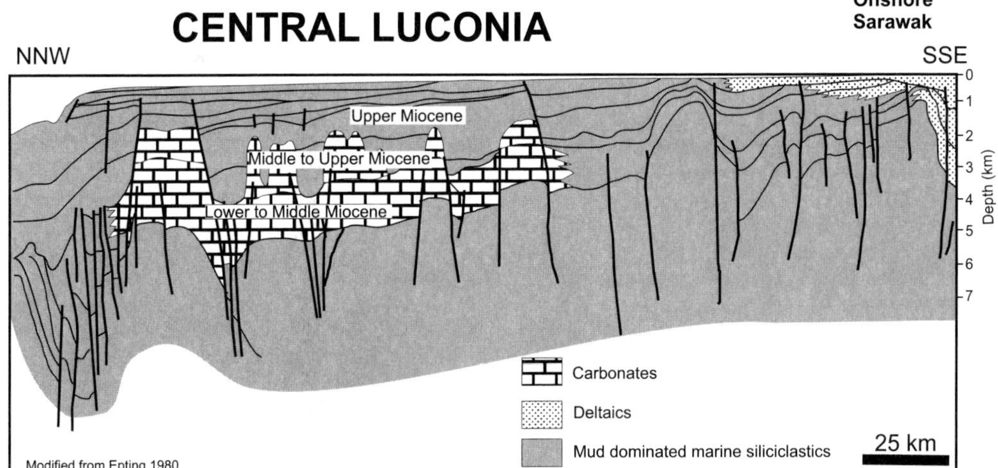

Fig. 13. Cross-section of the Sarawak Basin showing the location of carbonates in Central Luconia. Initial carbonate development was as a broad platform developed over a palaeotopographic high. Shoreline progradation from the southeast led to an increasingly siliciclastic-dominated deposystem and resulted in carbonates becoming restricted to patch and pinnacle reefs. Hydrocarbons are reservoired in these reefal build-ups. (Modified from Epting 1980.)

recrystallization and dolomitization are apparent. Initial leaching and cementation of primary pores is regarded as a freshwater stabilization process that resulted in a redistribution of calcium carbonate with little change in bulk volume. Dolomites are encountered throughout and appear to be associated with exposure surfaces (Fig. 14). Dolomitization occurred intermittently during carbonate build-up and build-out phases (but not build-in or bank phases) and mostly affected protected and reefoid environments (Epting 1980). Lagoonal mudstones and wackestones have been replaced by sucrosic dolomite with a resultant increase or retention of reservoir quality (Fig. 15). Elsewhere dolomitization is less extensive and is confined to matrix in grain-rich sediments. Stable isotope studies have yielded inconclusive results (average $\delta^{18}O$ (PDB) = –0.5‰; average $\delta^{13}C$ = +1.2‰, where PDB is the Peedee belemnite isotope standard). Three theories have been suggested for dolomitization in the Central Luconia Platform, namely:

- Meteoric–marine mixing related to subaerial exposure (Epting 1980).
- Refluxing brines with a high Mg/Ca ratio (Epting 1980).
- Evaporatic model with dolomitization occurring during sea-level lows and during dry seasons (Ali & Abolins 1999).

Climatic evidence suggests that SE Asia experienced an equatorial–subtropical humid climate during most of the Neogene (Frakes 1979). The humid climate suggests that dolomitization in the mixing zone was more likely than reflux or evaporitic mechanisms. However, drier monsoonal intervals are inferred, particularly for the late Miocene and Pliocene, based on pollen records (Morley 2000). Primary evaporitic minerals have not been reported in the area, but the reflux mechanism may have played a role in dolomitization during drier periods. In older (phase IV) carbonates tight zones occur, with intercrystalline porosity cemented by dolomite rhombs. On the basis of carbon isotope studies (average $\delta^{13}C$ (PDB) = –12‰) it has been suggested that bacterial degradation of organic material in argillaceous sediments may have supplied additional CO_2 for dolomite precipitation.

In SE Asia back-barrier/lagoonal carbonates are commonly highly micritic and tend towards poor porosity and permeability. However, in Central Luconia dolomitization of these facies is associated with retained or increased porosity and permeability, resulting in a net improvement in reservoir quality (Figure 15).

Tigapapan Unit, Sabah Basin

Carbonates are rare in the Sabah Basin, reflecting the enormous siliciclastic input during the

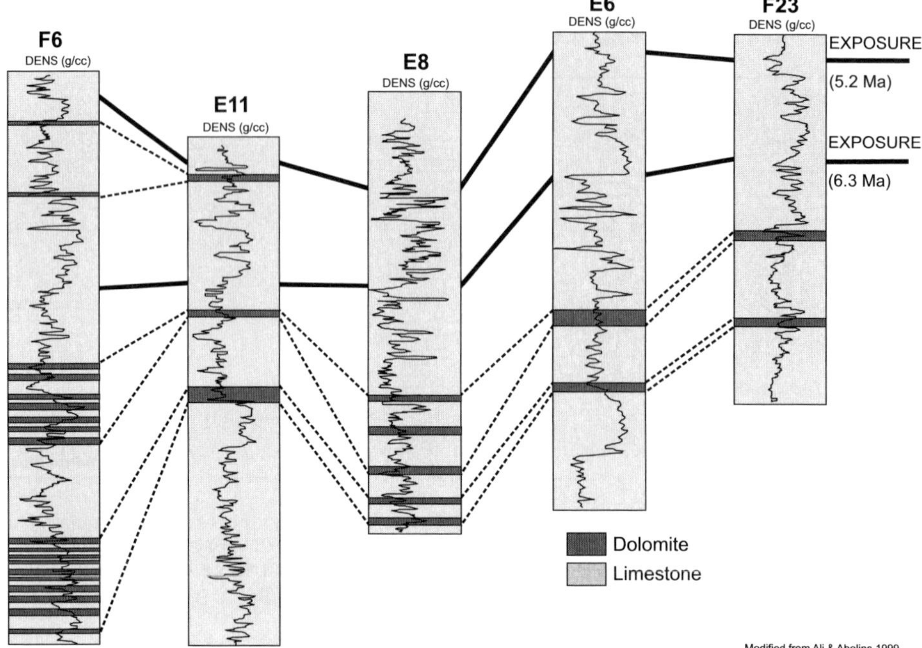

Fig. 14. Dolomitized intervals occur in every well in Luconia and are best developed below exposure surfaces. An association of dolomite with exposure has been noted elsewhere in SE Asia (e.g. Batu Raja Formation) with a rather variable influence on reservoir quality. (Modified from Ali & Abolins 1999.)

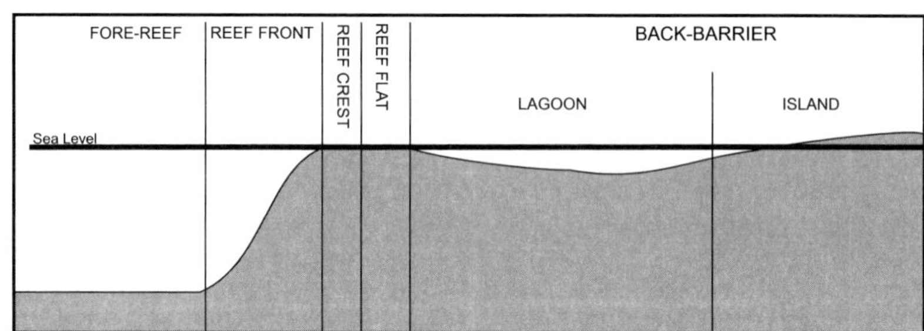

Setting	Fore-reef	Back-barrier/Reefal		Back-barrier
Dunham	Arg M/W	Grainstone/ Packstone	Wackestone/ Mudstone	Wackestone/ Mudstone
Process	Compaction/ Pressure Sol.	Leaching		Dolomitization
End product	Arg Lst	Mouldic Lst	Microporous Lst	Sucrosic Dol
Porosity (%)	2–8	20–40	10–20	15–30
Permeability (mD)	<1–5	10–400	5–50	10–500

Fig. 15. Schematic topographic profile of a typical Luconia Reef. Dolomitization is best developed in back barrier sediments, which are replaced by sucrosic dolomite leading to higher porosity and permeability values than might normally be expected from mudstone–wackestone sediments. (Data from Epting 1980.)

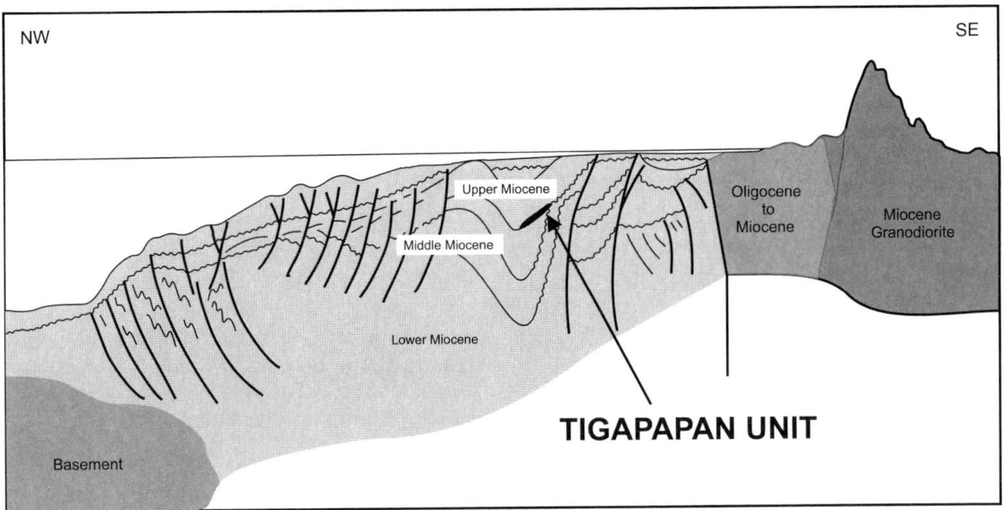

Fig. 16. The Tiga Papan Unit of Sabah, Malaysia, formed in the middle–late Miocene as storm-dominated shoals developed over palaeotopographic highs. Diagenesis is extensive with widespread calcite and dolomite developed which have significantly reduced an initially good reservoir quality. Not to scale. (Modified from Ali 1995.)

Tertiary. The Middle–Late Miocene Tigapapan unit comprises interbedded bioclastic sandstones and sandy limestones (Fig. 16). This mixed carbonate–siliciclastic unit was deposited as a series of progradational storm deposits on a palaeotopographic high (Ali 1995). Sediments are dominantly composed of siliciclastic (15–70%) and carbonate (3–55%) grains, with the latter comprising bryozoans, coralline algae, echinoderms, planktonic foraminifera, and small and larger benthonic foraminifera. The Tigapapan Unit has been extensively studied by Ali (1995), who concentrated on determining the relative ages of cement stages and their relationships to the palaeopore-fluid system, stratigraphy, tectonics and petroleum migration. Nine well-defined phases of diagenetic carbonate are recognizable; with four of these dolomitic (Fig. 17). The four phases of dolomite are thought to have had three distinct origins: methane associated, clay mineral associated and clay/organic interaction associated, with the last of these generating two stages of cement.

An initial phase of dolomite has limited distribution but, where present in the lower sandstone unit, it occurs as pervasive finely crystalline non-ferroan anhedral–euhedral crystals. Isotope analyses show strongly negative $\delta^{13}C$ PDB (–32.22 to –37.64‰) and positive $\delta^{18}O$ PDB (1.19–3.56‰), and it is inferred that the dolomitizing fluids were at least in part derived from the oxidation of methane. The source of this methane is unclear,

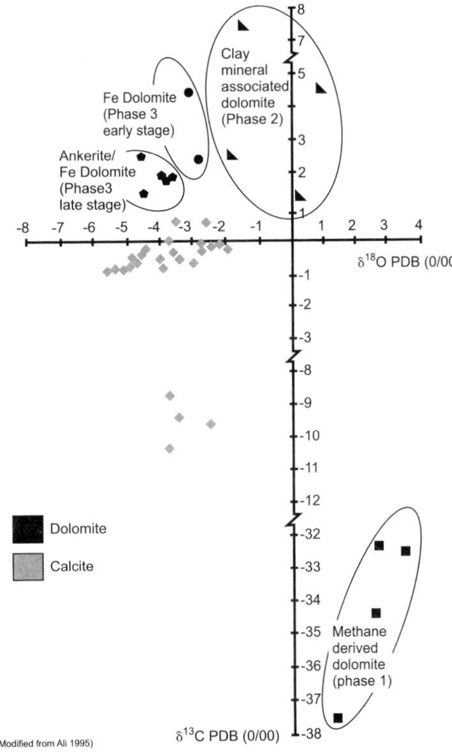

Fig. 17. Petrographic and diagenetic studies of the Tiga Papan Unit have shown the presence of three types of dolomite precipitated in three stages. (Modified from Ali 1995.)

but it is believed to reflect low-temperature biodegradation of organic material (Ali 1995).

The second-stage dolomite is a relatively late, widely developed, diagenetic phase post-dating the development of dissolution seams. It occurs as a sucrosic–finely disseminated rhombic ferroan microspar and is associated with terrigenous mud matrix. Isotope analyses show positive $\delta^{13}C$ PDB (2.99–7.22‰) and negative $\delta^{18}O$ PDB (–1.63 to –2.52‰), and this phase of dolomite is interpreted as being related to fluids derived from clay mineral reactions (smectite–illite transformation) and/or clay dewatering (Ali 1995).

The third and fourth iron-rich phases of dolomite also formed during late diagenesis, and are restricted to basal sandier units. Both are localized and are most common in fine-grained deposits with a high organic content. The third phase is a fine–medium crystalline ferroan dolomite developed as replacive rhombs. Isotope analyses show $\delta^{13}C$ PDB and $\delta^{18}O$ PDB values of 4.56‰ and –3.37‰, respectively. The fourth phase occurs as an anhedral–euhedral ferroan cement associated with ankerite. Isotope analyses of the dolomite show positive $\delta^{13}C$ PDB (1.70–3.23‰) and negative $\delta^{18}O$ PDB (–3.78 to –4.72‰). Both dolomites are inferred to have formed at intermediate burial depths, from fluids derived from organic-rich sediments in which smectite is transforming to illite.

Despite the presence of trace amounts of hydrocarbons, the reservoir potential of this formation is considered to be low. Initial porosity was good and comprised evenly distributed intergranular pores. But although there has been some leaching to generate secondary pores, successive phases of dolomite and calcite cementation have resulted in a progressive loss of porosity.

Batu Raja Limestone, South Sumatra, Sunda and Northwest Java basins

The Early Miocene Batu Raja Formation occurs in the South Sumatra, Sunda and Northwest Java basins where it forms highly productive reservoirs (e.g. Rama, Bima, Cinta, Krisna and Ramba fields). Stratigraphically, some parts of the Batu Raja Formation consist of upper and lower members (Park *et al.* 1995), although many authors treat it as a single unit. The Lower Member is generally considered to be platformal or reefal to near reefal, and was affected by a lowstand in sea level that resulted in much karstic porosity. The remainder of the Lower Member and all of the Upper Member were deposited within an overall transgressive systems tract. The Upper Member shows a series of minor shallowing-upwards cycles. Continuing transgression led to deposition of shales (Batu Raja Shale and Gumai Formation) that act as a seal to the carbonates beneath.

Dolomite is variably recorded in the Batu Raja Formation, and published data suggests it is best developed in the Sunda Basin. Here there is a tendency for the more argillaceous Upper Member to be more dolomitic than the Lower Member (Park *et al.* 1995). In the South Sumatra Basin, dolomite is rarely reported and is considered to be a trace or minor component associated with argillaceous carbonates (e.g. Longman *et al.* 1992). However, the apparent increase in dolomite content in the Sunda Basin may be a function of the available literature, with more detailed studies from this area.

In a study of the Batu Raja Formation (concentrating on the Krisna Field), Park *et al.* (1995) noted a complicated diagenetic history in which three phases of dolomite were recognized. Dolomite is most common as a microcrystalline replacive phase associated with more argillaceous sediment. It was derived from sulphate reduction via oxidation of organic material. In the Rama Field, Tonkin *et al.* (1992) record dolomite crystals of 5–25 μm diameter, most commonly developed in intervals containing greater than 7–10% terrigenous mud. Here, dolomite occurs as isolated to clustered rhombs that can make up to 27% of the bulk volume. This early phase of dolomite tends to have a limited effect on reservoir quality, although locally permeability is improved by increasing, or perhaps maintaining original, pore-throat diameters.

A second phase of dolomite in the Krisna Field occurs as a cement and is best developed in karstic pores (Park *et al.* 1995). This phase has $\delta^{18}O$ PDB values of –2 to –4‰, and $\delta^{13}C$ PDB values of +5–+6‰. Cathodoluminescence reveals that it is complexly zoned. It formed due to mixing-zone processes with a progressive increase in marine-derived fluids. The same phase is recorded in the Rama Field (Tonkin *et al.* 1992), where it is a minor porosity-reducing component. A third and final phase of dolomite in the Krisna Field is ferroan, forming coarse crystalline syntaxial overgrowths on existing crystals. This late phase has an isotopically distinct signature ($\delta^{18}O$ PDB –8 to –10‰ and $\delta^{13}C$ PDB +5–+6‰) interpreted to reflect mixing of expelled methanogenic basinal fluids and downward-percolating meteoric waters derived from the Sumatra hinterland. In some fields, such as Air Sedang, a later phase of saddle dolomite is patchily distributed (Longman *et al.* 1992).

In the Batu Raja Formation the presence of dolomite has either not affected or reduced reservoir quality compared with undolomitized carbonates. Dolomite is most common in argillaceous carbonates that have low reservoir quality. Localized improvements in permeability in such rocks reflect increased pore-throat diameters as a result of dolomitization. However, where present as a cement, dolomite causes a reduction in porosity and permeability by restricting pore throats.

Kais Formation, Salawati and Bintuni basins, New Guinea

The Middle–Late Miocene Kais Limestone is a significant reservoir unit in the Salawati and Bintuni basins of eastern Indonesia (e.g. Walio, Kasim and Cenderawasih fields). This formation is a good example of a reservoir rock in which dolomitization has had a mixed influence on quality (Gibson-Robinson *et al.* 1990). The Kais Formation consists of platform and reefal carbonates covering much of the Birds Head area of Irian Jaya (Fig. 18). To the south, more argillaceous sediments were deposited in deeper water (Klamogun Formation). Upwards, there is a marked increase in siliciclastic influx (Klasafet Formation), and carbonate-producing areas became progressively smaller. By the end of the Miocene, carbonates were restricted to isolated reefal complexes in which hydrocarbons have accumulated. These build-ups are surrounded by the argillaceous Klasafet Formation, which acts as both seal (lateral and vertical) and source rock.

Diagenesis strongly influenced the Kais Formation, with widespread leaching, cementation, recrystallization, dolomitization and fracturing reported (Hendarjo & Netherwood 1986; Gibson-Robinson *et al.* 1990; Livingstone *et al.* 1992; Nurzaman & Pujianto 1994). Initial diagenetic phases are commonly masked by later dolomitization. However, where evidence survives, early diagenetic events affecting the formation appear to have been similar to those of many other carbonates in the region with marine phreatic and freshwater phreatic processes operating. Dolomite is widespread and typically occurs as a microcrystalline–finely crystalline fabric-retentive phase, preferentially replacing the matrix (Fig. 19) (Livingstone *et al.* 1992). Locally there is also some replacement of skeletal grains in areas where dolomitization

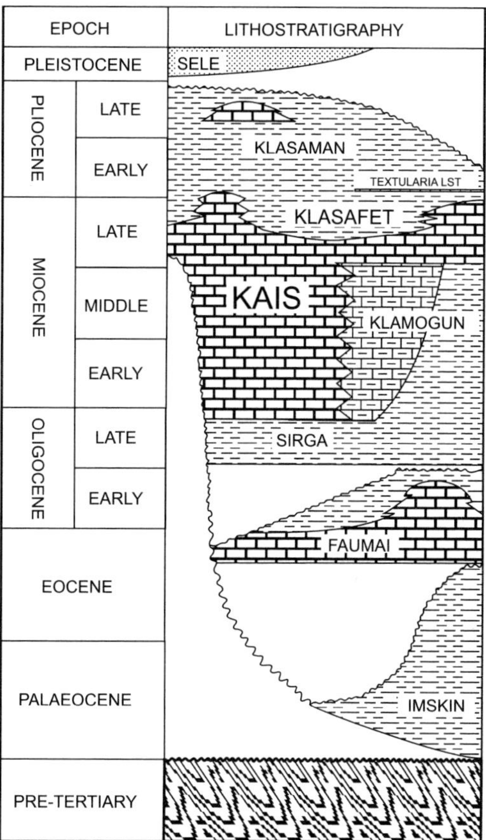

Fig. 18. Stratigraphic column of the Salawati Basin. The extensively dolomitized Kais Formation is the main reservoir, with hydrocarbons trapped in reefal build-ups. (Various sources.)

is more extensive. This replacive dolomite is interpreted to be a relatively early diagenetic phase reflecting mixing-zone processes (Hendarjo & Netherwood 1986). Gibson-Robinson *et al.* (1990) report the presence of celestine as an intercrystalline cement and this has been suggested as supporting evidence for an early diagenetic origin to the dolomite, as at this time strontium would have been liberated from aragonite.

Although dolomite is best developed as a replacement, it is also widespread as a cement, where it occludes residual primary pores and secondary moulds, vugs and fractures (Livingstone *et al.* 1992). The development of dolomite as a fracture cement suggests a second phase of precipitation.

Dolomitization is reported to be more extensive around the flanks of reefal build-ups and

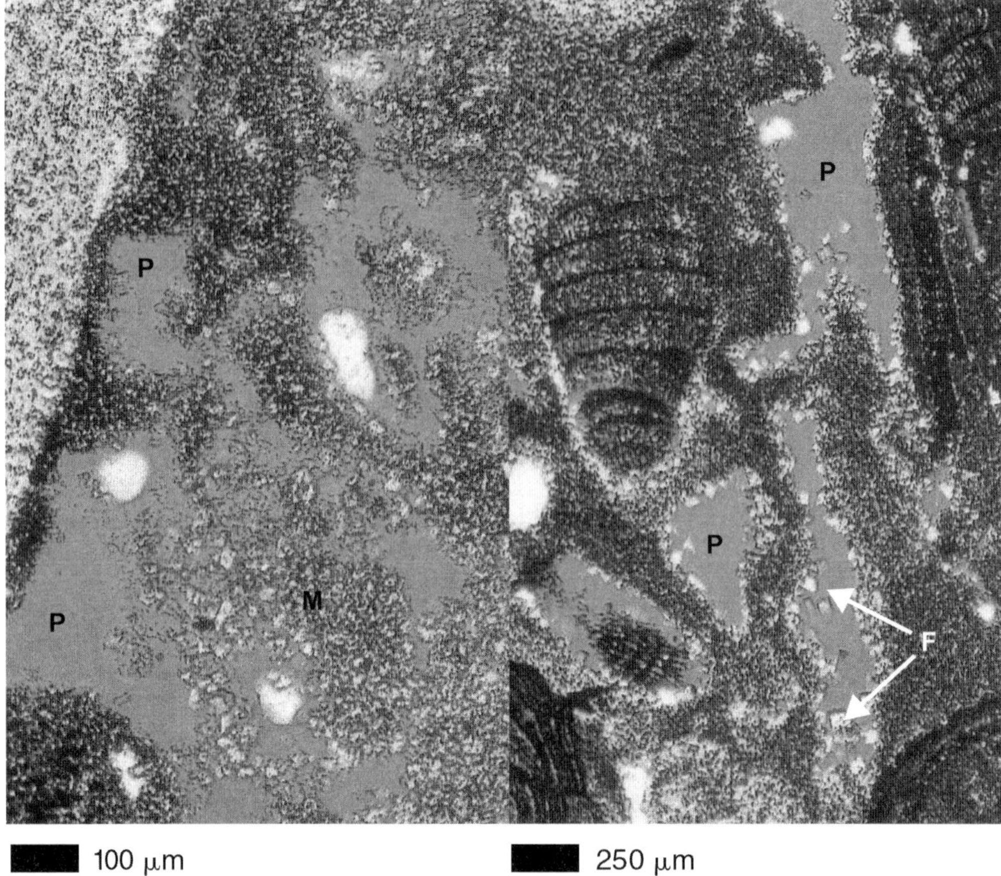

Fig. 19. Photomicrographs of dolomitized Kais limestone showing large pores (P) within dolomicrite (M). Dolomite is developed as fringing cements (F) in mouldic pores. Scale: left 1 cm = 100 μm; right 1 cm = 250 μm. (Modified from Livingstone *et al.* 1992.)

this has led to an interpretion of dolomitizing fluids, at least in part, derived from the surrounding clays (Livingstone *et al.* 1992). It is possible that the late phase of largely cementing dolomite is derived from waters expelled from clays (Dolan & Hermany 1988). Unfortunately, no detailed diagenetic studies have been published that might support this hypothesis. Minor amounts of dolomite are reported in association with stylolites.

In the Kais Formation dolomitized horizons often have good reservoir quality (Fig. 20), although where dolomite cements are present these may reduce porosity and permeability (e.g. Kasim Utara Field). The best reservoir quality has been reported where dolomites are present as a matrix-replacive phase. For example, 20–30% porosities, and permeabilities of tens to hundreds of millidarcies (mD) are reported in pervasively dolomitized cores (Hendarjo & Netherwood 1986; Livingstone *et al.* 1992).

CO_2 from thermal degradation of dolomite

Non-hydrocarbon gases (principally CO_2) are a significant feature of many SE Asian basins (e.g. the Bac Bo Yinggehai and North Sumatra basins) and are a major consideration in hydrocarbon exploration and development programmes. These gases are perhaps best known in the East Natuna–Sarawak and North Sumatra basins, and indeed CO_2 disposal has been the major consideration delaying development of the Natuna L-Alpha Field, which, although containing large hydrocarbon gas

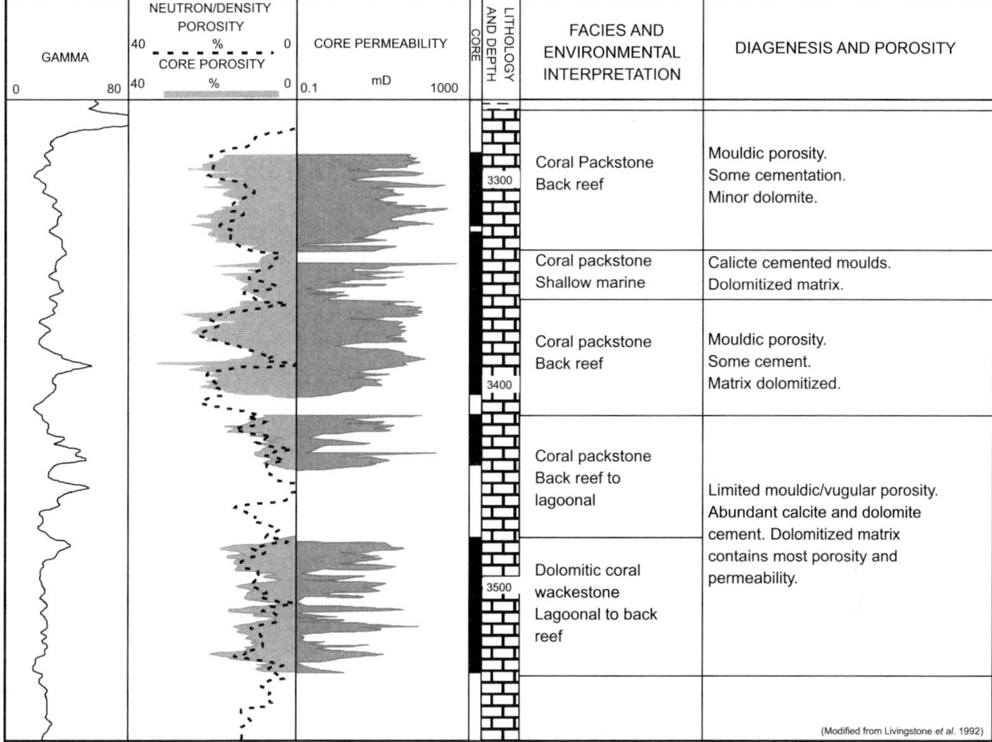

Fig. 20. Summary of Kasim log and core data from the Kasim-24 well. Widespread replacive dolomitization is noted with dolomite preferentially replacing the matrix. (Modified from Livingstone *et al.* 1992.)

reserves (67 TCF gas, where tcf is trillion cubic ft) also contains enormous volumes of CO_2 (200 TCF).

The origin and distribution of the carbon dioxide are perhaps best understood in the North Sumatra Basin (Fig. 21), where CO_2 contents can be as high as 85% (Table 5) and have long been recognized as a problem. However, it was only with the drilling of Kuala Langsa-1 well, where a potential supergiant gas discovery similar in size to the Arun Field was found to be dominated by CO_2, that the importance of understanding its distribution was recognized.

There are a number of mechanisms for generating this CO_2 (Hunt 1996; Cooper *et al.* 1997), which can be sourced from both inorganic and organic sources. Organically derived CO_2 can be generated by:

- thermal breakdown of kerogen;
- biogenic processes;
- thermochemical sulphate reaction.

Inorganic CO_2 can be derived from:

- mantle degassing;
- metamorphic mineral reactions;
- diagenetic mineral reactions (both of siliciclastic and carbonate deposits).

Isotope studies

Caughey & Wahyudi (1993) suggested that CO_2 was derived from an inorganic source, based on carbon isotope measurements ranging from $\delta^{13}C$ –0.95 to –3.97 ‰. Similarly, Reaves & Sulaeman (1994) give carbon isotope measurements ranging from $\delta^{13}C$ +2.0 to –4.0 ‰. These ranges are similar to those that might be expected from thermal decomposition of carbonates (typically $\delta^{13}C$ 0.0 to –10 ‰) and are in marked contrast to those derived from methane in the same reservoirs ($\delta^{13}C$ –32.76 to –37.16 ‰). This difference has been interpreted to reflect methane and CO_2 generation from different sources. The most likely source of

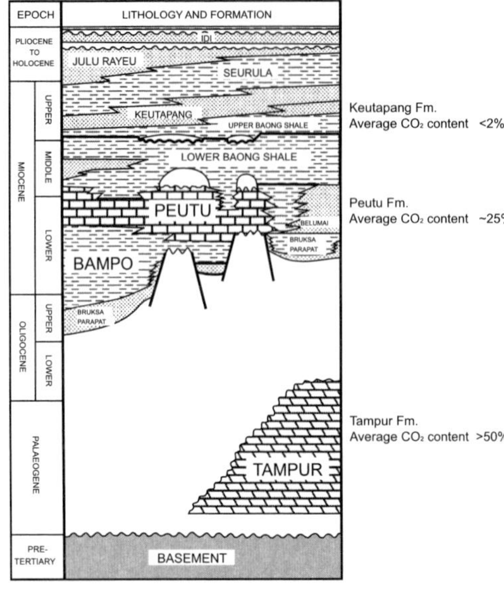

Fig. 21. Chronostratigraphic cross-section of the North Sumatra Basin. There is a marked upwards decrease in CO_2 content from the Tampur Formation to the Keutapang Formation.

CO_2 is from the basement (reflecting the tectonically active nature of the area) or from thermal decomposition of carbonates. Although mantle-derived carbon isotope measurements ($\delta^{13}C$ –4.7 to –8.0 ‰) are similar to those derived from thermally decomposing carbonates, a lack of obvious conduits from the basement and low helium isotope values (Cooper *et al.* 1997 give an $^3He/^4He$ isotope ratio of 0.132 relative to air) suggest that any CO_2 contribution from the basement is minimal, and that CO_2 is most probably derived from carbonates.

Distribution

In the North Sumatra Basin there are several potential sources for CO_2 generation. Kissin & Pakhamov (1967) have shown that carbonates containing impurities (such as magnesium) generate large volumes of CO_2 when heated to temperatures over 150 °C. Carbonates of the Peutu Formation and the Tampur Formation are considered to be the most likely CO_2 sources, and over much of the North Sumatra Basin are buried to temperatures in excess of 150 °C. Other potential carbonate sources for

Table 5. *CO_2 concentrations for North Sumatra Basin (from Caughey & Wahyudi 1993; Cooper* et al. *1997)*

Field	Formation	Substrate	Depth (m)	CO_2 (%)	Comments
Alur Siwah	Peutu	Tampur	2900	30	
Arun	Peutu	Bampo	3050	15	Average of field
Arun A1	Peutu	Bampo	2895	15	
Arun A2	Peutu	Bampo	2890	17	
Arun A3	Peutu	Bampo	3080	14	
Arun A5	Peutu	Bampo	3100	13	
Arun A6	Peutu	Bampo	3060	14	
Arun A-8	Peutu	Tampur	3078	25	Off Arun structure
Bata-1	Peutu	Belumai	2065	14	
Bata-1	Belumai	Tampur	2089	18	
Cunda	Peutu	Phyllite	3755	26	Average of field
Cunda A2	Peutu	Phyllite	3050	25	
Iee Tabeue-50	Peutu	Unknown	2386	18	
Jeuku-1	Bampo	Unknown	3300	36	Reported as up to 73%
Kuala Langsa-1	Peutu	Metased	3370	82	
Kuala Muku-1A	Peutu	Bampo	3220	7	
Lho Sukon S. A	Peutu	Tampur	2590	24	
NSB-A	Peutu	Unknown	1533	31	
NSB-J1	Peutu	Unknown	1488	28	
NSB-J2	Peutu	Unknown	1493	29	
NSB-R	Peutu	Unknown	1320	19	
Pase-C1	Peutu	Unknown		33	
Pergidatit-AB1	Belumai	Pre-Tertiary	2397	11	
Peudawa Rayeu	Belumai	Unknown	2979	10	
Peulalu-3	Peutu	Bampo	2311	22	
Peutouw-1	Peutu	Tampur	2764	64	

CO_2, such as those in the Woyla Group or the Baong Carbonate, are considered too localized to generate the large volumes of CO_2 seen over the greater part of the basin.

In the North Sumatra Basin, CO_2 distribution is by no means uniform and a number of trends can be identified. There is an overall relationship between CO_2 content and stratigraphic position (Fig. 21). In general, CO_2 contents are highest in the Tampur Formation (average 51%) and decrease with decreasing age (Peutu carbonates average 26% and Keutapang Formation average 1%). Superimposed on this stratigraphic trend is a tendency for increased CO_2 content with increased depth. For example, the Arun (15%), South Lho Sukon (24% CO_2) and Bata (14% CO_2) fields show lower CO_2 contents compared to the stratigraphically equivalent, but more deeply buried, Alur Siwah Field, which contains a CO_2 content of some 30%.

In addition to these trends there is a strong correlation of CO_2 content and proximity to dolomites of the Tampur Formation, suggesting that these may be a source for CO_2. For example, the Arun Field overlies Bampo shales and has an average CO_2 content of 15%, whereas the nearby Arun A8 well (to the east of the Arun Field) has a CO_2 level of 25.33% and the time-equivalent Miocene carbonates unconformably overlie the Tampur Formation. Similarly, the Kuala Langsa deposits, with some 82% CO_2, are either in direct or proximal faulted contact with the Tampur Formation. It is therefore considered that although the Peutu Formation, and time-equivalent carbonates in North Sumatra, are locally dolomitic (dolomites up to 15 m thick occur in the Arun Field) it is the more consistently dolomitic and deeper buried Tampur Formation that is the main source for CO_2 generation (Caughey & Wahyudi 1993). In most instances dolomites of the Tampur Formation are directly or indirectly associated with elevated CO_2 levels.

CO_2 source rocks

All evidence points towards thermal decomposition of dolomites of the Tampur Formation as the major source for CO_2 in the basin, with CO_2 migrating along faults and carrier beds away from kitchen areas and into reservoirs (Fig. 22). For example, the structural grain of the basin has favoured migration of CO_2 to the north onto the Malacca Platform where discoveries to date show elevated CO_2 levels. Knowledge of potential migration paths ('basin plumbing') is therefore essential to determine the likelihood of elevated CO_2 levels in any given prospect (Reaves & Sulaeman 1994). It is interesting to note that locally high CO_2 levels in the Malay Basin are associated with deep-seated faults that may penetrate to basement, which, at least locally, contains carbonates. Studies of onshore locations indicate that basement carbonates are likely to be strongly dolomitized. High heat flows are reported for the Malay Basin and it is possible that much of the CO_2 is derived from thermal decomposition of carbonates.

Conclusions

Dolomite distribution and habits

Although dolomite comprises a small part of the total volume of carbonates in SE Asia, it is spatially and temporally widespread, reflecting the distribution of the precursor limestones. On a formation and reservoir scale, dolomite content and distribution are highly varied. For example, within the Manusela Formation of Seram dolomitization may be absent or complete. Dolomitization within the Kais Formation of Irian Jaya also shows considerable variability from field to field.

Many workers record the presence of dolomite without commenting on habit, distribution or origins. Most information is available on Neogene carbonates, reflecting their economic importance to the region, but even here the knowledge and understanding of dolomite is patchy. For example, in the Arun Field porous and permeable dolomitized zones 3–15 m thick are reported without information on their origin, lateral extent or any variations in dolomite fabric (Graves & Weegar 1973). By contrast, the Tigapapan Unit of the Sabah Basin has been well studied with much associated discussion of dolomite habit, origin and distribution (Ali 1995).

Although there are significant shortfalls in the availability and quality of data it is clear that in SE Asia dolomite occurs in both cementing and replacive habits. On the basis of available data, dolomite appears to be variably distributed, but is commonly associated with: (a) particular facies; (b) specific parts, or surfaces within platforms or build-ups; and (c) secondary compaction, or tectonic-induced, features.

Facies associated. Overall, dolomite is most widely encountered as a replacive phase associated with argillaceous carbonates (Berry 1976). In formations such as the Batu Raja, Batu Putih and Vanda carbonates, it is most commonly

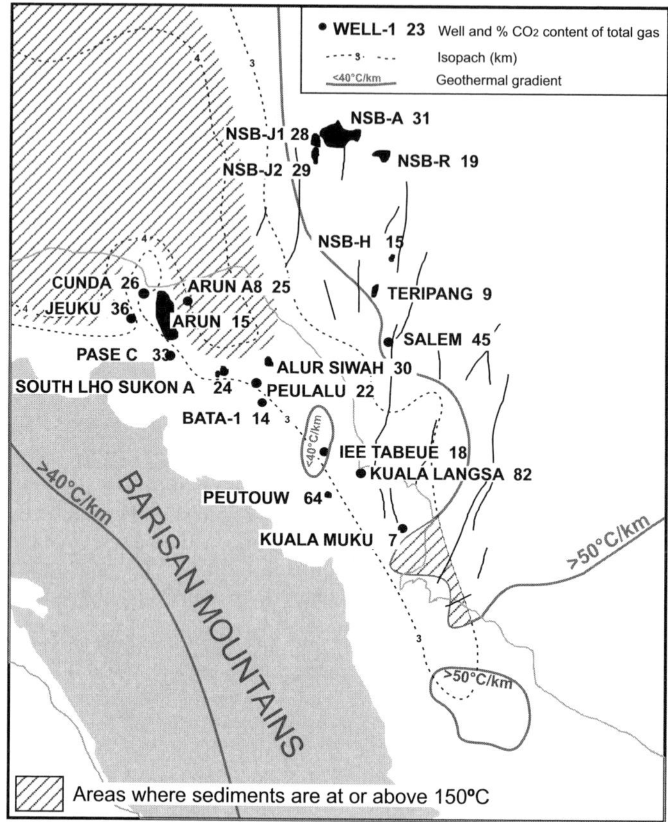

Fig. 22. A map of the North Sumatra Basin showing measured CO_2 distributions in the Peutu Formation together with areas where dolomites of the Tampur Formation are most likely to be generating CO_2.

reported in deposits containing greater than 7–10% terrigenous mud (Netherwood & Wight 1992; Tonkin *et al.* 1992). In these argillaceous deposits dolomite occurs as microcrystalline–finely crystalline rhombs, which locally form clusters. Although many authors record the presence of this dolomite, its paragenesis is generally not interpreted. Park *et al.* (1995) noted that this form of dolomite is locally partially enclosed in fringing cements developed in early formed voids, and interpreted it as a phase developed prior to significant compaction of mudstones. However, in other argillaceous deposits microcrystalline dolomites post-date the formation of dissolution seams and are interpreted as a burial feature (Ali 1995).

Where pure carbonates have been affected, dolomite typically replaces the micrite matrix first. This is a reflection of the large surface area to volume ratio and the high number of nucleation sites, with smaller particles more reactive than larger. Such replacement tends to be fabric-retentive and commonly contains high microporosity. Based on diagenetic relationships, it is generally inferred that this is a relatively early diagenetic phase, although later dolomitization is suggested locally (e.g. the Manusela Formation, Seram).

Associated with specific parts of, or surfaces within, platforms or build-ups. Many authors have noticed an association between dolomite and exposure surfaces, with dolomite developed as a cementing and replacive phase. Examples in which dolomites are well developed below exposure surfaces are seen in the Luconia Platform (Epting 1980), the Batu Raja Formation (Park *et al.* 1995) and the Tacipi Formation (Mayall & Cox 1988). Dolomite is developed in association with both syndepositional exposure and exposure related to post-burial uplift. In a number of examples, dolomite formation followed karstification, and dolomite cements precipitated in vugs or cavities (Mayall & Cox 1988; Park *et al.* 1995).

Sun & Esteban (1994) suggested that dolomites are commonly encountered in the basal parts and margins of SE Asian carbonate platforms. The occurrence of dolomites in the basal parts of platforms has only been reported from Miocene carbonates in northern Sumatra (Arun, Peutu and Peusangan). In contrast, the Batu Raja Formation of southern Sumatra is most commonly dolomitized in its upper part. However, in most carbonates in the region, dolomitization is not associated with either upper or lower parts of carbonate successions. Replacive dolomites are reported from the flanks of build-ups in the Berai (Saller & Vijaya 2002) and Kais formations (Livingstone *et al.* 1992). In many of these associations with specific parts of platforms, dolomite occurrences are in clay-rich facies (Livingstone *et al.* 1992; Park *et al.* 1995). Other occurrences in specific parts of platforms have been linked to fluids derived from underlying or adjacent compacting shales (Livingstone *et al.* 1992; Saller & Vijaya 2002).

Associated with secondary compaction, or tectonic-induced, features. Dolomite is commonly encountered as a late diagenetic stage associated with stylolites where it occurs as a replacement and as a cement. Replacive dolomite tends to form non-ferroan rhombs within and adjacent to the stylolites, but also occurs as a microcrystalline replacement of adjacent fabric. Dolomite cement occludes tension gashes and pores in the precursor fabric. Stylolite-associated dolomite is volumetrically minor and is typically overlooked in the literature.

Dolomite is widespread as a fracture-filling cement (e.g. Batu Raja Formation: Park *et al.* 1995), but also occurs as a replacive phase associated with fractures (e.g. Taballar Formation and Rajamandala Limestone). Dolomite cements associated with stylolites and fractures may be either planar or nonplanar saddle dolomites. The degree of replacement decreases away from fractures. It is inferred that fractures were important conduits, providing pathways for dolomitizing fluids to move into adjacent limestones.

Models for dolomite origin

Although dolomite is widely recognized in SE Asia, few studies provide data to enable the origins of dolomitizing fluids to be evaluated. As a consequence, the processes by which carbonates have been dolomitized remain poorly understood. Most published work suggests dolomitization has taken place in the mixing zone. In examples such as the Tacipi, BatuRaja and Luconia carbonates, this interpretation is supported with stratigraphic, textural and isotopic data. However, for a number of other formations no isotopic data are available and a mixing-zone origin has been suggested without substantiating evidence (Hendarjo & Netherwood 1986). The perceived, although unproven, dominance of mixing zone models in the Neogene is perhaps to be expected as most of the studied carbonates are reefal and by their very nature susceptible to mixing-zone processes. Dolomitization in response to exposure is widely hypothesized and in the past may have been underestimated or misidentified. The deposition of carbonates (particularly Neogene carbonates) close to sea-level makes them susceptible to exposure (by eustatic sea-level fluctuations) and karstic processes. Isotopic signatures in dolomite cements in karst cavities in the Batu Raja Formation record progressive re-emplacement of marine fluids (i.e. mixing-zone dolomitization) following karstification (Park *et al.* 1995)

Textural relationships and isotopic data suggest that microcrystalline replacive dolomite in organic-rich, argillaceous sediments has at least two origins. An early diagenetic phase recognized in the Batu Raja Formation has been related to sulphate reduction through the oxidation of organic material (Park *et al.* 1995). In a number of other argillaceous carbonates, such as the Tigapapan and Vanda limestones, microcrystalline dolomite is a late diagenetic feature formed at intermediate burial depths. During compaction the dewatering of shales, and the transformation of smectite–illite, are thought to have provided the fluids responsible (Ali 1995). Although these dolomites are widespread in the region, their occurrence has probably been under-reported as they are present in poorly studied, tight argillaceous limestones with low reservoir potential.

Methanogenic dolomites have been reported in SE Asia. Ali (1995) recorded dolomite cements in the Tigapapan unit of Sabah with light carbon isotope signatures and suggested that the carbon was derived from the action of methane-oxidizing bacteria. Calcite cements with a similar isotope signature have been reported in the Sunda Basin (Park *et al.* 1995), although here precipitation from fluids expelled from the Banuwati Formation source rock is envisaged. Dolomite (and calcite) precipitation from methane-derived fluids is potentially more widespread than has been reported to date and may merit further study.

Many authors favour an interpretation of dolomite formation from Mg-rich fluids derived from the clays that commonly encase the precursor limestone. Dolomites are very common in carbonates with a significant component of admixed clays, and this association has been taken to support the compaction of shales as generating dolomitizing fluids (Berry 1976; Ali 1995; Park *et al.* 1995). However, only a few carbonate successions that overlie or are laterally adjacent to shales have been dolomitized in their bases or margins (Hendardjo & Netherwood 1986; Sun & Esteban 1994; Saller & Vijaya 2002). Most Neogene carbonates developed as part of mixed carbonate–clastic deposystems and should therefore be prone to dolomitization by this process. However, this is clearly not the case and there is considerable variation in the extent of dolomitization within and between units. The Kais Formation, for example, is extensively dolomitized, whereas the Batu Raja Formation shows only limited dolomitization. Even within the Kais Formation there are significant variations in the degree of alteration, with the Kasim Utara Field more extensively dolomitized than the Walio Field (Livingstone *et al.* 1992). In the Berai and Kais limestones that are dolomitized along their flanks, the process was preceded by dissolution (Livingstone *et al.* 1992; Saller & Vijaya 2002). It may be that dolomitizing fluids cannot penetrate into carbonate formations unless fracturing and/or dissolution provide suitable conduits (Livingstone *et al.* 1992). In the case of the Berai Limestone, dissolution porosity is inferred to be related to acidic water compacting out of adjacent shales. It is possible that reduction of organic material within muds plays a significant role in dissolution during burial and/or dolomitization. During moderate–deep burial of organic-rich shales, acidic waters can be generated from the maturation of organic material (Saller & Vijaya 2002). The Klasafet Formation acts as both seal and source to the Kais Formation and has been suggested as a possible source of fluids (Livingstone *et al.* 1992). It is possible that the increased organic component to the Klasafet Formation gives it greater potential for dolomitization than muds that surround the Batu Raja Formation. Internal variations within the Kais and Klasafet formations may account for the varying degrees of dolomitization of Kais Formation fields. Thin foraminiferal carbonates associated with marine high stands in SE Asia are commonly dolomitized. These highstand mudstones show increased organic components and reducing conditions, suggesting that reduction of organic material may have a significant influence on dolomitization.

Many models for dolomite formation favour an evaporitic component, and some authors have suggested that this is a factor that can be applied to SE Asia. Sun & Esteban (1994) suggested an evaporitic origin for unspecified dolomites in SE Asia, with dolomitization occurring during sea-level lows and during dry seasons. Ali & Abolins (1999) suggested that this model might provide an alternative mechanism to mixing-zone dolomitization for the Luconia platforms. The isotopic data do not, however, support an evaporitic origin, and they cannot be used to distinguish between a reflux or mixing zone model. However, evaporitic models for dolomitization are very much the exception, with most workers considering conditions in the Tertiary to have been unsuitable for evaporitic processes to have operated. Evaporites are extremely rare in SE Asia. In the Pre-Tertiary, evaporites were apparently confined to the Cretaceous Maha Sakham Formation in the Khorat Plateau Basin in Thailand and the Triassic Pagno Formation in Northern Shan, Myanmar. The Puragi Limestone of Late Cretaceous–?Middle Eocene age in Irian Jaya is the only example in which dolomites are associated with possible primary evaporites. If the possible algal laminites, cryptoolitic structures and nodules of anhydrite described in this formation formed in a sabkha or evaporitic back-reef area they would have done so when Irian Jaya was in a more southerly latitudes, before it drifted into the humid tropics. In Tertiary rocks, primary gypsum has been reported in Miocene-aged Pegu Group (Taungtalon Sandstone) clastics in Central Myanmar (Khin & Myitta 1999). Its development here most probably reflects the unusual climatic conditions in Central Burma, which lies in the rain shadows of both the SW and NE monsoons. Elsewhere in the Tertiary of SE Asia, evaporite minerals are extremely rare or absent, with records attributable to misidentification or precipitation as a late diagenetic phase from fluids derived from basement and transported along faults. In addition to a lack of evaporite minerals, palaeoclimatic evidence suggests that throughout most of the Tertiary, with the possible exception of the late Miocene and Pliocene, conditions would have been too humid to allow the development of hypersalinity (Frakes 1979; Morley 2000). In summary, the interpretation of dolomitization of Tertiary carbonates by evaporitic processes is not supported by lithological, isotopic or climatic data for the region.

Although dolomitization of carbonates by evaporitic processes is considered unlikely in the Tertiary, such processes may have operated earlier. For example, dolomites in the Permian and Carboniferous replace limestones that were deposited on shallow platforms that show evidence of at least localized restriction. These carbonates were deposited on terranes translocated across Tethys from Gondwana and must have passed through more arid climatic zones (Ziegler 1990; Rees *et al.* 1999).

Association of dolomite with reservoir quality

Dolomite has a mixed relationship with reservoir quality, but overall is often associated with good porosity and permeability characteristics. Best reservoir quality tends to occur where dolomite is developed as a replacive phase (e.g. Central Luconia and Kais Limestone). It may be that early fabric retentive dolomites preserve original porosity by preventing later compaction. Replacive dolomites, such as those formed in argillaceous carbonates or sometimes those associated with late diagenetic events, often have low porosities and permeabilities. In argillaceous carbonates, dolomites are developed in essentially non-reservoir lithologies, and appear to have had a minimal effect on reservoir quality. In general, when compared with undolomitized limestones, replacive dolomites have a greater influence on permeability, which is commonly improved, than on porosity, which is commonly reduced. For example, James (1983) summarizing Indonesian and Malaysian carbonates reported average porosities of 10 and 20% for dolomite and limestone, respectively, but similar average permeabilities for both (100 mD).

Dolomite cements have a destructive influence on reservoir quality reducing both porosity and permeability. The extent of this reduction depends on the amount of cementation that has taken place. In the Kasim Utara Field (Kais Formation, Salawati Basin) an initial preservation or improvement in reservoir quality as a result of dolomitization has been offset by later widespread dolomite cementation.

In addition to its direct association with reservoir quality, dolomite also has an indirect influence on petroleum systems in the region by acting as a source of non-hydrocarbon gases. As can be seen in the North Sumatra Basin, the distribution of CO_2 can be a major factor influencing reservoir viability, and can make an otherwise attractive target uneconomic. As a consequence, knowledge of dolomite distribution, potential CO_2 kitchens and potential pathways ('basin plumbing') can have a considerable impact on exploration and production strategies.

We would like to thank the following organisations for their support and encouragement; Shell EP Solutions UK, the SE Asia Research Group at Royal Holloway College (London), Robertson Research International and the University of Durham. Two anonymous referees and the editor, C. Braithwaite, are thanked for reviewing the manuscript.

References

ADAMS, C.G. 1965. The foraminifera and stratigraphy of the Melinau Limestone, Sarawak, and its importance in Tertiary correlation. *Quarterly Journal of the Geological Society, London*, **121**, 283–338.

AKHMAD, F. 2002. Stratigraphy and structural analysis in the Gunung Bijih (Erstberg) mining district, Irian Jaya. *In*: *Proceedings ot the 28th Annual Convention*. Indonesian Petroleum Association, Jakarta, 335–344.

ALAM, H., PATERSON, D.W., CORBIN, S.G., SYARIFUDDIN, N. & BUSONO, I. 1999. Reservoir potential of carbonate rocks in Kutai Basin region, East Kalimantan, Indonesia. *Journal of Asian Earth Sciences*, **17**, 203–204.

ALI, M.Y. 1992. Carbonate cement stratigraphy and timing of hydrocarbon migration: an example from Tigapapan Unit, offshore Sabah. *Geological Society of Malaysia, Bulletin*, **32**, 185–211.

ALI, M.Y. 1995. Carbonate cement stratigraphy and timing of diagenesis in a Miocene mixed carbonate-clastic sequence, offshore Sabah, Malaysia: Constraints from cathodoluminescene, geochemistry and isotope studies. *Sedimentary Geology*, **99**, 191–214.

ALI, M.Y. & ABOLINS, P. 1999. Central Luconia Province. *In*: *The Petroleum Geology and Resources of Malaysia*. Petronas, Malaysia, 369–392.

AMPORNMAHA, A. 1995. Triassic carbonate rocks in the Phatthalung area, Peninsular Thailand. *Journal of Southeast Asian Earth Sciences*, **11**, 225–236.

ARPANDI, D. & PATMOSUKISMO, S. 1975. The Cibulakan Formation as one of the most prospective stratigraphic units in the north-west Java basinal area. *In*: *Proceedings of the 4th Annual Convention*. Indonesian Petroleum Association, Jakarta, 181–210.

ASCARIA, N.A., HARBURY, N.A. & WILSON, M.E.J. 1997. Hydrocarbon potential and development of Miocene knoll-reefs, South Sulawesi. *In*: HOWES, J.V.C. & NOBLE, R.A. (eds) *Proceedings of an International Conference on Petroleum Systems of SE Asia & Australasia*. Indonesian Petroleum Association, Jakarta, 569–584.

AUDLEY-CHARLES, M.G. 1968. *The Geology of Portuguese Timor*. Geological Society of London, Memoir, **4**.

BAIRD, A. & BOSENCE, D. 1993. The sedimentological and diagenetic evolution of the Ratburi Limestone, Peninsular Thailand. *Journal of Southeast Asian Earth Sciences*, **8**, 173–180.

BARKHAM, S.T. 1993. *The structure and stratigraphy of the Permo-Triassic carbonate formations of West Timor, Indonesia*. PhD thesis, University of London.

BARLIANA, A., BURGON, G. & CAUGHEY, C.A. 2000. Changing perceptions of a carbonate gas reservoir: Alur Siwah Field, Aceh Timur, Sumatra. *In*: *Proceedings of the 27th Annual Convention*. Indonesian Petroleum Association, Jakarta, 159–176.

BENDER, F. 1983. *Geology of Burma*. Gebrüder Borntraeger, Berlin.

BENNETT, J.D., BRIDGE, D.MCC. *ET AL.* 1981. *The Geology of the Langsa Quadrangle, Sumatra (Quadrangle 0420)*. Scale 1: 250 000. Geological Survey of Indonesia, Directorate of Mineral Resources, Geological Research and Development Centre, Bandung.

BERRY, J.T. 1976. Aspects of diagenesis and porosity in Indonesian Miocene carbonates. *In*: *Proceedings of the Carbonate Seminar, Special Volume*. Indonesian Petroleum Association, Jakarta, 109–111.

BOOTH, J. & SATTAYARAK, N. 2000. Multiple inversion of the Khorat Plateau Basin, NE Thailand, with the formation and preservation of a productive gas play. *AAPG Bulletin*, **84**, 1406.

BRASH, R.A., HENAGE, L.F., HARAHAP, B.H., MOFFAT, D.T. & TURNER, R.W. 1991. Stratigraphy and depositional history of the New Guinea Limestone Group, Lengguru, Irian Jaya. *In*: *Proceedings of the 20th Annual Convention*. Indonesian Petroleum Association, Jakarta, 67–84.

BUKHARI, T., KALDI, J.G., YAMAN, F., KAKUNG, H.P. & WILLIAMS, D.O. 1992. Parigi Carbonate Buildups: Northwest Java Sea. *In*: SIEMERS, C.T., LONGMAN, M.W., PARK, R.K. & KALDI, J.G. (eds) *Carbonate Rocks and Reservoirs of Indonesia: A Core Workshop*. Indonesian Petroleum Association Core Workshop Notes, **1**, 6-1–6-36.

CARNELL, A.J.H. 1996. *The Rajamandala Limestone of the Sukabumi Area of West Java*. Society of Petroleum Engineers, Indonesia Branch, Field Guide Book, 46.

CARNELL, A.J.H. 2000. The Rajamandala Limestone at Sukabumi; Can it be considered an analogue for the Batu Raja Limestone. *AAPG Bulletin*, **84**, 1409.

CAUGHEY, C.A. & WAHYUDI, T. 1993. Gas reservoirs in the Lower Miocene Peutu Formation, Aceh Timur, Sumatra. *In*: *Proceedings of the 22nd Annual Convention*, Volume 1. Indonesian Petroleum Association, Jakarta, 191–218.

CHINOROJE, O. 1993. Petrographic studies of Permian Carbonates in Southern Thailand. *Journal of Southeast Asian Earth Sciences*, **8**, 161–171.

COFFIELD, D.Q., BERGMAN, S.C., GARRARD, R.A., GURITNO, N., ROBINSON, N.M. & TALBOT, J. 1993. Tectonic and stratigraphic evolution of the Kalosi PSC area and associated development of a Tertiary petroleum system, South Sulawesi, Indonesia. *In*: *Proceedings of the 22nd Annual Convention*. Indonesian Petroleum Association, Jakarta, 679–706.

COLLINS, J.F., KRISTANTO, A.S., BON, J. & CAUGHEY, C.A. 1996. Sequence stratigraphic framework of Oligocene and Miocene carbonates, North Sumatra Basin, Indonesia. *In*: *Proceedings of the 25th Annual Convention*. Indonesian Petroleum Association, Jakarta, 267–269.

COOPER, B.A., RAVEN, M.J., SAMUEL, L., HARDJONO & SATOTO, W. 1997. Origin and geological controls on subsurface CO_2 distribution with examples from Western Indonesia. *In*: HOWES, J.V.C. & NOBLE, R.A. (eds) *Proceedings of the Conference on Petroleum Systems of SE Asia and Australasia, Jakarta, Indonesia*, 21–23 May Indonesian Petroleum Association, Jakarta, 877–892.

CROTTY, K.J. & ENGELHARDT, D.W. 1993. Larger foraminifera and palynomorphs of the upper Malawa and lower Tonasa Formations, southwestern Sulawesi Island, Indonesia. *In*: THANASUTHIPITAK, T. (ed.) *Symposium on Biostratigraphy of Mainland Southeast Asia: Facies and Paleontology, Chiang Mai, Thailand*, 71–82.

CRUMB, R.E. 1989. Petrophysical properties of the Bima Batu Raja carbonate reservoir, offshore N.W. Java. *In*: *Proceedings of the 18th Annual Convention*. Indonesian Petroleum Association, Jakarta, 161–208.

CUCCI, M.A. & CLARK, M.H. 1993. Sequence stratigraphy of a Miocene carbonate buildup, Java Sea. *In*: LOUCKS, R.G. & SARG, J.F. (eds) *Carbonate Sequence Stratigraphy, Recent Developments and Applications*. American Association of Petroleum Geologists, Memoir, **57**, 291–303.

CUCCI, M.A. & CLARK, M.H. 1995. Carbonate systems tracts of an asymmetric Miocene buildup near Kangean Island, E. Java Sea. *In*: CAUGHEY, C.A., CARTER, D.C., CLURE, J., GRESKO, M.J., LOWRY, P., PARK, R.K. & WONDERS, A. (eds) *Proceedings of the International Symposium on Sequence Stratigraphy in Southeast Asia*. Indonesian Petroleum Association, Jakarta, 231–251.

DAVIES, H.L. 1973. *Gazelle Peninsula, New Britain*. Australian Bureau of Mineral Resources, 1:250 000 Explanatory Notes, Sheet SB/56–2.

DAVIES, I.C. 1990. Geology and exploration review of the Tomori PSC, Eastern Indonesia. *In*: *Proceedings of the 19th Annual Convention*. Indonesian Petroleum Association, Jakarta, 41–68.

DE SMET, M.E.M. 1991. *A guide to the stratigraphy of Sumatra, Part 1: Pre-Tertiary*. London University, Research in SE Asia Internal Report, **101**.

DE SMET, M.E.M. 1992. *A guide to the stratigraphy of Sumatra, Part 2: Tertiary*. London University, Research in SE Asia Internal Report, **108**.

DJUANDA, H. 1985. Facies distribution in the Nurbani carbonate build-up, Sunda Basin. *In*: *Proceedings of the 14th Annual Convention*. Indonesian Petroleum Association, Jakarta, 507–533.

DOLAN, P.J. & HERMANY 1988. The geology of the Wiriagar field, Bintuni Basin, Irian Jaya. *In*: *Proceedings of the 17th Annual Convention*. Indonesian Petroleum Association, Jakarta, 53–87.

DOW, D.B., HARAHAP, B.H. & HAKIM, S.A. 1990. *Geology of the Enarotali Sheet Area, Irian Jaya*. Geological Survey of Indonesia, Directorate of Mineral Resources, Geological Research and Development Centre, Bandung.

DUNN, P.A., KOZAR, M.G. & BUDIYONO 1996. Application of geoscience technology in a geologic study of the Natuna gas field, Natuna Sea, Offshore Indonesia. *In*: *Proceedings of the 25th Annual Convention*. Indonesian Petroleum Association, Jakarta, 117–130.

EPTING, M. 1980. Sedimentology of Miocene carbonate buildups, Central Luconia, Offshore Sarawak. *Geological Society of Malaysia, Bulletin*, **12**, 17–30.

FONTAINE, H. 1992. Pre-Tertiary sediments found at the bottom of wells drilled in Malacca Straits. *CCOP Technical Bulletin*, **17**, 12–17.

FRAKES, L.A. 1979. *Climates Through Geologic Time*. Elsevier, Amsterdam.

FRANCIS, G. 1988. Stratigraphy of Manus Island, Western New Ireland Basin, Papua New Guinea. *In*: MARLOW, M.S., DADISMAN, S.V. & EXON, N.F. (eds) *Geology and Offshore Resources of Pacific Island Arcs – New Ireland and Manus Region, Papua New Guinea*. Circum-Pacific Council for Energy and Mineral Resources, Earth Science Series, **9**, 31–40.

GIBSON-ROBINSON, C., HENRY, N.M., THOMPSON, S.J. & RAHARJO, H.T. 1990. Kasim and Walio Fields, Indonesia: Salawati Basin, Irian Jaya. *In*: BEAUMONT, E.A. & FOSTER, N.H. (eds) *AAPG Treatise on Petroleum Geology Atlas, Oil and Gas Fields: Stratigraphic Traps I*, 257–295.

GOBBETT, D.J. 1973. Carboniferous and Permian correlation in Southeast Asia. *Geological Society of Malaysia, Bulletin*, **6**, 131–142.

GRAINGE, A.M. & DAVIES, K.G. 1983. Reef exploration in the East Sengkang basin, Sulawesi, Indonesia. *In*: *Proceedings of the 12th Annual Convention*. Indonesian Petroleum Association, Jakarta, 207–227.

GRAINGE, A.M. & DAVIES, K.G. 1985. Reef exploration in the East Sengkang basin, Sulawesi, Indonesia. *Marine and Petroleum Geology*, **2**, 142–155.

GRAVES, R.R. & WEEGAR, A.A. 1973. Geology of the Arun gas field (North Sumatra). *In*: *Proceedings of the 2nd Annual Convention*. Indonesian Petroleum Association, Jakarta, 53–72.

GRÖTSCH, J. & MERCADIER, C. 1999. Integrated 3-D reservoir modelling based on 3-D seismic: The Tertiary Malampaya and Camago buildups, offshore Palawan, Philippines. *AAPG Bulletin*, **83**, 1703–1728.

HENDARJO, K.S. & NETHERWOOD, R.E. 1986. Palaeoenvironmental and diagenetic history of the Kais Formation, K.B.S.A., Irian Jaya. *In*: *Proceedings of the 15th Annual Convention*, Volume 1. Indonesian Petroleum Association, Jakarta, 423–438.

HEWARD, A.P., CHUENBUNCHOM, S., MÄKEL, G., MARSLAND, D. & SPRING, L. 2000. Nang Nuan Field, B6/27, Gulf of Thailand: karst reservoirs of meteoric or deep burial origin? *Petroleum Geoscience*, **6**, 15–27.

HINTON, L.B., ATMADJA, W.S. & SUWITO, P.S. 1987. Peusangan–C1 reef interpretation with top reef transparent to seismic. *In*: *Proceedings of the 16th Annual Convention*. Indonesian Petroleum Association, Jakarta, 339–362.

HOUPT, J.R. & KERSTING, C.C. 1976. Arun reef, Bee Block, North Sumatra. *In*: *Proceedings of the Carbonate Seminar, Special Volume*. Indonesia Petroleum Association, Jakarta, 42–60.

HUNT, J.M. 1996. *Petroleum Geochemistry and Geology*, 2nd edn. W.H. Freeman, New York.

JAMES, S.J. 1983. Micritization and calichification: their recognition and implications. *In*: *Proceedings of the 12th Annual Convention*. Indonesian Petroleum Association, Jakarta, 81–106.

KEMP, G. 1992. The Manusela Formation – An example of a Jurassic carbonate unit of the Australian Plate from Seram, Eastern Indonesia. *In*: SIEMERS, C.T., LONGMAN, M.W., PARK, R.K. & KALDI, J.G. (eds) *Carbonate Rocks and Reservoirs of Indonesia: A Core Workshop*. Indonesian Petroleum Association Core Workshop Notes, **1**, 11-1–11-41.

KENYON, C.S. 1977. Distribution and morphology of early Miocene reefs, East Java Sea. *Proceedings of the 6th Annual Convention*. Indonesian Petroleum Association, Jakarta, 215–238.

KHIN, K. & MYITTA 1999. Marine transgression and regression in Miocene sequences of northern Pegu (Bago) Yoma, Central Myanmar. *Journal of Asian Earth Sciences*, **17**, 369–393.

KIESSLING, W. & FLUGEL, E. 2000. Late Paleozoic and late Triassic limestones from North Palawan block (Philippines): microfacies and paleogeographical implications. *Facies*, **43**, 39–78.

KISSIN, I.G. & PAKHAMOV, S.L. 1967. The possibility of carbon dioxide generation at depth at moderately high temperatures. *Doklady Akademii Nauk*, **174**, 451–454.

LIVINGSTONE, H.J., SINCOCK, B.W., SYARIEF, A.M., SRIWIDADI & WILSON, J.N. 1992. Comparison of Walio and Kasim Reefs, Salawati Basin, Western Irian Jaya, Indonesia. *In*: SIEMERS, C.T., LONGMAN, M.W., PARK, R.K. & KALDI, J.G. (eds) *Carbonate Rocks and Reservoirs of Indonesia: A Core Workshop*. Indonesian Petroleum Association Core Workshop Notes, **1**, 4-1–4-37.

LONGMAN, M.W. 1980. Carbonate petrology of the Nido B-3A core, offshore Palawan, Philippines *In*: HALLEY, R.B. & LOUCKS, R.G. (eds) *Carbonate Reservoir Rocks*. Society of Economic Paleonlotogists and Mineralogists Core Workshop, **1**, 161–183.

LONGMAN, M.W., MAXWELL, R.J., MASON, A.D.M. & BEDDOES, L.R., JR. 1987. Characteristics of a Miocene intrabank channel in Batu Raja Limestone, Ramba Field, South Sumatra, Indonesia. *AAPG Bulletin*, **71**, 1261–1273.

LONGMAN, M.W., SIEMERS, C.T. & SIWINDONO, T. 1992. Characteristics of low relief carbonate mudbank reservoir rocks, Baturaja Formation (Lower Miocene), Air Serdang and Mandala Fields South Sumatra Basin, Indonesia. *In*: SIEMERS, C.T., LONGMAN, M.W., PARK, R.K. & KALDI, J.G. (eds) *Carbonate Rocks and Reservoirs of Indonesia*. Indonesian Petroleum Association Core Workshop Notes, **1**, Chapter 9.

LUNT, P. & DJAAFAR, R. 1991. Aspects of the stratigraphy of western Irian Jaya and implications for the development of sandy facies. *In*: *Proceedings of the 20th Annual Convention*. Indonesian Petroleum Association, Jakarta, 107–124.

MALAIHOLLO, J.F.A. 1993. *The geology and tectonics of the Bacan region, eastern Indonesia*. PhD thesis, University of London.

MAY, J.A. & EYLES, D.R. 1985. Well log and seismic character of Tertiary Terumbu carbonate, South China Sea, Indonesia. *AAPG, Bulletin*, **69**, 1339–1358.

Mayall, M.J. & Cox, M. 1988. Deposition and diagenesis of Miocene limestones, Sengkang Basin, Sulawesi, Indonesia. *Sedimentary Petrology*, **59**, 77–92.

MAYALL, M.J., BENT, A. & ROBERTS, D.M. 1997. Miocene carbonate build-ups offshore Socialist Republic of Vietnam. *In*: FRASER, A.J., MATTHEWS, S.J. & MURPHY, R.W. (eds) *Petroleum Geology of Southeast Asia*. Geological Society, London, Special Publications, **126**, 117–120.

MCARTHUR, A.C. & HELM, R.B. 1982. Miocene carbonate buildups, offshore North Sumatra. *In*: *Proceedings of the 11th Annual Convention*. Indonesian Petroleum Association, Jakarta, 127–146.

METCALFE, I. 1990. Allochthonous terrane processes in Southeast Asia. *Philosophical Transactions of the Royal Society of London*, **331**, 625–640.

MORLEY, R.J. 2000. *Origin and evolution of Tropical Rain Forests*. John Wiley, Chichester.

MUNDT, P.A. 1982. Miocene reefs, offshore North Sumatra. *In*: *SEAPEX Proceedings*.

MUNDT, P.A. 1983. Miocene reefs, offshore North Sumatra. *In*: *SEAPEX Proceedings*, 1–9.

NETHERWOOD, R.E. & WIGHT, A.W.R. 1992. Structurally-controlled, linear reefs in a Pliocene delta-front setting, Tarakan Basin, Northeast Kalimantan. *In*: SIEMERS, C.T., LONGMAN, M.W., PARK, R.K. & KALDI, J.G. (eds) *Carbonate Rocks and Reservoirs of Indonesia: A Core Workshop*. Indonesian Petroleum Association Core Workshop Notes, **1**, 3-1–3-40.

NILANDAROE, N., MOGG, W. & BARRACLOUGH, R. 2002. Characteristics of the fractured carbonate reservoir of the Oseil Field, Seram Island, Indonesia. *In*: *Proceedings of the 28th Annual Convention*. Indonesian Petroleum Association, Jakarta, 439–456.

NURZAMAN, Z.Z. & PUJIANTO, A. 1994. Geology and reservoir characterisation of Wiriagar Field as a diagenetic facies for reservoir simulation. *In*: *Proceedings of the 23rd Annual Convention*, Volume 2. Indonesian Petroleum Association, Jakarta, 29–49.

PANGGABEAN, H. 1990. *Geology of the Omba Sheet Area, Irian Jaya*. Geological Survey of Indonesia, Directorate of Mineral Resources, Geological Research and Development Centre, Bandung.

PARK, R.K., MATTER, A. & TONKIN, P.C. 1995. Porosity evolution in the Batu Raja carbonates of the Sunda Basin – Windows of evolution. *In*: *Proceedings of the 24th Annual Convention*, Volume 1. Indonesian Petroleum Association, Jakarta, 163–184.

PIETERS, P.E., HAKIM, A.S. & ATMAWINATA, S. 1990. *Geologi lembar Ransiki, Irian Jaya*. Geological Survey of Indonesia, Directorate of Mineral Resources, Geological Research and Development Centre, Bandung.

PIETERS, P.E., PIGRAM, C.J., TRAIL, D.S., DOW, D.B., RATMAN, N. & SUKAMTO, R. 1983. The stratigraphy of western Irian Jaya. *In*: *Proceedings of the 12th Annual Convention*. Indonesian Petroleum Association, Jakarta, 229–261.

PIGRAM, C.J. & PANGGABEAN, H. 1989. *Geology of the Waghete Sheet Area, Irian Jaya*. Geological Survey of Indonesia, Directorate of Mineral Resources, Geological Research and Development Centre, Bandung.

PIGRAM, C.J. & SUKANTA, U. 1989. *Geologi lembar Taminabuan, Irian Jaya*. Geological Survey of Indonesia, Directorate of Mineral Resources, Geological Research and Development Centre, Bandung.

PRINGGO, H., SURYO, S.P. & ROSKAMIL 1977. The Kromong carbonate rocks and their relationship with the Cibulakan and Parigi Formations. *In*: *Proceedings of the 6th Annual Convention*. Indonesian Petroleum Association, Jakarta, 221–240.

REAVES, C.M. & SULAEMAN, A. 1994. Emprical models for predicting CO_2 concentrations in North Sumatra. *In*: *Proceedings of the 23rd Annual Convention*, Volume 1. Indonesian Petroleum Association, Jakarta, 33–43.

REES, P.M., GIBBS, M.T., ZIEGLER, A.M., KUTZBACH, J.E. & BEHLING, P.J. 1999. Permian climates: Evaluating model predictions using global paleobotanical data. *Geology*, **27**, 891–894.

ROBINSON, G.P., HARAHAP, B.H., SUPARMAN, M. & BLADON, G.M. 1990*a*. *Geologi lembar Fak Fak, Irian Jaya*. Geological Survey of Indonesia, Directorate of Mineral Resources, Geological Research and Development Centre, Bandung.

ROBINSON, G.P., RYBURN, R.J., HARAHAP, B.H., TOBING, S.L., BLADON, G.M. & PIETERS, P.E. 1990*b*. *Geology of the Kaimana Sheet area, Irian Jaya*. Geological Survey of Indonesia, Directorate of Mineral Resources, Geological Research and Development Centre, Bandung.

RORY, R. 1990. Geology of the South Lho Sukon 'A' Field North Sumatra, *In*: *Proceedings of the 19th Annual Convention*. Indonesian Petroleum Association, Jakarta, 1–40.

RUSMANA, E., HARTONO, U. & PIGRAM, C.J. 1989. *Geology of the Misool Sheet Area, Irian Jaya*. Geological Survey of Indonesia, Directorate of Mineral Resources, Geological Research and Development Centre, Bandung.

SALLER, A.H. & VIJAYA, S. 2002. Depositional and diagenetic history of the Kerendan carbonate platform, Oligocene, Central Kalimantan, Indonesia. *Journal of Petroleum Geology*, **25**, 123–150.

SARG, J.F., WEBER, L.J. *ET AL.* 1995. Carbonate sequence stratigraphy – a summary and perspective with case history, Neogene, Papua New Guinea. *In*: CAUGHEY, C.A., CARTER, D.C., CLURE, J., GRESKO, M.J., LOWRY, P., PARK, R.K. & WONDERS, A. (eds) *Proceedings of the International Symposium on Sequence Stratigraphy in Southeast Asia.* Indonesian Petroleum Association, Jakarta, 137–179.

SHI, G.R. & ARCHBOLD, N.W. 1998. Permian marine biogeography of SE Asia. *In*: HALL, R. & HOLLOWAY, J.D. (eds) *Biogeography and Geological Evolution of SE Asia.* Backhuys, Leiden, 57–72.

SIEMERS, C.T., DECKELMAN, J.A., BROWN, A.A. & WEST, E.R. 1992. Characteristics of the fractured Ngimbang Carbonate (Eocene), West Kangean-2 Well, Kangean PSC, East Java Sea, Indonesia. *In*: SIEMERS, C.T., LONGMAN, M.W., PARK, R.K. & KALDI, J.G. (eds) *Carbonate Rocks and Reservoirs of Indonesia: A Core Workshop.* Indonesian Petroleum Association Core Workshop Notes, **1**, 10-1–10-52.

SITUMORANG, R.L., SMIT, R. & VAN VESSEM, E.J. 1992. *Geology of the Jatirogo, Jawa. (Quadrangle 1509–2).* Scale 1:100 000. Geological Research and Development Centre, Bandung, Indonesia.

SUHERMAN, T. & SYAHBUDDIN, A. 1986. Exploration history of the MB field coastal area of northwest Java. *In*: *Proceedings of the 15th Annual Convention.* Indonesian Petroleum Association, Jakarta, 101–122.

SULITRA, M.D. 1991. NSB 'A' Field construction of geological model incorporating data from the anomalous well NSB-A7. *In*: *Proceedings of the 20th Annual Convention.* Indonesian Petroleum Association, Jakarta, 479–518.

SUN, S.Q. & ESTEBAN, M. 1994. Paleoclimatic controls on sedimentation, diagenesis and reservoir quality: Lessons from Miocene carbonates. *AAPG Bulletin*, **78**, 519–543.

SURONO 1994. A petrographic study of an oolitic limestone succession of the Eocene–Oligocene Tampakura Formation, South-east Sulawesi, Indonesia. *Jurnal Geologi dan Sumberdaya Mineral*, **5**, 2–11.

TJOKROSAPOETRA, S., BUDHITRISNA, T. & RUSNAMA, E. 1993. *Geology of the Buru Quadrangle, Maluku*, 1:250 000 scale. Geological Survey of Indonesia, Geological Research and Development Centre, Bandung, Indonesia.

TONKIN, P.C., TEMANSAJI, A. & PARK, R.K. 1992. Reef complex lithofacies and reservoir, Rama Field, Sunda Basin, Southeast Sumatra, Indonesia. *In*: SIEMERS, C.T., LONGMAN, M.W., PARK, R.K. & KALDI, J.G. (eds) *Carbonate Rocks and Reservoirs of Indonesia: A Core Workshop.* Indonesian Petroleum Association Core Workshop Notes, **1**, p. 7-1–7-46.

VAHRENKAMP, V.C. 2000. Sr-isotope stratigraphy of Miocene carbonates, Luconia Province, Sarawak, Malaysia: Implications for platform growth and demise and regional reservoir behaviour. *In*: CREVELLO, P.D., VAHRENKAMP, V. & SINGH, U. (eds) *Carbonate Depositional Systems and High-resolution Reservoir Stratigraphy: Methods and Applications for Building Integrated Reservoir Simulation Models.* American Association Petroleum Geologists Short Course, 8.2-1–8.2-4.

VINCELETTE, R.R. 1973. Reef exploration in Irian Jaya, Indonesia. *In*: *Proceedings of the 2nd Annual Convention.* Indonesian Petroleum Association, Jakarta, 243–277.

WICAKSONO, P., WIGHT, A.W.R., LODWICK, W.R., NETHERWOOD, R.E., BUDIARTO, B. & HANGGORO, D. 1995. Use of sequence stratigraphy in carbonate exploration: Sunda Basin, Java Sea, Indonesia. *In*: CAUGHY, C.A., CARTER, D.C., CLURE, J., GRESKO, M.J., LOWRY, P., PARK R.K. & WONDERS, A. (eds) *Proceedings of the International Symposium on Sequence Stratigraphy in Southeast Asia.* Indonesian Petroleum Association, Jakarta, 197–229.

WIEDICKE, M.C. 1987. Biostratigraphy, microfacies and diagenesis of Tertiary carbonates from the South China Sea (Dangerous Grounds – Palawan, Philippines). *Facies*, **16**, 195–302.

WIGHT, A.W.R. & HARDIAN, D. 1982. Importance of diagenesis in carbonate exploration and production, Lower Batu Raja Carbonates, Krisna Field, Java Sea. *In*: *Proceedings of the 11th Annual Convention.* Indonesian Petroleum Association, Jakarta, 211–236.

WILLMOTT, W.F. 1972. *Markham, Papua New Guinea (sheet SB/55-10) International Index.* Scale 1:250 000 Geological Series Explanatory Notes. Geological Survey of Papua New Guinea/Australian Bureau of Mineral Resources, Geological Map Series.

WILSON, M.E.J. 1995. *The Tonasa Limestone Formation, Sulawesi, Indonesia: Development of a Tertiary carbonate platform.* PhD thesis, University of London.

WILSON, M.E.J. 2000. Tropical carbonate evolution during the Cenozoic in SE Asia: Implications for hydrocarbon exploration. *AAPG Bulletin*, **84**, 1514.

WILSON, M.E.J. 2002. Cenozoic carbonates in SE Asia: Implications for equatorial carbonate development. *Sedimentary Geology*, **147**, 295–428.

WILSON, M.E.J. & BOSENCE, D.W.J. 1996. The Tertiary evolution of South Sulawesi: A record in redeposited carbonates of the Tonasa Limestone Formation. *In*: HALL, R. & BLUNDELL, D.J. (eds) *Tectonic Evolution of SE Asia.* Geological Society, London, Special Publications, **106**, 365–389.

WILSON, M.E.J. & EVANS, M.J. 2002. Sedimentology and diagenesis of Tertiary carbonates on the Mangkalihat Peninsula, Borneo: implications for subsurface reservoir quality. *Marine and Petroleum Geology*, **19**, 873–900.

WILSON, M.E.J. & LOKIER, S.J. 2002. Siliciclastic and

volcaniclastic influences on equatorial carbonates; insights from the Neogene of Indonesia. *Sedimentology*, **49**, 583–601.
WILSON, M.E.J. & ROSEN, B.R. 1998. Implications of paucity of corals in the Paleogene of SE Asia: Plate tectonics or centre of origin? *In*: HALL, R. & HOLLOWAY, J.D. (eds) *Biogeography and Geological Evolution of SE Asia*. Buckhuys, Leiden, pp. 165–195.
YAMAN, F., AMBISMAR, T. & BUKHARI, T. 1991. Gas exploration in Parigi and Pre-Parigi carbonate buildups, NW Java Sea. *In*: *Proceedings of the 20th Annual Convention*. Indonesian Petroleum Association, Jakarta, 319–346.
ZIEGLER, A.M. 1990. Phytogeographic patterns and continental configurations during the Permian Period. *In*: MCKERROW, W.S & SCOTESE, C.R. (eds) *Palaeozoic Palaeogeography and Biogeography*. Geological Society, London, Memoir, **12**, 363–377.

Patterned dolomites: microbial origins and clues to vanished evaporites in the Arab Formation, Upper Jurassic, Arabian Gulf

A. KIRKHAM

Ty Newydd, 5 Greys Hollow, Rickling Green, Saffron Walden, Essex CB11 3YB, UK (e-mail: kirkhama@compuserve.com)

Abstract: The common occurrence of patterned dolomites in the upper Arab Formation of the Arabian Gulf is highlighted. Their mottled appearance is due to concentrations of microcrystalline iron sulphide. The dolomites are typically devoid of bioclastic debris and are usually interbedded with stromatolites and thin carbonates, both of which commonly contain anhydrite nodules. Previous workers have interpreted the patterned appearance as originating from birds-eye porosity, burrowing activity or plant roots, but their mottled appearance, and possibly also the dolomite, probably originated as a by-product of the activities of sulphate-reducing bacteria in water-logged or subaqueous sediments on the floors of salinas or very highly saline lagoons from which dissolved sulphate and/or gypsum had been removed. This implies that the Arab Formation was originally locally even more gypsiferous than is evident from its existing bedded anhydrites. The related processes of sulphate removal and sulphide formation were penecontemporaneous, as shown by occurrences of the mottled fabrics as intraclasts. Contorted lamination within the patterned dolomites and their frequent involvement with sediment injection structures are evidence of fluidization. Reasons for the contortion and fluidization are considered with respect to compatible Holocene depositional and diagenetic environments. In addition to the comparatively simple depositional models of extensively prograding supratidal sabkhas, which are traditionally applied to facies associations of stromatolites and bedded anhydrites in the Arab Formation, the presence of patterned dolomites demands consideration of less continuous and static facies belts with reduced correlatability that imply higher levels of reservoir heterogeneity.

This paper discusses a very familiar type of dolostone characteristic of one of the world's great reservoirs, the Kimmeridgian–Tithonian Arab Formation of the Arabian Gulf. This formation is partly highly dolomitized, and includes a common and very distinctive dolofacies that may be overlooked because of difficulties in understanding its origins. It is suggested here that bacterially dominated diagenesis provides a plausible explanation for its occurrence. The paper by Wright & Wacey (2004) discusses the importance of Recent bacterially mediated dolomite from the Coorong Lakes. The present paper supports this hypothesis by highlighting ancient examples of this process and brings attention to what appears to be a relatively common type of dolofacies.

The Arab Formation is produced in several giant oil and gas fields of the Arabian Gulf states of Saudi Arabia, Qatar and Abu Dhabi (eg. Ghawar, Dukhan, Bunduq, Nasr and Umm Shaif). Its ramp carbonates were largely deposited as shallow-water sediments interbedded with anhydrites that may be several metres in thickness. The very massive Hith Anhydrite Formation directly overlies the Arab and forms the main cap rock. Anhydrites within the Arab Formation are interpreted as having a mixture of both sabkha and subaqueous origins (Warren & Kendall 1985; Al Silwadi *et al.* 1996). In fact, the anhydrites of the Arab Formation were some of the first to be compared as analogues (Wood & Wolfe 1969) of the Recent anhydrites discovered during the early 1960s beneath the coastal sabkhas of Abu Dhabi (Curtis *et al.* 1963; Shearman 1966; Evans *et al.* 1969). The Arab Formation is regionally subdivided into four carbonate units (A–D) separated usually by anhydrite units. Units A, B and C commonly contain dolomites that are generally regarded as originating from seepage reflux or evaporitic pumping. However, a large proportion have the distinctive characteristics of 'patterned carbonates', a term first introduced by Dixon (1976) to describe very distinctive limestones and dolomites in Cambro-Ordovician–Triassic rocks of Canada. Many experienced geologists are either unaware of, or fail to appreciate, their significance and this paper attempts to highlight this phenomenon that is too often overlooked.

From: BRAITHWAITE, C. J. R., RIZZI, G. & DARKE, G. (eds) 2004. *The Geometry and Petrogenesis of Dolomite Hydrocarbon Reservoirs*. Geological Society, London, Special Publications, **235**, 301–308. 0305-8719/$15.00

(a)

(b)

Fig. 1. Core slabs of the Arab Formation. (**a**) Well-laminated, patterned dolomite with elongated dark blebs of iron sulphide caused by sulphate-reduction. The arcuate patterns are artefacts of the core-sawing process. (**b**) Interbedding of patterned dolomicrites and crypto-microbial dolomicrites. Former lithologies appear to have undergone internal reorganization or slumping. The white patches are anhydrite nodules that post-date the sulphate reduction of the sediment in which the anhydrite nodules grew. Bar scales are 1 cm.

Description

The patterned dolomites of the Arab Formation are generally buff-coloured, partly due to oil staining, and are finely crystalline. Their distinction arises from their mottled appearance generated by concentrations of very finely crystalline iron sulphide. The mottles are generally blue-grey–black, although they frequently show a brown discoloration due to iron oxidation. Sizes of the mottles are varied and aspect ratios range from thin, subcontinuous laminae or 'dashes', that parallel seemingly undisturbed bedding (Fig. 1a), to irregular and somewhat erratically distributed 'clots' (Figs 1b and 2a & b), which locally accompany pseudocontorted and (internally?) slumped lamination. Microscopically, the mottles are seen to comprise aggregates, rather than nodules, of opaque, finely

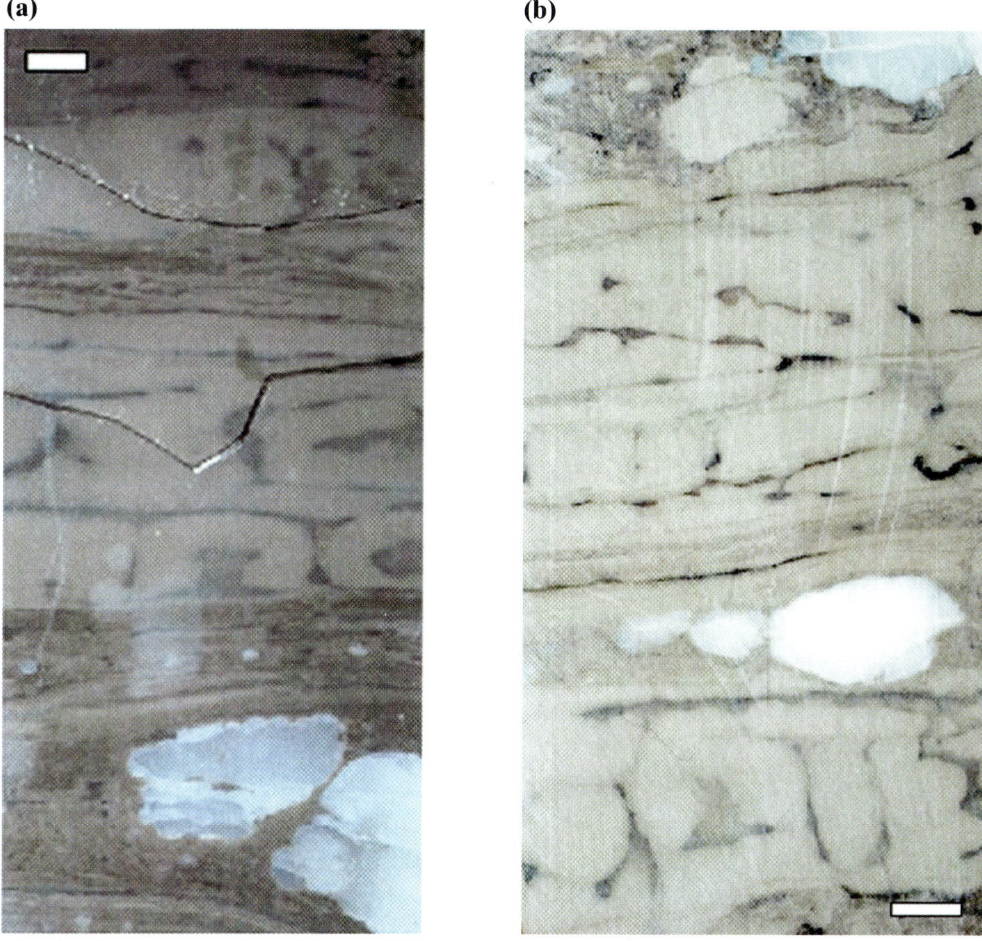

Fig. 2. Core slabs of Arab Formation showing interbedded, patterned dolomicrites and grain-supported dolostones. The dolomicrites appear to have experienced intraformational contortion perhaps by fluidization. The grain-supported dolostones contain white anhydrite nodules interpreted as post-dating sulphate reduction in the patterned dolomicrites, but predating compaction because laminae are draped around the nodules. An intraclastic horizon forms the top of the core in (b). Bar scales are 1 cm .

crystalline sulphide surrounded by the otherwise featureless dolomicrite (Fig. 3).

Patterned dolomites are usually interbedded with stromatolites, cryptomicrobial laminites, thin carbonate sands or massive anhydrite. Their thicknesses range from less than 3 cm to over 1 m. Tops and bases of the patterned dolomite beds are usually planar, although they also occur within small, upwardly and laterally intrusive synsedimentary pipes and sills, which commonly dislodged and incorporated adjacent lithologies. Occasionally, they formed firmgrounds or hardgrounds that were penecontemporaneously fractured and reworked (Fig. 4). The Arab Formation patterned dolomites are invariably devoid of bioclastic debris and isolated anhydrite nodules commonly developed within them or in adjacent cryptomicrobially laminated strata (Figs 1a and 2a & b).

Origins

Wood & Wolfe (1969) illustrated patterned dolomites in the Arab Formation and interpreted them as ‘birds-eyes’, pore types usually (but not exclusively) diagnostic of beach environments. However, these sulphide-rich

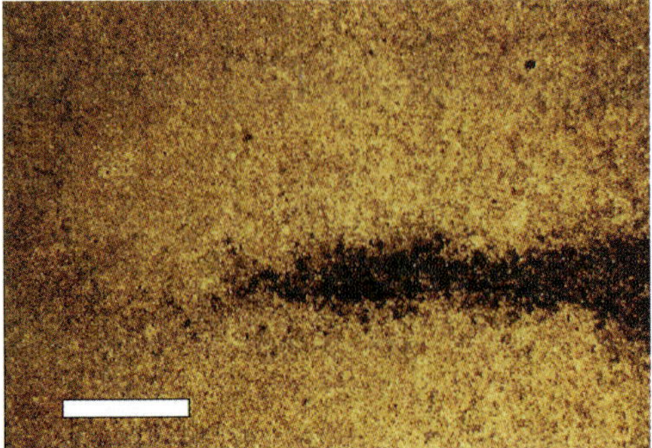

Fig. 3. Photomicrograph of dolomicrite from the Arab Formation showing details of one of the mottles comprising an aggregate of microcrystalline iron sulphide. Note the lack of pores or obvious pore-filling cement, although total rock porosity is 20.8%. Permeability is 0.35 milliDarcies (mD). Cross-polarized light. Bar scale is 1 mm.

patches show no visible porosity or pore/vug cementation. Other geologists have interpreted them as burrows (which they commonly resemble) or root pseudomorphs. While not denying that such features exist within the Arab Formation, the author concurs with Dixon (1976) and Kendall (1977) that such patterned carbonates are a by-product of sulphate-reducing bacteria. The sulphide is a result of sulphate reduction (Berner *et al.* 1979) acting either on sulphates dissolved in the circulating groundwaters or on discrete gypsum crystals. Kendall (1977) further observed that some fracture-filling examples developed from the expulsion of sulphurated water or hydrogen sulphide gas.

The very fine dolomite crystallinity indicates rapid crystallization and suggests formation under hypersaline conditions, probably produced by the high evaporation rates that would have been commensurate with depositional environments of the upper Arab Formation. As in the Coorong Lakes of

(a)

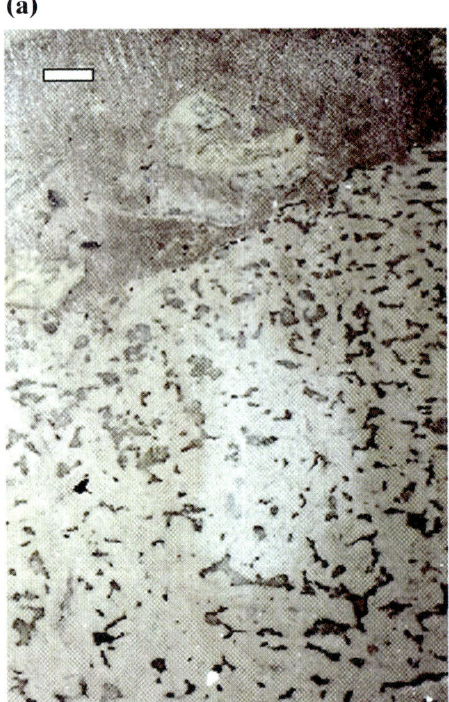

(b)

Fig. 4. Cores of patterned dolomites from the Arab Formation showing early lithification, i.e. firmgrounds or hardgrounds. (**a**) Shows rounded intraclasts of patterned dolomite. The pale patch in the centre of the dolomicrite merely indicates the former location of a core label. (**b**) Shows a fractured patterned dolomite layer that is lighter coloured than its adjacent rocks due to lack of oil staining. The fracture (former mud-crack?) was later filled with intraclasts and other carbonate debris. Bar scales are 1 cm.

(a)

(b)

Fig. 5. Cores from the Arab Formation composed entirely of patterned dolomicrites. (**a**) A few horizontal undisturbed layers are evident, but the bulk of the core appears to have experienced internal deformation. A small injection feature penetrated upwards through one of the 'layers' in the upper central part of the core. (**b**) Most of the left side of the core is well laminated, but much of the right side is chaotically bedded probably due to upward sediment injection. Bar scales are 1 cm.

southern Australia, the dolomite is considered by the author to have been a primary precipitate induced by sulphate-reducing conditions (Warren 1989; Vasconcelos & McKenzie 1997; Wright 1999; Wright & Wacey 2004). The absence of cross-cutting relationships with other facies strongly argues against a seepage-reflux origin for these dolostones. Dixon (1976) favoured an intertidal or supratidal environment of origin whilst preferring the latter, but Kendall (1977, 1979*a*, *b*) interpreted some patterned carbonates as having formed subaqueously.

The ferruginous mottles tend to conform to bedding contortions to enhance the 'patterned' appearance. Some contortions resemble stromatolitic features but many are convincingly post-depositional and caused by fluidization. Such sediments obviously experienced internal slumping or flow (Fig. 5a & b). Even the seemingly undisturbed example illustrated in Figure 1a may have experienced internal shearing. Lapointe (1992) recorded flame structures and load casts within the Arab Formation, although none have been observed in conjunction with patterned dolomites.

The fluidization would have been within water-logged sediments. If in the intertidal–supratidal zone it would probably have taken place beneath the water table. If it was subaqueous and at (or just beneath) the sediment–water interface, it probably occurred whilst the sediments were on the floors of salinas or highly saline lagoons where reworking would have been more likely than at a subterranian water table. The lack of fossils in the dolomites suggests a highly inhospitable environment. One would normally expect only microbes to survive highly evaporitic conditions. If, indeed, the convolutions formed subaqueously, the intimately associated stromatolites, cryptalgal laminites and bedded anhydrites may also be of subaqueous origin. Al-Silwadi *et al.* (1996) argued that some regionally correlatable anhydrites of the upper Arab Formation strata of offshore Abu Dhabi are replacements of subaqueous gypsum precursors. The patterned dolomites may simply be the tell-tale signs of vanished evaporites, if indeed the dolomite formed by sulphate reduction.

Mechanisms for fluidization

Hydrostatic head within shallow and relatively uncompacted sediments may have aided fluidization. Park (pers. comm.) and his co-workers experienced short-duration, artesian water fountains as they penetrated hardgrounds in observation pits on the modern coastal sabkha onshore of Abu Dhabi island. McKenzie *et al.* (1980) reported that the piezometric head was higher than the groundwater table on the mainland sabkha in the vicinity of Abu Dhabi Island. The hydrostatic head could have been supported by freshwater lenses in the high Hajar (Oman) Mountains over 100 km to the east. Patterson & Kinsman (1981) described the gradual inclination of the water table from these mountains towards the coastline and there are many reports of freshwater springs in the coastal lagoons along the Emirates shores, including many tales about the local population, even within living memory, being dependent on freshwater or brackish-water springs (onland and submarine) for survival in the inhospitable natural environments of the United Arab Emirates (UAE). Chafetz & Rush (1995) recorded similar freshwater migration over distances of 250 miles across Saudi Arabia from the Tuwaiq Mountains to the Arabian Gulf. In both cases, the freshwater is clearly migrating at shallow depths within or beneath thin Quaternary strata into offshore environments. The potable water salinities are probably kept low in the coastal regions by aquicludes such as impervious anhydrite beds, microbial mats or hardgrounds that prevent vertical mixing with seawater.

Whilst the contemporary hinterland of the Arab Formation is unknown, similar hydrological conditions involving long-distance water migration may also have functioned during its accumulation. Evidence for some freshwater influx, possibly associated with an elevated hydraulic head, is provided by a richly charophytic horizon in the Upper Jurassic (Al-Silwadi *et al.* 1996). Sediment fluidization could have been triggered by shocks created by seismic activity, storm waves or rapid sediment loading. Any or all of these could have induced spasmodic increases in fluid pressures leading to breakdown of the sediment packing, intrastratal reconfiguration, differential compaction and sediment flow.

However, it is significant that the patterned dolomites of the Arab Formation are confined to a specific, evaporite-dominated facies association indicating environments that were inhospitable to fauna and flora other than during brief episodic microbial mat development. Biogenic generation of methane and/or hydrogen sulphide is typical of microbially dominated deposition. There are many instances from the Abu Dhabi Holocene sabkhas of shallow gas accumulations causing short-duration water fountains due to local overpressuring. Typically, these reach about 1 m in height, but more spectacular gas-driven water fountains reaching 15 m have been witnessed personally. Such occurrences reflect overpressuring and fluidization in the types of sedimentary environments under discussion.

The favoured mechanism leading to the intrastratal contortions of the patterned dolomites is that proposed by Kendall (1977) that involves soft-sediment adjustments to fill spaces vacated by the removal of gypsum crystals or nodules by sulphate-reducing bacteria. Kendall's theory conveniently allows for sulphide production before or during the internal deformation, as required by the evidence of sulphide concentrations apparently being contorted by post-depositional sediment flow.

Modern examples

The SW coastline of the Arabian Gulf with its well-developed, evaporitic, peritidal deposits is frequently cited as an excellent analogue for the upper Arab Formation, and one could expect to find modern examples of patterned carbonates being actively formed. Reducing conditions are extremely common within the peritidal and lagoonal carbonates. This is evident from varying degrees of sediment blackening by sulphide impregnation, often within millimetres of the sediment surface.

Examples of patterned carbonates may conceivably be found in the sediments of Khor Odaid, SE Qatar, which is the only known present-day subaqueous environment of primary gypsum precipitation (Loreau & Purser 1973). With time, the present lagoons behind the Great Pearl Bank barrier along the Abu Dhabi coastline may also become more restricted and form environments conducive to extensive patterned dolomite formation.

Extensive work on the Holocene sediments of Abu Dhabi indicates that supratidal sabkhas and intertidal flats contain areas where dolomitization is interpreted to have occurred as a result of marine flood recharge/reflux seepage (Bush 1973) or evaporative pumping (McKenzie *et al.* 1980). The dolomite is difficult to identify in the field but patterned carbonate has been observed in the bank of the Mussafah channel – a 7 km man-made waterway excavated through the mainland sabkha, immediately west of Abu Dhabi island. Subaqueous salina gypsum probably formed at the same stratigraphic level in very close proximity to this occurrence (Kirkham 1998).

Reservoir implications

The realization that patterned dolomites represent a discrete lithofacies formed subaqueously within a salina or restricted lagoonal environment may significantly impact depositional modelling of a reservoir. They could provide supporting evidence that associated interbedded stromatolitic horizons and anhydrite (formerly gypsum) beds may also have been subaqueous rather than intertidal and supratidal. Rather than the depositional models of extensively prograding supratidal sabkhas traditionally applied to facies associations of stromatolites and bedded anhydrites in the Arab Formation (e.g. Wood & Wolfe 1969), the presence of patterned dolomites demands reconsideration of the sedimentary dynamics. These resulted in more static facies belts with reduced correlatability that implies higher levels of reservoir heterogeneity, than those discussed by Kirkham & Twombley (1995). Salinas generally aggrade rather than prograde (Warren & Kendall 1985) and so any long-distance correlation and facies continuity within the Arab Formation may need reassessment, although it is stressed that no size limitations are implied for such evaporitic environments.

Conclusions

Patterned dolomite, created partly by sulphate-reducing bacteria, is a relatively common lithology in the Arab Formation. Its sedimentary characteristics indicate that significant amounts of sulphate were removed from the system penecontemporaneously. Much of this could have been crystalline gypsum, and patterned dolomite may therefore be an indicator of vanished evaporites. The gypsum removed is likely to have formed as a primary precipitate in salinas or highly restricted lagoons, and intimately associated stromatolitic horizons may therefore also be subaqueous or subtidal rather than intertidal.

If these lithologies did, indeed, form subaqueously, the classical models of extensively prograding sabkha cycles commonly applied to the Arab Formation need to be fundamentally reassessed. Whilst not refuting all such models that have a well-established modern analogue in the coastal Holocene strata of Abu Dhabi, many may need revision to account for the presence of more static and transitory depositional elements. A reduction in lateral continuity implied by such a model may significantly impact on reservoir characterization and performance.

H. Mueller, D. Wright and C. Braithwaite are thanked for reviewing earlier drafts and proposing highly appreciated modifications.

References

AL-SILWADI, M.S., KIRKHAM, A., SIMMONS, M.D. & TWOMBLEY, B.N. 1996. New insights into regional correlation and sedimentology, Arab Formation (Upper Jurassic), offshore Abu Dhabi. *GeoArabia*, **1**, 6–27.

BERNER, R.A., BALDWIN, T. & HOLDREN, G.A., JR. 1979. Authigenic iron sulphides as paleaosalinity indicators. *Journal of Sedimentary Petrology*, **4**, 1345–1350.

BUSH, P.R. 1973. Some aspects of the diagenetic history of the Sabkha in Abu Dhabi, Persian Gulf. *In*: PURSER, B.H. (ed.) *The Persian Gulf, Holocene Carbonate Sedimentation in a Shallow Epicontinental Sea*. Springer, New York, 395–407.

CHAFETZ, H.S. & RUSH, P.F. 1995. Two-phase diagenesis of Quaternary carbonates, Arabian Gulf: insights from $\delta^{13}C$ and $\delta^{18}O$ data. *Journal of Sedimentary Research*, **65**, 294–305.

CURTIS, R., EVANS, G., KINSMAN, D.J.J. & SHEARMAN, D.J. 1963. Association of dolomite and anhydrite in the Recent sediments of the Persian Gulf. *Nature*, **197** (4868), 679–680.

DIXON, J. 1976. Patterned carbonate – a diagenetic feature. *Bulletin of Canadian Petroleum Geology*, **24**, 450–456.

EVANS, G., SCHMIDT, V., BUSH, P. & NELSON, H. 1969. Stratigraphy and geological history of the sabkha, Abu Dhabi, Persian Gulf. *Sedimentology*, **12**, 145–159.

KENDALL, A.C. 1977. Patterned carbonate – a diagenetic feature, by James Dixon. Discussion. *Bulletin of Canadian Petroleum Geology*, **25**, 695–697.

KENDALL, A.C. 1979*a*. Facies Models 13. Continental and supratidal (sabkha) evaporites. *In*: WALKER, R.G. (ed.) *Facies Models*. Geoscience Canada Reprint Series, **1**, 145–157.

KENDALL, A.C. 1979*b*. Facies Models 14. Subaqueous evaporites. *In*: WALKER, R.G. (ed.) *Facies Models*. Geoscience Canada Reprint Series, **1**, 159–174.

KIRKHAM, A. 1998. A Quaternary proximal foreland ramp and its continental fringe, Arabian Gulf, UAE. *In*: WRIGHT, V.P. & BURCHETTE, T.P. (eds) *Carbonate Ramps*. Geological Society, London, Special Publications, **149**, 15–41.

KIRKHAM, A. & TWOMBLEY, B.N. 1995. Heterogeneity and fluid saturation predictions in complex Jurassic carbonates of Abu Dhabi. *Geo '94, Middle East Petroleum Geosciences Conference*, Volume II, 605–614. Gulf Petrolink.

LAPOINTE, PH. 1992. *Fluid Transfer Structures Identification in the Upper Arab Formation of the Abu*

Dhabi Offshore Fields (UAE). Society of Petroleum Engineers, SPE Paper 24512 (ADSPE Paper 304).

LOREAU, J.-P. & PURSER, B.H. 1973. Distribution and ultrastructure of Holocene ooids in the Persian Gulf. *In*: PURSER, B.H. (ed.) *The Persian Gulf. Holocene Carbonate Sedimentation and Diagenesis in a Shallow Epicontinental Sea*. Springer, Berlin, 279–328.

MCKENZIE, J.N., HSU, K.J. & SCHNEIDER, J.F. 1980. Movement of subsurface waters under the sabkha, Abu Dhabi, UAE, and its relation to evaporative dolomite genesis. *In*: ZENGER, D.H., DUNHAM, J.B. & ETHINGTON, R.L. (eds) *Concepts and Models of Dolomitization*. Society of Economic Palaeontologists and Mineralogists, Special Publications, **28**, 11–30.

PATTERSON, R.J. & KINSMAN, D.J.J. 1981. Hydrologic framework of a Sabkha along the Arabian Gulf. *AAPG Bulletin*, **65**, 1457–1475.

SHEARMAN, D.J. 1966. Origin of marine evaporites by diagenesis. *Transactions of the Institute of Mineralogy and Metallurgy*, Section B, **75**, 208–215.

VASCONCELOS, C. & MCKENZIE, J.A. 1997. Microbial mediation of modern dolomite preciptation and diagenesis under anaoxic conditions (Lagoa Vermelha, Rio de Janeiro, Brazil). *Journal of Sedimentary Research*, **67**, 378–390.

WARREN, J.K. 1989. *Evaporite Sedimentology*. Prentice-Hall, Englewood Cliffs, NJ.

WARREN, J.K. & KENDALL, C.G.ST.C. 1985. Comparison of sequences formed in marine sabkha (subaerial) and salina (subaqueous) settings – modern and ancient. *AAPG Bulletin*, **69**, 1013–1023.

WOOD, G.V. & WOLFE, M.J. 1969. Sabkha cycles in the Arab/Darb Formation off the Trucial Coast of Arabia. *Sedimentology*, **12**, 165–191.

WRIGHT, D.T. 1999. The role of sulphate-reducing bacteria and cyanobacteria in dolomite formation in distal ephemeral lakes of the Coorong region, South Australia. *Sedimentary Geology*, **126**, 147–157.

WRIGHT, D.T. & WACEY, D. 2004. Sedimentary dolomite: a reality check. *In*: BRAITHWAITE, C.J.R., RIZZI, G. & DARKE, G. (eds) *The Geometry and Petrogenesis of Dolomite Hydrocarbon Reservoirs*. Geological Society, London, Special Publications, **235**, 65–74.

Palaeozoic dolomite reservoirs in the Permian Basin, SW USA: stratigraphic distribution, porosity, permeability and production

ARTHUR H. SALLER

Unocal Corporation, Sugar Land, TX 77478, USA

Abstract: Dolomite and dolomite reservoirs are common in Ordovician and middle to upper Permian carbonates in the 'Permian Basin' area of the SW United States. Scattered, but significant, dolomite reservoirs also occur in the Silurian and Devonian. In the middle to upper Permian, platform and shelf-top carbonates were dolomitized, while limestones in slope and basinal environments were not. Most dolomitization of Ordovician and Permian carbonates occurred in evaporated seawater shortly after deposition (reflux dolomitization). Most dolomitizing brines apparently formed in tidal flat and restricted lagoonal environments in the platform and shelf interiors, and on depositional highs near shelf and platform margins.

More than 26 billion barrels of oil have been produced from the Permian Basin, with most of that oil coming from Palaeozoic (mainly Permian) dolomites. Permian Basin dolomites have very heterogeneous porosity and permeability on a wide range of scales. Field-scale (km-scale) variations in porosity are commonly related to position in the original dolomitizing system. Porosity generally increases away from the apparent source of the dolomitizing brines because a greater volume of dolomite was precipitated in proximal parts of the dolomitizing system than in the distal parts; hence, porosity is greater in dolomites in the basinward parts of fields.

Most Permian Basin dolomite reservoirs are structural traps with stratigraphic enhancement of closure by loss of porosity and permeability towards the shelf or platform interior. Many traps were formed by compactional drape over the same features that created highs during deposition. Hence, the structurally highest parts of many fields have the poorest porosity and permeability because they coincide with proximal parts of the original dolomitizing system. The most porous, permeable and productive dolomites are on the basinward flanks of structures, and often near the oil–water contact.

Dolomite reservoirs in the Permian Basin are quite variable. Ultimate oil recoveries from these fields range from <1000 barrels to 2 billion barrels, with the largest fields in shallow middle Permian (San Andres/Grayburg) reservoirs. Reservoir depths range from 1500 to more than 14 000 ft (500–4300 m). Average porosities for fields are 1–21%, with porosities generally decreasing with depth. Average permeabilities are 1–1000 mD. Many deeper reservoirs have high permeability related to fractures in karstified Ordovician reservoirs. Recovery efficiencies are 10–65% of the original oil in place, with higher recovery efficiency associated with larger pores and higher permeabilities.

The Permian Basin of the United States is a major petroleum-producing area that is mainly in west Texas, but extends into SE New Mexico (Fig. 1). More than 26 billion barrels of oil have been produced in the basin since the 1920s. Almost all of this production has come from Palaeozoic strata, with most from middle Permian dolomites. However, large oil and gas fields also occur in Ordovician dolomites, and significant fields are present in Siluro-Devonian dolomites. There are also major fields in Pennsylvanian limestones of the Horseshoe Atoll, and oil and gas fields have been found in sandstones, limestones and cherts of a variety of ages within the basin (Galloway *et al.* 1983).

Dolomite reservoirs within the Permian Basin are complex and heterogeneous regardless of age. They are typically structural traps with stratigraphic enhancement of closure. More than 50 years after discovery, many reservoirs continue to produce substantial amounts of hydrocarbons with the help of waterfloods, infill wells, CO_2 injection and innovative reservoir management. The purposes of this paper are to: (1) show the stratigraphic distribution of Permian Basin dolomites; (2) discuss a model for Permian dolomitization; (3) discuss the distribution of porosity within Permian dolomites; (4) show a common relationship between structure and porosity; and (5) summarize major features of west Texas dolomite reservoirs.

From: BRAITHWAITE, C. J. R., RIZZI, G. & DARKE, G. (eds) 2004. *The Geometry and Petrogenesis of Dolomite Hydrocarbon Reservoirs*. Geological Society, London, Special Publications, **235**, 309–323. 0305-8719/$15.00

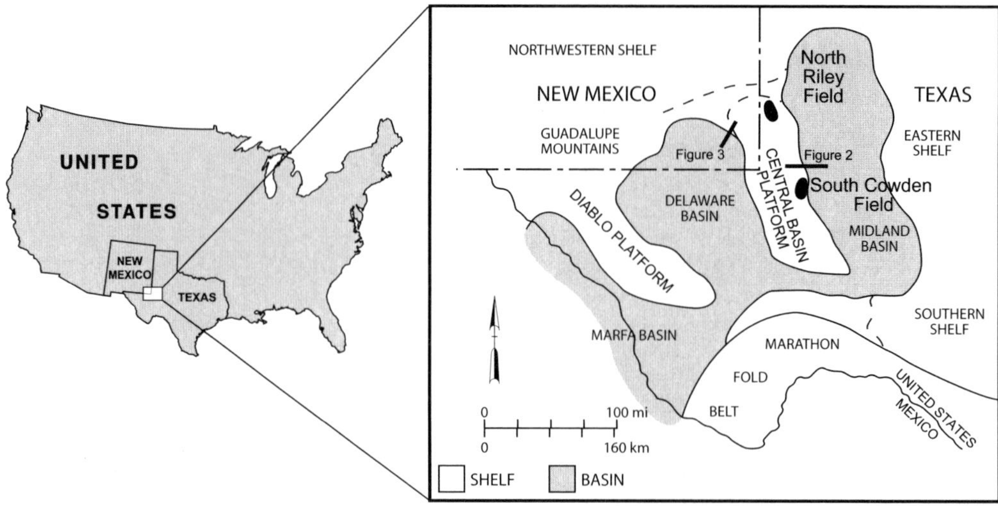

Fig. 1. Location of the Permian Basin and general Permian palaeogeography.

Geological setting

The geological history of the Permian Basin is dominated by Palaeozoic events. Cambrian sandstones and conglomerates occur above a crystalline pre-Cambrian basement. Early Palaeozoic deposition occurred over a broad, gently subsiding area in the SW part of the North American continent. The Lower Ordovician Ellenburger Group is a karstified dolomite that forms an important deep reservoir. Broad subsidence with an axis along the current 'Central Basin Platform' (Fig. 1) created the 'Tobosa Basin' during the Silurian and Devonian. Spiculitic chert was deposited in the deeper part of that basin during the early Devonian (Ruppel & Barnaby 2001).

Major tectonic activity occurred during the Mississippian and early Pennsylvanian (Carboniferous) when South America collided with North America. That collision resulted in folding, faulting, uplift, and some erosion of the middle and lower Palaeozoic section in the Permian Basin. Regional subsidence began during the middle Pennsylvanian and continued through the Permian. Areas of greater subsidence resulted in the Delaware and Midland basins, which were bounded by upper Pennsylvanian and Permian carbonate-shelf and platform margins (Fig. 1). Regional subsidence slowed during the late Permian allowing the Delaware and Midland basins to be filled. More than 10 000 ft (*c.* 3000 m) of Permian strata were deposited in the Delaware Basin, and 4000 ft (*c.* 1200 m) of Permian carbonates were deposited on the Central Basin platform (Fig. 2; Matchus & Jones 1984).

Little deposition has occurred in the Permian Basin area since the Permian. During most of the Mesozoic, the basin was tectonically stable. Triassic and Jurassic strata are thin or absent. A thin Cretaceous section suggests minor short-lived subsidence and/or eustatic sea-level rise. The western Permian Basin was at the eastern limits of the Laramide orogeny causing some uplift and tilting during the latest Cretaceous and early Tertiary. The basin is also just east of the 'Basin and Range' faulting that dominated the western United States during the middle and late Cenozoic.

Distribution of dolomite in Palaeozoic strata of the Permian Basin

The distribution of dolomite in the Palaeozoic of the Permian Basin displays some consistent patterns (Fig. 2). Dolomite is widespread in the Ordovician (Ellenburger Group). Limestones and dolomites occur in the Silurian and Devonian. Limestone is the dominant carbonate in the Mississippian, Pennsylvanian and lowest Permian. In the middle and upper Permian, dolomite is again widespread and dominates in platform and shelf-top strata, whereas limestone is common in contemporaneous shelf-margin and slope deposits (Figs 2 and 3). Dolomite generally increases upward in the Permian section, until evaporites dominate at the end of the Permian. Non-carbonate lithologies also

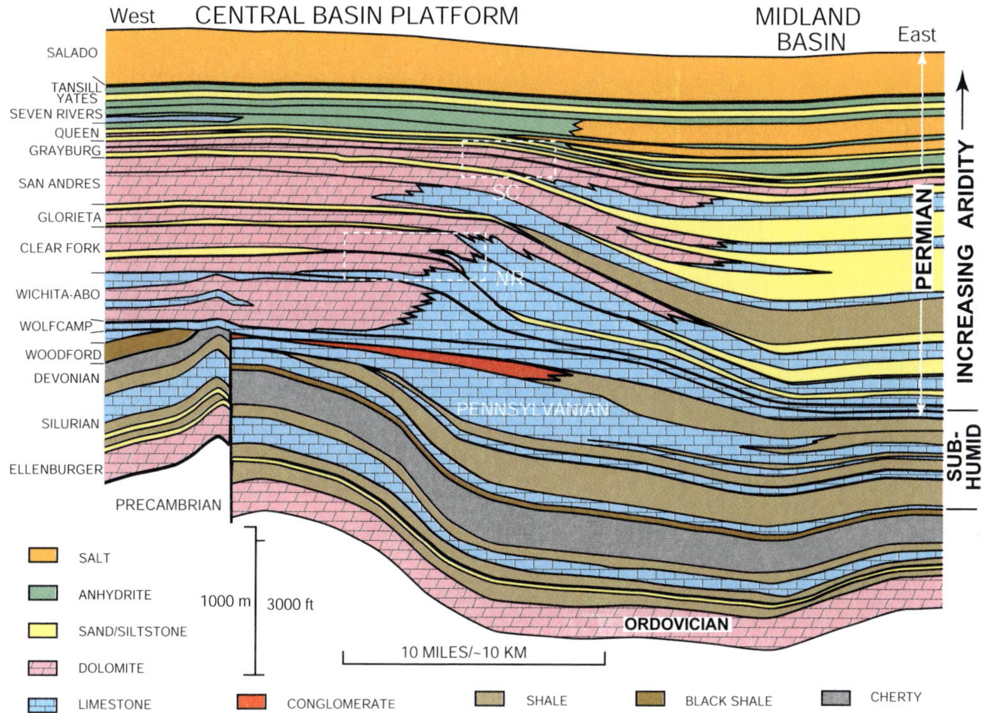

Fig. 2. Simplified structural cross-section at the eastern margin of the Central Basin Platform based on data in Matchus & Jones (1984). The approximate location is shown in Figure 1. Dolomite is most common in platform-top areas during arid times. Limestone is the dominant carbonate everywhere during subhumid times, and dominates in slope and basinal environments during more arid times. 'SC' indicates the approximate stratigraphic position of South Cowden Field. 'NR' indicates the approximate position of North Riley Field.

show systematic variations in the Palaeozoic (Table 1). Shales are common in the Devonian, Mississippian, Pennsylvanian and lowest Permian, whereas evaporites are extremely rare in those intervals. Evaporites become increasingly widespread up-section during the Permian, and clastics become dominated by very fine-grained sands and silts during the middle and upper Permian (Fig. 2).

The distribution of Palaeozoic lithologies in the Permian Basin indicates a correlation between climate and dolomite. The widespread occurrence of shale and lack of evaporites suggests that the Permian Basin area was humid to semi-humid during the Mississippian and Pennsylvanian (Carboniferous). Limestone is the dominant carbonate lithology during the Mississippian and Pennsylvanian (Carboniferous; Fig. 2 and Table 1). In contrast, dolomite is associated with sand and silt in the Ordovician, and with sands, silts and evaporites in the middle and upper Permian, suggesting dolomitization was associated with a semi-arid or arid climate. The preferential occurrence of dolomite in the interior of Permian platforms and shelves (Figs 2 and 3) suggests dolomitization in slightly to very restricted platform-interior environments.

Mechanisms of dolomitization

Several lines of evidence indicate that most dolomitization of Ordovician and Permian carbonates occurred shortly after deposition from evaporated seawater (i.e. reflux dolomitization of Adams & Rhodes 1960). As mentioned above, preferential dolomitization of platform and shelf-interior carbonates (Figs 2 and 3) supports dolomitization by brines formed in the interior of carbonate shelves and platforms (Fig. 4). The association of dolomite with semi-arid and arid climates during the Palaeozoic is consistent with dolomitization by evaporated seawater.

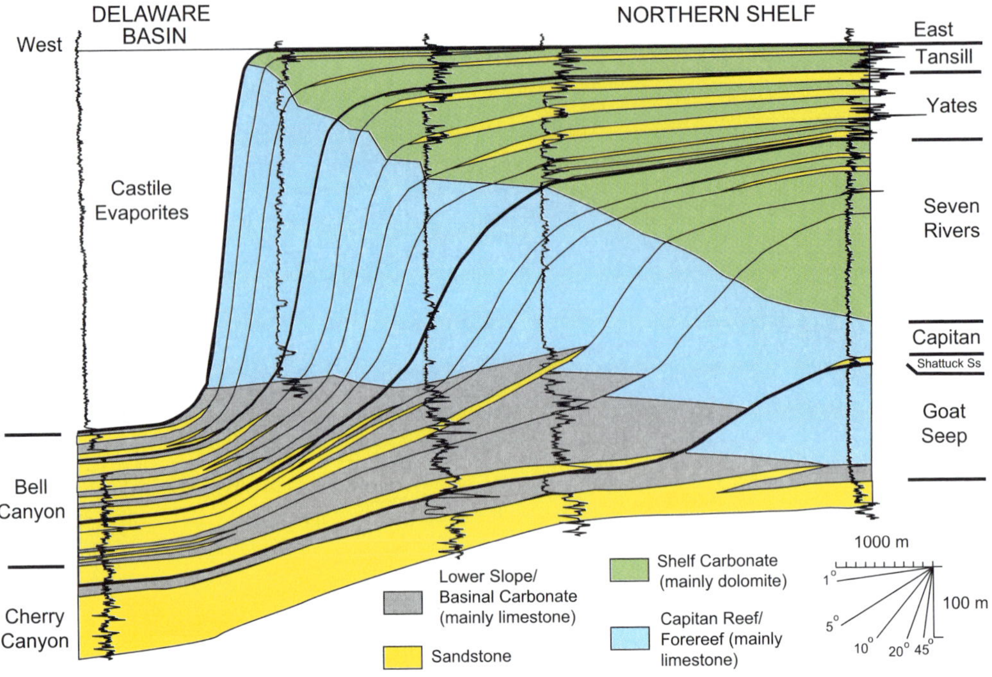

Fig. 3. Subsurface stratigraphic cross-section of the Capitan depositional system (upper Guadalupian) in the NE Delaware Basin. Datum is top of Tansill Formation and a horizon in the Castile. Dolomite is the dominant carbonate in platform-top environments; whereas limestone is the dominant carbonate in platform-margin and basinal environments. See Figure 1 for the location (modified from Harris & Saller 1999).

Table 1. *Lithologies in the Permian Basin*

Interval	Main carbonate	Main clastic	Evaporites	Inferred climate
Middle Permian	Dolomite	Sand/silt	Common	Arid
Wolfcampian–Pennsylvanian	Limestone	Shale	Rare	Subhumid
Mississippian	Limestone	Shale	None	Humid
Siluro-Devonian	Both	Shale	Rare	Subhumid–subarid
Ordovician	Dolomite	Sand and shale	None	Subhumid–subarid

Previous researchers have used other lines of evidence to argue for dolomitization of Palaeozoic carbonates in and around the Permian Basin by evaporated seawater. Widespread dolomitization of Ordovician carbonates (Ellenburger Group and outcrop equivalents) was attributed to evaporated seawater by Gao & Land (1991) and Kupecz & Land (1994) based on petrography and geochemistry. They also note that some early dolomite recrystallized and additional dolomite formed during deeper burial.

Permian dolomite has generally been interpreted as forming in evaporated seawater, based on its distribution, petrography and geochemistry (Adams & Rhodes 1960; Ruppel & Cander 1988; Major *et al.* 1990; Leary & Vogt 1986, 1990; Saller & Henderson 1998). In the Permian, dolomites show heavy carbon and oxygen isotopes with increasing enrichment in ^{13}C and ^{18}O up-section (Fig. 5), which also supports dolomitization by evaporated seawater and increasing aridity during the Permian.

In contrast, limestones are common in the Mississippian, Pennsylvanian and lower Permian (Wolfcamp) strata., These limestones were commonly exposed and subjected to intense freshwater diagenesis (Saller *et al.* 1999).

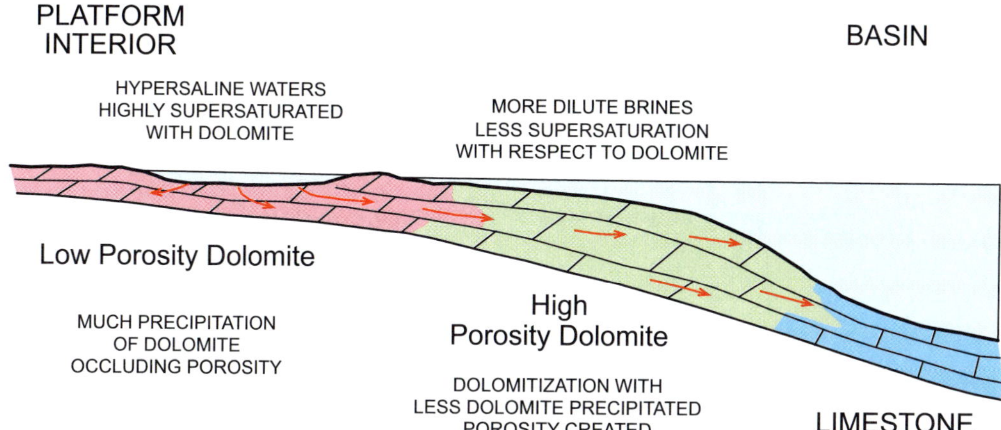

Fig. 4. Model for dolomitization for middle Permian carbonates showing that dolomitization is mainly by evaporated seawater generated in restricted platform-interior environments. Porosity is low in proximal parts of the dolomitizing system and high in distal parts of the system.

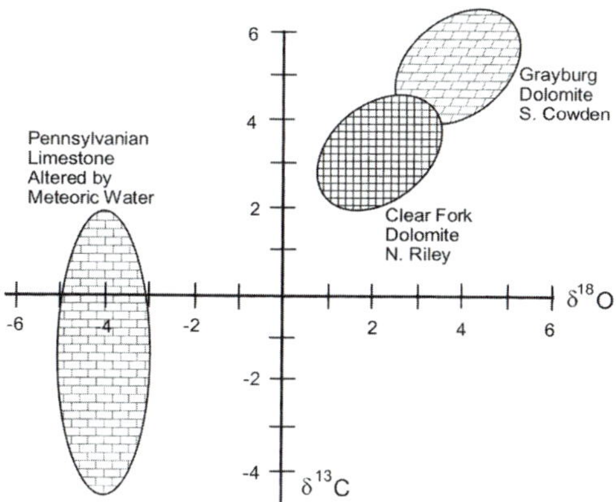

Fig. 5. Plot of stable carbon- and oxygen-isotope fields in Permian Basin strata. Stable isotopes become heavier upward in the section from Pennsylvanian to Clear Fork (lower middle Permian) to Grayburg (middle middle Permian) suggesting diagenesis, especially dolomitization, in more evaporitic waters.

Hence, mixing zones must have been common, but the lack of early dolomitization suggests that they were not significant environments for dolomitization.

Porosity distribution in Permian dolomite reservoirs

Permian dolomites in the Permian Basin have been repeatedly penetrated and cored in the subsurface, allowing insight into spatial trends in their porosity and permeability. Porosity and permeability in Permian dolomite reservoirs are very heterogeneous (Kerans *et al.* 1994, and many others); however, some distinct trends are present (Saller & Henderson 1998; Atchley *et al.* 1999). Dolomite reservoirs occur in shelf and platform-margin and shelf and platform-interior settings. Typical depositional facies are shown in Figure 6. Some trends in porosity and permeability are clearly related to original depositional facies (Kerans *et al.* 1994).

In reservoirs associated with shelf and platform margins, porosity in dolomite commonly increases from shelf and platform interior to shelf and platform margin independent of depositional facies. This is shown in the South Cowden and North Riley fields (Figs 1, 7A and 8A) where porosity is heterogeneous but generally increases basinward (Figs 7B and 8B). In both fields, fenestral wackestone/packstones are more common toward the platform interior and fusulinid wackestone/packstones are more common in basinward parts of the fields (Figs 7B and 8B), as would be expected given the depositional model in Figure 6. A given facies is generally more porous in basinward parts of the field (Figs 7B and 8B). For example, fusulinid wackestone/packstones in the western part of the South Cowden Field have less porosity than in the eastern, basinward part of the field (Fig. 7B). Although many pore types are present, intercrystalline porosity is dominant throughout the South Cowden Field. Many pore types, including intercrystalline,

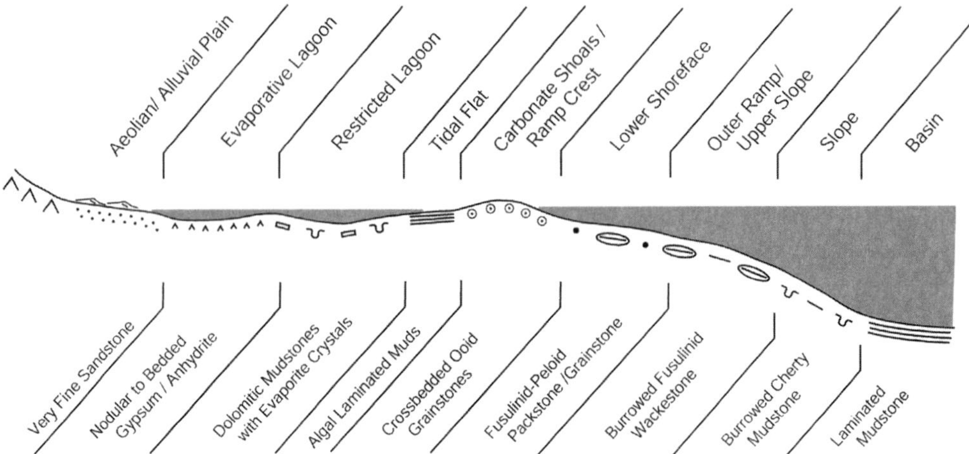

Fig. 6. Depositional model for middle Permian carbonates in the Permian Basin.

moulds and intergranular, are important in the North Riley Field. In both fields, pore size and, hence, matrix permeability generally increase basinward.

Many dolomitizing systems show a systematic increase in porosity away from the apparent source of the dolomitizing water (Lucia & Major 1994; Saller & Henderson 1998). This is commonly the case in the Permian Basin. As mentioned in the previous section, Permian dolomite is generally interpreted to have formed in seawater that was concentrated by evaporation in restricted lagoons and tidal flats. That evaporated seawater apparently moved down and basinward into more open marine facies that were dolomitized as a result (Fig. 4). The evaporation of seawater increases dolomite supersaturation, and precipitation of dolomite will decrease saturation. Hence, proximal dolomitization environments will have the most dolomite precipitated by the initial dolomitizing brines, and more distal environments will have lower supersaturations and, hence, less dolomite precipitated (Saller & Henderson 1998). Waters supersaturated with respect to dolomite should continue to flow through the initially dolomitized sediment, precipitating additional dolomite and further decreasing porosity (Figs 9 and 10). This precipitation of additional layers of dolomite, decreasing porosity, is similar to the 'overdolomite' of Lucia & Major (1994) and Sun (1995).

Relationship of structure to porosity and production

Most oil and gas fields in the Permian Basin are structural traps, although many Permian fields have stratigraphic enhancement of structural closure. Lower Palaeozoic strata were folded during the Mississippian and early Pennsylvanian. As a result, most Ordovician Ellenburger reservoirs are four-way dip closures (doubly plunging anticlines). The Permian Basin was relatively stable tectonically during the Permian. However, structures in many middle and upper Permian carbonates were created during regional subsidence by differential compaction (Fig. 11). Compaction structures

Fig. 7. (**A**) South Cowden map. Structural contours are in feet subsea. Note that the greatest oil production was from wells on the east (basinward) flank of the field where porosity was highest. Production data are per well recoveries from Lucia & Ruppel (1996). Red lines indicate the location of the cross-section in (B) (modified from Saller & Henderson 1998). (**B**) South Cowden stratigraphic section illustrating facies and porosity in the Upper Grayburg from core data. The section was pervasively dolomitized. Facies are generally more restricted to the west (left), and hence the sources of dolomitizing brines are interpreted to be mainly to the west. The section also thins to the west, due to differential compaction during deposition. Porosity >5% is shaded green. Note that porosity is best developed offstructure, in the basinward part of the Upper Grayburg. The higher porosity area would also correspond to more distal parts of the dolomitizing system. Porosity in the bottom 6–20 well has been increased by anhydrite dissolution. Datum is top of Grayburg (modified from Saller & Henderson 1998).

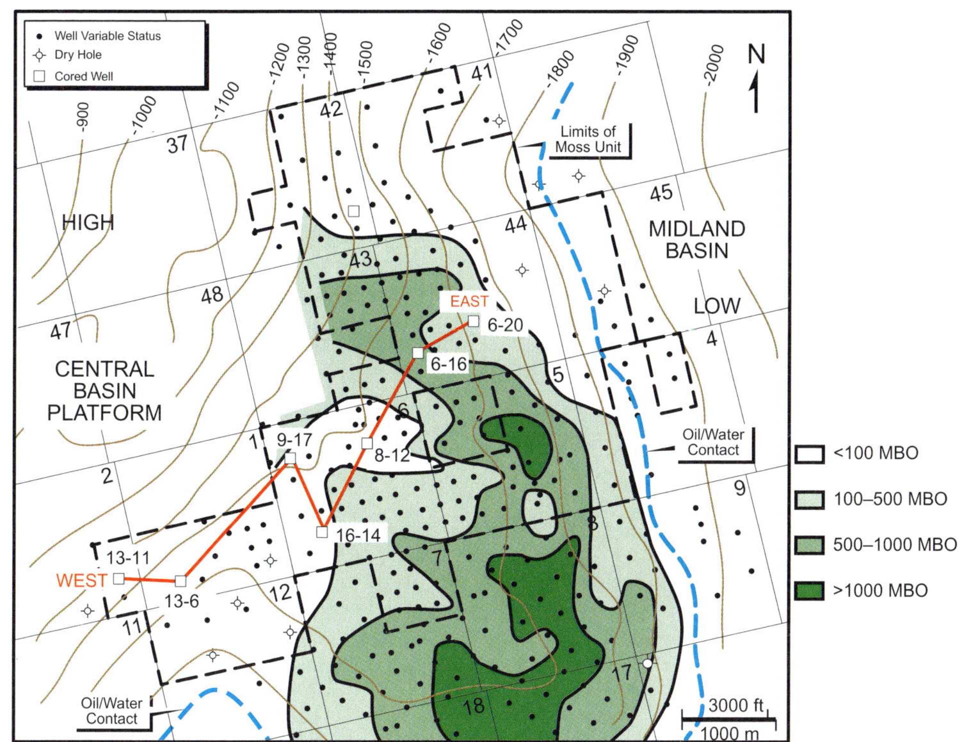
(A)
Well Variable Status
Dry Hole
Cored Well
HIGH
CENTRAL
BASIN
PLATFORM
MIDLAND
BASIN
LOW
Limits of
Moss Unit
Oil/Water
Contact
EAST
WEST
6-20
6-16
8-12
9-17
16-14
13-11
13-6
<100 MBO
100–500 MBO
500–1000 MBO
>1000 MBO
3000 ft
1000 m
N

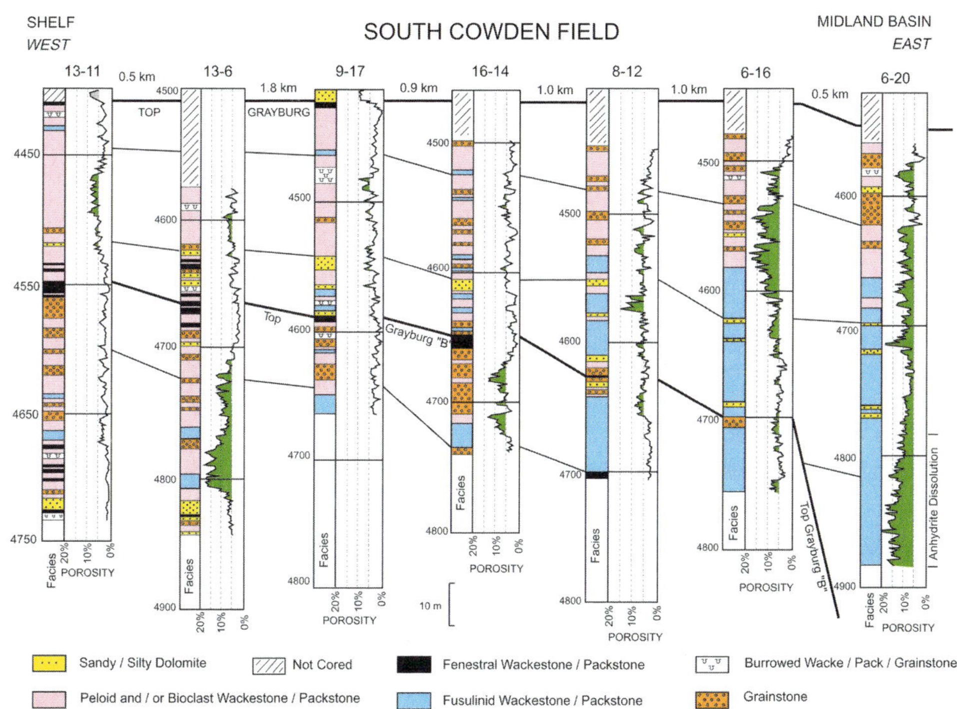
(B)
SHELF
WEST
SOUTH COWDEN FIELD
MIDLAND BASIN
EAST
13-11
13-6
9-17
16-14
8-12
6-16
6-20
0.5 km
1.8 km
0.9 km
1.0 km
1.0 km
0.5 km
TOP
GRAYBURG
Top Grayburg "B"
Anhydrite Dissolution
Facies
POROSITY
10 m
Sandy / Silty Dolomite
Not Cored
Fenestral Wackestone / Packstone
Burrowed Wacke / Pack / Grainstone
Peloid and / or Bioclast Wackestone / Packstone
Fusulinid Wackestone / Packstone
Grainstone

(A)

STRUCTURAL LOW
TOP CLEAR FORK
NORTH RILEY UNIT
1 mile
1 km
-3100
-3000
-2900
>300 MBO/ WELL
200-300 MBO/ WELL
100-200 MBO/ WELL
<100 MBO/ WELL
N
STRUCTURAL HIGH

(B)

PLATFORM INTERIOR
SOUTH
NORTH RILEY FIELD
BASINWARD
NORTH
NRU 357
NRU 337
NRU 294
NRU 272
NRU 44R
NRU 232
1.6 km
1.6 km
0.9 km
1.0 km
0.3 km
TOP OF MIDDLE CLEAR FORK
20%
10%
0%
POROSITY
5m
Grainstone Facies
Tidal Flat Facies
Peloid and Bioclast Wacke / Packstone Facies
Fusulinid Wacke / Packstone Facies

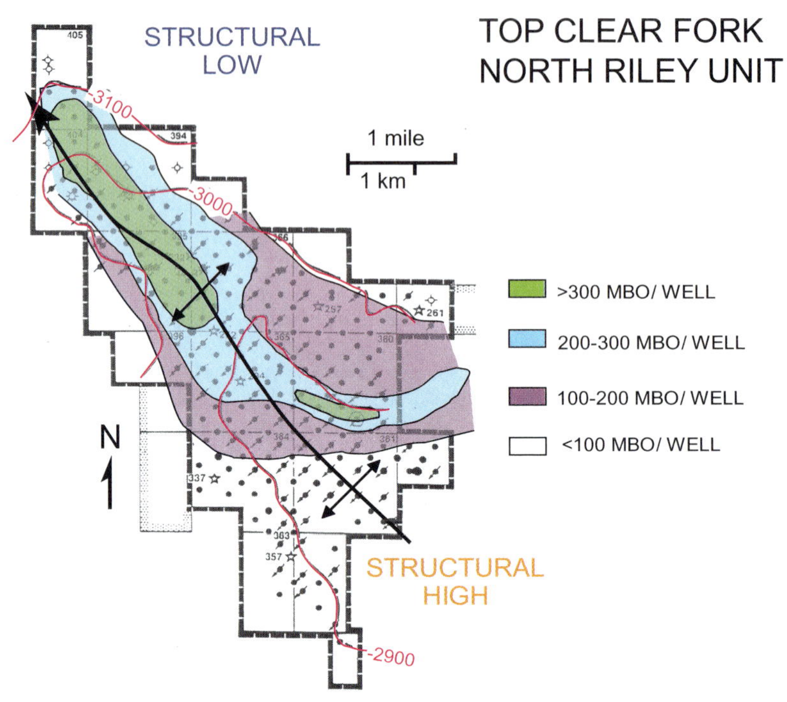

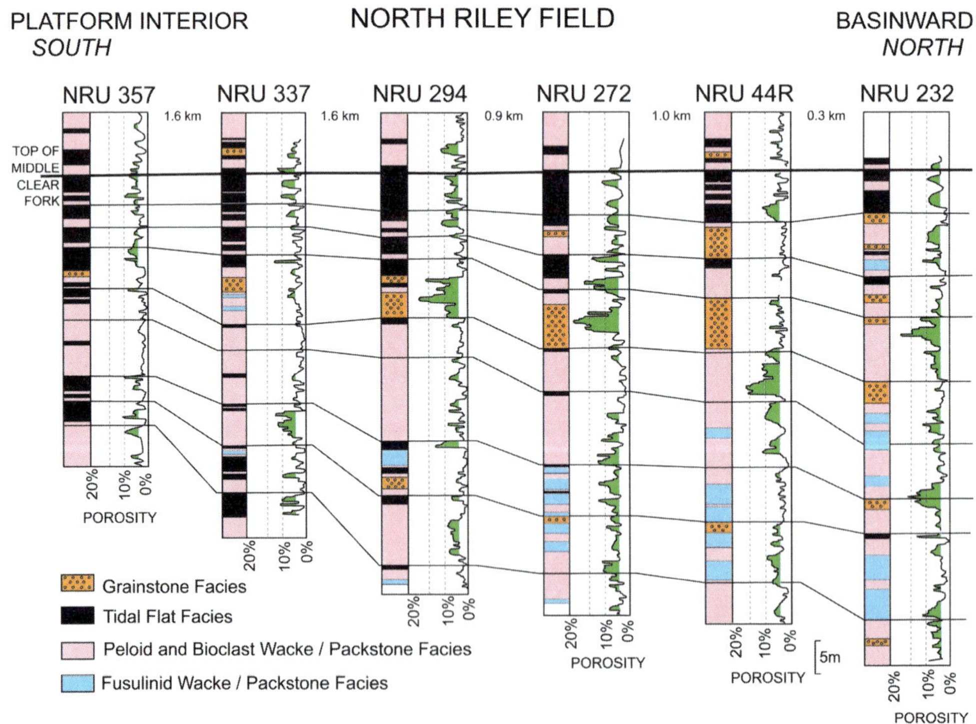

have originated in two main ways: (1) differential compaction of Pennsylvanian and Permian strata over tectonic highs (anticlines) formed during the Mississippian and Pennsylvanian; and (2) differential compaction of different Permian facies.

(1) Younger more compactable shales and carbonates were deposited adjacent to Lower Palaeozoic strata in anticlines. During the Permian, younger, offstructure strata compacted more than the Lower Palaeozoic material on anticlines causing Permian strata to have structural closure in the same location as older tectonic structures.

(2) In other locations, certain carbonate facies were more rigid than adjacent strata, causing differential compaction. For example, the lower San Andres Formation is generally transgressive and backstepping. Some grain-rich carbonate build-ups and bioherms formed during that transgression. Mud-rich carbonate sediments were deposited adjacent to these (Kerans *et al.* 1994) and compacted more during subsequent burial (Kumar & Foster 1982). Thus, structural closure was created in the upper San Andres and Grayburg formations as a result of less compaction of grain-rich and biohermal parts of the lower San Andres (Kumar & Foster 1982). Less compaction of Abo reefs (Leonardian; lower middle Permian) compared to adjacent strata also created closure in the San Andres (Guadalupian; upper middle Permian; Fig. 2) in some parts of the Permian Basin (Galloway *et al.* 1983).

Differential compaction created many local depositional highs during much of the Permian. Systematic porosity variations in the dolomitizing systems caused less porous dolomite on depositional highs (Saller & Henderson 1998). Differential compaction continued to localize structural highs in the same areas. As a result, low-porosity dolomite occurs in the structurally highest parts of many fields; whereas dolomites are most porous on the flanks of structures (Figs 7B, 8B and 11). Atchley *et al.* (1999) recognized a similar relationship of porosity to structure in middle Permian dolomites.

Very porous, sometimes vuggy, carbonates occur near the oil–water contact of many fields because of anhydrite dissolution (Lucia & Ruppel 1996) (Figs 7 and 8). Dissolution of anhydrite is thought to have been a result of middle–late Cenozoic meteoric waters flowing through the aquifer below the oil–water contact (Saller & Henderson 1998). The meteoric waters were derived from areas to the west uplifted during the Cenozoic (Stueber *et al.* 1998). Oil has apparently moved down into the zone of anhydrite dissolution at the edges of some fields. It is not clear whether the structure has shifted slightly or the oil zone has expanded down-dip.

As might be expected from this model, the most productive parts of fields are commonly on the basinward flanks of structural highs. This can be observed in both the South Cowden and North Riley fields. In the South Cowden Field many original wells (40-acre spacing) on the eastern (basinward) side of the field produced more than 500 000 barrels of oil, whereas structurally higher wells on the west side of the field produced less than 100 000 barrels of oil (Fig. 7A). A similar pattern is observed in the North Riley Field. The original 40-acre wells to the south on the crest of the structure produced less than 100 000 barrels of oil; whereas many in the northern (basinward) part of the field produced more than 300 000 barrels of oil (Fig. 8A).

General characteristics of west Texas dolomite reservoirs

Hundreds of thousands of wells have been drilled in west Texas, and over 120 fields have major reservoirs in Palaeozoic dolomites. Palaeozoic dolomite reservoirs in the Permian Basin can be divided into four major groups by age (from top down): (1) San Andres/Grayburg

Fig. 8. (**A**) North Riley map showing structure at top of Clear Fork. Structural contours are every 100 ft subsea. Production data are per well recoveries for original 40-acre spaced wells. The structure is a northward-plunging anticline. Oil production on the crest of the structure (south part of figure) is low (<100 000 barrels per well) because of low porosity in the reservoir dolomite. Production from down-dip wells is greater because of higher porosity and permeability in reservoir dolomites. (**B**) S–N stratigraphic section through the middle Clear Fork at North Riley Field. Facies and porosity are from core data. Porosity >5% is shaded green. The section was pervasively dolomitized. Facies indicate slightly deeper and more open marine environment to the north (offstructure). Although highly variable, porosity generally increases to the north. More restricted environments to the south are apparently the source of dolomitizing brines. Hence, porosity generally increases away from the source of the dolomitizing brines. The section thins to the south indicating that differential compaction kept the southern area relatively high during deposition. Datum is top of Middle Clear Fork (modified from Saller & Henderson 1998).

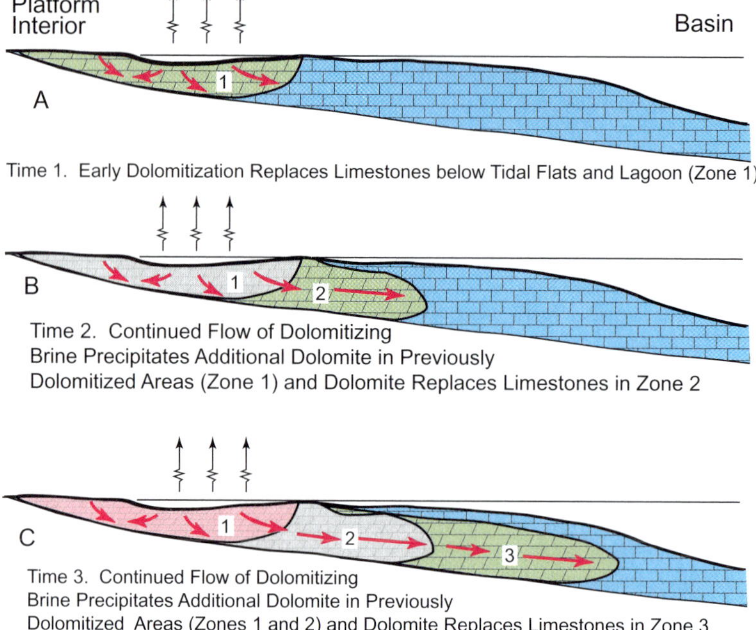

Fig. 9. Schematic cross-sections showing the evolution of dolomite porosity through time. Initial dolomitization of limestone by evaporated seawater results in dolomite with substantial porosity (green; zone 1 in A; zone 2 in B; zone 3 in C). Brine flowing through previously dolomitized areas decreases porosity by precipitating additional dolomite (zone 1 in B; zones 1 and 2 in C). See Figure 10 for small-scale evolution of pores.

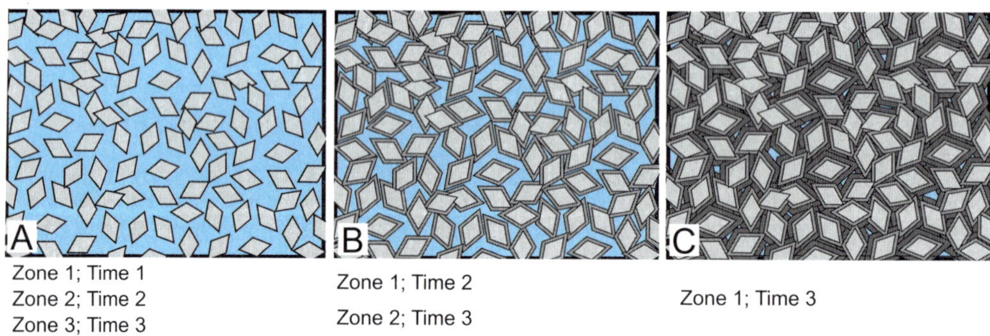

Fig. 10. Pore-scale evolution of dolomite porosity through time. Increasing time is to the right. Pore space is blue. (**A**) Initial dolomites have a high porosity, but circulation of additional brines results in precipitation of additional dolomite ('overdolomite') decreasing porosity (**B**) and (**C**). See Figure 9 for diagenetic environments (zones) and times.

(upper middle Permian); (2) Abo, Clear Fork, Glorietta (lower middle Permian); (3) Siluro-Devonian; and (4) Ellenburger (Ordovician).

The San Andres/Grayburg reservoirs include the largest in the Permian Basin. Three will recover in excess of a billion barrels of oil (Fig. 12). They are mainly Guadalupian in age (upper middle Permian). The fields are commonly structural or combination structure–stratigraphic traps with permeability pinching out in a shelf or platform-interior direction. Field size is generally related to trap size with many reservoirs apparently filled to spill point (see maps in Galloway *et al.* 1983). Dolomitization

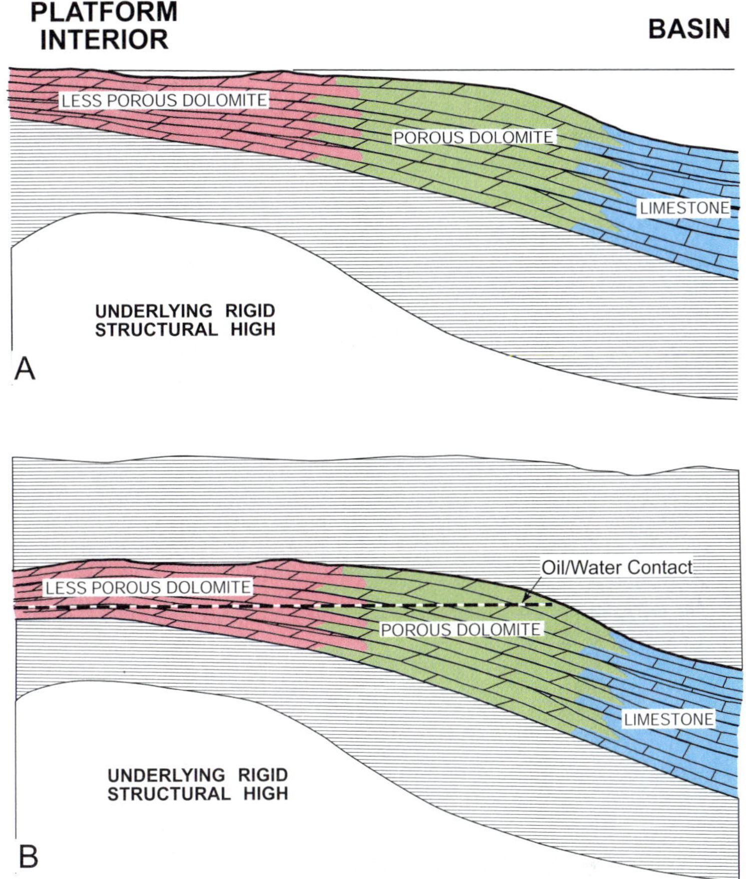

Fig. 11. Model for porosity and structure development for a Permian reservoir associated with differential compaction/compaction drape over an underlying high. Long-term differential compaction caused an area to be a depositional high in the Permian (**A**), and a structural high now (**B**).

extends well beyond the limits of the fields. Many different pore types are present, and pore networks are typically complex and heterogeneous. Fine matrix porosity (intercrystalline, mouldic and intracrystalline; Kerans *et al.* 1994; Lucia & Ruppel 1996; Saller & Henderson 1998) is dominant, but fractures and vugs cause local permeability streaks in some fields. These are commonly solution-gas drive reservoirs, although the margins of some fields have a weak water drive.

Leonardian (lower middle Permian) age reservoirs include the Abo, Clear Fork and Glorietta formations. These are also structural or combination structure–stratigraphic traps with porosity and permeability pinching out in a shelf or platform-interior direction. These reservoirs are a variety of sizes (10–220 million barrels of oil; Fig. 12). The largest Leonardian reservoirs are smaller than the largest San Andres/Grayburg reservoirs because Leonardian reservoirs cover much less area. Heterogeneous matrix porosity is dominant. Most of these reservoirs have an original solution-gas drive and commonly need waterfloods to recover more than 10% of the original oil in place.

The largest Silurian and Devonian dolomite reservoirs are smaller than the largest of other groups of dolomite reservoirs in the basin (Fig. 12). These reservoirs generally cover a smaller area and have a lower porosity than Permian reservoirs (Figs 12 and 13). Some of these are structural traps, and others have traps related to erosional truncation. They include burial dolomite as well as early dolomite.

Ellenburger (Ordovician) reservoirs are the

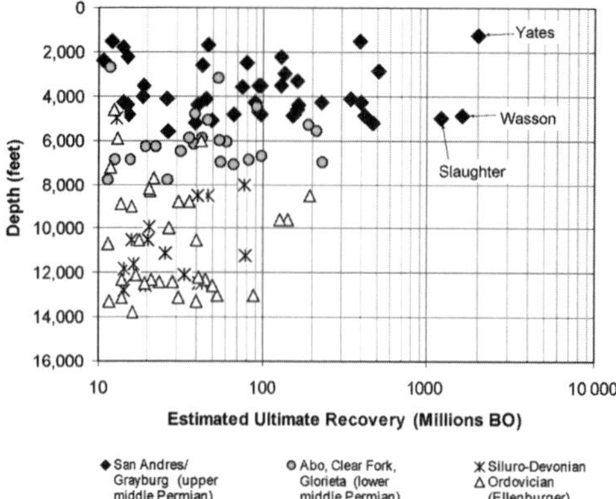

Fig. 12. Plot of estimated ultimate recovery v. depth for major west Texas dolomite oil fields. Note that the largest reservoirs are shallow San Andres/ Grayburg dolomite. 'BO' is barrels of oil (data from Galloway *et al.* 1983).

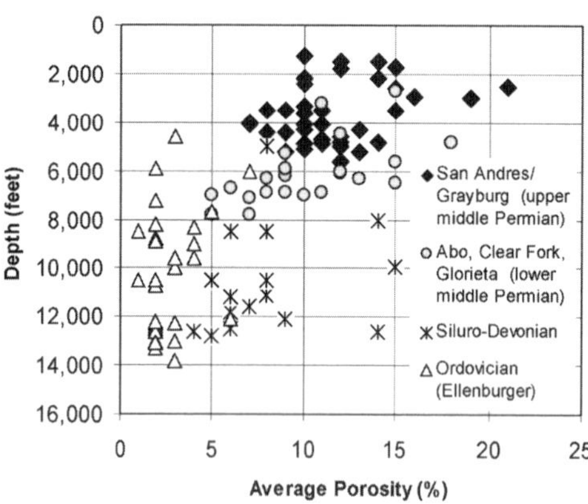

Fig. 13. Plot of average porosity v. depth for major west Texas dolomite oil fields. Note a general decrease in porosity with depth, although porosity is highly variable (data from Galloway *et al.* 1983).

oldest and deepest reservoirs in the Permian Basin (Fig. 12). They are typically structural, four-way dip closures formed during Mississippian and early Pennsylvanian tectonic activity. The Ellenburger was karst-modified during deposition, and the collapse of cave systems has resulted in widespread well-connected fracture porosity (Loucks 1999). These are commonly low-porosity, high-permeability reservoirs with a bottom or edge-water drive.

Galloway *et al.* (1983) compiled a list of the general characteristics of oil fields in west Texas. Their data show a number of interesting trends

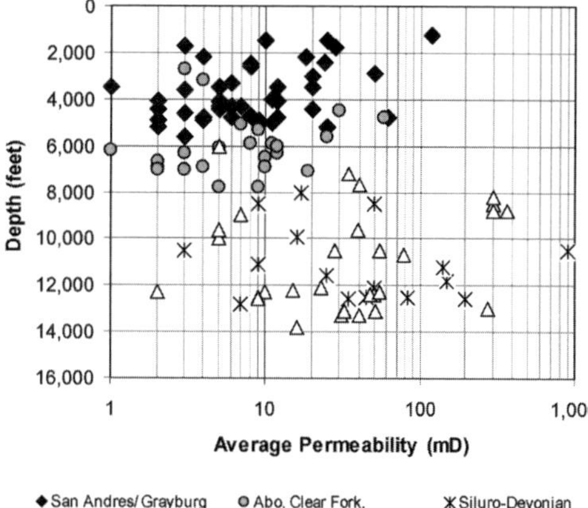

Fig. 14. Plot of average permeability v. depth for major west Texas dolomite oil fields. Shallow, low-permeability reservoirs are dominated by matrix porosity with small pore throats. Deep, high-permeability reservoirs (Siluro-Devonian and Ordovician) have fracture and vuggy pores (data from Galloway *et al.* 1983).

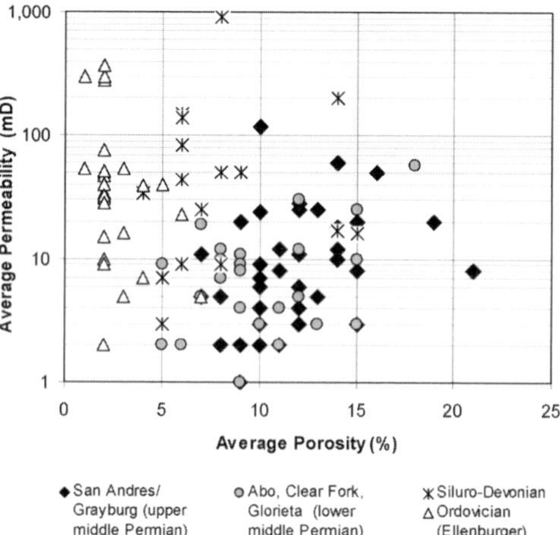

Fig. 15. Plot of average permeability v. average porosity for major west Texas dolomite oil fields. The poor correlation between porosity and permeability is caused by the variations in pore types and pore-throat sizes common in these dolomites. Permian reservoirs with high porosity and low permeability are generally dominated by fine–microcrystalline dolomite with intercrystalline porosity. Fractures and vugs are common in Siluro-Devonian reservoirs with low porosity and high permeability (data from Galloway *et al.* 1983).

in Permian Basin dolomite reservoirs. First, the largest are shallow and commonly within the San Andres or Grayburg formations (Figs 2 and

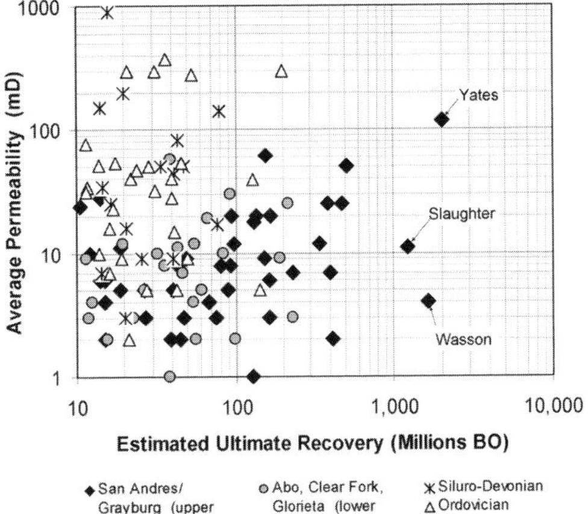

Fig. 16. Plot of estimated ultimate recovery v. average permeability for major west Texas dolomite oil fields. Note the lack of correlation. Some very large reservoirs have very low permeabilities (data from Galloway *et al.* 1983).

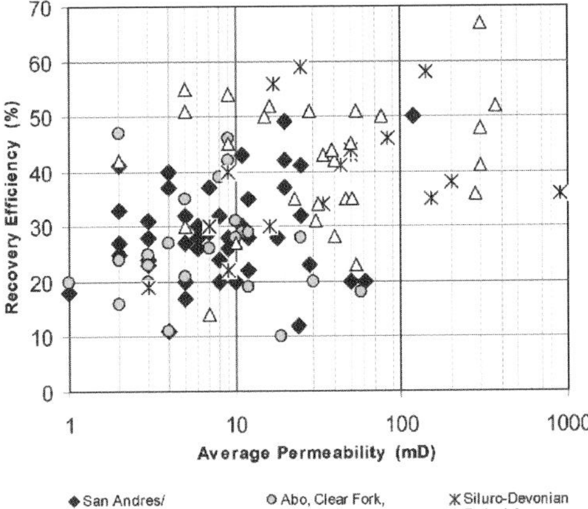

Fig. 17. Plot of recovery efficiency v. average permeability for major west Texas dolomite oil fields. Low-permeability Permian reservoirs tend to have low recovery efficiencies (10–40% of original oil in place will be recovered). Ordovician and Silurian reservoirs tend to have higher permeabilities and higher recovery efficiencies (30–60% of original oil in place will be recovered; data from Galloway *et al.* 1983).

12). Smaller fields occur in all formations. The extensive oil-gathering system in west Texas allows very small fields to be economic. The large San Andres and Grayburg fields cover vast areas indicating very large traps (for example, the Slaughter–Levelland Field is more than 45 km across). The largest fields in other formations cover distinctly smaller areas.

The average porosity of dolomite reservoirs generally decreases with depth (Fig. 13), but average permeability generally increases with depth (Fig. 14). The San Andres and Grayburg reservoirs are shallow and are characterized by high matrix porosity (7–21%) with low average permeability (most are 2–30 mD). In contrast, Ellenburger reservoirs are deep with low average porosity (most are 2–6%), but high average permeability (most are 10–200 mD; Figs 13–15) caused by open fractures (Fig. 15).

Permian Basin reservoirs show no correlation between average permeability and ultimate recovery (Fig. 16). For example, the Wasson Field is expected to recover more than 1.6 billion barrels of oil from a dolomite reservoir with an average permeability of 4 mD. The West Texas dolomite reservoirs have a rough correlation between average field permeability and recovery efficiency (Fig. 17). Recovery efficiencies in fractured pore systems of the Ellenburger are commonly 30–60% of the original oil in place. In contrast, Permian dolomite reservoirs with low matrix permeability commonly have recovery efficiencies of less than 40%.

Conclusions

(1) Major dolomite reservoirs occur in Ordovician, Silurian, Devonian and Permian strata in the Permian Basin.

(2) Dolomite was most widespread during times of relatively arid climate in the Ordovician and middle to late Permian. Most of that dolomitization occurred in evaporated seawater shortly after deposition (reflux dolomitization).

(3) Carbonates deposited in relatively humid climates (Carboniferous–Mississippian and Pennsylvanian) are generally limestone, suggesting that mixing-zone dolomitization was not widespread.

(4) Ordovician reservoirs are generally fractured dolomites that are producing in anticlines formed during Carboniferous (Mississippian and early Pennsylvanian) tectonism.

(5) The largest and most widespread reservoirs are Permian dolomites. They are commonly in anticlinal structures formed by differential compaction over underlying tectonic structures or over less compactable facies. Closure is commonly enhanced by up-dip pinchouts of porosity, resulting in many traps having a stratigraphic component.

(6) Porosity in Permian platform dolomites apparently increases away from the depositional highs that were the sources of the dolomitizing brines (evaporated seawater). Hence, porosity generally increases in a basinward direction until slope mudstones and/or limestones are encountered.
(7) Because differential compaction caused many Permian depositional highs to correspond to current structural highs, the structurally highest parts of dolomite fields commonly have poor porosity, poor permeability and poor production. The basinward flanks of dolomite reservoirs commonly have relatively high porosity, high permeability and good oil production.
(8) Permian Basin dolomite reservoirs are quite variable with respect to size, depth, porosity, permeability and recovery efficiency. Ultimate oil recovery of fields is generally related to trap size. Permeability and recovery efficiency vary with small- and large-scale pore geometries. Average porosity generally decreases with depth; however, permeability commonly increases with depth because deep Ordovician dolomite reservoirs frequently have fracture permeability.

Many people at Unocal have shared their insights with me in projects related to this paper. Those people include T. Elliott, B. Ball, N. Henderson, S. Frost, S. Boyd, G. Moore, A. Crawford, J. A. Dickson and S. Walden. C. Braithwaite and another reviewer provided many constructive comments that greatly improved this manuscript. T. Saller helped enter data used for this paper. I thank Unocal for letting me publish this.

References

Adams, J.E. & Rhodes, M.L. 1960. Dolomitization by seepage refluxion. *AAPG Bulletin*, **44**, 1912–1921.

Atchley, S.C., Kozar, M.G. & Yose, L.A. 1999. A predictive model for reservoir distribution in the Permian (Leonardian) Clear Fork and Glorieta formations, Robertson field area, west Texas. *AAPG Bulletin*, **83**, 1031–1055.

Galloway, W.E., Ewing, T.E., Garrett, C.M., Tyler, N. & Bebout, D.G. 1983. *Atlas of Major Texas Oil Reservoirs*. Texas Bureau of Economic Geology.

Gao, G. & Land, L.S. 1991. Early Ordovician Cool Creek dolomite, Middle Arbuckle Group, Slick Hills, SW Oklahoma: origin and modification. *Journal of Sedimentary Petrology*, **61**, 161–173.

Harris, P.M. & Saller, A.H. 1999. Subsurface expression of the Capitan depositional system and implications for hydrocarbon reservoirs, northeastern Delaware basin. *In*: Saller, A.H., Harris, P.M., Kirkland, B. & Mazzullo, S. (eds) *Geologic Framework of the Capitan Reef*. Society of Economic Paleontologists and Mineralogists, Special Publications, **65**, 37–49.

Kerans, C., Lucia, F.J. & Senger, R.K. 1994. Integrated characterization of carbonate ramp reservoirs using Permian San Andres Formation outcrop analogs. *AAPG Bulletin*, **78**, 181–216.

Kumar, N. & Foster, J.D. 1982. Effect of an underlying bioherm on the San Andres (Permian) reservoir and trap, Hanford Field, Gaines County, Texas. *AAPG Bulletin*, **12**, 2571–2583.

Kupecz, J.A. & Land, L.S. 1994. Progressive recrystallization and stabilization of early-stage dolomite, Lower Ordovician Ellenburger Group, west Texas. *In*: Purser, B., Tucker, M. & Zenger, D. (eds) *Dolomites: A Volume in Honour of Dolomieu*. International Association of Sedimentologists, Special Publications, **21**, 255–279.

Leary, D.A. & Vogt, J.N. 1986. Diagenesis of the Permian (Guadalupian) San Andres Formation reservoirs, Central Basin Platform, west Texas. *In*: Bebout, D.G. & Harris, P.M. (eds) *Hydrocarbon Reservoir Studies. San Andres/Grayburg Formations, Permian Basin*. Permian Basin Section, Society of Economic Paleontologist and Mineralogists, 67–68.

Leary, D.A. & Vogt, J.N. 1990. Diagenesis of the San Andres Formation (Guadalupian) reservoirs, University Lands, Central Basin Platform. *In*: Bebout, D.G. & Harris, P.M. (eds) *Geological and Engineering Approaches in Evaluation of San Andres/Grayburg Hydrocarbon Reservoirs – Permian Basin*. The University of Texas at Austin, Bureau of Economic Geology, 21–28.

Loucks, R.G. 1999. Paleocave carbonate reservoirs: origins, burial-depth modifications, spatial complexity, and reservoir implications. *AAPG Bulletin*, **83**, 1795–1834.

Lucia, F.J. & Major, R.P. 1994. Porosity evolution through hypersaline reflux dolomitization. *In*: Purser, B., Tucker, M. & Zenger, D. (eds) *Dolomites: A Volume in Honour of Dolomieu*. International Association of Sedimentologists, Special Publications, **21**, 325–341.

Lucia, F.J. & Ruppel, S.C. 1996. Characterization of diagenetically altered carbonate reservoirs, South Cowden Grayburg reservoir, west Texas. *In*: *Proceedings of the 1996 SPE Annual Technical Conference and Exhibition, Formation Evaluation and Reservoir Geology*. Society of Petroleum Engineers, SPE 36650, 883–893.

Major, R.P., Van der Stoep, G.W. & Holtz, M.H. 1990. *Delineation of Unrecovered Mobile Oil in a Mature Dolomite Reservoir: East Penwell San Andres Unit, University Lands, West Texas*. The University of Texas at Austin, Bureau of Economic Geology Report of Investigations, **194**.

Matchus, E.J. & Jones, T.S. 1984. *East–West Cross Section Through the Permian Basin of West Texas*. West Texas Geological Society Publication, 84–79.

RUPPEL, S.C. & BARNABY, R.J. 2001. Contrasting styles of reservoir development in proximal and distal chert facies: Devonian Thirtyone Formation, Texas. *AAPG Bulletin*, **85**, 7–34.

RUPPEL, S.C. & CANDER, H.S. 1988. Dolomitization of shallow water platform carbonates by seawater and seawater-derived brines: San Andres Formation (Guadalupian), west Texas. *In*: SHUKLA, V. & BAKER, P.A. (eds) *Sedimentology and Geochemistry of Dolostones*. Society of Economic Paleontologists and Mineralogists, Special Publications, **43**, 245–262.

SALLER, A.H. & HENDERSON, N. 1998. Distribution of porosity and permeability in platform dolomites: insight from the Permian of west Texas. *AAPG Bulletin*, **82**, 1528–1550.

SALLER, A.H., DICKSON, J.A.D. & MATSUDA, F. 1999. Evolution and distribution of porosity associated with subaerial exposure in upper Paleozoic platform limestones, west Texas. *AAPG Bulletin*, **83**, 1835–1854.

STUEBER, A.M., SALLER, A.H. & ISHIDA, H. 1998. Origin, migration, and mixing of brines in the Permian basin: Geochemical evidence from the eastern Central Basin Platform, Texas. *AAPG Bulletin*, **82**, 1652–1672.

SUN, S.Q. 1995. Dolomite reservoirs: porosity evolution and reservoir characteristics. *AAPG Bulletin*, **79**, 186–204.

Contrasting patterns of pore-system modification due to dolomitization and fracturing in Dinantian basin-margin carbonates from the UK

JON E. BOUCH, ANTONI E. MILODOWSKI & KEITH AMBROSE

British Geological Survey, Keyworth, Nottingham NG12 5GG, UK

(e-mail: jbouch@bgs.ac.uk)

Abstract: Two examples of Dinantian basin-margin carbonates from the United Kingdom provide potential analogues for fracture and porosity distributions in hydrocarbon reservoirs in dolostones. In both examples pore systems have been extensively modified by dolomitization, fracturing and related mineralization. However, the detailed processes and the end results show some significant differences, highlighting the importance of developing an understanding of the specific pore-system modifying processes when characterizing and modelling porosity distributions in these settings.

In the first example, predominantly low-porosity and low-permeability limestones in Lower Carboniferous inliers in Leicestershire (Cloud Hill) and south Derbyshire (Ticknall) had their pore systems further degraded by extensive dolomitization. Subsequent fracturing and related mineralization were responsible for significant porosity generation adjacent to fractures.

In contrast, in the Sellafield area (west Cumbria), dolomitization was strongly controlled by fractures that were also mineralized by sulphate-rich brines. In the north of the area fractures were filled with an assemblage of barite–fluorite–hematite–calcite that is resistant to corrosion by low-temperature meteoric groundwater. However, in the south of the area the fractures were cemented by anhydrite, which is readily corroded by saline, but sulphate-poor, groundwater formed by percolating meteoric recharge from the east. Progressive dissolution has been ongoing since Tertiary uplift, and has rejuvenated fracture porosity within the dolomitized limestones.

This paper describes the relationships between dolomitization, fracturing and porosity in Dinantian limestones from inliers in Leicestershire (Cloud Hill) and south Derbyshire (Ticknall; Fig. 1a & b), and compares these with Dinantian limestones from west Cumbria (Sellafield area; Fig. 1c). These examples represent situations where pore systems of limestones have been significantly modified by dolomitization, fracturing and related mineralization. In Derbyshire and Leicestershire the formation was dolomitized prior to fracturing. Corrosion in the dolomitic wall rocks associated with fracturing generated highly porous and permeable domains in an otherwise essentially non-, or very poorly, porous lithology. In Cumbria dolomitization was directly related to fracturing, and significant porosity developed due to the precipitation and corrosion of successive generations of mineral cements in domains constrained by the fractures. Both areas provide examples of fracture-related porosity generation or rejuvenation, and analogues for fracture and porosity distribution in hydrocarbon reservoirs in dolomitized limestones.

Geological settings

Leicestershire and south Derbyshire

The Dinantian inliers of Leicestershire and south Derbyshire have previously attracted scant attention (Hull 1860; Fox-Strangways 1905, 1907; Parsons 1918*a*; Mitchell & Stubblefield 1941; Spink 1965; Ford 1968; King 1968; Monteleone 1973). They lie on the Hathern Shelf, south of the Widmerpool half-graben and bounded to the north by the Normanton Hills Fault (Fig. 1b) (Carney *et al.* 2001*a*, *b*; Hoton Fault of Ebdon *et al.* 1990). These faults were active throughout the Dinantian and Namurian, until the early Westphalian (Carney *et al.* 2002*a*, *b*). Variscan inversion removed at least 800 m of Namurian and an unknown thickness of Westphalian strata. There is also evidence for Tertiary fault movements, with up to 95 m of displacement on the Normanton Hills Fault (Carney *et al.* 2001*a*, *b*; Green *et al.* 2001).

Recent mapping by the British Geological Survey (BGS) reappraised Monteleone's (1973) stratigraphy of the inliers, and divided the

From: BRAITHWAITE, C. J. R., RIZZI, G. & DARKE, G. (eds) 2004. *The Geometry and Petrogenesis of Dolomite Hydrocarbon Reservoirs*. Geological Society, London, Special Publications, **235**, 325–348. 0305-8719/$15.00

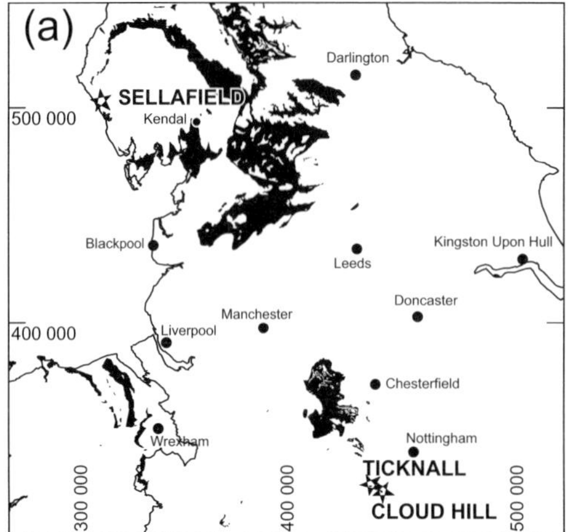

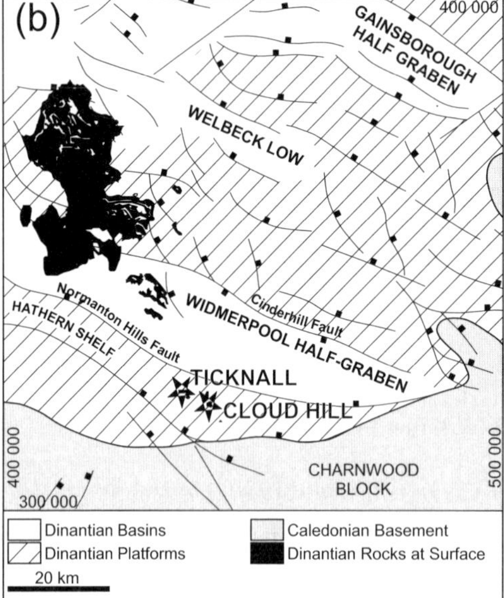

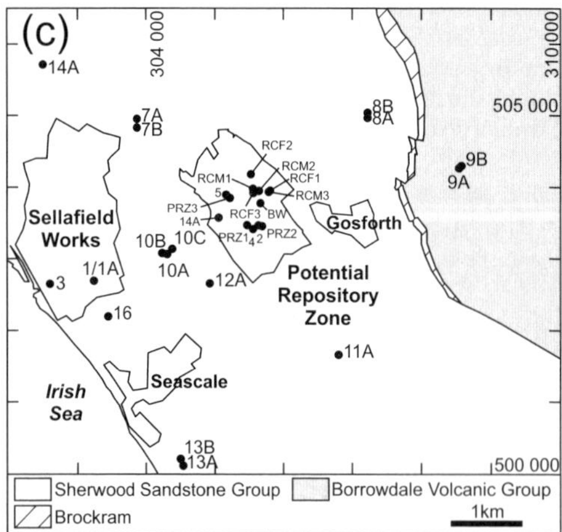

Fig. 1. Location maps showing the study sites. (**a**) The distribution of Lower Carboniferous outcrops in central and northern England, showing the locations of Cloud Hill (Leicestershire), Ticknall (south Derbyshire) and Sellafield (West Cumbria). (**b**) Map showing the positions of the Ticknall and Cloud Hill inliers in relation to blocks and basins in Central England during Dinantian times (fault positions from Ebdon *et al.* 1990). (**c**) Simplified geological map of the Sellafield area, west Cumbria, showing the locations of the deep boreholes drilled by Nirex in the area. The Carboniferous is unconformably overstepped by the Sherwood Sandstone Group, but is present at depth to the west of the 'Potential Repository Zone' (PRZ).

Dinantian into three formations (Ambrose & Carney 1997*a*, *b*; Carney *et al.* 2001*a*, *b*). The *Milldale Limestone Formation* (Early Chadian) only crops out in the Leicestershire inliers (400 m thick), where it is dominated by dolomitized, bedded, fine–coarse-grained, probably shallow-water platform or shelf-storm deposits and deeper-water ramp deposits, interbedded with thin siliciclastic mud- and siltstones deposited during quieter periods. Where undolomitized, it comprises bioclastic and locally oolitic and peloidal grainstones. However, due to obliterative dolomitization, it is typically poorly fossiliferous making biostratigraphical correlation difficult. Fenestral cavities are locally present. In the Breeden inlier, a thickness of up to 100 m of massive, structureless, fine-grained dolostone is interpreted as representing mud-mound (Waulsortian) reef on the basis of its occurrence within an otherwise well-bedded sequence, and on the presence of fabrics and textures comparable with mud-mound reefs described in Derbyshire (e.g. Lees & Miller 1985; Bridges & Chapman 1988). The *Holly Bush Member* (61 m, Cloud Hill inlier) comprises fine–medium-grained, locally pebbly dolomitic sandstone and sandy dolostone beds interbedded with dolostone.

The *Cloud Hill Dolostone Formation* (Early Asbian: Turner 1995; Riley 1997) is well developed in the Leicestershire and the south Derbyshire inliers (125 and 45 m thick, respectively). In Leicestershire, its base is defined by the Main Breeden Discontinuity, and the oldest proven rocks are of Asbian age. The formation is of similar character to the Milldale Limestone, with dolomitized, fossiliferous, bedded shallow-platform or shelf-storm deposits. Mud-mound reef deposits comprise fossiliferous, commonly mouldic pore-rich, finely to coarsely crystalline, massive dolostones. Bedding is rare in these, although crude stromatolitic lamination and

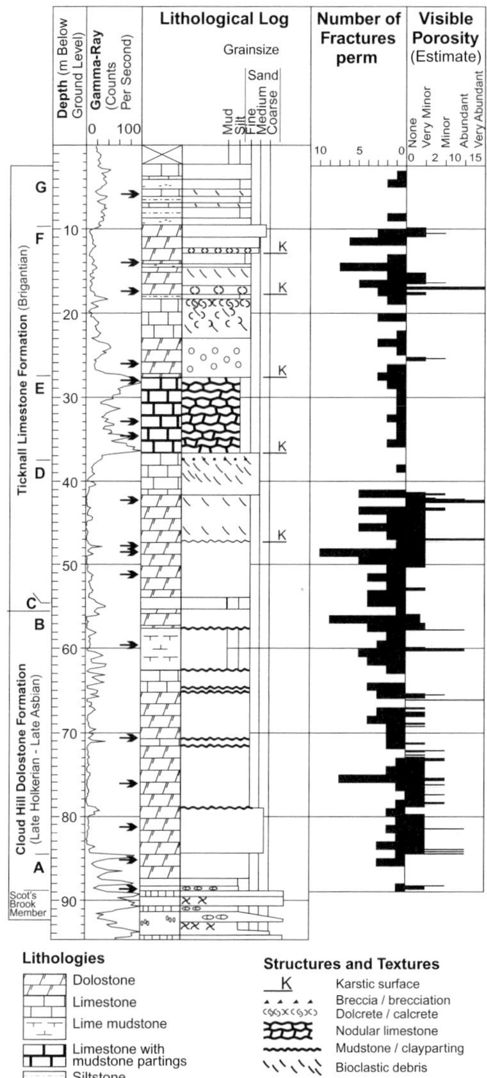

Fig. 2. Summary lithological, fracture and porosity log of the Cloud Hill Dolostone and Ticknall Limestone formations in the BGS Ticknall Borehole (lithological log modified from Ambrose & Carney 1997*a*). The porosity log is based on visual estimates of macro (i.e. visible) pore volume using a classification of none (no visible porosity), very minor (less than approximately 2%), minor (2–10%), abundant (10–15%) and very abundant (approximately 15%).

thin, elongate subparallel cavities filled with sparry calcite are locally preserved. The lowermost *Cloud Wood Member* (36 m thick, Cloud Hill inlier) comprises grey and green, locally stained red and orange, interbedded dolostones, siltstones and mudstones. The member is locally folded and mudstones are sheared and brecciated, suggesting possible earthquake-induced gravity slides at the basin margin.

In south Derbyshire, the BGS Ticknall Borehole (Fig. 2), proved the formation to comprise very-fine to medium-grained, bedded bioclastic and mouldic pore-rich limestones, lime mudstones and dolostones interpreted as storm deposits, with beds of sabkha–peritidal lime mudstones and birds-eye micrites. The *Scot's Brook Member* (3.4 m thick, Ticknall inlier) contains argillaceous sandstones, mudstones and lime mudstones, with at least two horizons of pedogenically modified mudstones and limestones suggesting periods of emergence. The carbonates were deposited in shallow peritidal conditions with the sandstones representing fluvial deposits.

The *Ticknall Limestone Formation* (Brigantian) occurs in the south Derbyshire inliers (55 m thick), and within Cloud Hill Quarry (Leicestershire; 14 m thick). It has previously been described by Hull (1860), Harrison (1877), Fox-Strangways (1905, 1907), Parsons (1918*a*), Mitchell & Stubblefield (1941), Monteleone (1973), Lott (1996) and Kemp (1997). The Ticknall Borehole recovered a complete sequence of variably dolomitized, muddy to coarse-grained, typically bioclastic, locally oolitic and peloidal grainstones with muddy and silty interbeds, representing shallow-water deposits. Periods of emergence and desiccation are recorded by very finely crystalline sabkha–peritidal deposits, and local rubbly palaeosols, with minor karstification.

Much of the sequence has been dolomitized, and in early work on the Cloud Hill and Breedon Hill inliers Parsons (1918*a*, *b*) described a lower, dense, 'yellow dolomite', considered to be contemporaneous with deposition and an upper unit of red, 'ferruginous dolomite', considered to be Triassic. However, Spink (1965) concluded that the dolomitization was probably Permian.

West Cumbria

This analysis is based on cores from deep boreholes (BHs) drilled as part of the United Kingdom Nirex Limited (Nirex) Sellafield site characterization programme of the 1990s (Barclay *et al.* 1994; Michie 1996; Akhurst *et al.* 1997) (Fig. 1c). The Carboniferous Limestone was only encountered in the western and southern parts of the area (BHs 3, 7A, 10A, 12A, 13A and 14A; Fig. 1c) and dies out eastwards. The Dinantian is largely confined to the subsurface, attaining a maximum thickness of 300 m to the north of Sellafield. The Dinantian rocks were mostly deposited in shallow tropical seas near the equator (Scotese *et al.* 1979) on the northern margin of a tilted fault-block ramp, bordered by the Lake District–Manx High to the east (Adams *et al.* 1990). The interplay between tectonism and eustacy produced alternating shallow-marine and subaerial conditions. A four-fold subdivision for the Dinantian rocks of the Sellafield district was proposed by Barclay *et al.* (1994), incorporating work by Dunham & Rose (1941), Rose & Dunham (1977) and Grayson & Oldham (1987).

The *Basal Beds* (maximum 2.5 m thick) comprise thin, impersistent, red–brown, poorly sorted sandstones and conglomerates, with a significant detrital component from the Borrowdale Volcanic Group. Subordinate lithologies include wackes, arenites, siltstones, sandy limestones and lime mudstones. No evidence of age has been found and variable relationships with overlying strata have been recorded. In some boreholes the Basal Beds are truncated by erosion surfaces, suggesting a Courceyan age, but elsewhere they pass conformably upwards into the Frizington Limestone, suggesting a Holkerian age.

The *Martin Limestone Formation* (Late Chadian; BH 3 only; 7.37 m thick) mainly comprises sandy packstones with subordinate sandy lime mudstones and a thin, fine-grained sandstone near the top.

The *Frizington Limestone Formation* (Holkerian; generally about 50 m thick, maximum 100 m) has been proved in the Nirex boreholes and is exposed in two quarries to the north of Sellafield. In the boreholes, the formation comprises a varied sequence that has been informally divided into two members. The *Lower Member* comprises mainly sandy limestones and sandstones with thin interbeds of mudstone and siltstone and some palaeosols, with a thin basal conglomerate. The *Upper Member* mostly comprises bioclastic and peloidal grainstones with thin interbeds of mudstone and siltstone, and subordinate palaeosols, lime mudstones and sandy beds. The uppermost bed has been dolomitized where it is overlain by Permian rocks. The entire member is dolomitized in BH 13A.

The *Urswick Limestone Formation* (Late Asbian; BH 3 only; 47 m thick) rests unconformably on the Frizington Limestone with all of the Arundian and Early Asbian missing. It mainly comprises bioclastic and peloidal grainstones, with subordinate pseudobreccias and numerous closely spaced palaeokarst surfaces and palaeosols indicating periods of emergence. The top of the formation is dolomitized beneath the sub-Permian unconformity.

The Carboniferous Limestone is overlain by the Permian Brockram Formation (breccia 1–>100 m thick), St Bees Evaporite Formation (dolomite and gypsum/anhydrite; 1–50 m), St Bees Shale Formation (mudstones, siltstones and fine sandstones; up to 200 m) and the Triassic Sherwood Sandstone Group that crops out over the Sellafield area.

Samples and methods

Samples were collected from Cloud Hill Quarry (Leicestershire), a series of disused quarries near Ticknall (south Derbyshire) and the Ticknall Borehole (Table 1). Samples from five Nirex deep boreholes in the western and southern part of the Sellafield area of west Cumbria, which had previously been examined as part of the Sellafield site investigation programme, were revisited for this study (BHs 3, 7A, 10A, 12A, 13A and 14A; Fig. 1c) (Nirex 1997*a–l*; Milodowski *et al.* 1998, 2002). Depths for samples from these boreholes are referenced in metres below Ordnance Datum (m bOD).

Unoriented core from the Ticknall Borehole was fracture logged, noting fracture positions, inclinations and any associated mineralization or porosity. Fracture logging was not attempted at Cloud Hill Quarry as the exposed sections contain numerous features induced and/or enhanced by the continuing blasting. Continuous core through the Carboniferous Limestone of the Nirex boreholes had previously been fracture logged, with attention paid to the relationships between fracturing and the hydrogeological properties of the rock mass (Nirex 1995, 1997*k*, *l*).

Petrographic and microchemical analyses were conducted on polished thin sections and dual-stained (Alizarin-red S and potassium-ferricyanide; Dickson 1966) thin sections using a variety of techniques. Back-scattered electron scanning electron microscope (BSEM) analyses were conducted using a LEO 435VP instrument, with a KE Developments back-scattered electron detector. Qualitative X-ray analyses were acquired during imaging using an Oxford Instruments ISIS300 energy dispersive X-ray analyser (EDXA). Some of the Sellafield materials were analysed using a Cambridge Instruments S250 SEM, with a Link Systems 860 EDXA system. Cathodoluminescence (CL) observations were made using a Technosyn

Table 1. *List of samples taken from Cloud Hill Quarrry and the Ticknall Borehole for this study*

Formation	Member	Locality	Sample number	Depth (m, below ground level)	Short description	Unit (Ticknall Borehole)	Petrography	Minipermeametry	SEM pore image analysis	Fluid-inclusion analysis	Lithology
Ticknall Limestone		**Ticknall Quarries**	MPLG279	–	bioclastic lime wackestone, with calcite veins	–	Y	Y	Y	–	
		(SK32SE	MPLG280	–	bioclastic lime wackestone	–	Y	Y	Y	–	
		Ticknall Borehole	MPLG317	5.6	bioclastic lime wackestone, with haematized/calcite cemented fracture	G	Y	Y	Y	–	
		(SK32SE	MPLG521	13.7	dolomite-sparstone, brecciated with calcite/haematite cement	F	Y	Y	Y	–	
		435910 323630;	MPLG522	17.2	sucrose dolomite-sparstone, with vugs	F	Y	–	Y	–	
		bore hole ID 103)	MPLG523	25.8	calcite cemented dolomite-sparstone breccia with dolocrete nodules	F	Y	–	Y	–	
			MPLG524	27.8	dolomite-sparstone with calcite- and galena-cemented fracture	E	Y	–	–	Y	
			MPLG525	32.7	quartzitic dolomite-sparstone	E	Y	Y	Y	–	
			MPLG526	34.5	bioclastic lime wackestone with calcite vein	E	Y	Y	Y	–	
			MPLG527	42.0	sucrose dolomite-sparstone	D	–	Y	–	–	
			MPLG528	47.4	fractured, locally sucrosic dolomite-sparstone	D	Y	Y	Y	–	
			MPLG529	48.3	dolomite-sparstone, with haematized microfractures	D	Y	Y	Y	–	
			MPLG530	50.9	partially brecciated, dolomite-sparstone with vugs	D	Y	Y	Y	–	
			MPLG531	59.4	dolomitic lime wackestone	C	Y	Y	Y	–	
			MPLG532	70.4	brecciated dolomite sparstone, with calcite cement	B	–	Y	–	–	
Cloud Hill Dolostone	Scot's Brook		MPLG533	75.9	dolomite-sparstone, with heamatized and calcite cemented fracture and vuggy porosity	B	Y	Y	Y	Y	
			MPLG534	81.0	dolomite-sparstone with partially calcite cemented, open fracture	B	Y	–	Y	–	
			MPLG535	84.9	dolomite-sparstone with fracture bounded sucrose porosity	A	Y	Y	Y	Y	
			MPLG536	88.2	dolomitic, peloidal, lime wackestone	A	Y	Y	Y	–	
		Cloud Hill Quarry (SK42SW 441120 321520)	MPLG275	–	dolomite-sparstone with fractures	–	Y	–	–	–	
Milldale Limestone	Holly Bush	**Cloud Hill Quarry**	MPLG285	–	sucrose dolomite-sparstone adjacent to calcite-cemented fracture	–	Y	Y	Y	–	
		(SK42SW	MPLG287	–	sucrose dolomite-sparstone	–	Y	Y	Y	Y	
		441310 321120)	MPLG289	–	calcite-cemented, variably sucrose dolomite-sparstone breccia	–	Y	–	–	–	
			MPLG290	–	sucrose dolomite-sparstone	–	Y	Y	Y	Y	
			MPLG293	–	brecciated sucrose dolomite-sparstone	–	Y	–	–	–	
			MPLG294	–	dedolomitized sucrose dolomite-sparstone	–	Y	Y	Y	–	
			MPLG295	–	brecciated sucrose dolomite-sparstone	–	Y	–	–	–	
			MPLG297	–	sucrose, clay-bearing dolomite-sparstone	–	Y	–	–	–	
			MPLG301	–	sucrose dolomite-sparstone	–	Y	–	–	–	
			MPLG302	–	brecciated dolomitic lime sparstone	–	Y	Y	–	–	
			MPLG303	–	sucrose dolomite-sparstone, adjacent to calcite-cemented fracture and calcite-filled vug	–	Y	–	–	–	
			MPLG304	–	sucrose dolomite-sparstone	–	Y	Y	Y	–	
			MPLG305	–	sucrose dolomite-sparstone, adjacent to fracture with calcite and galena cement	–	Y	–	–	Y	
			MPLG308	–	calcite-cemented, sucrose dolomite-sparstone breccia	–	Y	–	–	–	
			MPLG310	–	dolomitic, bioclastic, lime sparstone, with galena-cemented fracture	–	Y	Y	Y	–	
			MPLG311	–	lightly dolomitic, bioclastic lime sparstone	–	Y	Y	Y	–	
			MPLG309	–	dolomite-sparstone	–	Y	Y	Y	–	
		Cloud Hill Quarry	MPLG313	–	dedolomitized, calcite-cemented, sucrose dolomite-sparstone	–	Y	Y	Y	–	
		(SK42SW	MPLG316	–	dolomitic, bioclastic, lime-sparstone with cross-cutting calcite-cemented fractures and sulphide mineralization	–	Y	–	Y	–	
		441320 321120)									
		Cloud Hill Quarry	MPLG312	–	dolomite-sparstone	–	Y	–	Y	–	
		(SK42SW	MPLG314	–	dolomite-sparstone	–	Y	Y	Y	–	
		441340 321120)	MPLG315	–	dolomite-sparstone, with partially dolomite-cemented mouldic porosity	–	Y	Y	Y	–	
		Cloud Hill Quarry	MPLG277	–	bedded, dolomite-sparstone.	–	Y	–	Y	Y	
		(SK42SW 441320	MPLG278	–	dolomite-sparstone with calcite cemented fracture	–	Y	–	Y	–	
		321200)	MPLG281	–	dolomite-sparstone with calcite cemented fracture	–	Y	Y	Y	–	
			MPLG282	–	dolomite-sparstone with calcite cemented fracture	–	Y	Y	Y	Y	

Key to lithologies:
- dolomite-sparstone
- sucrose dolomite-sparstone
- limestone

Mark II apparatus. Limited wavelength-dispersive microchemical mapping was undertaken using a Cameca SX-50 electron microprobe (EPMA).

Permeabilities were measured on flat surfaces of the Cloud Hill and Ticknall samples using a Temco Inc. MP-402 mini-permeameter. Pore volumes were determined using image analysis of representative BSEM images.

Fluid-inclusion microthermometry was conducted on a Linkam MDS600 heating–freezing stage, controlled by Linksys software and a Linkam TP93 programmer. Calibration checks were performed using synthetic fluid-inclusion standards, and the controller reprogrammed if the measured temperatures differed from the known values by more than ±0.1 °C. Homogenization temperatures (T_h) were determined on 'heating runs' prior to the determination of first (T_{fm}) and final (T_{ice}) ice melting, and hydrohalite melting (T_{hyd}) measurements on 'freezing runs'. This prevented homogenization temperatures being artificially elevated due to inclusion 'stretching' during freezing. For

			TIME →
Early marine/meteoric Reducing(?)	MC1	Micrite neomorphism	
		Non-luminescent intercrystalline calcite	
		Luminescent intercrystalline calcite	
Dolomitization	MD1	Nodular, luminescent dolomite (pedogenic?)	
	MD2	Rhombic, non-/patchily luminescent dolomite	
Fracturing	Fracturing		
	Generation of significant intercrystalline porosity (saccharoidal dolostone)		
	Internal Sediment		
Fracture-filling dolomite	FD1	Non-ferroan, dully-luminescent dolomite	
	FD2	Non-ferroan, brightly-luminescent dolomite	
	Oxidation		
	FD3	Ferroan, non-luminescent dolomite	
Mineralization	Sulphides		
	Barite		
	Fluorite		
	Dolomite corrosion		?? ??
Calcite	FC1	Weakly ferroan, brightly-luminescent calcite	
	FC2	Strongly ferroan, dully-luminescent calcite	
	FC3	Brightly luminescent calcite, with high heavy metal concentrations	
Fresh-water calcite	FC4	Non-luminescent calcite, with luminescent bands	

Fig. 3. Summary paragenetic scheme for limestones, dolostones and fracture fills from Leicestershire and south Derbyshire.

the determination of all phase changes a 'cycling' protocol, compatible with the general methodology for the study of fluid inclusions in diagenetic cements outlined by Goldstein & Reynolds (1995), was applied. This exploits the fact that nearly all phase changes exhibit supercooling phenomena. The phase change end-point temperature was approached incrementally, followed by rapid cooling of the inclusion. If the end-point of the phase change had not been passed, then the 'disappearing' phase (e.g. vapour bubble or ice crystal) returned gradually. If the end point had been passed, the disappearing phase typically only returned after considerable supercooling. The errors in determination of T_h, T_{fm} T_{ice} and T_{hyd} are controlled by the temperature increment used during the approach to the end point of the phase change being measured, and are typically ±5 °C for T_h, ±0.1 °C for T_{ice}, ±5 °C for T_{fm} and ±1 °C for T_{hyd}. No pressure corrections have been applied to the homogenization temperatures. Owing to the small size of the inclusions, in most cases it proved impossible to acquire measurements of T_{fm} and T_{hyd}, therefore salinities were calculated based on freezing-point depression in the system NaCl–H_2O and are expressed in terms of wt% NaCl equivalent (Roedder 1984). Because of the small sizes of the inclusions, the turbid nature of much of the material and the general difficulties inherent in working with inclusions in carbonate minerals, it only proved possible to acquire limited data on carbonates within, or directly associated with, fractures (T_h n = 86; T_{ice} n = 50; T_{hyd} n = 7; T_{fm} n = 21).

Results

This section describes the petrographical characteristics of the limestones, dolostones and fractures from the Leicestershire and south Derbyshire inliers, and those from west Cumbria. These are described in the context of the overall paragenesis deduced for each area. Whilst these areas display similarities in the overall sequence of mineralization, the fluids responsible were sourced from different basins, and the mineralization events were not necessarily coeval. Therefore, a different notation for the description of diagenetic stages has been adopted for each area.

Leicestershire and south Derbyshire

In these inliers, limestones have been dolomitized (incipient to complete replacement) and subsequently fractured, with related alteration and mineralization. A generalized paragenetic scheme is given in Figure 3.

Limestone petrography. The limestones have varied textures, fabrics and degrees of diagenetic overprinting (Fig. 4). The *Milldale Limestone Formation* was only sampled in Cloud Hill Quarry, where it is extensively dolomitized (see sections below). However, partially dolomitized bioclastic wackestones and packstones contain significant neomorphic, partially fabric-retentive, non-ferroan, sparry calcite, which preserves the structures of micrite-walled foraminifera, but only the outlines of other bioclasts. Incipient dolomitization is typically

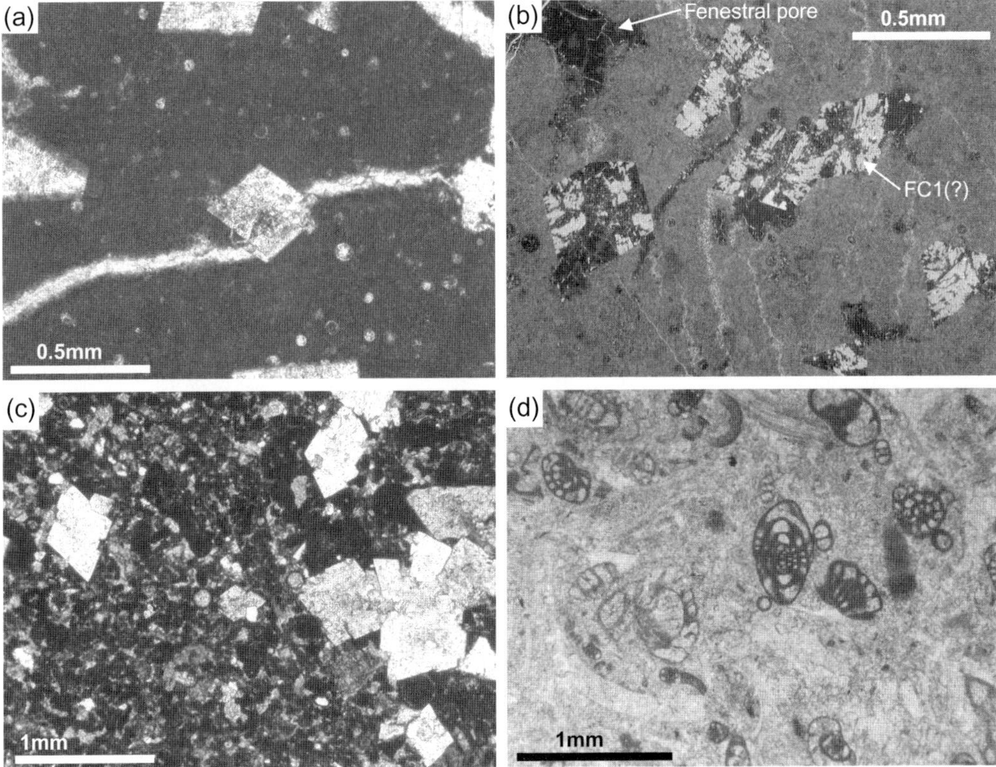

Fig. 4. Petrographical characteristics of non- and partially dolomitized limestones from Leicestershire and south Derbyshire. (**a**) and (**b**) Lightly dolomitized lime mudstone. The lime mud matrix is weakly luminescent and cut by brightly luminescent calcite microfractures. Fenestral pores are filled with non-luminescent calcite. Cemented pores and the lime mudstone matrix are replaced by rhombs of non-luminescent, weakly ferroan dolomite that are replaced in turn by brightly luminescent calcite, also seen in microfractures (sample MPLG531; Cloud Hill Dolostone; Ticknall Borehole; a – plane polarized light (PPL), b – CL). (**c**) Partially dolomitized peloidal grainstone, with non-ferroan and ferroan sparry calcite cement (sample MPLG536; Cloud Hill Dolostone; Ticknall Borehole; PPL). (**d**) Bioclastic wackestone–packstone with sparry calcite cement. Micrite-walled foramifera occur throughout (sample MPLG279; Ticknall Limestone; Ticknall Quarry; PPL). Images copyright BGS.

non-fabric-selective with isolated rhombs of non- to weakly ferroan dolomite dispersed throughout the limestones.

Two lightly dolomitized samples were available from the *Cloud Hill Dolostone Formation*. One sample of lime mudstone comprises non-ferroan and dully luminescent micrite, with minor calcispheres, forams and rare shelly and coral fragments (Fig. 4a & b). Rounded patches of sparry non- to weakly ferroan calcite may represent cemented mouldic or fenestral cavities. Dolomite rhombs (up to 1 mm in diameter; approximately 5–10%) follow irregular microfractures, possibly representing former fluid-flow pathways through the muddy matrix. However, the dolomite is partially replaced by weakly ferroan calcite, similar to that which fills minor fractures, although the mechanism of this replacement is uncertain. The other sample is a peloidal grainstone (Fig. 4c), with minor calcispheres, micrite-walled foraminifera and rare shelly fragments (approximately 5%), in a matrix of non-ferroan and ferroan sparry calcite that may represent either pore-filling cement or neomorphosed micrite matrix. Rhombs of weakly ferroan dolomite are preferentially developed within certain laminae, but the primary textural and mineralogical variation that has given rise to this variability is obscured by the dolomite.

The *Ticknall Limestone Formation* was only sampled in the Ticknall inlier, and where undolomitized facies comprise bioclastic wackestones and packstones (Fig. 4d). Most bioclasts are non-ferroan and retain their original structures, but some have been dissolved and the resulting secondary pores are filled by sparry, non-ferroan calcite or, more rarely, chert. Sparry, weakly ferroan, orange-luminescent calcite cement has developed within minor intraparticle and mouldic pores, and within micrite-poor intervals.

Paragenesis. Non- and partially dolomitized limestones are overprinted by neomorphism of micrite to sparite and by calcite cementation. Sparry calcite cement fills interparticle, intraparticle and local fenestral pores, and is predominantly non-luminescent (Fig. 4b–d), with minor dull orange-luminescence. Very thin, irregular bands of brightly luminescent calcite within the lime mudstones resemble stylolites, but their origin and timing relative to fenestral-pore-filling non-luminescent calcite is uncertain. However, they predate dolomitization (Fig. 4a & b).

Unaltered dolostone petrography. In all three formations, unaltered dolostones comprise mosaics of predominantly non-fabric-selective, planar–nonplanar crystals, typically <0.1–0.5 mm in diameter, with euhedral, inclusion-rich cores, limpid overgrowths and irregular compromise boundaries (Fig. 5a; 'sparstones' using the diagenetic classification of Wright 1992). The majority of the dolostones are clay-free; however, some clay-rich laminae are preserved towards bed tops in the bedded facies, and minor muddy dolostones, locally with pedogenic fabrics, occur in the Ticknall Borehole. In rare examples the outlines and structures of bioclasts are partially preserved (Fig. 5b), but in most cases the only evidence for the former presence of bioclasts is in mouldic macropores, commonly lined or filled by euhedral dolomite. The bulk of the dolomite is non-ferroan, with dull-red–yellow-orange luminescence.

Paragenesis. Dolomitization largely predates fracturing, and two principal generations of matrix dolomite are recognized (MD1 and MD2). The earliest (MD1) is only locally developed, has an irregular, nodular habit resembling that developed in palaeosols in the Ticknall Borehole, and is non- or dully luminescent (Fig. 5c). The MD2 dolomite is non-ferroan and comprises interlocking rhombs with inclusion-rich, typically weakly red-luminescent, cores and clearer non-luminescent rims.

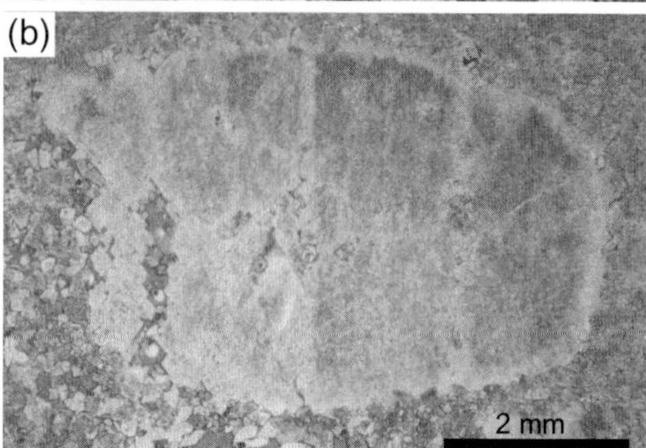

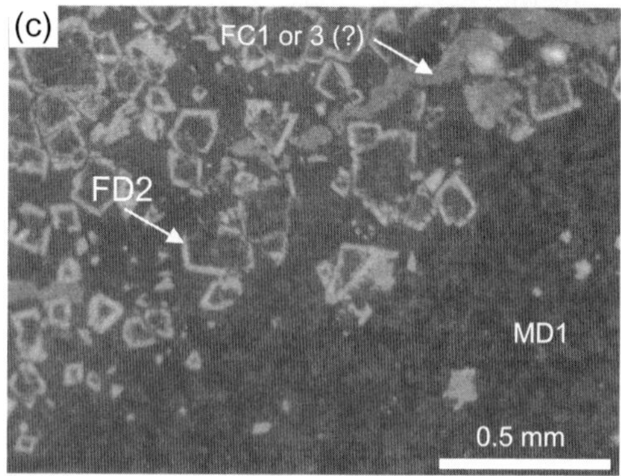

Fig. 5. Crystal fabrics of dolostones from Leicestershire and south Derbyshire. (**a**) Dolostone with a tightly interlocking mosaic of nonplanar dolomite crystals and negligible intercrystalline porosity cut by a partially calcite cemented fracture (sample MPLG277; Milldale Limestone; Cloud Hill Quarry; PPL). (**b**) A dolomitized crinoid ossicle, and associated relict mouldic porosity in an otherwise non-porous dolostone (sample MPLG315; Milldale Limestone; Cloud Hill Quarry; PPL). (**c**) Saccharoidal dolostone. In the top left of the image dully/non-luminescent dolomite rhombs are overgrown by brightly luminescent dolomite (FD2). In the bottom-right of the image, non-luminesent dolomite has no overgrowths. (sample MPLG525; Ticknall Limestone; Ticknall Borehole; CL). Images copyright BGS.

Fracture distribution. As no fracture logging was possible at Cloud Hill Quarry, there are no data to formally characterize fracture distributions in the Milldale Limestone Formation. Qualitatively, however, bedded dolostones contain abundant bedding-normal fractures 10–20 cm long and 0.1–1.0 cm wide that terminate at mudstone–siltstone partings, and contain dolomite, calcite and sulphide cements. At the time of sampling (January 2001) an area of at least 100 m^2 of extensively fractured and brecciated dolostone with well-developed calcite, dolomite and sulphides with wall-rock alteration was exposed on the lowermost level of the quarry. This feature has irregular margins subparallel to bedding.

Fracture logging of the Cloud Hill Dolostone and Ticknall Limestone Formations in the Ticknall Borehole (Fig. 2) indicates that:

- Average fracture density is 1.8 m^{-1} with marginally higher fracture densities in the Cloud Hill Dolostone relative to the Ticknall Limestone Formation.
- Fractures are typically high angle, with approximately 70% at >60° from the horizontal, 20% between 40° and 60°, and 10% at <40°.
- In the Cloud Hill Dolostone Formation, there is little difference in average fracture density or fracture length between dolostones and limestones, although the most intense fracturing occurs in dolostones (Fig. 2). In the Ticknall Limestone Formation, dolostones are notably more intensely fractured than limestones.
- Fracture widths range from micron- to cm-scale, with subequal proportions of open (non- or partially cemented) and closed (tightly cemented) fractures. Brecciated intervals may also be cemented. The cements and fracture fills are discussed below.

Fracture and wall-rock alteration petrography. In non- and partially dolomitized limestones, fractures tend to be relatively narrow (micron- to mm-scale, rarely up to cm-scale) and are completely cemented by multiple generations of calcite, with no apparent wall-rock alteration. In contrast, fractures in dolostones (Fig. 6a–d) are more variable, with apertures ranging from micron- to cm-scale, and variable degrees of fill by multiple generations of dolomite and calcite with minor sulphides, barite and fluorite. In addition, fractures are commonly associated with wall-rock dissolution and alteration with the development of vugs and enhancement of intercrystalline porosity.

Saccharoidal dolostones comprising floating to loosely grain-supported, and/or calcite-cemented networks of euhedral dolomite rhombs (up to 1 mm in diameter; Fig. 6a) with high volumes of intercrystalline porosity are locally developed adjacent to fractures, and may contain small volumes of washed-in clay and/or finely comminuted carbonate. The cores of the rhombs are of similar character to those of dolomite crystals in adjacent uncorroded dolostone mosaics, but they are commonly corroded and have variably ferroan overgrowths. Saccharoidal dolostones typically grade into non-porous, apparently uncorroded, mosaic dolomite, within 5 cm of fracture surfaces. Millimetre- to cm-scale vugs occur directly adjacent to fractures and are lined by dolomite, with later generations of calcite and minor sulphides that match the paragenesis seen in the fractures (see below; Fig. 6b).

Paragenesis. In the non- and partially dolomitized limestones, fracture-related wall-rock alteration is limited. However, in the dolostones, the presence of enhanced volumes of intercrystalline and vuggy porosity adjacent to fractures indicates dolomite corrosion during or after fracturing. Because the distribution of saccharoidal dolostones is controlled by fractures within otherwise non-porous, pervasively dolomitized material, it seems likely that these textures reflect corrosion of a dolostone rather than selective removal of calcite from a partially dolomitized limestone. Multiple episodes, or a protracted period of fracturing, are implied by the fact that early fracture-fills are themselves fractured and locally brecciated prior to further cementation. In the southern part of Cloud Hill Quarry, breccias containing angular–subrounded, cm- to dm-sized clasts of saccharoidal dolostone suggest localized collapse of the most heavily corroded material (Fig. 6d).

Fluid-inclusion microthermometric data are summarized in Figure 7. Most of these data are from dolomite and the earlier calcite generations, with later calcite generations (FC3 and FC4, see below) particularly inclusion-poor. Homogenization temperatures are typically 60–120 °C, with a broad maximum in the frequency distribution at 80–90 °C (Fig. 7b). Whether the highest homogenization temperatures (maximum 200 °C) are real or an artefact of post-trapping leakage and/or stretching of inclusions is uncertain. Salinities are typically 18–24 wt% NaCl eq., with a small population of very low salinity inclusions (0–2 wt%) largely

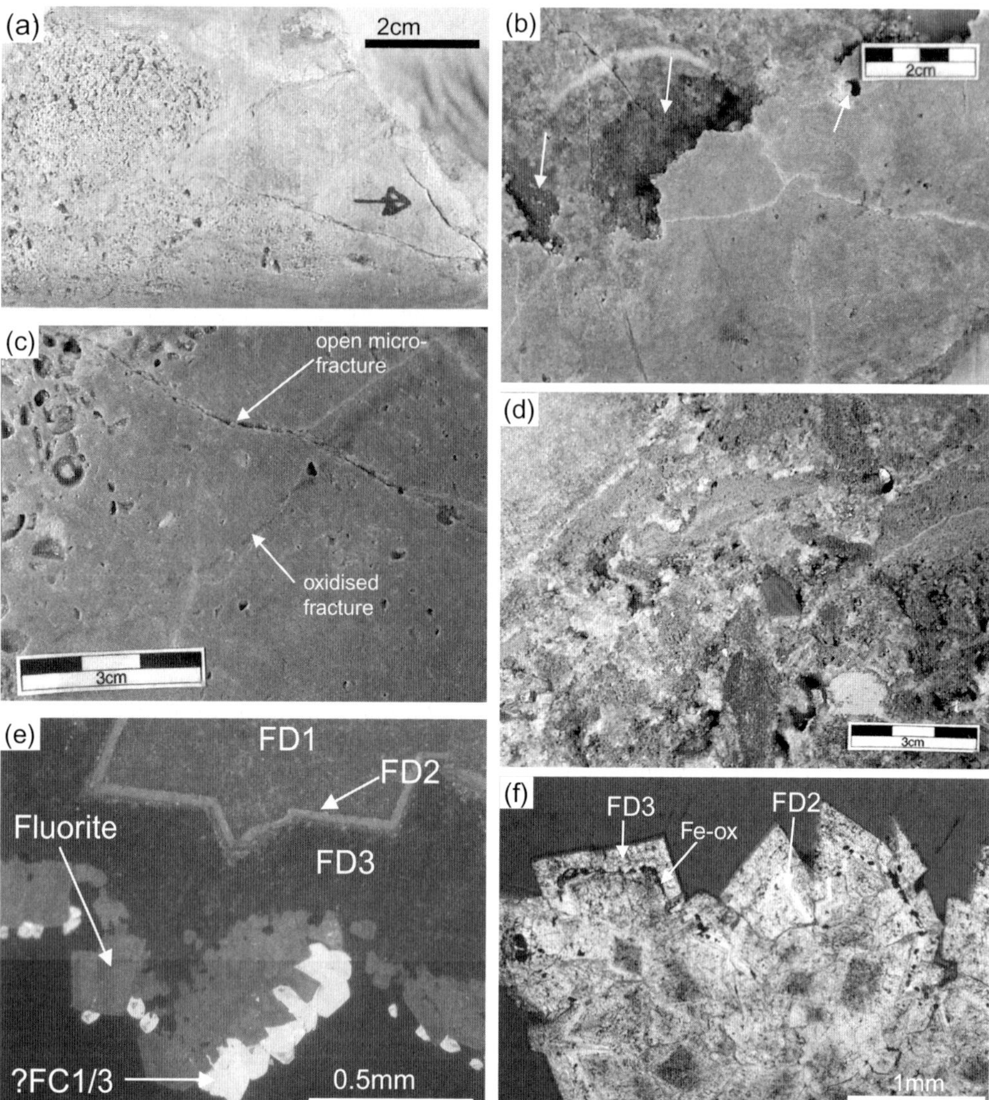

Fig. 6. Fractures, fracture-fills and related features from Leicestershire and south Derbyshire. (**a**) Low-porosity dolostone cut by mm-scale fractures with adjacent intercrystalline porosity (sample MPLG528; Ticknall Limestone; Ticknall Borehole). (**b**) Low-porosity dolostone cut by an open fracture with adjacent vuggy porosity (arrowed). The fracture and vuggy porosity are lined by ferroan dolomite (sample MPLG277; Milldale Limestone; Cloud Hill Quarry). (**c**) Low-porosity dolostone with well-developed mouldic porosity (top left of image), cut by a thin, open fracture (sample MPLG309; Milldale Limestone; Cloud Hill Quarry). (**d**) Calcite-cemented breccia with irregularly shaped clasts of saccharoidal dolostone, muddy dolostone and lime mudstone (sample MPLG308; Milldale Limestone; Cloud Hill Quarry). (**e**) and (**f**) Fracture-lining dolomite (FD1–FD3). In (e), the dolomite generations are readily differentiated by their luminescence characteristics, and are overgrown by finely crystalline fluorite and calcite (FC1 and FC3?; sample MPLG534; Cloud Hill Dolostone; Ticknall Borehole; CL). In (f), FD1 is difficult to distinguish optically from the dolostone wall rock. FD2 forms a thin, clear overgrowth associated with a band of Fe-oxides that separates FD1 and FD2 from FD3. FD3 is ferroan, non-luminescent and euhedral with curved crystal faces (sample MPLG278; Milldale Limestone; Cloud Hill Quarry; PPL). Images copyright BGS.

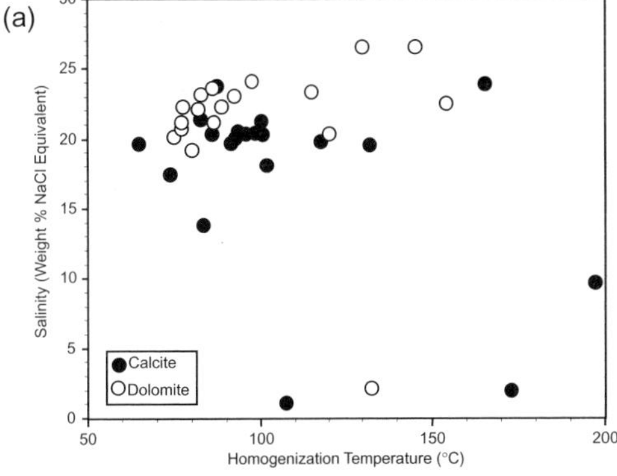

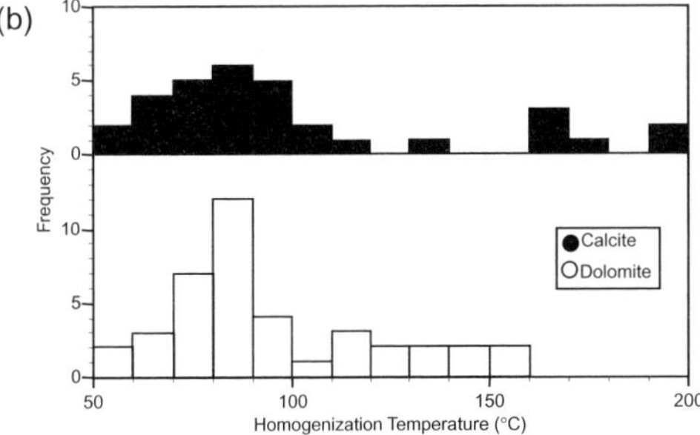

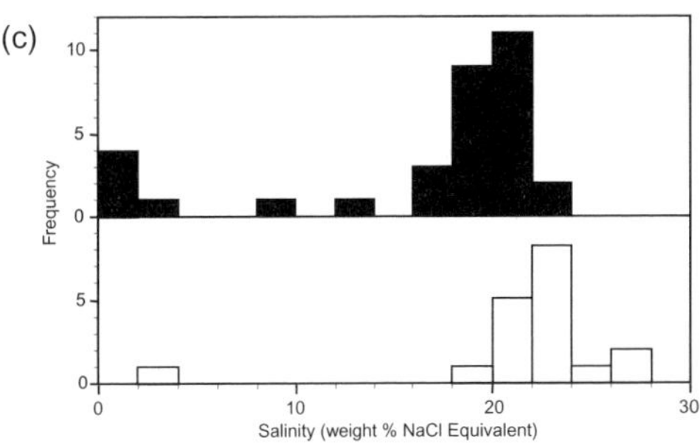

Fig. 7. Cross-plot (**a**) and histograms, showing variations in homogenization temperature (**b**) and salinity (**c**) for inclusions in samples of dolostones and limestones from the Milldale Limestone, Cloud Hill Dolostone and Ticknall Limestone formations from Leicestershire and south Derbyshire.

restricted to secondary inclusions in calcite of a single sample. No simple relationships between T_h and salinity are apparent (Fig. 7a). First melting temperatures are between –60 and –50 °C, suggesting a predominance of NaCl–$CaCl_2$–H_2O brines (Roedder 1984). No significant differences in T_h or salinity are observed between inclusions with different T_{fm}, and there is no apparent difference between the T_{fm} of inclusions in calcite and those in dolomite. Three generations of fracture-related dolomite (FD1–FD3) line and fill fractures, and form overgrowths in adjacent porous dolostones.

- FD1 is non-ferroan, with dull red luminescence. It primarily replaces dolomite in the wall rock adjacent to fractures (Fig. 6e & f) and forms overgrowths on corroded rhombs in saccharoidal dolostones.
- This is overlain by FD2, which comprises a thin (10 μm) zone of clear, non-ferroan, bright orange-luminescent, manganoan, non-ferroan dolomite, which occurs in fractures and forms thin overgrowths on mouldic-pore-lining dolomite away from the fractures.
- Between dolomite generations FD2 and FD3, a period of oxidizing conditions is recorded by localized bands of Fe-oxides, which impart a red or orange coloration to some fracture linings. The possibility that these represent oxidized diagenetic pyrite cannot be discounted. However, no sulphides older than those associated with the last stages of FD3 have been observed. Some contacts between FD2 and FD3 are corroded.
- FD3 is the most volumetrically significant fracture-related dolomite generation, and forms coatings up to 1 mm thick on fracture surfaces. It also occurs as overgrowths on rhombs in saccharoidal dolostones, and lining vuggy and mouldic pores. It is ferroan, non-luminescent and has slightly curved crystal faces.
- Fluid-inclusion data suggest that dolomites of all generations precipitated from saline (20–24 wt% NaCl eq.; Fig. 7) NaCl–$CaCl_2$–H_2O brines, with homogenization temperatures providing minimum estimates for precipitation temperatures of 70–100 °C.

Towards the close of dolomite FD3 precipitation, non-carbonate phases also began to precipitate. The most widespread of these are pyrite, galena and chalcopyrite, with minor sphalerite and barite. Fluorite is observed in the Ticknall Borehole and, to the best of our knowledge, has not previously been reported in the area (Fig. 6e). Prior to precipitation of the earliest fracture-filling calcite (FC1, see below) some dolomite corrosion occurred, leaving a residue of insoluble Mn-oxide that was engulfed by later calcite. Barite and fluorite precipitation were restricted to the period between the FD3 and FC1 precipitation. Four generations of fracture-filling calcite (FC1–FC4) are recognized:

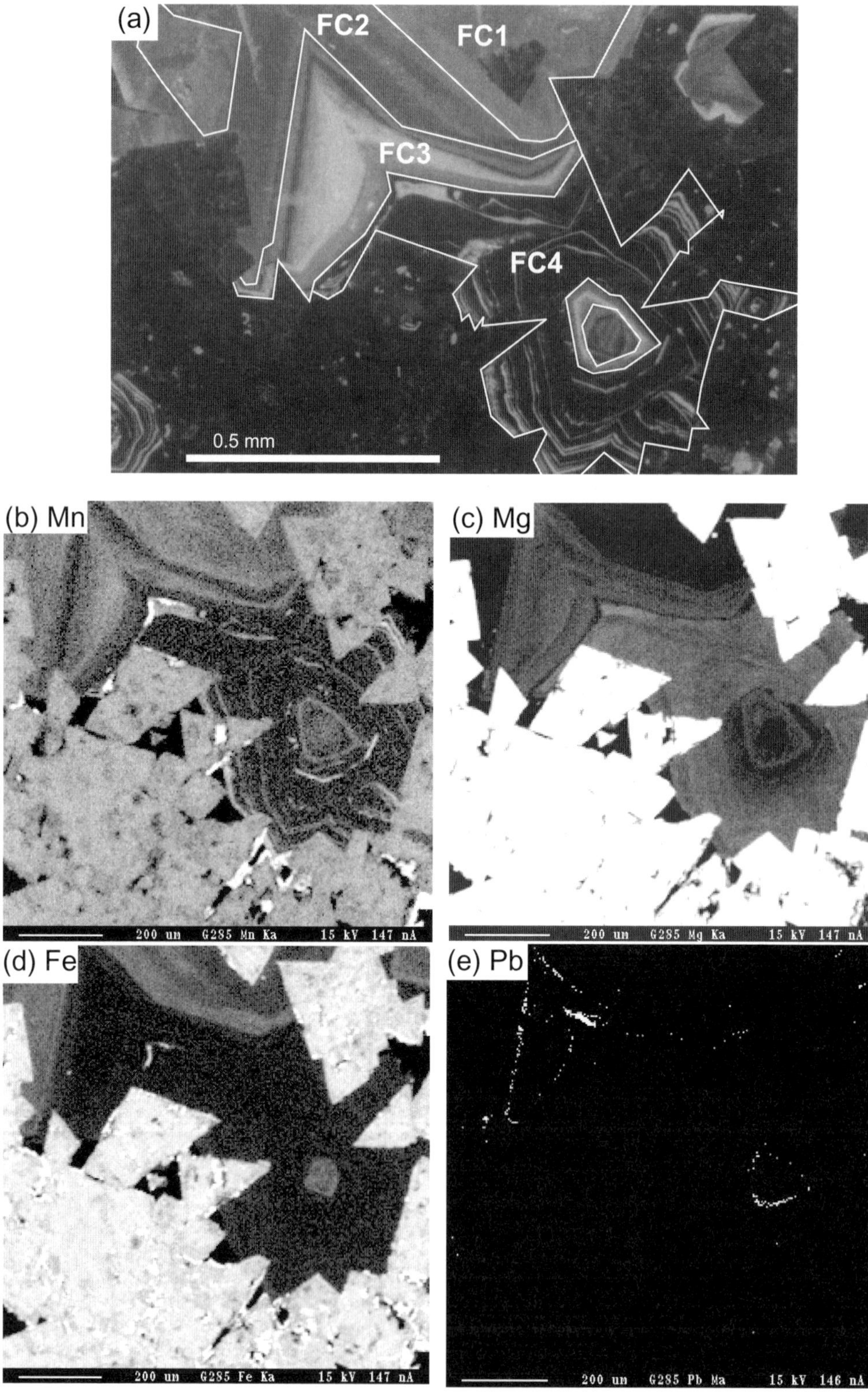

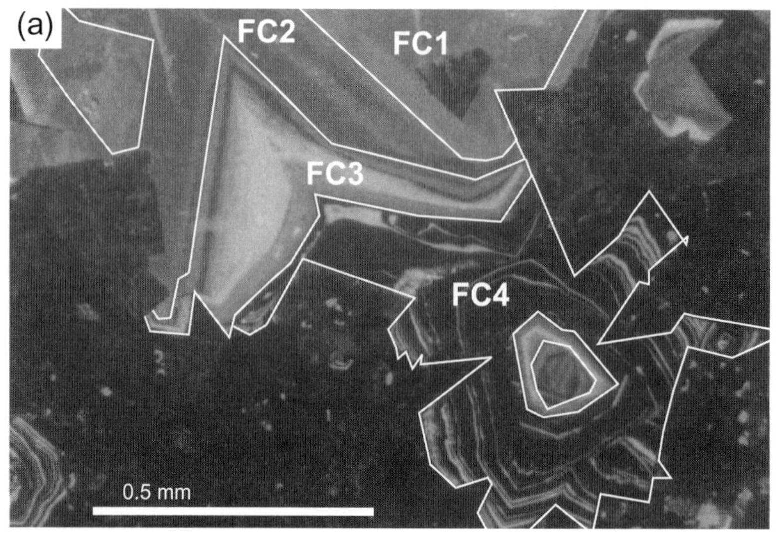
(a)
FC2
FC1
FC3
FC4
0.5 mm

(b) Mn
200 um G285 Mn Ka 15 kV 147 nA
(c) Mg
200 um G285 Mg Ka 15 kV 147 nA
(d) Fe
200 um G285 Fe Ka 15 kV 147 nA
(e) Pb
200 um G285 Pb Ma 15 kV 146 nA

- FC1 is the most volumetrically significant generation. It is weakly ferroan and stains pink, with orange–yellow-orange luminescence that highlights the presence of complex, irregular patterns of internal zoning (Fig. 8). It occurs both as cement and as replacing dolomite.
- FC2 forms relatively thin overgrowths on FC1, stains pale-blue due to marginally higher Fe-contents than in FC1 (EPMA microchemical mapping; Fig. 8), and has dull-orange luminescence that reveals concentric and sector zoning
- Most sulphide mineralization occurs within FC1 and FC2, with thin bands of pyrite and chalcopyrite inclusions present locally along growth zones.
- FC3 is discontinuous, but where present typically forms a thin (approximately 100 μm), luminescent zone with well-developed sector, concentric and irregular zonation, defined by variations in Mn-content. This has lower Fe-, but higher Mg-, content than the earlier calcite generations. This generation also contains micron-scale Cd- and/or Pb-rich zones (Fig. 8e; up to approximately 0.7 oxide wt%).
- FC4 forms localized overgrowths on earlier crystals, and is best developed in saccharoidal dolostones in the Ticknall Borehole, and in the extensively fractured area in Cloud Hill that is associated with groundwater flowing at the present day. It stains pink, is non-ferroan and predominantly non-luminescent, with micron-scale Mn-bearing brightly luminescent bands (Fig. 8). Like FC3, this generation has a low, but consistent, Mg-content. No sulphides are enclosed within FC4, but it locally engulfs patches of Cu-oxides (after chalcopyrite) suggesting that sulphide alteration had at least been initiated prior to the onset of FC4 precipitation.

FC1 and FC2 typically have euhedral, *c*-axis elongate (scalenodehral) morphologies. In contrast, FC3 and FC4 form dentate overgrowths on earlier calcites. Similar variations in calcite morphology have been reported in the Sellafield area by Bath *et al.* (2000), who related them to salinity changes during growth. The strong variations in CL response seen in generations FC3 and FC4 suggest variations in Mn- and/or Fe-content that may indicate fluctuating redox conditions during crystal growth.

Fluid-inclusion data from calcite are predominantly from generations FC1 and FC2, and suggest that most of the fracture-filling calcite precipitated from NaCl–$CaCl_2$–H_2O-dominated brines of marginally lower salinity (18–22 wt% NaCl eq.) than the fluid associated with dolomitization. Homogenization temperatures provide a minimum estimate for precipitation temperatures of 70–100 °C. The salinities and temperatures, and the association with base-metal ore minerals, are consistent with the fluid having been sourced from compacting Lower Carboniferous mudstones/shales in the Widmerpool Gulf.

West Cumbria

Fracture and wall-rock alteration petrography and paragenesis. An overview of the mineralogical and petrological characteristics of the fracture mineralization in Ordovician–Triassic strata in the Sellafield Area of west Cumbria is provided by Milodowski *et al.* (1998, 2002), who differentiated nine mineralization events, from ME1 (oldest) to ME9 (youngest). ME1–ME3 represent pre-Carboniferous high-temperature hydrothermal mineralization confined to Lower Palaeozoic strata, and are not discussed further. ME4–ME7 are Mesozoic, and ME8 and ME9 are interpreted as post-Tertiary uplift, telogenetic episodes. ME4–ME9 occur in all strata in the sequence from the Lower Palaeozoic to the Triassic and are summarized in Figure 9. The same paragenesis is recognized here, with the single exception that the age and relative paragenetic position of calcite-hematite mineralization ('Early ME6a' of Milodowski *et al.* 1998, 2002) is now revised to Late Carboniferous or Early Permian (identified as 'CME1' in Fig. 9).

Fracture mineralization varies systematically across the area, with fills dominated by calcite–dolomite–ankerite–hematite–fluorite–anhydrite–barite, with minor–trace amounts of

Fig. 8. Chemical variations in fracture-related calcites from Leicestershire. (**a**) CL photomicrograph of variably luminescent calcite filling intercrystalline porosity in a saccharoidal dolostone (dolomite is non-luminescent) with calcite of generations FC1–FC4 labelled. (**b**)–(**e**) EPMA microchemical maps of part of the area shown in (a), showing variations in Mn, Mg, Fe and Pb content, respectively. FC1 and FC2 are characterized by moderate Fe-, moderate Mn- and low Mg-contents. In contrast, FC3 and FC4 are relatively Mg-rich, Fe-poor and Mn-poor, with thin Mn-rich and Pb-rich zones (sample MPLG285; Milldale Limestone; Cloud Hill Quarry). Images copyright BGS.

Event	Code	Description	
Early hydrothermal calcite mineralization: [Late Carboniferous or Early Permian (?)]	CME1*	Micrite recrystallization to non-luminescent non-ferroan calcite Non-luminescent intercrystalline and fracture-filling non-ferroan calcite + specular hematite Luminescent intercrystalline and fracture-filling ferromanganoan calcite	
Uplift and erosion [Permian unconformity]		**Internal sediment geopetal fills by Permian silt and sand sediment**	
Anhydrite fracture mineralization [Remobilized Permian evaporite sulphate]	ME4	Monomineralic, creamy yellow-white bladed vein-filling anhydrite	
Major calcite + hematite fracture mineralization [Warm basinal brine]	'Late' ME6a	Finely banded non-ferroan to ferromanganoan calcite and colloform earthy hematite Replacement of ME4 anhydrite	
Major dolomite-anhydrite -hematite fracture miner. Wallrock dolomitization [Warm basinal brine]	ME6b	Non-ferroan,brightly-luminescent dolomite +/- hematite Ferroan,non-luminescent dolomite and ankerite Fibrous and bladed anhydrite Wallrock dolomitization	
Major calcite-hematite -fluorite-barite miner. [Warm basinal brine + mixing of two fluids (barite)]	ME6c	Ferromanganoan brightly and moderately luminescent calcite Hematite +/- barite +/- fluorite +/- sulphide	
Secondary porosity + Calcite mineralization [Post-Tertiary uplift-present day]	ME9	Dissolution of ME4 and ME6b anhydrite mineralization Precipitation of calcite (+/- pyrite, marcasite, barite or anhydrite)	

*Note: *Originally defined by Milodowski et al. (1998) as 'Early ME6a' Permo-Triassic mineralization but now recognized to be much earlier than the ME6 event.*

Fig. 9. Summary paragenetic scheme for the Lower Carboniferous in west Cumbria.

siderite, baritocelestite, celestite, quartz, pyrite, sphalerite and galena, and trace amounts of uraniferous minerals in the Carboniferous Limestone in BH 3. Calcite and dolomite are present throughout, and several generations of carbonate mineralization have been distinguished (discussed below). Hematite is present throughout and is more abundant further north, ultimately forming the West Cumbrian Ore Field (Rose & Dunham 1977; Shepherd & Goldring 1993; Akhurst *et al.* 1997). Barite and fluorite occur only in Carboniferous Limestone in the north of the area, but persist to the south and east in Permian, Permo-Triassic and Borrowdale Volcanic Group strata. Anhydrite has an antithetic relationship to barite and fluorite, and is absent in the north, but present to the south and east, and is particularly abundant in BH 3 to the west. The fracture mineralization affecting the Carboniferous Limestone has been broken down into the sequence described below.

Pre-Permian unconformity mineralization (CME1). The earliest mineralization identified in the Carboniferous Limestone from the Nirex boreholes is represented by fracture-lining euhedral calcite associated with specular hematite. The calcite has euhedral cores of dog-tooth to sub-equant non-luminescent non-ferroan and non-manganoan calcite. These are developed as syntaxial overgrowths on neomorphosed, non-luminescent calcite exposed on fracture surfaces, and overgrown by finely banded non- and brightly luminescent calcite (Fig. 10a). CME1 fracture mineralization appears to be coeval with limestone neomorphism with trace–minor amounts of brightly luminescent calcite forming thin overgrowths on non-luminescent neomorphosed calcite and filling residual intercrystalline pores in the wall rock.

This mineralization is found only in the Lower Palaeozoic basement and the Carboniferous Limestone, and was described by Milodowski *et al.* (1998, 2002) as the earliest part ('Early ME6a') of a complex sequence of fracture-filling minerals defined as 'ME6' (see below). However, Milodowski *et al.* (1998, 2002) were uncertain about its timing and relationships with later features, and suggested that it might reflect an earlier event – possibly related to Permian weathering. New observations demonstrate that CME1 mineralization in fissures in the Carboniferous Limestone beneath the Permian unconformity is overlain by a geopetal fill of sand and silt derived from the overlying Brockram and St Bees Shale formations. Therefore, the CME1 mineralization may be Late Carboniferous or Early Permian in age, probably coeval with the neomorphism of the Carboniferous Limestone. CME1 calcite is locally corroded and a coating of fine-grained hematite was precipitated on the corroded surface prior to precipitation of Late ME6a calcite (see below).

Early anhydrite veining (ME4). This is preserved in a few fractures in the upper part of the Carboniferous Limestone sequence in BH 3 (at around 1507 m bOD, Late Asbian–Urswick Limestone) in peloidal bioclastic wackestones interbedded with thin, silicified, red–brown and green, illitic mudstones (Nirex 1997*a*). This interval is faulted and within the competent limestone interbeds steep dilational fractures are filled by creamy- to pale-yellow bladed anhydrite crystals, which are isotopically distinct (+6.5–+9.5‰ $\delta^{34}S_{CDT}$; sulphur isotopes are reported relative to the Canyon Diablo Troilite (CDT) standard) from later anhydrite filling vuggy cavities in dolomite–calcite veins. There is no evidence of any earlier mineralization of fracture surfaces. The presence of bladed calcite pseudomorphs after anhydrite, and tiny corroded relicts within later carbonate minerals in other boreholes (Nirex 1997*b*; Milodowski *et al.* 1998, 2002), indicates that early anhydrite was originally widespread, but has subsequently been leached or replaced.

Milodowski *et al.* (1998, 2002) reported that the isotopic composition of the sulphur of this anhydrite is similar to that of primary bedded anhydrite in the overlying Permian St Bees Shale and St Bees Evaporite formations, and early pore-filling cements in the Brockram (+8–+10‰ $d^{34}S_{CDT}$). It differs from isotopically heavier ME6b fracture anhydrite (+13.4–+18.7‰ $\delta^{34}S_{CDT}$) reported by Milodowski *et al.* (1998, 2002) and Lower Carboniferous bedded anhydrite (+14–+21‰ $\delta^{34}S_{CDT}$) reported by Crowley *et al.* (1997). The $^{87}Sr/^{86}Sr$ ratio of the ME4 anhydrite (0.70935–0.70958) lies within the range (0.70845–0.70993) for the Basal Dolomite and bedded anhydrite of the overlying St Bees Evaporite (equivalent of the Magnesian Limestone; Milodowski *et al.* 1998, 2002). There is evidence for the remobilization of primary anhydrite into secondary veins and early cements in the overlying Permian strata (Milodowski *et al.* 1998, 2002), and the ME4 anhydrite is considered to have been derived locally from this source.

A kaolinite–illite mineralization (ME5), present in the Lower Palaeozoic and overlying Permo-Triassic rocks, is absent in the Carboniferous, probably reflecting a scarcity of reactive detrital aluminosilicates (Milodowski *et al.* 1998).

Calcite mineralization (ME6a). As noted above, mineralization previously attributed to an early phase of this event (Early ME6a; Milodowski *et al.* 1998, 2002) is now considered older (late Carboniferous; CME1). Late ME6a comprises widespread fracture-filling, white, anhedral calcite that encloses corroded CME1 calcite, and ME4 anhydrite. The calcite is weakly–moderately luminescent (Fig. 10a), and is commonly finely interbanded with colloform to earthy hematite. It fills fractures or forms an early fracture coating beneath dolomite, ankerite, anhydrite, barite, fluorite and hematite. Late ME6a mineralization is also associated with reddening of the adjacent limestone wall rock due to replacement of framboidal pyrite by fine-grained hematite, which also filled some intercrystalline porosity in the neomorphosed limestone.

Dolomite–anhydrite mineralization (ME6b). This is often extensive and rests on earlier calcite, or completely fills new fractures. Dolomite typically forms coarse euhedral crystals, often with curved faces (Fig. 10c), and partially replaces earlier calcite and anhydrite, with traces of these commonly present as corroded inclusions. The cores of dolomite crystals are generally non- or only weakly ferroan, but crystals are usually strongly growth zoned, becoming increasingly ferroan up to ankerite in later zones (Fig. 10c). Dolomite-cemented veins either contain tightly interlocking crystals with very little porosity, or are vuggy with large cavities (up to several centimetres in diameter) lined by dolomite. In BH 7A, and BH 14A vugs are filled by ferroan calcite, barite, fluorite and/or hematite.

Dolomitization is commonly limited or directed by fractures, with most intense dolomitization occurring within the fracture-damage envelope around large fault intersections (Fig. 10f). Minor dolomitization produces fine-grained saccharoidal dolomite with tightly interlocking rhombs similar to those from Leicestershire and south Derbyshire described above. More extensive dolomitization produces a highly porous coarse-grained dolostone, often with large cavities where the host rock has been dissolved between a skeletal framework of dense dolomite-cemented veins (Fig. 10g). Dissolution voids up to 1 m diameter in BH 3 correspond to major present-day groundwater flow zones and are commonly lined or filled by very coarse, clear pale blue–colourless bladed crystals of anhydrite (Fig. 10g), with later calcite or hematite (Fig. 10h). In BH 7A and BH 14A dissolution cavities are filled by major hematite, usually associated with calcite, fluorite and barite. Petrographical analysis, together with strontium and stable isotope analyses

(Milodowski *et al.* 1998, 2002 and references therein), show that the fracture-filling and limestone-replacive dolomites were coeval.

This episode was accompanied by anhydrite precipitation. In BH 3 and BH 13A anhydrite commonly fills residual pores between dolomite crystals (Fig. 10d & e). In BH 10A and BH 12A many dolomite veins contain mouldic pores

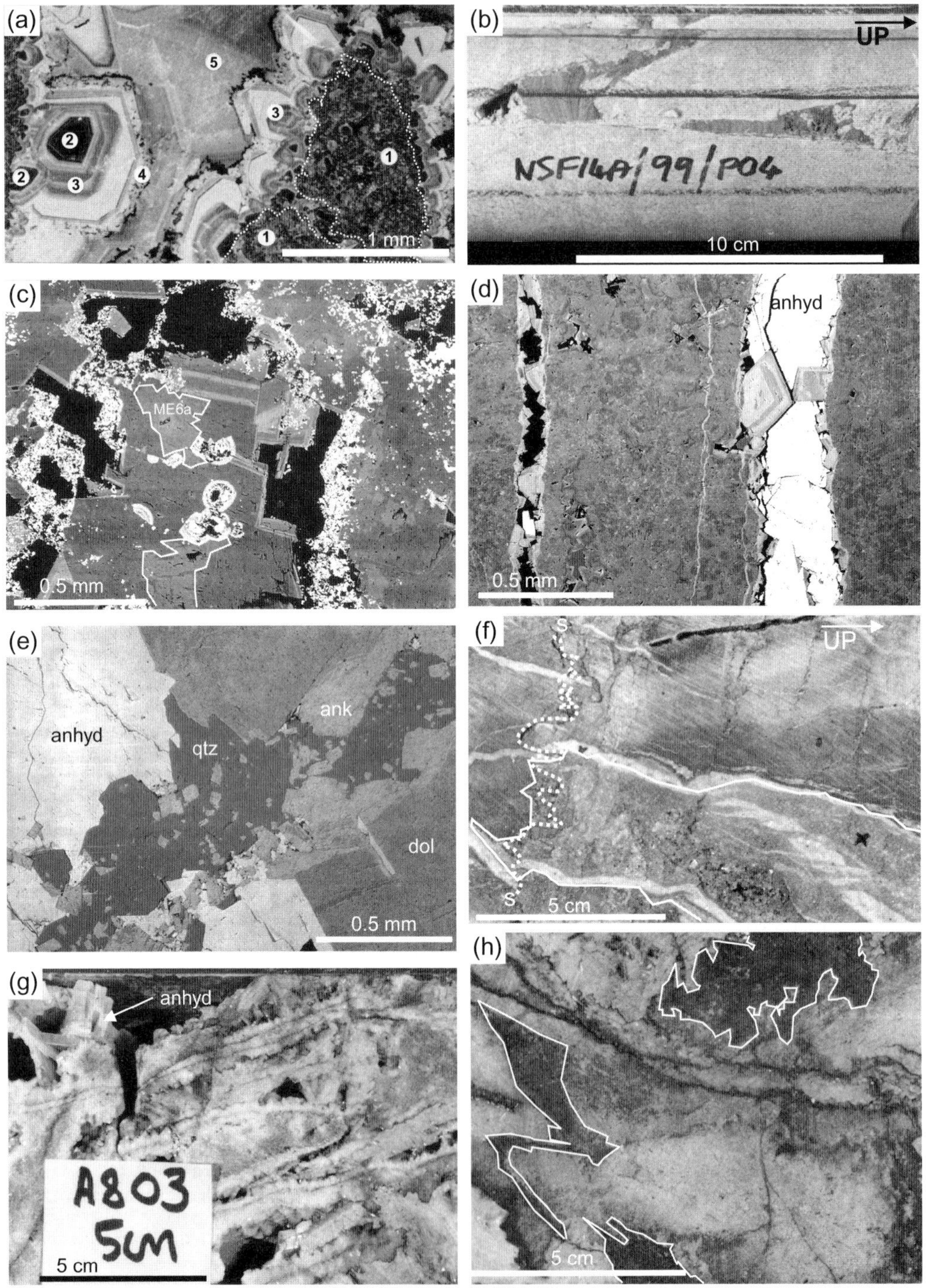

formed by the dissolution of coarse, bladed anhydrite, with rare anhydrite relicts preserved. These vuggy areas correspond closely to zones of present-day groundwater movement. As discussed above, ME6b anhydrite is isotopically heavier than the ME4 anhydrite. Milodowski *et al.* (1998) suggested that ME6b anhydrite was derived by remobilization of sulphates from either the Lower Carboniferous of the adjacent Solway Basin or from the overlying Triassic Mercia Mudstone Group that have similar isotopic compositions. Quartz and/or hematite are also present, resting on the dolomite–ferroan dolomite–ankerite (Fig. 10c).

Calcite–hematite mineralization (ME6c). This comprises major calcite, accompanied by major–minor amounts of quartz, barite, fluorite and hematite, with trace amounts of celestite, baritocelestite, sulphides and uranium minerals. In many veins, the calcite replaced ME4 and ME6b anhydrites (and to a lesser extent ME6b dolomite–ankerite), or lined or filled cavities formed by anhydrite dissolution. Quartz, barite, hematite, fluorite and other trace minerals formed during the early part of ME6c, commonly rest on corroded earlier carbonates. Two generations of ME6c calcite are separated by a corrosion surface marked by minor brecciation and locally coated by very-fine-grained sulphide, fluorite, barite, baritocelestite and celestite.

Fluid-inclusion data for ME6 (Milodowski *et al.* 1998, 2002) show that minerals crystallized from warm (T_h = 53–180 °C) highly saline brines (up to 25 wt% NaCl eq.). First melting temperatures of <–50 °C indicate that calcium was a significant component of the fluid. This was confirmed by SEM-EDXA observations that identified Na–Cl–Ca–SO_4 residues around decrepitated fluid inclusions (Milodowski *et al.* 1998). The mineralization temperature generally increases from ME6a calcite (T_h = 53–85 °C) to ME6b dolomite–ankerite (T_h = 85–164 °C). However, the temperature was reduced during quartz precipitation at the beginning of ME6c (T_h = 35–85 °C) before increasing again during fluorite (T_h = 91–121 °C) and later calcite precipitation (T_h = up to 180 °C). Carbon and oxygen stable isotope compositions vary systematically from ME6a to ME6c, with oxygen isotopes becoming progressively lighter (from –2 to –17‰ $\delta^{18}O_{PDB}$; relative to Pee Dee Belemnite) and carbon progressively heavier (from –11 to –1‰ $\delta^{13}C_{PDB}$). In the Nirex deep boreholes ME6 mineralization is correlated with major hematite–fluorite–dolomite–barite mineralization of the Carboniferous Limestone in the West Cumbrian Ore Field, and is considered to reflect precipitation from warm brines expelled from the East Irish Sea Basin during deep burial of the evaporite-bearing Permo-Triassic fill (Milodowski *et al.* 1998, 2002).

Fig. 10. Fractures and fracture fills from the Carboniferous Limestone of the Sellafield area. (**a**) CME1 euhedral dog-tooth crystals with cores of non-luminescent calcite (2), surrounded by finely banded bright and non-luminescent calcite (3), resting on weakly to non-luminescent neomorphosed calcite in the wall rock (1). The outer surface of the CME1 calcite is corroded (4) and coated by fine-grained hematite (black). This is enclosed by later moderately luminescent Late ME6a calcite (5; BH 3A, 1596.01–1596.37 mbOD; CL; plate from Nirex 1997*b*). (**b**) Subvertical fissues filled by horizontal-bedded, sandy sediment derived from the overlying Permo-Triassic strata. The fissure margins beneath the sediment fill are lined by fine white, CME1 calcite. Where present, the head-space above the sediment is filled or partially filled by dolomite and/or calcite, often containing mouldic cavities after anhydrite) (BH 14A; 574.5–575.48 mbOD). (**c**) Intergrown hematite (bright), and rhombic dolomite–ankerite crystals (grey). The dolomite is finely growth zoned with brighter zones of ankerite and ferroan dolomite in the outer zones. The cores of the dolomite crystals contain relict patches of earlier calcite that have been partially replaced (ME6a; outlined). Fragments of earlier (ME6a) hematite were included within the dolomite after replacement of the calcite (BH 7A, 502.87–503.40 mbOD; BSEM; from Nirex 1997*d*). (**d**) A fracture in dolostone is lined by rhombic crystals of darker-grey dolomite and lighter-grey ankerite, enclosed by later fracture-filling anhydrite (anhyd; BH 13A, 1572.46–1572.74 mbOD; BSEM; from Nirex 1997*i*). (**e**) A veinlet of quartz (qtz) cuts earlier rhombs of fracture-filling ankerite (ank), ferrroan dolomite (dol) and fracture-filling anhydrite (anhyd). Rhombs of dolomite are included in the quartz (BH 13A, 1590.07–1590.61 mbOD; BSEM; from Nirex 1997*i*). (**f**) Slabbed core showing limestone cut by a narrow zone of closely-spaced, steep, en echelon, white dolomite-cemented veins. These are cut by and slightly offset by a stylolite (s-s). Dolomitization of the host limestone is constrained by the dolomite veins (outlined), and the rock outside this zone is undolomitized (BH 3, 1578.11–1578.35 mbOD; from Nirex 1997*b*). (**g**) Slabbed core showing intensely fractured and porous interval within dolomitized limestone. A skeletal network of close-spaced dolomite-cemented veins (white) has been produced as a result of the dissolution of the host rock. Coarse crystals of anhydrite partially fill the large dissolution cavities between the veins (BH 3, 1339.37–1539.91 mbOD; from Nirex 1997*b*). (**h**) Slabbed core showing complex mineralization within a highly dolomitized limestone. Large cavities produced by dissolution of the host rock between dolomite veins are filled by hematite (outlined; BH 3, 1530.04–1530.24 mbOD; from Nirex 1997*b*).

The development of barite mineralization to the north and east was attributed by Milodowski *et al.* (1998, 2002) to the mixing of this sulphate-rich saline brine with a local, less saline, low sulphate fluid carrying Ba during the early stages of ME6c. This is supported by fluid-inclusion data that indicate that barite precipitated at a lower temperature (T_h <80 °C) from a fluid of about 18 wt% NaCl eq. But the barite could have formed directly by reaction between ME6b anhydrite and a Ba-rich fluid. However, although the sulphur isotopic compositions of the ME6c barite and ME6b anhydrite overlap, most of the barite is significantly lighter (+3.9–+8.1‰ $\delta^{34}S_{CDT}$) suggesting a different source. The $^{87}Sr/^{86}Sr$ isotopic composition of the barite (Milodowski *et al.* 1998, 2002) is anomalously less radiogenic than either ME6b or ME6c minerals (that show a trend towards progressively more radiogenic compositions), lending further support to the involvement of a different fluid during ME6c barite mineralization.

Stylolites are very common throughout the area, and show a complex relationship to the dolomite–ankerite veins. In some cases they cut and displace the latter, together with associated wall-rock dolomitization (Fig. 10f). Elsewhere, similar veins cut stylolites, indicating either that there may be more than one generation of stylolites or that stylolitization was continuous during multiple episodes of fracturing.

Illite mineralization (ME7). ME7 is associated with faulting in both Palaeozoic and Permian strata, but is absent from the Carboniferous Limestone, probably due its low aluminosilicate content. However, faulting and shearing of interbedded mudstones, and reactivation and brecciation of ME6 veins, is possibly associated with ME7.

Post-Tertiary uplift (telogenetic) mineralization (ME8 and ME9). Milodowski *et al.* (1998, 2002) identified very late-stage mineralization, dominated by iron and manganese oxides and oxyhydroxides (ME8) near-surface, and by calcite at depth (typically >200 m), sometimes also accompanied by pyrite, barite or anhydrite (ME9). Milodowski *et al.* (1998), supported by Bath *et al.* (2000), demonstrated a very close relationship between the distribution of ME8 and ME9, and variation in crystal form, fluid-inclusion characteristics and $^{87}Sr/^{86}Sr$ composition (ME9 only), and the flow paths and geochemistry of present-day groundwaters to which they attributed mineralization. Limited radiometric dating (Milodowski *et al.* 1998, 2002) pointed to a post-Tertiary uplift and/or Quaternary age for this mineralization.

Discussion

Pore-system evolution and reservoir quality in Leicestershire and south Derbyshire

Dolomitization, fracturing and fracture-related alteration of the Dinantian rocks in the Leicestershire and south Derbyshire inliers has had a profound impact on the volume, size and connectivity of their pores. A number of pore types are recognized and summarized in Figure 11, with differences observed between the pore systems of limestones, dolostones and altered dolostones.

The non- and partially dolomitized limestones contain no significant macroporosity, with any fenestral vugs, interparticle, intraparticle and mouldic pores occluded by sparry calcite cement. In addition, interstitial micrite has been neomorphosed to sparite. Consequently, pore systems are dominated by 2–10 µm diameter *interparticle micropores* that represent up to 8.5% of the rock volume (Figs 11h and 12b). Despite their small sizes, these pores are relatively well connected through micron-scale pore throats, and the limestones have permeabilities of 0.6–14 mD (geometric mean 5 mD; Fig. 12c).

In unaltered dolostones, two distinct pore-types are recognized:

- Dolomitization typically appears to have been accompanied by a reduction in porosity, with the replacement of the micron-scale interparticle pores of the limestones by very poorly connected *intercrystalline pores* between tightly packed dolomite crystals (Fig. 11g; porosities <0.1–6.3%; mean 1%; Fig. 12). These dolostones display a comparable range in permeability (1–14 mD), but have a lower geometric mean permeability (3.2 mD) than the limestones.
- Within discrete beds, *mouldic porosity* may account for up to 12% of rock volume (Fig. 12). Mouldic pores are typically 0.5–5 mm in diameter, and are typically isolated features that are unlikely to significantly enhance larger-scale fluid flow (Fig. 11e). However, within small intervals they are sufficiently abundant to support permeabilities in excess of 100 mD (Fig. 12).

Fracturing in the limestones does not appear to have been accompanied by wall-rock

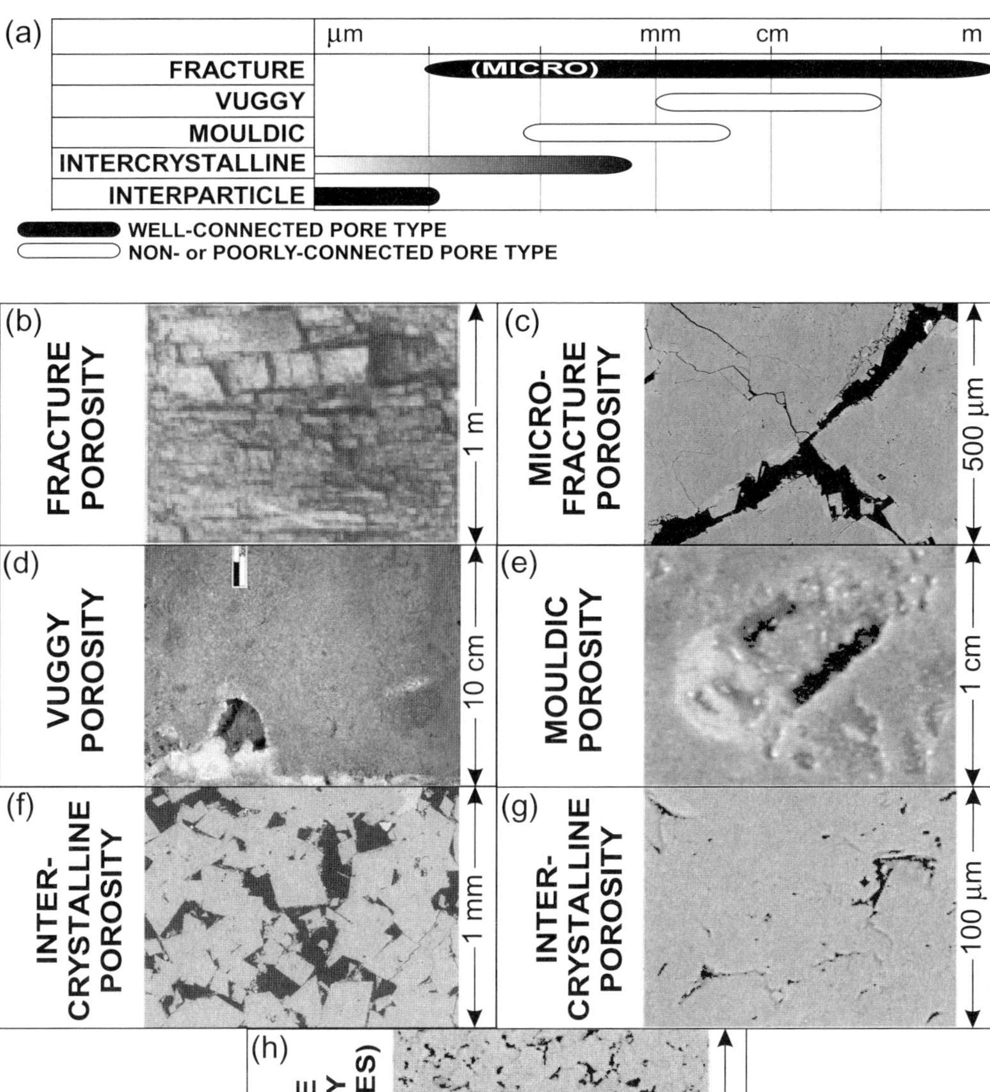

Fig. 11. Illustration of the different pore types encountered in the dolomitized limestones from Leicestershire and south Derbyshire, and the scales over which they occur.

alteration, and the fractures themselves are typically filled with calcite cement. Therefore, in limestones, the fractures are more likely to act as permeability baffles/barriers than as flow conduits. However, in the dolostones fracturing has been accompanied by localized porosity enhancement:

- Typically cm-scale, *vuggy pores* are locally well connected through the partially

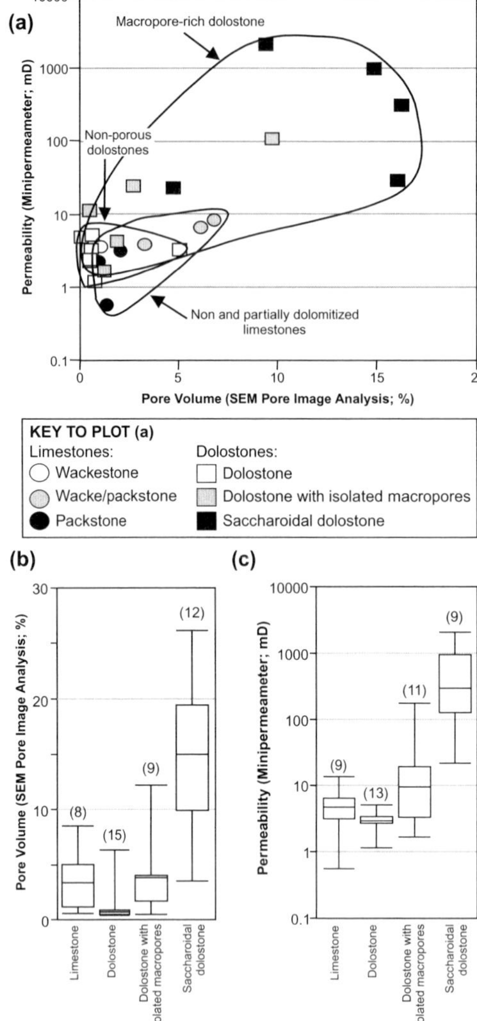

Fig. 12. Cross-plot (**a**) and box-plots showing variations in porosity (**b**) and permeability (**c**) for dolostones and limestones from the Milldale Limestone, Cloud Hill Dolostone and Ticknall Limestone Formations from the Leicestershire and south Derbyshire. Analyses are grouped into the major lithological types sampled. The box-plots show the range of the data, the interquartile range (the boxed interval) and the mean (geometric mean for permeability data) for the various rock types.

cemented fracture network (Fig. 11d). Owing to the scale and irregular distribution of these features, no attempt has been made to quantify their porosity–permeability characteristics.

- Corrosion of dolostones adjacent to fractures has locally created enhanced volumes of *intercrystalline porosity*, with the development of saccharoidal dolostones (Fig. 11f). Intercrystalline porosity may be very high (up to 26%; Fig. 12) and well connected, supporting high permeabilities (in excess of 2000 mD). However, these pores are locally re-cemented by calcite (related to the calcite in fracture-fills) that degrades pore volume and connectivity. Qualitative porosity logging in the Ticknall Borehole indicates that the Cloud Hill Dolostone and Ticknall Limestone Formations typically contain little or no macroporosity (94% of the core has <2% porosity), and that cm- to dm-thick intervals with >2% macroporosity almost exclusively comprise fracture-related intercrystalline porosity.
- *Fracture porosity* (Fig. 11b & c): relative to rock volume, fracture pore volume is likely to be relatively low. During and post-fracturing, dolomite, sulphide and calcite cements formed within fractures and adjacent porosity causing significant reduction of fracture apertures. In spite of this, however, the majority of fractures in dolostones remain open with good connectivity. Consequently, fractures may be anticipated to have a significant influence on larger-scale fluid flow, and this is supported by the fact that groundwater can be seen flowing through fractures in Cloud Hill Quarry. The irregular distribution of fractures and their occurrence over a wide range of scales precluded any attempt to quantify their porosity–permeability characteristics.

Dolomitization, wall-rock alteration and pore-system evolution in west Cumbria

Any significant porosity originally present in the Carboniferous Limestone was lost during recrystallization and neomorphism prior to or during CME1 mineralization. Virtually all of the present-day porosity is secondary vuggy fracture porosity in ME6a–ME6b veins, or large dissolution cavities (length up to 0.5 m) and intercrystalline porosity (10–100 μm) in the matrix of coarse saccharoidal dolomitized limestone. Dolomitization was spatially and temporally associated with faulting and ME6b dolomite–ankerite vein mineralization.

Much of the fracture porosity in the area was filled by ME4 and ME6b anhydrite. In the north of the area (particularly in BH-7A) anhydrite was removed by ME6c mineralization, which also sealed fractures with an essentially insoluble assemblage of fluorite, barite, calcite and hematite that prevented subsequent porosity rejuvenation. As a result, in this area, limestones and dolomitized limestones have retained the generally low fracture and matrix permeability seen in BH-7A. In the south of the area, ME6c mineralization was less intense and anhydrite was consequently preserved in the Carboniferous Limestone until the onset of progressive (E–W) dissolution in groundwaters recharged in the Lake District. This probably began during ME9 mineralization following Tertiary uplift (Milodowski *et al.* 1998, 2002; Bath *et al.* 2000). Where this process was advanced, large vuggy cavities have developed between the dolomite-cemented components of fractures.

Comparison of Leicestershire and south Derbyshire inliers with west Cumbria

The Lower Carboniferous limestones of Leicestershire, south Derbyshire and west Cumbria were all deposited on basin margins, and have pore systems modified by dolomitization, fracturing and mineralization by basinal fluids. In all cases, the predominantly fracture-controlled pore systems are similar. However, this state was reached by different mechanisms in each area.

In Leicestershire and south Derbyshire, large-scale dolomitization occurred prior to the onset of fracturing, and converted the micron-scale pore systems in the fine-grained matrices of lime mudstones to very poorly connected intercrystalline porosity in dolostones, locally supplemented by mouldic porosity. The mineralizing fluids associated with subsequent fracturing were warm, high-salinity brines probably derived from the adjacent Widmerpool Gulf, and were responsible for the corrosion of dolomite adjacent to fractures, locally generating highly porous and permeable saccharoidal dolostones.

By contrast, in west Cumbria, large-scale dolomitization of the Lower Carboniferous occurred during and after fracturing, and dolomite distribution is strongly controlled by the distribution of fractures and faults along which the mineralizing fluids migrated. The dolomite here is possibly related to warm, sulphate-rich, brines sourced by the dewatering of evaporitic Permo-Triassic sediments in the adjacent East Irish Sea Basin. In the north of the area, fractures were ultimately sealed by barite, fluorite, calcite and hematite, resistant to corrosion by low-temperature, neutral groundwater from meteoric recharge. In the south, fractures were sealed by anhydrite, which is readily hydrated to gypsum, and corroded by saline, but sulphate-poor, groundwater formed from meteoric recharge from the east, rejuvenating fracture porosity.

Conclusions

Two examples of dolomitized, basin-margin, Lower Carboniferous limestones that represent potential analogues for dolomitized and fractured limestone hydrocarbon reservoirs have been described. In both cases the pore systems of the rock masses are strongly controlled by fracturing and related mineralization. However, the relationships between dolomitization, fracturing and porosity differ.

In Leicestershire and south Derbyshire significant pore-system modification due to a combination of dolomitization, fracturing and associated cementation has occurred:

- In the non- and partially dolomitized limestones, pore systems are characterized by small volumes of microporosity (micron-scale), typically associated with the lime mudstone matrix (Fig. 11h). Porosities and permeabilities tend to be relatively low (<8.5% and 0.6–13 mD, respectively; Fig. 12).
- Throughout the inliers, dolomitization has modified pore systems. The dolostones contain low volumes (<1%) of micron-scale, very poorly connected, intercrystalline pores, and permeabilities are typically low (<5 mD). Mouldic porosity after bioclasts is typically poorly connected. However, some mouldic-pore-rich (up to 12% of rock volume) intervals have permeabilities in excess of 100 mD.
- Evidence from the Ticknall Borehole suggests that the dolostones are more prone to fracturing than the non- or partially dolomitized limestones. Fractures are associated with three generations of dolomite cement (FD1–FD3), followed by barite, fluorite and sulphide mineralization, and four generations of calcite cement (FC1–FC4). Sulphide mineralization continued during the first three of the calcite generations. FC3 contains notable Pb- and Cd-rich zones. FC4 potentially precipitated from modern groundwaters.
- Fluid-inclusion evidence suggests that the dolomite, sulphides, barite, fluorite and earlier calcite generations formed at temperatures around 70–90 °C, from highly saline $NaCl–CaCl_2–H_2O$ brines (18–24 wt% NaCl eq.), consistent with fluids sourced from the Lower Carboniferous basin of the Widmerpool Gulf.

- As well as forming a well-connected fluid-flow system in their own right, fractures in the dolostones are associated with elevated volumes of very-well-connected, intercrystalline porosity in adjacent dolostones (up to 26%) that support high permeabilities (in excess of 2000 mD).

In west Cumbria, there was a different pattern of mineralization and pore-system evolution, with the sulphate-rich mineralizing fluid sourced from the Permo-Triassic evaporitic sediments of the East Irish Sea Basin. Dolomitization is strongly fracture controlled, and sulphate-rich mineralization has sealed faults and fractures with barite–fluorite–hematite–calcite in the north of the area, and anhydrite in the south. In the south of the area, meteoric recharge is progressively dissolving the anhydrite leading to the rejuvenation of porosity within the fracture system.

The staff and management of Ennstone Breeden at Cloud Hill, and Breeden on the Hill Quarries, are thanked for access to the sites and for permission to take samples. This paper is published with the approval of the Director of the British Geological Survey (Natural Environment Research Council) and with that of United Kingdom Nirex Limited (Nirex). Most of the west Cumbria data were acquired for the Nirex Sellafield site investigation study and Nirex are thanked for permission to reproduce some images from this work. Cores and related information from the Nirex investigations are available for research at the British Geological Survey, Keyworth, UK. J. McKervey (BGS) is thanked for the EPMA microchemical maps used in Figure 8. Two anonymous reviewers are also thanked for their comments, which significantly improved the final version of this paper.

References

ADAMS, A.E., HORBURY, A.D. & ABDEL AZIZ, A.A. 1990. Controls on Dinantian sedimentation in south Cumbria and surrounding areas of northwest England. *Proceedings of the Geologists' Association*, **101**, 19–30.

AKHURST, M.C., CHADWICK, R.A., HOLLIDAY, D.W., MCCORMAC, M., MCMILLAN, A.A., MILLWARD, D. & YOUNG, B. 1997. *Geology of the West Cumbria District; Memoir for the 1:50 000 Geological Sheets 28 Whittehaven, 37 Gosforth and 47 Bootle (England and Wales)*. British Geological Survey, Nottingham.

AMBROSE, K. & CARNEY, J.N. 1997*a*. *Geology of the Calke Abbey area, 1:10 000 Sheet SK32SE: Part of 1:50 000 Sheet 141 (Loughborough)*. British Geological Survey Technical Report, **WA/97/17**.

AMBROSE, K. & CARNEY, J.N. 1997*b*. *Geology of the Breedon on the Hill area, 1:10 000 Sheet SK42SW: Part of 1:50 000 Sheet 141 (Loughborough)*. British Geological Survey Technical Report, **WA/97/42**.

BARCLAY, W.J., RILEY, N.J. & STRONG, G.E. 1994. The Dinantian rocks of the Sellafield area, west Cumbria. *Proceedings of the Yorkshire Geological Society*, **50**, 37–49.

BATH, A., MILODOWSKI, A., RUOTSALAINEN, P., TULLBORG, E.-L., CORTÉS RUIZ, A. & ARANYOSSY, J.-F. 2000. *Evidence from mineralogy and geochemistry for the evolution of groundwater systems during the Quaternary for use in radioactive waste repository safety assessment (EQUIP project)*. European Commission, Nuclear Science and Technology Report, **EUR 19613 EN**.

BRIDGES, P.H. & CHAPMAN, A.J. 1988. The anatomy of a deep water mud-mound complex to the south west of the Dinantian platform in Derbyshire UK. *Sedimentology*, **35**, 139–162.

CARNEY, J.N., AMBROSE, K. & BRANDON, A. 2001*a*. *Geology of the Loughborough District – A Brief Explanation of the Geological Map Sheet 141 (Loughborough)*. Sheet Explanation of the British Geological Survey, 1:50 000 Series Sheet 141 Loughborough (England and Wales).

CARNEY, J.N., AMBROSE, K. & BRANDON, A. 2001*b*. *Geology of the Country Between Loughborough, Burton and Derby: Sheet Description of the 1:50 000 Series Sheet 141 Loughborough (England and Wales)*. Sheet Description of the British Geological Survey.

CARNEY, J.N., AMBROSE, K. & BRANDON, A. 2002*a*. *Geology of the Melton Mowbray District: A Brief Explanation of Geological Sheet 142*. Sheet Explanation of the British Geological Survey, 1:50 000 Series Sheet 142 Melton Mowbray (England and Wales).

CARNEY, J.N., AMBROSE, K. & BRANDON, A. 2002*b*. *Geology of the Country Around Melton Mowbray*. Sheet Description of the British Geological Survey, 1:50 000 Series Sheet 142 Melton Mowbray (England and Wales).

CROWLEY, S.F., BOTTRELL, S.H., MCCARTHY, M.D.B., WARD, J. & YOUNG, B. 1997. $\delta^{34}S$ of Lower Carboniferous anhydrite, Cumbria and its implications for barite mineralization in the northern Pennines. *Journal of the Geological Society, London*, **154**, 597–600.

DICKSON, J.A.D. 1966. Carbonate identification and genesis as revealed by staining. *Journal of Sedimentary Petrography*, **36**, 491–505.

DUNHAM, K.C. & ROSE, W.H.C. 1941. *Geology of the Iron-ore Field of South Cumberland and Furness*. Wartime pamphlet of the Geological Survey of Great Britain, **16**.

EBDON, C.C., FRASER, A.J., HIGGINS, A.C., MITCHENER, B.C. & STRANK, A.R.E. 1990. The Dinantian stratigraphy of the East Midlands: a seismotectonic approach. *Journal of the Geological Society, London*, **147**, 519–537.

FORD, T.D. 1968. The Carboniferous Limestone. *In*: SYLVESTER-BRADLEY, P.C. & FORD, T.D. (eds) *The Geology of the East Midlands*. Leicester University Press, Leicester, 59–81.

FOX-STRANGWAYS, C. 1905. *The Geology of the Country Between Derby, Burton-on-Trent, Ashby-de-la-Zouch and Loughborough*. Memoir of the Geological Survey of Great Britain, Sheet 141 (England and Wales).

FOX-STRANGWAYS, C. 1907. *The Geology of the Leicestershire and South Derbyshire Coalfields*. Coalfield Memoir of the Geological Survey of Great Britain (England and Wales).

GOLDSTEIN, R.H. & REYNOLDS, T.K. 1995. *Systematics of Fluid Inclusions in Diagenetic Minerals*. Society of Economic Palaeontologists and Mineralogists, Short Course, **31**.

GRAYSON, R.F. & OLDHAM, L. 1987. A new structural framework for the northern British Dinantian as a basis for oil, gas and mineral exploration. *In*: MILLER, J., ADAMS, A.E. & WRIGHT, V.P. (eds) *European Dinantian Environments*. John Wiley, Chichester, 33–59.

GREEN, P.F., THOMSON, K. & HUDSON, J.D. 2001. Recognition of tectonic events in undeformed regions: contrasting results from the Midland Platform and East Midlands Shelf. *Journal of the Geological Society, London*, **158**, 59–73.

HARRISON, W.J. 1877. *A Sketch of the Geology of Leicestershire and Rutland*. W. White, Sheffield.

HULL, E. 1860. *The Geology of the Leicestershire Coalfield and the Country Around Ashby de la Zouch*. Coalfield Memoir of the Geological Survey of England and Wales.

KEMP, S.J. 1997. *The nature and maturity of mudstone intervals from the Ticknall Borehole, Derbyshire*. British Geological Survey Technical Report, **WG/97/18**.

KING, R.J. 1968. Mineralization. *In*: SYLVESTER-BRADLEY, P.C. & FORD, T.D. (eds) *The Geology of the East Midlands*. Leicester University Press, Leicester, 112–137.

LEES, A. & MILLER, J. 1985. Facies variations in Waulsortian buildups, part 2; Mid Dinantian buildups from Europe and North America. *Geological Journal*, **20**, 159–180.

LOTT, G.K. 1996. *Thin section petrography of selected samples from the Ticknall Borehole, Loughborough Sheet (141)*. British Geological Survey Technical Report, **WH/96/95R**.

MICHIE, U. 1996. The geological framework of the Sellafield area and its relationship to hydrogeology. *Quarterly Journal of Engineering Geology*, **29**, S13–S27.

MILODOWSKI, A.E., FORTEY, N.J., GILLESPIE, M.R., PEARCE, J.M. & HYSLOP, E.K. 2002. *Synthesis report on the mineralogical characteristics of fractures from the Nirex boreholes in the Sellafield area*. British Geological Survey, Technical Report, **WG/98/8**.

MILODOWSKI, A.E., GILLESPIE, M.R., NADEN, J., FORTEY, N.J., SHEPHERD, T.J., PEARCE, J.M. & METCALFE, R. 1998. The petrology and paragenesis of fracture mineralization in the Sellafield area, west Cumbria. *Proceedings of the Yorkshire Geological Society*, **52**, 215–241.

MITCHELL, G.H. & STUBBLEFIELD, C.J. 1941. The Carboniferous Limestone of Breedon Cloud, Leicestershire, and the associated inliers. *Geological Magazine*, **78**, 201–219.

MONTELEONE, P.H. 1973. *The Carboniferous limestone of Leicestershire and south Derbyshire*. PhD thesis, University of Leicester.

NIREX. 1995. *Flow zone characterisation: mineralogical and fracture orientation characteristics in the PRZ and Fleming Hall Fault Zone area boreholes, Sellafield*. Nirex Science Report, **SA/95/001**.

NIREX. 1997*a*. *Offsite core characterisation programme: the petrographic, mineralogical and lithogeochemical characteristics of Permo-Triassic and Carboniferous rocks from Sellafield BH3*. Nirex Science Report, **733**.

NIREX. 1997*b*. *Offsite core characterisation programme: the petrology of fractures and fracture mineralisation in Sellafield Borehole 3*. Nirex Science Report, **641**.

NIREX. 1997*c*. *Offsite core characterisation programme: the petrographic, mineralogical and lithogeochemical characteristics of Permo-Triassic and Carboniferous rocks from Sellafield BH7A and 7B*. Nirex Science Report, **536**.

NIREX. 1997*d*. *Offsite core characterisation programme: the petrology of fractures and fracture mineralisation in Sellafield BH7A and 7B*. Nirex Science Report, **743**.

NIREX. 1997*e*. *Offsite core characterisation programme: the petrographic, mineralogical and lithogeochemical characteristics of Permo-Triassic and Carboniferous rocks from Sellafield BH10A*. Nirex Science Report, **735**.

NIREX. 1997*f*. *Offsite core characterisation programme: the petrology of fractures and fracture mineralisation in Sellafield Borehole 10A*. Nirex Science Report, **752**.

NIREX. 1997*g*. *Offsite core characterisation programme: the petrology of fractures and fracture mineralisation in Sellafield BH12A*. Nirex Science Report, **737**.

NIREX. 1997*h*. *Offsite core characterisation programme: the petrographic, mineralogical and lithogeochemical characteristics of Permo-Triassic and Carboniferous rocks from Sellafield BH13A/B*. Nirex Science Report, **748**.

NIREX. 1997*i*. *Offsite core characterisation programme: the petrology of flow zones and potential flowing features in Sellafield Borehole 13A*. Nirex Science Report, **751**.

NIREX. 1997*j*. *Offsite core characterisation programme: the petrographic, mineralogical and lithogeochemical characteristics of Carboniferous and Permo-Triassic rocks from Sellafield BH14A*. Nirex Science Report, **750**.

NIREX. 1997*k*. *Offsite core characterisation programme: the petrology of flow zones and potential flowing features in Sellafield Borehole 14A*. Nirex Science Report, **755**.

NIREX. 1997*l*. *Sellafield geological and hydrogeological investigations: structural atlas of the Sellafield boreholes: structural domains and statistics illustrated by graphic display of discontinuity data. Section 1, PRZ area boreholes*. Nirex Science Report, **SA/96/005**.

PARSONS, L.M. 1918*a*. The Carboniferous Limestone bordering the Leicestershire Coalfield. *Journal of the Geological Society, London*, **73**, 84–110.

PARSONS, L.M. 1918*b*. Dolomitization and the Leicestershire dolomites. *Geological Magazine*, **55**, 246–258.

RILEY, N.J. 1997. *Foraminiferal biostratigraphy of the Carboniferous interval in Ticknall (Calke Abbey) Borehole*. British Geological Survey Technical Report, **WH/97/39R**.

ROEDDER, E. 1984. *Fluid Inclusions*. Mineralogical Society of America, Reviews in Mineralogy, **12**.

ROSE, W.H.C. & DUNHAM, K.C. 1977. *Geology and Hematite Deposits of South Cumbria*. Economic Memoir of the Geological Survey of Great Britain for 1:50 000 Geological Sheet 58 and Southern Part of Sheet 48.

SCOTESE, C.R., BAMBACH, R.K., BARTON, C., VAN DER VOO, R. & ZIEGLER, A.M. 1979. Palaeozoic base maps. *Journal of Geology*, **87**, 217–277.

SHEPHERD, T.J. & GOLDRING, D.C. 1993. Cumbrian hematite deposits, northwest England. *In*: PATRICK, R.A.D. & POLYA, D. (eds) *Mineralization in the British Isles*. Chapman & Hall, London, 419–443.

SPINK, K. 1965. Coalfield geology of Leicestershire and South Derbyshire: The exposed coalfields. *Transactions of the Leicester Literary and Philosophical Society*, **59**, 41–98.

TURNER, N. 1995. *Carboniferous palynology of the BGS Ticknall Borehole; 86.51–141.17 m*. British Geological Survey Technical Report, **WH/95/239R**.

WRIGHT, V.P. 1992. A revised classification of limestones. *Sedimentary Geology*, **76**, 177–185.

Geometry and origin of dolomudstone reservoirs: Pekisko Formation (Lower Carboniferous), western Canada

JOHN C. HOPKINS

Department of Geology and Geophysics, University of Calgary, Calgary, Alberta, Canada T2N 1N4 (e-mail: jchopkin@ucalgary.ca)

Abstract: Dolomudstones of the Pekisko Formation in western Canada form small but important oil and gas reservoirs. The reservoirs are irregularly shaped bodies 1 km or so wide and commonly 5–8 m thick. Porosity development within the dolomudstones is a complex function of sedimentation, early facies-selective dolomitization and later telogenetic leaching of calcareous components.

The carbonate sediment precursor of the dolomudstone, interpreted from relict textures preserved in chert nodules, was a microwackestone with abundant silt-sized skeletal fragments. Dolomudstone reservoirs are comprised of dolomudstones, calcareous dolomudstones, and subordinate interbedded dolowackestones and dolograinstones. Some dolomudstone reservoirs are contained entirely within grainstones. Others are capped by tight fenestral lime mudstone that has been dolomitized locally. Dolomitization has been most intense within the centres of these reservoirs, and dolomudstones grade laterally into calcareous dolomudstones.

The association of facies indicates that microwackestones were deposited in subtidal intershoal and lagoonal environments on an inner ramp. Grainstone shoals provided a broad barrier that absorbed wave energy seaward of the lagoon. Fenestral lime mudstones accumulated in peritidal environments in restricted areas of the inner ramp, landward of the lagoon. Dolomitization is interpreted to have been early and selective to the microwackestone facies because it retained permeability or was reactive during early burial. Dolomitizing fluids were most probably derived from overlying formations and made their way downwards through spatially separated conduits. The Pekisko Formation was exposed and sculptured at several Jurassic–Early Cretaceous unconformities. During these times, sandstones and shales were deposited in solution cavities developed within the dolomudstones. Concomitant leaching of calcite increased porosity of the dolomudstone reservoirs.

In many carbonate formations porosity is related to grain size so that interparticle porosity is preserved in grainstones, whereas lime mudstones have neomorphosed forming microcrystalline rocks with low porosity. By contrast, in other formations, grainstones are tightly cemented, whereas lime mudstone facies have become preferentially dolomitized to form porous dolomudstones. Various reasons have been put forward for these different diagenetic modifications including depositional environment, fabric of the host sediment, permeability, original mineralogy and the nature of dolomitizing solutions. First recognized by Murray & Lucia (1967), the process of dolomitization of lime mudstones was described as facies-selective and considered to be evidence for early dolomitization where surrounding grainstones had been rendered tight by early cementation. Other examples of facies-selective dolomitization of carbonate mudstones have since been described, particularly from peritidal to shallow inner-ramp successions (e.g. Choquette & Steinen 1980, 1985; Longman *et al.* 1983; Wegelin 1987; Johnson 1994).

Dolomudstones are an important reservoir group and form subtle traps. Principally because so few examples have been described, it is difficult to characterize the reservoirs. In examples that have been well documented, fine-crystalline–microcrystalline porous dolomudstones with a microsucrosic texture predominate, the shapes of the reservoirs are intricate and consequently recoveries are tenuous. Because they are so poorly known, dolomudstone reservoirs may have been overlooked in mature basins.

Dolomudstone reservoirs occur in the Pekisko Formation in the Western Canada Sedimentary Basin where they are present as unconformity-associated traps along an extensive subcrop edge (Podruski *et al.* 1988; Hopkins 1999). The purpose of this paper is to describe the sedimentary and diagenetic characteristics

From: BRAITHWAITE, C. J. R., RIZZI, G. & DARKE, G. (eds) 2004. *The Geometry and Petrogenesis of Dolomite Hydrocarbon Reservoirs*. Geological Society, London, Special Publications, **235**, 349–366. 0305-8719/$15.00

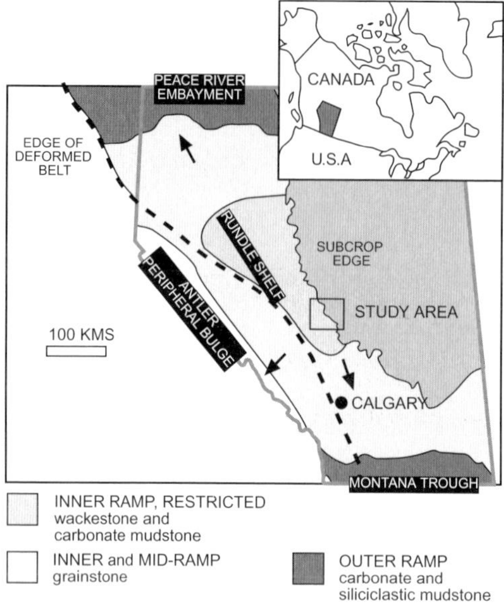

Fig. 1. Regional palaeogeography of Pekisko Formation compiled from Richards (1989) and Richards *et al.* (1994). The extent of the area of restricted sedimentation on the Rundle Shelf was determined from the occurrence of evaporites and shallow-water carbonate mudstones.

of these rocks, the relationship between dolomitization and porosity–permeability, and the geometries of several dolomudstone reservoirs. In this way the Pekisko reservoirs may serve as a template for the discovery and development of similar facies-selective dolomudstone reservoirs in other basins.

Units and symbols

Metric units are used throughout this paper except for depths in wells drilled prior to July 1978 when depths were recorded in feet (ft) on well logs and in core files. Well locations are given as Dominion Land Survey coordinates that are the legal designations for wells in Alberta. Core analyses are commercial analyses of whole-core samples from well data files available from Alberta Energy Utilities Board Information Services in Calgary or through commercial databases. Permeability is air permeability measured in millidarcies (mD) and reported as maximum horizontal permeability (K_{max}), permeability measured at 90° to the maximum permeability (K_{90}), and vertical permeability (K_v).

Palaeogeography and stratigraphy

Lower Carboniferous (Mississippian) sedimentation in western North America took place on a broad westward-facing ramp developed along the craton margin (Sando *et al.* 1990). In south and central Alberta (Fig. 1) fine-grained siliciclastics, carbonates and evaporites were deposited over a broad stable shallow area some 300 km long and 200 km wide termed the Rundle Shelf by Richards *et al.* (1994). To the north, shelf sediments grade into slope sediments in the subsiding Peace River Embayment; to the west, a peripheral bulge of low elevation associated with Antler deformation restricted sedimentation; to the south and west, the water deepened gradually into a re-entrant in the cratonic platform called the Montana Trough.

The Lower Carboniferous ramp in western Canada was interpreted to be thermally stratified by Martindale & Boreen (1997) who developed a facies model for sedimentation following principles outlined by Burchette & Wright (1992). Deeper-water carbonate mudstones–grainstones of the outer- and mid-ramp divisions were deposited under cool-water conditions influenced by oceanic upwelling. Inner-ramp grainstones and mudstones accumulated in warmer marine waters (Fig. 2). Fluctuations of relative sea-level commonly observed throughout Lower Carboniferous successions (Ross & Ross 1988) are represented by a number of transgressive–regressive cycles (sequences) across the ramp (Richards 1989). One of these is the Pekisko Formation formed in response to the latest middle Tournaisian transgression.

The Pekisko Formation is a depositional sequence at the third-order level (Richards 1989). Regional transgressive erosional surfaces separate the Pekisko Formation from the Banff Formation beneath (Richards 1989; Hopkins 1999) and the overlying Shunda Formation (Stoakes 1999). The Pekisko Formation in the study area comprises an upward-shallowing succession about 40 m thick (Fig. 3). The lower part is grainstone, dominated by crinoid grainstone interbedded with fine-grained grainstone and locally dolomudstone. The upper part is a dolomudstone–fenestral lime mudstone succession reflecting deposition in subtidal and intertidal environments, respectively.

Dolomudstone and associated facies

The dolomudstone facies is characterized by fine-crystalline porous dolostone, but dolomudstone reservoirs include calcareous

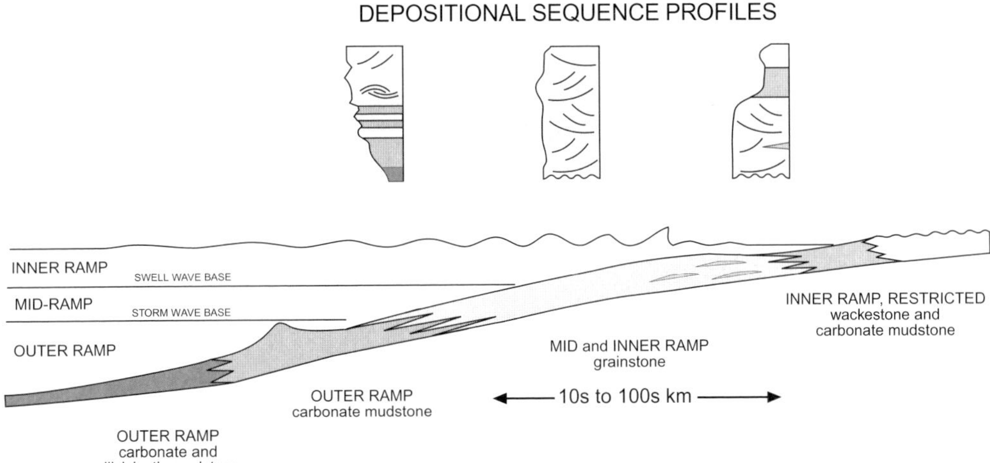

Fig. 2. Lower Carboniferous depositional model modified after Martindale & Boreen (1997) by the addition of depositional sequence profiles.

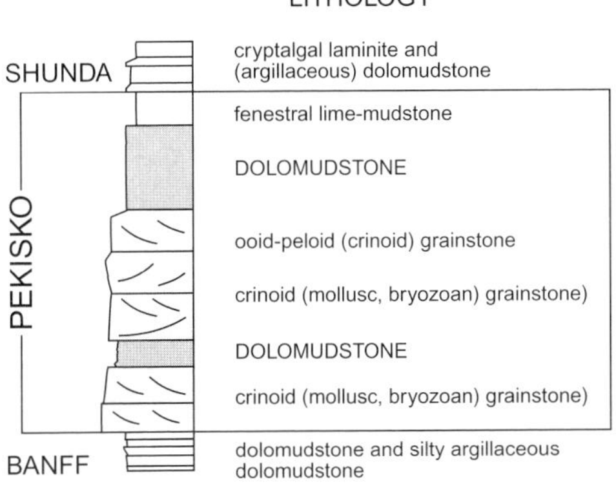

Fig. 3. General lithology of the Pekisko Formation in the study area after Hopkins (1999).

dolomudstones and subordinate interbedded dolograinstones and dolowackestones. These lithologies are described below. In general, limestones surrounding the dolomudstones have not been dolomitized.

Dolomudstone is a massive, current-laminated or mottled (bioturbated) fine-crystalline carbonate rock (Fig. 4) composed principally of 10–20 μm dolomite crystals and is generally stained brown by oil. Locally, discrete silica nodules up to 20 cm across (Fig. 4a) are present in the dolomudstones. Within these nodules (Fig. 5a & b) silicified ghosts reveal a distinctive fabric of the original calcareous host sediment with scattered thin-shelled fossils (mostly ostracods) and numerous silt-sized skeletal fragments set in a fine-grained matrix. The sediment is informally referred to as a microwackestone throughout this paper.

Dolowackestone is characterized by moulds of fenestrate bryozoa and whole-shell brachiopods set in fine-crystalline dolomite. Scattered angular equant moulds a few millimetres across probably represent leached crinoid plates. The matrix is a mosaic of euhedral 10–100 μm dolomite rhombs with intercrystalline porosity. Dolowackestone usually occurs in discrete beds within dolomudstone, although in a few places they grade one into the other. Typically, the matrix of the dolowackestone is more coarsely crystalline than the dolomudstone and fossil moulds are more numerous.

Dolograinstone commonly preserves grain lamination and/or cross-stratification with aligned angular skeletal fragments. Equant or elongate angular moulds a few millimetres across were skeletal grains, principally crinoids. Dolograinstones are generally interbedded with dolomudstones. Although the dolomudstone facies overlies or is laterally equivalent to grainstones, those immediately adjacent have been dolomitized only locally (see below).

Sandstone and shale fills dissolution cavities within the Pekisko Formation. These siliciclastic sediments are palaeokarst deposits superimposed on the Pekisko Formation as a result of later subaerial exposure (Hopkins 1999). Although they are not contemporaneous deposits of the dolomudstones, they are included here as their presence will later be

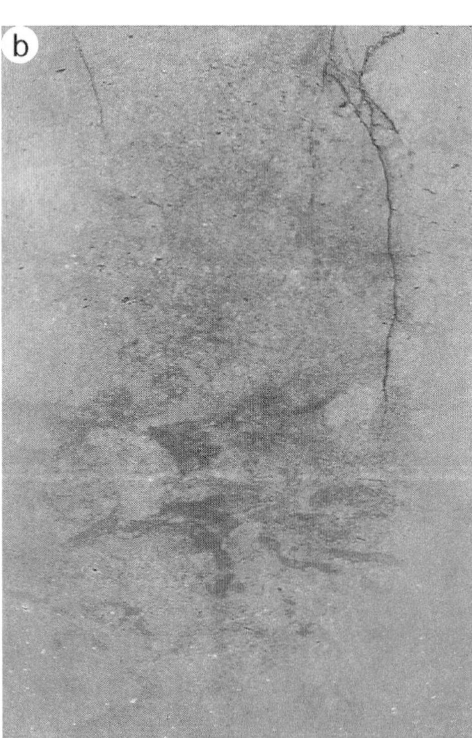

interpreted to have some bearing on porosity development within the dolomudstone facies.

Petrography

A variably porous mosaic of fine-crystalline dolomite rhombs dominates the dolomudstone fabric. Crystal size and fabric varies irregularly on the mm-scale following bioturbation or current lamination structures (Fig. 5c). Euhedral crystals (planar-e fabric of Sibley & Gregg 1987) dominate the porous areas, whereas subhedral–anhedral crystals are widespread throughout less porous areas (planar-s fabric of Sibley & Gregg 1987). Individual crystals are limpid and generally free of inclusions (Fig. 5d). They have unit extinction under crossed polars. The small crystal size and relatively high porosity combine to form a microsucrosic fabric (Fig. 5d & e). Calcareous dolomudstones are formed of a subhedral–anhedral mosaic of 10–30 μm calcite crystals and euhedral dolomite rhombs. Porosity is loosely organized around patches of euhedral dolomite crystals resulting in a general decrease in porosity in the dolomudstone with increasing calcite content. Locally, calcite rhombs have been etched (Fig. 5f).

Petrophysics

Porosity and grain density values from whole-core analyses of seven cores through dolomudstones are given in Figure 6. The most porous samples are pure dolostones and have grain density of about 2840 kg m^{-3}, and the least porous samples are limestones with a grain density of about 2710 kg m^{-3}. In general, porosity is only weakly related to grain density and calcite content (correlation coefficient r^2 is 0.388) and calcareous dolomudstones have a wide range in porosities for a given amount of calcite. For example, of 22 samples with grain density of 2800 kg m^{-3} (corresponding to 30% calcite, 70% dolomite) porosity values range from 5.7 to 30.1%; 10 samples have less than 10% porosity and five samples have more than 20% porosity. Petrographic evidence (Fig. 5c

Fig. 4. Core slabs of dolomudstones. (**a**) Massive (churned) and bioturbated dolomudstone with chert nodules. 14-08-039-03W5 7202.0 ft. Grain density 2830 kg m^{-3}, K_{max} 56 mD, K_{90} 53 mD, K_v 35 mD, porosity 26.1%. (**b**) Bioturbated calcareous dolomudstone. 14-08-039-03W5 7229.4 ft. Grain density 2750 kg m^{-3}, K_{max} 20 mD, K_{90} 13 mD, K_v 6.3 mD, porosity 18.5%.

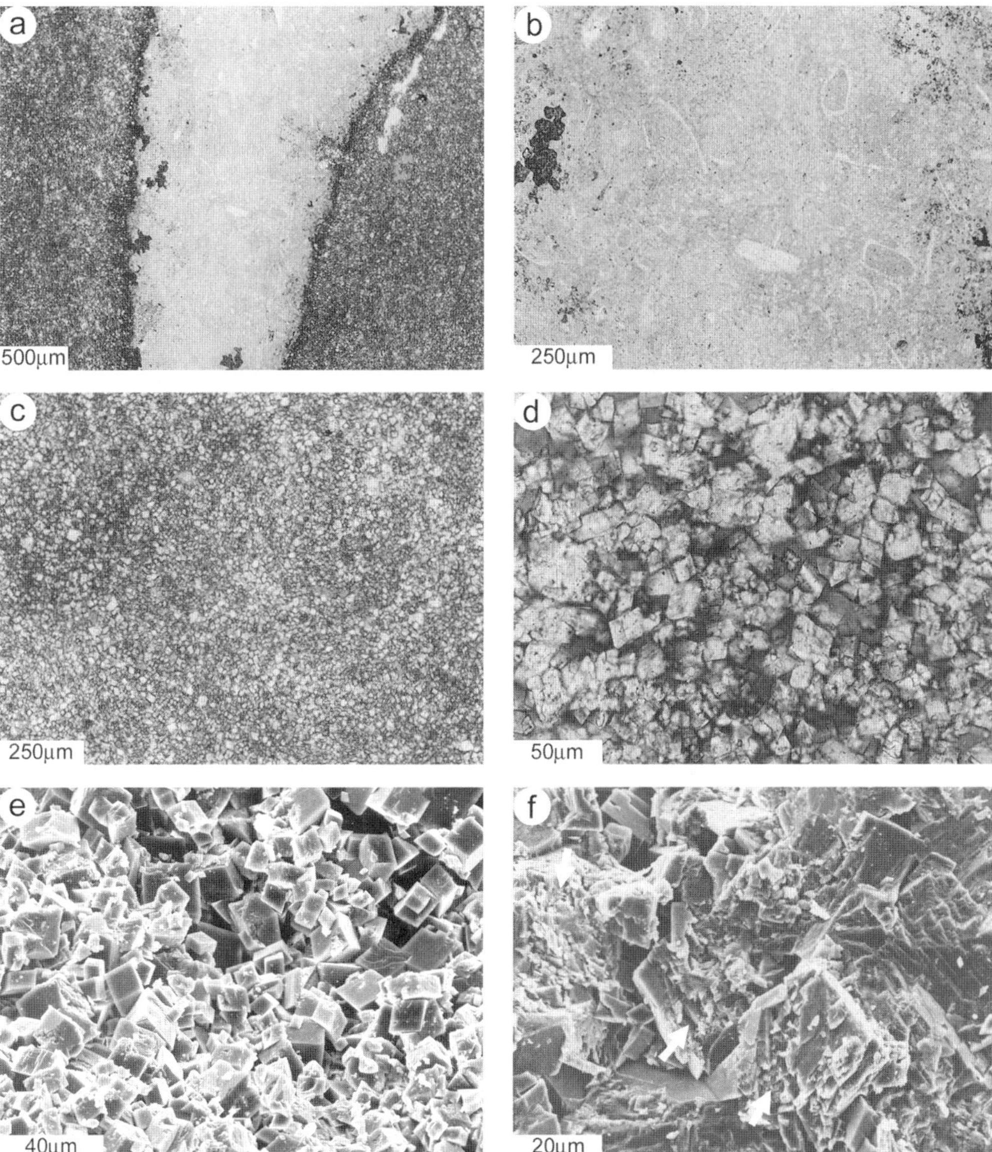

Fig. 5. (**a**) Photomicrograph of chert nodule in dolomudstone. 06–08–039–03W5 7210.3 ft. Grain density 2800 kg m^{-3}, K_{max} 25 mD, K_{90} 7.9 mD, K_v 26 mD, porosity 24.6%. (**b**) Details of chert nodule in (a) revealing ghosts of finely comminuted shell fragments and scattered ostracod shells. (**c**) Photomicrograph of bioturbated dolomudstone impregnated with dark-tinted epoxy resin. Irregular light-coloured areas are relatively low-porosity mosaics of anhedral dolomite crystals. Irregular darker areas are high-porosity mosaics dominated by euhedral dolomite crystals. 06-08-039-03W5 7206.0 ft. Grain density 2840 kg m^{-3}, K_{max} 128 mD, K_{90} 126 mD, K_v 113 mD, porosity 33.1%. (**d**) Details of the porous mosaic in (c) showing limpid euhedral rhombs of dolomite with few inclusions; porosity is dark grey. (**e**) SEM micrograph of dolomudstone with low porosity in anhedral mosaic (lower left) and high porosity in euhedral mosaic (upper right) of dolomite crystals. 04-08-039-03W5 7225 ft. Grain density 2840 kg m^{-3}, K_{max} 55 mD, K_{90} 55 mD, K_v 39 mD, porosity 28.7%. (**f**) Etched calcite rhombs (arrows) in calcitic dolomudstone. 04-08-039-03W5 7207 ft. Grain density 2750 kg m^{-3}, K_{max} 0.6 mD, K_{90} 0.42 mD, K_v <0.01 mD, porosity 9.6%.

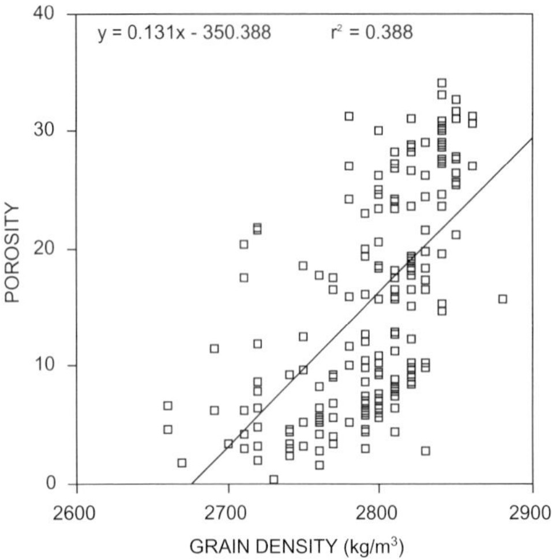

Fig. 6. Cross-plot of grain density v. porosity for 210 whole-core samples from seven wells in Medicine River Pekisko E Pool. Grain density for calcite is 2710 kg m^{-3}; grain density for dolomite is 2840 kg m^{-3}.

& e) shows that areas of anhedral dolomite in a sample may have low porosity, whereas areas of euhedral dolomite form a microsucrosic texture that is porous and permeable. Porosity may have been enhanced (and bulk grain density increased) by the leaching of calcite from the fabric; however, the fundamental determinative of high porosity is the presence of a crystal-supported fabric of fine-crystalline euhedral dolomite.

Carbon and oxygen stable isotopes

Thirty-one samples of dolomudstones and associated facies were analysed in the Stable Isotope Laboratory at the Department of Physics and Astronomy, University of Calgary. For samples with significant calcite, calcite and dolomite were analysed separately by selective acid leaching. Data for samples are presented in Table 1 and Figure 7.

The bulk of the dolomite samples (20 samples contained within the dotted circle in Fig. 7)

Table 1. *Stable isotope values (‰ PDB) for dolomudstones and associated lithologies of the Pekisko Formation*

Well	Depth	$\delta^{13}C$ calcite	$\delta^{18}O$ calcite	$\delta^{13}C$ dolomite	$\delta^{18}O$ dolomite	Lithology
02-08-39-03W5	7182.4 ft	0.9	–3.7			Fenestral lime mudstone
02-08-39-03W5	7190.0 ft	1.4	–3.3			Fenestral lime mudstone
02-08-39-03W5	7195.8 ft	–1.44	–9.96	2.57	–1.19	Dolomudstone
02-08-39-03W5	7195.8 ft			0.6	–5.1	Dolomudstone
02-08-39-03W5	7201.2 ft	0.55	–6.94	3.25	–0.89	Dolomudstone
02-08-39-03W5	7201.2 ft			2.5	–1.7	Dolomudstone
03-08-39-03W5	2189.79 ft	0.66	–4.34	1.9	–0.27	Dolomudstone
03-08-39-03W5	2192.89 m			2.92	–1.27	Dolomudstone
03-08-39-03W5	2195.54 m			1.29	–0.57	Dolowackestone
03-08-39-03W5	2196.30 m			2.54	–0.27	Dolomudstone
06-08-39-03W5	7139.5 ft	0.82	–4.32	1.6	–0.94	Dolomudstone
06-08-39-03W5	7151.0 ft	1.93	–1.73	1.55	2.39	Dolomudstone
06-08-39-03W5	7158.4 ft	–0.53	–5.62	–0.86	–5.18	Dolomudstone
06-08-39-03W5	7169.0 ft	0.44	–3.43	0.31	–2.9	Dolomudstone
06-08-39-03W5	7203.5 ft	2.72	–3.88	4.44	–1.06	Dolomudstone
06-08-39-03W5	7206.5 ft	3.65	–3.86	4.42	–1.71	Dolomudstone
06-21-39-03W5	7044.8 ft	2.18	–4.13	3.11	–0.38	Dolomudstone
08-16-39-03W5	6998.5 ft	0.3	–9.09	2.58	–2.25	Dolomudstone
2/06-08-39-03W5	2193.48 m	1.05	–5.64	3.06	–0.89	Dolomudstone
2/06-08-39-03W5	2195.45 m	–3.61	–12.02	0.08	–5.95	Dolomudstone
2/06-08-39-03W5	2196.00 m	–3.7	–11.44	–0.1	–5.69	Dolomudstone
2/06-08-39-03W5	2197.00 m	1.02	–6.9	3.26	–3.46	Dolomudstone
14-08-39-03W5	7177.0 ft			3.7	–0.3	Dolomudstone
14-08-39-03W5	7184.8 ft			2.6	–0.2	Dolomudstone
14-08-39-03W5	7195.6 ft			2.2	0.6	Dolomudstone
14-08-39-03W5	7203.1 ft			3.7	–1.3	Dolomudstone
14-08-39-03W5	7223.2 ft			3.2	3.0	Dolograinstone
14-08-39-03W5	7233.4 ft			4.0	–1.1	Dolowackestone
16-32-39-03W5	7030.0 ft	1.12	–4.73	2.83	0.54	Dolomudstone
16-32-39-03W5	7046.0 ft			3.93	–0.19	Dolomudstone
16-32-39-03W5	7062.0 ft			3.85	3.03	Dolomudstone

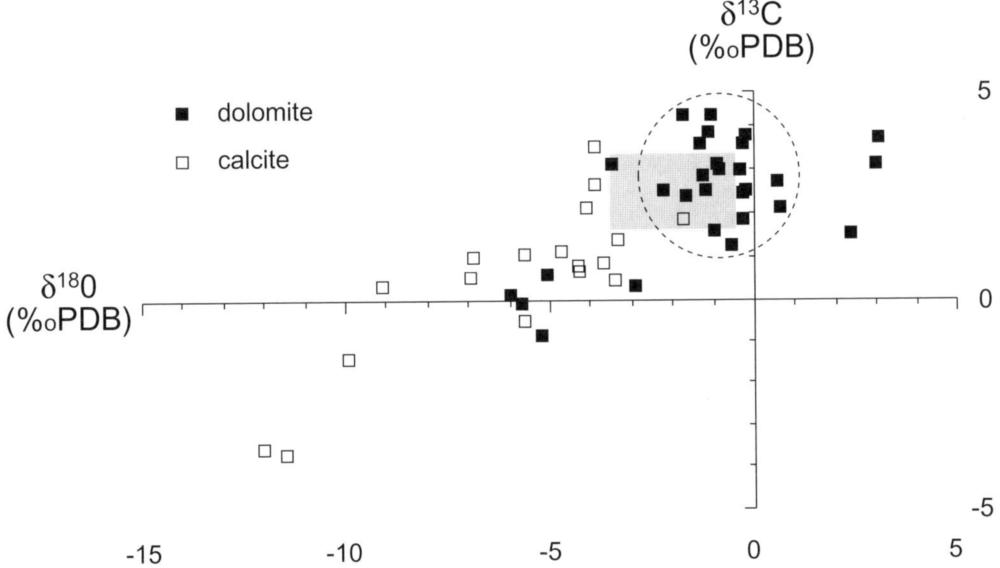

Fig. 7. Carbon and oxygen stable isotope values for 31 dolomudstones and associated lithologies from the Pekisko Formation (location of samples is giving in Table 1). Shaded area contains the field for Tournaisian marine carbonate from Mii *et al.* (1999).

cluster about a mean of $\delta^{13}C$ = 3.0‰ and $\delta^{18}O$ = –0.8‰. These values lie close to the field for Lower Carboniferous (Tournaisian) marine carbonate (Mii *et al.* 1999). Three samples relatively enriched in oxygen are closer to postulated values for dolomite precipitated from evaporated Early Carboniferous seawater (cf. Smith & Dorobek 1993; Durocher & Al-Aasm 1997). Six samples are relatively depleted in oxygen, and five of these are depleted in both carbon and oxygen compared to the clustered group.

Both carbon and oxygen isotope values for calcite in 18 samples of calcareous dolomudstones are nearly all depleted with respect to Tournaisian marine carbonates. One sample lies within the Tournaisian carbonate field, four others on the fringe of this field have a negative shift in oxygen of 1–2‰. The remaining 13 calcite samples are significantly depleted in both oxygen and carbon.

Discussion

Geological evidence for early dolomitization of peritidal carbonates has been presented by many workers, and a strong linkage between dolomitization of peritidal carbonates and eustatic cycles has been demonstrated by Montanez & Read (1992). Thus, dolomitization of Lower Carboniferous dolomudstones in western Canada and the western United States has been linked to the passage of evaporite brines from overlying sabkhas (e.g. Illing 1959; Murray & Lucia 1967; Wegelen 1987; Smith & Dorobek 1993). More recently, geochemical evidence has been used to detail fluid composition and dolomitization by evaporated seawater (Smith & Dorobek 1993; Al-Aasm & Packard 2000) or modified seawater in shallow coastal plain aquifers (Smith & Dorobek 1993; Al-Aasm & Lu 1994).

The bulk of the dolomudstones of the Pekisko Formation have isotope values similar to those of fine-crystalline dolomites in the Debolt Formation (Pekisko equivalent in NW Alberta – Al-Aasm & Lu 1994; Al Aasm & Packard 2000) and in the Mission Canyon Formation (Pekisko equivalent in Montana – Smith & Dorobek 1993). In each of these areas, many dolomites from dolomudstones have values close to those expected for precipitation from Early Carboniferous seawater or evaporated seawater. Other samples are more depleted in oxygen and, to a lesser extent, carbon. Explanations for isotope depletion have included alteration by meteoric waters at unconformities and deep burial diagenesis.

In the Debolt Formation (Al-Aasm & Packard 2000) dolomudstones locally retain fabrics and compositions consistent with dolomite precipitation from Early Carboniferous evaporated seawater or normal seawater. For these samples, it appears that subsequent neomorphism of the dolomudstones was arrested or contained within relatively closed systems despite burial to depths of several kilometres. Other Lower Carboniferous dolomudstones (Durocher & Al-Aasm 1997; Al-Aasm & Packard 2000) have depleted isotopic signatures suggesting later deep burial with continuing reactions between the dolomudstones and basinal fluids.

In the Mission Canyon Formation (Smith & Dorobek 1993) dolomudstones from two different geographic areas were precipitated from evaporated and normal seawater, respectively. Both groups were subject to variable degrees of alteration by meteoric waters beneath a Late Carboniferous unconformity. Stable isotopes of oxygen show a wide range of values up to –12‰; carbon isotopes are less depleted and generally lie within the range of 0–4‰.

Dolomudstones of the Pekisko Formation in the study area also have been exposed to meteoric fluids from the Late Carboniferous to the Early Cretaceous. About 120 Ma is represented along the unconformity that separates Lower Carboniferous and Lower Jurassic strata along the subcrop edge of the Pekisko Formation in the study area. The actual amount of time the Pekisko Formation was exposed is not known; however, late Palaeozoic and Triassic sediments were either not deposited or were very thin and have been removed by erosion across the Rundle Shelf (Kent 1994). Further exposure occurred across several unconformities in the Jurassic and Early Cretaceous (Hopkins *et al.* 1998). The presence of sandstone and shale-filled dissolution cavities, sinkholes and valleys testifies to extensive telogenetic modification of the Pekisko Formation along its subcrop edge (Hopkins 1999). It has already been noted that leaching of calcite within the dolomudstones is one of the factors critical to the development of porous reservoirs. The associated sandstone and shale-filled cavities simply represent local more intense leaching. Thus, it is likely that isotopically depleted meteoric fluids reset some of the isotopes in the dolomudstones. The general high depletion of both carbon and oxygen in calcite compared to the values in dolomite from calcareous dolomudstones (Fig. 7) is probably related to the greater solubility of calcite.

Individual dolomite crystals within the dolomudstones of the Pekisko Formation appear not to have suffered the etching and calcitization that commonly follows meteoric alteration of metastable dolomite (James *et al.* 1993; Smith & Dorobek 1993). One explanation is that these Early Carboniferous dolomudstones had already been mineralogically stabilized by the time they were exposed to Jurassic and Early Cretaceous meteoric leaching and alteration.

Distribution and geometry of dolomudstones

The Pekisko Formation has been removed by erosion from the eastern part of the study area (Fig. 8) and has been variously sculptured across the western part of the area. Sinkholes and incised river valleys up to 150 m deep filled with Jurassic and Lower Cretaceous sediments dominate the palaeotopography (Hopkins *et al.* 1998). The irregular distribution of dolomudstones along the subcrop belt is therefore partly due to erosion; but it is also apparent that the dolomudstone bodies are rarely contiguous for distances of more than a few kilometres (Fig. 9).

Dolomudstones correspond with relatively high porosity zones on geophysical logs (Fig. 9), an observation confirmed from cores and cuttings in older wells. Occurrences of dolomudstones are easily inferred from modern neutron-density and photoelectric (PEF) logs (e.g. Fig. 9c), where their presence is coincident with gamma radiation values a few API units higher than those of the limestones. However, geophysical logs do not readily distinguish dolomudstones from interbedded dolowackestones and dolograinstones. They define the body of dolomudstone and associated facies rather than the dolomudstone lithology.

Cross-sections through four different pools across the study area (Fig. 9), a map of porosity distribution in the Medicine River Pekisko E pool (Fig. 10) and two sets of closely spaced cores from two different pools (Fig. 11) are used to infer the geometry of the reservoirs.

Medicine River Pekisko E Pool (Fig. 9A–A′)

The Medicine River E Pool is one of the larger and more productive dolomudstone reservoirs in the study area (Hopkins 1999). The dolostone body is truncated to the NW by a Lower Jurassic unconformity (or terminates before the unconformity edge) and is limited to

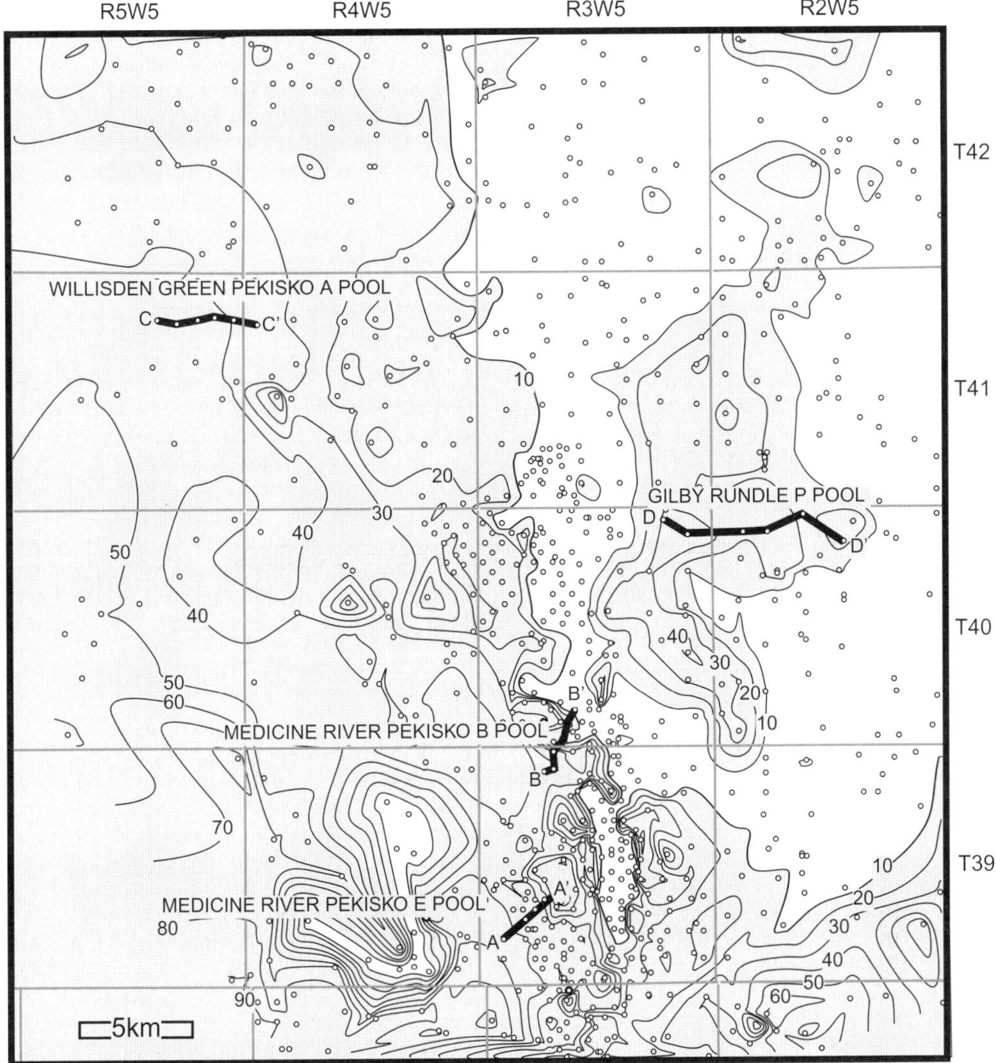

Fig. 8. Banff to top Mississippian isopach map (Pekisko and Shunda Formations) with 10 m-contour intervals. Small circles are well locations. Pekisko Formation is present within the shaded areas. A–A′, B–B′, C–C′ and D–D′ are locations of cross-sections in Figure 9.

the SW by a facies change into tight limestones (Figs 9A–A′ and 10). Cores of the porous intervals in 04-08-039-03W5 and 06-08-039-03W5 along the line of section are comprised of dolomudstone. Dolomudstones sharply overlie grainstones and grade sharply upwards into fenestral lime mudstones. A second, lower horizon of dolomudstone is present beneath the main porous body in 04-08-039-03W5 and is contained within grainstones.

The E Pool has the highest density of wells and cores of any of the dolomudstone pools in the study area so the shape of the porous body that hosts the pool can be inferred from well data. Figure 10 is an isolith map of dolostone with more than 10% porosity. In three dimensions this must represent an irregular flattened amoeboid body with irregularly distributed thicker intervals of more than 10% porosity.

Two other cored wells, 02-08-039-03W5 and 14-08-039-03W5 (Figs 10 and 11), contrast the

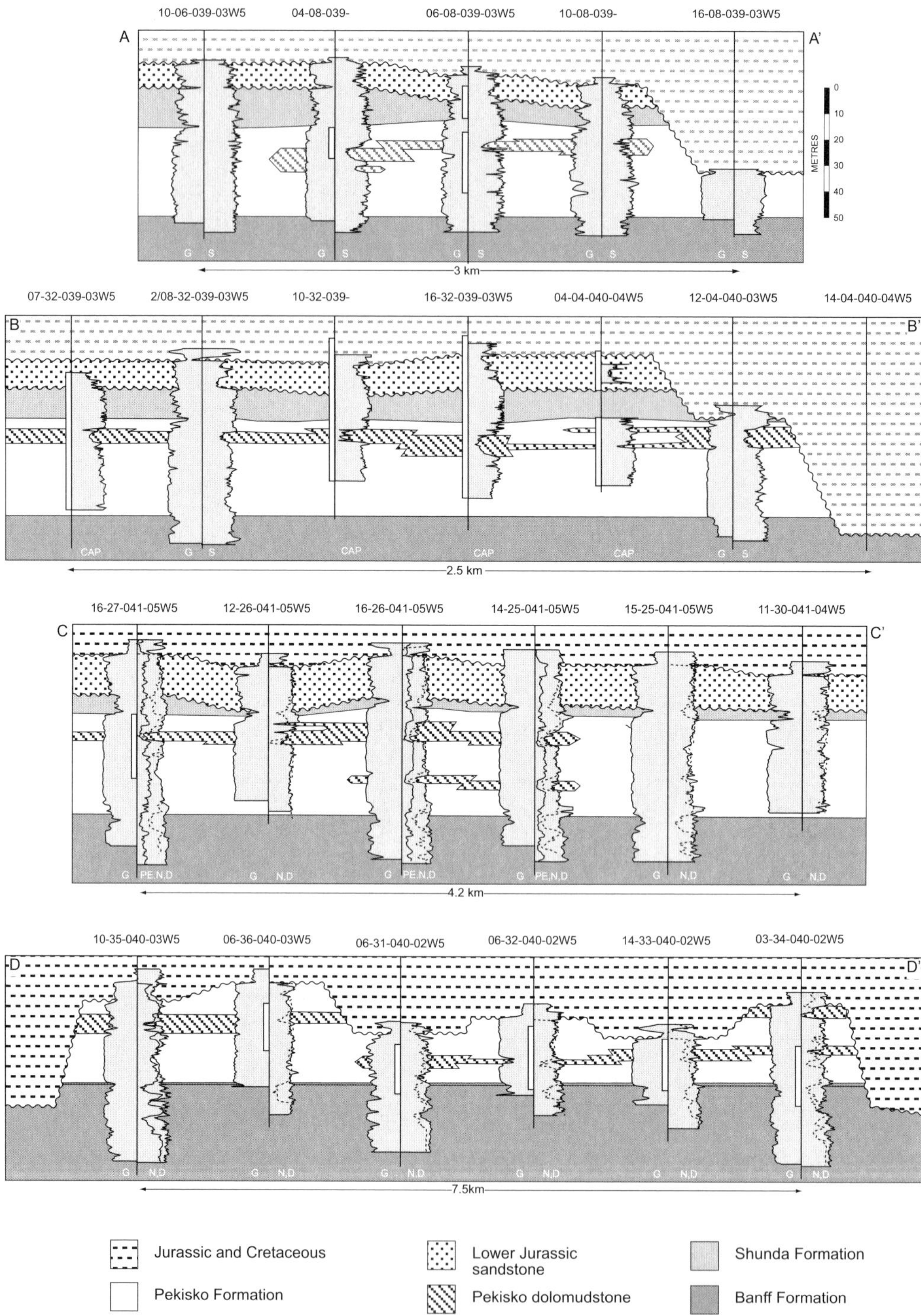

Fig. 9. Cross-sections based on well logs or core analyses in the study area. Datum is the top of the Banff Formation. Log traces are designated as G, gamma, S, sonic, D, density, N, neutron, PEF, photoelectric, CAP, porosity from core analyses. The vertical white-filled bars on the log axes represent cored intervals. Porous dolomudstone facies of the Pekisko Formation is highlighted by cross-hatching.

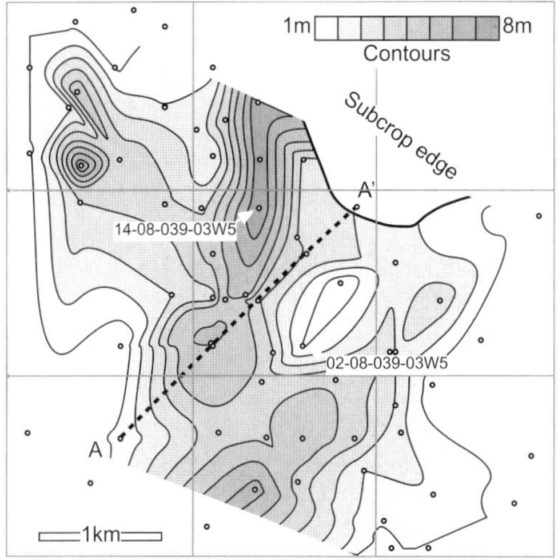

Fig. 10. Isopach map of porosity in Medicine River Pekisko E Pool centred around Section 8 Township 39 Range 3W5. Contours record the thickness in metres of porosity greater than 10% in each well. A–A′ is the same cross-section as in Figure 8.

extremes of dolomitization and porosity development. The two wells are 940 m apart; 02-08-039-03W5 is a dry hole, whereas 14-08-039-03W5 has produced in excess of 200 000 m^3 of oil. In both wells, dolomudstones overlie grainstones but the contacts are obscured by sandstone and shale-filled dissolution cavities: in 02-08-039-03W5 dolomudstone grades upwards into fenestral lime mudstones over a few decimetres and in 14-08-039-03W5 similar lime mudstones have been dolomitized to produce microcrystalline dolostones that lack significant porosity.

The dolostone body in 02-08-039-03W5 contains a single interval of variably calcareous dolomudstone with chert nodules. In 14-08-039-03W5 the dolostone body is composite: two horizons of dolomudstone are present and are separated by dolograinstone and dolowackestone (Fig. 11a). Porosity is markedly higher in the dolomudstones compared to the enclosed dolograinstones and dolowackestones; however, the permeabilities of all three lithologies are similar. The Pekisko Formation in 02-08-039-03W5 is 37 m thick, in 14-08-39-03W5 it is 35 m thick. Using either the top or the base of the Pekisko Formation in these two wells as a datum places the lower dolomudstone in 14-08-039-03W5 as the stratigraphic equivalent of grainstones in 02-08-039-03W5.

Medicine River Pekisko B Pool (Fig. 9B–B′)

The Medicine River Pekisko B Pool was developed in the early 1950s and, although a number of wells were cored and the cores analysed, few gamma-porosity logs are available. Several porosity profiles used in construction of a cross-section (Fig. 9B–B′) are derived from core analyses, with their stratigraphic position within the Pekisko Formation determined from electric logs. Despite marked changes in thickness from well to well, the porous dolomudstone is continuous along the line of section and the wells have a common pressure regime (one of the criteria used to define an oil pool). In general, contacts between the dolomudstone and adjacent lithologies are sharp or grade over a few centimetres; it overlies grainstone and is overlain by fenestral micrite.

There are pronounced facies changes within the dolomudstone horizon between two cored wells 16-32-039-03W5 and 04-04-040-03W5 that are just 600 m apart (Fig. 11b). In 16-32-039-03W5 the dolomudstone is 8 m thick, and represents one of the thickest occurrences of the lithology in the area. In well 04-04-040-03W5 two thin dolomudstone intervals are separated by undolomitized crinoidal grainstone. Within the cross-section between these wells a grainstone tongue must pass laterally into dolomudstone. For comparison, a grainstone tongue between two dolomudstone beds has been dolomitized in well 14-08-039-03W5 (Fig. 11a).

The upper and lower contacts of the dolostone body are different in the two wells. In 16–32–039–03W5 the overlying fenestral lime mudstone is thin and has been partly dolomitized to form low-porosity microcrystalline dolostone, distinct from the porous fine-crystalline dolomudstone beneath. The lower contact of the dolomudstone body in 16-32-039-03W5 is contained within a thin interval of lost core, but from the electric log appears to be a sharp boundary with undolomitized grainstone. In 04-04-040-03W5 the contact between the dolomudstones and the grainstones beneath is also sharp, but here the grainstones have been dolomitized for several metres below the contact and grade laterally into those below the dolomudstone in 16-32-039-03W5.

Willisden Green Pekisko A Pool (Fig. 9C–C′)

In the NW part of the study area (Fig. 8) dolomudstone is present in the upper part of the

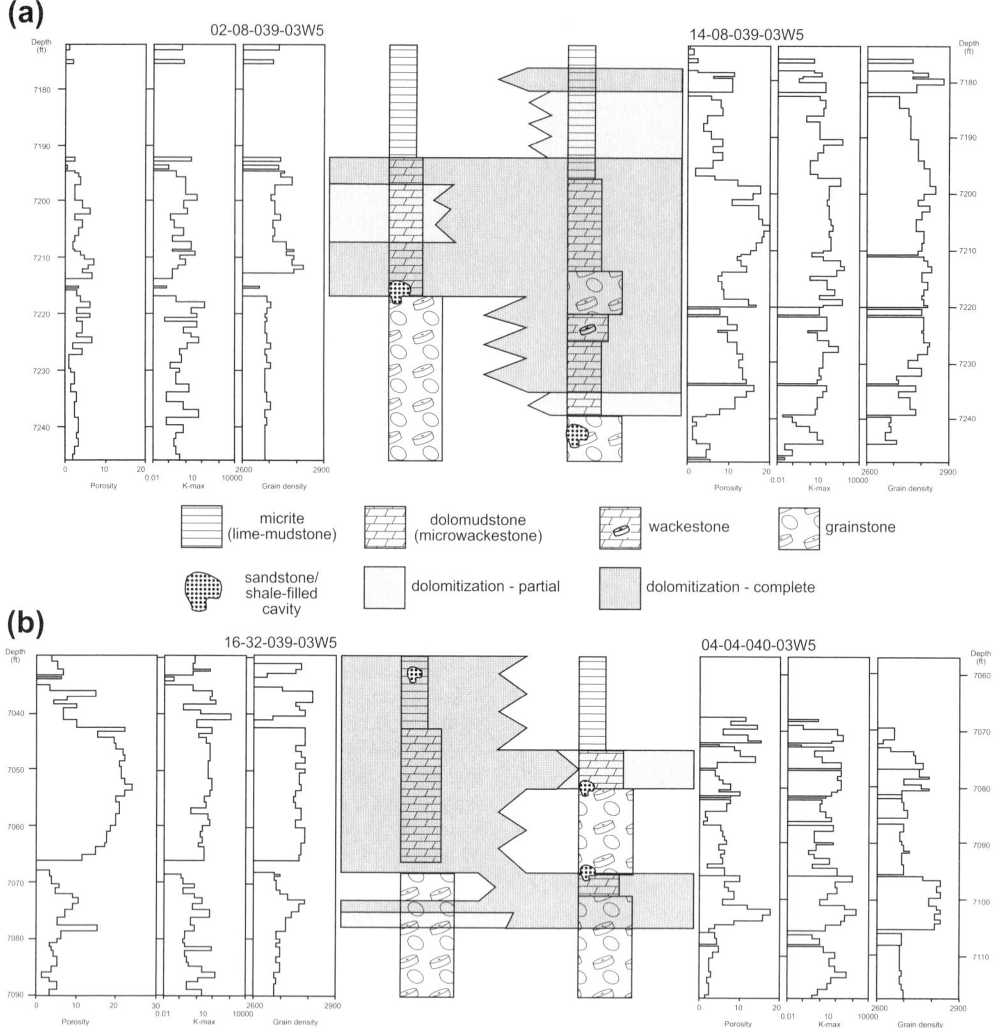

Fig. 11. Core logs and petrophysical data for cored dolomudstone intervals in closely spaced wells. Permeability values (K_{max}) are maximum horizontal permeability measured in millidarcies. Grain density values are in kg m^{-3}. Grain density for calcite is 2710 kg m^{-3}; grain density for dolomite is 2840 kg m^{-3}. Gaps in the petrophysical data are intervals that were not analysed. (**a**) Wells 02-08-039-03W5 and 14-08-039-03W5 from Medicine River Pekisko E Pool. (**b**) Wells 16-32-039-03W5 and 04-04-040-03W5 from Medicine River Pekisko B Pool.

Pekisko Formation at the same stratigraphic position (determined with respect to the base and top of the formation) as that to the south in the Medicine River area. From PEF and porosity logs (Fig. 9C–C′) it is evident that the lower part of the dolostone body is porous and the upper part generally tight. A core from well 16-27-041-05W5 shows that porous dolomudstones sharply overlie grainstones and grade up through low-porosity calcareous dolomudstones into tight fenestral lime mudstones.

Dolomudstone is also present at a lower stratigraphic horizon in wells 16-26-041-05W5 and 14-25-041-05W5. Although no cores have been cut, a dolomudstone body can be recognized from PEF logs and the slightly higher radiation on gamma ray logs; and its presence is confirmed from cuttings. It has a limited lateral distribution and apparently grades laterally into grainstones of the lower part of the Pekisko Formation.

Gilby Rundle P Pool (Fig. 9D–D′)

The Gilby Rundle P Pool (the name stems from the Rundle Group that contains the Pekisko Formation) is developed within dolomudstones in an isolated erosional outlier of the Pekisko Formation (Fig. 8). This contains the most eastern occurrences of dolomudstone along the subcrop edge in the study area. Two intervals of dolomudstone are present in the Pekisko Formation. Because the area is gas bearing, it is more difficult to interpret the dolostone lithology from the neutron-density logs. The characteristic high apparent neutron porosity for dolostones on the limestone scale is masked by low neutron absorption over porous gas-bearing intervals. However, most of the wells have been cored and the dolomudstones are continuous for several kilometres or more. It is not possible to establish their exact position within the Pekisko Formation, as the upper part of the formation has been removed by erosion. From their position with respect to the top of the Banff Formation they are roughly equivalent to the middle of the Pekisko Formation in the other pools. The lower dolomudstone unit occurs within the lower half of the Pekisko Formation, more or less in the same position as the lower dolomudstone in Willisden Green (Fig. 9C–C′) and grainstones to the south at Medicine River (Fig. 9A–A′).

The following generalizations are possible from cross-sections, logs and cores throughout the study area:

(1) Dolomudstones occur throughout the Pekisko Formation. However, in the lower part of the formation they are absent to the south, and are discontinuous to the NW where they are enclosed within grainstones. To the NE they are more numerous and more continuous. They are also more widespread and continuous in the upper part of the formation.
(2) Porous dolomudstone bodies are generally 5–8 m thick and 1 km or so wide. Judging from the two-dimensional (2-D) well data in cross-sections and 3-D interpretation of porosity in the Medicine River Pekisko E Pool, they are also very irregular in shape.
(3) The dolomudstone facies is the locus of porosity, permeability and dolomitization.

Depositional environments and sequence stratigraphy

The dolomudstones described are not part of a tidal-flat facies association as they lack associated stromatolites, cryptalgal laminites, desiccation structures and moulds of evaporite minerals. The presence of bioturbation, current-laminated structures, and interbedded fossiliferous wackestones and grainstones suggests subtidal deposition in seawater of normal or near-normal salinity. The texture of the microwackestones revealed in chert nodules hints that these sediments formed largely by mechanical and biological comminution of skeletal material. Bioturbated microwackestones with thin-shelled fossils and comminuted skeletal fragments are interpreted to have formed in a relatively low-energy subtidal marine environment. In this respect they are similar to subtidal dolomudstones in other Lower Carboniferous formations (cf. Choquette & Steinen 1980; Wegelin 1987).

Dolomudstone bodies within grainstone successions imply accumulation in low-energy environments between grainstone shoals. In the upper part of the formation they were also deposited in low-energy environments but are more widely distributed. They are interpreted to have formed in shallow protected lagoons between the wave-agitated grainstone belt on the outer margin of the inner ramp and the peritidal belt of the restricted inner margin.

The Pekisko Formation is a depositional sequence that in the study area comprises a grainstone–dolomudstone–fenestral lime mudstone succession deposited on the inner ramp (Fig. 12). This implies that the inner part of the Pekisko ramp included a peritidal belt that has been eroded below the sub-Cretaceous unconformity. Parasequences within the depositional sequence (in the broadest sense of van Waggoner *et al.* 1990) are comprised of grainstone–dolomudstone successions in which the base of the grainstone lies on a scoured surface, the dolomudstones accumulated in more restricted environments. Grainstones contain the transgressive part of each parasequence; dolomudstones represent the regressive part of the parasequences. Thicker grainstone successions in the Pekisko Formation, especially those that lie to the south and west (seaward direction, Figs 2 and 12), may also contain parasequences, but the regressive portion is either represented by grainstones or has been eroded beneath the scoured transgressive surface of the overlying parasequence. Similar parasequences have been described from Lower Carboniferous sequences in the Illinois Basin by Smith & Read (1999). Here, the grainstones are tidal deposits formed under high-energy conditions imposed by increased accommodation through eustatic sea-level rise; carbonate mudstones reflect

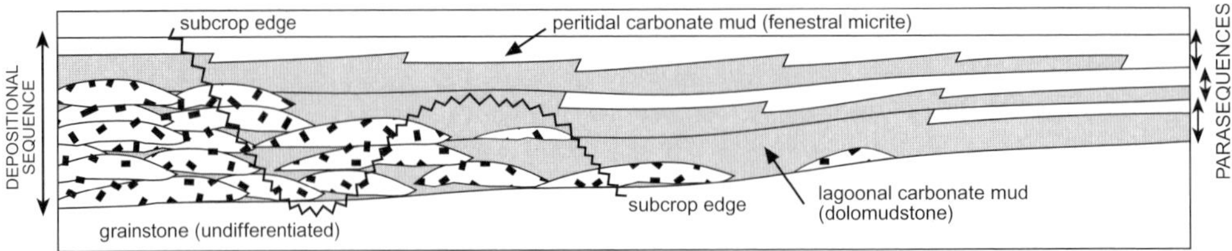

Fig. 12. Schematic sequence stratigraphic interpretation for the Pekisko depositional sequence in the study area. Projection of peritidal sediments and parasequences to the east (right-hand side of diagram) beyond the subcrop edge is hypothetical. Not to scale.

decreased tidal energy as sedimentation filled accommodation.

The lateral continuity of grainstone–dolomudstone parasequences in closely spaced wells in the study area (Fig. 9) is of the order of a few kilometres at most. In the lower part of the Pekisko Formation, some dolomudstone bodies are contained within grainstones. Even in the upper part, where the dolomudstone facies formed during the regressive phase of the sequence, it locally interfingers with grainstones (Figs 9 and 11). Intrinsic controls imposed by the vagaries of the depositional system appear to have had an important role in determining the facies distribution.

The Pekisko Formation fits well with the general model for Lower Carboniferous ramp sedimentation in western Canada proposed by Martindale & Boreen (1997). Dolomudstone was deposited between grainstone shoals and in the adjacent subtidal lagoon, seaward of the peritidal belt of carbonate mudstones and on the landward side of the broad grainstone belt that comprised the outer part of the inner ramp (Fig. 2).

Dolomitization

Porous dolomudstones of the Pekisko Formation lie beneath lime mudstones or are contained within grainstones that are not dolomitized, and are either tight or have very little porosity (Fig. 9). Stable isotopes suggest precipitation of the bulk of the dolomite from Early Carboniferous seawater or modified seawater. These aspects constrain both the timing and access of dolomitizing fluids.

There is no evidence that dolomudstone bodies were subaerially exposed at parasequence tops and, thus, altered by evaporated seawater or freshwater. The locus of dolomitization was the dolomudstone facies and the containing facies are dolomitized only where in close contact. Dolomitizing fluids were directed into the microwackestones and there effected dolomitization. It is possible that these had a mineralogy (aragonite or Mg calcite skeletal fragments) that was particularly susceptible to dolomitization. This may have aided dolomitization, but fluid access (permeability) would still have been the paramount control.

Permeability has been invoked as a controlling mechanism for the dolomitization of lime mudstones by Murray & Lucia (1967) who cited evidence for early cementation of Carboniferous grainstones in western Canada. On the basis of grain size and texture, it is probable that crinoidal carbonate sands were more permeable at the time of deposition than microwackestones that were, in turn, more permeable than fenestral lime mudstones. Substantial early cementation of grainstones and neomorphism of lime mudstones could have reversed this permeability order so that the microwackestones became by default the relatively more permeable lithology. However, early cementation of grainstones and neomorphism of lime mudstones would also have had the effect of isolating the dolomudstone bodies from circulating fluids.

The importance of fluid circulation through grainstones to dolomitize adjacent wackestones in the Lower Carboniferous Ste. Genevieve Formation has been emphasized by Choquette & Steinen (1980). In these, freshwater flowed through porous grainstones, mixing with connate seawater in the underlying lower permeability wackestones and effected dolomitization. The Ste. Genevieve dolostones are very similar to the dolomudstones of the Pekisko Formation with 5–20 μm euhedral crystals and up to 40% porosity. A similar mechanism applied to the Pekisko Formation may explain the marine isotopic composition of the dolomudstones. However, freshwater flow through grainstones adjacent to or beneath the dolomudstones would have to be invoked in many pools because the dolomudstones are overlain by lime mudstones. Presumably the lime mudstones would escape dolomitization because of their low permeability. The principle objection to this hypothesis is the difficulty of establishing a mechanism of freshwater recharge into the grainstones.

Local dolomitization of lime mudstones in the upper part of the Pekisko Formation and restricted dolomitization of facies associated with the dolomudstones suggest that the dolomitizing fluids may have descended from overlying formations. The most likely source of the fluids is the overlying peritidal sediments of the Shunda Formation. Dolomudstone reservoirs of limited dimensions (pods of dolomite 1 km or so wide and tens of metres thick) have been described from peritidal carbonates of the Ordovician Red River Formation by Longman *et al.* (1983). The unusual aspect of the Red River dolomudstones is that they occur only locally within laterally extensive sheets of lime mudstone capped by evaporites. Longman *et al.* (1983) attributed localization to restricted access of dolomitizing fluids from overlying sabkhas downwards through conduits produced by fracturing of the evaporites, or possibly associated with the escape of compaction waters from below. Kendall (1984) acknowledged the localized geometry of the reservoirs but invoked groundwater flow upwards through giant polygonal cracks developed in the overlying evaporites. Similar access of dolomitizing waters to the dolomudstones from overlying formations following early shallow burial of the Pekisko Formation could explain both the localization of dolomitization in the dolomudstones and their isotopic signatures.

Finally, there is the possibility that dolomitization in some way relates to a much later time along the Pekisko subcrop. Dolomudstones are not confined to the subcrop, however. They were deposited over a much wider area, and their isotopic composition indicates derivation from Early Carboniferous seawater (or modified seawater). Petrographic evidence points to dissolution of calcite along the subcrop edge, not to the precipitation of dolomite.

In the light of these observations, the dolomudstones of the Pekisko Formation are interpreted to have formed through facies-selective dolomitization of a lime mudstone facies comprised of microwackestone. Dolomitization may have occurred early, perhaps by the mixing of marine and meteoric freshwaters that flowed laterally though grainstones. Alternatively, fluids descended through fractures in the relatively impermeable lime mudstones above and spread laterally into the relatively more permeable (and possibly more reactive) microwackestones. Later exposure of the dolomudstones at unconformities caused dissolution of calcite and minor neomorphism of the dolomites, and partial isotopic equilibration with meteoric waters.

Conclusions

The steps in making a dolomudstone reservoir are presented diagrammatically in Figure 13 and include deposition of a microwackestone facies, early facies-selective dolomitization of this and adjacent lithologies, truncation and leaching of the calcitic residue. Each step has contributed to increasing the geometric complexity of the dolomudstone reservoirs.

(1) Dolomudstone reservoirs represent facies-selective dolomitization of microwackestone, a sediment comprised of silt-size skeletal material and lime mud. This accumulated in a broad shallow protected lagoon that lay between an open marine grainstone belt and peritidal facies belt on the inner part of a ramp.
(2) Dolomitization of the microwackestone and some closely associated facies occurred where dolomitizing solutions descended from overlying low-permeability carbonate mudstones through selective pathways.
(3) The Pekisko Formation was exposed at various times from the Late Carboniferous to Early Jurassic, and was deeply incised at times during the Jurassic–Early Cretaceous. Meteoric water entered porous parts of the formation, particularly the dolomudstones, and further enhanced porosity by dissolution. Neomorphism of dolomite and calcite to partial equilibration with isotopically depleted meteoric waters caused a shift towards more depleted values for the oxygen isotope and a lesser shift for the carbon isotope. Calcites were more affected that dolomite due to their greater solubility.
(4) Exposure and leaching at Late Carboniferous–Early Cretaceous unconformities formed dissolution cavities within both limestones and porous dolomudstones. Infiltration of siliciclastic sandstones and shales into these cavities, and associated local collapse of caves in dolomudstones, added additional complexities to these already geometrically complicated reservoirs.

1. Deposition of microwackestone facies within a shoaling grainstone to lime-mudstone succession

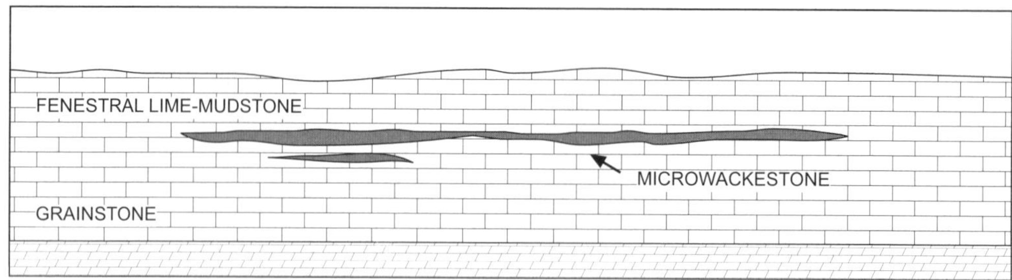

2. Burial, early facies-selective dolomitization of microwackestone and some adjacent limestones

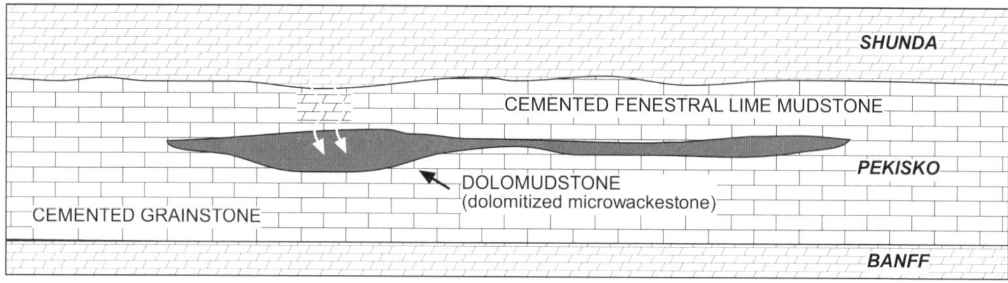

3. Erosion, exposure, formation of aquifer, leaching of calcite from dolomudstone

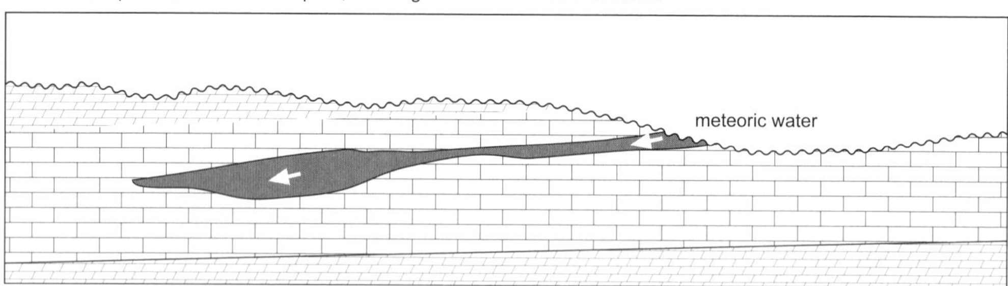

4. Cave formation, sandstone and shale deposition, local collapse of caves

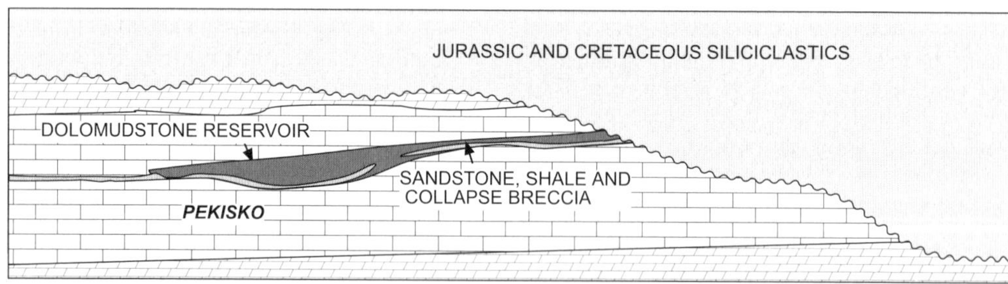

Fig. 13. Interpreted stages of formation of a dolomudstone reservoir in the Pekisko Formation.

References

AL-AASM, I.S. & LU, F. 1994. Multistage dolomitization of the Mississippian Turner Valley Formation, Quirk Field, Alberta: chemical and petrologic evidence. *In*: EMBRY, A.F., BEAUCHAMP, B. & GLASS, D.J. (eds) *Pangea: Global Environments and Resources*. Canadian Society of Petroleum Geologists, Memoirs, **17**, 657–675.

AL-AASM, I. & PACKARD, J. 2000. Stabilization of early-formed dolomite: a tale of divergence from two Mississippian dolomites. *Sedimentary Geology*, **131**, 97–108.

BURCHETTE, T.P. & WRIGHT, V.P. 1992. Carbonate ramp depositional systems. *Sedimentary Geology*, **79**, 3–57.

CHOQUETTE, P.W. & STEINEN, R.P. 1980. Mississippian non-supratidal dolomite, Ste. Genevieve Limestone, Illinois Basin; evidence for mixed-water dolomitization. *In*: ZENGER, D.H., DUNHAM, J.B. & ETHINGTON, R.L. (eds) *Concepts and Models of Dolomitization*. Society of Economic Paleontologists and Mineralogists, Special Publications, **28**, 163–196.

CHOQUETTE, P.W. & STEINEN, R.P. 1985. Mississippian oolite and non-supratidal dolomite reservoirs in the Ste. Genevieve Formation, North Bridgeport Field, Illinois Basin. *In*: ROEHL, P.O. & CHOQUETTE, P.W. (eds) *Carbonate Petroleum Reservoirs*. Springer, New York, 207–225.

DUROCHER, S. & AL-AASM, I.S. 1997. Dolomitization and neomorphism of Mississippian (Visean) Upper Debolt Formation, Blueberry Field, northeastern British Columbia: geologic, petrologic, and chemical evidence. *AAPG Bulletin*, **81**, 954–977.

HOPKINS, J.C. 1999. Characterization of reservoir lithologies within subunconformity pools: Pekisko Formation, Medicine River Field, Alberta, Canada. *AAPG Bulletin*, **83**, 1855–1870.

HOPKINS, J.C., CUPIDO, P. & HANDCOCK, P. 1998. Reservoir development in a marine valley-fill complex: Medicine River Jurassic 'D' Pool. *In*: HOGG, J.R. (ed.) *Oil and Gas Pools of the Western Canada Sedimentary Basin*. Canadian Society of Petroleum Geologists, Special Publications, **S-51**, 39–49.

ILLING, L.V. 1959. Deposition and diagenesis of some upper Palaeozoic carbonate sediments in western Canada. *In*: *Proceedings of the World Petroleum Congress, New York 1959*. Wiley, Chichester, 391–409.

JAMES, N.P., BONE, Y. & KSYER, T.K. 1993. Shallow burial dolomitization and dedolomitization of Mid-Cenozoic, cool-water, calcitic, deep-shelf limestones, southern Australia. *Journal of Sedimentary Petrology*, **63**, 528–538.

JOHNSON, R.A. 1994. Distribution and architecture of subunconformity carbonate reservoirs; lower Meramecian (Mississippian) subcrop trend, western Kansas. *In*: DOLSON, J.C., HENDRICKS, M.L. & WESCOTT, W.A. (eds) *Unconformity-related Hydrocarbons in Sedimentary Sequences; Guidebook For Petroleum Exploration and Exploitation in Clastic and Carbonate Sediments*. Rocky Mountain Association of Geologists, Denver, CO, 231–244.

KENDALL, A.C. 1984. Origin and geometry of Red River dolomite reservoirs, western Williston Basin: Discussion. *AAPG Bulletin*, **68**, 776–779.

KENT, D.M. 1994. Paleogeographic evolution of the cratonic platform – Cambrian to Triassic. *In*: MOSSOP, G.D. & SHETSEN, I. (eds) *Geological Atlas of the Western Canada Sedimentary Basin*. Canadian Society of Petroleum Geologists and Alberta Research Council, Calgary, 69–85.

LONGMAN, M.W., FERTAL, T.G. & GLENNIE, J.S. 1983. Origin and geometry of Red River dolomite reservoirs, western Williston Basin. *AAPG Bulletin*, **67**, 744–771.

MARTINDALE, W. & BOREEN, T.D. 1997. Temperature-stratified Mississippian carbonates as hydrocarbon reservoirs – examples from the foothills of the Canadian Rockies. *In*: JAMES, N.P. & CLARKE, J.A.D. (eds) *Cool-water Carbonates*. Society of Economic Paleontologists and Mineralogists, Special Publications, **56**, 391–409.

MII, H.-S., GROSSMAN, E.L. & YANCEY, T.E. 1999. Carboniferous isotope stratigraphies of North America; implications for Carboniferous paleoceanography and Mississippian glaciation. *Geological Society of America Bulletin*, **111**, 960–973.

MONTANEZ, I.P. & READ, J.F. 1992. Eustatic control on dolomitization of cyclic peritidal carbonates: Evidence from Early Ordovician Knox Group, Appalachians. *Geological Society of American Bulletin*, **104**, 872–886.

MURRAY, R.C. & LUCIA, F.J. 1967. Cause and control of dolomite distribution by rock selectivity. *Geological Society of America Bulletin*, **78**, 21–35.

PODRUSKI, J.A., BARCLAY, J.E. *ET AL.* 1988. *Conventional Oil Resources of Western Canada*. Geological Survey of Canada, Paper, **87–26**.

RICHARDS, B.C. 1989. Upper Kaskaskia Sequence: Uppermost Devonian and Lower Carboniferous. *In*: RICKETTS, B.D. (ed.) *Western Canada Sedimentary Basin, A Case History*. Canadian Society of Petroleum Geologists, Calgary, 165–201.

RICHARDS, B.C., BARCLAY, J.E., BRYAN, D., HARTLING, A., HENDERSON, C.M. & HINDS, R.C. 1994. Carboniferous strata of the western Canada sedimentary basin. *In*: MOSSOP, G.D. & SHETSEN, I. (eds) *Geological Atlas of the Western Canada Sedimentary Basin*. Canadian Society of Petroleum Geologists and Alberta Research Council, Calgary, 221–250.

ROSS, C.A. & ROSS, J.R.P. 1988. Late Paleozoic transgressive-regressive deposition. *In*: WILGUS, C.K., HASTINGS, B. S., ROSS, C.A., POSAMENTIER, H., VAN WAGGONER, W J. & KENDALL, C.G.S.C. (eds) *Sea-level Changes; An Integrated Approach*. Society of Economic Paleontologists and Mineralogists, Special Publications, **42**, 227–247.

SANDO, W.J., BAMBER, E.W. & RICHARDS, B.C. 1990. The rugose coral *Ankhelasma*; index to Visean (Lower Carboniferous) shelf margin in the Western Interior of North America. *US Geological Survey Bulletin*, **1895**, B1–B29.

SIBLEY, D.F. & GREGG, J.M. 1987. Classification of dolomite rock textures. *Journal of Sedimentary Petrology*, **57**, 967–975.

SMITH, L.B., JR. & READ, J.F. 1999. Application of high-resolution sequence stratigraphy to tidally influenced Upper Mississippian carbonates, Illinois Basin. *In*: HARRIS, P.M., SALLER, A.H. & SIMO, J.A. (eds) *Advances in Carbonate Sequence*

Stratigraphy; Application to Reservoirs, Outcrops and Models. Society of Economic Paleontologists and Mineralogists, Special Publications, **63**, 107–126.

SMITH, T.M. & DOROBEK, S.L. 1993. Alteration of early-formed dolomite during shallow to deep burial: Mississippian Mission Canyon Formation, central to southwestern Montana. *Geological Society of America Bulletin*, **105**, 1389–1399.

STOAKES, F.A. 1999. Sedimentary response to Early Carboniferous tectonics: Paleogeographic fluctuations in the Pekisko, Shunda and Turner Valley Formations of the west-central plains and foothills. *In*: *Program with Abstracts: XIV International Congress on the Carboniferous and Permian*. Canadian Society of Petroleum Geologists, Calgary, 142–143.

VAN WAGGONER, J.C., MITCHUM, R.M., CAMPION, K.M. & RAHMANIAN, V.D. 1990. *Siliciclastic Sequence Stratigraphy in Well Logs, Cores, and Outcrops; Concepts for High-resolution Correlation of Time and Facies*. American Association of Petroleum Geologists, Methods in Exploration Series, **7**.

WEGELIN, A. 1987. Reservoir characteristics of the Weyburn Field, southeastern Saskatchewan. *Journal of Canadian Petroleum Technology*, **26**, 60–66.

Early dolomitization and fluid migration through the Lower Carboniferous carbonate platform in the SE Irish Midlands: implications for reservoir attributes

ZS. R. NAGY[1], J. M. GREGG[2], K. L. SHELTON[3], S. P. BECKER[2], I. D. SOMERVILLE[4] & A. W. JOHNSON[3]

[1]*Schlumberger Data and Consulting Services, 1325 S. Dairy Ashford Road, Houston, TX 77077, USA (e-mail: znagy@houston.oilfield.slb.com)*

[2]*Department of Geology & Geophysics, University of Missouri-Rolla, Rolla, MO 65401, USA*

[3]*Department of Geological Sciences, University of Missouri-Columbia, Columbia, MO 65211, USA*

[4]*Department of Geology, University College Dublin, Belfield, Dublin 4, Ireland*

Abstract: Shallow-marine, Lower Carboniferous carbonate sequences of the SE Irish Midlands, close to the Leinster Massif, are intensely dolomitized. Fine-crystalline (<50 μm), planar-s (subhedral) dolomite is associated with evidence for evaporites, typical of arid peritidal sequences. However, stable isotope data suggest a diagenetic overprint. Volumetrically more important medium-crystalline (50–200 μm), planar-s and minor planar-e (euhedral) dolomites were precipitated from slightly modified Lower Carboniferous seawater. These dolomites replace open-marine intraclastic and bioclastic packstones and grainstones.

Length-slow, fibrous quartz partially replaces crinoids and fills dissolution cavities beneath peritidal strata. Associated dolomites are gradually enriched in ^{18}O downward through the underlying strata, suggesting vertical brine migration. The widespread occurrence of skeletal material replaced by chalcedony in open-marine wackestones and grainstones further to the west, within the Rathdowney Trend, suggests evaporite cementation in the Zn–Pb mineralized area.

Base-metal mineralization in the fractured Waulsortian 'reservoir' is associated with chloride-enriched brines (beyond that expected from seawater evaporation alone). The presence of evaporites in the Leinster Massif area suggests a possible source of the excess chloride. The dolomitizing brine may have contributed to the overall chemistry of the Zn–Pb mineralizing fluid and also to the distribution of porosity within the carbonate platform. An Arundian or younger age is suggested for the mineralization, based on the timing of evaporite cement emplacement, and this is compatible with numerical fluid-flow models of brine movement through the carbonate platform.

Dolomitization of Lower Carboniferous carbonate rocks of the Irish Midlands is comparable to diagenetic histories of several important dolomite petroleum reservoirs. This study provides an example that may be applied to petroleum exploration in similar geological settings.

Lower Carboniferous carbonate rocks of the Irish Midlands have undergone multiple dolomitization events (Gregg *et al.* 2001; Wright 2001 and references therein). Early diagenetic dolomitization largely governed the distribution of later nonplanar dolomite cements by enhancing the intercrystal porosity exploited by later fluids (Wright 2001). Nonplanar dolomite cements, commonly precipitating in vugs, fracture and breccia porosity, are associated with economically important metal sulphide (Zn, Pb, Ba and Ag) deposits in the Rathdowney Trend (Hitzman 1995*a*, *b*; Hitzman & Beaty 1996). Studies focusing on the origin of the dolomite have concentrated on the Waulsortian Limestone for two reasons: (1) most of the Zn–Pb ore deposits in the Irish Midlands are hosted by the Waulsortian Limestone or its equivalent; and (2) the Zn–Pb mineralization has been proposed to be syn-depositional with the Waulsortian Limestone (late Courceyan, *c.* 350 Ma; Wilkinson *et al.* 2003). Sevastopulo & Redmond (1999) and Peace & Wallace (2000), however, suggested a

From: BRAITHWAITE, C. J. R., RIZZI, G. & DARKE, G. (eds) 2004. *The Geometry and Petrogenesis of Dolomite Hydrocarbon Reservoirs*. Geological Society, London, Special Publications, **235**, 367–392. 0305-8719/$15.00

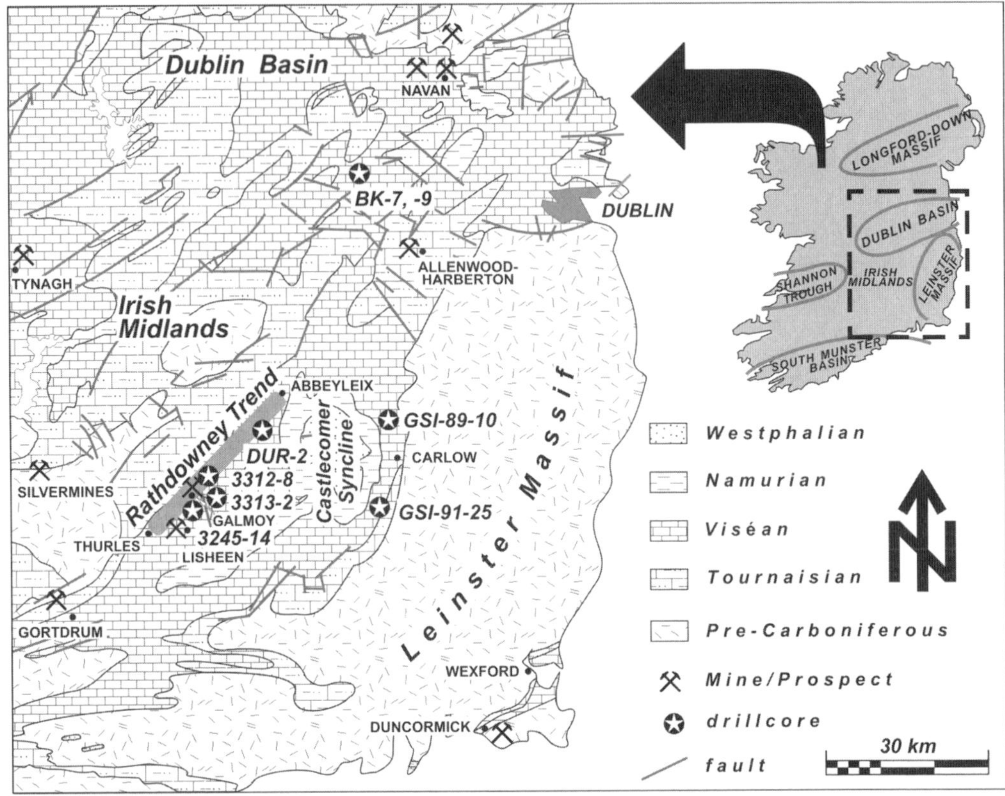

Fig. 1. Simplified geological map of SE Ireland showing the location of cores sampled for this study.

later, post-Arundian (*c.* 340 Ma) event. If the latter is correct, fluids causing Zn–Pb mineralization were probably in contact with lower–upper Viséan rocks in the SE Irish Midlands, proximal to the Leinster Massif. Consequently, the Chadian–Brigantian strata may bear a more significant geological importance to Zn–Pb mineralization than previously suspected. This study focuses on: (1) petrographic description and spatial distribution of early diagenetic dolomites in order to identify different diagenetic environments; and (2) characterization of the dolomitizing fluid(s) to establish the possible relationship between early diagenetic dolomitization and base-metal mineralization.

Stratigraphic framework

The areal distribution of Carboniferous sedimentary rocks in SE Ireland is presented in Figure 1, and a schematic cross-section of the Lower Carboniferous (Dinantian) stratigraphy of the Irish Midlands is shown in Figure 2. Exposures of Viséan rocks are scattered sparsely within the Irish Midlands. Mining exploration cores, however, have penetrated most of the Lower Carboniferous succession (Fig. 1).

Courceyan rocks of the study area conformably overlie Old Red Sandstone facies and consist of mixed siliciclastic–carbonate sedimentary rocks (Lower Limestone Group and Ballysteen Formation in the southern Midlands; Navan Group and Malahide Limestone Formation in the Dublin Basin) formed on gently dipping broad ramps or platforms (Philcox 1984; Strogen *et al.* 1996). These units are time-transgressive northward (Hitzman 1995*a*) and are overlain by the Waulsortian Limestone throughout the Irish Midlands (Fig. 2). The latter limestone was deposited as mud mounds and is characterized by facies associations such as core, flank and cover facies (Lees 1964; Hitzman 1995*a*; Somerville 2003).

Sedimentation of the Chadian–Brigantian sequence occurred with strong structural control over facies development. In the Dublin Basin, bioturbated mudstones and limestone turbidites were deposited (the Tober Colleen Mudstone and Lucan 'Calp' Limestone

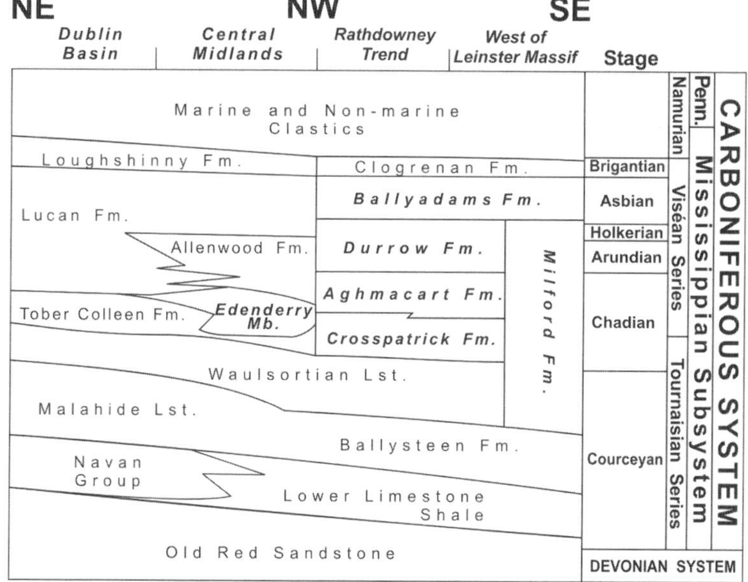

Fig. 2. Lithostratigraphic chart showing the Lower Carboniferous strata and facies relationship of southeast Ireland, after Hitzman (1995*a*) and Wright (2001). Formations involved in this study are marked with bold italics. Fm., Formation; Mb., Member; Lst., Limestone; Penn., Pennyslvanian.

formations; Strogen *et al.* 1996). Platform conditions were established, during the Chadian, around the margins of the Dublin Basin and in parts of the central and southern Midlands west of the Leinster Massif (Fig. 1). In the Leinster Massif region these rocks consist of argillaceous, cherty, bryozoan-crinoidal wackestone/packstones with horizons rich in sponge spicules (Crosspatrick Formation). Very shallow-marine oolitic shoals (Allenwood Formation, Edenderry Oolite Member) formed on NE–SW-trending blocks (McConnell *et al.* 1994; Hitzman 1995*a*). Rocks overlying the Crosspatrick Formation are typically fine-grained peloidal, oncoidal or micritic limestones with a restricted fauna of gastropods and ostracodes that probably reflects a very shallow-water lagoonal environment (Aghmacart Formation). In the upper part of the formation, fenestral fabrics, rhizoliths and desiccation cracks indicate periodic subaerial exposure. Oolitic or peloidal cross-bedded units, representing channel or barrier grainstones, occur interbedded with the finer-grained lithologies and cyclic shales (Nagy *et al.* 2001, 2005; Gatley *et al.* 2004). The Durrow Formation, marking the first appearance of Arundian foraminiferal assemblages (Tietzsch-Tyler *et al.* 1994; Gatley *et al.* 2004), is a more open-marine, but intermittently shaly, fossiliferous limestone. Its strata consist of an alternation of coral-rich grainstones, cross-bedded oolites, fossiliferous shales and rare birds-eye micrites of Arundian–Holkerian age. The Aghmacart and Durrow formations are termed the Milford Formation in County Carlow, SE Ireland (Fig. 2) (Gatley *et al.* 2004). The overlying units (Ballyadams and Clogrenan formations) consist of alternating pale-grey and dark-grey, thick-bedded, colonial-coral bearing (cerioid *Lithostrotion* and fasciculate *Siphonodendron*), bioclastic shelf limestones containing palaeokarsts of Asbian–Brigantian age (Cózar & Somerville 2005). The younger units are notably more cherty. The Upper Carboniferous (Namurian–Westphalian Series) consists mainly of regressive terrigenous rocks deposited in response to progressive glaciation coupled with regional uplift associated with the Hercynian (Variscan) orogeny. The Hercynian orogeny resulted in complex thrusting and folding in the southern Midlands with locally inverted stratigraphic successions (cf. Corfield *et al.* 1996; Hitzman 1999).

Methods

A total of 996 samples of Chadian–Brigantian-aged limestones and dolomites were examined in this study. Samples were collected from drill-core: Dur-2 (Durrow), 267 samples; 3312-8, 59 samples; 3313-2, 15 samples; 3245-14, 18 samples; GSI-91-25 (Milford), 126 samples;

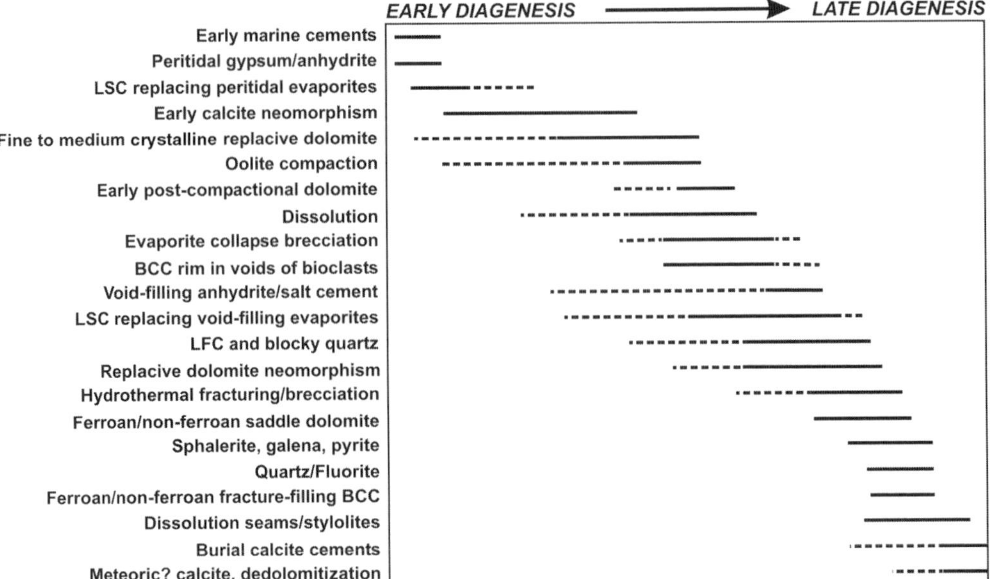

Fig. 3. Summary diagram of diagenetic sequence of the studied formations in the Irish Midlands. LSC, length-slow chalcedony; LFC, length-fast chalcedony; BCC, blocky calcite cement.

GSI-89-10 (Tankardstown), 76 samples; BK-7 and BK-9 (Ballykane), 91 samples; and from surface outcrops (344 samples; Fig. 1). Petrographic analysis was conducted using hand specimens and thin sections stained with alizarin-red S and potassium ferricyanide (Dickson 1966). Cathodoluminescence (CL) properties of selected polished thin sections were examined using a Technosyn 8200 MkII apparatus operated at an accelerating voltage of approximately 12 kV with beam currents of between 150 and 250 μA, and a vacuum of 0.05 torr.

Powdered carbonate samples of 0.2–0.5 mg were drilled from rock chips, from which thin sections had been prepared, for carbon and oxygen isotope analysis at the University of Missouri Stable Isotope Laboratory. The 26 calcite samples from brachiopod and crinoid shells, and nine samples of early micrite geopetal sediments and blocky calcites, were checked carefully in standard thin sections stained with alizarin-red S and were then analysed by CL microscopy for any evidence of recrystallization and/or dolomitization. Dolomite samples analysed include 29 from Dur-2, 45 from GSI-91-25, 10 from the Galmoy and Lisheen area, and five from the BK-9 drillcore. Samples were reacted at 25 °C with 103% phosphoric acid in an automated Kiel device and were analysed with a Finnigan MAT Delta Plus gas-source mass spectrometer. The $\delta^{18}O$ values of dolomites and calcites were corrected for differences in acid fractionation of oxygen isotopes (McCrea 1950, Friedman & O'Neil 1977). Data are reported in standard δ notation relative to PDB for both C and O. The standard error of each analysis is less than ±0.08‰.

Sr isotope analysis on 12 calcite and dolomite samples was performed at the University of Kansas Isotope Geochemistry Laboratory. Samples were digested in 500 μl of 3.5N HNO_3, and 10 ml of KU #11 ^{84}Sr spike was added. Strontium was separated using standard cation-exchange columns filled with a strontium-specific resin. The Sr was then concentrated onto a drop of H_3PO_4, placed onto rhenium single filaments and analysed using thermal ionization mass spectrometry with a VG Sector mass spectrometer. The reproducibility of the Sr standard (NBS 987) is $^{87}Sr/^{86}Sr = 0.710235 \pm 0.000025$ (2σ, $n = 33$) and the fractionation was corrected to $^{87}Sr/^{86}Sr = 0.710250$.

Petrography and spatial distribution

Limestone diagenesis

Limestone diagenetic events were identified to help constrain the relative timing of dolomitization within the paragenetic sequence (Fig. 3). Evaporite precipitation and diagenesis seem to have had important roles in Zn–Pb mineralization in the Irish Midlands that will be described here and addressed in the discussion.

Early diagenesis of the peritidal sequence adjacent to the Leinster Massif (Aghmacart Formation) was characterized by anhydrite–gypsum precipitation in an arid climate (Nagy *et al.* 2005). Evaporite mineral growth was displacive and replaced shortly after formation by length-slow fibrous chalcedony (quartzine) and megaquartz spherules (Fig. 4A & B). Coarse-crystalline, blocky quartz displaying sweeping extinction and containing abundant anhydrite inclusions is also common, forming pseudomorphs after anhydrite. These types of replacements are well known from the literature (Folk & Pitman 1971; Siedlecka 1972, fig. 4; Chowns & Elkins 1974; Milliken 1979, fig. 9b; Elorza & Rodríquez-Lazaro 1984, fig. 8; Swennen & Viaene 1986; Swennen *et al.* 1990) and are interpreted as reflecting early diagenetic alteration (Siedlecka 1976; Tucker 1976).

Gypsum–anhydrite precipitation and replacement by quartzine were found to predate the formation of both fine- and medium-crystalline planar dolomites by observing corrosion of quartz by carbonate (Fig. 4B). Quartzine and coarse blocky quartz cements were also commonly observed in dissolution voids and intercrystal spaces (Fig. 4C), and as partial void fillings in crinoids in the Crosspatrick Formation in the GSI-91–25 core. Here, the quartz phase replaces anhydrite cement, indicated by small relict anhydrite inclusions in the quartz. Both anhydrite and quartz were found to post-date fine- and medium-crystalline planar dolomites in this part of the section. This interpretation is based on the observation that abundant dolomite crystals were found in the quartz (Fig. 4C) as well as euhedral growth of the host dolomite inward from the edge of the cavity (planar-c type after Gregg & Sibley 1984).

At several horizons, large (3–5 cm) spherical cavities were observed with stylolitic argillaceous halos around them containing abundant fine-grained (*c.* 50 μm) angular quartz with diffuse or corroded outer crystal faces (Nagy *et al.* 2005, fig. 4c). The cavities, from edge to centre, contain 'rip-up' clasts of the halo, coarse-crystalline (100 μm–2 mm) euhedral blocky quartz with sweeping extinction, and coarse-crystalline (100 μm–3 mm) saddle dolomite. Geopetal features are common where 'rip-ups' of the halo and euhedral quartz characterize the bottom half of the cavity and the upper half is cemented by saddle dolomite. Similar cavities were described from the Lower Viséan of the Vesdre Formation, Belgium by Swennen & Viaene (1986, plate 2H) and were interpreted as replacements of 'cauliflower'-shape anhydrite nodules. Saddle dolomite cavity-fills in the Irish Midlands indicate, however, that the process was a result of sulphate dissolution and subsequent open-space filling.

Early diagenesis of shallow, open-marine carbonate sediments probably started at the sea floor, as indicated by abundant micritic envelopes, micro- and macro-borings, various crusts and geopetal fills (Nagy 2003). Rarely, vague isopachous cement is observed fringing ooids in grainstones. Oolitic sediments underwent early compaction, shown by stylolites and dissolution seams associated with dolomite (see the subsection on 'Early post-compactional replacement dolomite' later). The predominant cement type in grainstones is equant blocky calcite displaying unzoned or multiple dull or bright orange and non-luminescent CL zones, and occluding the remaining porosity.

The skeletal grains (crinoids, brachiopods and corals) in open-marine limestones observed in the Dur-2 and 3312-8 drillcores were subjected to partial dissolution. Small cavities are rimmed with calcite cement followed by quartzine (length-slow chalcedony; Fig. 4E). The calcite displays multiple bright or dull orange CL zones; the quartzine is non-luminescent (Fig. 4F). Precipitation of quartzine caused corrosion of calcite (Fig. 4F), indicating an origin from chemically different fluids. The main occurrence of skeletal replacement by quartzine is limited to Arundian strata in the Dur-2 drillcore. Crinoids in the Chadian-age Crosspatrick Formation in the 3312-8, 3313-2 and 3245-25 drillcores were commonly replaced by quartzine. Length-slow, fibrous chalcedony frequently contains 5–15 μm size cubic inclusions indicating halite (or other salt *isotype*) as the precursor cement (Fig. 4G & H). Quartzine replacing skeletal components has also been described from the Permian Spirifer Limestone of Svalbard, Norway (Siedlecka 1972) and from the Upper Cretaceous Cueva Formation of Spain (Elorza & Garcia-Garmilla 1993), and interpreted as the product of migrating hypersaline brines.

Various horizons of the Chadian–Brigantian strata were replaced by length-fast chalcedony, which post-dates fine- and medium-crystalline dolomites. The quartz is composed of microcrystalline (<20 μm), equant mosaics, and/or void- and fracture-filling fibrous chalcedony. The precipitation of this quartz caused corrosion of earlier dolomites.

The late diagenetic events represented by nonplanar dolomite, and precipitation of

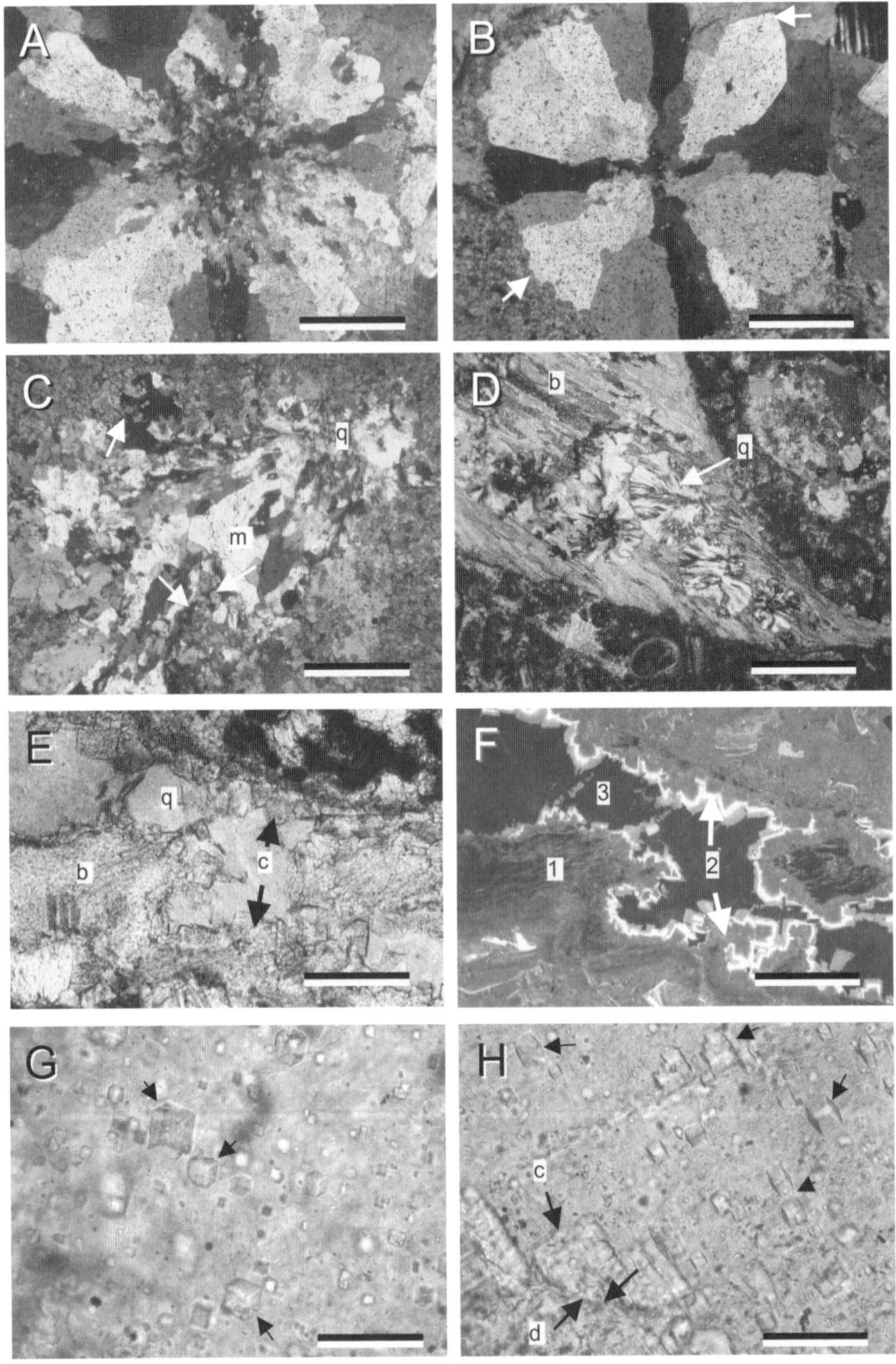
A
B
C
q
m
D
b
q
E
q
c
b
F
3
1
2
G
H
c
d

open-space-filling saddle dolomite, calcite, quartz and fluorite, are associated with intense fracturing and brecciation of the host rock (Everett *et al.* 1999; Hitzman 1999; Becker *et al.* 2002). This event predates burial diagenesis (Fig. 3), as indicated by the cross-cutting relationship between hydrothermal fractures and compactional stylolites. Burial cements are commonly found as syntaxial overgrowths and saw-tooth calcite on crinoids, and as drusy calcite. Coarse-crystalline, inclusion-free blocky calcite, displaying multiple bright or dull orange and non-luminescent CL zones, is commonly associated with dedolomitization of earlier saddle dolomite cements (Fig. 3). This calcite is also of late diagenetic origin.

Fine-crystalline dolomite

Lithological correlation and spatial distribution of replacive dolomite in the study area are shown in Figure 5. Fine-crystalline, planar dolomite comprises a minor proportion of early diagenetic replacive dolomitization. This type of dolomite was found predominantly in the successions proximal to the Leinster Massif and also in isolated outcrops in counties Tipperary and Kilkenny, where it is restricted to a few metre-thick sections in the Dur-2 core (Fig. 5).

Fine-crystalline planar dolomite is composed generally of subhedral–anhedral crystals, ranging from 5 to 25 μm. It is uniformly dark red or orange in CL and displays no compositional zoning. In the lower section of the GSI-91-25 core (Crosspatrick Formation), the size of the crystals ranges from 25 to 50 μm. Dolomitization is non-fabric-destructive, as the original sedimentary structures and the majority of the components are still identifiable. It is associated with rocks displaying sedimentary fabrics typical of peritidal sequences, such as former microbial mats, fenestral structures (Fig. 6A), desiccation cracks, vermiform gastropods, root casts and synsedimentary breccias. In the lower part of the GSI-91-25 core (Crosspatrick Formation) dolomite replaces the micritic matrix of bioclastic limestone (Fig. 6B); bioclasts and other components are replaced or filled by fibrous chalcedony or saddle dolomite cements, respectively. Fine-crystalline dolomite is also associated with pseudomorphs of former evaporites (see 'Limestone diagenesis'), such as calcitized lath- or lozenge-shape gypsum casts (Aghmacart Formation), quartz spherules, blocky quartz and length-slow fibrous chalcedony (Crosspatrick Formation).

Medium-crystalline dolomite

Medium-crystalline planar dolomite is the most abundant replacive dolomite encountered in the study. It occurs mainly in the NW part of the study area, including borehole sections within the Rathdowney Trend. It is very common at the top of and immediately above the Waulsortian Limestone, thus causing difficulties in distinguishing the Waulsortian Limestone from the overlying Crosspatrick Formation. It also occurs to a lesser extent in the upper part of the sections (Durrow and Ballyadams formations) (Fig. 5).

Medium-crystalline dolomite consists of euhedral–subhedral crystals, ranging in size from 50 to 200 μm (Fig. 7A). The crystals display unit extinction and predate stylolites and late saddle dolomite cement (Fig. 3). Dolomitization is generally texture-destructive, but in some cases preserves limestone textures as relicts (Fig. 7B). The dolomite preferentially replaces calcite cement in oolitic grainstones, high-magnesium calcite skeletal fragments (e.g. dasycladacean algae) or lime mud in

Fig. 4. Silica after former anhydrite. (**A**) Replacement spherule composed of two phases: quartzine in centre grading outward into megaquartz. 91-25/91, 210.65 m, Aghmacart Formation, scale = 750 μm, cross-polarized light (= cpl). (**B**) Coarse-crystalline blocky quartz replacing anhydrite nodule. Note the corrosion of quartz at the bottom left (arrow), the prismatic crystal habit on the top right (arrow) and the abundant small inclusions. 91-25/91, 210.65 m, Aghmacart Formation, scale = 750 μm, cpl. (**C**) Dissolution cavity filled by fibrous quartzine (q) grades into megaquartz (m) in fine-crystalline replacive dolomite. Note the abundant dolomite crystals in quartz (arrows). 91/25-126, 274.34 m, Crosspatrick Formation, scale = 750 μm, cpl. (**D**) Quartzine fans (q) filling dissolution void in a brachiopod shell (b), Dur-2/105, 239.74 m, Durrow Formation, scale = 500 μm, cpl. (**E**) Skeletal packstone containing large brachiopod shell (b) with secondary dissolution voids filled with calcite rim cement (c) and quartzine (q), Dur-2/164, 383.5 m, plane-polarized light (= ppl), scale = 400 μm. (**F**) CL image of (E) with dark red and non-luminescent brachiopod shell (1), cavity-filling calcite displaying multiple bright orange/yellow zones in calcite cement (2), followed by non-luminescent quartzine (3). Note the corrosion of calcite by quartz, scale = 400 μm. (**G**) Small solid halite (or other isotype) inclusions (arrows) in quartzine displaying cube form. 3312-8/3, 387.49 m, scale – 30 μm, ppl. (**H**) Similar inclusions (arrows) to (G) in void-filling quartzine. Note the dissolution margin (d) of void and the incipient calcite cement (c). 3312-8/34, 254.29 m, scale = 30 μm, ppl.

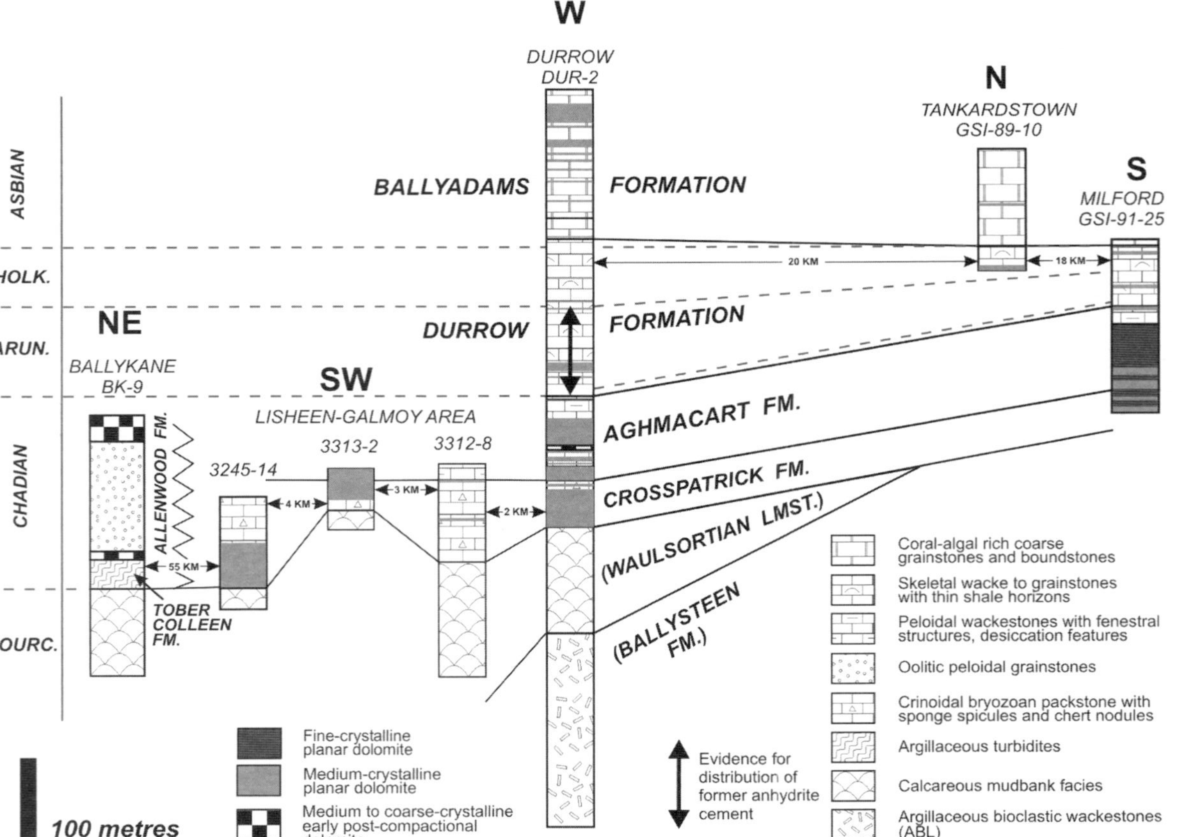

Fig. 5. Lithofacies correlation and spatial distribution of dolomite types in the studied cores. Solid lines represent lithofacies correlation; dashed lines represent equivalent time lines; zigzag line represents facies transition. Dolomite distribution is not shown in the strata below the Crosspatrick Formation. Note the intense dolomitization within the Crosspatrick Formation, and the relatively small amount of fine-crystalline dolomite in the upper part of the Aghmacart Formation. Note also the appearance of evaporite cement in the Arundian shelf carbonates (arrowed interval in the Dur-2 core). Courc., Courceyan; Arun., Arundian; Holk., Holkerian; Fm., Formation; Lmst., Limestone.

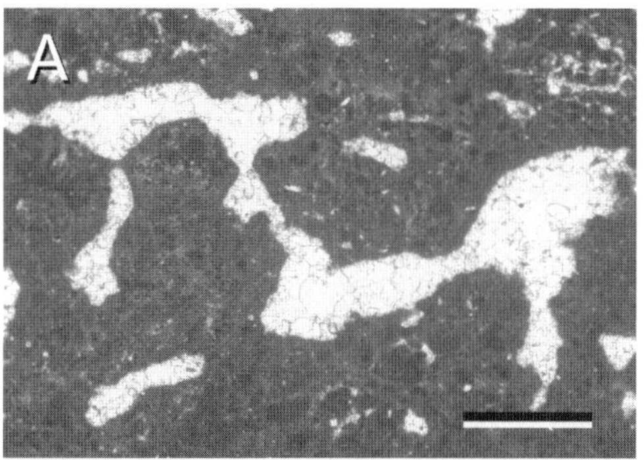

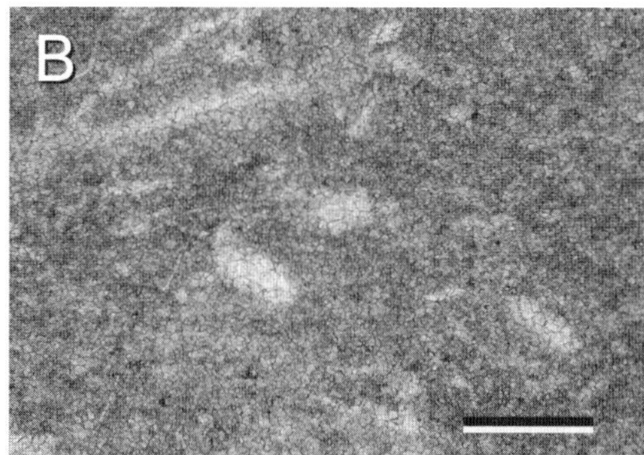

Fig. 6. Photomicrographs of fine-crystalline dolomite. (**A**) Fenestral structures in fine-crystalline peloidal dolomite, GSI 91-25/84, 194.45 m, Aghmacart Formation, scale = 750 μm, ppl. (**B**) Fine-crystalline dolomite replacing argillaceous bioclastic limestone in the Ballysteen Formation, GSI 91-25/126, 274.34 m, scale = 500 μm, ppl.

wackestones and packstones. Medium-crystalline planar dolomite crystals frequently display cloudy centres and clear rims (Fig. 7A), indicating replacement of original limestone rather than recrystallization of an earlier dolomite. Late neomorphism of medium-crystalline planar dolomite in the Irish Midlands (cf. Wright *et al.* 2003) is indicated by: (1) planar crystal boundaries gradually becoming curved and/or lobate (Fig. 7C), leading to fewer preserved crystal-face junctions (Gregg & Sibley 1984); and (2) increasing crystal size towards open-space-filling saddle dolomite cement (Fig. 7D), leading to energetically more favourable crystal size by reducing the surface energy (Gregg & Shelton 1990).

Medium-crystalline dolomite commonly displays uniform dark or dull red CL, however, euhedral crystals infrequently show multiple (bright red or dull red or non-luminescent) oscillatory zoning (Fig. 7E & F). Dolomite with transitional planar to nonplanar texture displays bright or mottled red CL, similar to late diagenetic replacive nonplanar dolomite.

Early post-compactional replacement dolomite

Volumetrically less important, medium–very-coarse-crystalline, planar-e dolomite rhombs or dolomite clusters, ranging in size from 100 to 450 μm (Fig. 8A–C), are limited to the oolitic grainstone strata found in great thickness in BK-7 and BK-9 drillcores (Allenwood Formation). Smaller thicknesses are found in the Dur-2 drillcore (Aghmacart Formation) (Fig. 5). The dolomitization occurs in isolated and thin, discontinuous horizons within boreholes. Dolomite crystals are generally compositionally zoned and display multiple CL zones (Fig. 8D). The crystals in the upper part of the BK-9 drillcore display uniform bright red CL with a bright orange–yellow dedolomitized outer zone.

The scattered dolomite rhombs are of interest in terms of the timing of diagenetic events. They are commonly associated with stylolites and cross-cut pressure-solution contacts (Fig. 8A & C). The crystals have nucleated at grain margins and grown into open spaces and into grains, with relicts of the latter retained as inclusions (Fig. 8B &C). The timing of dolomite precipitation may be evident from this inclusion pattern (Hird & Tucker 1988; Tucker & Wright 1990). In an example from the Lower Carboniferous Gully Oolite of South Wales, Hird & Tucker (1988) showed that as the result of grain-to-grain compaction, the grain margin preserved within the rhombs does not coincide with the grain–pore boundary outside the rhomb and they therefore interpreted the rhombs as pre-compactional. This is not, however, the case in the Irish Midlands. Grain margins preserved within the rhombs coincide with grain–pore boundaries outside the rhombs, and thus dolomite rhombs post-date ooid compaction. Intense compaction is found where early pore-filling cements are scarce. In such cases, sutures that were cross-cut by dolomite crystals can also be used as indicators (Fig. 8B & C). In areas where early calcite cements including isopachous fringes are more common, the cementation prevented any significant compaction (Hird & Tucker 1988; cf. Searl 1989). The dolomite rhombs are partially

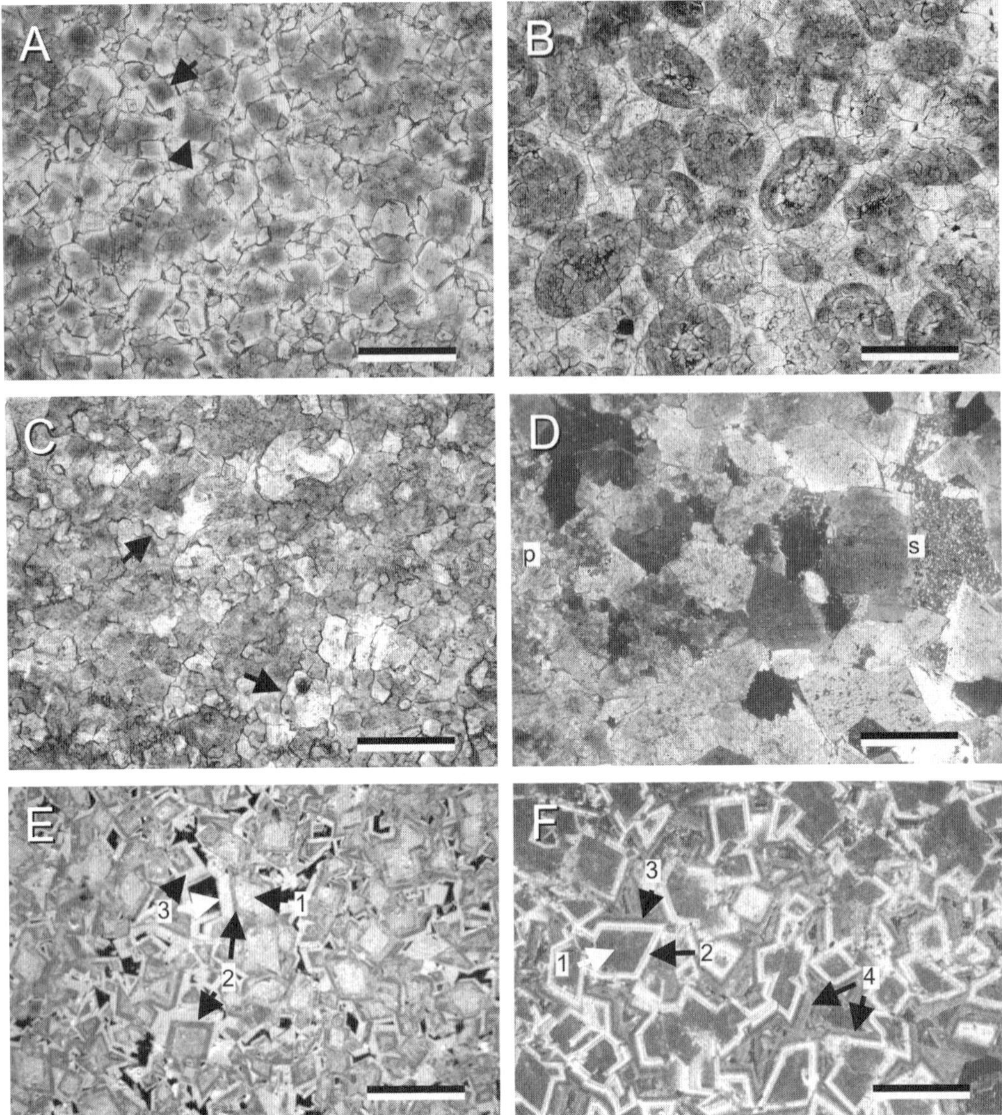

Fig. 7. Medium-crystalline dolomite textures occurring within the Chadian–Brigantian strata. (**A**) Medium-crystalline, planar-e and planar-s dolomite replacing limestone. Note the inclusion-rich centres and clear rims of crystals (arrows). 3245-14/3, 121.12 m, Crosspatrick Formation, scale = 500 μm, ppl. (**B**) Medium-crystalline, planar-s dolomite, oolitic grainstone (preserved as relicts). GSI-91-25/99, 228.0 m, Aghmacart Formation, scale = 500 μm, ppl. (**C**) Transitional dolomite from planar-s to nonplanar, note the curved boundaries (arrows) of nonplanar crystals. Dur-2 /148, 344.15 m, Durrow Formation, scale = 500 μm, ppl. (**D**) Texture variation from medium-crystalline, planar-s dolomite (p) to nonplanar dolomite (centre field), to fracture-filling, coarse-crystalline saddle dolomite (s). Crosspatrick Formation, 3313-2/7, 34.97 m, scale = 500 μm, cpl. (**E**) CL image of a medium-crystalline, planar-e dolomite displaying multiple CL zones: (1) bright red core; (2) thin dull red zone; (3) thick bright red zone. GSI-91-25/68, 167.0 m, Aghmacart Formation, scale = 300 μm. (**F**) CL image of a medium-crystalline, planar-s dolomite with multiple CL zones: (1) dark red core; (2) bright red zone; (3) dark red zone; and (4) dull red zone. Dur-2/212, 493.25 m, Aghmacart Formation, scale = 300 μm.

replaced by calcite cement, where: (1) spar intrudes the dolomite phase, causing an etched surface (Fig. 8A); and (2) the outer zone of dolomite crystals is dedolomitized, commonly leading to enhanced secondary porosity. A similar example of post-compactional replacive

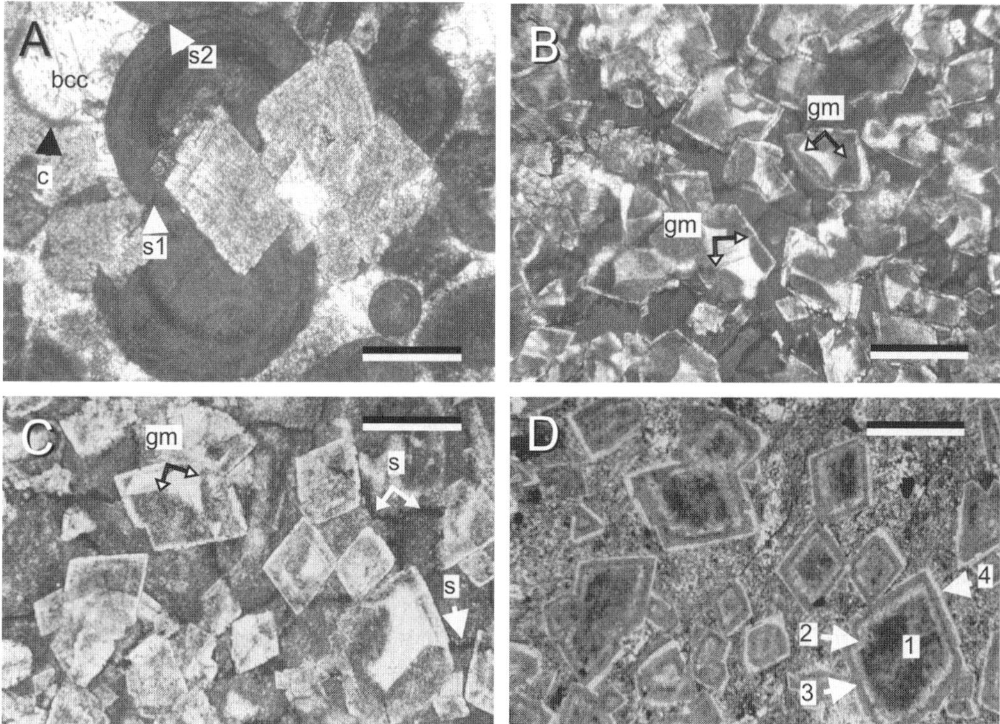

Fig. 8. Photomicrographs of early post-compactional dolomite textures. (**A**) A cluster of coarse-crystalline, planar-e dolomites cross-cut an ooid suture (s1). Note that ooid suture (s2) terminates at blocky calcite spar cement (bcc); and calcite spar intruded the dolomite, forming an etched contact (c). BK-9/53, 113.6 m, ppl, scale = 300 μm. (**B**) Intensely sutured and compacted oolite with abundant post-compactional dolomite rhombs. Note the grain margins (gm) preserved within the rhombs shows no compaction displacement. Dur-2/213, 497.0 m, ppl, scale = 500 μm. (**C**) Nearly completely compacted oolite with coarse-crystalline, post-compactional dolomite. Note that dolomite crystals cross-cut sutures (s) and grain margins (gm) do not show compactional displacement. Dur-2/214, 497.25 m, ppl, scale = 300 μm. (**D**) CL image of (C) displaying coarse dolomite with multiple CL zones: (1) dark mottled red core; (2) thin medium red; (3) thin dull red; and (4) thin medium red, which may be corroded. Host rock displays medium dull speckled orange CL, scale = 300 μm.

dolomite was described from the Upper Jurassic Smackover Formation by Moore *et al.* (1988) and interpreted as having been formed during shallow burial at elevated temperature (120 °C) and/or changing water composition.

Isotope geochemistry

Results

Plots of $\delta^{13}C$ and $\delta^{18}O$ values (‰ PDB) of Chadian–Brigantian-hosted calcites and dolomites are shown in Figure 9A, and isotopic data for individual samples are listed in Tables 1 and 2. The following data are presented in paragenetic order, from earliest to latest phases.

Brachiopods and crinoids have $\delta^{13}C$ values ranging from –4.57 to +4.04‰ (mean = +0.64‰, n = 26) and $\delta^{18}O$ values ranging from –7.39 to –3.44‰ (mean = –5.87‰, n = 26). Early blocky calcites and micrites have $\delta^{13}C$ values of 0.49 to 3.63‰ (mean = 2.6‰, n = 9) and $\delta^{18}O$ values of –8.92 to –4.14‰ (mean = –6.60‰, n = 9). Planar replacive dolomite from the Dur-2 core displays $\delta^{13}C$ values of –0.25 to 4.81‰ (mean = +3.89‰, n = 27) and $\delta^{18}O$ values of –7.49 to –3.87‰ (mean = –6.46‰, n = 27). Planar replacive dolomite from the Lisheen–Galmoy area has $\delta^{13}C$ values of 2.72 to 4.82‰ (mean = 3.72‰, n = 9) and $\delta^{18}O$ values of –8.11 to –3.42‰ (mean = –6.48‰, n = 9). Planar replacive dolomite from the GSI-91-25 core (fine and medium crystalline) has $\delta^{13}C$ values of –0.44 to +5.09‰ (mean = +2.44‰, n = 45) and $\delta^{18}O$ values of –13.81 to –0.25‰ (mean = –4.72‰, n = 44). Early post-compactional replacive dolomite from the BK-9 and Dur-2 drillcores yielded $\delta^{13}C$ values of 1.22 to 3.72‰

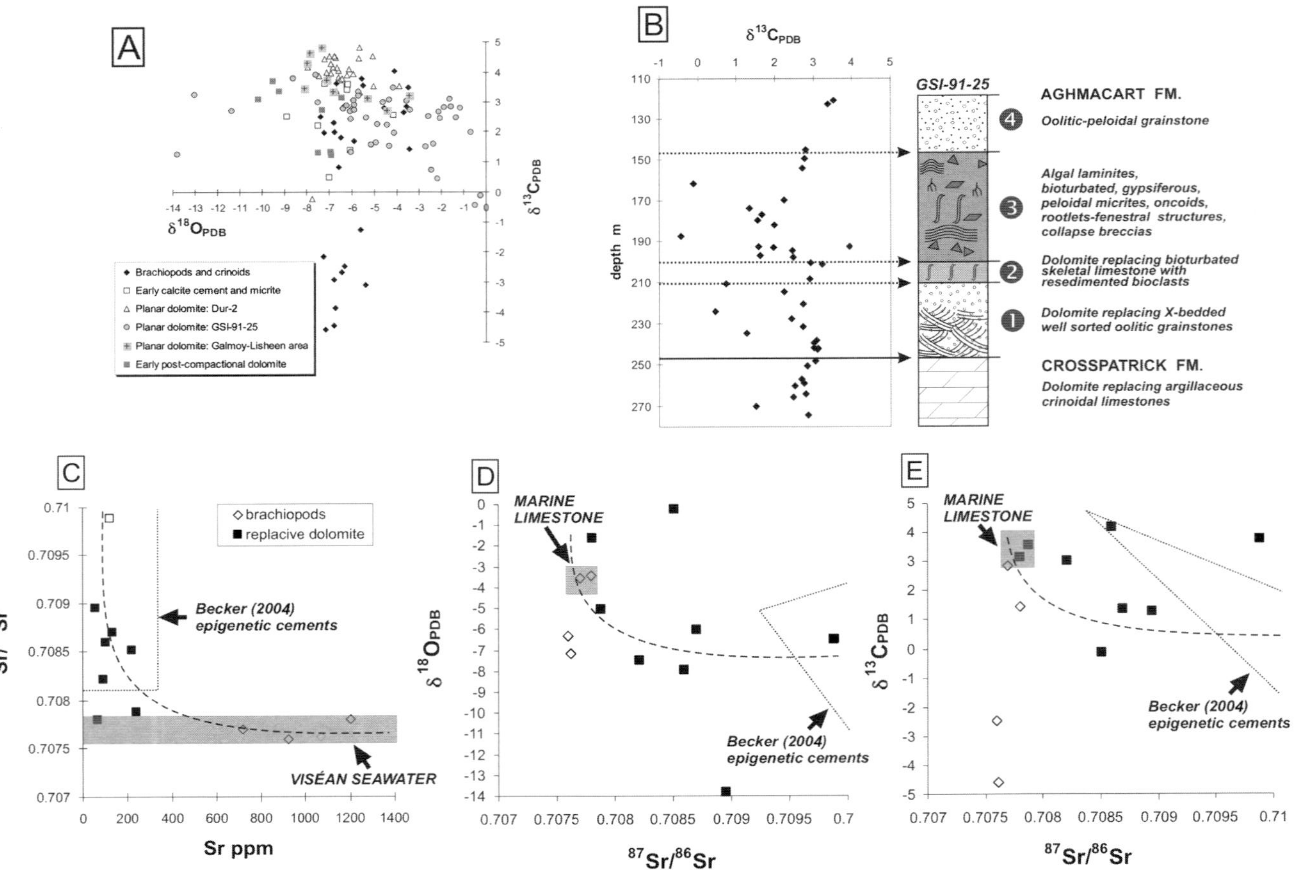

Fig. 9. (**A**) Plot of $\delta^{13}C$ and $\delta^{18}O$ values (‰, PDB) for the Chadian–Brigantian formations of SE Ireland. (**B**) Stratal variation of $\delta^{13}C$ within the Aghmacart and Crosspatrick formations in the lower part of the GSI-91-25 drillcore. Units are numbered and based on Nagy *et al.* (2005). (**C**)–(**E**) Covariation of $^{87}Sr/^{86}Sr$ ratios with Sr concentration, $\delta^{18}O$ and $\delta^{13}C$ values of replacive dolomite. Curves (dashed lines) illustrate fluid–rock interaction pathway (tentative, not calculated). Epigenetic dolomite cements measured from the Irish Midlands are from Becker (2004).

Table 1. *Summary of carbon and oxygen isotope values for calcite and dolomite from the Chadian–Brigantian formations of the study area*

Type	Crystal size	Textural and CL properties	$\delta^{13}C$ (‰, PDB)	$\delta^{18}O$ (‰, PDB)
Brachiopods and crinoids	N/A	Brachiopods: varies from non-luminescent to intensely recrystallized bright orange. Crinoids: bright orange	–4.57 to 4.04 mean = 0.64 (n = 26)	–7.39 to –3.44 mean = –5.87 (n = 26)
Early BCC and micrite	BCC: 150–500 μm; Micrite: <4 μm	BCC: unzoned dull orange or non-luminescent. Micrite: dull/dark orange	0.49 to 3.63 mean = 2.6 (n = 9)	–8.92 to –4.14 mean = –6.60 (n = 9)
Dur-2: Medium xln planar dolomite	50–200 μm	Subhedral–euhedral, fabric-destructive, unzoned medium red and/or multiple zoned dull/bright red/non-luminescent CL	–0.25 to 4.81 mean = 3.89 (n = 27)	–7.94 to –3.87 mean = –6.46 (n = 27)
GSI-91-25: Fine–medium xln planar dolomite	Fine: 5–25 μm and 25–50 μm Medium: 50–200 μm	Fine: anhedral, fabric-preserving; unzoned, non-luminescent or dark red CL. Medium: unzoned medium red; multiple zoned dull/bright red/non-luminescent CL	–0.44 to 5.09 mean = 2.44 (n = 45)	–13.81 to –0.25 mean = –4.72 (n = 45)
Galmoy-Lisheen: Medium xln planar dolomite	50–250 μm	Subhedral–euhedral, fabric-destructive, unzoned medium/dark red; dark core with bright outer zone in CL	2.72 to 4.82 mean = 3.72 (n = 9)	–8.11 to –3.42 mean = –6.48 (n = 9)
Early post-compaction replacive dolomite	100–450 μm	Euhedral–subhedral clusters, multiple CL zones in Dur-2, bright red with bright orange dedolomite outer zone in BK-9	1.22 to 3.72 mean = 2.5 (n = 8)	–10.19 to –6.45 mean = –8.02 (n = 8)

CL refers to cathodoluminescence properties of the dolomite. BCC, blocky calcite cement; xln, crystalline; N/A, not applicable.

(mean = 2.5‰, n = 8) and $\delta^{18}O$ values of –10.19 to –6.45‰ (mean = –8.02‰, n = 8).

Strontium isotope ($^{87}Sr/^{86}Sr$) analyses of representative brachiopods and replacive dolomites from Chadian–Asbian strata yield values ranging from 0.707600 to 0.707799 (mean = 0.707379, n = 4) and from 0.707805 to 0.709881 (mean = 0.708568, n = 8), respectively (Fig. 9C and Table 3). The brachiopod values are within the range for Viséan seawater (0.70752–0.70800) published by Bruckschen *et al.* (1999).

Characteristics and comparison

Calcites drilled from brachiopods and crinoids display similar $\delta^{18}O$ values, but based on different $\delta^{13}C$ values, two separate clusters can be distinguished (Fig. 9A; 1.43 to 4.04‰ and –1.27 to –4.44‰). Low $\delta^{13}C$ values for brachiopods, similar to those of our study, have been published for breccia matrix and palisade calcites of the Belle Roche Formation, Lower Carboniferous of Belgium, by Peeters *et al.* (1992). Nielsen *et al.* (2000) also published low $\delta^{13}C$ values from palaeosols, matrix crusts and columnar calcites for the Terwagne Formation, Dinantian of Belgium. These low $\delta^{13}C$ values were attributed to interference of near-surface meteoric diagenetic and soil-related processes. Early calcite cement and micrite values display a wide scatter and overlap the majority of the GSI-91-25 data.

Planar dolomites from the Dur-2 core display a tight clustering of values with similar $\delta^{18}O$ values and slightly higher $\delta^{13}C$ values than the skeletal components. Planar dolomite from the GSI-91-25 core displays a wide range of $\delta^{18}O$ and $\delta^{13}C$ values, possibly indicating a complex diagenetic history. Three samples of dolomite that replace ooids and are closely associated with late saddle dolomite cement have very low $\delta^{18}O$ values (ranging from –11.39 to –13.89‰), which probably reflect recrystallization in the presence of low ^{18}O late diagenetic fluids. Planar dolomites from the Galmoy–Lisheen area (3313-2, 3312-8 and 3245-14 drillcores) generally overlap the $\delta^{13}C$ and $\delta^{18}O$ values of dolomite from the Dur-2 drillcore. Early post-compactional replacive dolomites display spatial variations: (1) the dolomite values measured from the Dur-2 drillcore are more depleted in ^{18}O and slightly more enriched in ^{13}C; (2) the dolomites from the BK-9 drillcore display a narrow range of $\delta^{18}O$ values similar to those of the Durrow and Lisheen–Galmoy dolomites, but are more depleted in ^{13}C.

Planar dolomites from the Dur-2 core have values (see above) similar to the 'pure dolomite' values ($\delta^{13}C$ = 2.0 to 4.32‰, mean = 3.17‰; $\delta^{18}O$ = –9.83 to –4.9‰, mean = –7.05‰; n = 7) of Wright (2001) measured from

Table 2. *Carbon and oxygen isotope values of limestones and dolomites in the studied area*

Sample	Depth (m)	Formations	Description	$\delta^{13}C_{PDB}$ (‰)	$\delta^{18}O_{PDB}$ (‰)
Calcites from brachiopods and crinoids (n = 26)					
89-10/19	47.93	Ballyadams	B – 70% recrystallized dark red, 30% non-luminescent	2.49	–7.39
89-10/32	79.41	Ballyadams	B – 70–80% recrystallized bright orange, silicified	1.79	–6.48
89-10/45	107.22	Ballyadams	B – non-luminescent 50–70%, recrystallized bright orange	0.83	–6.59
89-10/67	157.76	Durrow	B – 95% non-luminescent, 5% recrystallized dark red	–4.57	–7.16
91-25/5	15.25	Ballyadams	B – 100% non-luminescent, 30% recrystallized bright orange	–2.15	–7.27
91-25/15	35.86	Durrow*	B – 70–80% recrystallized bright orange, 20% non-luminescent	1.70	–5.89
91-25/27	72.14	Durrow*	B – 70% recrystallized dark red, 30% non-luminescent	1.96	–7.23
Dur-2/6	22.08	Ballyadams	B – 85% medium orange recrystallized, 15% non-luminescent	–2.91	–6.79
Dur-2/46	117.55	Ballyadams	B – 60% non-luminescent, 40% very dark red	4.04	–4.09
Dur-2/82	193.40	Ballyadams	B – 100% non-luminescent with minor silicification	3.57	–5.50
Dur-2/99	230.88	Durrow	B – 80–90% non-luminescent and partially dolomitized by thin fractures	3.79	–5.55
Dur-2/115	265.55	Durrow	B – 100% non-luminescent	–2.47	–6.34
Dur-2/176	418.65	Durrow	B – 95% non-luminescent, 5% dark orange cross-cutting BCC	–4.45	–6.80
Dur-2/191	453.80	Durrow	B – 99% non-luminescent, 1% recrystallized dark red	–3.11	–5.40
6-20/38	n.a.	Ballyadams	B – 70% non-luminescent, 30% recrystallized dark red	–3.85	–6.73
BM-3	n.a.	Ballyadams	B – 90% non-luminescent, 10% dull orange on margins	–2.68	–6.45
6-22/29	n.a.	Durrow	B – 89–90% recrystallized bright orange	2.30	–6.80
6-23/1	n.a.	Ballyadams	B – 95% non-luminescent, 5% recrystallized dark orange	–1.27	–5.62
6-23/17	n.a.	Ballyadams	B – 95% non-luminescent, 5% recrystallized dark orange	1.43	–3.44
3312-8/34	254.29	Crosspatrick	B – 95% non-luminescent with 5% silicification	2.85	–3.55
89-10/24	60.22	Ballyadams	C – heterogeneous bright orange	3.63	–6.66
6-20/1	n.a.	Crosspatrick	C – dark/dull orange same as host rock	2.65	–3.69
6-23/22	n.a.	Clogrenan	C – dull/dark orange with some bright orange spots	1.97	–6.77
3312-8/4	385.46	Crosspatrick	C – very bright orange	2.81	–4.55
3312-8/15	339.38	Crosspatrick	C – heterogeneous bright orange	3.50	–3.46
3312-8/33	264.01	Crosspatrick	C – heterogeneous bright orange	2.87	–6.32
Early calcite cement and micrite (n = 9)					
91-25/9	28.60	Durrow*	clotted micrite in calcrete horizon	3.56	–6.19
91-25/10	29.63	Durrow*	calcimicrite matrix in gastropod rich wackestone	3.42	–6.19
91-25/12	31.90	Durrow*	coarse BCC in calcrete containing alveolar septal structure	3.62	–6.18
91-25/22	62.45	Durrow*	calcimicrite in ostracod rich fenestral wackestone	1.40	–6.09
91-25/51	132.41	Aghmacart*	pale-grey microsparite containing clasts of reworked pedogenic clasts	0.49	–7.01
Dur-2/42	110.67	Ballyadams	cross-laminated calcimicrite with scarce reworked skeletal fragments	2.51	–8.92
Dur-2/105	239.74	Durrow	calcimicrite geopetal fill in large reworked brachiopod shell	2.57	–4.14
Dur-2/201	474.62	Aghmacart	coarse BCC in very coarse peloidal bioclastic grainstone	3.63	–7.19
Dur-2/206	484.75	Aghmacart	coarse BCC in very coarse oncoidal grst	2.21	–7.53
Planar dolomite from Dur-2 (n = 27)					
Dur-2/5	19.02	Ballyadams	dull/dark red or non-luminescent mxpd	4.53	–6.99
Dur-2/13	38.00	Ballyadams	dull/dark red or non-luminescent mxpd-pntd with faint bright red zones	3.84	–6.59
Dur-2/34	97.33	Ballyadams	mxpd-pntd replacing skeletal limestone	4.06	–6.57
Dur-2/45	116.65	Ballyadams	unzoned medium red mxpd with rare thin bright red outer zone	4.10	–6.10
Dur-2/50	122.30	Ballyadams	mxpd replacing algal packstone	3.97	–6.93
Dur-2/57	136.60	Ballyadams	medium red mostly unzoned mxpd-pntd, some bright orange dedolomite	3.74	–6.52
Dur-2/62	150.60	Ballyadams	unzoned dark red mxpd-pntd, if euhedral, thin bright red outer zone	4.15	–6.64
Dur-2/67	155.85	Ballyadams	medium red mxpd-pntd with thin bright red outer zone	4.17	–7.94
Dur-2/75	175.10	Ballyadams	medium red mxpd-pntd with thin bright red outer zone	3.89	–7.45
Dur-2/100	234.10	Durrow	dark red or non-luminescent mxpd-pntd	3.91	–6.22
Dur-2/103	236.35	Durrow	mxpd replacing peloidal skeletal limestone	3.94	–5.92
Dur-2/106	242.70	Durrow	mxpd replacing peloidal skeletal pkst	3.55	–5.03
Dur-2/126	297.25	Durrow	medium/dark red speckled mxpd-pntd	3.35	–5.66
Dur-2/148	344.15	Durrow	medium/dark red mxpd-pntd	4.46	–5.69
Dur-2/178	425.70	Durrow	red/orange mxpd replacing skeletal wackstone/packstone	4.81	–5.64
Dur-2/181	427.90	Durrow	unzoned bright red fine to mxpd with intercrystal organics	4.43	–7.20
Dur-2/212	493.25	Aghmacart	multiple zoned mxpd with intercrystal organics	3.55	–3.87
Dur-2/217	498.40	Aghmacart	multiple zoned mxpd replacing peloidal detrital quartz rich limestone	4.17	–6.84
Dur-2/220	518.90	Aghmacart	dark red dolosilt with intercrystal organics, cross-cutting fractures	4.52	–5.05
Dur-2/226	549.20	Aghmacart	multiple zoned mxpd-pntd towards voids	3.79	–7.12
Dur-2/228	554.58	Aghmacart	very dark/non-luminescent inclusion rich mxpd	–0.25	–7.77
Dur-2/237	565.16	Crosspatrick	mxpd replacing crinoidal limestone	3.99	–7.12

Table 2. *Continued*

Sample	Depth (m)	Formations	Description	$\delta^{13}C_{PDB}$ (‰)	$\delta^{18}O_{PDB}$ (‰)
Dur-2/243	572.23	Crosspatrick	dark red/non-luminescent mxpd replacing crinoidal limestone	4.28	–6.92
Dur-2/251	585.00	Crosspatrick	dark red/non-luminescent mxpd with reworked crinoids	4.49	–6.71
Dur-2/252	585.55	Crosspatrick	bright red/orange fine to mxpd, dedolomite	4.52	–6.78
Dur-2/260	598.10	Crosspatrick	medium/dark red mxpd replacing crinoidal lmst	3.80	–6.43
Dur-2/266	623.64	Crosspatrick	fine to mxpd-pntd with large cavity filled epigenetic cements	3.38	–6.67
Planar dolomite from west of Leinster Massif (n = 45)					
91-25/15	35.86	Durrow*	fine–medium xln neomorphic planar dolomite	5.09	–5.92
91-25/30a	82.00	Durrow*	medium xln planar-s dolomite, partially dedolomitized	3.29	–13.02
91-25/30b	82.00	Durrow*	fine–medium xln planar-s dolomite	3.81	–8.65
91-25/46	120.71	Aghmacart*	medium xln planar-e, -s dolomite replacing peloidal/intraclastic pkst	3.51	–4.16
91-25/48	122.82	Aghmacart*	fine and medium xln planar dolomite replacing oolite	3.37	–5.70
91-25/58	145.36	Aghmacart*	fine–medium xln planar-s dolomite with nonplanar moulds	2.79	–6.41
91-25/61a	149.44	Aghmacart*	fine–medium xln planar-e, s dolomite, zoned	2.77	–5.99
91-25/64a	154.20	Aghmacart*	fine xln planar dolomite replacing peloidal micrite with serpulid horizon	2.70	–6.09
91-25/66a	161.90	Aghmacart*	fine xln planar dolomite replacing peloidal micrite	–0.13	–0.25
91-25/69	169.55	Aghmacart*	very fine xln planar dolomite associated with stromatolite	2.23	–4.44
91-25/72	173.71	Aghmacart*	very fine xln planar dolomite associated with stromatolite and gypsum laths	1.33	–6.06
91-25/73	176.71	Aghmacart*	medium xln planar dolomite replacing laminated lmst with dark clasts	1.65	–4.93
91-25/74a	179.46	Aghmacart*	very fine xln planar dolomite with gypsum laths	1.54	–4.34
91-25/76	182.00	Aghmacart*	very fine xln dolomite with quartz silt	1.97	–0.69
91-25/79	187.40	Aghmacart*	very fine xln planar dolomite with patches of medium xln dolomite	–0.44	–0.51
91-25/81a	192.46	Aghmacart*	medium xln planar-s dolomite infiltration in geopetal fashion	3.93	–7.60
91-25/81b	192.46	Aghmacart*	fine–medium xln planar-s dolomite breccia clasts	1.58	–5.15
91-25/82	193.10	Aghmacart*	fine xln dolomite replacing algal-stromatolitic bindstone	1.96	–4.12
91-25/83	194.65	Aghmacart*	dolosilt, planar dolomite replacing laminated lmst	2.46	–2.07
91-25/84	197.17	Aghmacart*	dolomicrite associated with fenestral, birds-eye structures	1.61	–2.37
91-25/85	197.80	Aghmacart*	medium xln planar-s dolomite replacing laminated lmst	2.47	–5.37
91-25/86b	200.35	Aghmacart*	dolosilt, planar dolomite replacing bioturbated lmst	2.91	–5.89
91-28/87	201.35	Aghmacart*	medium xln planar-s dolomite with detrital qtz and organic matter	3.23	–5.71
91-25/89	208.49	Aghmacart*	medium xln planar-e, -s dolomite replacing well-sorted oolite	2.90	–6.23
91-25/91	210.65	Aghmacart*	fine–medium xln planar dolomite breccia clast	0.73	–2.46
91-25/92a	214.73	Aghmacart*	medium xln planar-s dolomite with vague ooid remnants	2.25	–4.87
91-25/95c	220.65	Aghmacart*	mxpd replacing well-sorted oolites	2.74	–5.88
91-25/97	224.20	Aghmacart*	very fine xln dolomite associated with rhizoliths and desiccation cracks	0.45	–2.18
91-25/99a	228.00	Aghmacart*	medium to coarse xln planar-s dolomite replacing well-sorted oolite	2.43	–6.06
91-25/101	231.95	Aghmacart*	medium xln planar-s dolomite replacing oolitic/peloidal lmst	2.72	–11.39
91-25/103c	235.04	Aghmacart*	medium xln planar-s, -e dolomite replacing oolitic lmst	1.27	–13.81
91-25/105	238.33	Aghmacart*	medium xln planar-s dolomite replacing skeletal lmst	3.08	–5.91
91-25/106b	239.53	Aghmacart*	fine xln planar dolomite associated with late saddle dolomite	3.01	–7.51
91-25/108	242.00	Aghmacart*	medium xln planar-s dolomite with large cavity filling saddle dolomite	3.01	–4.61
91-25/109	242.55	Aghmacart*	vaguely laminated fine xln dolomite	3.11	–1.62
91-25/110	242.83	Aghmacart*	fine to mxpd clasts replacing skeletal lmst in breccia	3.10	–4.34
91-25/112a	248.18	Crosspatrick*	mxpd clasts in breccia	3.05	–3.56
91-25/113	250.69	Crosspatrick*	dolosilt replacing crinoidal lmst	2.83	–1.88
91-25/116	257.00	Crosspatrick*	dolosilt, planar-s dolomite with organic matter	2.68	–2.12
91-25/118	258.88	Crosspatrick*	dolosilt replacing skeletal lmst	2.75	–3.39
91-25/119a	260.25	Crosspatrick*	fine to mxpd with interxln organic matter	2.52	–2.60
91-25/121	264.38	Crosspatrick*	dolosilt, planar-s dolomite, intensely fractures	2.80	–1.14
91-25/122	265.67	Crosspatrick*	fine to mxpd	2.46	–1.33
91-25/124d	270.06	Crosspatrick*	dolosilt, planar-s dolomite replacing crinoidal lmst	1.50	–2.70
91-25/126b	274.34	Crosspatrick*	dolosilt, planar-s dolomite replacing crinoidal lmst	2.85	–1.56
Planar dolomite from the Galmoy-Lisheen area (n = 9)					
3313-2/6	37.39	Crosspatrick	bright/dark mottled red mxpd-pntd with cavity filled late cements	3.48	–8.11
3313-2/14	9.20	Aghmacart	zoned (PK1) mxpd geopetal fill in small fracture	3.36	–6.83
3245-14/7	93.43	Crosspatrick	multiple zoned mxpd-pntd replacing crinoidal bryozoan limestone	3.15	–5.29
3245-14/13	59.50	Crosspatrick	mxpd replacing crinoidal skeletal limestone	3.24	–3.42
3312-8/12	360.78	Crosspatrick	unzoned medium red or multiple zoned euhedral mxpd	3.74	–7.12
3312-8/15	339.38	Crosspatrick	mxpd replacing very coarse crinoidal bryozoan wackestone	2.72	–4.42
3312-8/41	212.20	Crosspatrick	dull/dark red mxpd with small bright spots graded to pntd	4.82	–7.33
3312-8/43	211.95	Crosspatrick	dull/dark red mxpd with bright red orange outer zone	4.64	–7.85
3312-8/49	206.60	Crosspatrick	dark/dull red mxpd-pntd	4.30	–7.98

Table 2. *Continued*

Sample	Depth (m)	Formations	Description	$\delta^{13}C_{PDB}$ (‰)	$\delta^{18}O_{PDB}$ (‰)
Early post-compactional dolomite (EPC) (n = 8)					
Dur-2/211	492.45	Aghmacart	vaguely zoned mxpd-cxpd coalesced rhombs in oolitic limestone	3.72	–9.54
Dur-2/213	497.00	Aghmacart	multiple zoned mxpd-cxpld rhombs in oolitic limestone	3.13	–10.19
BK-9/53	113.60	Allenwood	zoned mxpd with dull orange dedolomite outer zone	3.16	–6.45
BK-9/69	57.81	Allenwood	corroded mxpd (dedolomite) on ooid margins	1.34	–6.95
BK-9/70	52.42	Allenwood	oolitic grainstone with sutured contacts and mxpd	2.75	–7.32
BK-9/72	43.50	Allenwood	oolitic grainstone with sutured contacts and mxpd, dedolomite	1.22	–6.92
BK-9/75	27.90	Allenwood	dark red mxpd with corroded dedolomitized outer zone	1.30	–7.50
3312-8/54	193.93	Aghmacart	unzoned medium/dark red mxpd with corroded bright red outer zone	3.39	–9.26

*Refers to samples from the former Milford Formation by McConnell *et al.* (1994). B, brachiopods; C, crinoids; BCC, blocky calcite cement; mxpd, medium-crystalline planar dolomite; pntd, planar to nonplanar transitional dolomite; cxpd, coarse-crystalline planar dolomite.

Chadian–Brigantian strata of six other drill-cores from the Irish Midlands. Isotope values of planar dolomite from the GSI-91-25 core (see above) are similar to those of the 'pervasive dolomite' ($\delta^{13}C$ = 1.81 to 3.85‰, mean = 2.41‰; $\delta^{18}O$ = –5.46 to –0.31‰, mean = –2.43‰; n = 8) from the Lower Carboniferous of SE Wales described by Hird *et al.* (1987), although their 'peritidal dolomicrite' ($\delta^{13}C$ = –2.27 to +0.13‰, mean = –1.03‰; $\delta^{18}O$ = –1.72 to –0.46‰, mean = –0.87‰; n = 9) is more depleted in both ^{18}O and ^{13}C than our samples.

Variations of $\delta^{13}C$ values with depth in the Aghmacart Formation

The $\delta^{13}C$ values from the GSI-91–25 drillcore display a variation with stratigraphic depth in the Aghmacart Formation of late Chadian age (Fig. 9B). The formation is composed of four sedimentary units. Unit 1, well-sorted, cross-bedded, oolitic grainstone replaced by medium-crystalline planar dolomite, displays a progressive decrease in $\delta^{13}C$ values up-section from 3.08 to 2.25‰ with little variation. Shifts to lower values include a strongly recrystallized oolite ($\delta^{13}C$ = 1.27‰; $\delta^{18}O$ –13.81‰), an interbedded peritidal micrite ($\delta^{13}C$ = 0.45‰; $\delta^{18}O$ –2.18‰) and a microbial mat breccia ($\delta^{13}C$ = 0.73‰; $\delta^{18}O$ –2.46‰). Unit 2 consists of bioturbated skeletal limestone with resedimented fragments replaced by fine- to medium-crystalline planar dolomite and displays a shift in $\delta^{13}C$ values to 2.9‰. The overlying Unit 3 is composed of rocks with features typical of arid tidal-flat environments and displays a progressive decrease in $\delta^{13}C$ values up-section towards –0.44‰ at 187.4 m and again of –0.13‰ at 161.9 m. The upper part of Unit 3 and Unit 4 (an oolitic peloidal barrier shoal deposit) displays a progressive ^{13}C enrichment up section to a $\delta^{13}C$ value of 3.51‰ (Fig. 9B).

Discussion

Early diagenesis and dolomite formation

Planar textures, which characterize much of the replacement dolomite in the Chadian–Brigantian sequence, are consistent with an early diagenetic origin (Gregg & Sibley 1984; Sibley & Gregg 1987). Medium-crystalline dolomites are more common in the distal, open-marine sequence, where they alternate with skeletal wackestones and grainstones. Fine-crystalline dolomites, comprising a minor proportion of the entire sequence, are restricted to peritidal facies found predominantly in the proximal shelf area adjacent to the Leinster Massif (Fig. 5). This dolomite was probably formed where capillary evaporation and evaporative pumping were prevalent (cf. McKenzie *et al.* 1980). Similar fine-crystalline dolomite was described from the peritidal sequence in the coastal sabkha of Abu Dhabi (Illing *et al.* 1965; McKenzie 1981; Patterson & Kinsman 1982) and in the hypersaline environment of Bonaire, Netherlands Antilles (Deffeyes *et al.* 1965). It may also be analogous to the dolomite crust found in the tidal flat environment of the Bahamas (Shinn *et al.* 1965). However, evaporite minerals are not preserved there.

An estimate of the isotopic composition of Early Carboniferous seawater is possible if a value for the temperature of the seawater is assumed. Ireland, including the SE Irish Midlands, was located within 30° of the palaeo-equator during the Carboniferous (Leeder 1987). Using $\delta^{18}O$ values measured from brachiopods and crinoids (mean = –5.87‰ PDB, Table 1) and a temperature of 25 °C, the

Table 3. *Strontium isotope ratios and Sr concentrations of limestones and dolomites in the studied area*

Sample	Depth (m)	Formations	Description	$^{87}Sr/^{86}Sr$	Sr (ppm)
Calcites from brachiopods and crinoids (n = 4)					
89-10/67	157.76	Durrow	B – 95% non-luminescent, 5% recrystallized dark red	0.70762	1070.12
Dur-2/115	265.55	Durrow	B – 100% non-luminescent	0.70760	920.26
6-23/17	n.a.	Ballyadams	B – 95% non-luminescent, 5% recrystallized dark orange	0.70779	1202.11
3312-8/34	254.29	Crosspatrick	B – 95% non-luminescent with 5% silicification	0.70769	718.24
			$^{87}Sr/^{86}Sr$: 0.70760 to 0.70779, mean = 0.70767 Sr: 718.24 to 1202.11, mean = 977.68		
Planar dolomite from Dur-2 and GSI-91-25 drillcores (n = 8)					
Dur-2/57	136.60	Ballyadams	medium red mostly unzoned mxpd-pntd, some bright orange dedolomite	0.70988	122.27
Dur-2/67	155.85	Ballyadams	medium red mxpd-pntd with thin bright red outer zone	0.70859	101.84
Dur-2/106	242.70	Durrow	mxpd replacing peloidal skeletal pkst	0.70788	238.91
91-25/66a	161.90	Aghmacart*	fine xln planar dolomite replacing peloidal micrite	0.70851	219.05
91-25/72	173.71	Aghmacart*	very fine xln planar dolomite associated with stromatolite and gypsum laths	0.70869	130.13
91-25/103c	235.04	Aghmacart*	medium xln planar-s, -e dolomite replacing oolitic lmst	0.70895	57.00
91-25/106b	239.53	Aghmacart*	fine xln planar dolomite associated with late saddle dolomite	0.70821	92.88
91-25/109	242.55	Aghmacart*	vaguely laminated fine xln dolomite	0.70780	68.02
			$^{87}Sr/^{86}Sr$: 0.70780 to 0.70988, mean = 0.70856 Sr: 57.00 to 130.13, mean = 87.01		

*Refers to samples from the former Milford Formation by McConnell *et al.* (1994). B, brachiopods; mxpd, medium-crystalline planar dolomite; pntd, planar–nonplanar transitional dolomite; xln, crystalline; n.a., not applicable.

calculated $\delta^{18}O$ value for Lower Carboniferous seawater is –3.90‰ (Standard Mean Ocean Water) (SMOW)) using the Friedman & O'Neil (1977) fractionation equation. The large variation in the $\delta^{13}C$ values of the brachiopods (with mean = 0.64‰) suggests later diagenetic alteration. There is no consistency between the observations made with transmitted light or CL petrography to detect diagenetic alteration, and the stable isotope results from the brachiopods and crinoids (Table 2). This discrepancy raises the suspicion that non-luminescent, and apparently unaltered, brachiopods may not provide entirely reliable estimates of the original isotope value of Early Carboniferous seawater (Brand 1982; Popp *et al.* 1986; Veizer *et al.* 1986; Rush & Chafetz 1990; Banner & Kaufman 1994).

The $^{87}Sr/^{86}Sr$ ratios of the brachiopods are within the values for Viséan seawater (Bruckschen *et al.* 1999) (Fig. 9C). Two of the four brachiopods displaying ≥95% non-luminescence in CL (Tables 2 and 3) yielded the highest $\delta^{18}O$ values measured in this study (3.44 and 3.55‰, Fig. 9A). They yield an estimated $\delta^{18}O$ value of seawater of –1.46‰ (SMOW). This is comparable with the estimated $\delta^{18}O_{SMOW}$ value of –2.7±0.5‰ for Viséan seawater based on a larger sample set analysed by Bruckschen *et al.* (1999) for the European Carboniferous.

Five other brachiopods displaying ≥95% non-luminescence in CL (Table 2) also yielded low $\delta^{13}C$ carbon isotope values (Fig. 9A; ranging from –4.57 to –1.27‰). Several scenarios may be responsible for the diagenetic alteration of the brachiopod shells. The overall variation of $\delta^{13}C$ values for calcite is relatively insensitive to changes in temperature, but is commonly governed by the contribution of carbon from various inorganic and organic sources (Anderson & Arthur 1983). Soil weathering and dissolution of marine limestones with lower $\delta^{13}C$ will generally result in fluids with moderately low $\delta^{13}C$ values (Allen & Matthews 1982; James & Choquette 1984). The carbon isotopic composition of meteoric water can thus be modified rapidly as the water percolates through the soil (Lohmann 1988). Coarse-crystalline, inclusion-free, blocky calcite observed in the Dur-2 drillcore, displaying multiple bright or dull orange and non-luminescent CL zones, yielded $\delta^{13}C$ values ranging from –7.79 to –3.47‰ (mean = –5.81‰, n = 18; Nagy 2003). The precipitation of calcite frequently caused dedolomitisation of earlier saddle dolomite cements. This low ^{13}C, calcite-precipitating and dedolomitizing fluid may also have caused the observed recrystallization and isotope re-equilibration of brachiopods (Fig. 10A).

Chemistry of the dolomitizing fluid

Calculations can also be made for the early diagenetic planar dolomites in order to estimate

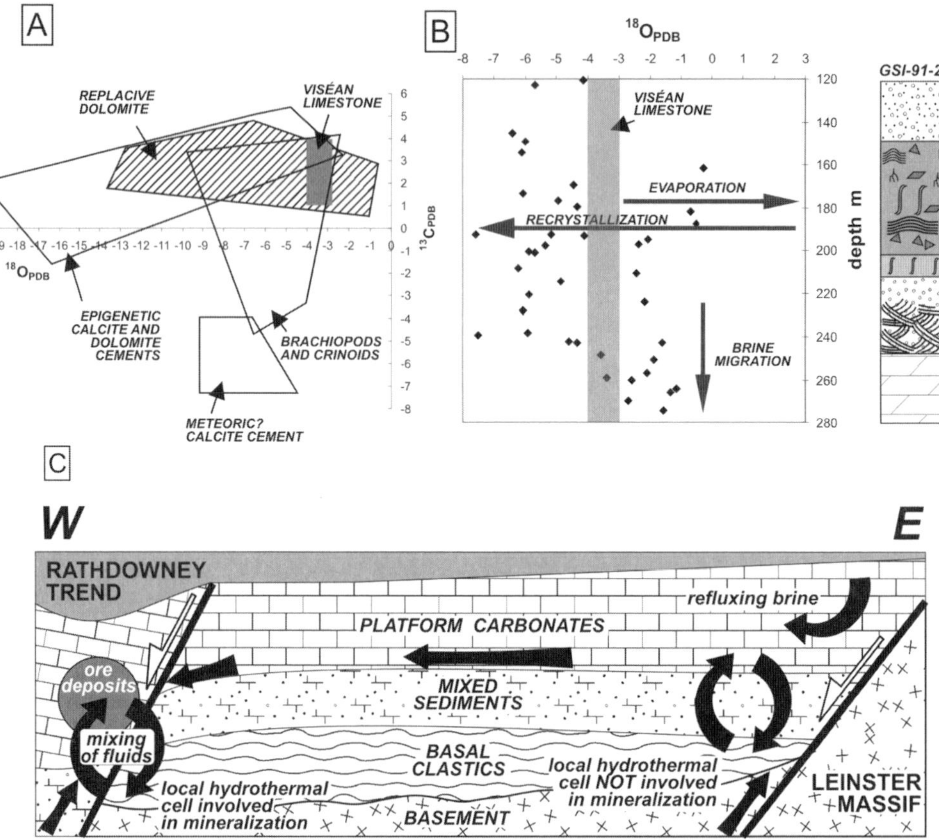

Fig. 10. (**A**) Simplified stable isotope summary plot of diagenetic carbonates and skeletal grains of the studied area. Epigenetic calcite and dolomite cement values from Becker (2004), Becker *et al.* (2002); meteoric? calcite cement values from Nagy (2003). (**B**) Relationship between $\delta^{18}O$ values and depth from the GSI-91-25 core, west of the Leinster Massif. The initial $\delta^{18}O$ values of the replacive dolomites in the tidal-flat strata of the Aghmacart Formation were controlled by evaporative fluids (evaporation arrow pointing to right) resulting in enrichment in ^{18}O. Later diagenetic alteration caused significant recrystallization with oxygen isotope exchange with lower ^{18}O fluid (recrystallization arrow pointing to left). Unit 1 and the underlying strata display higher $\delta^{18}O$ values indicating interaction with hypersaline pore waters during the formation of replacive dolomite (brine migration pointing arrow). Simplified lithological column for the lower part of the core is shown, see Figure 9B for explanation. Range of Viséan limestone was calculated in this study. (**C**) Conceptual fluid-flow model for the SE Irish Midlands, see text for discussion.

the $\delta^{18}O$ values of the waters from which they precipitated. At 25 °C, the dolomites from the Dur-2 and GSI-91-25 cores yield $\delta^{18}O_{water}$ values of –9.75 to –5.69‰ (mean = –8.27‰; SMOW) and –9.40 to –2.06‰ (mean = –5.85‰; SMOW), respectively, based on Land's (1985) equation. (Samples with $\delta^{18}O$ values ≤–8‰ were not used for calculation. Proximity of these samples to saddle dolomite cements increases the likelihood that they may reflect oxygen isotope exchange with late diagenetic fluids.) Using the oxygen isotope fractionation estimates of Land (1983) the difference ($\Delta^{18}_{dolomite-calcite}$) between co-existing dolomite and calcite ranges from about 3 to 6‰ at 25 °C. Therefore, our early diagenetic planar dolomites must have formed from, or been affected by, water depleted in ^{18}O with respect to seawater, or at elevated temperature. For example, the most ^{18}O-enriched dolomite ($\delta^{18}O$ = –0.24‰, PDB) measured from a typical evaporitic sequence in the GSI-91-25 core yields a $\delta^{18}O_{water}$ value of –2.06‰ (SMOW), about 0.5‰ less than our estimated value for Viséan seawater. Although there is textural evidence for evaporative conditions during initial carbonate precipitation, the later dolomite does not reflect estimated *c.* 3.2‰ enrichment in ^{18}O expected for dolomite in equilibrium with calcite found by McKenzie (1981) for modern sabkha dolomite. Thus, even our highest $\delta^{18}O$ dolomite reflects exchange with low ^{18}O fluids. The case is even more pronounced for planar dolomite from the Dur-2 core that yield $\delta^{18}O_{water}$ values (mean = −8.27‰, SMOW) approximately 7‰ lower than the calculated seawater (−1.46‰, SMOW). Thus, early replacement of evaporites (Fig. 4A & B) and late diagenetic saddle dolomite and calcite in the sequence (Fig. 7C & D) indicate the presence of multiple diagenetic fluids that probably overprinted the early diagenetic isotope signal.

Implication of carbon isotope variations in the Aghmacart Formation

Carbon isotope variations found in the GSI-91-25 drillcore (Fig. 9B) illustrate early freshwater diagenesis, where ^{12}C-enriched limestones suggest the presence of subaerial exposure surfaces (Allen & Matthews 1982; Humphrey *et al.* 1986). This variation is most prominent at the bottom of 'Unit 2', where the $\delta^{13}C$ value of the underlying limestone displays a strong decrease at the contact, from 2.9 to 0.73‰, followed by a progressive increase to higher values down-section. This pattern is also characteristic of isotope variations proximal to a subaerial exposure (Allen & Matthews 1982; Humphrey *et al.* 1986). Nagy *et al.* (2005) described a microbial mat horizon at the bottom of 'Unit 2', which confirms the results based on isotope chemistry.

Neomorphism of planar dolomite

Variations in the Sr isotopic composition of ancient limestones and dolomites can be used as indicators of diagenetic fluid compositions and the mechanisms of diagenesis (Banner 1995). Dolomites may be modified subsequent to deposition due to neomorphism (Land 1980; Gregg & Shelton 1990; Gregg *et al.* 1992). The $^{87}Sr/^{86}Sr$ isotope ratios measured from replacive dolomites of the Chadian–Asbian strata display elevated values with respect to estimated Viséan seawater (Fig. 9C). This variation indicates a diagenetic fluid with extraformational Sr derived from interaction with crystalline basement and/or siliciclastic sedimentary basins (Banner 1995). Oxygen and strontium isotopic variations of the replacement dolomite (Fig. 9D) suggest neomorphism by a fluid with similar $\delta^{18}O$ values but different $^{87}Sr/^{86}Sr$ ratios. The diagenetic history of the replacement dolomite can be understood by the 'model curve', which represents mixing of mineral end members (dashed lines in Fig. 9C–E). During neomorphism of the replacive dolomite, the original planar dolomite crystals were progressively dissolved and reprecipitated in the presence of the diagenetic fluid. Significant departure in $^{87}Sr/^{86}Sr$ ratios (0.00029–0.00236) from the estimated original marine $^{87}Sr/^{86}Sr$ value (*c.* 0.70752) was caused by addition of the extraformational Sr contained in the diagenetic fluid. Neomorphism of dolomite at elevated temperature (>60 °C) by the end-member diagenetic fluid can produce nonplanar textures (Gregg & Sibley 1984), which were observed in the studied formation (Fig. 7C & D).

Meteoric waters and brines that interact with marine carbonates may inherit Sr isotope compositions that are diagnostic of the age and/or composition of the carbonates (Müller *et al.* 1991; Banner *et al.* 1994). According to Morrison & Brand (1986) and Brand (1989), modern brachiopods contain 200–1500 ppm Sr, and well-preserved shells in ancient limestone may have a range of 300–3400 ppm, but are mostly in excess of 500 ppm. Fluid–rock interaction between meteoric water and brine (low $\delta^{13}C$ and $\delta^{18}O$ values, high $^{87}Sr/^{86}Sr$ ratios, low Sr concentration) and brachiopods would shift $\delta^{13}C$ and/or $\delta^{18}O$ values of the brachiopod calcite before any significant change in $^{87}Sr/^{86}Sr$ occurred (e.g. Banner & Kaufman 1994). This could account for the stable isotope exchange of brachiopods observed in this study (Fig 10A).

Wright (2001) and Wright *et al.* (2004) modelled the isotope and trace-element chemistry of early planar dolomites (similar to medium-crystalline dolomite in our study) from the Waulsortian Limestone in the Irish Midlands. They concluded that the origin of the dolomite is more compatible with formation from modified seawater at relatively low temperatures (50–70 °C). Localized depositional and palaeoenvironmental differences between the late Chadian–Brigantian units and the underlying Waulsortian, however, indicate potential variation in dolomitizing mechanisms.

Hypersaline brine migration

During the late Chadian, peritidal deposition was widespread proximal to the Leinster Massif (McConnell *et al.* 1994; Nagy *et al.* 2005). Similar algal-flat, intertidal, lagoon and sabkha facies were also characteristic of other regions adjacent to topographic highs during the Lower Carboniferous (West *et al.* 1968; Hird *et al.* 1987; Swennen *et al.* 1990; Peeters *et al.* 1992; Somerville *et al.* 2001). We speculate that hypersaline brines generated on peritidal flats descended into the underlying Crosspatrick Formation through porous oolitic sediments (Fig. 7B) resulting in evaporite cement precipitation in voids of bioclasts such as crinoids. An evaporite precursor is suggested by the occurrence of length-slow fibrous chalcedony pseudomorphs. The original host-rock texture was destroyed due to replacement by fine-crystalline dolomite (Figs 4C and 6B).

The hypothesis of descending brines is supported by the stratal trend of $\delta^{18}O$ values in the GSI-91-25 borehole (Fig. 10B). However, it is likely that the values were further reset by later diagenetic fluids (Fig. 10A) associated with fractures within the peritidal sequence. Figure 10B shows the possible effect of early and late diagenetic alteration.

Comparison to brine migration in the sequences of dolomite hydrocarbon reservoirs

Hypersaline brine migration has resulted in the development of important dolomite reservoirs. Vertical flow was an important hydrodynamic mechanism for the dolomitization of Jurassic strata of the Smackover Formation in east Texas and west Arkansas, USA (Moore *et al.* 1988). Considerable thicknesses of evaporites (Buckner Formation) were deposited above porous oolitic grainstones (Smackover Formation). Refluxing brines mixed with an active meteoric-water system and caused extensive dolomitization. Melim & Scholle (2002) suggested that mesosaline brines from carbonate lagoons were also capable of dolomitizing reef and forereef facies in the Permian Basin. The reflux dolomitization model may therefore work in areas where no well-developed evaporite facies is indicated.

Basin-wide diagenetic environments similar to those in the Irish Lower Carboniferous were described from the Capitan shelf margin of west Texas and SE New Mexico, USA. There are some major depositional and diagenetic differences in this portion of the Permian Basin, such as the occurrence of shelf siltstones and sandstones and a different biota in the shelf-edge 'barrier' facies (Sarg 1981; Ward *et al.* 1986; Borer & Harris 1991; Andreason 1992). The most important difference though, is that the rocks host hydrocarbons instead of base metals. Nevertheless, the overall early diagenetic environment was probably very similar. It involved hypersaline fluids causing both dolomitization and evaporite cementation in surrounding carbonate sediments (Adams & Rhodes 1960; Garber *et al.* 1990; Scholle *et al.* 1992; Melim & Scholle 2002) and meteoric diagenesis forming early calcite cement and leaching skeletal fragments (Given & Lohmann 1986). The location of evaporite precipitation and cementation had a major control on petroleum distribution in the Guadalupian strata, as hydrocarbons are found primarily at the up-dip transition from porous to evaporite-cemented shelf deposits (Ward *et al.* 1986; Harris & Saller 1999). During the interval between the termination of Guadalupian carbonate sedimentation and the onlap of Ochoan evaporites, the Capitan strata underwent significant subaerial alteration and developed large-scale, solution-enlarged fracture systems (Melim & Scholle 1989). These systems, coupled with synsedimentary fracturing of the massive reef facies, provided conduits through which hypersaline brines later flowed into more permeable reef and slope deposits, causing local dolomitization and evaporite cementation (Scholle *et al.* 1992; Ulmer-Scholle *et al.* 1993). Garber *et al.* (1989) suggested that evaporite formation was related to one of three mechanisms: (1) synsedimentary recycling of brines from Capitan-equivalent shelf evaporites; (2) lateral flow of brines from the immediately post-Capitan (lower Ochoan) Castile Formation evaporites of the Delaware Basin; or (3) reflux of brines from upper Ochoan Salado Formation evaporites forming on the Northwest Shelf. It should be noted, however, that no equivalent overlying or basinal evaporites have been recorded in the SE Irish Midlands or the Dublin Basin.

Implication for mineralization in the Irish Midlands

The Zn–Pb mineralization in the Irish Midlands (Fig. 10C) occurred adjacent to normal fault zones (Hitzman 1995*b*), where limestone host rocks of the Rathdowney Trend experienced secondary porosity enhancement by tectonic brecciation and dissolution during late dolomitization.

It is proposed that local hydrothermal convection flow occurred in the areas adjacent to the Leinster Massif and in the Rathdowney Trend with different characteristic features (Fig. 10C). In the Rathdowney Trend, extensional faults served as conduits for high-temperature (probably basement-derived) fluids to reach the Waulsortian Limestone and its equivalent (Russell 1986; Hitzman 1995*b*), where they subsequently mixed with lower temperature brines (Gregg *et al.* 2001; Johnson *et al.* 2001; Wright 2001). It is speculated that the refluxing hypersaline brine migrated westward from the Leinster Massif area through the carbonate platform and mixed with basement-derived, presumably metal-bearing ore fluids in the Rathdowney Trend (Fig. 10C). Fluid-inclusion studies of ore and gangue minerals have shown that evaporated Carboniferous seawater was one of the major fluid components involved in ore mineralization (Gleeson *et al.* 1999; Banks *et al.* 2002; Johnson 2003).

In the area adjacent to the Leinster Massif, early diagenetic replacive dolomite was overprinted by brines that had interacted with local basement rocks (Fig. 10B & C). These brines had distinct geochemical characteristics (higher $^{87}Sr/^{87}Sr$ ratios, higher temperatures: Becker *et al.* 2002; Becker 2004) from those present in the ore-mineralized Rathdowney Trend. Although refluxing brines may have interacted with fluids that circulated through local basement rocks of the Leinster Massif, we interpret the absence of ore deposits in the area adjacent to this as indicating that the local, chemically distinct, basement-involved fluids did not scavenge sufficient metals from the Massif (Fig. 10C).

Evaporite cementation of the shelf carbonate strata, in both SE Ireland and the Permian Basin, was completed within a few million years (Scholle *et al.* 1992; Nagy 2003; Nagy *et al.* 2005). Former evaporite cements in the Dur-2 core are restricted to Arundian shelf strata, implying that the migration of hypersaline brine from the upper Chadian peritidal sequence to the shelf lasted approximately 3–5 Ma. This is in agreement with numerical modelling of brine migration through a carbonate platform that has shown that refluxing brines will continue to sink, even when brine-generating conditions cease (Jones *et al.* 2002; Whitaker *et al.* 2004).

Another important implication for this study is the timing of Zn–Pb mineralization in the Rathdowney Trend. It could be as early as Arundian, if the excess concentration of chloride found in the fluid inclusions of ores (Gleeson *et al.* 1999; Banks *et al.* 2002; Johnson *et al.* 2002; Johnson 2003), indicative of halite dissolution, originated from evaporites proximal to the Leinster Massif.

A significant difference between the two dolomitized areas, SE Ireland and the Permian Basin, is the timing of silica replacement of evaporite cements. This is likely to have occurred between the latest Permian and the early Tertiary in the Permian Basin (Ulmer-Scholle *et al.* 1993). Silicification was coeval with, or post-dated, hydrocarbon migration based on abundant mixed hydrocarbon–brine inclusions in the silica (Scholle *et al.* 1992; Ulmer-Scholle *et al.* 1993). The evaporites in the peritidal sequence of the SE Irish Midlands were replaced before the formation of medium-crystalline planar dolomite, and also before the late diagenetic dolomite cementation and mineralization of the shelf sequence (Fig. 3) (Nagy 2003). Gregg *et al.* (2001) showed that marine cementation reduced primary porosity to less than 1% in much of the Waulsortian Limestone. If early diagenetic dolomitization enhanced porosity in the strata overlying the Waulsortian, this porosity was largely occluded by subsequent evaporite (and silica) cement. Consequently, there was no open pore space for further fluid migration, and late diagenetic flow was restricted to fractures.

Conclusions

Fabric-preserving, fine-crystalline, planar-s replacive dolomite predominates in the proximal shelf area of the Irish Midlands, and is associated with sedimentological and petrographical features typical of an arid peritidal sequence. Post-compactional dolomites were formed in oolitic grainstones during early diagenesis of the sediments. Volumetrically more significant, medium-crystalline, planar-e and planar-s dolomites are fabric-destructive and replace skeletal wackestones–grainstones.

Carbon and oxygen isotope compositions of the dolomites of the study area indicate formation from, or later alteration under the influence of, water depleted in ^{18}O with respect to seawater or at elevated temperature. It is likely that the brines generated on the peritidal flats of the proximal shelf became enriched in ^{18}O during evaporation, but subsequent multiple-stage diagenetic fluids also geochemically overprinted the rocks. Low $\delta^{13}C$ values preserved in the proximal sequence of the Aghmacart Formation indicate multiple subaerial exposure surfaces.

Based on petrographic and stable isotope evidence, we propose that hypersaline brines migrated westward through the porous platform carbonates, causing partial dolomitization and evaporite cementation, and changed the porosity distribution of the carbonate platform. These brines may have contributed excess chloride in the Rathdowney Trend ore system recognized in high-temperature mineralizing fluids.

An Arundian or younger age for Zn–Pb mineralization in the Rathdowney Trend is possible, based on the interpreted age of emplacement of evaporite cements in the post-Waulsortian strata. This timing is compatible with numerical fluid-flow modelling of hypersaline brine migration through the carbonate platform.

We acknowledge support from the American Association of Petroleum Geologists and the Jefferson Smurfit Corporation to Zs. R. Nagy. The National Science Foundation (NSF-INT-9729653 and NSF-EAR-0106388) and the donors of the Petroleum Research Fund, administered by the American Chemical Society (PRF 35893-AC8) provided support to J. M. Gregg and K. L. Shetton. The University of Missouri Research Board provided support to K. L. Shelton. We thank A. Bowden (ARCON Exploration plc) and A. Sleeman (Geological Survey of Ireland) for allowing us access to sample subsurface drillcores and to publish the results. We thank W. Wright (Robertson Research) for field assistance, T. Culligan, J. Kennedy and A. Keogh (UCD), and M. Roberson (UMR) for help in thin section preparation, photography and computer graphics, and T. McIntyre (GSI) for logistical help in sampling core.

An earlier version of this paper greatly benefited from the kind reviews by A. M. M. Aqrawi and an anonymous author.

References

ADAMS, J.E. & RHODES, M.L. 1960. Dolomitization by seepage reflux. *AAPG Bulletin*, **44**, 1912–1920.

ALLEN, J.R. & MATTHEWS, R.K. 1982. Isotope signatures associated with early meteoric diagenesis. *Sedimentology*, **29**, 797–817.

ANDERSON, T.F. & ARTHUR, M.A. 1983. Stable isotopes of oxygen and carbon and their application to sedimentologic and paleoenvironmental problems. *In*: ARTHUR, M.A. (ed.) *Stable Isotopes in Sedimentary Geology*. Society of Economic Paleontologists and Mineralogists, Short Course, **10**, 1.1–1.151.

ANDREASON, M.W. 1992. Coastal siliciclastic sabkhas and related evaporite environments of the Permian Yates Formation, North Ward-Estes Field, Wars County, Texas. *AAPG Bulletin*, **76**, 1735–1759.

BANKS, D.A., BOYCE, A.J. & SAMSON, I.M. 2002. Constraints on the origin of fluids forming Irish Zn–Pb–Ba deposits: Evidence from composition of fluid inclusions. *Economic Geology*, **97**, 471–480.

BANNER, J.L. 1995. Application of trace element and isotope geochemistry of strontium to studies of carbonate diagenesis. *Sedimentology*, **42**, 805–824.

BANNER, J.L. & KAUFMAN, J. 1994. The isotopic record of ocean chemistry and diagenesis preserved in non-luminescent brachiopods from Mississippian carbonate rocks, Illinois and Missouri. *Geological Society of America Bulletin*, **106**, 1074–1082.

BANNER, J.L., MUSGROVE, M. & CAPO, R. 1994. Tracing ground-water evolution in a limestone aquifer using Sr isotopes: effects of multiple sources of dissolved ion and mineral-solution reactions. *Geology*, **22**, 687–690.

BECKER, S.P. 2004. *Brine migration in Chadian–Brigantian carbonate strata of southeast Ireland*. MS thesis, University of Missouri-Rolla, MO.

BECKER, S.P., NAGY, ZS. R., JOHNSON, A.W., GREGG, J.M., SHELTON, K.L., SOMERVILLE, I.D. & WRIGHT, W.R. 2002. Relationship of evaporites to brine migration and carbonate hosted Zn–Pb mineralization in the Lower Carboniferous of Southeastern Ireland. *Geological Society of America, Abstracts with Programs*, **34**, 6, A338.

BORER, J.M. & HARRIS, P.M. 1991. Lithofacies and cyclicity of the Yates Formation, Permian Basin: Implications for reservoir heterogeneity. *AAPG Bulletin*, **75**, 726–779.

BRAND, U. 1982. The oxygen and carbon isotope composition of Carboniferous fossil components: seawater effect. *Sedimentology*, **29**, 139–147.

BRAND, U. 1989. Biochemistry of Palaeozoic North American brachiopods and secular variation of seawater composition. *Biogeochemistry*, **7**, 159–193.

BRUCKSCHEN, P., OESMANN, S. & VEIZER, J. 1999. Isotope stratigraphy of the European Carboniferous: proxy signals for ocean chemistry, climate and tectonics. *Chemical Geology*, **161**, 127–163.

CHOWNS, T.M. & ELKINS, J.E. 1974. The origin of quartz geodes and cauliflower cherts through the silicification of anhydrite nodules. *Journal of Sedimentary Petrology*, **3**, 885–903.

CORFIELD, S.M., GAWTHORPE, R.L., GAGE, M., FRASER, A.J. & BESLY, B.M. 1996. Inversion tectonics of the Variscan foreland of the British Isles. *Journal of the Geological Society, London*, **153**, 17–32.

CÓZAR P. & SOMERVILLE, I.D. 2005. Stratigraphy of Upper Viséan rocks in the Carlow area, southeast Ireland. *Geological Journal*, **40**(1).

DEFFEYES, K.S., LUCIA, F.J. & WEYL, P.K. 1965. Dolomitization of Recent and Plio-Pleistocene sediments by marine evaporite waters on Bonaire, Netherlands Antilles. *In*: PRAY, L.C. & MURRAY, R.C. (eds) *Dolomitization and Limestone Diagenesis, A Symposium*. Society of Economic Paleontologists and Mineralogists, Special Publications, **13**, 89–111.

DICKSON, J.A.D. 1966. Carbonate identification and genesis revealed by staining. *Journal of Sedimentary Petrology*, **36**, 491–505.

ELORZA, J.J. & GARCIA-GARMILLA, F. 1993. Chert appearance in the Cueva-Bedón carbonate platform (upper Cretaceous, northern Spain). *Geological Magazine*, **130**, 805–816.

ELORZA, J.J. & RODRIGUEZ-LAZARO, J. 1984. Late Cretaceous quartz geodes after anhydrite from

Burgos, Spain. *Geological Magazine*, **121**, 107–113.

EVERETT, C.E., WILKINSON, J.J. & RYE, D.M. 1999. Fracture-controlled fluid flow in the Lower Palaeozoic basement rocks of Ireland: implications for the genesis of Irish-type Zn–Pb deposits. *In*: MCCAFFREY, K.J., LONERGAN, L. & WILKINSON, J.J. (eds) *Fractures, Fluid Flow and Mineralization*. Geological Society, London, Special Publications, **155**, 247–276.

FOLK, R.L. & PITTMAN, J.S. 1971. Length-slow chalcedony: a new testament for vanished evaporites. *Journal of Sedimentary Petrology*, **41**, 1045–1058.

FRIEDMAN, I. & O'NEIL, J.R. 1977. *Composition of Stable Isotope Fractionation Factors of Geochemical Interest*. US Geological Survey, Professional Paper, 440-KK.

GARBER, R.A., GROVER, G.A. & HARRIS, P.M. 1989. Geology of the Capitan shelf margin – subsurface data from the northern Delaware Basin. *In*: HARRIS, P.M. & GROVER, G.A. (eds) *Subsurface and Outcrop Examination of the Capitan Shelf Margin, Northern Delaware Basin*. Society of Economic Paleontologists and Mineralogists, Core Workshop, **13**, 3–269.

GARBER, R.A., HARRIS, P.M. & BORER, J.M. 1990. Occurrence and significance of magnesite in Upper Permian (Guadalupian) Tansill and Yates Formations, Delaware Basin, New Mexico. *AAPG Bulletin*, **74**, 119–134.

GATLEY, S., SOMERVILLE, I.D., MORRIS, J.H., SLEEMAN, A.G. & EMO, G. 2004. *Geology of Galway – Offaly, and Adjacent Parts of Westmeath, Tipperary, Laois, Clare and Roscommon: A Geological Description to Accompany the Bedrock Geology 1:100 000 Scale Map Series, Sheet 15, Galway – Offaly*. Geological Survey of Ireland, in press.

GIVEN, R.K. & LOHMANN, K.C. 1986. Isotopic evidence for the early meteoric diagenesis of the reef facies, Permian Reef Complex of west Texas and New Mexico. *Journal of Sedimentary Petrology*, **56**, 183–193.

GLEESON, S.A., BANKS, D.A., EVERETT, C.E., WILKINSON, J.J. & SAMSON, I.M. 1999. Origin of mineralising fluids in Irish-type deposits: Constraints from halogens. *In*: STANLEY, C.J. ET AL. (eds) *Mineral Deposits: Processes to Processing. 5th Biennial SGA Meeting, Proceedings*, Volume 2. Balkema, Rotterdam, 857–860.

GREGG, J.M. & SHELTON, K.L. 1990. Dolomitization and dolomite neomorphism in the back reef facies of the Bonneterre and Davis Formations (Cambrian), southeastern Missouri. *Journal of Sedimentary Petrology*, **60**, 549–562.

GREGG, J.M. & SIBLEY, D.F. 1984. Epigenetic dolomitization and the origin of xenotopic dolomite texture. *Journal of Sedimentary Petrology*, **54**, 908–931.

GREGG, J.M., HOWARD, S.A. & MAZZULLO, S.J. 1992. Early diagenetic recrystallization of Holocene (<3000 years old) peritidal dolomites, Ambergis Cay, Belize. *Sedimentology*, **39**, 143–160.

GREGG, J.M., SHELTON, K.L., JOHNSON, A.W., SOMERVILLE, I.D. & WRIGHT, W.R. 2001. Dolomitization of the Waulsortian Limestone (Lower Carboniferous) in the Irish Midlands. *Sedimentology*, **48**, 745–766.

HARRIS, P.M. & SALLER, A.H. 1999. Subsurface expression of the Capitan depositional system and implications for hydrocarbon reservoirs, northeastern Delaware Basin. *In*: SALLER, A.H., HARRIS, P.M., KIRKLAND, B.L. & MAZZULO, S.J. (eds) *Geologic Framework of the Capitan Reef*. Society of Economic Paleontologists and Mineralogists, Special Publications, **65**, 37–49.

HIRD, K. & TUCKER, M.E. 1988. Contrasting diagenesis of two Carboniferous oolites from South Wales: a tale of climatic influence. *Sedimentology*, **35**, 587–602.

HIRD, K., TUCKER, M.E. & WATERS, R.A. 1987. Petrography, geochemistry and origin of Dinantian dolomites from South-east Wales. *In*: MILLER, J., ADAMS, A.E. & WRIGHT, V.P. (eds) *European Dinantian Environments*. Wiley, Chichester, 359–377.

HITZMAN, M.W. 1995*a*. Geologic setting of the Irish Zn–Pb–(Ba–Ag) orefield. *In*: ANDERSON, K., ASHTON, J., EARLS, G., HITZMAN, M.W. & TEAR, S. (eds) *Irish Carbonate-hosted Zn–Pb Deposits*. Guidebook Series, Irish Association of Economic Geology and Society of Economic Geology, Dublin, **21**, 3–24.

HITZMAN, M.W. 1995*b*. Mineralization in the Irish Zn–Pb–(Ba–Ag) orefield. *In*: ANDERSON, K., ASHTON, J., EARLS, G., HITZMAN, M.W. & TEAR S. (eds) *Irish Carbonate-hosted Zn–Pb Deposits*. Guidebook Series, Irish Association of Economic Geology and Society of Economic Geology, Dublin, **21**, 25–61.

HITZMAN, M.W. 1999. Extensional faults that localize Irish syndiagenetic Zn–Pb deposits and their reactivation during Variscan compression. *In*: MCCAFFREY, K.J., LONERGAN, L. & WILKINSON, J.J. (eds) *Fractures, Fluid Flow and Mineralization*. Geological Society, London, Special Publications, **155**, 233–245.

HITZMAN, M.W. & BEATY, D.W. 1996. The Irish Zn–Pb–(Ba) orefield. *In*: SANGSTER, D.F. (ed.) *Carbonate Hosted Lead–Zinc Deposits*. Society of Economic Geologists, Special Publications, **4**, 112–143.

HUMPHREY, J.D., RANSOM, K.L. & MATTHEWS, R.K. 1986. Early meteoric diagenetic control of Upper Smackover Production, Oaks Field, Louisiana. *AAPG Bulletin*, **70**, 70–85.

ILLING, L.V., WELLS, A.J. & TAYLOR, J.C.M. 1965. Penecontemporary dolomite in the Persian Gulf. *In*: PRAY, L.C. & MURRAY, R.C. (eds) *Dolomitization and Limestone Diagenesis, A Symposium*. Society of Economic Paleontologists and Mineralogists, Special Publications, **13**, 89–111.

JAMES, N.P. & CHOQUETTE, P.W. 1984. Diagenesis 9 – Limestones – The meteoric diagenetic environment. *Geoscience Canada*, **11**, 161–194.

JOHNSON, A.W. 2003. *Regional-scale geochemical analysis of carbonate cements: Reconstructing multiple fluid interaction related dolomitization and mineralization in Lower Carboniferous rocks of the Irish Midlands*. PhD thesis, University of Missouri-Columbia, MO.

JOHNSON, A.W., SHELTON, K.L., GREGG, J.M., SOMERVILLE, I.D. & WRIGHT, W.R. 2001. Fluid inclusion evidence for the presence of multiple fluids in Zn–Pb hosting Carboniferous carbonate rocks in the Irish Midlands: Initial findings. *In*: HAGNI, R.D. (ed.) *Studies on Ore Deposits, Mineral Economics, and Applied Mineralogy: With Emphasis on Mississippi-type Base Metal and Carbonatite-related Ore Deposits*. University of Missouri-Rolla, MO 18–30.

JOHNSON, A.W., SHELTON, K.L., GREGG, J.M., NAGY, ZS. R., BECKER, S.P., SOMERVILLE, I.D. & WRIGHT, W.R. 2002. Regional-scale isotopic, geochemical, and petrologic studies of fluid migration and dolomitization of Dinantian rocks of the Irish Midlands: Relationship to base-metal ore formation. *In*: *Workshop on Geodynamic Setting of Major Basin-hosted Lead–Zinc Mineral Provinces, Barcelona*, Abstract Volume, 29–32.

JONES, G.D., WHITAKER, F.F., SMART, P.L. & SANFORD, W.E. 2002. Fate of reflux brines in carbonate platforms. *Geology*, **30**, 371–374.

LAND, L.S. 1980. The isotopic and trace element geochemistry of dolomite: The state of the art. *In*: ZENGER, D.H., DUNHAM, J.B. & ETHINGTON, R.L. (eds) *Concepts and Models of Dolomitization*. Society of Economic Paleontologists and Mineralogists, Special Publications, **28**, 87–110.

LAND, L.S. 1983. The applications of stable isotopes to studies of the origin of dolomite and to problems of diagenesis of clastic sediments. *In*: ARTHUR, M.A. & ANDERSON, T.F. (eds) *Stable Isotopes in Sedimentary Geology*. Society of Economic Paleontologists and Mineralogists, Short Course, **10**, 4.1–4.22.

LAND, L.S. 1985. The origin of massive dolomite. *Journal of Geological Education*, **33**, 112–125.

LEEDER, M.R. 1987. Tectonic and palaeogeographic models for Lower Carboniferous Europe. *In*: MILLER, J., ADAMS, A.E. & WRIGHT, V.P. (eds) *European Dinantian Environments*. Wiley, Chichester, 1–20.

LEES, A. 1964. The structure and origin of the Waulsortian (Lower Carboniferous) reefs of west-central Eire. *Philosophical Transactions of the Royal Society of London*, **B247**, 483–531.

LOHMANN, K.C. 1988. Geochemical patterns of meteoric diagenetic systems and their application to studies of paleokarst. *In*: JAMES, N.P. & CHOQUETTE, P.W. (eds) *Paleokarst*. Springer, New York, 58–80.

MCCONNELL, B., PHILCOX, M.E., SLEEMAN, A.G., STANLEY, G., FLEGG, A.M., DALY, E.P. & WARREN, W.P. 1994. *Geology of Kildare – Wicklow. A Geological Description to Accompany the Bedrock Geology 1:100 000 Map Series, Sheet 16, Kildare–Wicklow*. Geological Survey of Ireland.

MCCREA, J.M. 1950. The isotope chemistry of carbonates and a paleotemperature scale. *Journal of Chemical Physics*, **18**, 849–857.

MCKENZIE, J.A. 1981. Holocene dolomitization of calcium carbonate sediments from the coastal sabkhas of Abu Dhabi, U.A.E.: A stable isotope study. *Journal of Geology*, **89**, 185–198.

MCKENZIE, J.A., HSÜ, K.J. & SCHNEIDER, J.F. 1980. Movement of subsurface waters under the sabkha, Abu Dhabi, UAE, and its relationship to evaporative dolomite genesis. *In*: ZENGER, D.H. & ETHINGTON, R.L. (eds) *Concepts and Models of Dolomitization*. Society of Economic Paleontologist and Mineralogists, Special Publications, **28**, 11–30.

MELIM, L.A. & SCHOLLE, P.A. 1989. Dolomitization model for the forereef of the Permian Capitan Formation, Guadalupe Mountains, Texas-New Mexico. *In*: HARRIS, P.M. & GROVER, G.A. (eds) *Subsurface and Outcrop Examination of the Capitan Shelf Margin, Northern Delaware Basin*. Society of Economic Paleontologists and Mineralogists, Core Workshop, **13**, 407–413.

MELIM, L.A. & SCHOLLE, P.A. 2002. Dolomitization of the Capitan Formation forereef facies (Permian, west Texas and New Mexico): seepage reflux revisited. *Sedimentology*, **49**, 1207–1227.

MILLIKEN, K.L. 1979. The silicified evaporite syndrome – two aspects of silicification history of former evaporite nodules from Southern Kentucky and Northern Tennessee. *Journal of Sedimentary Petrology*, **49**, 245–256.

MOORE, C.H., CHOWDHURY, A. & CHAN, L. 1988. Upper Jurassic Smackover platform dolomitization, northwestern Gulf of Mexico: A tale of two waters. *In*: SHUKLA, V. & BAKER, P.A. (eds) *Sedimentology and Geochemistry of Dolostones*. Society of Economic Paleontologists and Mineralogists, Special Publications, **43**, 175–189.

MORRISON, J.O. & BRAND, U. 1986. Paleocene: 5. Geochemistry of recent marine invertebrates. *Geoscience Canada*, **13**, 237–254.

MÜLLER, D.W., MCKENZIE, J.A. & MUELLER, P.A. 1991. Abu Dhabi Sabkha, Persian Gulf, revisited: Application of strontium isotopes to test an early dolomitization model. *Geology*, **18**, 618–621.

NAGY, ZS. R. 2003. *The Lower Carboniferous (Chadian–Brigantian) geology of shallow marine carbonate rocks in the central and southeast Irish Midlands and the Wexford Outlier*. PhD thesis, University of Missouri-Rolla, MO.

NAGY, ZS. R., SOMERVILLE, I.D., BECKER, S., GREGG, J.M., WRIGHT, W., JOHNSON, A.W. & SHELTON, K.L. 2001. Petrology and biostratigraphy of carbonate rocks from the ?lower to upper Viséan (Lower Carboniferous) Milford Formation, Co. Carlow, Ireland. *Geological Society of America, Abstracts with Programs*, **33**, 6, A-443.

NAGY, ZS. R., SOMERVILLE, I.D., GREGG, J.M., BECKER, S.P. & SHELTON, K.L. 2005. Lower Carboniferous peritidal carbonates and associated evaporites adjacent to the Leinster Massif, southeast Irish Midlands. *Geological Journal*, **40**.

NIELSEN, P, FLIPKENS, V., GROESSENS, E. & SWENNEN, R. 2000. Sedimentology and diagenesis of the Dinantian succession in the Vinalmont borehole. *Geologica Belgica*, **3**, 369–393.

PATTERSON, R.J. & KINSMAN, D.J.J. 1982. Formation of diagenetic dolomite in coastal sabkha along Arabian (Persian) Gulf. *AAPG Bulletin*, **66**, 28–43.

PEACE, W.P. & WALLACE, M.W. 2000. Timing of mineralization at the Navan Zn–Pb deposit: a post-Arundian age for mineralization. *Geology*, **28**, 711–714.

PEETERS, C., SWENNEN, R., NIELSEN, P. & MUCHEZ, PH. 1992. Sedimentology and diagenesis of the Viséan carbonates in the Vesdre area (Verviers synclinorum, E. Belgium). *Zentralblatt für Geologie und Paläontologie, Teil 1*, **H5**, 519–547.

PHILCOX, M.E. 1984. *Lower Carboniferous Lithostratigraphy of the Irish Midlands*. Irish Association for Economic Geology, Dublin.

POPP, B.N., ANDERSON, T.F. & SANDBERG, P.A. 1986. Brachiopods as indicators of original isotopic composition in some Paleozoic limestones. *Geological Society of America Bulletin*, **97**, 1262–1269.

RUSH, P.F. & CHAFETZ, H.S. 1990. Fabric-retentive, non-luminescent brachiopods as indicators of original $\delta^{13}C$ and $\delta^{18}O$ composition: A test. *Journal of Sedimentary Petrology*, **60**, 968–981.

RUSSELL, M.J. 1986. Extension and convection: a genetic model for the Irish Carboniferous base metal and barite deposits. *In*: ANDREW, C.J., CROWE, R.W., FINLAY, S., PENNEL, W.M. & PYNE, J.F. (eds) *Geology and Genesis of Mineral Deposits in Ireland*. Irish Association for Economic Geology, Dublin, 545–554.

SARG, J.F. 1981. Petrology of the carbonate–evaporite facies transition of the Seven Rivers Formation (Guadalupian, Permian) southeast New Mexico. *Journal of Sedimentary Petrology*, **51**, 73–95.

SCHOLLE, P.A., ULMER, D.S. & MELIM, L.A. 1992. Late-stage calcite in the Permian Capitan Formation and its equivalents, Delaware Basin margin, west Texas and New Mexico: evidence for replacement of precursor evaporites. *Sedimentology*, **39**, 207–234.

SEARL, A. 1989. Diagenesis of the Gully Oolite (Lower Carboniferous), South Wales. *Geological Journal*, **24**, 275–293.

SEVASTOPULO, G.D. & REDMOND, P. 1999. Age of mineralization of carbonate-hosted, base metal deposits in the Rathdowney Trend, Ireland. *In*: MCCAFFREY, K.J.W., LONERGAN, L. & WILKINSON, J.J. (eds) *Fractures, Fluid Flow and Mineralization*. Geological Society, London, Special Publications, **151**, 967–975.

SHINN, E.A., GINSBURG, R.N. & LLOYD, R.M. 1965. Recent supratidal dolomite from Andros Island, Bahamas. *In*: PRAY, LL. C. & MURRAY, R.C. (eds) *Dolomitization and Limestone Diagenesis, A Symposium*. Society of Economic Paleontologists and Mineralogists, Special Publications, **13**, 112–123.

SIBLEY, D.F. & GREGG, J.M. 1987. Classification of dolomite rock textures. *Journal of Sedimentary Petrology*, **57**, 967–975.

SIEDLECKA, A. 1972. Length-slow chalcedony and relicts of sulphates – evidences of evaporitic environments in the Upper Carboniferous and Permian beds of Bear Island, Svalbard. *Journal of Sedimentary Petrology*, **42**, 812–816.

SIEDLECKA, A. 1976. Silicified Precambrian evaporite nodules from northern Norway: a preliminary report. *Sedimentary Geology*, **16**, 161–175.

SOMERVILLE, I.D. 2003. Review of Irish Lower Carboniferous (Mississippian) mud-mounds: depositional setting, biota, facies and evolution. *In*: AHR, W., HARRIS, A.P., MORGAN, W.A. & SOMERVILLE, I.D. (eds) *Permo-Carboniferous Carbonate Platforms and Reefs*. Society of Economic Paleontologists and Mineralogists, Special Publications, **78**; AAPG Memoirs, **83**, 239–252.

SOMERVILLE, I.D., STROGEN, P., MITCHELL, W.I., SOMERVILLE, H.E.A. & HIGGS, K.T. 2001. Stratigraphy of Dinantian rocks in WB3 borehole from Co. Armagh, N. Ireland. *Irish Journal of Earth Sciences*, **19**, 51–78.

STROGEN, P., SOMERVILLE, I.D., PICKARD, N.A.H., JONES, G.LL. & FLEMING, M. 1996. Controls on ramp, platform and basin sedimentation in the Dinantian of the Dublin Basin and Shannon Trough, Ireland. *In*: STROGEN, P., SOMERVILLE, I.D. & JONES, G.LL. (eds) *Recent Advances in Lower Carboniferous Geology*. Geological Society, London, Special Publications, **107**, 263–279.

SWENNEN, R. & VIAENE, W. 1986. Occurrence of pseudomorphosed anhydrite nodules in the Lower Viséan (Lower Moliniacian of the Verviers Synclinorium, E. Belgium). *Bulletin de la Société Belge de Géologie*, **95**, 89–99.

SWENNEN, R., VIAENE, W. & CORNELISSEN, C. 1990. Petrography and geochemistry of the Belle Roche breccia (lower Viséan, Belgium): evidence for brecciation by evaporite dissolution. *Sedimentology*, **37**, 859–878.

TIETZSCH-TYLER, D., SLEEMAN, A.G., MCCONNELL, B., DALY, E.P., FLEGG, A.M., O'CONNOR, P.J., PHILCOX, M.E. & WARREN, W.P. 1994. *Geology of Carlow – Wexford. A Geological Description to Accompany the Bedrock Geology 1:100 000 Scale Map Series, Sheet 19, Carlow – Wexford*. Geological Survey of Ireland.

TUCKER, M.E. 1976. Replaced evaporites from the Late Precambrian of Finnmark, Arctic Norway. *Sedimentary Geology*, **16**, 193–204.

TUCKER, M.E. & WRIGHT, V.P. 1990. *Carbonate Sedimentology*. Blackwell Scientific, Oxford.

ULMER-SCHOLLE, D.S., SCHOLLE, P.A. & BRADY, P.V. 1993. Silicification of evaporites in Permian (Guadalupian) back-reef carbonates of the Delaware Basin, west Texas and New Mexico. *Journal of Sedimentary Petrology*, **63**, 955–965.

VEIZER, J., FRITZ, P. & JONES, B. 1986. Geochemistry of brachiopods: Oxygen and carbon isotopic record of Paleozoic oceans. *Geochimica et Cosmochimica Acta*, **50**, 1679–1696.

WARD, R.F, KENDALL, C.G.ST. C. & HARRIS, P.M. 1986. Upper Permian (Guadalupian) facies and their association with hydrocarbons – Permian Basin, west Texas and New Mexico. *AAPG Bulletin*, **70**, 239–262.

West, I., Branson, A. & Smith, M. 1968. A tidal flat evaporitic facies in the Viséan of Ireland. *Journal of Sedimentary Petrology*, **38**, 1079–1093.

Whitaker, F.F., Smart, P.L. & Jones, G.D. 2004. Dolomitization: from conceptual to numerical models. *In*: Braithwaite, C.J.R., Rizzi, G. & Darke, G. (eds) *The Geometry and Petrogenesis of Dolomite Hydrocarbon Reservoirs*. Geological Society, London, Special Publications, **235**, 99–139.

Wilkinson, J.J., Boyce, A.J., Everett, C.E. & Lee, M.J. 2003. Timing and depth of mineralization in the Irish Zn–Pb orefield. *In*: Kelly, J.G., Andrew, C.J., Ashton, J.H., Boland, M.B., Earls, G., Fusciardi, L. & Stanley, G. (eds) *Europe's Major Base Metal Deposits*. Irish Association for Economic Geology, Dublin, 483–497.

Wright, W.R. 2001. *Dolomitization, fluid-flow and mineralization of the Lower Carboniferous rocks of the Irish Midlands and Dublin Basin Regions*. PhD thesis, University College Dublin.

Wright, W.R., Somerville, I.D., Gregg, J.M., Johnson, A.W. & Shelton, K.L. 2003. Dolomitization and neomorphism of Irish Lower Carboniferous (Early Mississippian) limestones: evidence from petrographic and isotopic data. *In*: Ahr, W., Harris, A.P., Morgan, W.A. & Somerville, I.D. (eds) *Permo-Carboniferous Carbonate Platforms and Reefs*. Society of Economic Paleontologists and Mineralogists, Special Publications, **78**; AAPG Memoirs, **83**, 395–408.

Wright, W.R., Somerville, I.D., Gregg, J.M., Shelton, K.L. & Johnson, A.W. 2004. The Petrogenesis of dolomite, regional patterns of dolomitisation and fluid flow in the Lower Carboniferous of the Irish Midlands and Dublin Basin. *In*: Braithwaite, C.J.R., Rizzi, G. & Darke, G. (eds) *The Geometry and Petrogenesis of Dolomite Hydrocarbon Reservoirs*. Geological Society, London, Special Publications, **235**, 75–97.

Genesis of some Carboniferous dolomites in the south and east of Ireland

CLAIRE M. MULHALL & GEORGE D. SEVASTOPULO
Department of Geology, Museum Building, Trinity College, Dublin 2, Ireland
(e-mail: mulhallc@tcd.ie)

Abstract: Dolomitized Tournaisian rocks occur widely in Ireland. They host the major Zn–Pb ore bodies, at Lisheen and Galmoy. Despite significant research and exploration into fluid flow related to both dolomitization and base-metal mineralization, the evolution of the porosity and permeability of the dolomites and precursor limestones remains poorly understood. Study areas, both within and outside the main mineralized zones, have been chosen in an effort to unravel the evolution of the porosity and permeability of the Tournaisian successions. In each area the earliest phase of dolomitization of the limestone lithosome occurred after all of the primary porosity had been occluded. Fluid inclusions from dolomite and authigenic quartz indicate relatively high temperatures (>105 °C) of formation of all phases of dolomite. The fluid-inclusion data demonstrate that there has been no re-equilibration due to later Variscan heating. Laterally restricted bodies of dolomite at Newcastle West and Subulter may have formed under hydrothermal or geothermal conditions. The geometry of the dolomite body at Newcastle West suggests that the dolomitizing fluids moved within a steeply dipping zone. The laterally extensive Regional Dolomite of SE Ireland, examined at Lisheen and in Co. Carlow, is likely to have had a geothermal origin and to have formed at depths of more than 2.5 km. The dolomites are moderately porous with the notable development of vugs; but their permeability is low. The vugs are not related to primary cavities within the limestone.

The fluid flow related to base-metal mineralization hosted within Carboniferous carbonate rocks in Ireland has been the subject of recent study and debate. The search for a model to explain the origin of the major Zn–Pb (Ba) deposits has stimulated research on the sources (Everett *et al.* 1999; Gleeson *et al.* 1999; Banks *et al.* 2002), timing (Sevastopulo & Redmond 1999; Peace & Wallace 2000; Blakeman *et al.* 2002; Boyce *et al.* 2003), and the pathways and mechanisms driving flow of mineralizing fluids (Russell 1978, 1986; Lydon 1986; Hitzman *et al.* 1998). Numerical models (Hazlett 1997) have been used to evaluate competing hypotheses of patterns of flow within the Irish Carboniferous basins. Of the parameters for which values are required for input into these models, the porosity and permeability of the host rocks at the time of mineralization are the least well constrained. As part of a research project on the evolution of the porosity and permeability of the host rocks, sample areas both within and outside the main mineralized zones were chosen for study. Samples of the dolomite and limestone were selected to evaluate whether the permeability of the host carbonate was sufficient to permit significant fluid flow at the time of dolomitization and mineralization. The study areas are at Lisheen, Co. Tipperary, where dolomites host mineralization, and four locations without significant mineralization: Newcastle West, Co. Limerick; Subulter, Co. Cork (Hitzman & Beaty 1996); and Quinagh and Clonmelsh, Co. Carlow (Fig. 1).

Geological setting

Regional geology of the Irish Midlands

Sevastopulo & Wyse-Jackson (2001) provided an overview of the stratigraphy of the Carboniferous rocks of the Irish Midlands. The Tournaisian succession, in which the dolomites under discussion occur, consists of an overall deepening sequence from continental red beds (Old Red Sandstone) through shallow-water marine sandstones, shales and limestones, to deeper water limestones (Ballysteen Formation) deposited on a southward dipping ramp. The culmination of the deepening trend is in the overlying late Tournaisian Waulsortian carbonate mudbank limestones. These are succeeded in the areas discussed by skeletal packstones and grainstones (Crosspatrick Formation; Fig. 2) of late Tournaisian and probably early Viséan age. Younger Viséan rocks in SE Ireland are all of carbonate shelf facies. They are capped by Namurian and

From: BRAITHWAITE, C. J. R., RIZZI, G. & DARKE, G. (eds) 2004. *The Geometry and Petrogenesis of Dolomite Hydrocarbon Reservoirs*. Geological Society, London, Special Publications, **235**, 393–406. 0305-8719/$15.00

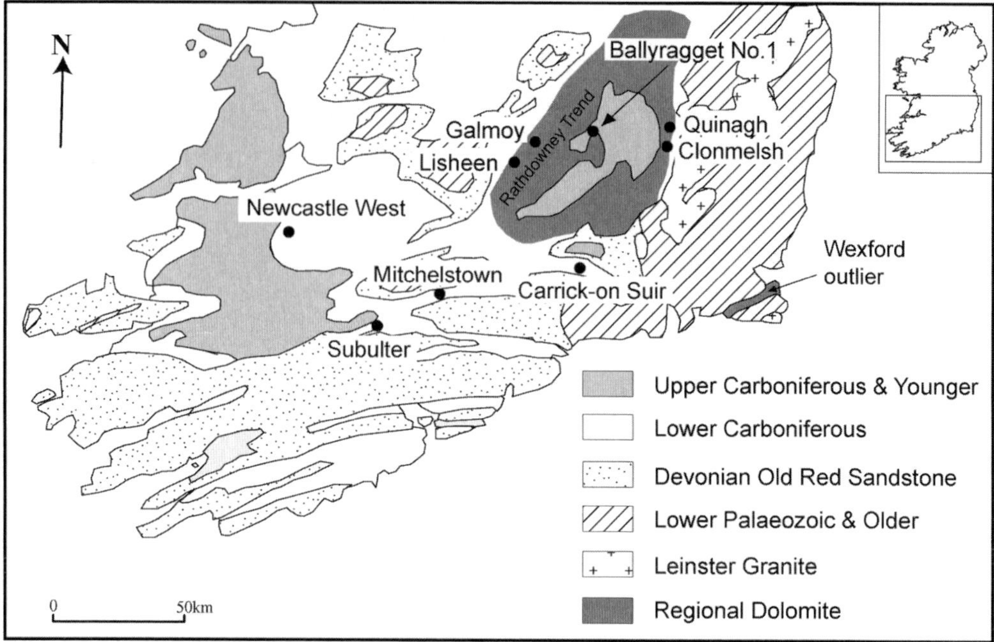

Fig. 1. Geological map of southern Ireland showing the known extent of the Regional Dolomite, the Rathdowney Trend and places cited in the text.

Westphalian siliciclastic rocks. In SE Ireland the Waulsortian Limestone has been pervasively dolomitized on a regional scale. This so-called 'Regional Dolomite' (Hitzman *et al.*, 1992) is the host for the ore bodies currently exploited at Lisheen and Galmoy (Fig. 1).

Description of the Regional Dolomite

The Regional Dolomite consists of two main components: a buff–pale grey, fine-grained, replacive dolomite; and a white, coarse-grained dolomite that later partially, or completely filled open spaces in the grey replacive dolomite (Hitzman *et al.* 1998). The grey replacive dolomite consists of euhedral or subhedral crystals, averaging 100–500 μm in maximum dimension. Their grain size distribution, at the scale of a thin section, is generally unimodal. The white, coarse-grained dolomite has a polymodal crystal size distribution with euhedral–subhedral grains up to 5 mm in size. The crystals have curved faces and exhibit the sweeping extinction characteristic of saddle dolomite (Radke & Mathis 1980). Many are turbid with inclusions. The dolomite is commonly vuggy, with vugs typically 1–3 cm and rarely up to 20 cm in maximum dimension. In thin section intergrain porosity within the white coarse-grained dolomite is commonplace. However, Hitzman *et al.* (2002) reported maximum permeabilities of dolomite from the Rathdowney Trend (see below) of approximately 0.5 millidarcies (mD), except where fracture-related permeability was dominant.

Mineralization and the distribution of the Regional Dolomite

The Irish ore field is the richest in the world per square kilometre (Singer 1995) and all of the significant Zn–Pb mineralization is contained within carbonates of Tournaisian age (Phillips & Sevastopulo 1986). Sphalerite and galena are the principle sulphides developed, and are associated with variable amounts of both barite and pyrite. The deposits at Lisheen and Galmoy are both hosted by Regional Dolomite. They are located within the Rathdowney Trend, a structurally controlled belt of mineralization that extends NE–SW for approximately 40 km (Fig. 1). The Rathdowney Trend is itself part of a broader zone of Regional Dolomite that extends westwards from the Leinster granite and its Lower Palaeozoic envelope to the Lisheen mine area. Within this zone the

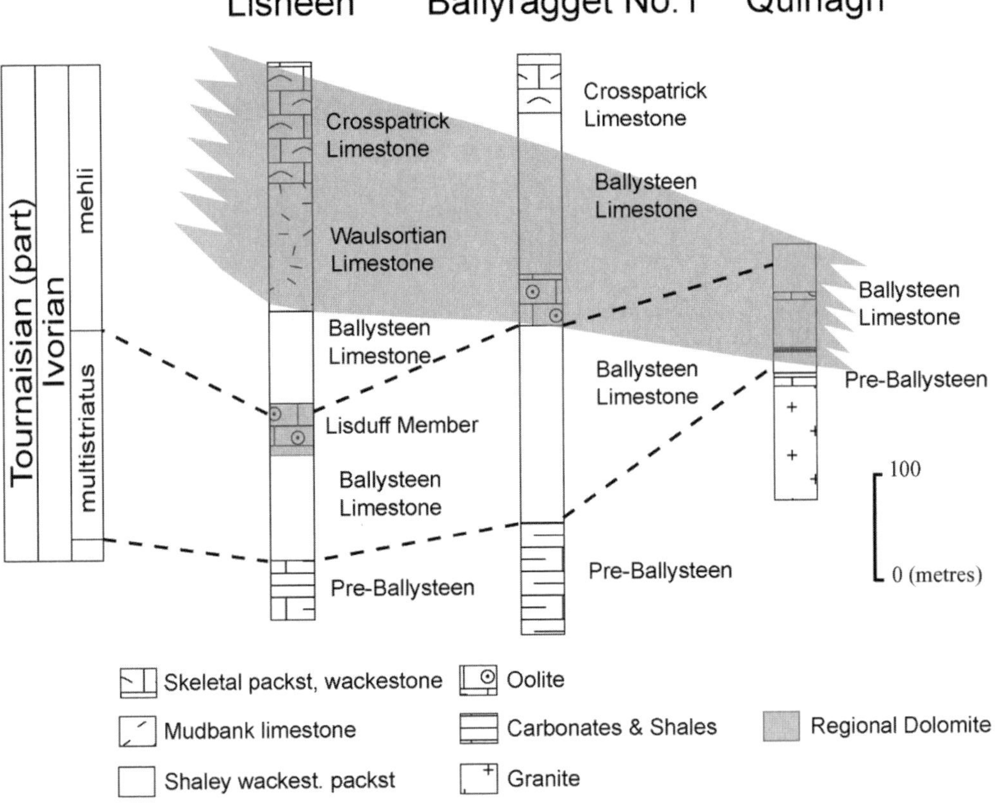

Fig. 2. Schematic view of the stratigraphical distribution of the Regional Dolomite moving from Lisheen in the west to Quinagh in the east using data from Tietzsch-Tyler & Sleeman (1994) for Quinagh and from Sheridan (1977) for the Ballyragget No. 1 borehole.

Waulsortian Limestone is pervasively dolomitized, but around its margins dolomitization is more patchy. This is clearly seen around Lisheen, where drillcore and outcrop include both dolomitized and undolomitized representatives. Mapping around Lisheen suggests that the contact is gradational over a distance of approximately 1 km (Hitzman *et al.* 1992). Across this transition zone the upper part of the Waulsortian tends to be dolomitized, whereas the lower part is less so. Within the Rathdowney Trend, the Ballysteen Limestone below the Waulsortian is generally not dolomitized, except for the oolitic Lisduff Member (Fig. 2). However, in the east close to the Leinster Granite, the Ballysteen Limestone is dolomitized in the Quinagh borehole (Fig. 2) and Clonmelsh Quarry, Co. Carlow (Fig 1), and in the Wexford outlier (the Lower Dolomite: Carter & Wilbur 1986) (Fig. 1). Several other occurrences of dolomitized Waulsortian Limestone have been documented outside this broad zone, for example in the Carrick-on-Suir syncline (Keeley 1980) and in the Mitchelstown syncline (Shearley 1988) (Fig. 1).

Methods

Petrographic analyses were conducted using hand specimens and thin sections that were routinely stained with alizarin red S and potassium ferricyanide (Dickson 1966). Cathodoluminescence (CL) microscopy was carried out using a Reliotron cold cathode stage, mounted on a Nikon Labophot microscope, with a 12–15 kV beam, a current intensity of 0.5 mA and a vacuum of 0.05 torr.

Microthermometric analyses were carried out using a Linkam THMG600 heating–freezing stage mounted on a Nikon Optiphot microscope equipped with ×40, ×60 and ×100 long working distance objectives. Stage calibration was carried out using the techniques described by Shepherd *et al.* (1985). Accuracy is estimated to be ±0.2 °C in the –100 to 30 °C range and 0.5 °C at higher temperatures. The precision of the measurements, determined from repeated analyses on synthetic inclusions, is 0.1 °C. Homogenization temperatures (T_h) and, where possible, first melting temperatures (T_{fm}), ice melting temperatures ($T_{m_{ice}}$) and hydrate melting temperatures (T_{m_h}) were measured for each inclusion. To limit any error in estimating the salinities, they are calculated as combined wt% NaCl and $CaCl_2$ equivalent using a numerical algorithm developed by Naden (1996). This was done because the eutectic temperatures of the inclusions are consistent with being part of the NaCl–$CaCl_2$–H_2O ternary system.

Dolomitized Waulsortian Limestone

Samples have been selected across the transition (less then 1 m wide) from weakly dolomitized to pervasively dolomitized Waulsortian Limestone in drillcore at Newcastle West; and from both drillcore and quarry exposures at Lisheen and Subulter (where the transition zone may be up to 1 km wide). These provide evidence of the cementation history of the original limestone and its relationship to dolomitization. The assumption is made that the paragenetic sequence observed in partially dolomitized Waulsortian Limestone is the same as that in the adjoining pervasively dolomitized limestone, where all primary textures have been destroyed.

Original calcite cements

The Waulsortian facies in Ireland has been reviewed in detail by Lees & Miller (1995). Using CL microscopy, the early void-filling calcite cements of the Waulsortian Limestone at Lisheen, Newcastle West and Subulter can be separated into four distinct stages (stages A–D; terminology of Lees & Miller 1995). Stage A are cryptofibrous cements (CFC), and are interpreted as neomorphosed, early marine calcite cements. Under CL, they have a characteristic blotchy appearance, partly non-luminescent and partly brightly luminescent (Fig. 3a). The brightly luminescent parts are microveins and partial replacement of the CFC by later cement generations. Stage B cements are usually non-luminescent, although they may have thin, bright zones within them. They were followed by Stage C cements, which are brightly luminescent, and Stage D, which usually have a complex zoning pattern with many substages, all of which are generally dully luminescent. Stage D cements occluded any remaining porosity (Fig. 3a).

The sequence of cements from black (B) to brightly luminescent (C) to dully luminescent (D) has been observed in original cavities in Waulsortian Limestone throughout Ireland and England (King & Meyers 1985; Gillies 1987; de Brit 1988; Lees & Miller 1995). This pattern is believed to reflect differences in the trace-element chemistry (especially in Mn^{2+} and Fe^{2+}) in the cements, related to the interaction with increasingly reduced pore fluids as a direct consequence of burial (Machel 1985; Lees & Miller 1995). Lees & Miller (1995, p. 249, fig. 46) suggested that cement stages A–D formed at relatively shallow depths of burial (within approximately 500 m of the sea floor). However, Reed & Wallace (2001) argued that at Courtbrown, Co. Limerick (approximately 25 km NNE of Newcastle West) cement stages C and D formed at substantially greater depths because of their timing relative to stylolite formation. Evidence from elsewhere in Ireland (de Brit 1988; Lee & Wilkinson 2002) suggests shallow depths of burial, possibly even shallower than envisaged by Lees & Miller (1995).

Dolomitization of the Waulsortian Limestone at Lisheen, Newcastle West and Subulter occurred after the occlusion of all of the primary porosity (Fig. 3b–d). In partially dolomitized samples the earliest (fine-grained) dolomite preferentially replaced bryozoan fronds (Fig. 3c & d) and micrite (Fig. 3e & f). Cryptofibrous calcite cements and larger, more robust, fossil fragments, such as crinoid ossicles, remained undolomitized, indicating that they were amongst the last components to be replaced (Fig. 3e & f). Small clusters of fine-grained dolomite rhombs occur along dissolution seams that post-date the latest calcite cement (Fig. 4a & b). This implies that during the initial phase of dolomitization the Mg-rich fluids exploited the dissolution seams as conduits for fluid flow.

Minimum estimates of the temperature of formation of the coarse dolomite at Lisheen have been made from primary fluid inclusions. Hitzman *et al.* (1992) reported T_h values ranging from 105 to 233 °C and salinity data ranging from 10 to 13 wt% NaCl equivalent. The higher homogenization temperatures were interpreted as a result of re-equilibration due to a later regional heating event. Four fluid inclusions measured in the current study had T_h values of 173–236 °C (Table 1). One reproducible salinity measurement of 25.6 combined wt% NaCl and $CaCl_2$ equivalent was also recorded.

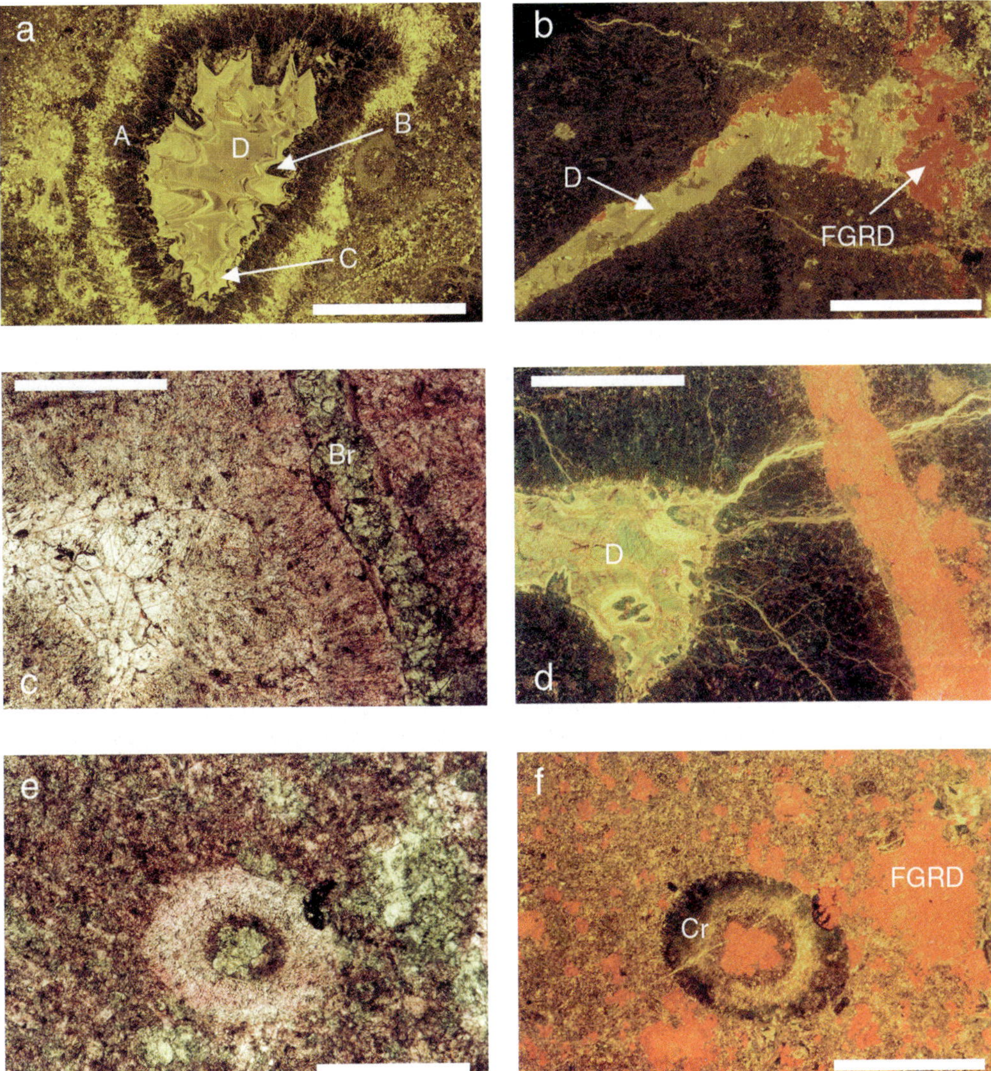

Fig. 3. Photomicrographs illustrating the paragenesis of the Waulsortian Limestone. (**a**) CL view of a cavity within the Waulsortian Limestone at Newcastle West with the early calcite cement stages (A–D) labelled. The last cement stage, Stage D, occludes any remaining porosity within the cavity. (**b**) CL view of a vein of Stage D cement, from within the same sample as (a), illustrating that the fine-grained replacive dolomite (FGRD) post-dates porosity occlusion within this rock. (**c**) PPL view of the selective dolomitization of a bryozoan (Br) at Lisheen. (**d**) Same view in CL. Dolomite clearly post-dates the porosity occluding Stage D cement. (**e**) PPL view of a crinoid ossicle from Lisheen within a micrite-rich matrix. (**f**) Same view in CL (Cr, crinoid ossicle) . Note the selective dolomitization of the micrite that is both surrounding and within the lumen of the ossicle, and the lack of replacement by dolomite of the crinoid ossicle itself. All scale bars are 500 μm.

The dolomite at Newcastle West is indistinguishable from the Regional Dolomite at Lisheen and elsewhere in the Rathdowney Trend. However, it is restricted to a narrow, steeply dipping zone trending NE–SW. It is not possible from the limited subsurface information available to determine whether the zone is structurally controlled. Euhedral quartz crystals, up to 5 mm in length, occur within the micritic parts of the limestone (Fig. 4c) and increase in both size and abundance towards the limestone–dolomite contact. CL petrography

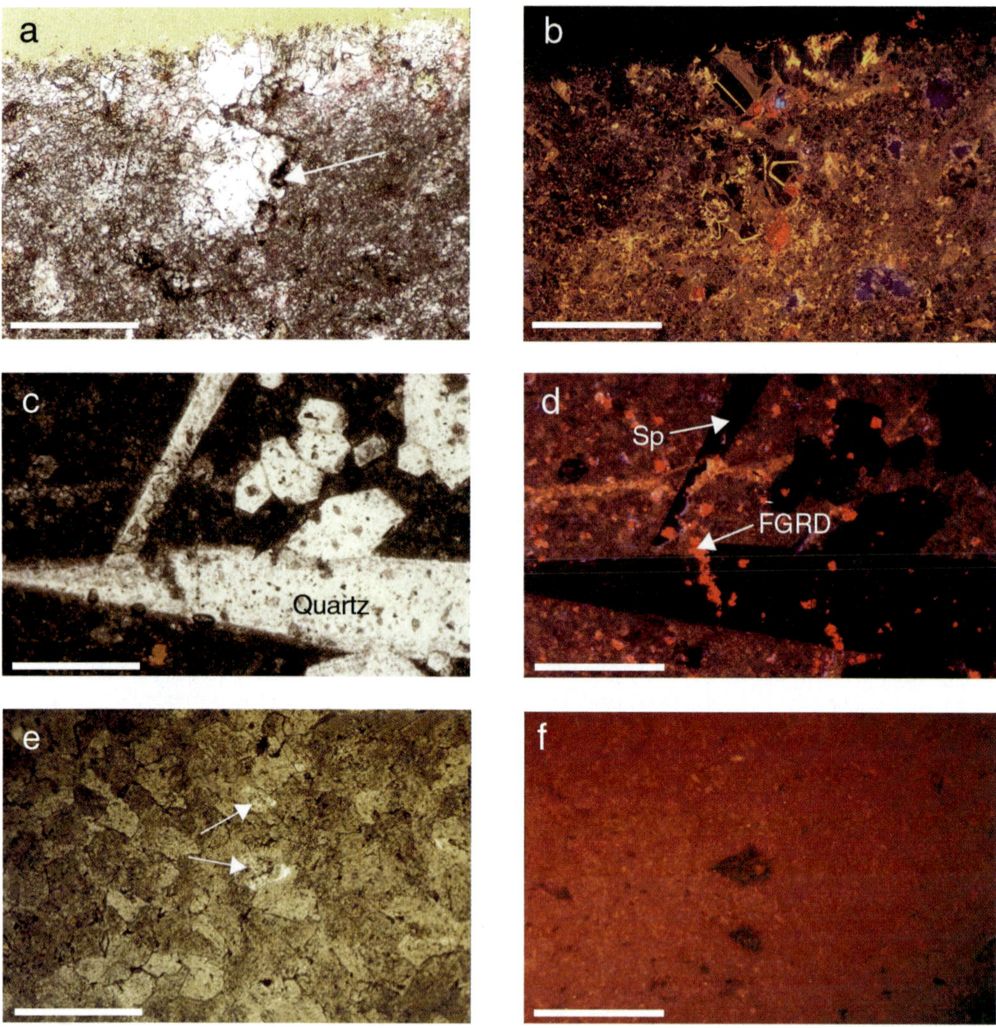

Fig. 4. (**a**) PPL view of a small cement-filled cavity that has been cut by a dissolution seam (arrow) at Newcastle West. The yellowish substance to the top if the image is epoxy resin. (**b**) Same view in CL. The dissolution seam post-dates the early calcite cements. Dolomite rhombs occur along the seam implying that it was exploited as a conduit. (**c**) PPL view of euhedral quartz crystals within micrite of the Waulsortian Limestone from Newcastle West. (**d**) The same view in CL illustrating that the fine-grained replacive dolomite (FGRD) post-dates or is contemporaneous with the precipitation of the quartz. Note the selective dolomitization of the micritic components of the background limestone. Sp indicates a mould of a sponge spicule infilled with silica. (**e**) PPL view of the relict pre-dolomite quartz crystals (arrows) at Subulter. (**f**) The same view under CL. All scale bars are 500 μm.

shows that quartz precipitation predated or was contemporaneous with dolomitization (Fig. 4d). T_h values of fluid inclusions in the quartz range between 131 and 187 °C and in the dolomite between 125 and 199 °C (Table 1). Salinity data for both have an average of 25.3 combined wt% NaCl and $CaCl_2$ equivalent.

As there is evidence of substantial late Carboniferous heating (Clayton *et al.* 1989), some authors have expressed concerns over the validity of fluid-inclusion data measured in Irish Carboniferous rocks (Hitzman *et al.* 1992; Peace 1999; Reed & Wallace 2001). Bodnar (2003) suggested that the most important factor that determines the ease with which fluid inclusions re-equilibrate is the mineralogy of the host

Table 1. *Microthermometric data measured from the coarse dolomite and associated quartz phases in the areas studied*

Sample	Phase	Size (μm)	ψ_{liquid}	T_h	T_{fm}	$T_{m_{ice}}$	T_{m_h}	Salinity*
Newcastle West								
5.1.1	CD	20	0.9	161.6	–49	–22.5	–11.3	24.8
5.2.2	CD	4	0.9	188.5	–52	–22	–11	24.6
5.2.3	CD	8	0.85	170.5	–54.2	–23.5	–11.5	25.7
5.3.4a	CD	5	0.85	159.5				
5.3.4b	CD	5	0.85	162.0				
5.4.5	CD	10	0.9	171.7	–55	–23	–10.1	25.4
5.4.6a	CD	4	0.85	149	–52	–24	–10.6	26.1
5.4.6b	CD	3	0.85	148.4				
5.4.6c	CD	2	0.85	195.1				
5.4.6d	CD	2	0.85	198.7				
M1b	CD	15	0.9	181	–54.4	–21.9	–11.9	24.5
M1a	CD	10	0.9	175.9	–53.8			
M2a	CD	6	0.9	124.8				
M2b	CD	8	0.9	136.8				
M2c	CD	3	0.9	166.8				
M3a	CD	10	0.9	195.8				
M3b	CD	4	0.9	166.2				
1.5.7	Qtz	4	0.8	187.4				
3.1.7a	Qtz	4	0.8	147.7	–52	–24	–23.4	24.8
3.1.7b	Qtz	3	0.8	145.5				
3.1.7c	Qtz	3	0.8	166.2				
11.C1	Qtz	4	0.9	171.6	–52.3	–24.5	–24.1	25.1
11.C2	Qtz	4	0.9	130.7	–50.7	–22.8	–22.4	23.9
11.C3	Qtz	4	0.9	159.3	–51.2	–25.8	–24.9	25.6
11.C4	Qtz	4	0.9	157.2	–54.2	–26.1	–25.7	25.6
11.C5	Qtz	3	0.9	155.1	–53.5	–26	–25.6	25.6
11.C6	Qtz	3	0.9	155.5	–52.3	–28.7	–25.8	26.1
11.C7	Qtz	3	0.9	159.8	–51.6	–25.8	–24.9	25.6
11.C8	Qtz	3	0.9	161.4	–50.9	–27.7	–24.9	26.1
11.C9	Qtz	3	0.9	161.3				
11.C10	Qtz	3	0.9	154.3				
11.C11	Qtz	5	0.95	166.1	–50.3	–26.3	–25	25.8
11.C12	Qtz	5	0.9	166.5				
Subulter								
A 1	CD	8	0.8	134.7	–50.5	<–23.1		
B 3.2.1	CD	4	0.8	137.4				
B 3.2.2	CD	10	0.8	155.2				
D 3.2.1	CD	6	0.9	140.4	<–49.2			
D 3.2.2	CD	6	0.9	160.1				
D 3.2.3	CD	4	0.9	155				
E 5.2.1	CD	20	0.8	141.5	–55	–27.2		
E 5.2.2	CD	5	0.8	126.2				
E 5.2.3	CD	4	0.8	130.7				
E 5.2.4	CD	6	0.8	128.2				
E 5.2.5	CD	4	0.8	126				
F 3.1	CD	12	0.8	124	–54	–22.5	–26.7	22.9
G 3.1	CD	20	0.9	125	–54.8	–23.6	–26.5	23.6
E1.1	CD	2	0.85	134.8				
E 1.2	CD	2.5	0.9	123.1				
D 1.1	CD	2.5	0.9	127.7		–26.2		
D 1.2	CD	1.5	0.9	120.9				
D 1.3	CD	1.5	0.9	134.7				
D 2.1	CD	2	0.9	123.1				
D 2.2	CD	8	0.9	129.8	–54.6			

Table 1. *Continued*

Sample	Phase	Size (μm)	ψ_{liquid}	T_h	T_{fm}	$T_{m_{ice}}$	T_{m_h}	Salinity*
D 2.3	CD	4	0.9	127				
G 1.1	CD	20	0.85	119.2	–54.3		-27.5	
G 1.2.1	CD	2	0.9	132.4				
G 1.2.2	CD	2	0.9	124.6				
G 1.2.3	CD	2	0.9	120.5				
G 2.1	CD	4	0.9	115.9				
G 3.1b	CD	5	0.9	123.5	–54.3	–25.4	–26.2	24.7
G 4.1	CD	10	0.85	109.2	–54.4	–24.6	–27.2	23.7
G 4.2.1	CD	20	0.9	129.3	–54.7	–22.6	–23.4	24.0
G 4.2.2	CD	8	0.9	127.8	–54.7	–23.4	–24.2	24.0
D.2	Pre-qtz	1	0.85	123.5				
D.4	Pre-qtz	2	0.8	149				
D.5	Pre-qtz	2	0.85	134.8				
D.6	Pre-qtz	2	0.85	133.9				
D.B1	Pre-qtz	13	0.85	125.2	–54	–26.1		
D.B3	Pre-qtz	6	0.85	105	–54.2	–26.2		
F.A.1	Pre-qtz	5	0.9	135.4	–54.4	–27.6	–30.2	25.0
F.A.2.1	Pre-qtz	8	0.9	127.4	–54.7	–27.9		
F.A.2.2	Pre-qtz	10	0.9	134.5	–54.8	–27.9		
F.A.2.3	Pre-qtz	22	0.9	134	–54.6	–30.3		
F.A.3	Pre qtz	6	0.9	122.7	–54.7	–27.4		
A 1.2	Post-qtz	15	0.9	82	–54.2	–16.5	–26.2	19.2
A 1.3	Post-qtz	35	0.9	83.4	–53	–16.5		
B 1e	Post-qtz	6	0.9	78.3				
A 1.2	Post-qtz	12	0.9	87.4	–55	–17		
C 3	Post-qtz	15	0.9	78				
A 1.1	Post-qtz	8	0.9	82.1	–55	–16.4		
Quinagh								
A1	CD	10	0.8	264.7	–48	–13.4	–25	17.0
A2	CD	5	0.8	234	–50.1	–14.8	–24.5	18.1
A3	CD	5	0.8	265	–54	–14.5	–22.9	18.0
B1	CD	5	0.85	192.7				
C1.1	CD	2	0.8	232.5				
C1.2	CD	3	0.8	219.5				
D1	CD	4	0.85	267				
E1.1	CD	4	0.8	230.6				
E1.2	CD	7	0.8	228.1				
E1.3	CD	3	0.8	229.8				
E1.4	CD	3	0.8	211				
E1.5	CD	3	0.8	212.6				
E1.6	CD	10	0.75	243				
E1.7	CD	3	0.8	211				
E1.8	CD	3	0.8	224.4				
E2.1	CD	2	0.8	257.6				
E2.2	CD	5	0.8	226				
E2.3	CD	2	0.8	253.4				
E2.4	CD	3	0.8	253.2				
E3.1	CD	3	0.8	251.8				
E3.2	CD	4	0.8	248.1				
E3.3	CD	4	0.8	246.5				
E3.4	CD	3	0.8	238.6				
E3.5	CD	4	0.8	249.5				
E3.6	CD	4	0.8	239.2				
E3.7	CD	4	0.8	262.2				
E4.1	CD	5	0.8	252.6	–49.8			

Table 1. *Continued*

Sample	Phase	Size (μm)	ψ_{liquid}	T_h	T_{fm}	$T_{m_{ice}}$	T_{m_h}	Salinity*
A1	Post-qtz	50	0.8	214.7	–54.6	–22.7	–16.5	24.4
A2	Post-qtz	25	0.85	174	–54.8	–25	–16.5	26.2
A3.1	Post-qtz	10	0.85	185				
A3.2	Post-qtz	35	0.85	193	–54.2	–25.5		
A3.3	Post-qtz	15	0.85	275	–54.1			
A3.4	Post-qtz	20	0.85	187.2	–54.9	–25.4		
A3.5	Post-qtz	15	0.85	195	–54			
A4	Post-qtz	50	0.9	196.1	–53.3	–24.2	–22.9	25.1
A5.1	Post-qtz	25	0.9	187.3	–53.2			
A5.2	Post-qtz	25	0.9	195.4	–52			
A5.3	Post-qtz	25	0.9	198.3	–53.3	–25.8	–22.8	25.9
A5.4	Post-qtz	10	0.9	190	–52.7	–24.6	–23.9	25.2
Lisheen								
J1.a	CD	7	0.85	177		–6.8		
J1.c	CD	8	0.85	183.4				
J1.d	CD	6	0.9	236.2	<-47.7			
J2.a	CD	9	0.9	172.6	-53.2	–23.3	–11	25.6

*Salinity given in combined NaCl and $CaCl_2$ equivalent.
CD, coarse dolomite; Post-qtz, post-dolomite quartz; Pre-qtz, pre-dolomite quartz.
ψ_{liquid}, is defined as the % occupied by the liquid at room temperature.

phase. Fluid inclusions in dolomite, which has a Mohs hardness of 3.5–4, are more likely to stretch than fluid inclusions in quartz, which has a hardness of 7 and is therefore a more robust mineral (Bodnar 2003, p. 8.3, fig. 8.2). The overlap in the homogenization temperature data from both the quartz and coarse dolomite at Newcastle West suggests that there has been no stretching as a result of Variscan heating. This means that the fluid-inclusion data from Newcastle West can be considered as being reliable.

The dolomitized Waulsortian Limestone at Subulter is notably vuggy. It contains isolated, relict crystals of a pre-dolomite quartz generation (Fig. 4e). Although relatively uncommon and difficult to locate in plane-polarized light (PPL), this pre-dolomite quartz generation is more easily observed under CL (Fig. 4f). It consists of crystals that are usually less than 1 mm long and have ragged outlines due to replacement by dolomite. Fluid-inclusion data indicate that the pre-dolomite quartz and dolomite formed at similar temperatures (T_h of 105–149 °C and 109–160 °C, respectively). It should be noted that that the lower end of these temperature ranges are the same as those estimated by Hitzman *et al.* (1998) from oxygen isotope values. Salinity measurements for both pre-dolomite quartz and dolomite range from 22.9 to 25 combined wt% NaCl and $CaCl_2$ equivalent.

Dolomitized Ballysteen Limestone

Petrographic studies of the Ballysteen Limestone have been based on drillcore from the Geological Survey of Ireland borehole in the townland of Quinagh, Co. Carlow (grid reference: 2732 1744, Fig. 1) and an extensive quarry section at Clonmelsh (grid reference: 2724 1702), approximately 6 km south of Quinagh. The fossiliferous packstones and grainstones of the Ballysteen Limestone in the Quinagh area are pervasively dolomitized, and undolomitized or partially dolomitized limestones are rare.

The calcite cements within the partially dolomitized limestones consist of an early stage that is zoned under CL and a later unzoned, dully luminescent stage that totally occluded the primary porosity. The zoned cement predates compactional features, while the unzoned cement is later than compactional features. Stylolites and dissolution surfaces are common, especially in the more argillaceous parts of the succession. Silicification of bioclasts, particularly brachiopods and crinoid ossicles, is also common.

Throughout the Quinagh borehole, the

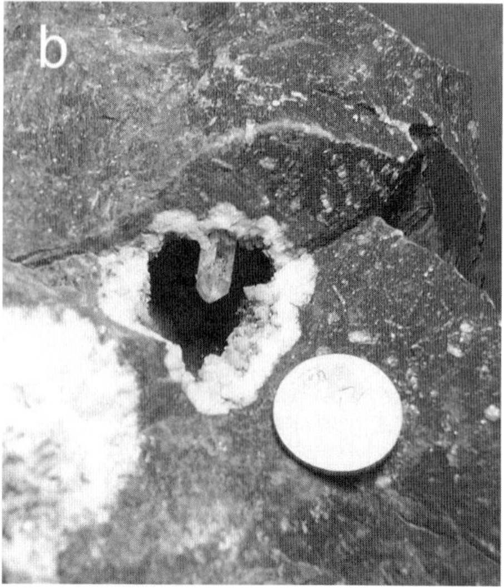

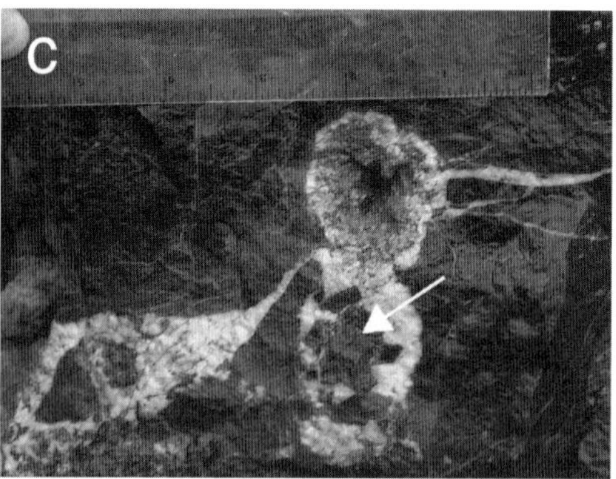

Fig. 5. (**a**) Distribution of vugs along the quarry face at Clonmelsh. (**b**) Detailed view of the vugs within the regionally dolomitized Ballysteen Limestone. The vug in the bottom left of the image has been completely infilled by the late coarse white dolomite, whereas the vug in the centre of the image is lined by the coarse white dolomite. The centre of this vug remains as open space into which quartz has grown. The coin is 22 mm. (**c**) The vug is almost completely filled with coarse white dolomite with some bed-parallel veining and minor brecciation (arrow). (**d**) Hydraulic fracturing and brecciation of regionally dolomitized Ballysteen Limestone by coarse white dolomite. The coin is 22 mm.

Ballysteen Limestone is partially–completely replaced by dark-grey dolomite that can be shown to be later than the calcite cements, silicification of bioclasts, and many of the stylolites and dissolution seams. Therefore, as in the Waulsortian Limestone, dolomitization occurred after the primary porosity had been occluded. However, in contrast, the calcite cements were preferentially dolomitized and fine-grained skeletal components in partially dolomitized limestones were not completely replaced.

A striking feature of both the pervasively dolomitized parts of the Quinagh drillcore and of the quarry exposure at Clonmelsh is large open vugs. These range from less than 1 cm to about 20 cm at maximum dimension and appear to be preferentially developed within particular beds (Fig. 5a). The vugs are rimmed, or less commonly completely filled, by coarse white dolomite, that is clearly later than the dark-grey dolomite. This also fills veins parallel and normal to bedding that commonly connect to the dolomite in the vugs. The veins form the matrix of breccias containing fragments of the dark-grey dolomite (Fig. 5c) that display refittable fabrics, which are likely to be the products of hydraulic fracturing (Fig. 5d). This style of brecciation is very similar to the white matrix breccia described from the Lisheen deposit (Hitzman *et al.* 2002). The white matrix breccia occurs in zones formed by a stockwork of coarse veins cross-cutting the regionally dolomitized Waulsortian Limestone and forming zones of brecciation. T_h data from the coarse white dolomite at Quinagh range from 193 to 267 °C, similar to the range of 170–230 °C recorded from white matrix breccia at Lisheen by Eyre *et al.* (1996). Salinity data range from 17 to 18.1 combined wt% NaCl and $CaCl_2$ equivalent. In most vugs at Quinagh, quartz crystals, some as large as 40 mm in length, have grown into open spaces from the surfaces of coarse white dolomite crystals (Fig. 5b). T_h data from fluid inclusions in quartz range from 174 to 215 °C (with one outlier at 275 °C) and indicate growth from a fluid at a temperature similar to that at which the coarse white dolomite formed. Salinity data indicate a range of 24.4–26.2 combined wt% NaCl and $CaCl_2$ equivalent.

Vug formation

The vugs in the dolomites at Quinagh and Clonmelsh must have developed after calcite cementation of the Ballysteen Limestone and were probably contemporaneous with, or followed, the formation of the dark-grey dolomite. They formed before the hydraulic fracturing and brecciation of the dark-grey dolomite and prior to the emplacement of the coarse white dolomite. There is no evidence to suggest that they represent the amplification by dolomitization of pre-existing, primary cavities within the limestone lithosome, as there are no 'stromatactis-like' cavities within the Ballysteen Limestone resembling those developed in the Waulsortian Limestone. The depth at which vug formation took place can therefore be bracketed between the depth of formation of the dark-grey dolomite and that of the coarse white dolomite.

Discussion

In all study locations petrographic analysis indicates that dolomitization of the Waulsortian Limestone and the underlying Ballysteen Limestone occurred after the complete occlusion of primary porosity by calcite cements. This raises questions regarding the depth at which dolomitization occurred and the mechanism by which Mg-rich fluids were able to infiltrate the low-permeability limestone lithosome.

Based on oxygen isotope data, Gregg *et al.* (2001) interpreted fine-grained dolomite found in the Irish Midlands and Dublin Basin (their pk1 dolomite) as a low-temperature (<50 °C), diagenetic replacement of limestone by seawater. They speculated that the fine-grained dolomite provided the necessary porosity and permeability within the Waulsortian Limestone to enable flow of later dolomitizing fluids. However, in this study the paragenetic relationship between the fine-grained dolomite and the pre-dolomite quartz at Newcastle West and Subulter can be used to infer the temperature of formation of the fine-grained dolomite. As the fine-grained dolomite was precipitated after (or during) quartz precipitation, and before coarse white dolomite precipitation, it follows that the temperature of formation of the fine-grained dolomite lies between that of the quartz and the coarse dolomite, and is therefore >131 °C at Newcastle West and >105 °C at Subulter. It remains to be established whether the fine-grained dolomite found within the Irish Midlands has an origin different from that of areas further south.

Evidence regarding the depth at which dolomitization occurred has been based on the relationship of dolomite to compactional features (Peace 1999; Reed & Wallace 2001). These authors have used an estimate of 800 m burial for the formation of 'macrostylolites', which they have shown predate dolomitization. However, the relationship between pressure solution and burial depth is complex (Railsback 1993*a*, *b*).

The inference of depth of formation of the dolomites in this study from microthermometric analysis depends on whether they were hydrothermal or geothermal (Machel & Lonnee 2002). As the dolomite bodies at Newcastle West and Subulter are isolated and small, they may have formed under either hydrothermal or geothermal conditions. Therefore, estimation of their depths of occurrence from microthermometric data would be speculative. In contrast, the Lisheen and Co. Carlow occurrences form part of a very large body of dolomite (Regional Dolomite), which has been conservatively estimated to have been more than >50 km^3 (Sevastopulo & Redmond 1999). It is difficult to envisage how the enormous volume of fluid needed to dolomitize such a large volume of limestone could remain out of thermal equilibrium with the country rocks. It seems probable, therefore, that the Regional Dolomite was geothermal in origin. Based on the minimum T_h measured from Lisheen and a palaeogeothermal gradient of 60 °C km^{-1}, the minimum depth of formation of the Regional Dolomite would have been greater than 2.5 km.

The mechanisms by which dolomitizing fluids infiltrated the limestone lithosome are still not clear. The geometry of the dolomite body at Newcastle West suggests that the Mg-rich fluids moved within a steeply dipping zone. Although Lower Carboniferous faulting has been documented at Lisheen (Hitzman *et al.* 2002), the Regional Dolomite in the Rathdowney Trend is not restricted to known fault zones. Dolomitizing fluids may have moved vertically into the Waulsortian and then laterally. Evidence from Newcastle West suggests that, at least during the earliest phase of dolomitization, fluids may have exploited dissolution seams as conduits. As suggested by Gregg *et al.* (2001), the fine-grained dolomite may have created sufficient secondary porosity and permeability for further infiltration of dolomitizing fluids.

Conclusions

The primary porosity of the Waulsortian Limestone and Ballysteen Limestone formations was totally occluded by calcite cements within several hundred metres of burial, prior to the onset of dolomitization. Dolomitizing fluids may have exploited steep fracture zones but also spread laterally. Movement through the tightly cemented limestones probably occurred, at least initially, along dissolution seams. The early fine-grained dolomite provided permeability for further infiltration of dolomitizing solutions.

The overlap in T_h data measured in dolomite and quartz from both Newcastle West and Subulter indicate that fluid inclusions have not been re-equilibrated as a result of Variscan heating, and that the data from the fluid inclusions are reliable. Fine-grained dolomite at Newcastle West and Subulter formed at a minimum of 131 and 105 °C, respectively. T_h data suggest that the Regional Dolomite at Lisheen and in Co. Carlow formed at a temperature in excess of 173 °C. If the Regional Dolomite is geothermal in origin its formation was at a depth of at least 2.5 km.

The large vugs that occur in dolomitized units do not result from the amplification of pre-existing primary cavities within the host limestone. The coarse white dolomite, which lines vugs and is contained within veins associated with vugs, formed at high temperatures (>193 °C).

The work contained within this paper forms part of the PhD research project of C. M. Mulhall, funded by a Basic Research Grant from Enterprise Ireland. The authors would like to thank A. Boyce and T. Fallick of the Scottish Universities Research and Environmental Centre (SUERC), and S. Gleeson of the University of Alberta for many discussions and advice on various aspects of this project.

References

BANKS, D.A., BOYCE, A.J. & SAMSON, I.M. 2002. Constraints on the origins of fluids forming Irish Zn–Pb–Ba deposits: Evidence from the composition of fluid inclusions. *Economic Geology*, **97**, 471–480.

BLAKEMAN, R.J., ASHTON, J.H., BOYCE, A.J., FALLICK, A.E. & RUSSELL, M.J. 2002. Timing of interplay between hydrothermal and surface fluids in the Navan Zn + Pb orebody, Ireland: Evidence from metal distribution trends, mineral textures and $\delta^{34}S$ analyses. *Economic Geology*, **97**, 73–91.

BODNAR, R.J. 2003. Re-equilibration of fluid inclusions. *In*: SAMSON, I.M., ANDERSON, A. & MARSHALL, D. (eds) *The Analysis and Interpretation of Fluid Inclusions*. Mineralogical Association of Canada, Short Course, **32**, 8.1–8.16.

BOYCE, A.J., LITTLE, C.T.S. & RUSSELL, M.J. 2003. A new fossil vent biota in the Ballynoe baryte deposit, Silvermines, Ireland: evidence for intracratonic sea-floor hydrothermal activity about 352 Ma. *Economic Geology*, **98**, 649–656.

CARTER, J.S. & WILBUR, D.G. 1986. The geological setting of Zn mineralisation near Duncormick in the Wexford Permo-Carboniferous outlier. *In*: ANDREW, C.J., CROWE, R.W.A., FINLAY, S., PENNELL, W.M. & PYNE, J.F. (eds) *Geology and Genesis of Mineral Deposits in Ireland*. Irish Association for Economic Geology, Dublin, 471–474.

CLAYTON, G., HAUGHEY, N., SEVASTOPULO, G.D. & BURNETT, R. 1989. *Thermal Maturation Levels in the Devonian and Carboniferous Rocks of Ireland. Geological Survey of Ireland.*

DE BRIT, T.J. 1988. *The diagenetic and structural relationships of veins from Lower Carboniferous Limestones, Counties Longford and Westmeath, Ireland*. PhD thesis, University of Dublin.

DICKSON, J.A.D. 1966. Carbonate identification and genesis as revealed by staining. *Journal of Sedimentary Petrology*, **36**, 491–505.

EVERETT, C.E., RYE, D.M., WILKINSON, J.J., BOYCE, A.J., ELLAM, R.M., FALLICK, A.E. & GLEESON, S.A. 1999. The genesis of Irish-type Zn–Pb deposits: Characterisation and origin of the principle ore fluid. *In*: STANLEY, C.J. (ed.) *Mineral Deposits: Processes and Processing*. Balkema, Rotterdam, 854–848.

EYRE, S.L., WILKINSON, J.J., STANLEY, C.J. & BOYCE, A.J. 1996. Geochemistry of dolomitisation and zinc-lead mineralisation in the Rathdowney Trend, Ireland. *Geological Society of America, Abstracts with Programs*, **28**, A210–A211.

GILLIES, D. 1987. *Diagenetic history of the Waulsortian buildups (Dinantian), Craven Basin, N.W. England*. PhD thesis, University of Edinburgh.

GLEESON, S.A., BANKS, D.A., EVERETT, C.E., WILKINSON, J.J., SAMSON, I.M. & BOYCE, A.J. 1999. Origin of mineralising fluids in Irish-type deposits: Constraints from halogens. *In*: STANLEY, C.J. (ed.) *Mineral Deposits: Processes and Processing*. Balkema, Rotterdam, 857–860.

GREGG, J.A., SHELTON, K.L., JOHNSON, A.W., SOMERVILLE, I.D. & WRIGHT, W.R. 2001. Dolomitisation of the Waulsortian Limestone (Lower Carboniferous) in the Irish Midlands. *Sedimentology*, **48**, 745–766.

HAZLETT, T.J. 1997. *A hydrogeologic analysis of the Irish carbonate-hosted lead–zinc ore deposits*. Unpublished PhD thesis, Johns Hopkins University.

HITZMAN, M.W. & BEATY, D.W. 1996. *The Irish Zn–Pb–(Ba) Orefield*. Society of Economic Geologists, Special Publications, **4**, 112–143.

HITZMAN, M.W., ALLAN, J.R. & BEATY, D.W. 1998. Regional dolomitisation of the Waulsortian Limestone in southeastern Ireland; evidence of large scale fluid flow driven by the Hercynian Orogeny. *Geology*, **26**, 547–550.

HITZMAN, M.W., CONNOR, P.O.O., SHEARLEY, E., SCHAFFALITZKY, C., BEATY, D.W., ALLAN, J.R. & THOMPSON, T. 1992. Discovery and geology of the Lisheen Zn–Pb–Ag prospect, Rathdowney Trend, Ireland. *In*: BOWDEN, A., EARLS, G., O'CONNOR, P. & PYNE, J. (eds) *The Irish Minerals Industry 1980–1990*. Irish Association of Economic Geology, Dublin, 227–246.

HITZMAN, M.W., REDMOND, P. & BEATY, D.W. 2002. The carbonate-hosted Lisheen Zn–Pb–Ag deposit, County Tipperary, Ireland. *Economic Geology*, **97**, 1627–1655.

KEELEY, M.L. 1980. *The Carboniferous geology of the Carrick-on Suir syncline, southern Ireland*. PhD thesis, University of Dublin.

KING, D.E. & MEYERS, W.J. 1985. Regional cement stratigraphy and diagenetic history of Waulsortian Limestones, eastern Midlands, Republic of Ireland. (Abs.) *AAPG Bulletin*, **69**, 273.

LEE, M.J. & WILKINSON, J.J. 2002. Cementation, hydrothermal alteration and Zn–Pb mineralization of carbonate breccias in the Irish Midlands: Textural evidence from the Cooleen Zone, near Silvermines, County Tipperary. *Economic Geology*, **97**, 653–662.

LEES, A. & MILLER, J. 1995. Waulsortian banks. *In*: MONTY, C.L.V., D.W.J. BOSENCE, BRIDGES, P.H. & PRATT, P.R. (eds) *Carbonate Mud-mounds – Their Origin and Evolution*. International Association of Sedimentologists, Special Publications, **23**, 191–271.

LYDON, J.W. 1986. Models for the generation of metalliferous hydrothermal systems within sedimentary rocks, and their applicability to Irish Carboniferous Zn–Pb deposits. *In*: ANDREW, C.J., CROWE, R.W.A., FINLAY, S., PENNEL, W.M. & PYNE, J.F. (eds) *Geology and Genesis of Minerals Deposits in Ireland*. Irish Association of Economic Geology, Dublin, 555–577.

MACHEL, H.G. 1985. Cathodoluminescence in calcite and dolomite and its chemical interpretation. *Geoscience Canada*, **12**, 139–147.

MACHEL, H.G. & LONNEE, J. 2002. Hydrothermal dolomite – a product of poor definition and imagination. *Sedimentary Geology*, **152**, 163–171.

NADEN, J. 1996. CalcicBrine 1.5: a Microsoft Excel 5.0 Add-in for calculating salinities from microthermometric data in the system NaCl–$CaCl_2$–H_2O, *In*: *PACROFI VI, Madison, Wisconsin, May 27–June 1, Program with Abstracts*, 97–98.

PEACE, W. 1999. *Carbonate-hosted Zn–Pb mineralisation, Upper Pale Beds, Navan, Ireland*. PhD thesis, University of Melbourne.

PEACE, W.M. & WALLACE, M.W. 2000. Timing of mineralization at the Navan Zn–Pb deposit: A post Arundian age for Irish mineralization. *Geology*, **28**, 711–714.

PHILLIPS, W.E.A. & SEVASTOPULO, G.D. 1986. The stratigraphic and structural setting of the Irish mineral deposits. *In*: ANDREW, C.J., CROWE, R.W.A., FINLAY, S., PENNELL, W. M. & PYNE, J.F. (eds) *Geology and Genesis of Mineral Deposits in Ireland*. Irish Association for Economic Geology, Dublin, 1–30.

RADKE, B.M. & MATHIS, R.L. 1980. On the formation and occurrence of saddle dolomite. *Journal of Sedimentary Petrology*, **50**, 1149–1168.

RAILSBACK, L.B. 1993*a*. Contrasting styles of chemical compaction in the Upper Pennsylvanian Dennis Limestone in the midcontinent region, USA. *Journal of Sedimentary Petrology*, **63**, 61–72.

RAILSBACK, L.B. 1993*b*. Intergranular pressure dissolution in a Plio-Pleistocene grainstone buried no more than 30 metres: Shoofy oolite, southwestern Idaho. *Carbonates and Evaporites*, **8**, 163–169.

REED, C.P. & WALLACE, M.W. 2001. Diagenetic evidence for an epigenetic origin of the Courtbrown Zn–Pb deposit, Ireland. *Mineralium Deposita*, **36**, 428–441.

RUSSELL, M.J. 1978. Downward-excavating hydrothermal cells and Irish type ore deposits: importance of an underlying thick Caledonian prism. *Transactions of the Institution of Mining and Metallurgy*, **87**, B168–B171.

RUSSELL, M.J. 1986. Extension and convection: a genetic model for the Irish Carboniferous base metal and barite deposits. *In*: ANDREW, C.J., CROWE, R.W.A., FINLAY, S., PENNELL, W.M. & PYNE, J.F. (eds) *Geology and Genesis of Mineral Deposits in Ireland*. Irish Association for Economic Geology, Dublin, 545–554.

SEVASTOPULO, G.D. & REDMOND, P. 1999. Age of mineralisation of carbonate-hosted, base metal deposits in the Rathdowney Trend, Ireland. *In*: MCCAFFREY, K.J.W., LONERGAN, L. & WILKINSON, J.J. (eds) *Fractures, Fluid Flow and Mineralisation*. Geological Society, London, Special Publications, **155**, 303–311.

SEVASTOPULO, G.D. & WYSE-JACKSON, P.N. 2001.

Carboniferous (Dinantian). *In*: HOLLAND, C.H. (ed.) *The Geology of Ireland*. Dunedin Academic, Edinburgh, 241–289.

SHEARLEY, E.P. 1988. *The Carboniferous geology of the Fermoy and Mitchelstown syncline, southern Ireland*. PhD thesis, University of Dublin.

SHEPHERD, T., RANKIN, A.H. & ALDERTON, D.H.M. 1985. *A Practical Guide to Fluid Inclusion Studies*. Blackie, Glasgow.

SHERIDAN, D.J. 1977. The hydrocarbons and mineralisation proved in the Carboniferous strata of deep boreholes in Ireland. *In*: GARRARD, P. (ed.) *Proceedings of the Forum on Oil and Ore in Sediments*. Geology Department, Imperial College, London, 113–144.

SINGER, D.A. 1995. World class base and precious metal deposits – a quantitative analysis. *Economic Geology*, **90**, 88–104.

TIETZSCH-TYLER, D. & SLEEMAN, A.G. 1994. *Geology of Carlow–Wexford; A Geological Description to Accompany the Bedrock Geology 1:100 000 Scale Map Series, Sheet 19, Carlow–Wexford with Contributions by B.J. McConnell, E.P. Daly, A.M. Flegg, P.J. O'Connor and W.P. Warren*. Geological Survey of Ireland, Dublin.

Index

Page numbers in *italic* refer to figures, page numbers in **bold** refer to tables